# SPRING NIGHT SKY

S0-BZH-239

## When to Use This Star Map

| | |
|---|---|
| Early March: | 2 a.m. |
| Late March: | 1 a.m. |
| Early April: | midnight |
| Late April: | 11 p.m. |
| Early May: | 10 p.m. |
| Late May: | Dusk |

This star chart is most accurate if used within an hour or so of the times listed and is plotted for observers located between 30° and 50° north latitude. All times are standard time; if daylight-saving time is in effect, add one hour.

To use this chart, hold it in front of you and rotate it so that the yellow label corresponding to the direction you are facing is positioned at the bottom, right-side up. The stars in the sky should match those depicted on the chart. The center of the chart is the zenith, the point in the sky directly overhead.

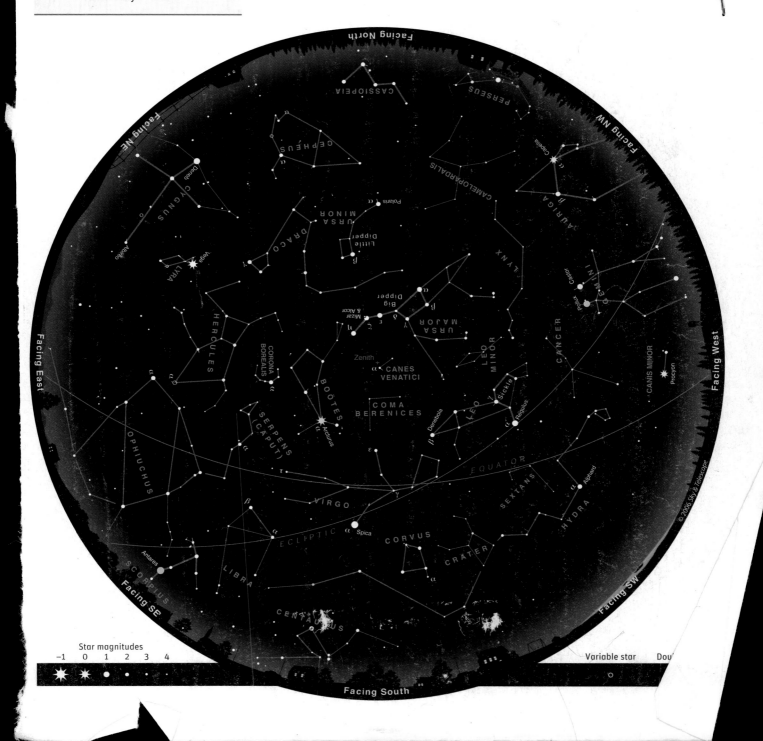

Star magnitudes
−1   0   1   2   3   4

# SUMMER NIGHT SKY

## When to Use This Star Map

| | |
|---|---|
| Early June: | 1 a.m. |
| Late June: | midnight |
| Early July: | 11 p.m. |
| Late July: | 10 p.m. |
| Early August: | 9 p.m. |
| Late August: | Dusk |

This star chart is most accurate if used within an hour or so of the times listed and is plotted for observers located between 30° and 50° north latitude. All times are standard time; if daylight-saving time is in effect, add one hour.

To use this chart, hold it in front of you and rotate it so that the yellow label corresponding to the direction you are facing is positioned at the bottom, right-side up. The stars in the sky should match those depicted on the chart. Ignore all the parts of the map above horizons you are not facing.

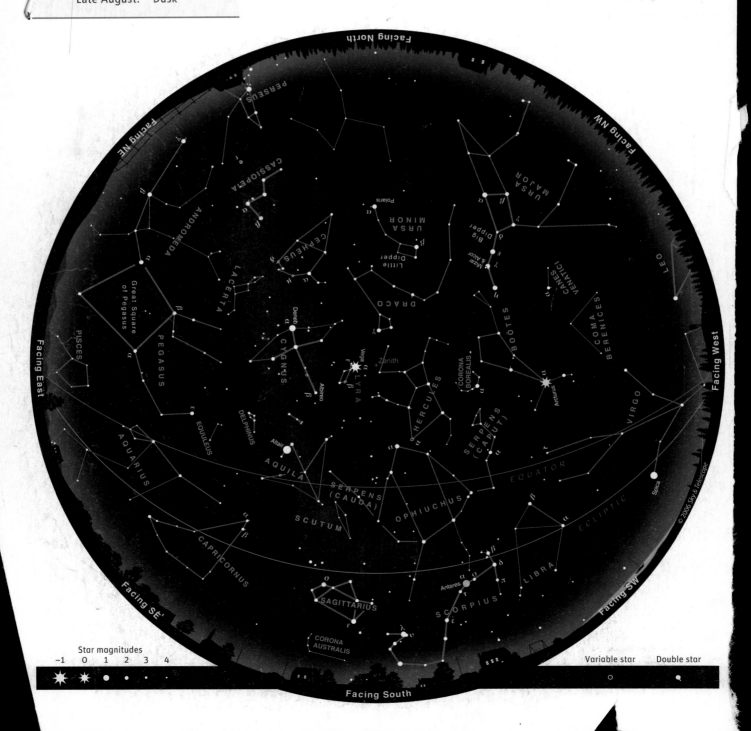

## When to Use This Star Map

| | |
|---|---|
| Early September: | midnight |
| Late September: | 11 p.m. |
| Early October: | 10 p.m. |
| Late October: | 9 p.m. |
| Early November: | 8 p.m. |
| Late November: | 7 p.m. |

This star chart is most accurate if used within an hour or so of the times listed and is plotted for observers located between 30° and 50° north latitude. All times are standard time: if daylight-saving time is in effect, add one hour.

To use this chart, hold it in front of you and rotate it so that the yellow label corresponding to the direction you are facing is positioned at the bottom, right-side up. The stars in the sky should match those depicted on the chart. The farther up from the map's edge they appear, the higher they'll be shining in your sky.

Star magnitudes
−1  0  1  2  3  4

Variable star     Double star

# WINTER NIGHT SKY

## When to Use This Star Map

| | |
|---|---|
| Early December: | midnight |
| Late December: | 11 p.m. |
| Early January: | 10 p.m. |
| Late January: | 9 p.m. |
| Early February: | 8 p.m. |
| Late February: | Dusk |

This star chart is most accurate if used within an hour or so of the times listed and is plotted for observers located between 30° and 50° north latitude. All times are standard time.

To use this chart, hold it in front of you and rotate it so that the yellow label corresponding to the direction you are facing is positioned at the bottom, right-side up. The stars in the sky should match those depicted on the chart. Remember that star patterns will look much larger in the sky than they do here on paper.

# Explorations

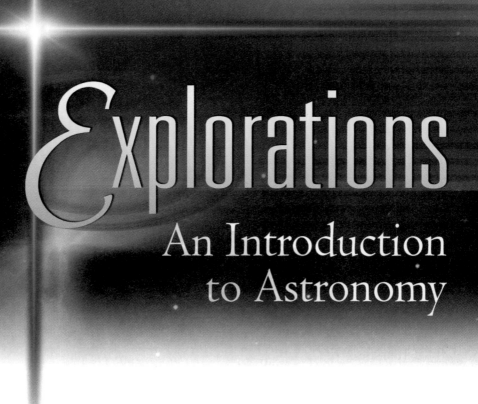

# Explorations

## An Introduction to Astronomy

**Thomas T. Arny**

Professor Emeritus
Department of Astronomy
University of Massachusetts, Amherst

**Stephen E. Schneider**

Professor of Astronomy
University of Massachusetts, Amherst

 **McGraw-Hill**
**Higher Education**

Boston   Burr Ridge, IL   Dubuque, IA   New York   San Francisco   St. Louis
Bangkok   Bogotá   Caracas   Kuala Lumpur   Lisbon   London   Madrid   Mexico City
Milan   Montreal   New Delhi   Santiago   Seoul   Singapore   Sydney   Taipei   Toronto

The McGraw·Hill Companies

# McGraw-Hill
# Higher Education

EXPLORATIONS: AN INTRODUCTION TO ASTRONOMY, FIFTH EDITION

Published by McGraw-Hill, a business unit of The McGraw-Hill Companies, Inc., 1221 Avenue of the Americas, New York, NY 10020. Copyright © 2008 by The McGraw-Hill Companies, Inc. All rights reserved. Previous editions 2006, 2004, 2002, 2000 (revised second edition), 1998, and 1996. No part of this publication may be reproduced or distributed in any form or by any means, or stored in a database or retrieval system, without the prior written consent of The McGraw-Hill Companies, Inc., including, but not limited to, in any network or other electronic storage or transmission, or broadcast for distance learning.

Some ancillaries, including electronic and print components, may not be available to customers outside the United States.

This book is printed on acid-free paper.

1 2 3 4 5 6 7 8 9 0 QPV/QPV 0 9 8 7

ISBN 978–0–07–294360–3 (softcover)
MHID 0–07–294360–2

ISBN 978–0–07–331598–0 (hardcover)
MHID 0–07–331598–2

Publisher: *Thomas D. Timp*
Sponsoring Editor: *Debra B. Hash*
Developmental Editor: *Liz Recker*
Senior Marketing Manager: *Lisa Nicks*
Senior Project Manager: *Gloria G. Schiesl*
Senior Production Supervisor: *Sherry L. Kane*
Lead Media Project Manager: *Judi David*
Media Producer: *Daniel M. Wallace*
Designer: *John Joran*
Cover/Interior Designer: *Rokusek Design, Inc.*
Senior Photo Research Coordinator: *Carrie K. Burger*
Photo Research: *Mary Reeg*
Supplement Producer: *Mary Jane Lampe*
Compositor: *ICC Macmillan*
Typeface: 10/12 Minion
Printer: *Quebecor World Versailles, KY*

The credits section for this book begins on page 577 and is considered an extension of the copyright page.

**Library of Congress Cataloging-in-Publication Data**

Arny, Thomas.
    Explorations : an introduction to astronomy. — 5th ed. / Thomas T. Arny, Stephen E. Schneider.
      p. cm.
    Includes index.
    ISBN  978–0–07–294360–3 — ISBN  0–07–294360–2 (hard copy : alk. paper)
      1. Astronomy—Textbooks.  I. Schneider, Stephen E. (Stephen Ewing), 1957-  II. Title.
    QB45.2.A76 2008
    520—dc22
                                 2007021192

www.mhhe.com

# BRIEF CONTENTS

When I began *Explorations: An Introduction to Astronomy,* many people asked me why I was writing an astronomy book when so many already existed. In answering such questions, I found that I had many reasons, ranging from personal to pedagogic. Part of my motivation came from my sense of wonder about the Universe. The Universe is an amazing place, and I wanted to convey some of my own wonderment to my students.

As I began writing this book, I began to understand why the existing textbooks were very long and overwhelming in detail. I kept finding things that I didn't want to leave out—such as calendars and the history of astronomy. Moreover, reviewers kept suggesting things that I must include, but rarely did they suggest things I could omit. This led me to rethink how I wanted to organize *Explorations.* I covered some topics, such as timekeeping and exobiology, in essays that the instructor might choose to skip. Moreover, small duplications in material make it possible to jump directly to some of the later chapters without having to work through the details of all the earlier chapters.

Perhaps my main goal was to give simple explanations of *why* things happen. Many astronomy books today seem to simply say "this is how it is." I wanted instead to offer explanations, drawing as much as possible on simple, everyday effects that students can see around them in the world. For example, why do some stars pulsate? A simple analogy of steam building up pressure under the lid of a pan offers a model of this phenomena that is easy to understand and reasonably accurate. I have also tried to link complex physical processes to simple everyday experiences. For example, you can see the effects of diffraction by looking at a bright light through a lock of your hair pulled over your eyes or through glasses that you have fogged with your breath. When we can thus link physical principles to everyday observations, many of the more abstract and remote ideas become more familiar. I have also used analogies heavily throughout the book, and I have designed the illustrations to make those analogies more concrete.

Knowing the facts about astronomical objects is important, but it is almost equally important to understand how astronomers deduce those facts. Thus, an additional aim throughout this text is to explain *how* astronomers have come to their understanding of our Universe. The Science at Work boxes illustrate how observations may force astronomers to revise their ideas of how a given process occurs. As part of showing how science works, I have set many of the modern discoveries in their historical context to illustrate that science is a dynamic process and subject to controversy—many ideas are not immediately accepted, even if they ultimately prove to be "correct." I hope that by seeing the arguments for and against various ideas, students will have a better understanding of how science works. I myself am amazed at how many widely accepted ideas have such flimsy underpinnings and how many widely quoted values for astronomical quantities are *very* imperfectly known.

I would be remiss not to mention the role of mathematics in science. Math is essential for understanding a number of methods used by astronomers, and I have therefore included some mathematics in a number of places. However, because math is so intimidating to so many students, I begin these discussions by introducing the essence of the calculation in everyday language so that the basic idea can be understood without understanding the mathematics. For example, Wien's law relates the temperature of a hot object to its color by a mathematical law. However, illustrations of the law can be seen in everyday life when we estimate how hot an electric stove burner is by the color of its glow. Where I present the mathematics, I work through it step by step, explaining where terms must be cross-multiplied and so forth.

## ABOUT THE FIFTH EDITION

### Content Updates and Additions

In this fifth edition of *Explorations*, I have updated the areas in which substantial changes in our knowledge have occurred, where different ideas are emerging, or where users have requested fuller coverage of a topic.

#### New Topics

- Discussion of newly defined "dwarf planets" in the Solar System (chapters 7 and 10)
- Discussion of the effects of Special Relativity in Space Travel (Essay 4)

#### Updated and Revised Topics

- Surface of Titan
- Further information on the importance of interactions and mergers in the origin and evolution of galaxies (chapter 16)
- Revised discussion of the evidence for dark energy (chapter 17).
- Tables of nearby bright and local group galaxies.
- New star charts
- Revised Periodic Table illustrating origin of chemical elements.

#### New and Updated Images

- Titan's Surface
- H-R diagrams
- Hubbe Tuning Fork diagram
- Galaxy and Galaxy Cluster images
- Map of Dark Matter
- Stellar Evolution

**New! Coauthor** In addition to the new topics and artwork in *Explorations*, this fifth edition now has a second author, Professor Stephen E. Schneider. Steve contributed advice and information to many of the earlier editions, and his participation in this book is now official. Besides the essay on relativity (mostly his work) and numerous revisions throughout the text, much of the new artwork bears his stamp. For these and many other reasons I am delighted to have Steve as coauthor. We both hope you will enjoy learning about the Universe from this book.

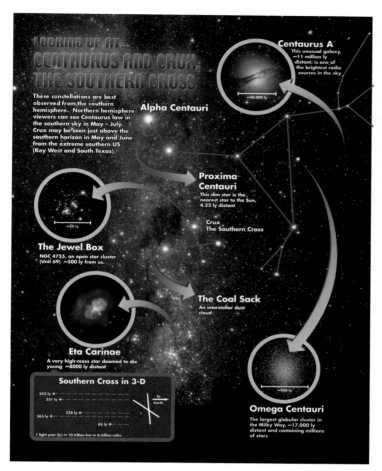

**Looking Up Features.** The eight Looking Up features— each one a full-page art piece—are designed to show students how some of the astronomical objects discussed in the text connect with the real sky that they can see overhead at night. In particular, they show where a variety of the frequently mentioned and important astronomical objects can be seen. Each Looking Up feature presents a photograph of one or more constellations in which nebulas, star clusters, and so forth are identified and illustrated. These latter illustrations include scale factors to help students visualize how even immense objects many light-years across can appear as mere dots in the sky. Along with the illustrated objects, most of the Looking Up features include a small insert to show how the constellation's stars are arranged in space.

The following are the Looking Up features

- Looking Up at Orion, chapter 1, page 26
- Looking Up at the Summer Triangle, Essay 1, page 65
- Looking Up at Taurus, chapter 4, page 136
- Looking Up at the Northern Circumpolar Constellations, chapter 5, page 172
- New! Looking Up at Centaurus and Crux, the Southern Cross, chapter 12, page 351
- Looking Up at Ursa Major, chapter 12, page 368
- Looking Up at Sagittarius, chapter 15, page 438
- Looking Up at M31 & Perseus, chapter 16, page 469

In constructing the Looking Up features, we have had to draw upon many different sources for the distances and sizes of the objects presented. For all but the very nearest of these, distances (and therefore sizes) are poorly known. Thus, none of the numbers appearing in these Looking Up features should be assumed to be precise.

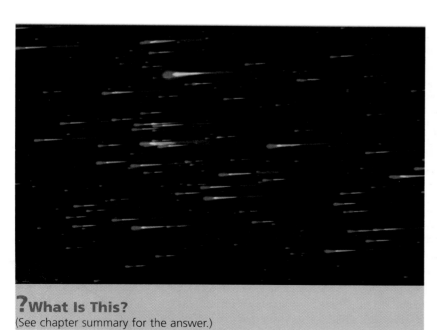

**?What Is This?**
(See chapter summary for the answer.)

**"What Is This?" Photographs.** At the beginning of each chapter, students are presented with a photo of an astronomical object and asked to guess what it is. After reading the chapter, they should have some idea of what is shown in the photo. The answer is provided at the end of each chapter summary under the heading "What It Is."

**Planetarium Activities.** Activities to be integrated with the Starry Night planetarium software are now available at the end of most chapters.

**Foldout: Scale of the Universe.** This two-page foldout shows the Earth, Sun, Moon, and their orbits to scale, as well as a "model" of some of the galaxies in the Local Group to illustrate their sizes and separations. The other side of the foldout depicts deep-space objects to scale. This foldout is constructed so students can easily leave it open and refer to it while reading the text to gain a better sense of scale.

**Electronic Media Integration** To help better grasp key concepts, these icons have been placed near figures and selections where students can gain additional understanding through the interactives and animations on the Explorations ARIS website Center.

## CUSTOM OPTIONS

### Primis Online

Customize this text in print or in electronic format to meet your exact course needs. Through McGraw-Hill's Primis Online, simply select your desired chapters and preferred sequence, and supplement them with the many science items on our database. Visit www.mhhe.com/primis/online to begin today.

### Volumes

Now one of the best-selling astronomy textbooks is available in smaller versions. Great for one-semester courses, these textbooks (volume 1: *Solar System,* and volume 2: *Stars and Galaxies*) contain the same great writing, the same great pedagogical tools, and the same great technology package.

## SUPPLEMENTS

McGraw-Hill offers various tools and technology products to support *Explorations: An Introduction to Astronomy,* Fifth Edition. Instructors can obtain teaching aids by calling the Customer Service Department at 800-338-3987 or contacting your local McGraw-Hill sales representative.

### NEW McGraw-Hill's ARIS—Assessment, Review, and Instruction System for *Explorations*
### http://www.mhhe.com/arny

This is a complete, online tutorial, electronic homework, and course management system designed for greater ease of use than any other system available. The ARIS website for *Explorations* allows instructors to create and share course materials and assignments with colleagues with a few clicks of the mouse. All PowerPoint® lectures, assignments, quizzes, tutorials, animations, and interactives are directly tied to text-specific materials in *Explorations,* but instructors can also edit questions, import their own content, and create announcements and due dates for assignments. ARIS has automatic grading and reporting of easy-to-assign homework, quizzing, and testing. All student activity within McGraw-Hill's ARIS is automatically recorded and available to the instructor through a fully integrated grade book that can be downloaded to Excel®.

## NEW ARIS Presentation Center

*Build instructional materials wherever, whenever; and however you want!*
ARIS Presentation Center is an online digital library containing assets such as photos, artworks, animations, PowerPoint presentations, and other media types that can be used to create customized lectures, visually enhanced tests and quizzes, compelling course websites, or attractive printed support materials.

*Access to your book, access to all books!*
Illustrations from Explorations are available as digital files. In addition, the Presentation Center Library includes thousands of assets from many McGraw-Hill titles. This ever growing resource gives instructors the power to utilize assets specific to an adopted textbook as well as content from all other books in the library.

*Nothing could be easier!*
Accessed from the instructor side of your textbook's ARIS website, Presentation Center's dynamic search engine allows you to explore by discipline, course, textbook chapter, asset type, or key word. Simply browse, select, and download the files you need to build engaging course materials. All assets are copyrighted by McGraw-Hill Higher Education but can be used by instructors for classroom purposes.

## Interactives

McGraw-Hill is proud to bring you an assortment of 23 out-standing Interactives like no other. Each Interactive is programmed in Flash for a stronger visual appeal. These Interactives offer a fresh and dynamic method for teaching the astronomy basics. Each Interactive allows users to manipulate parameters and gain a better understanding of topics such as Blackbody Radiation, The Bohr Model, a Solar System Builder, Retrograde Motion, Cosmology, and the H-R diagram by watching the effect of these manipulations. Each Interactive includes an analysis tool (interactive model), a tutorial describing its function, content describing its principal themes, related exercises, and solutions to the exercises. Users can jump between these exercises and analysis tools with just the click of the mouse.

## Starry Night CD

This planetarium software is available with every new text. It allows users to manipulate and take control of an accurate visual representation of the sky. Users become active observers and gain a far better understanding of how the sky works.

*New*-30 exercises that use the Starry Night Software are now available through the ARIS Website for *Explorations*.

## Instructor's Testing and Resource CD

McGraw-Hill's EZ Test is a flexible and easy-to-use electronic testing program. The program allows instructors to create tests from book specific items. It accommodates a wide range of question types and instructors may add their own questions. Multiple versions of the test can be created and any test can be exported for use with course management systems such as WebCT, BlackBoard, or PageOut. The program is available for Windows and Macintosh environments.

## Instructor's Manual

The Instructor's Manual is found on the Instructor's side of the Explorations ARIS website and on the Instructor's Testing and Resource CD.

## Classroom Performance System and Questions

The Classroom Performance System (CPS) brings interactivity into the classroom or lecture hall. CPS is a wireless response system that gives an instructor immediate feedback from every student in the class. Each CPS unit comes with up to 512 individual response pads and an appropriate number of corresponding receiver units. The wireless response pads are essentially remotes that are easy to use and engage students. The CPS system allows instructors to create their own questions or use the astronomy questions provided by McGraw-Hill.

## PowerWeb

Harness the assets of the Web to keep your course current with PowerWeb! This online resource provides high-quality, peer-reviewed content, including up-to-date articles from leading periodicals and journals, current news, weekly updates with assessment, interactive exercises, a Web research guide, study tips, and much more! PowerWeb is available packaged with a McGraw-Hill text or for online purchase from the website http://www.dushkin.com/powerweb.

## How to Study with This Book

Learning anything requires a certain amount of work. You certainly don't expect to be able to pick up a guitar and play it without practice, nor do you expect to be able to jog 5 miles without working out regularly. Learning astronomy also requires some work. The steps below may help you learn the material better and more easily.

In reading any assignment, begin by looking at the *pictures*. Turn the pages of the chapter and familiarize yourself with what the objects you will be reading about look like. Then read the introduction. Next, jump to the summary. Finally, start again and read the assigned material through. As you read, make notes of things you don't understand and ask your instructor or teaching assistant for clarification. For example, if you are puzzled about why eclipses don't happen every month, make a note. I would urge you *not* to highlight as you read. Making a few short notes is much more effective than highlighting whole paragraphs.

Look carefully at the pictures and diagrams. If the figure caption has a question in it, try to answer it. Make your own sketch of diagrams to be sure you understand what they represent.

In a first reading of a chapter, I'd suggest that if you are troubled by math, you should simply skip it for the time being. Be sure, however, to read the material leading into the math so you at least understand the basic idea. When you encounter a mathematical expression of a physical law, express the law in words. For example, the law of gravity relates the force of gravity to the mass of the objects and their distance from each other.

If you encounter words or terms as you read that you don't know, look them up in the glossary, the index, or a dictionary or encyclopedia. You are just wasting your time if you read a description of some object and you don't know what it is.

When you finish the assignment, try to answer the review questions. They are short and are designed to show you whether you have assimilated the basic factual material of the assignment. Try to do this without looking back into the chapter, but if you can't remember, look it up rather than skip over the question. You might find it helpful to write out short answers to the questions.

Having read the material once, go back and try to work through the math parts. Then try a practice problem to see if you can work through the material on your own.

If you get stuck at any point, see your teaching assistant or professor for help. Don't be shy about asking questions. I wish someone had made me understand this when I was a student. Learning is a thousand times easier if you ask questions when you get stuck.

Seeing a clear night sky spangled with stars is for me a nearly religious experience. And yet the beauty that I see and my sense of wonder are enriched even more by an appreciation of the complex processes that make the Universe work. I hope this book will similarly increase your appreciation of our Universe's wonders.

If you find mistakes or have suggestions about how to make this book better, please contact one of us. Write T. Arny at P.O. Box 545 Patagonia, AZ 85624, or by e-mail at tarny@theriver.com; or S. Schneider at Department of Physics and Astronomy, 536 Lederle Tower, Amherst, MA 01003, or by email at schneider@astro.umass.edu.

## REVIEWERS OF THIS EDITION

Special thanks and appreciation go out to reviewers of the fourth edition. Their contributions, constructive suggestions, new ideas, and invaluable advice played an important role in the development of this fifth edition and its supplements. These reviewers include:

C. Armendariz-Picon *Syracuse University*
Luca Bombelli *University of Mississippi*
David S. Curtis *University of West Georgia*
Pat Durrell *Youngstown State University*
Robert B. Friedfeld *Idaho State University*
Martin Hackworth *Idaho State University*
Clint D. Harper *Moorpark College*
Ronald A. Hartman *Mount San Antonio College*
Eric R. Hedin *Ball State University*
Harry L. F. Houpis *Sierra College*
Patrick Koehn *Eastern Michigan University*
N. A. Levenson *University of Kentucky*
Loris Magnani *University of Georgia*
Larry Sessions *Metropolitan State College of Denver*
James Webb *Florida International University*
David A. Wood, Jr. *San Antonio College*

The author and McGraw-Hill would also like to extend a special thank you to those who contributed to the supplements of this edition. These include:

Ronald H. Kaitchuk *Ball State University*
Patrick Koehn *Eastern Michigan University*
Nilakshi Veerabathina *University of Texas at Arlington*

## ACKNOWLEDGMENTS

Many people have played an important role in bringing this book into being. For this fifth edition of *Explorations.* We are very grateful to many people at McGraw-Hill, but especially Daryl Bruflodt, Debra Hash, Liz Recker, Gloria Schiesl, Carrie Burger, John Joran, Lisa Nicks, Sherry Kane, Dan Wallace, Judi David, and Mary Reeg for their help and patience.

We also want to thank a number of special reviewers and users who have sent comments. Linda Arny read carefully and commented thoughtfully on the entire manuscript. Bradley E. Schaefer and Roger E. Thayer also made valuable and helpful comments.

## REVIEWERS OF PREVIOUS EDITIONS

This revision of *Explorations* has also been made possible by the many users and reviewers of its previous editions. The author and publisher are grateful to the following reviewers of previous editions for their critical reviews, comments, and suggestions.

Parviz Ansari  *Seton Hall University*
Bruce Balick  *University of Washington, Seattle*
Tom Balonek  *Colgate University*
Wendy Hagen Bauer  *Wellesley College*
Peter A. Becker  *George Mason University*
Timothy Beers  *Michigan State University*
Ralph Benbow  *Northern Illinois University*
Daniel C. Boice  *University of Texas at San Antonio*
Luca Bombelli  *University of Mississippi*
Bernard Bopp  *University of Toledo*
Michael J. Bozack  *Auburn University*
Brother Guy Consolmagno, SJ  *Vatican Observatory*
Jeffrey A. Brown  *Washington State University*
Eugene Capriotti  *Michigan State University*
Tai L. Chow  *California State University, Stanislaus*
Thomas Christensen  *University of Colorado at Colorado Springs*
John J. Cowan  *University of Oklahoma*
Benjamin de Mayo  *West Georgia College*
Alexander Dickson  *Seminole Community College*
Jess Dowdy  *Abilene Christian University*
Mark Edwards  *Hofstra University*
Robert A. Egler  *North Carolina State University*
Heinrich Eichhorn  *University of Florida*
Don P. Engelberg  *Queensborough Community College*
Eric Feigelson  *The Pennsylvania State University*
Donald Foster  *Wichita State University*
Aaron Galonsky  *Michigan State University*
Harold Geller  *George Mason University*
Robert Geller  *University of California*
David J. Griffiths  *Oregon State University*
Bruce Gronich  *University of Texas–El Paso*
Bruno Gruber  *Southern Illinois University*
Alexander Gurshtein  *Mesa State College*
Heidi Hammel  *Massachusetts Institute of Technology*

Andrew Harris  *University of Maryland*
Harold M. Hastings  *Hofstra University*
John F. Hawley  *University of Virginia*
David Hedin  *Northern Illinois University*
Philip L. Hegenderfer  *University of Akron*
Paul Hintzen  *California State University–Long Beach*
Thomas Hockey  *University of Northern Iowa*
John Greg Hoessel  *University of Wisconsin*
Harry Houpis  *Sierra College*
James Christopher Hunt,  *Prince George's Community College*
Douglas R. Ingram  *Texas Christian University*
Scott B. Johnson  *Idaho State University*
Terry Jay Jones  *University of Minnesota*
William Keel  *University of Alabama*
Douglas M. Kelly  *University of Arizona*
Marvin D. Kemple  *Indiana University/Purdue University/Indianapolis*
Yong H. Kim  *Saddleback College*
Robbie F. Kouri  *Our Lady of the Lake University*
Joan P. S. Kowalski  *George Mason University*
Robert Kren  *University of Michigan*
David Kriegler  *University of Nebraska at Omaha*
Jeffrey Kuhn,  *Michigan State University*
Claud Lacy,  *University of Arkansas*
Kristine Larsen  *Central Connecticut State University*
John K. Lawrence  *California State University, Northridge*
W. R. Luebke  *Modesto Junior College*
Jonathan I. Lunine,  *University of Arizona*
James Mackey,  *Harding University*
Carolyn Mallory  *Moorpark College*
Norman Markworth  *Stephen F. Austin State University*
Nancy D. McDonald  *Palm Beach Community College*
Rahul Mehta  *University of Central Arkansas*
Chris Mihos  *Case Western Reserve University*
Milan Mijic  *California State University, Los Angeles*
Jordi Miralda-Escude  *Ohio State University*
Gerald Newsom  *Ohio State University*
Paul Nienaber  *Marquette University*
Andrew P. Odell  *Northern Arizona University*
John Oliver  *University of Florida*
Melvyn Oremland  *Pace University*
Peter O'Shull Jr.  *Oklahoma State University*
Robert Page  *University of Maine at Augusta*
Brian M. Patten  *Harvard-Smithsonian Center for Astrophysics*
Jon Pedicino  *College of the Redwoods*
James N. Pierce  *Minnesota State University at Mankato*
Mike Reed  *Southwest Missouri State University*
Richard Rees  *Westfield State College*
David D. Reid  *Eastern Michigan University*
Carl Rosenzweig  *Syracuse University*
Dwight Russell  *Baylor University*
John L. Safko Sr.  *University of South Carolina*
Edward Schaub  *Baylor University*
James Schombert  *University of Oregon*
William Seeley  *Massachusetts College of Liberal Arts*
Larry Sessions  *Metropolitan State College of Denver*
J. Scott Shaw  *University of Georgia*
Edward M. Sion  *Villanova University*

Alex G. Smith  *University of Florida*
George F. Spagna Jr.  *Randolph-Macon College*
Norman Sperling  *Chabot Observatory and Science Center*
Roger Stanley Jr.  *San Antonio College*
Michael Stewart  *San Antonia College*
Jack W. Sulentic  *University of Alabama*
Donald Terndrup  *Ohio State University*
Walter Wesley  *Moorehead State University*
Dan Wilkins  *University of Nebraska*
Richard Williamon  *Emory University*
Suzanne Willis  *Northern Illinois University*
George Wolf  *Southwest Missouri State University*
W. C. Woods  *Glassboro State College*
Jon K. Wooley  *Eastern Michigan University*
John Wayne Wooten  *Pensacola Junior College*

## Acknowledgments for Interactives

McGraw-Hill would like to thank Adam Frank, professor at the University of Rochester and president of Truth-N-Beauty, as well as the other employees of Truth-N-Beauty, especially Ted Pawlicki and Carol Latta.

# The Cosmic Landscape

"The Ancient of Days" takes the measure of the Universe in an etching by William Blake (c. 1794).

## Concepts and Skills to Remember

- The Universe shows structure from the very small to the very large.
- The Earth (our planet) orbits the Sun along with seven sister planets.
- The Sun (our star) orbits within the Milky Way along with some 100 billion other stars.
- The Milky Way (our galaxy) orbits with a few dozen other galaxies in the Local Group of galaxies.
- The Universe is made up of energy and matter.
- The ordinary matter we see around us is composed of atoms.
- The structure that we see in the Universe—from the very large to the very small—results from forces. Gravity is the force that gives structure to most astronomical objects and systems.
- To measure the immense sizes of astronomical systems, astronomers use special units—for example, the astronomical unit (based on the Earth–Sun distance) and the light-year (the distance light travels in one year).
- Scientific notation offers a concise way to express the very large numbers encountered in describing astronomical objects.

**?What Is This?**
(See chapter summary for the answer.)

Astronomy is the study of the heavens, the realm extending from beyond the Earth's atmosphere to the most distant reaches of the Universe. Within this vast space we discover an amazing diversity of planets, stars, and galaxies. That creatures as tiny as ourselves can not only contemplate but also understand such diversity and immensity is amazing. But even more amazing are the objects themselves: planets with dead volcanos whose summits dwarf Mount Everest, stars a hundred times the diameter of the Sun, and galaxies—slowly whirling clouds of stars—so vast that they make the Earth seem like a grain of sand in comparison. All this is the cosmic landscape in which we live, a landscape we will briefly explore now to familiarize ourselves with its features and to gain an appreciation of its vast scale.

## P.1    THE EARTH, OUR HOME

We begin with the Earth, our home **planet** (fig. P.1). This spinning sphere of rock and iron circling the Sun is huge by human standards, but it is one of the smallest bodies in the cosmic landscape. Nevertheless, it is an appropriate place to start because, as the base from which we view the Universe, it influences what we can see. We cannot travel from object to object in our quest to understand the Universe. Instead, we are like children who know their neighborhood well but for whom the larger world is still a mystery, known only from books and television.

**FIGURE P.1**
The planet Earth, our home, with blue oceans, white clouds, and multihued continents.

But just as children use knowledge of their neighborhood to build their image of the world, so astronomers use their knowledge of Earth as a guide to more exotic worlds. For example, we can deduce from the glowing lava of an erupting volcano and the boiling water shooting from a geyser that the interior of our planet is extremely hot. That heat creates motion inside the Earth, much the way heat makes soup in a pot bubble and churn. Although the motions inside Earth are far slower than those we see in bubbling soup, over millions of years they buckle the seemingly firm rock of our planet's crust to heave up mountains and volcanoes. Deeper inside Earth, similar motions generate magnetic forces that extend through the surface and into space. On Earth's surface these forces tug on the needle of a compass so that it points approximately north–south. High in our atmosphere, these same magnetic forces shape the northern lights.

Looking outward to our planetary neighbors, we find landscapes on Venus and Mars that bear evidence of many of the same processes that sculpt our planet and create its diversity. Likewise, when we look at the atmospheres of other planets, we see many of the same features that occur in our atmosphere. For example, winds in the thin envelope of gas that shelters us swirl around our planet much as similar winds sweep the alien landscapes of Venus and Mars.

## P.2 THE MOON

The Moon is our nearest neighbor in space, a **satellite** that orbits the Earth some quarter million miles (384,000 km) away. Held in tow by the Earth's gravity, the Moon is much smaller than Earth (only about one-quarter our planet's diameter).

With the naked eye, and certainly with a pair of binoculars, we can clearly see that its surface is totally unlike Earth's. Instead of white whirling clouds, green-covered hills, and blue oceans, we see an airless, pitted ball of rock that shows us the same face night after night (fig. P.2).

Why are the Earth and Moon so different? Their different surfaces arise in large part from their great difference in mass. The Moon's mass is only about 1/80 the Earth's, and it was therefore unable to retain an atmosphere. Because of its smaller bulk it was also less able to retain heat. Without that strong internal heat, no crustal motions—so important in shaping Earth—have altered the Moon's surface. In fact, the Moon has changed so little over the past 3 billion years that its surface may hold important clues to what planets may have been like billions of years ago. In addition to this scientific importance, the Moon has symbolic significance for us—it is the farthest place from Earth that humans have explored.

FIGURE P.2
The Moon, as we see it with unaided eyes.

# P.3 THE PLANETS

Beyond the Moon, circling the Sun as the Earth does, are seven other planets, sister bodies of Earth. In the order of their average distance from the Sun, working outward, the eight planets are Mercury, Venus, Earth, Mars, Jupiter, Saturn, Uranus, and Neptune. These worlds have dramatically different sizes and landscapes. For example:

- Ancient craters blasted out by asteroid impacts scar the airless surface of Mercury.
- Dense clouds of sulfuric acid droplets completely shroud Venus.
- White clouds, blue oceans, green jungles, and red deserts tint Earth.
- Huge canyons and deserts spread across the ruddy face of Mars, where possibly, long ago, lakes and even oceans may have reflected a pink sky.
- Immense atmospheric storms with lightning sweep across Jupiter.
- Trillions of icy fragments orbit Saturn, forming its bright rings.
- Dark rings girdle Uranus, its spin tipped by some cosmic catastrophe.
- Choking methane clouds whirl in the deep blue atmosphere of Neptune.

Mercury, Venus, Mars, Jupiter, and Saturn are visible to the naked eye as bright points of light, much like stars. But whereas stars do not noticeably change their positions relative to one another, the planets, because of their orbital motion around the Sun, drift slowly and regularly against the pattern of the background stars. This regular motion gave the planets a special significance to people in ancient times—a significance that

**FIGURE P.3**
Portraits of the eight planets along with silhouettes show their correct relative size. Also shown are three dwarf planets.

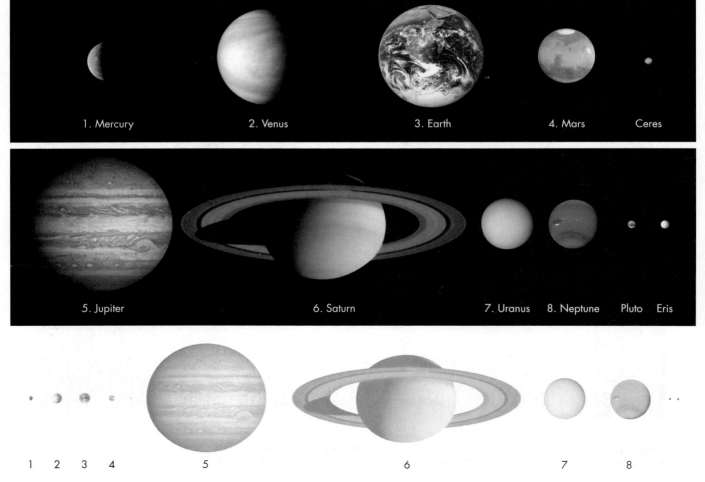

1. Mercury  2. Venus  3. Earth  4. Mars  Ceres

5. Jupiter  6. Saturn  7. Uranus  8. Neptune  Pluto  Eris

1  2  3  4  5  6  7  8

has been carried forward to today in the names of many of the days of the week. For example, in English, Saturday gets its name from Saturn, while in Spanish, miércoles gets its name from Mercury.

The planets are significant to us today because with modern telescopes and spacecraft we can see that they are truly other worlds. Figure P.3 shows pictures of these eight distinctive bodies and reveals something of their relative size and appearance. Some are far smaller and others vastly larger than Earth, but all are dwarfed by the Sun, whose immense gravity holds them in orbit.

## P.4    THE SUN

The Sun is a **star,** a huge ball of gas more than 100 times the diameter of the Earth and more than 300,000 times more massive: if the Sun were a volleyball, the Earth would be about the size of a pinhead, and Jupiter roughly the size of a nickel (fig. P.4). The Sun contains about 1000 times more matter than all of the Solar System's planets combined. The Sun, of course, differs from the planets in more than just size: it generates energy in its core by nuclear reactions that convert hydrogen into helium. From the core, the energy flows to the Sun's surface, and from there it pours into space to illuminate and warm the planets.

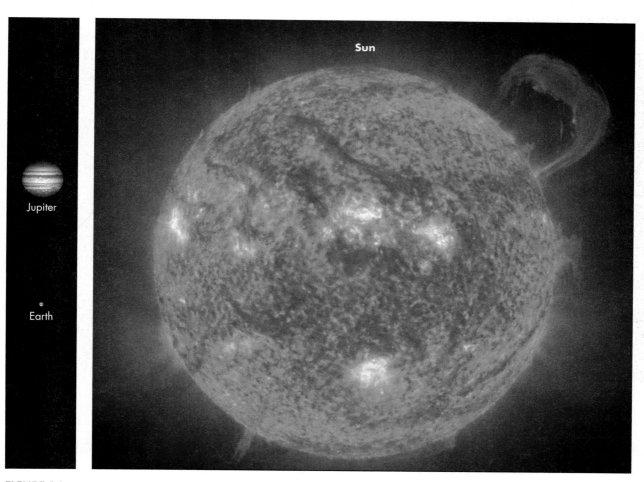

Jupiter

Earth

Sun

**FIGURE P.4**
The Sun as viewed through a filter that allows its hot outer gases to be seen. The Earth and Jupiter are shown to scale beside it for comparison. (The filter allows astronomers to see extremely hot helium gas.)

## P.5    THE SOLAR SYSTEM

The Sun and the eight planets orbiting it form the **Solar System.** But other objects orbit the Sun as well. These include the recently defined dwarf planets,* such as Pluto and Eris, as well as even smaller objects such as the asteroids and comets. Most asteroids orbit between Mars and Jupiter in the so-called asteroid belt, home also to the dwarf planet Ceres. Comets orbit mainly in the outermost fringes of the Solar System, where several other dwarf planets are found too.

## P.6    A SENSE OF SCALE

If the paths that the planets follow around the Sun were visible, we would see that the Solar System is like a huge set of nested, nearly circular rings, centered approximately on the Sun and extending about 4 billion miles outward to the dwarf planet Pluto's orbit (fig. P.5). It is hard to imagine such immense distances measured in miles. In fact, it is as foolish to use miles to measure the size of the Solar System as it is to use inches to measure the distance between New York and Tokyo. Whenever possible, we try to use units appropriate to the scale of what we seek to measure. For example, in earlier times people used units that were quite literally at hand, such as finger widths or the spread of a hand to measure a piece of cloth and paces to measure the size of a field. In the same tradition, although on a different scale, astronomers use distance scales related to familiar objects, such as the Earth. As we shall see in later chapters, the Earth's radius is a convenient unit for measuring the size of the other planets. Likewise, the Earth's distance from the Sun is a good unit for measuring the scale of the Solar System.

**FIGURE P.5**
Sketch of the orbits of the planets in our Solar System. To show the orbits to scale, the inner and outer Solar System are shown separately.

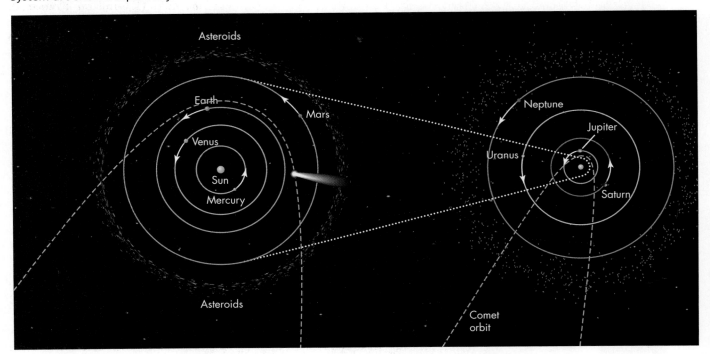

---

*In mid-2006, the International Astronomical Union voted to define a planet in such a way as to make Pluto no longer a full-fledged planet. By the new definition (described in more detail in chapter 7), Pluto and several other objects in our Solar System are now officially known as "dwarf planets."

**FIGURE P.6**

(A) This view of the Solar System is based on a series of real images made by the *Voyager I* spacecraft. The craft was about 40 AU from the Sun and about 20 AU above Neptune's orbit. The images of the planets (mere dots because of their immense distance) and the Sun have been made a little bigger and brighter in this view to allow you to see them more clearly. Mercury is lost in the Sun's glare and Mars happened to lie nearly in front of the Sun at the time the image was made, so it too is invisible. (B) A sketch of the orbits of the planets, showing approximately where each object was located at the time the image was made (February 1990). Notice the flatness of the system and the immense, empty spaces between the planets.

A

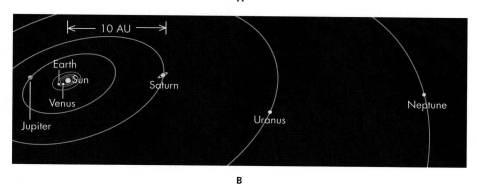

B

## P.7  THE ASTRONOMICAL UNIT

The **astronomical unit,** abbreviated as *AU,* is defined by the distance from the Earth to the Sun.* This translates into about 93 million miles (150 million kilometers). If we use the AU to measure the scale of the Solar System, Mercury turns out to be 0.4 AU from the Sun, while the dwarf planet Pluto is about 40 AU (fig. P.5). Figure P.6 shows a picture of the Solar System made by the spacecraft *Voyager I.* Notice how *empty* space is.

The Solar System extends far beyond the planets. For example, small icy objects, such as comets, drift along immense orbits that stretch 100,000 AU or so away from the Sun. The *Voyager* spacecraft has penetrated a small way into this remote region. The Solar System still remains the limit to our direct exploration of the Universe, but telescopes extend our view far beyond the Solar System to reveal that just as the Earth is but one of many planets orbiting the Sun, so too the Sun is but one of a vast swarm of stars orbiting the center of our galaxy, the Milky Way.

## P.8  THE MILKY WAY GALAXY

The **Milky Way galaxy** is a cloud of several hundred billion stars with a flattened shape like the Solar System (fig. P.7). The Sun and other stars orbit the Milky Way at some 140 miles per second (220 kilometers per second), but so vast is our galaxy that it still takes the Sun about 240 million years to complete one trip around this immense disk. The Milky Way's myriad stars come in many varieties, some hundreds of times larger than the Sun, others hundreds of times smaller. Some stars are much hotter than the Sun and shine a dazzling blue-white, while others are cooler and glow a deep red.

---

*Because the Earth's orbit is actually an ellipse, the AU is technically defined slightly differently, a point we will discuss further when we consider planetary orbits.

**FIGURE P.7**
The Milky Way galaxy. The top picture is a side view made by plotting stars in the 2MASS star catalog. The bottom picture is an artist's depiction of how our galaxy might look to an observer seeing it from above.

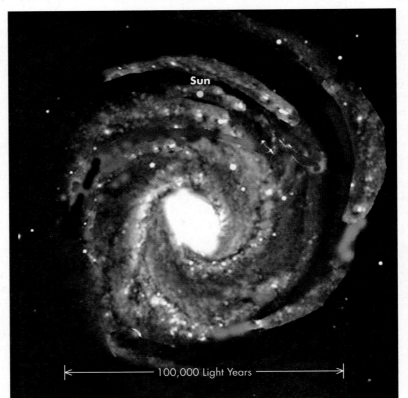

**FIGURE P.8 (Below)**
Photographs of interstellar clouds in the Milky Way taken with an Earthbound telescope. On the scale of these pictures, the Solar System out to Pluto is about 1000 times smaller than the period ending this sentence. (A) A cold, dark cloud and a star cluster beside it. Dust in the cloud blocks our view of the stars behind it. (B) A group of clouds heated by young stars. Glowing hydrogen in the clouds creates their red color.

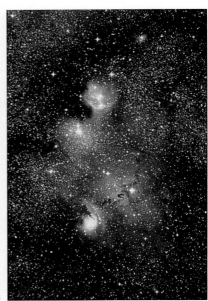

A

B

In the Milky Way, as in many other galaxies, stars intermingle with immense clouds of gas and dust. These clouds, enormously larger than the Solar System, are the sites of stellar birth and death. Deep within their cold, dark gas, gravity draws their matter into dense clumps that eventually turn into new stars, lighting the gas and dust around them. Stars eventually burn themselves out and explode, spraying matter outward to mix with the surrounding clouds. This matter from exploded stars is ultimately recycled into new stars (fig. P.8).

In this huge swarm of stars and clouds, the Solar System is all but lost—like a single grain of sand on a vast beach—forcing us again to grapple with the problem of scale. Stars are almost unimaginably remote: the nearest one to the Sun is over 25 trillion miles away. Such distances are so immense that analogy is often the only way to grasp them. For example, if we think of the Sun as a pinhead, the nearest star would be another pinhead about 35 miles away and the space between them would be nearly empty. In fact, distances between stars are so immense that even astronomical units are inappropriately small, and so we again choose a new unit of length—the light-year.

## P.9    THE LIGHT-YEAR

Measuring a distance in terms of a time may at first sound peculiar, but we do it often. We may say, for example, that our town is a 2-hour drive from the city, or our dorm is a 5-minute walk from the library, but expressing a distance in this fashion implies that we have a standard speed.

Astronomers are fortunate to have a superb speed standard: the speed of light in empty space, which is a constant of nature and equal to 299,792,458 meters per second (about 186,000 miles per second). Moving at this constant and universal speed, light in 1 year travels a distance defined to be 1 **light-year,** abbreviated as *ly.* As we will show below, this works out to be about 6 trillion miles (10 trillion kilometers). To demonstrate that, however, is cumbersome because it involves multiplying a series of large numbers. We therefore will use a more concise way to write them called scientific notation.

In **scientific notation** (also sometimes called powers-of-ten notation), we write numbers using ten to an exponent, or power. Thus we write $100 = 10 \times 10 = 10^2$ and 1 million (1,000,000) as $10 \times 10 \times 10 \times 10 \times 10 \times 10 = 10^6$. Instead of writing out all the zeros, therefore, we use the exponent to tell us the number of zeros. A number like the speed of light (186,000 miles per second) may also be written in scientific notation, becoming $1.86 \times 10^5$ miles per second. Likewise, the astronomical unit (150 million kilometers) can be written as $1.5 \times 10^8$ km.

One reason to use scientific notation is that multiplying and dividing becomes enormously easier. For example, to multiply two powers of ten we just add the exponents, and to divide we subtract them. Thus $10^2 \times 10^5 = 10^7$, and $10^8/10^3 = 10^5$. More details on using scientific notation are given in the appendix, but as an illustration of its usefulness, let us now calculate the number of miles in a light-year.

To find how far light travels in a year, we multiply its speed by the travel time. One year is about 31,600,000 (about $3.16 \times 10^7$) seconds.* Multiplying this by the speed of light we get $3.16 \times 10^7$ seconds $\times\ 1.86 \times 10^5$ miles/second $= 3.16 \times 1.86 \times 10^{12}$ seconds $\times$ miles/second $= 5.88 \times 10^{12}$ miles, or about 6 trillion miles ($10^{13}$ kilometers), as we previously claimed. In these units, the star nearest the Sun is 4.2 light-years away. Although we achieve a major convenience in adopting such a huge distance for our scale unit, we should not lose sight of how truly immense such distances are. For example, if we were to count off the miles in a light-year, one every second, it would take us about 186,000 years!

---

*Numbers here and elsewhere are rounded off.

We can now use the light-year for setting the scale of the Milky Way galaxy. In light-years, our galaxy is about 100,000 light-years across, with the Sun orbiting roughly 30,000 light-years from the center. Within the Milky Way's disk, stars are separated by a few light-years. The gas clouds, such as those in figure P.8, which we mentioned earlier, range in size from a fraction of a light-year to hundreds of light-years.

## P.10    GALAXY CLUSTERS AND THE UNIVERSE

Having gained some sense of scale for the Solar System and the Milky Way, we now resume our exploration of the cosmic landscape, pushing out to the realm of other galaxies. Here we find that just as stars assemble into galaxies, so galaxies themselves assemble into **galaxy clusters.**

The cluster of galaxies to which the Milky Way belongs is called the **Local Group.** It is "local," of course, because it is the one we inhabit. It is termed a "group" because it is small as galaxy clusters go, containing only about 40 galaxies as members, but it is still about 5 million light-years in diameter. Yet despite such vast dimensions, the Local Group is itself part of a still larger assemblage of galaxies known as the **Local Supercluster.** Figure P.9 puts this in perspective for us.

Our supercluster consists of a few dozen galaxy groups and clusters, spread over some 100 million light-years, but it is perhaps itself part of an even larger structure known as the Great Attractor, a cluster of superclusters, probably more than 300 million light-years across. Structures of such vast size are about the largest objects we can see before we take the final jump in scale to the **Universe** itself.

The visible Universe is the largest astronomical structure of which we have any knowledge. From the observations presently available to them, astronomers have deduced that the Universe is about 14 billion years old. This limits the distance we can see, even in principle, to about 14 billion light-years, a number often taken as the approximate radius of the *observable* Universe. However, that does not mean the Universe ends at such a distance. Rather, it means we cannot see what lies beyond. But regardless of our uncertainty about the known Universe's size, we can observe that its structure is similar throughout the visible Universe. Small objects are clustered into larger systems, which are themselves clustered: planets around stars, stars in galaxies,

**FIGURE P.9**

(A) A sketch of the central region of the Local Group.* Galaxies are shown at about twice their correct size relative to their separation. Only a few of the numerous dwarf members are shown. Outlying members are scattered throughout a region about twice the size of the region shown here. Distances are uncertain for many of these galaxies, even relatively nearby ones. (B) A sketch of the Local Supercluster. Only the larger groups of galaxies are labeled. Clusters are shown at approximately correct size relative to their separation, but here, too, sizes and distances are very uncertain.

A

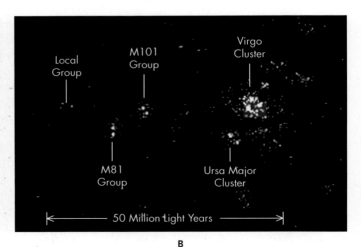

B

*The names of the galaxies M31 and M33 in the Local Group stand for Messier 31 and 33. They get their names from a list of galaxies and other astronomical objects that was compiled in the late 1700s by French astronomer Charles Messier ("Mess-yay"). We will encounter many other Messier objects later in our exploration of the Universe.

galaxies in clusters, clusters in superclusters, and superclusters into even larger groups. Although astronomers do not yet understand completely how this orderly structure originated, they do know that gravity plays a crucial role.

## P.11   GRAVITY

**Gravity** gives the Universe structure because it creates a force of attraction between *all* objects. You experience gravity's attraction in everyday life. For example, if you drop a book, the Earth's gravitational force makes the book fall. Moreover, that same force spans the vast distance between the Earth and the Moon to hold our satellite in its orbit. Similarly, gravity holds our planet in its orbit around the Sun and the Sun in its orbit around the Milky Way.

Although gravity dominates the large-scale structure of the Universe, other forces dominate on smaller scales. To understand these forces, we need to look briefly at the small-scale structure of matter.

## P.12   ATOMS AND OTHER FORCES

Matter is composed of submicroscopic particles called **atoms.** Atoms are incredibly small. For example, a hydrogen atom is about one ten-billionth of a meter ($10^{-10}$ m) in diameter. Ten million hydrogen atoms could be put in a line across the diameter of the period at the end of this sentence. But despite this tiny size, atoms themselves have structure. Every atom has a central core called the **nucleus** that is orbited by smaller particles called **electrons** (fig. P.10). The nucleus is in turn composed of two other kinds of particles called **protons** and **neutrons.**

Although the particles in an atom exert a gravitational attraction on one another, atoms are not held together by gravity. Instead, an electric force gives them their structure. That force arises because protons and electrons have a property called **electric charge.** A proton has a positive electric charge, and an electron has a negative electric charge. A neutron, as its name suggests, has no charge.

The **electric force** can either attract or repel, depending on the charges. Opposite charges attract, and like charges repel. Thus, two electrons (both negative) repel each other, while an electron and a proton (negative and positive, respectively) attract each other. That attraction is what holds the electrons in their orbits around the nucleus of an atom.

You can see the electric force at work in many ways. For example, the static electric charges generated when a clothes drier tumbles your laundry creates an attraction that

**FIGURE P.10**
An atom consists of a nucleus around which electrons orbit. The nucleus is itself composed of particles called protons and neutrons. (A) A classical view of an atom. (B) A more modern view with the electrons orbiting as "clouds."

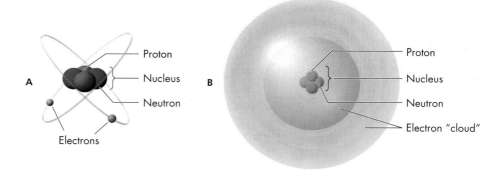

may make clothes cling together. The crackling sound you hear as you pull fuzzy socks away from a shirt is the electric charges jumping and making tiny sparks.

The electric force is closely linked with the magnetic force that makes a compass work or holds the little magnets to the door of your refrigerator. In fact, the theory of relativity demonstrates that electric and magnetic forces are fundamentally the same, and scientists generally refer to them jointly as the electromagnetic force.

Yet another force plays a critical role in atoms. This force, called the **strong force,** binds the protons and neutrons together to form the atom's nucleus. Although the effects of the strong force cannot be seen directly in everyday life, without it the nuclei of atoms would disintegrate and, with them, our familiar world.

In addition, a fourth force, known as the **weak force,** operates on the atomic scale and plays a role in radioactive decay.*

## P.13    THE STILL-UNKNOWN UNIVERSE

Our quick trip from Earth outward has shown us a Universe of planets, stars, and galaxies. But astronomers today have evidence that, although these objects are the ones we notice, the bulk of the Universe must consist of something completely different. That evidence comes from many sources, the most convincing of which are that (1) stars within galaxies, and (2) galaxies within clusters of galaxies, experience a far stronger gravitational force than can be explained by the directly observable matter. That is, both galaxies and galaxy clusters appear to contain huge amounts of what astronomers call **dark matter.**

Dark matter is so named because it emits no as-yet-observed radiation. But from its gravitational effects, astronomers deduce that it outweighs luminous matter by about a factor of ten to one. What is the dark matter? Astronomers do not know, but that is only part of the mystery. When astronomers study the expansion of the Universe, they find evidence for an as yet unidentified kind of energy. Because this energy is, like dark matter, of an unknown nature, astronomers call it dark energy. If we compare the mass of the dark matter and the dark energy with the mass that we directly detect (such as stars, galaxies, and gas clouds), those luminous objects amount to a mere 1% of the Universe's total mass. What we see of the Universe is therefore much like the footprints of an invisible man: a being who leaves tracks, but whose build and nature we do not yet know.

## P.14    THE SCIENTIFIC METHOD

Our scientific understanding of the Universe has not come easily. It has grown out of the work of thousands of men and women over thousands of years. Their work is part of the broad field that we call science.

By "science" we mean the systematic study of things and the search for the underlying principles that govern them, be they living things, matter, or, in our case, the astronomical universe. An essential part of that study is the rigorous testing of ideas. We call the process of such testing the **scientific method.** In using the scientific method, a scientist typically proposes an idea—a hypothesis—about some property of the Universe and then tests that hypothesis by experiment. In fact, whether an idea is "scientific" depends to some extent on whether it can be verified by either a real or an imagined experiment. Ideally the experiment either confirms the hypothesis or refutes it.

---

*The weak force is linked to the electromagnetic force, and their combination is known technically as the electroweak force.

If refuted, the hypothesis is rejected. On the other hand, if the experiment confirms the hypothesis, the scientist may then go on to develop related hypotheses or perhaps to make predictions about some as-yet-undiscovered aspect of the subject.

Once a set of ideas has been thoroughly tested and verified, the ideas may be incorporated into a theory. When we use the term *theory* here, we do *not* mean that the ideas are unproven or tentative. Rather, we mean that they have achieved wide acceptance by successful testing. For example, scientists have subjected the quantum theory of atomic structure and the theory of relativity to numerous tests, and these theories have passed all such tests with high precision.

Astronomers face a special difficulty in applying the scientific method because usually they cannot experiment with their subject matter directly: in virtually all cases, they can only passively observe. Nevertheless, they try—like all scientists—to use the scientific method. You will find some specific examples of this method in later chapters, where *Science at Work* boxes show how this process has led to new ideas and revision of old ones.

Application of the scientific method is no guarantee that its results will be believed. For example, we will see in chapter 1 that even before 300 B.C., the Greek philosopher Aristotle taught that the Earth is a sphere. Yet despite the proofs he offered to support that hypothesis, many people continued to believe the Earth to be flat. Today, too, some scientific hypotheses might be rejected despite their experimental verification, and others might be accepted though untrue. For example, one astronomer might find evidence supporting some hypothesis, but another astronomer might claim that the experiment was done incorrectly or the data were analyzed improperly.* Therefore, throughout this book, whenever we discuss our knowledge of a given topic, we must keep in mind the fact that such knowledge is not always proved or even universally accepted. This is especially true of topics at the frontiers of our understanding, such as the origin and structure of the Universe or the properties of black holes. Therefore, keep in mind that some of what we discuss in this book will be proved wrong in the future. That is not a failing of science, however. It is its strength.

# SUMMARY

The Earth is one of eight planets orbiting the Sun, and the Sun is one of about a hundred billion stars that make up the Milky Way galaxy. The Milky Way and about 40 other galaxies compose the Local Group, which in turn is part of the Local Supercluster of galaxies. Superclusters seem to be grouped into even larger systems that fill the visible Universe (fig. P.11). We can speak with some certainty about the size and properties of objects in our immediate neighborhood, but the farther we move from Earth, the less certain we become.

Astronomers use the astronomical unit (AU) and light-year (ly) to measure the immense sizes and distances of astronomical systems. The AU is defined by the average distance between the Earth and the Sun, and the light-year is defined as the distance light travels in a year, which is about 6 trillion miles. Using these units, we can see the immense scale of the Universe in table P.1 and figure P.11. The latter shows a series of drawings to help you visualize how enormous the Universe is. To further help you visualize the immensity of the Universe also look at the foldout (following page 18).

Matter is made up of atoms in which charged particles called electrons orbit a nucleus. The nucleus is itself composed of smaller particles called protons and neutrons.

Four forces give the Universe its structure: gravity on the cosmic scale and the electromagnetic, strong, and weak forces on the scale of atoms.

## ?What It Is.

These are galaxies, but *very* distant ones. This image was made by the Hubble Space Telescope. It spans a piece of sky that you could cover with a grain of sand held at arms length. The light from these galaxies left them long before the Solar System formed and has been traveling for more than 10 billion years to reach us.

---

*In fact, some historians of science have argued that old ideas are discarded not so much because they are proved wrong, but rather because their supporters have died.

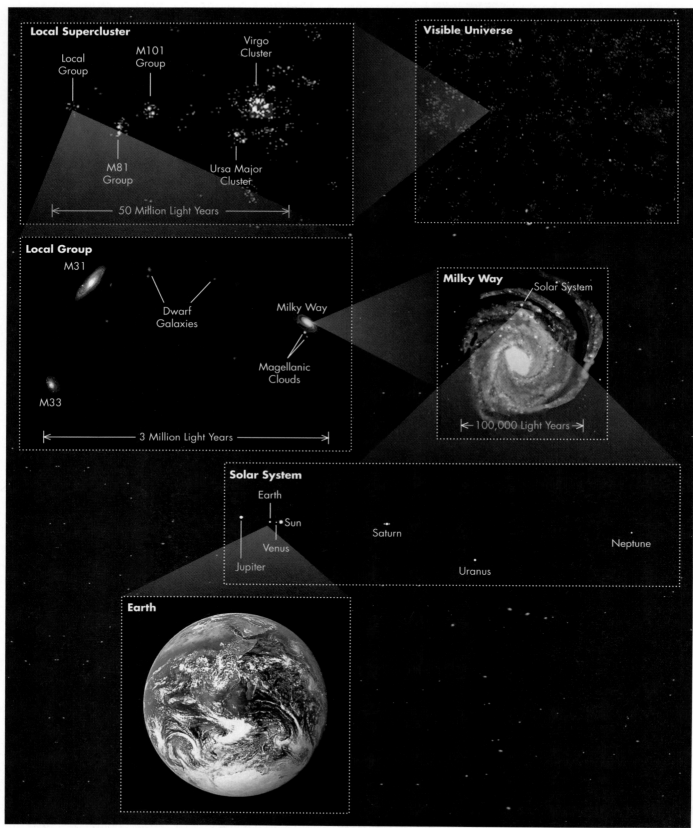

**FIGURE P.11**
How the Earth fits into the Universe.

### TABLE P.1

**The Scale of the Universe**

| Object | Approximate Radius |
|---|---|
| Earth | 6400 km<br>(about 4000 miles) |
| Sun | 700,000 km<br>(about 100 times radius of the Earth) |
| Earth's orbit | 150 million km<br>(about 200 times radius of Sun) = 1 AU |
| Solar System out to the dwarf Planet Pluto | 40 AU<br>(about 8600 times radius of the Sun) |
| Milky Way galaxy | 50,000 ly<br>(about $10^8$ times radius of the Solar System) |
| Local Group | 2.5 million ly<br>(about 50 times radius of the Milky Way) |
| Local Supercluster | 50 million ly<br>(about 20 times radius of the Local Group) |
| Visible Universe | 14 billion ly<br>(about 300 times radius of the Local Supercluster) |

## QUESTIONS FOR REVIEW

1. To what systems, in increasing order of size, does the Earth belong?

2. What force holds the different astronomical systems described in this section together?

3. About how much bigger in radius is the Sun than the Earth?

4. Roughly how big across is the Milky Way galaxy?

5. How big is an astronomical unit?

6. How is a light-year defined?

7. What particles make up an atom?

8. What force holds the electrons to an atom's nucleus?

9. What is meant by the scientific method?

## THOUGHT QUESTION

Propose a hypothesis about something you can experiment with in everyday life and try to verify or disprove the hypothesis. For example, what kind of surfaces will the little magnetic note holders people use on refrigerators stick to? Any smooth surface? Any metal surface?

## PROBLEMS

1. The radius of the Sun is $7 \times 10^5$ kilometers, and that of the Earth is about $6.4 \times 10^3$ kilometers. Use scientific notation to show that the Sun's radius is about 100 times the Earth's radius.

2. Given that an astronomical unit is $1.5 \times 10^8$ kilometers and a light-year is about $10^{13}$ kilometers, how many AU are in a light-year?

3. If the Earth were the size of a B-B (with a radius of about 0.2 centimeter), how big would the Sun be? How big would the Milky Way be?

4. Imagine building a model of the Solar System on your campus. Work out the diameter and spacings of the planets in millimeters and meters respectively.

5. If the Milky Way were the size of a nickel (about 2 centimeters), how big would the Local Group be? How big would the Local Supercluster be? How big would the visible Universe be? The data in table P.1 may help you here.

6. Using scientific notation, numerically evaluate $(4 \times 10^8)^3/(5 \times 10^{-6})^2$.

7. Calculate approximately how long it takes light to travel from the Sun to the dwarf planet Pluto.

8. A typical bacterium has a diameter of about $10^{-6}$ meters. A hydrogen atom has a diameter of about $10^{-10}$ meters. How many times smaller than a bacterium is a hydrogen atom?

9. Use scientific notation to numerically evaluate $(3 \times 10^4)^2/(4 \times 10^{-6})^{1/2}$.

10. Suppose two galaxies move away from each other at 6000 km/s and are 300 million ($3 \times 10^8$) light-years apart. If their speed has remained constant, how long has it taken them to move that far apart? Express your answer in years.

## TEST YOURSELF

1. The light-year is a unit of
   (a) time.
   (b) distance.
   (c) speed.
   (d) weight.
   (e) age.

2. You write your home address in the order of street, town, state, and so on. Suppose you were writing your cosmic address in a similar manner. Which of the following is the correct order?
   (a) Earth, Milky Way, Solar System, Local Group
   (b) Earth, Solar System, Local Group, Milky Way
   (c) Earth, Solar System, Milky Way, Local Group
   (d) Solar System, Earth, Local Group, Milky Way
   (e) Solar System, Local Group, Milky Way, Earth

3. Which of the following astronomical system is/are held together by gravity?
   (a) The Sun
   (b) The Solar System
   (c) The Milky Way
   (d) The Local Group
   (e) All of them are.

4. Judging from the lower part of figure P.3, about how much larger is Jupiter's diameter than the Earth's?
   (a) 2 times
   (b) 5 times
   (c) 10 times
   (d) 25 times
   (e) 100 times

5. Which of the following statements can be tested for correctness using the scientific method?
   (a) The Sun's diameter is about 100 times larger than the Earth's diameter.
   (b) The sky is sometimes blue.
   (c) An astronaut cannot survive on the Moon without life-support systems.
   (d) The Moon is an uglier place than the Earth.
   (e) All of the above except (d).

## FURTHER EXPLORATIONS

Boorstin, Daniel J. *The Discoverers.* New York: Random House, 1983.

Bronowski, Jacob. *A Sense of the Future.* Cambridge, Mass.: MIT Press, 1977.

Morrison, Philip, and Phylis Morrison. *Powers of Ten.* New York: W. H. Freeman, 1982.

Pine, Ronald C. *Science and the Human Prospect.* Belmont, Calif.: Wadsworth Publishing, 1989.

Sagan, Carl. *Cosmos.* New York: Random House, 1980.

**Website**

Please visit the *Explorations* website at **http://www.mhhe.com/arny** for additional online resources on these topics.

## VIDEO

*Powers of Ten.* Pyramid Film and Video, Santa Monica, Calif., 1989.

http://www.mhhe.com/arny

## KEY TERMS

| | | | | |
|---|---|---|---|---|
| astronomical unit, 7 | electron, 11 | Local Supercluster, 10 | proton, 11 | star, 5 |
| atom, 11 | galaxy cluster, 10 | Milky Way galaxy, 7 | satellite, 3 | strong force, 12 |
| dark matter, 12 | gravity, 11 | neutron, 11 | scientific method, 12 | Universe, 10 |
| electric charge, 11 | light-year, 9 | nucleus, 11 | scientific notation, 9 | weak force, 12 |
| electric force, 11 | Local Group, 10 | planet, 2 | Solar System, 6 | |

## PROJECTS

1. Get a long roll of paper and draw a scale model of the Earth, Sun, and Moon on it. If you make the Earth a blue dot about 1 millimeter (the size of a poppyseed, or about 1/32 inch) in diameter, the Moon should be about $1\frac{3}{16}$ inches away from it. The Sun should be a circle about 4 inches in diameter and about 38 feet away from it. Shelf paper will work fine, and art stores have very long rolls of paper for use on children's easels. A roll of paper towel would work too.

2. Make a scale model of the Solar System and the Local Group. Use the information given on the foldout and/or table P.1 to determine how far apart the various objects should be. To make the model, you will need to gather a variety of objects, such as a pin to represent the Earth for the Solar System; a poppy-seed; metal bottle cap (for the size of the Moon's orbit); a grapefruit, etc. For the Local Group you might use coins to represent the galaxies. Set up your model on an athletic field—perhaps the football field, if it is accessible. You might make a video of the project to share with students in a local elementary school or invite them to watch you set up the model.

## FOLD OUT ▶

The front of the following foldout shows scale drawings of the Earth, Moon, and Sun, and the orbits of the inner planets. The backside shows "models" that depict the scale of the entire Solar System (top) and the inner parts of the Local Group of galaxies (bottom). The purpose of the foldout is to help you appreciate the immensity of astronomical systems.

### FRONT—Earth, Moon, and Sun to Scale

The right-hand part of the figure shows the orbits of the inner planets (dashed circles) and the disk of the Sun to correct relative scale. Note that the diameter of the Earth's orbit is approximately 100 times the diameter of the Sun. The orbit of the Moon is approximately the thickness of the dashed line in this figure. The planets themselves on this scale would be invisible—far smaller than the thickness of the dashed lines. To have made the planets visible would have required making the foldout more than 10 feet long.

The left-hand part of the figure shows a "magnified" view (by about a factor of 250) of the Sun and the Earth–Moon system. These magnified views show the Earth, Moon, and Sun, and the size of the Moon's orbit, to correct relative size. Notice how the Sun dwarfs the Earth and Moon, and is even larger than the orbit of the Moon around the Earth.

### BACK—Models of the Solar System and Local Group of Galaxies

Here you see two "models." The top one depicts the Solar System and the bottom one depicts the Local Group of galaxies.

In the top model of the Solar System, the scale is set by imagining the Earth to be the size of a poppy seed on a football field. With the Earth shown at that size, the Sun is about the size of a grapefruit located about 10 yards from the poppy seed. Notice that the dwarf planet Pluto is positioned approximately 4 football fields away from the grapefruit.

The bottom model starts with the orbit of Pluto being the size of a pinhead. The nearest star would be about 10 yards away. (The Sun and planets in such a model would be far too small to see.) The Milky Way would be the size of a major metropolitan area (about 50 miles across). The nearest large galaxy to the Milky Way (M31) would lie more than 1000 miles away.

*Note:* The numbers quoted above have been rounded for simplicity.

## Scale Model of the Solar System

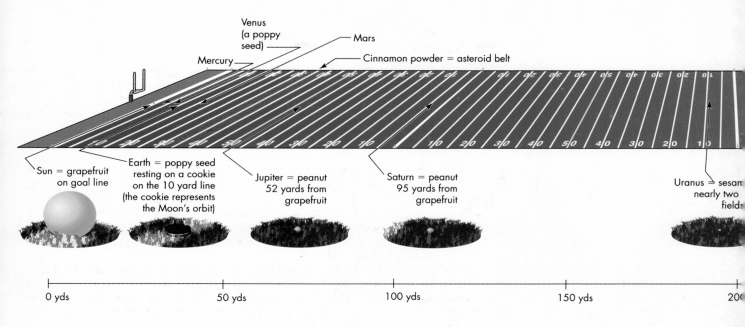

Venus (a poppy seed)

Mars

Mercury

Cinnamon powder = asteroid belt

Sun = grapefruit on goal line

Earth = poppy seed resting on a cookie on the 10 yard line (the cookie represents the Moon's orbit)

Jupiter = peanut 52 yards from grapefruit

Saturn = peanut 95 yards from grapefruit

Uranus = sesam... nearly two fields...

0 yds          50 yds          100 yds          150 yds          200

## Scale Model of the Local Group of Galaxies

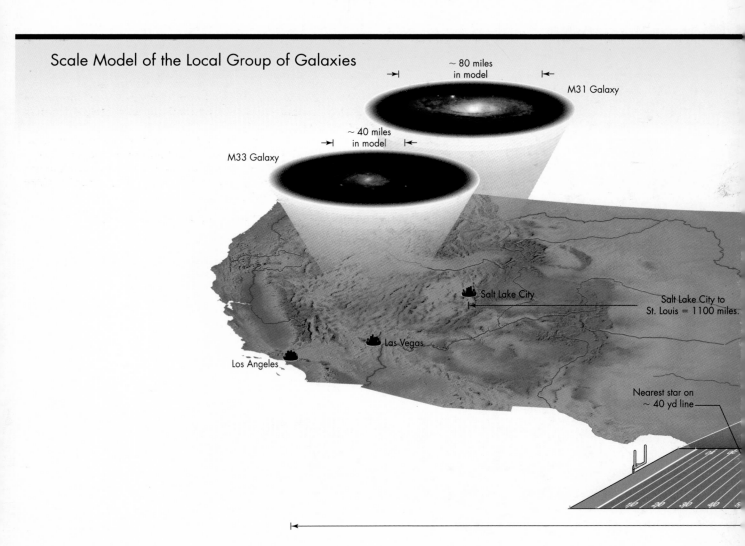

~ 80 miles in model

M31 Galaxy

~ 40 miles in model

M33 Galaxy

Salt Lake City

Salt Lake City to St. Louis = 1100 miles.

Las Vegas

Los Angeles

Nearest star on ~ 40 yd line

# Sun and orbits of three inner planets to correct relative size

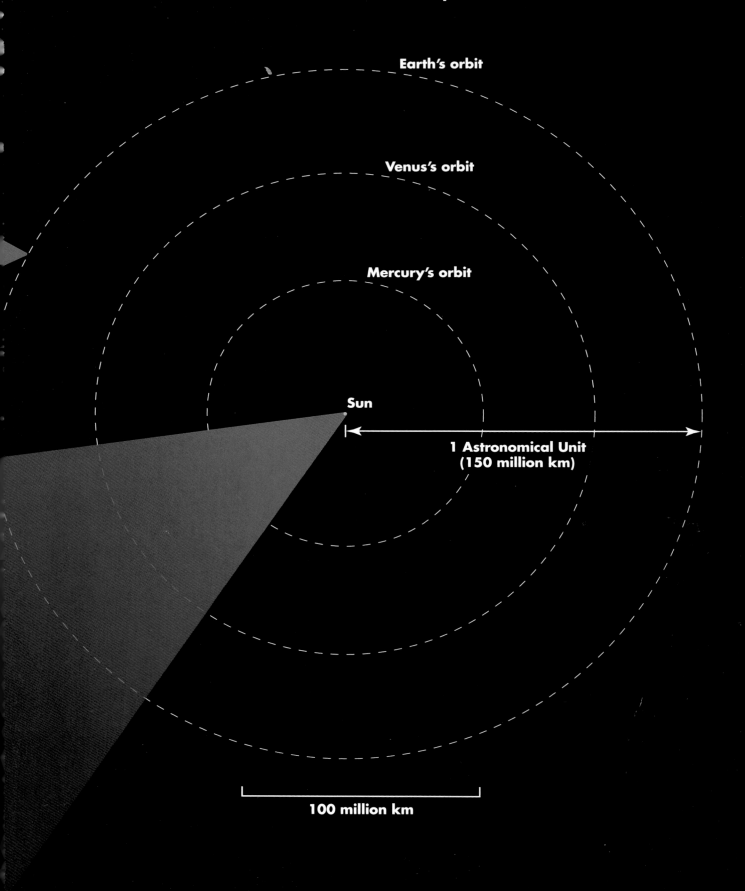

Earth's orbit

Venus's orbit

Mercury's orbit

Sun

1 Astronomical Unit
(150 million km)

100 million km

# Sun, Earth, Moon, and Moon's orbit to correct relative size

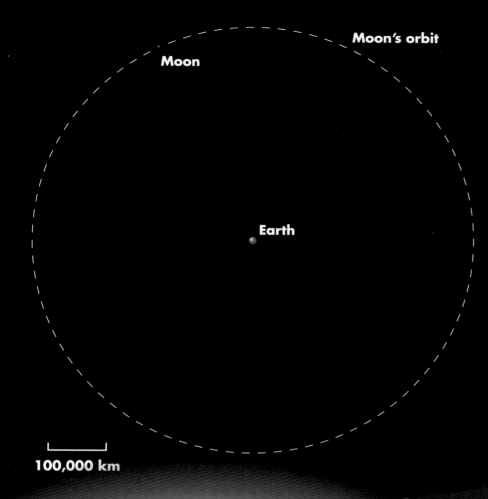

Moon

Moon's orbit

Earth

100,000 km

Sun

Sugar sprinkled 300 to 500 yards out = Kuiper belt

...seed
...otball
...way

Neptune = sesame seed
three football fields away

Pluto = a grain of salt
nearly four football
fields away

...yds          250 yds          300 yds          350 yds          400 yds

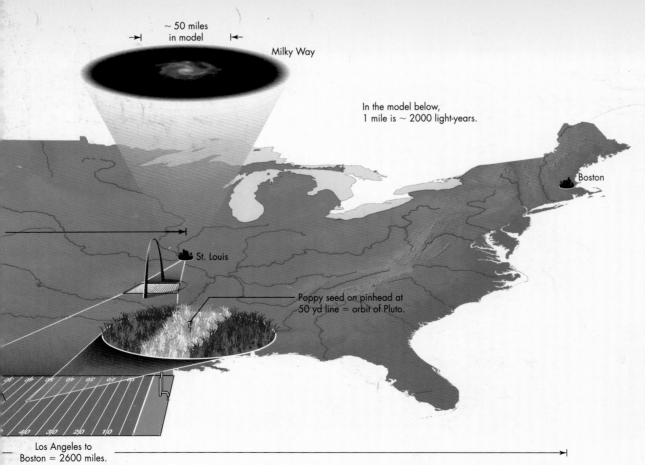

~ 50 miles
in model

Milky Way

In the model below,
1 mile is ~ 2000 light-years.

Boston

St. Louis

Poppy seed on pinhead at
50 yd line = orbit of Pluto.

Los Angeles to
Boston = 2600 miles.

# History of Astronomy

1

Stonehenge—Summer Sunrise

## Concepts and Things to Remember

- We see the Sun, Moon, and stars rise along the eastern horizon, move across the sky, and set along the western horizon.
- The stars form patterns (constellations) whose shapes do not change.
- We see different constellations at different times of the year—Orion in January, Scorpius in August, for example.
- The shape of the lit portion of the Moon changes during the month.
- Cycles of the Moon, Sun, and stars are the basis for keeping time and our calendar.
  - The day is based on the Earth's spin (rotation about its own axis).
  - The month is based on the Moon's orbital motion around the Earth.
  - The year is based on the Earth's orbital motion around the Sun.
- These motions also lead to seasons and eclipses.
- Astronomers of antiquity deduced the shape and size of the Earth.
- Starting around A.D. 1500, scientists such as Copernicus, Galileo, Kepler, and Newton deduced laws to describe and explain planetary motion.

**?What Is This?**
(See chapter summary for the answer.)

ur picture of the Universe has been assembled bit by bit from many separate discoveries—discoveries made by scientists from many parts of the world, at many times in the past, and in many disciplines. How those discoveries led to our current knowledge is the subject of this chapter. For convenience in our discussion, we will divide the history of Western astronomy into four main periods that can be characterized as follows:

1. Prehistoric (before 500 B.C.):* In the prehistoric era, people observed the daily and seasonal motions of the Sun, Moon, and stars and learned to use their cyclic motions for keeping time and determining direction.
2. Classical (500 B.C.–A.D. 1400): In the classical period, scientist-philosophers began to make measurements of the heavens and, with their knowledge of geometry, constructed idealized models that could account for the motion of heavenly bodies.
3. Renaissance (1400–1650): In the Renaissance period, those geometrical models were reassessed and found inadequate. Astronomers therefore devised new models that took into account a much greater body of data based on observational records accumulated over centuries. Astronomers also benefited from a technological advance that allowed them to observe even more—the telescope.
4. Modern (1650 to the present): Finally, in the modern period, scientists began the search for the physical laws (such as the law of gravity) that underlie the observed movements in the heavens. Other important contributions to our understanding of the Universe came from technological advances (for example, in optics, electronics, and computers) and better mathematical techniques (such as calculus). Such factors continue to be important today.

## 1.1    PREHISTORIC ASTRONOMY

We do not know when people of antiquity first began studying the heavens, but it was certainly many thousands of years ago. Astronomical observations are part of virtually every culture and include the obvious events that anyone who watches the sky can see without any equipment, such as the rising of the Sun in the eastern sky and its setting toward the west; the changing appearance of the Moon at different times of the month from new moon to crescent to full and back again; and eclipses. Other discoveries, such as the discovery that the planets are a class of heavenly bodies different from the stars, required more careful observation but were also probably made by prehistoric people.

For many prehistoric people, observations of the heavens had more than just curiosity value. Because so many astronomical phenomena are cyclic—that is, they repeat day after day and year after year—they can serve as time keepers. For example, when is it safe to set out on a sea voyage? When is it time to harvest crops? When will an eclipse occur? Moreover, the cyclic behavior of the heavens implies that many events seen in the sky are predictable. Thus, some of the motivation for studying the heavens probably came from the desire to foretell future events in the sky, and it may have led early cultures to build monumental stone structures such as Stonehenge (fig. 1.1), which dates to approximately 3000 B.C.

Ironically, many of the astronomical phenomena well known to ancient people are not nearly so familiar to people living today because the smog and bright lights of cities make it hard to see the sky and its rhythms. Perhaps more important, we no

---

*A few written records of astronomical observations made by astronomers in ancient China, India, Korea, and Japan predate 500 B.C.

North

To Summer Sunrise

A

B

**FIGURE 1.1**

(A) Stonehenge, a stone monument built by the ancient Britons on Salisbury Plain, England. Its orientation marks the seasonal rising and setting points of the Sun. (B) Sketch showing how the stones are aligned with respect to the solstice.

longer rely upon direct astronomical observations to tell us what season it is, when to plant, and so on. Therefore, if we are to appreciate the growth of astronomical ideas, we need to first understand what our distant ancestors knew and what we ourselves can learn by watching the sky over the course of a year.

One of nature's spectacles is the night sky seen from a clear, dark location with the stars scattered across the vault of the heavens. The night sky is a particularly appropriate place to begin our study of the history of astronomy because its array of stars has changed very little over the last several thousand years. The night sky affords us a direct link with what our remote ancestors viewed as they tried to understand the nature of the heavens. Thus, our first goal is to familiarize ourselves with some general aspects of the sky at night.

In the following discussion, you might want to imagine yourself as a shepherd in the Middle East, a hunter-gatherer on the African plains, a trader sailing along the coast of the Mediterranean, or even a ship's navigator in the early twentieth century. Whichever role you choose to assume, try to get out and actually look at the sky.

## The Celestial Sphere

The thousands of stars visible on a clear night are at vastly different distances from us. For example, the nearest star is about 4 light-years away, but others are more than a thousand times farther away. Such huge distances prevent us from getting any sense of their true three-dimensional arrangement in space. For purposes of naked-eye observations, we can therefore treat all stars as if they are at the same distance from the Earth, imagining that they lie on the inside of a gigantic dome that stretches overhead. Astronomers describe this dome as part of the **celestial sphere** and picture it completely surrounding the Earth, with the Earth at its center. Thus, if you were suspended in space far from Earth, you would see the entire celestial sphere surrounding you, as depicted in figure 1.2.* However, if you are on the Earth and look around, you will see that the ground blocks your view of approximately half the celestial sphere. The line where the sky meets the ground, and below which your view of the celestial sphere is blocked, is called the **horizon** (fig 1.2B).

---

*In figure 1.2 and many others throughout the book, distances and sizes of astronomical bodies are exaggerated for clarity.

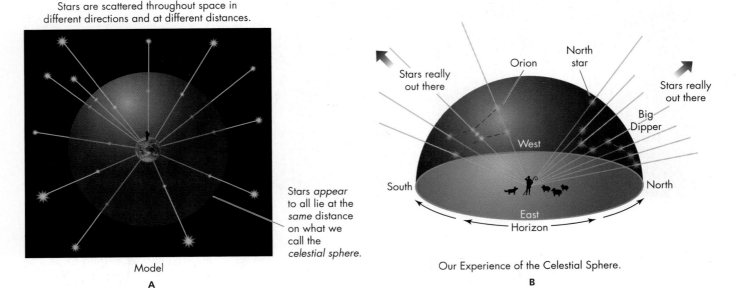

FIGURE 1.2

(A) Although in reality the stars are scattered through space at very different distances, they *appear* to lie at the same distance from us on what we call the celestial sphere. Note: Sizes and distances are drastically exaggerated. For Earth at the size shown, the *nearest* star would really be 6000 miles away. (B) The celestial sphere meets the ground at the horizon.

The celestial sphere has no physical reality, but it serves as a **model** of the heavens—an easy way to visualize the arrangement and motions of celestial bodies. We use the term *model* to mean a representation of some aspect of the Universe. That is, the celestial sphere represents a way of thinking about or viewing the location and motions of stars and planets. The celestial sphere is the first of many models we will encounter that humans have used to describe the known Universe. In later chapters, we will use models to enhance our understanding whenever the size or other properties of what we study fall outside the range of our everyday experience. Thus, we will speak of models of atoms, models of stars, and models of the Universe itself.

## Constellations

As human beings, we seek order in what we look at. Thus, when ancient people looked at the night sky, they noticed that the stars form fixed patterns on the celestial sphere, what we today call **constellations.** Some of these constellations resemble animals if we use a little imagination. For example, the pattern of stars in Leo looks a little like a lion, whereas that of Cygnus looks like a swan in flight, as depicted in figure 1.3. However, you will discover, as you learn to identify the constellations, that many have shapes that bear little resemblance to their namesakes. Also, keep in mind that stars in a constellation generally have no physical relation to one another. They simply happen to be in more or less the same direction in the sky.

All stars move, but as seen from Earth, their positions change very slowly, taking tens of thousands of years to make any noticeable shift. Thus, we see today virtually the same pattern of stars that was seen by ancient peoples. A shepherd who lived 5000 years ago in the Middle East would have no trouble recognizing the star patterns of the night sky we see and might even call them by the same names.

We do not know how the constellation names were chosen, but most of them date back thousands of years. In fact, we don't know when the names were first given to the constellations or why, although there is some evidence they served as mnemonic

**FIGURE 1.3**

The two constellations Leo (A) and Cygnus (B) with figures sketched in to help you visualize the animals they represent.

devices for keeping track of the seasons and for navigating. For example, the beginning of the stormy winter months, when sailing was dangerous and ships were often wrecked, was foretold by the Sun's appearance in the constellations Pisces and Aquarius, the water constellations. Likewise, the harvest time was indicated by the Sun's appearance in Virgo, a constellation often depicted as the goddess Proserpine, holding a sheaf of grain.

## Motions of the Sun and Stars

**Daily or Diurnal Motion.** Take a look at the night sky, and you will see stars rise along the eastern horizon, move across the sky, and set along the western horizon, exactly as the Sun does. You can verify this by watching the night sky for as little as 10 minutes. A star seen just above the eastern horizon will have risen noticeably higher and stars near the western horizon will have sunk lower or disappeared. Likewise, if you look at a constellation, you see its stars rise as a fixed pattern in the eastern sky, move across the sky, and set in the western sky. In terms of our model of the heavens based on the celestial sphere, we therefore explain the rising and setting of stars as rotation of the celestial sphere around us (fig. 1.4). For ancient people, that rotation was far easier to believe than that the Earth moved. Thus, they attributed all celestial motion—that of the Sun, Moon, stars, and planets—to a vast sphere slowly turning overhead. Today, of course, we know that it is the Earth's rotation that makes the Sun, Moon, and stars rise and move westward across the sky. It is not the celestial sphere that spins but the Earth.

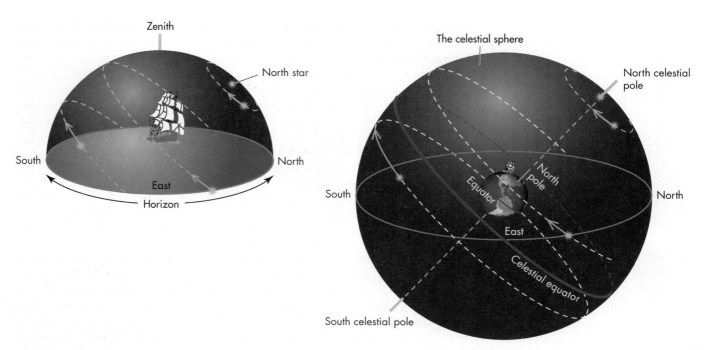

**FIGURE 1.4**

Stars appear to rise and set as the celestial sphere rotates overhead. Also shown are the celestial equator and poles and where they lie on the celestial sphere with respect to the Earth's equator and poles.

If you look at the celestial sphere turning overhead, two points on it do not move, as you can see in figure 1.4. These points are defined as the north and south **celestial poles.** The celestial poles lie exactly above the North and South Poles of the Earth, and just as our planet turns about a line running from its North to South Poles, so the celestial sphere rotates around the celestial poles. Because it lies directly above the Earth's North Pole, the north celestial pole always marks the direction of true north. Therefore, the north celestial pole is an important guide for travelers on land and sea and was widely used by early peoples for this purpose. Even today, the celestial poles serve as reference marks for navigators. Moreover, the celestial poles are used by astronomers to locate stars on the sky just as mapmakers use the Earth's North and South Poles as reference marks for locating places on our planet.

Another useful sky marker frequently used by astronomers is the **celestial equator.** The celestial equator lies directly above the Earth's equator, just as the celestial poles lie above the Earth's poles, as figure 1.4 shows.

At the same time that the Earth's spin causes the apparent daily motion of the Sun and stars across the sky, the Earth's orbital motion around the Sun also causes changes in the sky. However, this motion is much slower and harder to observe without making records of star positions. Nevertheless, astronomers today think that many ancient peoples built monuments to keep track of such motions. Because these movements repeat on a yearly cycle, they are called *annual motions.*

In discussing the motion of astronomical objects, astronomers often find it helpful to use different terms to distinguish an object's spinning motion from its orbital motion. For example, we say that the Earth rotates on its axis (spins) at the same time as it revolves around the Sun (moves along its orbit). Likewise we use the words *rotation* and *revolution* to distinguish between spin motion and orbital motion.

**Annual Motion.** If you watch the sky over a period of time, say, several months, you will discover that new constellations appear in the eastern sky and old ones disappear from the western sky. For example, on an early July evening across most of North

America, Europe, and Asia, the constellation Scorpius will be visible in the southern half of the sky. On December evenings, Scorpius is not visible. Instead, you will see the brilliant constellation Orion, the hunter. "Looking Up at Orion" (see page 26) shows Orion and some of the astronomical objects in the constellation, such as the Orion Nebula, that we will mention later in the book.

The discovery that different stars are visible at different times of the year was extremely important to early people because it provided a way to measure the passage of time other than by simply counting days. Moreover, the stars demonstrated that many celestial events are predictable and that they may be used to order our lives on Earth. For example, using the constellations mentioned above, if the evening sky shows Leo, it will soon be time to plant. If the sky shows Scorpius, it must be summer, and if it shows Orion, it must be winter. Determining the season by using the stars may sound strange to people living in climates with strong temperature differences between summer and winter. But for the ancient civilizations that were based in semitropical climates, such temperature differences are much smaller.

This change of the constellations with the seasons is caused by the Earth's motion around the Sun. As the Earth moves around the Sun, the Sun's glare blocks our view of the part of the celestial sphere that lies toward the Sun, making the stars that lie beyond the Sun invisible, as figure 1.5 shows. For example, in early June, a line from the Earth to the Sun points toward the constellation Taurus, and so its stars are lost in the Sun's glare. After sunset, however, we can see the neighboring constellation, Gemini, just above the western horizon. By early August, the Earth has moved to a new position in its orbit. At that time of year, the Sun lies in the direction of Cancer, causing this constellation to disappear in the Sun's glare. Looking to the west just after sunset, we now see Leo just above the horizon.

Month by month, the Sun covers one constellation after another. It is like sitting around a campfire and not being able to see the faces of the people on the far side. But if we get up and walk around the fire, we can see faces that were previously hidden. Similarly the Earth's motion allows us to see stars previously hidden in the Sun's glare. Our planet's motion also makes a given star rise 3 minutes and 56 seconds earlier each night. That 3 minutes and 56 seconds, when added up each night over an entire year, amounts to 24 hours. Thus a year later, when the Earth returns to the same spot in its orbit, the sky looks the same again.

**The Ecliptic.**   If we could mark on the celestial sphere the path traced by the Sun as it moves through the constellations, we would see a line that runs around the celestial sphere, as illustrated in figure 1.6. Astronomers call the line that the Sun traces across the celestial sphere the **ecliptic.** The name *ecliptic* arises because only when the new or full moon crosses this line can an eclipse occur, as you will learn later in this chapter. If you look at figure 1.6, you can see that the ecliptic is the extension of the Earth's orbit onto the celestial sphere, just as the celestial equator is the extension of the Earth's equator onto the celestial sphere.

It is not easy to track the Sun's position relative to the stars because the Sun's glare makes it impossible to see the stars. However, if you look just before sunrise or just after sunset, you can identify the stars near the Sun on any given date. By watching over several weeks, you will be able to see that the Sun changes its position with respect to the stars.*

Another complication we will discover when trying to track the Sun relative to the stars is that the Sun's path does not neatly align with the path of the stars. Instead, the path is tipped with respect to the celestial equator. As a result, the Sun lies north of the celestial equator for half the year and south of the celestial equator for the other

*Plotting the Sun's path through the stars can be done more precisely if you measure its position at noon and examine that position on the midnight sky half a year later.

# LOOKING UP AT... ORION

Orion is easy to identify because of the three bright stars of his "belt." You can see Orion in the evening sky from November to April, and before dawn from August through September.

Betelgeuse is a Red Giant star that has swelled to a size that is as large as our Solar System out to Mars. Its red color indicates that it is relatively cool for a star, 3400 degrees Kelvin.

## Betelgeuse

Sun

Mars' orbit

3 ly

## Horsehead Nebula

The Horsehead Nebula is an active star-forming region rich with dust and gas.

1.8 ly

3 ly

1 ly

## Orion Nebula

The Orion Nebula is an active star-forming region rich with dust and gas.

Rigel

1 ly

## Orion in 3-D

Betelgeuse
430 ly
240 ly

920 ly
1300 ly
820 ly
1500 ly

To Earth

770 ly
720 ly

Rigel

1 light year (ly) = ~6 trillion miles

SOUTH

Sun

Pluto's orbit

## A protoplanetary disk

This is the beginning of a star; our early Solar System may have looked like this!

**FIGURE 1.5**

The Sun hides from our view stars that lie beyond it. As we move around the Sun, those stars become visible, and the ones previously seen are hidden. Thus, the constellations change with the seasons.

Evening
August twilight – looking westward

Leo

Sun

Evening
June twilight – looking westward

Gemini

Sun

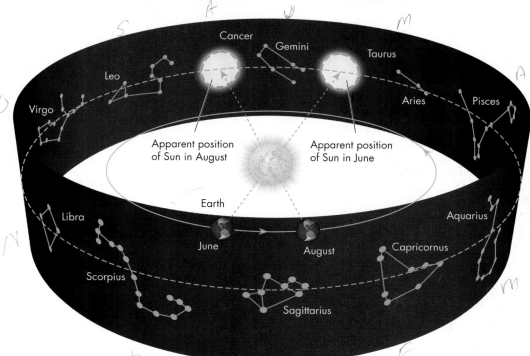

Cancer

Gemini

Taurus

Leo

Aries

Pisces

Virgo

Apparent position
of Sun in August

Apparent position
of Sun in June

Earth

Libra

Aquarius

June

August

Capricornus

Scorpius

Sagittarius

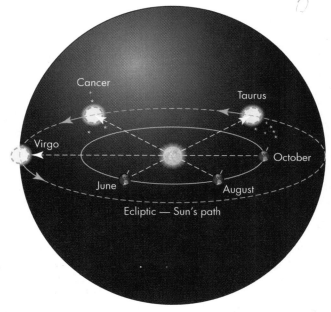

Cancer

Taurus

Virgo

October

June

August

Ecliptic — Sun's path

**FIGURE 1.6**

As the Earth orbits the Sun, the Sun appears to move around the celestial sphere through the background stars. The Sun's path is called the *ecliptic.* The Sun appears to lie in Taurus in June, in Cancer during August, in Virgo during October, and so forth. Note that the ecliptic is also where the Earth's orbital plane cuts the celestial sphere. Sizes and distances of objects are *not* to scale.

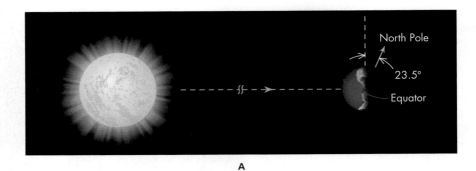

**A**

**FIGURE 1.7**
(A) The Earth's rotation axis is tilted 23.5° to its orbit around the Sun.
(B) The Earth's rotation axis keeps nearly the same tilt and direction as it moves around the Sun. Sizes and distances are *not* to scale.

**B**

*The Sun's motion north and south in the sky as the seasons change*

*The Earth's rotation axis*

**FIGURE 1.8**
This tendency of a spinning object to keep its orientation is called "conservation of angular momentum," and it is the principle on which gyroscopic compasses operate and the reason a quarterback puts "spin" on a football.

half of the year. Why is the Sun's path tipped in this way? To understand, it will help if we first discuss another change that happens with a yearly cycle: the seasons.

## The Seasons

Many people believe that we have seasons because the Earth's orbit is elliptical. They suppose that summer occurs when we are closest to the Sun and winter when we are farthest away. It turns out, however, that the Earth is nearest the Sun in early January, when the Northern Hemisphere is coldest. Thus, seasons must have some other cause.

To see what *does* cause seasons, we need to look at how our planet is oriented in space. As the Earth orbits the Sun, our planet also spins. That spin is around an imaginary line—the **rotation axis**—that runs through the Earth from its North Pole to its South Pole. The Earth's rotation axis is *not* perpendicular to its orbit around the Sun. Rather, it is tipped by 23.5° from the vertical, as shown in figure 1.7A.

As our planet moves along its orbit, its rotation axis maintains nearly exactly the same tilt and direction,* as figure 1.7B shows. That is, the Earth behaves much like a giant spinning top. This tendency of the Earth to preserve its tilt is shared by all spinning objects. For example, it is what keeps a rolling coin upright and a thrown football pointed properly (fig. 1.8). Moreover, you can easily feel this tendency of a spinning object to resist changes in its orientation by lifting a spinning bicycle wheel off the ground and trying to twist it from side to side.

Because the Earth's tilt remains nearly constant as we move around the Sun, sunlight falls more directly on the Northern Hemisphere for part of the year and more directly on the Southern Hemisphere for another part of the year, as illustrated in figure 1.9. This changes the amount of heat we receive from the Sun.

---

*At this time in history, the axis happens to point approximately toward the Star Polaris, which is therefore known as the North Star.

**FIGURE 1.9**
Because the Earth's rotation axis keeps the same tilt as we orbit the Sun, sunlight falls more directly on the Northern Hemisphere during part of the year and on the Southern Hemisphere during the other part of the year. Sizes and distances are *not* to scale.

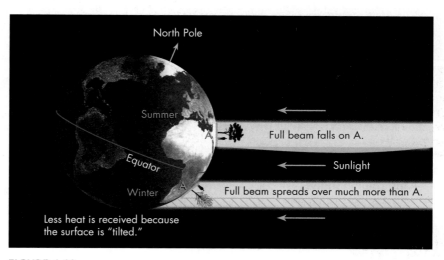

**FIGURE 1.10**
A flat surface directly facing the Sun receives more light (and thus more heat) than a tilted surface.

*Seasons*

A flat surface area presented directly to the Sun intercepts a larger portion of the Sun's radiation, and hence its heat, than the same surface area when tilted, as figure 1.10 illustrates. You take advantage of this effect instinctively when you warm your hands at a fire by holding your palms flat toward the fire, not edgewise. Similarly, the Northern Hemisphere receives its greatest heating at the time of year when the Sun shines most directly on it, making it summer. Six months later, the Northern Hemisphere receives its sunlight least directly, and so it is colder and therefore winter. This heating difference is enhanced because the Earth's tilt leads to many more hours of daylight in the summer than in winter. As a result, we not only receive the Sun's light more directly, we receive it for a longer time. Thus,

**the seasons are caused by the tilt of the Earth's rotation axis.**

Figure 1.10 also illustrates that this makes the seasons reversed between the Northern and Southern Hemispheres; when it is summer in one, it is winter in the other.

**The Ecliptic's Tilt.** The tilt of the Earth's rotation axis not only causes seasons, it also is why the Sun's path across the celestial sphere—the ecliptic—is tilted with respect to the celestial equator. As you can see in figure 1.9, in June, the Northern Hemisphere is tipped

Earth's position in its orbit
at different times of year.

Sun's position on celestial
sphere at start of each season.

**FIGURE 1.11**

The left-hand column shows
schematically the Earth's tilt and
position in its orbit at different
times of year. You can see that
in June, the direction to the Sun
(indicated by the arrow) lies north
of the Earth's equator. On the other
hand, in December, the direction
to the Sun lies south of the Earth's
equator. If we mark the Sun's
position on the celestial sphere
throughout the year, we get the
ecliptic, which, as you can see, is
tilted with respect to the celestial
equator. The Sun crosses the celestial
equator in March and September.
The times of these crossings define
the equinoxes.

toward the Sun. As a result, the Sun lies north of the celestial equator. On the other hand,
in December, when the Earth is on the other side of the Sun, the Sun lies south of the
celestial equator. The consequence of such north–south motion is that the Sun's path—
the ecliptic—must cross the celestial equator and therefore the ecliptic must be tilted with
respect to that line, as the sequence of sketches in figure 1.11 shows. This motion of the
Sun north and south in the sky over the course of the year is also why the Sun is so high
in the summer sky at noon and so low in the winter sky at noon (fig. 1.12). For example,
in June at mid-northern latitudes (40°), the noon Sun is about 74° above the horizon. In
December, on the other hand, it is only about 27° above the horizon.

**Solstices and Equinoxes.** The tilt of the ecliptic with respect to the celestial
equator means that during the year, the points on the horizon where we see the Sun
rise and set are not in general due east and west. Rather, over a year, the direction
to the rising and setting position of the Sun changes. From about March 21 until
about June 21, the Sun's rising point shifts a little farther northward each day, as
illustrated in figure 1.13A. From about June 21 to about September 23, the rising
position shifts southward each day, rising due east again about September 23. The
rising position then continues its southward shift until about December 21 (again,
see figure 1.13A). On that date it stops its southward drift and begins to move north
again. A similar shift occurs in the Sun's setting point to the northwest in June and
the southwest in December.

Astronomers give special names to the Sun's position at the times when the seasons
change. For example, spring and autumn begin when the Sun crosses the celestial equator. At these times of year, the days and nights are of equal length (approximately), and
so they are called the **equinoxes.** The spring or vernal equinox occurs near March 21,
but the fall or autumnal equinox occurs near September 23. Likewise, the beginning
of summer and winter are given special names. At these times of year, the Sun pauses
in its steady north–south motion and changes direction. Accordingly, these times are

The summer and winter beams carry the same amount of energy, but spread that energy over very different amounts of ground.

**FIGURE 1.12**

Why the Sun at noon is high in the sky in summer and low in the sky in winter.

**FIGURE 1.13**
(A) The direction of the rising and setting Sun changes throughout the year. At the equinoxes, the rising and setting points are due east and west. The sunrise direction shifts slowly northeast from March to the summer solstice, whereupon it shifts back, reaching due east at the autumn equinox. The sunrise direction continues moving southeast until the winter solstice. The sunset point similarly shifts north and south. (B) Sunrise on the summer solstice at Stonehenge.

Although the seasons begin on the solstices and equinoxes, the hottest and coldest times of year occur roughly 6 weeks after the solstices. The delay, known as the lag of the seasons, results from the oceans and land being slow to warm up in summer and slow to cool down in winter.

called the **solstices,** meaning the Sun (sol) has stopped its motion north or south and is *static* and about to reverse direction. The dates of the solstices (summer and winter) also change slightly from one year to the next, but they are always close to June 21 and December 21.

Just as the changing position of the Sun against the constellations could be used as an indicator of the seasons, so too could the position of the rising and setting Sun. One well-known example is Stonehenge, the ancient stone circle in England, illustrated in figure 1.1. Although we cannot know for certain how this ancient monument was used, it seems likely that it was laid out so that such seasonal changes in the Sun's position could be observed by noting through which stone arches the Sun rose or set. For example, figure 1.13B (a photograph taken on the summer solstice) shows that on that day, an observer standing at the center of this circle of immense standing stones would see the rising Sun framed by an arch. Similarly, some ancient Egyptian temples and pyramids have astronomical alignments, such as the Temple of Amen-Ra at Karnak, whose main chamber points toward the position of sunrise at the winter solstice.

Astronomical alignment of buildings occurs in many other places as well. For example, the Maya, native peoples of Central America, and their neighbors built pyramids from the summits of which they could get a clear view of the sky over the surrounding rain forest. The Mayans devised a precise and complex calendar based on the planet Venus, which they believed determined the course of their history. Through such interest in the sky, the Mayans and other ancient cultures learned to recognize another type of celestial motion: that of the planets.

## The Planets and the Zodiac

All cultures whose astronomy has been studied have noted that some bright objects that looked like stars did not follow the same cyclic motion associated with the rest of the stars in the sky. We do not know when humans first recognized the special motion

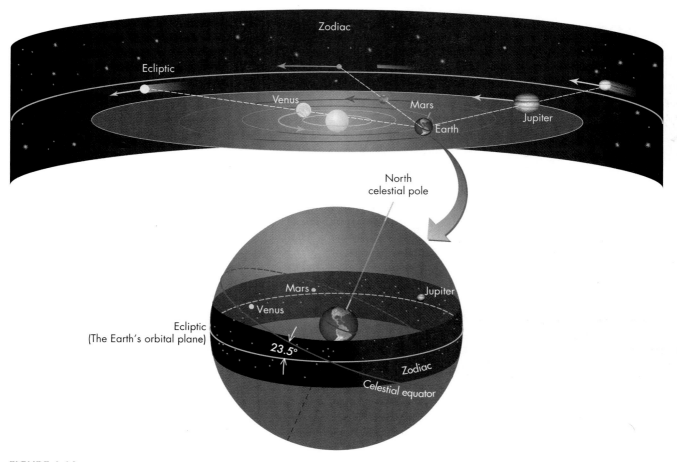

**FIGURE 1.14**
The zodiac is tilted 23.5° with respect to the celestial equator because Earth's rotation axis is tilted by that amount with respect to its orbit.

of these objects, but the Greeks are credited with giving them the name *planētai* meaning "wanderers," from which our word *planet* comes.

Planets move across the background stars because of a combination of the Earth's and their own orbital motion around the Sun. One of the more striking features of this motion is that the planets *always* move within a very narrow band on the celestial sphere. We call this band the **zodiac.** The motion of the planets lies in this narrow zone because their orbits, including that of the Earth, all lie in nearly the same plane.

The zodiac circles the heavens following the ecliptic, which lies exactly along its centerline, and both are tilted by 23.5° with respect to the celestial equator, as shown in figure 1.14. This tilt arises (as we discussed just previously) because the Earth's rotation axis is tilted with respect to our orbit around the Sun.

The word *zodiac* is from Greek *zoidion*, "little animal," leading to "animal sign." That is, zodiac refers to a circle of animals, which for the most part, its constellations represent. The names of these constellations are Aries, Taurus, Gemini, Cancer, Leo, Virgo, Libra, Scorpius, Sagittarius, Capricornus, Aquarius, and Pisces.*

---

*The zodiac also passes through part of the modern boundaries of the constellation Ophiuchus, the Serpent Bearer. Astronomers nevertheless do not consider that one of the traditional constellations of the zodiac.

**FIGURE 1.15**
A planet's eastward drift against the background stars plotted on the celestial sphere. Note: Star maps usually have east on the left and west on the right, so that they depict the sky when looking south.

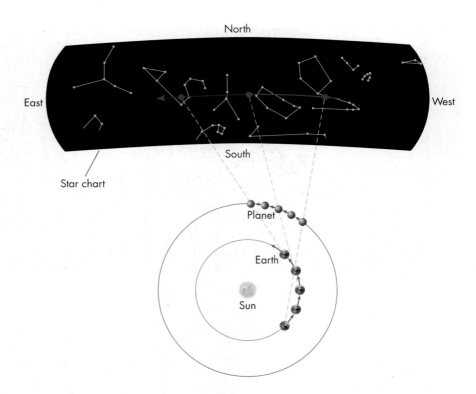

The motion of the planets along the zodiac can be seen easily by marking off the position of a planet on the celestial sphere over a period of a week or so. Figure 1.15 illustrates such a plot and shows that planets normally move eastward through the stars as a result of their orbital motion around the Sun.

Although the apparent motion is usually from west to east through the stars, this does *not* mean that the planet rises in the west and sets in the east. As seen from Earth, planets *always* rise in the east and set in the west because they are carried across the sky—just as the stars are—by the Earth's rotation. However, the motion of the planets is usually slower than that of the stars because their orbital motion partly offsets the rotation of the Earth that causes the apparent motion of the stars. Thus, if a star and planet rise side by side, at some later time the planet will not be as far above the horizon as the star. Therefore, *with respect to the stars,* the planet has moved to the east because of its orbital motion around the Sun.

This simple pattern of movement is sometimes interrupted. Occasionally a planet will move west with respect to the stars, a condition known as **retrograde motion** and shown in figure 1.16. The word *retrograde* means "backward," and when a planet is in retrograde motion, its path through the stars may bend backward for a few days or it may even form a loop (refer to fig. 1.16). All planets undergo retrograde motion for at least some portion of their paths around the sky. This motion greatly complicates the otherwise straightforward idea that the celestial sphere and its bodies rotate around the Earth. In fact, the need for a simple, plausible explanation of retrograde motion was a major reason that astronomers ultimately rejected models of the Solar System with the Earth at the center.

## The Moon

Like all celestial objects, the Moon rises in the east and sets in the west. Also, like the Sun, the Moon shifts its position across the background stars from west to east. You can verify this motion by observing the Moon at the same time each evening and noticing its change in position with respect to nearby stars. In fact, if the Moon happens to lie close to a bright star, its motion may be seen in only a few minutes, because in 1 hour the Moon moves a distance, on the sky, that is approximately equal to its own apparent diameter.

*Retrograde motion*

*Lunar phases*

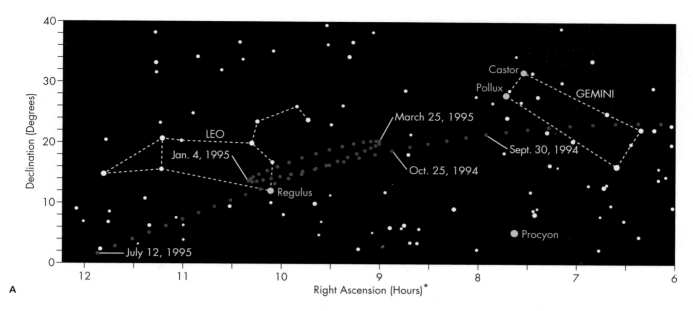

**FIGURE 1.16**
(A) The position of Mars marked out on the background stars and showing its retrograde motion. In what constellation is Mars in October 1994? (Use the star charts at the back of this book to identify the constellations.) (B) Images of Mars as it underwent retrograde motion in late 2005. The pictures were taken roughly a week apart.

Q* Why does the brightness of Mars change in the image? (Hint: Draw a sketch of the positions of Mars and the Earth as Mars undergoes retrograde motion.)

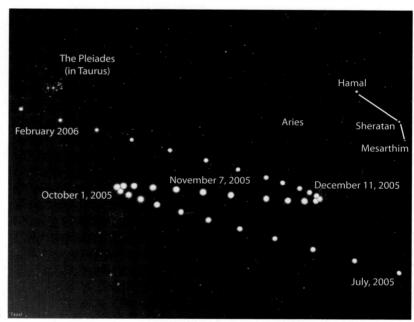

Perhaps the most striking feature of the Moon is that, unlike the Sun, its shape seems to change throughout the month in what is called the "cycle of lunar **phases.**" That is, the Moon appears alternately as a thin crescent, a fully illuminated disk, and then as a crescent again. During a period that is approximately 30 days, the Moon goes through a complete set of phases from invisibility (new) to fully lit (full) and back to invisibility. This is the origin of the month as a time period and also the source of the name "month," which was derived from the word *moon.*

The cycle of the phases and the Moon's changing position against the stars are caused by the Moon's orbital motion around the Earth. A new moon occurs when the Moon lies approximately between us and the Sun. A full moon occurs when the Moon is on the other side of the Earth from the Sun, opposite it in the sky. Many people

---

*See page 69 for the definition of right ascension.

**FIGURE 1.17**
The cycle of the phases of the Moon from new to full and back again. The phases are caused by our seeing different amounts of its illuminated surface. Sizes and distances of objects are *not* to scale.

believe these changes in shape are caused by the Earth's shadow falling on the Moon. However, that is *not* the explanation, as you can deduce from the fact that crescent phases occur when the Moon and Sun lie approximately in the same direction in the sky and the Earth's shadow must therefore point *away* from the Moon. In fact, we see the Moon's shape change because, as it moves around us, we see different amounts of its illuminated half. For example, when the Moon lies approximately opposite the Sun in the sky, the side of the Moon toward the Earth is fully lit. On the other hand, when the Moon lies approximately between us and the Sun, its fully lit side is turned nearly completely away from us, and therefore we glimpse only a sliver of its illuminated side, as illustrated in figure 1.17. This figure also shows the cycle of the lunar phases and the names used to describe the Moon's appearance.

The Moon's motion around the Earth has other effects as well. For example, because the Moon shifts eastward through the stars, the Earth itself must rotate eastward a little extra each evening to bring the Moon above the horizon. This extra rotation takes about an hour (actually 50 minutes, on the average) each day, and so moonrise is later each night by about that amount. The most beautiful and dramatic effect of the Moon's motion, however, is that it leads to eclipses.

# EXTENDING OUR REACH

## VISIBILITY OF THE MOON

If you watch the Moon go through a cycle of its phases, you will notice that the times at which it is visible change. For example, shortly after the new phase, you can see the Moon low in the western sky after sunset. But a few hours later that same night it will have set and become invisible. On the other hand, when the Moon is full, it rises at about sunset and doesn't set until dawn. Thus, the full moon is visible throughout the night. At some of its phases, you can see the Moon even during the day, if you know where to look. These differences in when the Moon is visible are the theme of this box. But to understand these differences, it will help if we review what determines the Moon's phases.

A

(1) The center figure shows the orbit of the Moon and its position on 7 days of the lunar cycle (days 4, 8, 11, 15, 19, 22, and 25). A figure on the Earth shows the position of an observer at the four times of day (midmorning, early evening, midnight, and predawn).

(2) The four surrounding pictures show what an observer at each of those times of day would see as the Moon goes through its phases.

**BOX FIGURE 1.1**

When the Moon Is Visible. This figure shows where to look for the Moon and how it appears at different times of day as it goes through its monthly cycle of phases. The central sketch shows the Earth and Moon and where the Moon is in its orbit at different phases, indicated here by the number of days since the new Moon. The yellow arrow points toward the Sun.

*(continued)*

## EXTENDING OUR REACH (CONTINUED)

Earlier in this section we described how the Moon's phase depends on where the Moon lies in its orbit around the Earth with respect to the Sun (see figure 1.17). For example, you can see in that figure that when the Moon is full, it lies in the sky approximately opposite to where the Sun lies. Likewise, when the Moon is a thin crescent, it lies in the sky more or less toward where the Sun is. These connections between the Moon's phase and its position with respect to the Sun are the key to understanding when the Moon is visible from Earth. They are also a key you can use to tell the approximate time at night from the Moon's phase and position in the sky. For example, because the full Moon must be more or less opposite the Sun, it *rises* above the eastern horizon at about the same time that the Sun *sets* below the western horizon. Likewise, the full Moon *sets* at about the time the Sun *rises*.

Thus, the full Moon is visible all night and, if we take sunrise to be about 6 A.M. and sunset to be about 6 P.M., we can then deduce that the full Moon rises at about 6 P.M. Furthermore, we can deduce that it will be highest in the sky near midnight.

On the other hand, the crescent moon is *not* visible during most of the night. This is because it lies in nearly the same direction as the Sun and so, once the Sun is well below the horizon, the crescent Moon is below the horizon too. Moreover, the crescent Moon is hard to see during the day because it is only a sliver of light, so it is lost in the brightness of the daytime sky. Therefore, the best times to see it are shortly after sunset (when the Moon is just past new) or shortly before sunrise (when the Moon is near to being "new" again). Thus, when the Moon is a few days past its new phase and is a thin crescent, you can see it low in the western sky at sunset. This crescent moon will set shortly after the Sun and not be visible again until after sunrise the next day.

As a further example, let's look at the Moon when it is at first quarter (half lit). At this phase (as you can see in Box figure 1.1), it is about one-quarter of the way around the sky and eastward of the Sun by about 90° on the sky. Thus the Earth must turn about an additional 90° to bring the Moon above the horizon. How long will it take the Earth to rotate those extra 90°? We can easily figure that out by noting that it takes the Earth 24 hours to rotate once (360°). Thus, to rotate 90° degrees (one-quarter of 360°) takes one-quarter of 24 hours, or 6 hours. Thus, the first quarter Moon rises about 6 hours after the Sun rises. Moreover, because it is about 90° from the Sun, it will be at its highest point in the sky at sunset and by midnight it will have moved to the western horizon and be setting.

With similar arguments, you can figure out when and where it is easiest to see the Moon during the day. For example, near first quarter, look for the Moon eastward of the Sun in the afternoon. Near third quarter, look for it westward of the Sun in the morning.

You can see these and other connections between the Moon's phase and position in the sky at a given time (sunset, midnight, dawn, etc.) in Box figure 1.1. This figure shows where the Moon lies in the sky, for a given time of day or night, as its phase changes. Note that at a given time, the Moon lies a little farther east in the sky each night and also that its phase changes. This shift to the east from night to night is simply the result of the Moon's orbital motion around the Earth. You may also find it helpful for understanding these features of the Moon's phases and its visibility to look at Box table 1.1, which lists the approximate time of moonrise.

### BOX TABLE 1.1

**Moonrise Times (Approximate) for Different Phases**

| Phase | Moonrise |
| --- | --- |
| Crescent | Morning—after sunrise |
| Ist Quarter | About noon |
| Gibbous | Midafternoon |
| Full | Sunset |
| Gibbous | Midevening |
| 3rd Quarter | About midnight |
| Crescent | Early morning—predawn |
| New | Sunrise |

**INTERACTIVE**

*Eclipses*

## Eclipses

An eclipse occurs when the Moon happens to lie exactly between the Earth and the Sun, or when the Earth lies exactly between the Sun and the Moon so that all three bodies lie on a straight line. Thus, there are two types of eclipses: solar and lunar.

A **solar eclipse** occurs whenever the Moon passes directly between the Sun and the Earth and blocks our view of the Sun, as depicted in figure 1.18. A **lunar eclipse** occurs when the Earth passes between the Sun and the Moon and casts its shadow on the Moon, as shown in figure 1.19.

Given that the Moon orbits the Earth approximately once every 30 days, you may wonder why we do not have eclipses every month. The answer is that the Moon's orbit is tipped with respect to the Earth's orbit (fig. 1.20). Because of this tip, even if

**FIGURE 1.18**
A solar eclipse occurs when the Moon passes between the Sun and the Earth so that the Moon's shadow touches the Earth. The photo inset shows what the eclipse looks like from Earth.

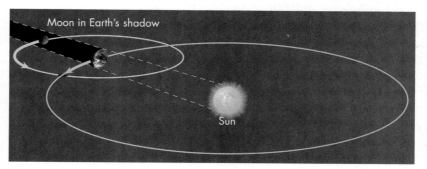

**FIGURE 1.19**
A lunar eclipse occurs when the Earth passes between the Sun and Moon, causing the Earth's shadow to fall on the Moon. Some sunlight leaks through the Earth's atmosphere, casting a deep reddish light on the Moon. The photo inset shows what the eclipse looks like from Earth.

**FIGURE 1.20**
The Moon's orbit (seen here edge on) is tilted with respect to the Earth's orbit by about 5°. The Earth and Moon are drawn to correct relative size and separation. Note how thin their shadows are.

the Moon is new so it lies on the part of its orbit between us and the Sun, the Moon's shadow may pass above or below Earth, as you can see in figure 1.20. As a result, no eclipse occurs. Similarly, when the Moon is full, the Earth's shadow may pass above or below the Moon so that again no eclipse occurs. Only a nearly exact alignment of the Earth, Moon, and Sun leads to eclipses, a point that is easier to appreciate if you look at figure 1.20, which shows the Earth and Moon drawn to scale.*

---

*We will discuss conditions for eclipses in greater detail in chapter 6.

Eclipses can be amazing sights. During a solar eclipse, the midday sky may become dark as night for a few minutes. During a lunar eclipse, the Moon may become a dull red or disappear altogether for over an hour. Therefore, it is not surprising that early people recorded such events and sought (successfully) to predict them.

The astronomical phenomena that we have discussed so far (rising and setting of Sun, Moon, and stars; the constellations; annual motion of Sun; motion of planets through the zodiac; phases of the Moon and eclipses) were the basis of ancient knowledge of the heavens. With these observations, we can now describe the attempts of early people to explain the heavens. We will see that many of their conclusions were incorrect, just as we today are probably in error about some aspects of astronomy. We study ancient ideas of the heavens not so much for what they tell us about the heavens but to learn how observation and reasoning can lead us to an understanding of the Universe.

## 1.2    EARLY IDEAS OF THE HEAVENS: CLASSICAL ASTRONOMY

As far as we know, the ancient Greek astronomers of classical times were the first people to try to explain the workings of the heavens in a careful, systematic manner, using models and observations. Given the limitations of naked-eye observation, these astronomers were extraordinarily successful, and their use of logic and mathematics (geometry) as tools of inquiry created a method for studying the world around us that we continue to use even today. This method is in many ways as important as the discoveries themselves.

### The Shape of the Earth

The ancient Greeks knew that the Earth is round. As long ago as about 500 B.C., the mathematician Pythagoras (about 560–480 B.C.) was teaching that the Earth is spherical, but the reason for his belief was as much mystical as rational. He, like many of the ancient philosophers, believed that the sphere was the perfect shape and that the gods would therefore have utilized that perfect form in the creation of the Earth.

By 300 B.C., however, Aristotle (384–322 B.C.) was presenting arguments for the Earth's spherical shape that were based on simple naked-eye observations that anyone could make. Such reliance on careful, firsthand observation was the first step toward acquiring scientifically valid knowledge of the contents and workings of the Universe. For instance, Aristotle noted that if you look at an eclipse of the Moon when the Earth's shadow falls upon the Moon, the shadow can be clearly seen as curved, as figure 1.21A shows. He wrote in his treatise "On the Heavens":

> **The shapes that the Moon itself each month shows are of every kind—straight, gibbous, and concave—but in eclipses the outline is always curved: and, since it is the interposition of the Earth that makes the eclipse, the form of this line will be caused by the form of the Earth's surface, which is therefore spherical.**

Another of Aristotle's arguments that the Earth is spherical was based on the observation that a traveler who moves south will see stars that were previously hidden below the southern horizon, as illustrated in figure 1.21B. For example, the bright star Canopus is easily seen in Miami but is invisible in Boston. This could not happen on a flat Earth.

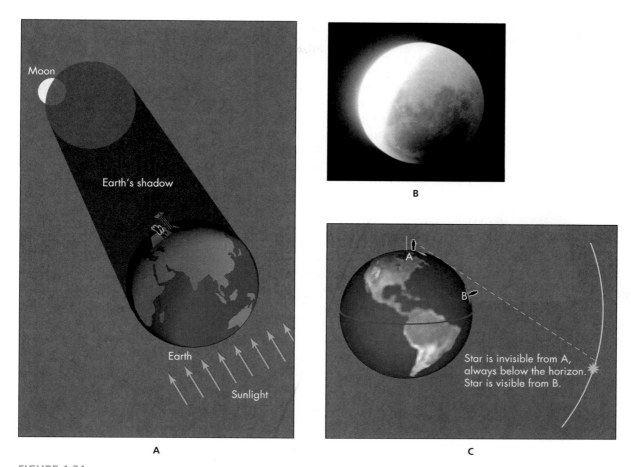

**FIGURE 1.21**

(A) During a lunar eclipse, we see that the Earth's shadow on the Moon is curved. Thus the Earth must be round. (B) Photograph of a partial lunar eclipse. (C) As a traveler moves from north to south on the Earth, the stars that are visible change. Some disappear below the northern horizon, whereas others, previously hidden, become visible above the southern horizon. This variation would not occur on a flat Earth.

## The Size of the Earth

Knowing that the Earth is spherical, the Greeks wanted to know its size. Eratosthenes (276–195 B.C.), head of the famous Library at Alexandria in Egypt, made the first measurement of the Earth's size. He obtained a value for its circumference of about 25,000 miles, which is roughly the presently known value. Eratosthenes's demonstration is one of the most beautiful ever performed. Because it so superbly illustrates how science links observation and logic, the demonstration is worth describing in some detail.

You must realize that at this point in time, astronomers were very well acquainted with the yearly movement of the Sun and could predict accurately the times of the solstices and equinoxes. The summer solstice marked the day of the year in Alexandria when the Sun would reach its highest point in the noonday sky. However, despite its height, the Sun still cast a shadow at noon. Eratosthenes, a geographer as well as an astronomer, heard that lying to the south, in the Egyptian town of Syene (the present city of Aswân), the Sun would be directly overhead at noon and cast no shadow. Proof of this was the fact that at that time the Sun shone exactly down a well near there. Knowing the distance between Alexandria and Syene and appreciating

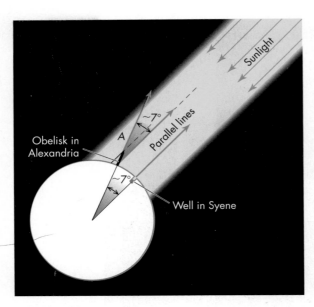

**FIGURE 1.22**

Eratosthenes's calculation of the circumference of the Earth. The Sun is directly overhead at local noon on the summer solstice at Syene, in southern Egypt. On that same day, and time, Eratosthenes found the Sun to be 1/50 of a circle (about 7°) from the vertical in Alexandria, in northern Egypt. Eratosthenes deduced that the angle between two verticals placed in northern and southern Egypt must be 1/50 of the circumference of the Earth.

the power of geometry, Eratosthenes realized he could deduce the size (circumference) of the Earth. He analyzed the problem as follows: Because the Sun is far away from the Earth, its light travels in parallel rays toward the Earth. Thus, two rays of sunlight, one hitting Alexandria and the other shining down the well, are parallel lines, as depicted in figure 1.22.

Now imagine drawing a straight line from the center of the Earth outward so that it passes vertically through the Earth's surface in Alexandria. The angle, call it A, between that line and the Sun's rays in Alexandria is the same as the angle between that line and the line from the center of the Earth up through the well in southern Egypt (see fig. 1.22). The reason is that the corresponding angles formed where a single line crosses two parallel lines are equal (a geometric theorem).

The angle A can be measured with sticks and a protractor (or its ancient equivalent) and is the angle between the direction to the Sun and the vertical to the ground (see fig. 1.22). Eratosthenes found this angle to be about 1/50th of a circle. Therefore the angle formed by a line from Alexandria to the Earth's center and a line from the well to the Earth's center must also be 1/50th of a circle, as deduced from the parallel line–equal angle theorem.

To find the circumference of the Earth, all that is needed is to note that the distance between Alexandria and the well thus represents 1/50th of the distance around the Earth. Because Eratosthenes knew the distance between Alexandria and the well to be 5000 stadia (where a stadium is about 0.1 mile), the distance around the entire Earth is 5000 stadia times 50, or 250,000, stadia. When expressed in miles, this is roughly 25,000 miles, which is approximately the circumference of the Earth as we know it today.

By modern standards, there is absolutely nothing wrong with this technique. You can use it yourself to measure the size of the Earth, as described in the projects at the end of this chapter. Eratosthenes's success was a triumph of logic and the scientific technique, but it suffered from one weakness. The method required that he assume the Sun is so far away that its light reaches Earth along parallel lines. That assumption, however, was supported by another set of measurements made by the ancient Greeks—namely, a rough measurement of the relative diameters and distances of the Sun and Moon.

## Distance and Size of the Sun and Moon

About 75 years before Eratosthenes measured the Earth's size, Aristarchus of Samos (an island in the Mediterranean) had estimated the relative size of the Earth, Moon, and Sun, and the relative distances to the Moon and Sun. His values for these numbers gave at least the correct sense of their proportionate sizes and their relative distances from Earth. For example, by comparing the size of the Earth's shadow on the Moon during a lunar eclipse to the size of the Moon's disk, illustrated in figure 1.23, Aristarchus calculated that the Moon's diameter is about one-third of the Earth's.

Aristarchus also calculated the Sun to be about 20 times farther away from the Earth than the Moon is. He did this calculation by measuring the angle between the Sun and the Moon when the Moon is exactly half lit, as shown in figure 1.24. From the apparent size in the sky of the Sun and Moon and their relative distances, and his calculation of the Moon's diameter relative to the Earth's, Aristarchus was able to deduce that the Sun's diameter is about 7 times that of the Earth's, far too small as we know today (it is actually about 100 times greater), but, nevertheless, evidence that the Sun is bigger than the Earth.

INTERACTIVE

*Stellar parallax*

Although Aristarchus's determination of the Moon's relative size was quite accurate (he found the Moon's diameter to be about 0.35 that of the Earth, whereas the correct ratio is about 0.27), his distance to the Sun and consequently its size were too small by a factor of about 20. Nevertheless, he was the first person to establish that the Sun is not only very far away but also much larger than the Earth.

**FIGURE 1.23**
Aristarchus used the size of the Earth's shadow on the Moon during a lunar eclipse to estimate the relative size of the Earth and Moon.

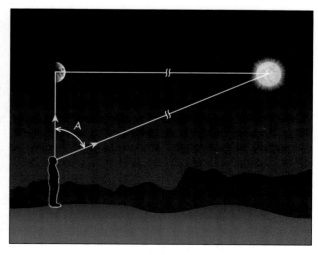

**FIGURE 1.24**
Aristarchus estimated the relative distance of the Sun and Moon by observing the angle *A* between the Sun and the Moon when the Moon is exactly half lit. Angle *B* must be 90° for the Moon to be half lit. Knowing the angle *A*, he could then set the scale of the triangle and thus the relative lengths of the sides. (Sizes and distances not to scale.)

# EXTENDING OUR REACH

## MEASURING THE DIAMETER OF ASTRONOMICAL OBJECTS

One of the most basic properties of an astronomical body is its diameter. But there is no astronomical body for which this fundamental observation can be made directly. Since we cannot stretch a tape measure across the disk of the Sun or Moon, how do we know the size of such heavenly bodies?

The basic method for measuring the size of a distant object was worked out long ago and is still used today. This method involves first measuring how big an object looks, a quantity called its **angular size.** We can measure an object's angular size by drawing imaginary lines to each side of it, as shown in Box figure 1.2, and then measuring the angle between the lines using a protractor. For example, we can measure the angular diameter of the Moon with two straight sticks connected at one end with a nut and bolt. We sight along the edge of one stick to one edge of the Moon and then sight along the other stick to the Moon's opposite edge. Measuring the angle between the sticks, we will find that the Moon's angular diameter is about one-half degree.

We can find an astronomical body's true diameter from its angular diameter if we know its distance. We need the distance because a body's angular size changes with distance. For example, a building looks big when we are near it and small when we are far away, as shown in Box figure 1.3. Furthermore, it is easy to verify that the angular size of a distant object changes inversely with the object's distance. That is, if we double the distance to an object, its angular size is halved. These basic features of angular size were known in antiquity and were in fact used by Aristarchus to determine the size of the Moon and Sun compared to the Earth.

To find an object's true diameter from its angular diameter and distance, imagine we are at the center of a circle passing through the object, as illustrated in Box figure 1.4. Let $L$ be the diameter of the body and $D$ the distance to the body, the radius of the circle. Next draw lines from the center to each end of $L$, letting the angle between the lines be $A$, the object's angular diameter.

We now determine the object's true size, $L$, by forming the following proportion: $L$ is to the circumference of the circle as $A$ is to the total number of degrees around the circle, which we know is 360.

Thus,

$$\frac{\text{Object's diameter}}{\text{Circumference}} = \frac{\text{Angle between lines}}{360}$$

$$\frac{L}{\text{Circumference}} = \frac{A}{360}$$

**BOX FIGURE 1.2**
Definition of angular size.

**BOX FIGURE 1.3**
How angular size varies with distance.

$$\frac{L}{\text{Circumference}} = \frac{L}{2\pi D} = \frac{A}{360°}$$

therefore, $L = \dfrac{2\pi D A}{360}$

**BOX FIGURE 1.4**
How to determine linear size from angular size.

However, we know from geometry that the circle's circumference is $2\pi D$. Thus,

$$\frac{L}{2\pi D} = \frac{A}{360}$$

We can now solve for $L$ and find that

$$L = \frac{2\pi DA}{360}$$

Thus, given a body's angular diameter and distance, we can calculate its true diameter. For example, suppose we apply this method to measure the Moon's diameter. We stated previously that the Moon's angular diameter is about 1/2°. Its distance is about 384,000 kilometers. Thus, its real diameter is

$$L = \frac{2\pi(384,000)(0.5)}{360}$$

$$= \text{about } 3350 \text{ km}$$

Suppose we try another example. The angular diameter of the Sun also turns out to be about 1/2°, an interesting coincidence. The Sun's distance is 150 million kilometers. The Sun's diameter must therefore be $2\pi$ (150,000,000) (0.5)/360, or about 1,310,000 kilometers.

An easy way to get at this same answer is to recognize that because the Sun and Moon have the same angular size, their true sizes must be in direct proportion to their distances. The Sun is about (150,000,000/384,000) = 391 times farther away. Therefore, it must be about 391 times larger than the Moon.

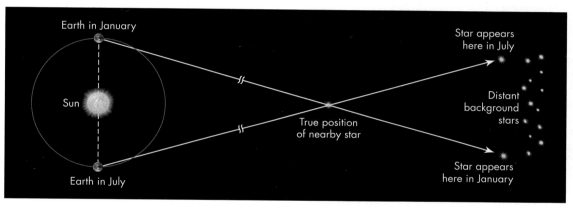

**FIGURE 1.25**

Motion of the Earth around the Sun causes stellar parallax—a shift in where in the sky a star appears to lie. Because the stars are so remote, this shift is too small to be seen by the naked eye. This lack of observable parallax led the ancient Greeks incorrectly to deduce that the Sun could not be the center of the Solar System. (Sizes and distances are not to scale.)

It was perhaps his recognition of the vast size of the Sun that led Aristarchus to the revolutionary idea that the Sun, and not the Earth, is the center of the heavens. Aristarchus was, of course, correct, but his idea was too revolutionary, and another 2000 years passed before scientists became convinced of its correctness.

However, this was not mere pigheadedness. There was a good reason for not believing that the Earth moves around the Sun. If it did, the positions of stars should change during the course of the year. Looking at figure 1.25, you can see that the nearby star should appear to lie in a different position in January and in July.

This shift in a star's apparent position resulting from Earth's motion is called **parallax,** and Aristarchus's critics were absolutely right in supposing that it should occur. So, when they could observe no parallax caused by the Earth's motion, they concluded that Aristarchus's Sun-centered system must be wrong. But Aristarchus's critics failed to appreciate how tiny parallax is. In particular, they did not realize that stars are so enormously far away that parallax is far too small to be detected without a powerful telescope. As a result, in Aristarchus's time (about 2000 years before the telescope was invented) there was no hope of detecting the parallax of stars. It was not until 1838 that astronomers had telescopes of sufficient power and accuracy to measure this parallax. Thus, an idea may be rejected for reasons that are logically correct but are based on data of limited accuracy.

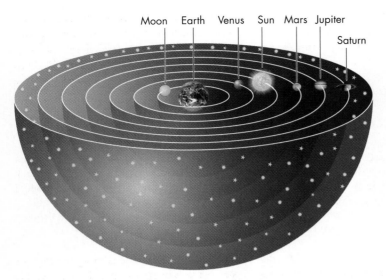

Moon   Earth   Venus   Sun   Mars   Jupiter   Saturn

**FIGURE 1.26**
A cutaway view of the geocentric model of the Solar System according to Eudoxus. (Mercury's sphere is omitted for clarity.)

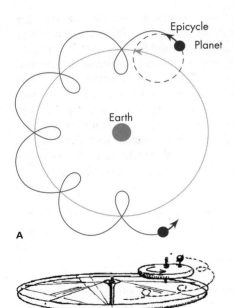

Epicycle
Planet
Earth

A

B

**FIGURE 1.27**
Epicycles are a bit like a bicycle wheel with a Frisbee bolted onto its rim.

**ANIMATION**

*Ptolemy's model of motion of a planet*

## Explaining the Motion of the Planets

Following the basic discoveries about the size and distance of the Sun and Moon, the main thread of astronomical research for almost the next 2000 years centered on the motion of the planets. From earliest Greek times, one simple model formed the basis for understanding those motions. It was based on the observation that as one looks at the sky, everything seems to move around the Earth from east to west. This led naturally to the theory that the Earth was at the center of the Universe and the planets and stars moved around it. Models of the Universe of this type are called **geocentric theories.**

Figure 1.26 shows a typical geocentric model based on the work of the Greek astronomer Eudoxus, who lived about 400–347 B.C. In this model, the Sun, Moon, and planets all revolve around the Earth. The bodies that move fastest across the sky are those that are nearest to the Earth. Thus, the Moon, whose path through the stars takes only about 28 days, is nearest to the Earth, whereas Saturn, whose path through the stars takes roughly 29 years, is located the farthest out of the planets known then. By assuming that each body was mounted on its own revolving transparent (crystalline) sphere and by tipping the spheres slightly with respect to one another, Eudoxus was able to give a quite satisfactory explanation of the motions of the heavenly bodies at any given time.

Unfortunately, such models do not work well if they are used to explain retrograde motion, unless one believes that the planets sometimes stop moving, reverse direction, pause, and then resume their original motion. This idea is clumsy and unappealing. Eudoxus was able to explain retrograde motion only by requiring that each planet moved on *two* spheres, one inside the other. By adjusting their rotation rates and sizes, he was able to get reasonable agreement with the observed positions of the planets as they shifted across the sky. However, about A.D. 150, far more elaborate models, which could be used to predict with much better accuracy, were developed by Ptolemy, the great astronomer of Greco-Roman times.

## Ptolemy

Ptolemy lived in Alexandria, Egypt, which at that time was one of the intellectual centers of the world, in part because of its magnificent library. Ptolemy fashioned a model of planetary motions in which each planet moved on one small circle, which in turn moved on a larger one (fig. 1.27). The small circle, called an **epicycle,** was supposed to be carried along on the large circle like a Frisbee spinning on the rim of a bicycle wheel.*

According to Ptolemy's model, the motion of the planet from east to west across the night sky is caused by the rotation of the large circle (the bicycle wheel in our analogy). Retrograde motion occurs when the epicycle carries the planet in a reverse direction (caused by the rotation of the Frisbee, in our model). Thus, with epicycles, it is possible to account for retrograde motion, and Ptolemy's model was able to predict planetary motions with fair precision.

Unfortunately, discrepancies remained between the predicted and true positions of the planets. This led to further modifications of the model, each of which led to slightly better agreement but at the cost of adding much greater complexity.

---

*Ptolemy seems to have got the idea of epicycles from the writings of Hipparchus, who lived about 150 B.C. Hipparchus is best known to astronomers for his invention of the magnitude system (see chapter 12) for measuring stellar brightness and for his discovery of precession (see chapter 5). The latter was made possible by his meticulous observations of star positions and the care with which he compared his data to those of his predecessors.

Nevertheless, Ptolemy's theory survived until the 1500s and ultimately collapsed from both its failure to adequately predict where planets should be and also a loss of faith in its complexity. It had become far too complex to be plausible, and simplicity is an important element of scientific theory. As the medieval British philosopher William of Ockham wrote in the 1300s, "Entities must not be unnecessarily multiplied," a principle known as "Ockham's razor."

Ptolemy's era was one of decay and general political instability for the Greco-Roman civilization, which accounts for our uncertainty about the year of his birth or death. We know of him mainly through his great book, *Almagest,* a compendium of the astronomical knowledge of the ancient Greeks. The book includes tables of star positions and brightnesses. In fact, much of what we know of Ptolemy (and of Greek and Roman civilization in general) we owe to the Islamic civilization that flourished around the southern edge of the Mediterranean from about 700 to 1200.

### Islamic Contributions

Islamic civilization, like so many others, relied on celestial phenomena to set its religious calendar. For example, Islam's influence is very evident in astronomy through Arabic words such as *zenith* and the names of nearly all the bright stars—Betelgeuse, Aldebaran, and so on. In addition, Islamic scholars revolutionized mathematical techniques through innovations such as algebra (another Arabic word) and Arabic numerals.

### Asian Contributions

The early people of Asia, like their contemporaries to the west, studied the heavens. They too devised constellations, but based on their own mythologies, and they too made maps of the sky. Although the ancient astronomers of the East did not devise elaborate geometric models of the heavens, their careful observations of celestial events nevertheless prove useful to astronomers even today. For example, Chinese, Japanese, and Korean astronomers kept detailed records of unusual celestial events, such as eclipses, comets, and exploding stars, and they even noted dark spots on the Sun (sunspots) that they could occasionally see with the naked eye when the Sun was low in the sky and its glare was dimmed by dust or haze. Based on such records, Chinese astronomers devised ways to predict eclipses. Their records of exploding stars are extremely, valuable and allow today's astronomers to determine the dates of many of these celestial outbursts.

## 1.3   ASTRONOMY IN THE RENAISSANCE

### Copernicus

The man who began the demolition of the geocentric model and the revolution in astronomical ideas that continues to this day was a Polish physician and lawyer by the name of Copernicus (1473–1543). Copernicus (fig. 1.28) made many attempts to reconcile the centuries of data on planetary positions that had been collected since Ptolemy's geocentric model, but all such attempts failed. Thus, he was led to reconsider Aristarchus's ancient idea that the Earth moves around the Sun.

You will remember that **heliocentric models** in which the Sun (*hēlios,* in Greek) was the center of the system had been proposed nearly 2000 years earlier by Aristarchus but had been rejected partly because the observational tools available at that time were inadequate to detect stellar parallax. Nevertheless, such models offer an enormously simpler explanation of retrograde motion. In fact, if the planets orbit the Sun, retrograde motion becomes a simple consequence of one planet on a smaller orbit overtaking and passing another on a larger orbit, as Copernicus was able to show.

**FIGURE 1.28**
Copernicus

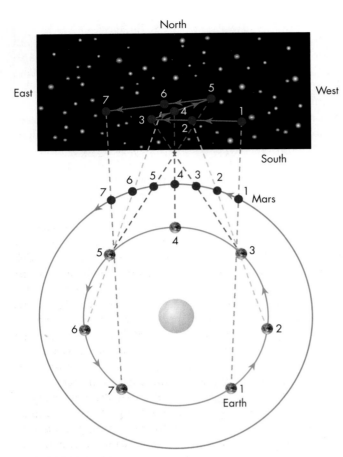

**FIGURE 1.29**
Why we see retrograde motion.
(Object sizes, positions, and distances
are exaggerated for clarity.)

**ANIMATION**

*The retrograde motion of Mars according to
the heliocentric model*

To see why retrograde motion occurs, look at figure 1.29. Here we see the Earth and Mars moving around the Sun. The Earth completes its orbit, circling the Sun in 1 year, whereas Mars takes 1.88 years to complete an orbit, with the Earth overtaking and passing Mars every 780 days.

If we draw lines from the Earth to Mars, we see that Mars will appear to change its direction of motion against the background stars each time the Earth overtakes it. A very similar phenomenon happens when you drive on a highway and pass a slower car. Both cars are, of course, moving in the same direction. However, as you pass the slower car, it *looks* as if it is shifting backward against the stationary objects behind it.

With his heliocentric model, Copernicus not only could give a simple explanation of retrograde motion but also could determine each planet's distance from the Sun by a geometric construction, as shown in figure 1.30. The distances found in this manner must be expressed in terms of the Earth's distance from the Sun (whose value was not known until several hundred years later), but table 1.1 illustrates that they agree well with modern values. These distances became an important starting place for those who came later.

Copernicus described his model of a Sun-centered universe in one of the most influential scientific books of all time, *De revolutionibus orbium coelestium (On the Revolutions of the Celestial Orbs)*. Because his ideas were counter to the teaching of the Christian church, they were met with hostility and skepticism. The book itself was not published until shortly before Copernicus's death (which was perhaps just as well), and he saw the first copy while on his deathbed.

**A**

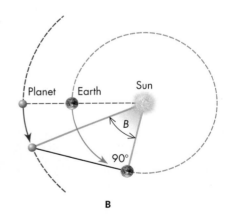

**B**

**FIGURE 1.30**
How Copernicus calculated the distance to the planets. (A) When an inner planet appears in the sky at its farthest point from the Sun, the planet's angle on the sky away from the Sun, *A*, can be measured. You can see from the figure that at the same time angle *B* is 90°. The planet's distance from the Sun can then be calculated with geometry, if one knows the measured value of angle *A* and the fact that the Earth–Sun distance is 1 AU. (B) Finding the distance to an outer planet requires determining how long it takes the planet to move from being opposite the Sun in the sky (the planet rises at sunset) to when the Sun–Earth–planet angle is 90° (the planet rises at noon or midnight). Knowing that time interval, one then calculates what fraction of their orbits the Earth and planet moved in that time. Multiplying those fractions by 360° gives the angles the planet and Earth moved. The difference between those angles gives angle *B*. Finally, using geometry and the value of angle *B* as just determined, the planet's distance from the Sun can be calculated.

**TABLE 1.1**

## Planetary Distances According to Copernicus

| Planet | Distance in AU According to Copernicus | Actual Distance |
|---|---|---|
| Mercury | 0.38 | 0.39 |
| Venus | 0.72 | 0.72 |
| Earth | 1.00 | 1.00 |
| Mars | 1.52 | 1.52 |
| Jupiter | 5.22 | 5.20 |
| Saturn | 9.17 | 9.54 |

Ironically, some of the criticism of Copernicus's work was justified. Although his model was basically correct, it did not account for the observed positions of the planets any more accurately than did Ptolemy's more complicated but incorrect model. This lack of complete agreement between model and observation arose at least in part because Copernicus insisted that the planetary orbits were circles. Furthermore, his model again raised the question of why no stellar parallax could be seen. Finally, his views of planetary motion ran counter to the teachings of Aristotle, views supported both by "common sense" and by the Catholic church at that time. After all, when we observe the sky, it *looks* as if it moves around us. Moreover, we do not detect any sensations caused by the Earth's motion—it feels at rest. This mixture of rational and irrational objections made even scientists slow to accept the Copernican view.

However, by this time, there was a growing recognition of the immensity of the Universe. Astronomers such as the Englishman Thomas Digges and the Italian Giordano Bruno went so far as to claim that the stars were other suns, perhaps with other worlds around them. This new scientific open-mindedness coupled with the aesthetic appeal of the simpler system led to a growing belief in the Copernican system.

## Tycho and Kepler

Copernicus's model, although not the only stimulus, marked the opening of a new era in the history of astronomy. Conditions were favorable for new ideas: the cultural renaissance in Europe was at its height; the Protestant reformation had just begun; the new world was being settled. In such an intellectually stimulating environment, new ideas flourished and found, at least among scientists, a more receptive climate than in earlier times.

One scientist whose ideas flourished in this more intellectually open environment was the sixteenth-century Danish astronomer Tycho Brahe (1546–1601). Born into the Danish nobility, Tycho (fig. 1.31) utilized his position and wealth to indulge his passion for study of the heavens, a passion based in part on his professed belief that God placed the planets in the heavens to be used as signs to mankind of events on Earth. Driven by this interest in the skies, Tycho designed and had built instruments of far greater accuracy than any yet devised in Europe. Tycho then used these devices to make precise measurements of planetary positions. His meticulous observations turned out to be crucial not only for distinguishing the superiority of the heliocentric over the geocentric system but also for revealing the true shape of planetary orbits.

Tycho was more than just a recorder of planetary positions; he recognized opportunity when he saw it. In 1572, when an exploding star (what we would now call a supernova) became visible, Tycho demonstrated from its lack of motion with respect to the other stars that it was far beyond the supposed spheres on which the planets move. Likewise, when a bright comet appeared in 1577, he showed that it lay far beyond the Moon, not within the Earth's atmosphere, as taught by the ancients. These

**FIGURE 1.31**
Tycho Brahe

**FIGURE 1.32**
Johannes Kepler

observations suggested that the heavens were both changeable and more complex than was previously believed.

Although Tycho could see the virtues of the simplicity offered by the Copernican model, he was also unconvinced of its validity because he could find no evidence for stellar parallax. Therefore, he offered a compromise model in which all of the planets except the Earth went around the Sun, while the Sun, as in earlier models, circled the Earth. Tycho was the last of the great astronomers to hold that the Earth was at the center of the Universe.

Upon Tycho's death, his observational data were passed to his young assistant, Johannes Kepler. Kepler (1571–1630) was able to derive from this huge set of precise information a detailed picture of the path of the planet Mars. Whereas all previous investigators had struggled to fit the planetary paths to circles, Kepler (fig. 1.32), by using Tycho's superb data, was able to show that the path of Mars was not circular but elliptical.

The shape of an **ellipse** is described by its long and short dimensions, called its major and minor axes, respectively. An ellipse can be drawn with a pencil inserted in a loop of string that is hooked around two thumbtacks. If you move the pencil while keeping it tight against the string, as shown in figure 1.33A, you will draw an ellipse. Each point marked by a tack is called a **focus** of the ellipse.

Having calculated from the observations that the orbit of Mars was elliptical, Kepler noticed that the Sun was located at a spot that was *not* the center of the ellipse but was off center at a focus. With the elliptical shape of the orbit now established, he was able to obtain excellent agreement between the calculated and the observed position not only of Mars but also of the other planets. Thus, Kepler's seemingly simple proposal that planetary orbits are ellipses and not circles was a critical step in understanding planetary motion.

Along with discovering the shape of planetary orbits, Kepler also measured the relative sizes of the orbits. Because the orbits are elliptical, he could not describe them by a single radius. He therefore used the orbit's **semimajor axis**—half the major axis, as shown in figure 1.33B—as a measure of its size. When he compared these orbital sizes with how long it takes the planets to orbit the Sun, their periods, Kepler noticed the relation illustrated in table 1.2: the square of the orbital period was proportional to the cube of the orbital size, as measured by the semimajor axis.

**FIGURE 1.33**
(A) Drawing an ellipse. (B) The Sun lies at one focus of the ellipse.

### TABLE 1.2

**Table Illustrating Kepler's Third Law for the Planets Known at His Time**

| Planet | Distance from Sun (*a*) (in Astronomical Units) | Orbital Period (*P*) (in Years) | $a^3$ | $P^2$ |
|---|---|---|---|---|
| Mercury | 0.387 | 0.241 | 0.058 | 0.058 |
| Venus | 0.723 | 0.615 | 0.378 | 0.378 |
| Earth | 1.0 | 1.0 | 1.0 | 1.0 |
| Mars | 1.524 | 1.881 | 3.54 | 3.54 |
| Jupiter | 5.20 | 11.86 | 141.0 | 141.0 |
| Saturn | 9.54 | 29.46 | 868.0 | 868.0 |

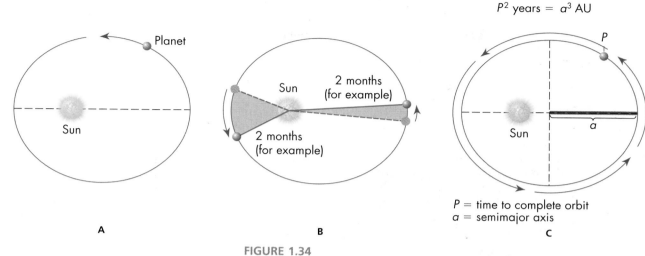

$P^2$ years $= a^3$ AU

$P =$ time to complete orbit
$a =$ semimajor axis

**FIGURE 1.34**

Kepler's three laws. (A) A planet moves in an elliptical orbit with the Sun at one focus. (B) A planet moves so that a line from it to the Sun sweeps out equal areas in equal times. Thus, the planet moves fastest when nearest the Sun. For purposes of the drawings a 2-month interval is chosen. (C) The square of a planet's orbital period (in years) equals the cube of the semimajor axis of its orbit (in AU), the planet's distance from the Sun if the orbit is a circle.

*Kepler's second law*

*Kepler's third law*

Kepler's discoveries of the nature of planetary motions are expressed in what are known today as **Kepler's three laws:**

I. Planets move in *elliptical* orbits with the Sun at one focus of the ellipse (see fig. 1.34A).
II. The orbital speed of a planet varies so that a line joining the Sun and the planet will sweep over equal areas in equal time intervals (see fig. 1.34B).
III. The amount of time a planet takes to orbit the Sun is related to its orbit's size, such that the period, $P$, squared is proportional to the semimajor axis, $a$, cubed (fig. 1.34C). Mathematically,

$$P^2 = a^3$$

where $P$ is measured in years and $a$ is measured in astronomical units.

These three laws describe the essential features of planetary motion around our Sun. They describe not only the shape of a planet's path but also its speed and distance from the Sun. For example, the second law—in its statement that a line from the planet to the Sun sweeps out equal areas in equal times—implies that when a planet is near the Sun, it moves more rapidly than when it is farther away. We can see this by considering the shaded areas in figure 1.34B. For the areas to be equal, the distance traveled along the orbit in a given time must be larger when the planet is near the Sun. Thus, according to Kepler's second law, as a planet moves along its elliptical orbit, its speed changes, increasing as it nears the Sun and decreasing as it moves away from the Sun.

The third law also has implications for planetary speeds, but it deals with the relative speeds of planets whose orbits are at different distances from the Sun, not the speed of a given planet. Because the third law states that $P^2 = a^3$, a planet far from the Sun (larger $a$) has a longer orbital period ($P$) than one near the Sun (see table 1.2). For example, the Earth takes 1 year to complete its orbit, but Jupiter, whose distance from the Sun is slightly more than 5 times Earth's distance, takes about 12 years. Thus, a planet orbiting near the Sun overtakes and passes a planet orbiting farther out, leading to the phenomenon of retrograde motion, as discussed earlier.

*Kepler's laws*

Kepler's third law has other implications. For example, as we shall see in section 2.7, the law gives information about the nature of the force holding the planets in orbit. Also, it implies that a planet close to the Sun moves along its orbit faster than a planet far from the Sun. Finally, the third law allows us to calculate the distance from the Sun of any body orbiting it if we measure the body's orbital period. The distance we obtain will be relative only to the Earth's distance, but the law thereby gives us at least the relative scale of the Solar System.

For example, suppose we wish to determine how far Pluto is from the Sun compared with the Earth's distance from the Sun. We measure Pluto's motion against the background stars to determine how long Pluto takes to circle the Sun, finding from such observations that its orbital period, $P$, is 248 years. Kepler's third law tells us that $P^2 = a^3$. If $P = 248$ yrs, then $(248)^2 = a^3$. Squaring 248 and then taking the cube root gives us $a = 39.5$ astronomical units. That is, Pluto is about 40 times farther from the Sun than the Earth is.

Apart from such astronomical applications, Kepler's laws have an additional significance. Kepler's laws are the first mathematical formulas to describe the heavens correctly, and as such they revolutionized our way of thinking about the Universe. Without such mathematical formulations of physical laws, much of our technological society would be impossible. These laws are therefore a major breakthrough in our quest to understand the world around us.

It is perhaps ironic that such mathematical laws should come from Kepler because so much of his work is tinged with mysticism. For example, as a young man, he sought to explain the spacing of the planets as described in Copernicus's work in terms of nested geometrical figures, the sphere, the cube, and so on. In fact, it was Tycho's notice of this work that led to his association with Kepler. Moreover, Kepler's third law evolved from his attempts to link planetary motion to music, using the mathematical relations known to exist between different notes of the musical scale. Kepler even attempted to compose "music of the spheres" based upon such a supposed link. Nevertheless, despite such excursions into these nonastronomical matters, Kepler's discoveries remain the foundation for our understanding of how planets move. As a final remark about his contributions, we owe credit to Kepler for the word *satellite*. When he saw the moons of Jupiter with a small telescope, their motion around the planet made him think of attendants or bodyguards—*satellēs*, in Latin.

## Galileo

At about the same time that Tycho Brahe and Kepler were striving to understand the motion of heavenly bodies, the Italian scientist Galileo Galilei (1564–1642) was likewise trying to understand the heavens. However, his approach was dramatically different.

Galileo (fig. 1.35) was interested not just in celestial motion but in all aspects of motion. He studied falling bodies, swinging weights hung on strings, and so on. In addition, he used the newly invented telescope to study astronomical objects. Galileo did not invent the telescope himself. That invention seems to have been the work of the Dutch spectacle maker Johannes (Hans) Lippershey. However, Galileo was the first person we know of who used the telescope to study the heavens and interpret his findings, and what he found was astonishing.*

In looking at the Moon, Galileo saw that its surface had mountains and was in that sense similar to the surface of the Earth. Therefore, he concluded that the Moon was not some mysterious ethereal body but a ball of rock. He looked (without taking adequate precaution) at the Sun and saw dark spots (now known as sunspots) on its surface. He noticed that the position of the spots changed from day to day, showing that the Sun not only had blemishes and was not a perfect celestial orb but that it also

**FIGURE 1.35**
Portrait of Galileo

http://www.mhhe.com/arny

*Thomas Harriott (1560–1621), an English mathematician–scientist, appears to have used a telescope to study the heavens a little before Galileo. He too saw sunspots and the moons of Jupiter, but he failed to publish his discoveries at the time.

changed. Both these observations were in disagreement with previously held conceptions of the heavens as perfect and unchangeable. In fact, by observing the changing position of the spots from day to day, Galileo deduced that the Sun rotated.

Galileo looked at Jupiter and saw four smaller objects orbiting it, which he concluded were moons of the planet. These bodies, known today in his honor as the Galilean satellites, proved unambiguously that there were at least *some* bodies in the heavens that did not orbit the Earth. But perhaps even more important, the motion of these bodies raised the crucial problem of what held them in orbit. He looked at Saturn and discovered that it did not appear as a perfectly round disk but that it had blobs off the edge. However, his telescope was too small and too crudely made (inferior to inexpensive binoculars of today) to show these as rings, a discovery that had to wait until 1656, when they were first seen by the Dutch scientist Christiaan Huygens as features that were detached from the planet.

Galileo also examined the Milky Way and saw that it was populated with an uncountable number of stars. This single observation, by demonstrating that there were far more stars than previously thought, shook the complacency of those who believed in the simple Earth-centered universe.

Galileo observed that Venus went through a cycle of phases, like the Moon, as illustrated in figure 1.36. The relation between the phase of the planet and its position

*The phases of Venus according to the Ptolemaic and Copernican Systems*

A

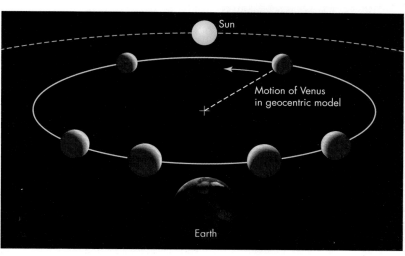

B

**FIGURE 1.36**

As Venus orbits the Sun, it goes through a cycle of phases (A). The phase and its position with respect to the Sun show conclusively that Venus cannot be orbiting the Earth. The gibbous phases Galileo observed occur for the heliocentric model but cannot happen in the Earth-centered Ptolemaic model (B), where Venus is shown on its epicycle.

with respect to the Sun left absolutely no doubt that Venus must be in orbit around the Sun exactly as the Moon is in orbit around the Earth. Perhaps more than any other observation, this one dealt the death blow to the old geocentric model of planetary motion.

Galileo's contributions to science would be honored even had he not made all these important observational discoveries, for he is often credited with originating the experimental method for studying scientific problems. From his experiments on the manner in which bodies move and fall, Galileo deduced the first correct "laws of motion," laws that ultimately led Newton to his explanation of why the planets obey the laws of planetary motion that Kepler discovered.

Galileo's probings into the laws of nature led him into trouble with religious "law." He was a vocal supporter of the Copernican view of a Sun-centered universe and wrote and circulated his views widely and somewhat tactlessly. His exposition followed the style of Plato, presenting his arguments as a dialog between a wise teacher (patterned after himself) and an unbeliever in the Copernican system whom he called Simplicio and who, according to his detractors, was patterned after the pope. Although the pope was actually a friend of Galileo, more conservative churchmen urged that Galileo be brought before the Inquisition because his views that the Earth moved were counter to the teachings of the Catholic Church. Considering that his trial took place at a time when the papacy was attempting to stamp out heresy, Galileo escaped lightly. He was made to recant his "heresy" and was put under house arrest for the remainder of his life. Only in 1992 did the Catholic Church admit it had erred in condemning Galileo for his ideas.

## 1.4     ISAAC NEWTON AND THE BIRTH OF ASTROPHYSICS

Isaac Newton (1642–1727), who was born the year that Galileo died, is arguably the greatest scientist of all time. Newton's contributions span mathematics, physics, and astronomy. Moreover, Newton (fig. 1.37) pioneered the modern studies of motion, optics, and gravity. In his attempts to understand the motion of the Moon, Newton not only deduced the law of gravity but also discovered that he needed mathematical methods for calculating the gravitational force of a spherical body and that no such methods were then available. This realization led him to invent what we now know as calculus.

What is especially remarkable about Newton's work is that the discoveries he made in the seventeenth century still form the core for most of our understanding of gravity and the motion of bodies, discoveries we will discuss in more detail in chapter 2. In chapter 3, we will discuss some of Newton's ideas and discoveries about light, ideas that also are still in use.

Newton was a fascinating individual. He came from very modest origins and rose to high positions not only in academia but also in the government. He was Warden of the Mint and is alleged to have invented milling, the process whereby grooves are cut in the edge of coins to detect metal being pared off them, which would debase their value. He was also a deeply religious man and wrote prolifically on theological matters as well as science.

Newton's laws of motion, when combined with his law of gravity, were successfully applied for the next 200 years to essentially all problems of the motion of astronomical bodies. Only at the beginning of the twentieth century, with the discovery of a tiny discrepancy in the motion of Mercury as calculated using Newton's work, did scientists recognize that Newton's laws may not have been the last word on planetary motion. For example, his descriptions of motion require modification if we are to correctly describe motion at speeds near that of light or where gravitational fields are very intense. These modifications are incorporated in Einstein's theories of relativity.

**FIGURE 1.37**
Isaac Newton

## 1.5   THE GROWTH OF ASTROPHYSICS

Newton's enormous contributions tend to overshadow other advances in astronomy during the eighteenth and nineteenth centuries. That period began with observational discoveries that increased astronomers' faith in using physical laws to understand the structure and workings of astronomical bodies. However, by the end of the period, newly found physical laws gave astronomers totally new tools for studying the heavens. In fact, the increasing use of the word *astrophysics* describes that shift well.

### New Discoveries

Unexpected discoveries play a major role today in expanding our knowledge of the heavens, no less so than in the time immediately after Newton's death. For example, in 1781 the English astronomer Sir William Herschel discovered the planet Uranus. He also discovered that some stars have companion stars in orbit around them. The motion of such double stars offered additional tests of Newton's laws, but the most striking triumph of these laws of motion was their explanation of irregularities in the orbital motion of Uranus. Such irregularities hinted that another body was exerting a gravitational force on Uranus, and from Newton's laws, astronomers could calculate the position of the unseen body. As we will discuss further in chapter 9, a search of the sky near the calculated position revealed the planet Neptune.

### New Technologies

New technologies also played an important role during this period. For example, improvements in optics allowed astronomers to build bigger telescopes and thereby observe much fainter objects. Among these objects, astronomers found dim, fuzzy patches of light—the so-called nebulas. Some of these were gas clouds within the Milky Way; others turned out to be external star systems similar to the Milky Way.

Another important technological advance was the application of photography to astronomy, starting in the middle of the nineteenth century. Photographic film gave astronomers permanent records of what they saw, and because film could store light during long exposures, astronomers were now able to detect objects much fainter than the eye could see in a single moment.

The scientific and technical advances described above have a direct bearing on astronomy, but scientific discoveries often influence totally unconnected areas. For example, during the eighteenth and nineteenth centuries, many scientists were studying the nature of matter and heat. The study of heat was prompted, at least in part, by a desire to improve the newly invented steam engine.

### The Nature of Matter and Heat

Like so many of our ideas about the nature of the Universe, our ideas of matter date back to the ancient Greeks. For example, Leucippus, who lived about the fifth century B.C. in Greece, and his student Democritus taught that matter was composed of tiny indivisible particles. They called these particles *atoms,* which means "uncuttable" in Greek. Our current model for the nature of atoms dates back to the early 1900s, with the work of the British physicist Ernest Rutherford. Rutherford showed with a series of experiments that atoms have a tiny core, the nucleus, around which yet smaller particles, called electrons, orbit. We will discuss this model of the atom more fully as we describe the production of light (chapter 3) and as we discuss the source of energy for stars (chapter 11).

The recognition of a link between energy and atoms began in the eighteenth and nineteenth centuries as scientists attempted to understand the behavior of gases. For example, why does heating a gas raise its pressure? The answer comes from the recognition (discussed more fully in chapter 11) that heating increases the motion of a gas's molecules and that pressure arises from that motion. One of the most important contributors to the understanding of heat and molecular motion was the English physicist Lord Kelvin.

## The Kelvin Temperature Scale

Kelvin studied numerous problems in physics and astronomy, ranging from the motion of fluids to the properties of gases. Much of this latter work was motivated by his attempts to improve the energy efficiency of steam engines. In the course of studying the energy content of gases, Kelvin devised a temperature scale that is used today in virtually all the physical sciences. The reason for this wide usage is that on the Kelvin scale, a body's temperature is directly related to its energy content and to the speed of its molecular motion. That is, the greater a body's Kelvin temperature, the more rapidly its atoms move and the more energy it possesses. Similarly, if the body is cooled toward a temperature of zero on the Kelvin scale, molecular motion within it slows to a virtual halt and its energy approaches zero. Partly as a result of this, the Kelvin scale has no negative temperatures.

Temperatures on the Kelvin scale are not given in degrees but are simply called "Kelvin." For example, the freezing and boiling points of water are very nearly 273 and 373 Kelvin, respectively. Room temperature is about 300 Kelvin. Relatively simple formulas allow conversion between Kelvin and the more familiar Fahrenheit and Celsius scales. Celsius temperatures are simply Kelvin temperatures minus 273. Fahrenheit temperatures, $F$, can be calculated by using the formula $F = 9/5\ K - 459.4$, where $K$ is the Kelvin temperature. Figure 1.38 shows how the Kelvin, Celsius, and Fahrenheit temperature scales compare. Because of its direct relation to so many physical processes, we will use the Kelvin scale in most of the remainder of this book.

**FIGURE 1.38**
Temperatures in Kelvin (K) and on the Celsius and Fahrenheit scales.

# SUMMARY

We create a mental model of the night sky overhead as a giant dome, forming part of the celestial sphere. Star patterns on the celestial sphere are called *constellations*. Using this model, stars rise in the east and set in the west as the celestial sphere rotates overhead. This apparent motion is actually caused by the Earth's spin. The points about which the celestial sphere appears to turn are called the *celestial poles*.

At any given time of year, the Sun's glare hides those stars lying near it on the celestial sphere. However, as the Earth moves around the Sun, it looks to us as if the Sun changes its position with respect to the stars. Thus, stars previously visible are lost in the Sun's glare, whereas stars previously hidden become visible. Therefore, different constellations are visible at different times of year. We call the path that the Sun follows around the celestial sphere the *ecliptic*.

Planets also move with respect to the constellations but always within the narrow band of the zodiac. The zodiac follows the Sun's path around the sky, but it is slightly tilted with respect to the celestial equator, an imaginary line on the celestial sphere that lies directly above the Earth's equator. The usual direction of planetary motion is from west to east with respect to the stars. However, during part of the year, each planet shifts in the other direction, undergoing apparent retrograde motion.

The zodiac and the Sun's path are tipped at an angle of 23.5° to the celestial equator because the Earth's rotation axis is tipped by that amount with respect to its orbit. The solstices and equinoxes mark when the Sun reaches its maximum distance from the celestial equator and when it crosses the equator, respectively. These points define the onsets of the seasons.

The Earth's spin keeps its rotation axis pointing in nearly a fixed direction as we orbit the Sun. Because the axis is tipped, the Sun shines more directly on the Northern Hemisphere for half the year and on the Southern Hemisphere for the other half of the year. This difference in exposure to the Sun's light and warmth creates the seasons.

Ancient peoples noted the basic patterns of the night sky, but the Greeks appear to have been the first to give explanations of planetary motion, based on a combination of observational and geometric analysis. The Greeks pictured the planets, Sun, and Moon all orbiting the Earth on crystalline spheres.

Through the work of Aristotle and Eratosthenes respectively, the Greeks knew the shape and size of the Earth. Aristarchus measured the relative size and distance of the Moon and Sun and proposed about 300 B.C. that the Earth orbited the Sun. However, his model was rejected because the expected shift in star positions (parallax) was unobservable at that time.

Ptolemy (about A.D. 150) developed a complex model of planetary motion with the Earth at the center (geocentric) and with retrograde motion explained by planets moving on epicycles.

The geocentric model began to crumble in the 1500s with the revival of the heliocentric model by Copernicus. Better observations by Tycho Brahe and detailed mathematical models by Kepler based on those observations placed the heliocentric model on a firmer basis. Galileo's observations with the recently invented telescope helped prove the heliocentric model.

Newton's discovery in the 1700s of the law of gravity and the laws of motion allowed him to explain why Kepler's laws worked, thereby completing the understanding of planetary motions.

## ?What It Is.

Chapter opener: This is the Moon's shadow on the Earth's surface. The shadow is usually a few hundred kilometers across. People within the region of the shadow would be able to see a total solar eclipse.

# QUESTIONS FOR REVIEW

1. What is the celestial sphere?
2. What is a constellation?
3. What is the zodiac?
4. What is the ecliptic?
5. What causes the seasons?
6. When it is winter in Australia, what season is it in New York? Is it then April in Paris?
7. Where on the celestial sphere would you look for the planets?
8. How long does it take the Moon to go through a cycle of phases?
9. Sketch the path on the sky that a planet makes when undergoing retrograde motion.
10. Will a planet in retrograde motion rise in the east or west?
11. If you see a bright "star" in the sky, how could you tell whether it is a star and not, for example, Venus?
12. Describe the major astronomical contribution(s) of the following in a sentence or two for each: Aristotle, Aristarchus, Eratosthenes, Ptolemy, Copernicus, Tycho, Kepler, Galileo, and Newton.
13. What is meant by the phrase *angular diameter*?
14. If you triple your distance from an object, what happens to its angular size?

## THOUGHT QUESTIONS

1. If you were standing on the Earth's equator, where would you look to see the north celestial pole? Could you see this pole from Australia?

2. Draw a sketch of the Earth and persuade yourself that your latitude is the angle of the north celestial pole above the northern horizon.

3. Can you think of an astronomical reason why the zodiac may have been divided into 12 signs rather than 8 or 16?

4. Some people still believe the Earth is flat. What "proof" would you offer them that it is round? Could you persuade them?

5. If the shape of the Earth's orbit was unaltered but its rotation axis was shifted so that it had no tilt with respect to the orbit, how would seasons be affected?

6. Why does the position of sunrise along the eastern horizon change during the year?

7. Suppose the stars were very much closer than they really are. How might that have made it easier for Aristarchus to persuade people that the Earth moves around the Sun?

8. Tycho argued that the Sun orbited the Earth but that the other planets orbited the Sun. Could Tycho's model explain the phases of Venus as observed by Galileo? Why?

## PROBLEMS

1. Suppose you were an alien living on the fictitious warlike planet Myrmidon and you wanted to measure its size. The Sun is shining directly down a missile silo 1000 miles to your south, while at your location, the Sun is 36° from straight overhead. What is the circumference of Myrmidon? What is its radius?

2. Make a table listing the astronomers named in the review questions above and then add the approximate dates of their births and deaths. Then add a few historic events of each period, as well as names of famous artists, writers, musicians, or politicians who lived about the same time.

3. Suppose a planet were found with an orbital period of 64 years. How might you estimate its distance from the Sun? If its orbit is circular, what is its radius?

4. Suppose you received a message from aliens living on a planet orbiting a star identical to our Sun. They say they live 4 times farther from their star than the Earth is from the Sun. What is the length of their year compared to ours?

5. The great galaxy in Andromeda has an angular diameter along its long axis of about 5°. Its distance is about 2.2 million light-years. What is its linear diameter?

6. A shell of gas blown out of a star has an angular diameter of 0.1° and a linear diameter of 1 light-year. How far away is it?

7. A small robot probe is exploring a spherical asteroid. As the probe creeps over the surface, it drills holes to take soil samples. At one hole, scientists on Earth notice that the Sun shines straight down the hole. At the same time, 10 kilometers due "north," the shadow of the vertical antenna on the main landing craft allows the scientists to deduce that the Sun is 15° from directly overhead. What is the radius of the asteroid? How many times smaller or bigger is its radius than Earth's?

8. A planet is discovered orbiting a nearby star once every 125 years. If the star is identical to the Sun, how could you find the planet's distance from its star? If the planet's orbit is a perfect circle, how far from the star is the planet in AU?

9. Suppose that, in the future, observations with some new telescope reveal a planet about 16 AU from a star whose mass is the same as our Sun's mass. How long does it take the planet to orbit the star?

## TEST YOURSELF

1. If you are standing at the Earth's North Pole, which of the following will be directly overhead?
   (a) The celestial equator
   (b) The ecliptic
   (c) The zodiac
   (d) The north celestial pole
   (e) The Sun

2. A planet in retrograde motion
   (a) rises in the west and sets in the east.
   (b) shifts westward with respect to the stars.
   (c) shifts eastward with respect to the stars.
   (d) will be at the north celestial pole.
   (e) will be exactly overhead no matter where you are on Earth.

3. The circular shape of the Earth's shadow on the Moon led early astronomers to conclude that
   (a) the Earth is a sphere.
   (b) the Earth is at the center of the Solar System.
   (c) the Earth must be at rest.
   (d) the Moon must orbit the Sun.
   (e) the Moon is a sphere.

4. Galileo used his observations of the changing phases of Venus to demonstrate that

   (a) the Sun moves around the Earth.

   (b) the Universe is infinite in size.

   (c) the Earth is a sphere.

   (d) the Moon orbits the Earth.

   (e) Venus follows an orbit around the Sun rather than around the Earth.

5. Kepler's third law

   (a) relates a planet's orbital period to the size of its orbit around the Sun.

   (b) relates a body's mass to its gravitational attraction.

   (c) allowed him to predict when eclipses occur.

   (d) allowed him to measure the distance to nearby stars.

   (e) showed that the Sun is much farther away than the Moon.

## FURTHER EXPLORATIONS

Ahmad, Imad A. "The Science of Knowing God: Astronomy in the Golden Era of Islam." *Mercury* 24 (March/April 1995): 28.

Aveni, Anthony F. "Emissaries to the Stars: The Astronomers of the Ancient Maya." *Mercury* 24 (January/February 1995): 15.

Aveni, Anthony F. "Native American Astronomy." *Physics Today* 37:24, 1984.

Farrell, Charlotte. "The Ninth-Century Renaissance in Astronomy." *Physics Teacher* 34 (May 1996): 268.

Gingerich, Owen. "Astronomy in the Age of Columbus." *Scientific American* 267 (November 1992): 100.

Gingerich, Owen. "Islamic Astronomy." *Scientific American* 254 (April 1986): 74.

Gingerich, Owen. "Copernicus and Tycho." *Scientific American* 229 (December 1973): 86.

Gingerich, Owen. *The Great Copernicus Chase.* New York: Cambridge University Press, 1992.

Gould, Stephen J. "The Persistently Flat Earth." *Natural History* 103 (March 1994): 12.

Hoskin, Michael, ed. *The Cambridge Illustrated History of Astronomy.* New York: Cambridge University Press, 1997.

Krupp, E. C. *In Search of Ancient Astronomies.* New York: McGraw-Hill, 1978.

Krupp, E. C. *Echoes of the Ancient Skies: The Astronomy of Lost Civilizations.* New York: Harper & Row, 1983.

Schaefer, Bradley E. "The Origin of the Greek Constellations." *Scientific American* 295 (November 2006): 96.

### Website

Please visit the *Explorations* website at **http://www.mhhe.com/arny** for additional online resources on these topics.

## PLANETARIUM EXERCISES

All of the planetarium exercises in this text book make use of the Starry Night planetarium software provided with your text. You should install the program and when the program asks you to enter a password, simply click OK to run Starry Night.

Familiarize yourself with the various parameters and controls before trying these exercises. Set the location from which you are viewing to the nearest large city, or your latitude and longitude. Do this by clicking on the location box on the toolbar. Verify that the local time is correct. Set the field of view to 60 degrees on the toolbar. Set the magnitude limit to 5.0 by going to the heading Settings and finding Star Magnitudes.

1. Use the Starry Night program to determine what planets are currently visible in your night sky as well as where they are and when they can be best viewed. After setting up the program as described above, center on each of the visible planets and Uranus (visible with a good pair of binoculars on a dark night). In each case, find the planets visible in the night sky for your location by clicking the Find option on the horizontal toolbar (it looks like a magnifying glass). Move the time forward and back to determine each planet's rising and setting times as well as the constellation it is closest to (click on Guides in the toolbar, then select Constellations). Record your data. Go outside and verify your data by locating at least one of the planets.

2. *Star motions looking east and west.* Set the time step to 30 minutes with the Time option on the toolbar. Set the program so that it displays stars and illustrations for the constellations by using the heading Settings, then Constellations, Illustrations. Set your view to due East. Now step the program forward in 30-minute increments using the Time option in the toolbar. Also try this while facing due West, due South, and due North. What does this star motion tell you about the rotation of Earth on its axis? If stars rise in the eastern part of the sky and set in the western part of the sky (like the Sun), is Earth rotating from west to east or east to west?

3. *Motion of constellations across the sky.* Pick a constellation, such as Orion, that is easily visible. Center on it using the heading Settings and then using the Find option. Set the time step to half-hour intervals and step ahead a few hours. Now go outside and find that same constellation in the sky and follow its motion in half-hour intervals just like in the program. Does this

constellation follow the path that you expected after seeing it on the planetarium program? What, if anything, seems different?

4. *Motion of a planet across the sky.* Set the time step to 7 days. Find a planet that is high in the night sky at sunset, and set the field of view to 60 degrees using the toolbar. Step by step, advance the time forward and observe the position of the planet with respect to the background stars and constellations. In what direction does the planet consistently move with respect to the stars? Explain this motion in a heliocentric model of the solar system.

## KEY TERMS

| | | | | |
|---|---|---|---|---|
| angular size, 44 | ecliptic, 25 | geocentric theories, 46 | model, 22 | semimajor axis, 50 |
| celestial equator, 24 | ellipse, 50 | heliocentric models, 47 | parallax, 45 | solar eclipse, 38 |
| celestial poles, 24 | epicycle, 46 | horizon, 21 | phases, 35 | solstices, 32 |
| celestial sphere, 21 | equinoxes, 30 | Kepler's three laws, 51 | retrograde motion, 34 | zodiac, 33 |
| constellations, 22 | focus, 50 | lunar eclipse, 38 | rotation axis, 28 | |

## PROJECTS

**1. Motion of the Moon and planets.** Obtain a star map (Sky Publishing Company's SC1 available from Sky Publishing Co., P.O. Box 9111, Belmont, MA 02178-9118, is excellent and inexpensive). Determine from the calendar when the Moon is a few days past new so that it will be visible in the early evening. Go outside shortly after sunset and look for the Moon in the west near where the Sun went down. Next check the dates along the bottom of the star map to find what region of the sky is visible at that date and time. As the sky darkens, locate the brighter stars near the Moon and mark on the chart where the Moon is with respect to those stars. Finally, make a sketch of the Moon's shape.

Repeat this process for the next four or five nights. The Moon will set a little later each evening, so you need to adjust your observing time accordingly. After watching a few nights, mark out the Moon's path on the star map. Ideally, you might want to follow the Moon's track for about 2 weeks, although as the Moon reaches third quarter (3 weeks after new moon), you'll have to stay up pretty late because the third-quarter moon does not rise until midnight. If you are really ambitious, get up before dawn and watch the crescent moon shrink as it approaches the new phase.

You can also use this method to study the motion of the planets. Venus is a good choice because it is bright and moves rapidly across the sky. However, Venus may lie too close to the Sun to be easily visible.

To locate the planets, you can look in an almanac, which will tell you what constellation the planet is in on the day you are observing. You may also find this information in the local newspaper, or you can ask your instructor to mark off the planet's approximate position on the star chart to get you started. Because the outer planets move relatively slowly across the sky, you should space out your observations, perhaps marking positions once a week rather than every night.

As you progress in your project, you might ask whether the planets really follow the ecliptic? Can you see retrograde motion?

**2. Motion of the stars across the sky.** Many people are startled when they are told that the stars move across the sky the same way that the Sun does. However, it is easy to show that they do.

Get a stick that you can poke into the ground so it will stand upright. Get a second smaller stick that you can tape or affix to the upright in some manner. A ruler taped to a camera tripod would be ideal, as sketched in Project figure 1.1.

Find a bright star and sight along the smaller stick toward the star. If you now wait 5 or so minutes and again sight along the stick, you will see that it no longer points to the star. That is, the star has moved so that it now lies west of where the stick is pointing. You could do this experiment indoors if you have a window on which you can put a small mark. Set up a chair by the window so that you can watch a star through the glass. While you remain seated, have a friend place a mark on the glass with a grease pencil or piece of tape where the star appears to be. Again, wait a few minutes. The star's motion will be clearly visible.

**3. Observe the changing location of sunset.** Find a spot, perhaps a window facing west, where you can see the western horizon in the evening. Make a sketch of the horizon, noting hills, buildings, or trees that might serve as reference marks. From your chosen viewing spot, watch the sunset and mark on your sketch where the Sun goes down. Label the date and time. Make an observation each week over a period of a month or so. Does the Sun set at the same point each night? If not, which direction along the horizon is the Sun moving? Does the Sun go down at the same

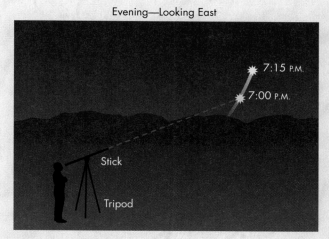

Evening—Looking East

7:15 P.M.

7:00 P.M.

Stick

Tripod

**PROJECT FIGURE 1.1**
Sketch illustrating how to observe the motion of the stars across the sky by sighting along a stick.

time each night? Does it set earlier or later? If you enjoy photography, you might try taking a photograph or videotaping the sunset or sunrise.

**4. Measure the diameter of the Earth.** This project duplicates the method used by Eratosthenes; he used the length of a shadow at two different locations to determine the Earth's size. Thus, you need either to collaborate with someone relatively far away (at least several hundred miles north or south of you) or be able to travel such a large distance yourself. The larger the distance, the better.

Begin by setting up a vertical stick where you live. Tape down a piece of paper beside the stick and mark off on it the length of the stick's shadow over a few hours around noontime.

Record the length of the stick and the smallest shadow.

On the same day as you do your measurement, have your collaborator in the other location do the same. If you are going to travel yourself—for example, during a school break—make your measurements at the two different locations as close together in time as possible.

From a road atlas, measure the straight-line, north–south distance between the two places where you made the measurements. On a piece of paper, make a careful scale drawing of the stick and shadow at each location. A sample pair of drawings is shown in Project figure 1.2A. From these drawings, measure the angles A and B at the top of the triangles.

Now look at Project figure 1.2B. This shows a beam of sunlight striking the Earth and the sticks you used in your experiment at the two different locations. Lines 1, 2, and 3 represent rays of light from the Sun, with lines 1 and 2 just cutting the top of the stick at the two locations, and line 3 passing through the center of the Earth. Notice that the difference in latitude between the two locations is just the angle $AB$. To see why this is so, notice that a line from the center of Earth to $A$ makes equal angles with lines 1 and 3 because these lines are parallel. Similarly, a line from the center to $B$ makes equal angles with lines 2 and 3. If we let $l$ be the difference in latitude between the two locations, then $l = A - B$. You could, of course, just look up that latitude difference in an atlas, but there were no atlases when Eratosthenes lived.

Once you know the angle of the difference in latitude, $l$, you can find the circumference of the Earth. A little geometry will show you that $360/l =$ the Earth's circumference$/D$, where $D$ is the distance in miles or kilometers between the two places at which you made the observations. Thus, you can solve for the Earth's circumference. With the Earth's circumference known, you can then find its radius from the formula Circumference $= 2\pi R$. How does your answer compare with the value you find in the appendix? What might explain the difference? How might you conduct the experiment more accurately?

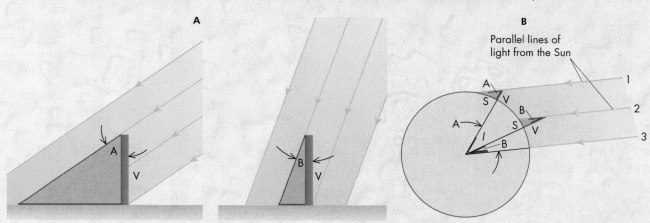

A

B

Parallel lines of light from the Sun

A
S V
B
A
S V
l
B

1

2

3

A

V

B

V

**PROJECT FIGURE 1.2**
Sketch illustrating a method for finding the radius of the Earth.

# Backyard Astronomy

You can learn many of the same things classical astronomers did by simply watching the night sky. But there is a bonus as well. Backyard astronomy is just plain fun, as evidenced by the many thousands of amateur astronomers who in their spare time pursue activities ranging from simply stargazing to searching for new comets.

This section is intended to give you some hints on how to become a backyard astronomer, beginning with learning the constellations and some of the stories associated with them. We will then briefly discuss small telescopes and star charts, introducing some of the terms used to describe the location of the planets. We will conclude with a description of some of the physical changes in your eye when you are observing in very dim light.

## LEARNING THE CONSTELLATIONS

One of the best ways to get started as a backyard astronomer is to learn the constellations. All it takes is a star chart, a dim flashlight, and a place that is dark and has an unobstructed view of the night sky. The star chart will generally tell you how to hold it so that it matches the sky for the date and time that you are observing.

Start by determining which way is north, using a compass if necessary. Then try to locate a few of the brighter stars, matching them up with the chart. This will give you some sense of how big a piece of the sky the chart corresponds to. Next, try to identify a few of the constellations. Focus at first on just a few of the brighter ones. For example, if you live at midlatitude in the Northern Hemisphere, the Big Dipper—part of the constellation Ursa Major—is a good group to start with because it is always in the northern part of the sky.

As you attempt to find and identify stars, your spread hand held at arm's length makes a useful scale. For most people, a fully spread hand at arm's length covers about 20° of sky, or about the length of the Big Dipper from tip of handle to bowl, as shown in figure E1.1. For smaller distances, you can use finger widths (a few degrees).

This scaling of sky distances with your hand makes it easy to point out stars to other people. For example, you can say that a star is two hands away from the Moon and at the 4 o'clock position, as illustrated in figure E1.2.

Once you have come to recognize a few of the constellations, you may be interested in the stories associated with them, that is, star lore.

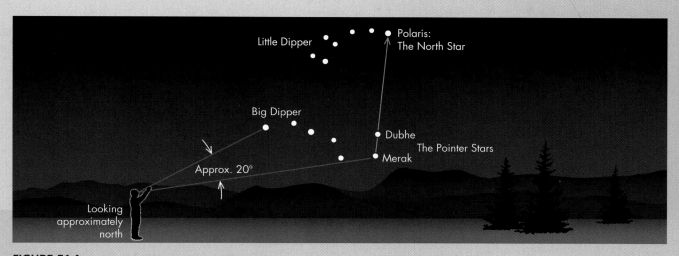

**FIGURE E1.1**
The Big Dipper, part of the constellation Ursa Major, the Great Bear. A line through the two pointer stars points toward Polaris. The Big Dipper spans about 20° of the sky, about a spread hand width at arm's length for most people. The sky is shown approximately as it looks in early October at about 8 P.M.

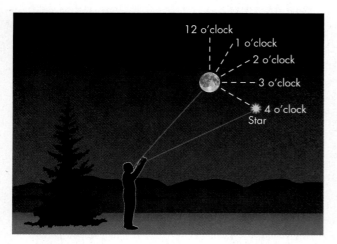

**FIGURE E1.2**
Describing the location of stars by clock position. The star is two hands from the Moon and at 4 o'clock position.

## STAR LORE

Star lore is part of virtually all cultures. The ancient Greeks, the Pawnee tribes of the American Midwest, the Australian aborigines, for example, all created stories about the star groupings they saw in the sky.

Because the star groupings do not change except on time scales of tens of thousands of years, the night sky we see is essentially the same night sky that ancient peoples saw. Star lore can therefore link us to the remote past. Furthermore, star lore can help us to remember constellations. In fact, it has been suggested that many of the stories were created as aids to memory, especially important in a time when familiarity with the stars could be literally a matter of life or death to a farmer or a navigator. Scientists have shown that baby birds learn to recognize star patterns and movements and use them to navigate safely—unguided by their parents—across thousands of miles of ocean to their winter homes. Perhaps we too have such instinctive faculties that help us learn the stars.

Probably the most familiar star grouping is the Big Dipper. It is not a constellation but rather is called an **asterism**. An asterism is an easily recognized grouping of stars that may be part of one constellation or may incorporate pieces of several. For example, the Big Dipper is part of the constellation Ursa Major, the Great Bear. The asterism of the Summer Triangle, on the other hand, spans three constellations. It consists of the three bright stars conspicuous in the summer evening: Deneb (in Cygnus, the Swan), Altair (in Aquila, the Eagle), and Vega (in Lyra, the Harp), shown in figure E1.3. "Looking Up at the Summer Triangle" (p. 65) shows a photograph of the Summer Triangle and some of the other interesting objects that lie near it, which are discussed later in the text.

The Big Dipper is not only easy to spot, but it is also an excellent signpost to other asterisms and stars. For example, two of the stars in its "bowl" (see fig. E1.1) are called the "pointers" because they point, roughly, to the North Star, Polaris.

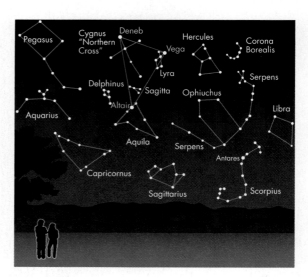

**FIGURE E1.3**
Dominating the night sky in July, August, and September are the three bright stars Vega, Altair, and Deneb, which form the summer triangle. This sketch shows how the sky looks looking south (from mid-northern latitudes) at about 8 P.M. in early September.

Polaris lies almost exactly above the Earth's North Pole, and because of its position there, it is the only star in the northern sky that shows, to the naked eye, no obvious motion during the night. Its relatively fixed position is illustrated by the time exposure in figure E1.4, showing the other stars rotating around it. Because Polaris always lies nearly true north, it is useful in orienting yourself to compass directions.

Polaris marks the end of the handle of the Little Dipper, another asterism and part of the constellation Ursa Minor, the Little Bear. The Little Dipper is not easy to see because its stars are dim, but it curves back toward the Big Dipper. Look for these two star groupings near the north celestial pole.

The native inhabitants of North America had a story about the Big Dipper. Its bowl represented a huge bear, and the handle represented three warriors in pursuit of the bear. They had wounded it, and it was bleeding. The red color of the leaves in autumn was said to be caused by the bear's blood dripping on them when the constellation lies low in the sky during the evening hours of the autumn months.

If you follow the pointer stars in the Big Dipper past the Little Dipper and Polaris, you will come to a set of constellations tied together by an ancient Greek myth, the story of Perseus and Andromeda. The cast includes a king (Cepheus), a queen (Cassiopeia), the hero (Perseus), the princess (Andromeda), a sea monster (Cetus), and the winged horse (Pegasus). The constellations are shown in figure E1.5, and their story goes as follows:

In ancient days there lived a queen of Ethiopia, Cassiopeia, who was very beautiful but also very vain. She and king Cepheus, her husband, and their daughter, Andromeda, lived happily until one day the queen boasted that she was more

# M57 The Ring Nebula

This planetary nebula is about 2000 ly away from us. It is 7000 years old and has a white dwarf at its center.

1 ly

The Summer Triangle consists of the three bright stars Deneb, Vega, and Altair. These stars, the brightest to our eyes in the constellations Cygnus, Lyra, and Aquila, respectively, rise in the east shortly after sunset in late June and are visible throughout the northern summer and into late October (when they set in the west in the early evening). Deneb is intrinsically the brightest of the three, although Vega looks the brightest to us. Deneb looks dim only because it is so much farther from us than Vega and Altair.

Vega

Lyra

## Epsilon Lyra

A double, double star

Cygnus

Deneb

Altair

## Alberio

This star pair (easily seen in a small telescope) shows a strong color contrast (orange and blue). Astronomers disagree about whether they orbit each other or just happen to lie in the same direction in the sky.

## M27 The Dumbbell Nebula

~2.5 ly

Another planetary nebula, The Dumbbell is about 900 ly distant and is about 2.5 ly in diameter.

## The Summer Triangle in 3-D

25 ly

Vega

3000 ly

Deneb

To Earth

17 ly

Altair

1 light year (ly) = 6 trillion miles

Polaris

**FIGURE E1.4**
A time exposure showing how Polaris remains essentially fixed while the sky pivots around it.

Pegasus

Cepheus

Andromeda

Cetus

Cassiopeia

Perseus

Pleiades

**FIGURE E1.5**
Perseus, Andromeda, Cassiopeia, Cepheus, Cetus, and Pegasus. The sky is drawn as it looks in early November at about 8 P.M., looking approximately northeast.

☆ Can you spot the Big Dipper?

☆ It is just above the tree.

beautiful than the daughters of Nereus, a sea god. In punishment for such pride, the sea-god Neptune sent a sea monster, Cetus, to ravage the kingdom. To save his people and appease the gods, Cepheus was instructed to tie his daughter, Andromeda, to a rock for the monster to devour. Meanwhile, Perseus was returning home from a quest in which he slew the snake-haired Gorgon, Medusa. On her death, her blood dripped into the sea and turned into the flying winged horse, Pegasus. Medieval versions of the story have Perseus riding Pegasus, but the classical myth has him borne on winged sandals. Regardless of his means of travel, Perseus saw the maiden's peril and landed, slaying the monster and delivering the kingdom. They all lived as happily ever after as most mythological families, and their astronomical depiction is most easily seen in the late autumn, when the brighter of its constellations are high in the sky.

Stories are also told about stars in other parts of the sky. For example, in the winter sky, you can see the sad legend of the Hunter, Orion, and the maiden who refused to fall in love with him. The story also involves Orion's hunting dogs (Canis Major and Canis Minor), a bull (Taurus), a rabbit (Lepus), the maiden's sisters (the Pleiades—a cluster of stars in the constellation Taurus), and a scorpion (Scorpius).

The king of the island Chios had a lovely daughter, Merope. His island was filled with savage beasts, and to rid his kingdom of these dangerous animals, the king called upon Orion to kill the beasts and make his kingdom safe. When the task was done, Orion met Merope and made unwelcome advances. In punishment, he was blinded by the king, but after doing penance, he had his sight restored. After reaching an old age, however, Orion one day stepped on a scorpion, which stung and killed him. On his death, the gods placed him in the sky with his faithful dogs (one of whom chases Lepus, the rabbit), forever attacking the wild bull, Taurus. Beyond the bull, Merope and her sisters (the Pleiades) run from the hunter, who pursues them each night across the sky. The scorpion was also placed in the sky, but on the other side of the heavens so that Orion would never again be threatened by it (Orion is visible in the evening only in the winter, whereas Scorpius is visible in the evening only in the summer).

The Orion myth has several versions, but the one described here fits together many of the astronomical references. Another myth with several versions involves the late summer constellations Boötes and Virgo. They are linked by the gloomy story of Icarius, the first cultivator of grapes for wine. According to legend, his recompense for sharing this knowledge was to be killed by drunken peasants, who buried his body under a tree. His dog, Maera, led Erigone, his daughter, to the spot, where, on discovering her father's body, she killed herself out of grief. Icarius was placed in the sky as Boötes, and Erigone as Virgo. The dog Maera became the star Procyon.

There are many other stories about constellations, but the above may give you some sense of the ones that have been handed down for thousands of years of written and oral history. Pass them on.

## AMATEUR ASTRONOMY

Anyone with access to even very modest equipment such as binoculars or low-powered telescopes has better equipment than Galileo ever had. With such equipment and access to a dark sky, that person can become an amateur astronomer. The pleasures of the hobby can range from the aesthetic satisfaction of taking a lovely photograph (fig. E1.6) to the thrill of discovering a new comet or an exploding star.

Many amateur astronomers now use digital cameras for astrophotography. With a middle- to upper-price-range camera mounted on a tripod you can make surprisingly good images of constellations. If you use still more expensive, special-purpose CCDs,* you can take pictures with a backyard telescope that rival those made 40 years ago with a 200-inch telescope.

To take astronomy pictures with a digital camera, you generally must change the camera's settings from automatic. For example, because you will typically need an exposure time of 10 or so seconds, the camera should be mounted on a tripod. You should also set the lens aperture as wide open as possible and the focus distance at infinity. In addition, if you can change the ISO setting, make it at least 400, preferably higher (higher ISO values allow you to image dimmer objects). Your camera's instruction guide will tell you how to make these changes. You can also find helpful suggestions at a number of websites that describe how to do astrophotography. A cable release will help you expose without shaking the camera but is not essential.

If you expose for more than about 15 seconds, the Earth's rotation will smear the star image into a streak. Deliberately allowing the smearing to occur can produce dramatic pictures of what are called "star trails" (see fig. E1.4). To make star-trail pictures, leave the shutter open for 20 minutes or so.

To take untrailed long exposures or to use a telephoto lens, you will need a way to compensate for the Earth's rotation. You can make a simple device to do this from hinged boards with a carriage bolt through them. Mount the camera on one board and set the other on a firm surface. Slowly turning the bolt will shift the camera, allowing you to take exposures of half an hour or so. For best results, however, you need a motor drive. Check your local library or the references at the end of the chapter for books or articles on drives and astrophotography. You may also find in such sources suggestions for scientifically useful projects, such as variable star observing or comet hunting. Projects like these are ideal for amateur astronomers because the results are not only valuable but also require basically only patience and care without the need for a big telescope.

## SMALL TELESCOPES

A small telescope will greatly increase the number and interest of objects you can observe. Such telescopes come in a wide range of styles and prices, but selecting the best one for your

---

*CCD = "charge-coupled device." See chapter 4 for more details on these electronic cameras.

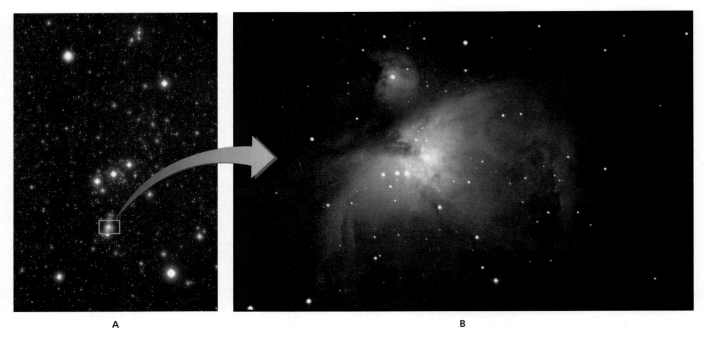

**FIGURE E1.6**
(A) Photograph of the constellation Orion made with an ordinary camera. (B) Picture of the Orion Nebula taken with a small backyard telescope.

Q★ What causes the red glow in the Orion Nebula?

A★ Hydrogen gas being heated by young stars in the vicinity of the nebula cause a characteristic red glow.

needs can be confusing. Moreover, it is a sad truth that you generally get what you pay for. Many amateur astronomers begin with a 3- to 4-inch reflecting telescope. Such a telescope uses a mirror to collect and focus the light; thus, its name, "reflecting." The numbers refer to the diameter of the mirror, important because a larger mirror collects more light, allowing you to see fainter objects. A larger mirror also permits seeing finer details with other things being equal (see chapter 4 for how a telescope works). With such a telescope, you can easily see the moons of Jupiter, the rings of Saturn, and many lovely star clusters and galaxies. However, these latter objects will not look like the pictures in books because your eyes—unlike a photograph or electronic detector—cannot store up light. Moreover, your eyes are far less sensitive to colors in very dim light.

Notice that we have said nothing of magnifying power. For most amateur telescopes, the maximum useful power is limited by distortions to the light as it passes through our atmosphere. These distortions make a magnification of about 100 to 200 the useful limit. Beyond that, the distortions dominate and higher power gives no increase in clarity.

Whatever type of telescope you choose, be sure to get a sturdy mounting for it. Even at 100 power, tiny vibrations of the telescope caused by wind or the touch of your hand will make the image jiggle, hopelessly blurring it.

Before you actually buy a telescope, you might want to talk with your instructor or a local amateur astronomer.

Such people often belong to an astronomy club, some of whose members may have second-hand telescopes they are willing to sell at reduced prices. For information about new telescopes, browse through the magazines *Astronomy* or *Sky and Telescope*— publications widely read by both amateur and professional astronomers—which contain many advertisements for small telescopes.

## STAR CHARTS

One of the pleasures of using a telescope is to look at objects too faint to be seen by the naked eye, such as most galaxies, faint star clusters, and remnants of dying stars. To find many of these dim objects you will need a good star chart.

Astronomers use star charts to find objects in the sky much as navigators use charts to find places on Earth. A typical star chart (fig. E1.7) shows the location of the constellations, the stars, and other objects. It also gives some indication of the relative brightness of the stars. Finally, many charts also have information about the season and time of night at which the stars are visible.

Star charts are designed much like maps of the Earth. For example, both represent on a flat surface a map of something curved—in one case the celestial sphere; in the other the surface of the Earth. Also, both use a coordinate grid.

Nowadays, star charts are as likely to be on a computer as on paper. These offer tremendous ease of use, and many are

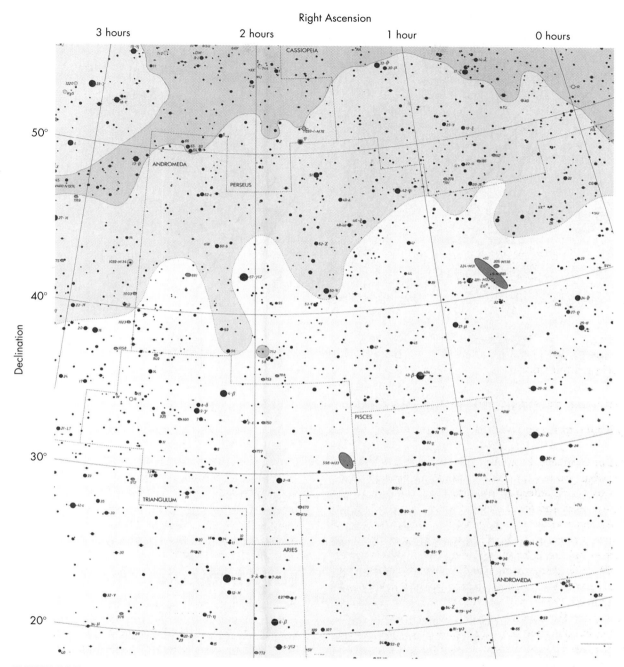

**FIGURE E1.7**
A modern star chart showing stars, galaxies, and coordinates. Black circles are stars. Their size indicates their brightness—larger are brighter. Red ellipses are galaxies. The shaded blue area is the Milky Way.

very beautifully done with many additional features, such as the charts in the Starry Night software that accompanies the text.

## CELESTIAL COORDINATES

The coordinate grid used by astronomers is similar to that used by navigators. The grid consists of one set of lines running east–west on the celestial sphere, parallel to the celestial equator, and another set running north–south, connecting one celestial pole to the other. The east–west lines play the same role as latitude on the Earth, but to avoid confusion with terrestrial coordinates, they are called lines of **declination,** or "dec" for short. The north–south lines play the same role as longitude on the Earth and are called lines of **right ascension,** or "RA" for short. Declination values run from +90° to −90° (the north and south celestial poles), with 0° being the celestial

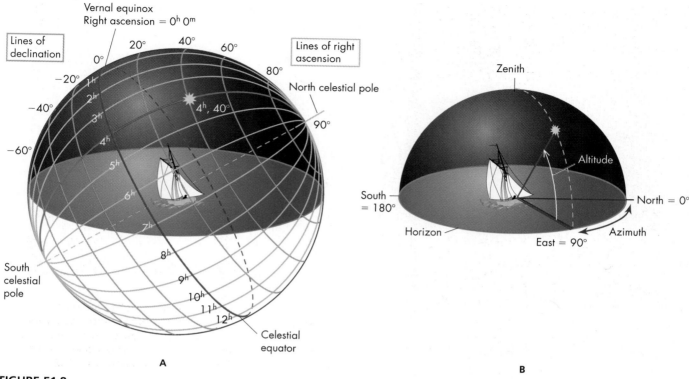

**FIGURE E1.8**
(A) Locating a star according to right ascension and declination. (B) Locating a star according to altitude and azimuth.

equator. Right-ascension lines divide the celestial sphere into 24 equal zones that are labeled not in degrees but in units of time. Thus, the right ascension of an object is given in hours (h), minutes (m), and seconds (s). Because the 360° around the sky is divided into 24 segments, each hour of RA equals 15°; that is, 360° divided by 24 = 15°. The point $0^h\ 0^m\ 0^s$ of RA is arbitrarily chosen to be where the Sun's path, the ecliptic, crosses the celestial equator as the Sun moves north (fig. E1.8A).

Right ascension and declination are not the only way to locate objects on the sky. For example, a celestial object may be located by its altitude and azimuth, as illustrated in figure E1.8B. In this so-called horizon system, **altitude** is an object's angle above the horizon. **Azimuth** is generally defined as the angle measured eastward along the horizon from North to the point directly below the object, although sometimes it is measured from South. Whichever convention is used for azimuth, these coordinates are useful for pointing out objects or tracking their motion. However, they have a serious drawback compared to the otherwise seemingly cumbersome right ascension and declination: an astronomical body's altitude and azimuth constantly change as it moves across the sky, whereas, because they are defined with respect to the rotating celestial sphere, a body's right ascension and declination remain the same as a body rises and sets.

With a set of coordinate lines established, we can now locate astronomical objects on the sky the same way we can locate places on the Earth. For example, M31, a galaxy in the Local Group, is at right ascension 0 hours 42.7 minutes ($0^h\ 42.7^m$) declination +41°16′, as you can see in figure E1.7.

## PLANETARY CONFIGURATIONS

Because planets move across the stellar background, astronomers have invented some terms to help describe where they are located at any given time. These terms describe a planet's position with respect to the Earth and the Sun—planetary configurations—and are shown in figure E1.9. Understanding these terms when they are used can help you find planets.

If a planet lies in the sky in the same direction as the Sun, it is said to be at **conjunction.** If it lies approximately between us and the Sun, it is at **inferior conjunction.** If it is on the other side of the Sun, it is at **superior conjunction.**

Planets are very hard to see at either conjunction because they are hidden in the Sun's glare. On some very rare occasions, a planet may pass directly between us and the Sun. We may then see it silhouetted against the Sun's bright disk, as shown in figure E1.10. Such an event is called a **transit.** Only Mercury and Venus can transit the Sun as seen from Earth, but we can imagine talking with an astronaut on Mars who

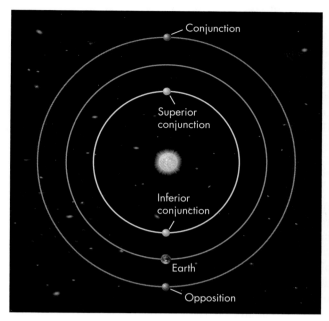

**FIGURE E1.9**
Planetary configurations: opposition, superior conjunction, and inferior conjunction.

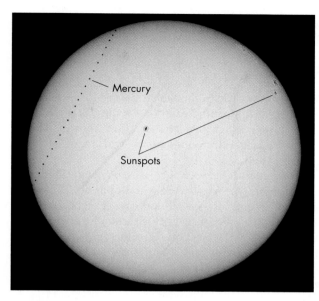

**FIGURE E1.10**
Transit of Mercury, May 7, 2003, photographed by Dominique Dierick of Belgium. Twenty-three separate images, taken about 15 minutes apart, were combined to form this picture. The next three transits will occur on May 9, 2016; Nov. 11, 2019; and Nov. 13, 2032.

View from Space

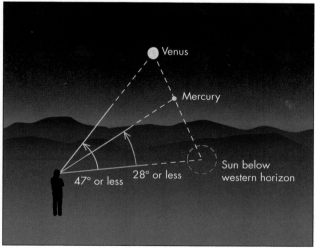

View from Earth

**FIGURE E1.11**
The greatest elongations of Mercury and Venus and the Evening Star phenomenon. The left-hand diagram also shows that Mercury and Venus can *never* appear more than 28° and 47°, respectively, from the Sun.

has just witnessed the Earth transiting the Sun. This would occur when Mars is directly opposite the Sun in the sky, or at what seen from Earth is called **opposition.**

A planet's configuration strongly affects how easily it can be viewed from Earth. For example, when an outer planet is at opposition, it is at its nearest to the Earth. Other things being equal, the planet is therefore also at its brightest. Being opposite the Sun, a planet at opposition rises at sunset. Inner planets, on the other hand, are easiest to see when they lie far from the Sun in our sky, so that they are not lost in the Sun's glare. However, there is a limit to how far from the Sun in our sky an inner planet can be. As figure E1.11 shows, Mercury can never be more than 28° from the Sun and Venus never more than 47° as seen on the sky. It is for this reason that Mercury and Venus are usually visible only in the morning or evening sky when the Sun is just below the horizon.

ANIMATION

*Mercury and Venus as the morning star and the evening star*

A planet seen close to the Sun at dawn or dusk is sometimes called the **Morning Star** or **Evening Star.** When a planet is at its greatest angular separation from the Sun, it is said to be at **greatest elongation**—which can be either western or eastern.

The time interval between successive planetary configurations of the same type is called the **synodic period.** The synodic period differs from the planet's orbital period because both the Earth and the other planets move around the Sun. Thus, the interval between oppositions is neither an Earth year nor another planet's orbital period. For example, the Earth takes about 2 years to catch up to and overtake Mars after an opposition. The Earth overtakes the slower moving, more distant planets more quickly, and the interval between oppositions is close to a year. Thus, the Martian synodic period is about 780 days, whereas the Saturnian synodic period is 378 days.

## YOUR EYES AT NIGHT

You will soon discover that the longer you stay outside in dim light, the more sensitive your eyes will become and the fainter the stars you will be able to see. This is the result of physiological changes in your eye referred to as **dark adaption.**

The simplest change in your eye occurring in dim light is that the pupil opens wider. This is easy to verify by looking at yourself in a mirror in a dimly lit room. In full sunlight, your pupil normally has a diameter of about 2 millimeters, but in total darkness, its diameter may expand to about 7 or 8 millimeters, thereby allowing more light to enter your eye. The same principle is used in cameras when you adjust the aperture to match the available light.

Your eyes undergo another change in the dark. Chemical changes make the dark-adapted retina about 1 million times more sensitive to light than that under full daylight conditions. The process takes about 20 minutes to get well established but is undone by even a few seconds' exposure to bright light. Thus, once you are dark-adapted, you should stay away from bright lights for as long as you intend to observe.

In addition to becoming more sensitive to light, your eye also changes its sensitivity to color slightly, a phenomenon known as the "Purkinje effect." In full daylight, the eye responds best to greenish colors. At low light levels, it responds best to slightly bluer colors. This is probably the result of natural selection because starlight is much bluer than sunlight and eyes responsive to blue will therefore aid survival. It is certainly the case that night-flying insects see blue light better than yellow. That is the reason bug-zappers use a blue light to attract insects and why a yellow light bulb is often used for outdoor night lighting to be less attractive to insects.

You may also notice that it is easier to see very faint objects if you don't look directly at them but instead look a little to one side. The greater sensitivity you enjoy from this so-called **averted vision** arises because the center of your field of view is densely packed with receptors designed to allow you to see fine details. These receptors are better at showing fine structure than at seeing faint light. Thus, looking slightly to one side of a faint object makes it easier to see.

## SUMMARY

Looking at the night sky is not only fun, it will help you understand some of the phenomena described in chapter 1. Star maps will help you identify constellations and bright stars, and by learning the mythology of the stars, you will be able to find your way around the night sky more easily. In addition, you will forge a link to distant and ancient cultures. For many people, backyard astronomy—even with simple equipment—is an enjoyable and exciting hobby. Perhaps you will discover a comet and have it named for you!

## QUESTIONS FOR REVIEW

1. Approximately where would you look for Mercury in the sky about sunset?
2. What is meant by "Morning Star"?
3. If a planet is at opposition and you see it high in the sky, about what time of night must it be?
4. Can you see Mercury in the western sky at dawn?
5. What is meant by "dark adaption"?
6. Why does the pupil of your eye get bigger in dim light?
7. What is averted vision?
8. What is right ascension?
9. What is declination?
10. What is azimuth?

## TEST YOURSELF

1. As a star rises and moves across the sky, which of the following change?
   (a) Its right ascension
   (b) Its declination
   (c) Its azimuth
   (d) Both (a) and (b)
   (e) None of the above
2. A planet is at inferior conjunction. It therefore rises at approximately
   (a) sunset.
   (b) sunrise.

(c) midnight.

(d) 2 hours before the Sun.

(e) You can't tell from the available information.

3. If Mercury is at greatest elongation to the west of the Sun, it will be easiest to see

(a) just before dawn.

(b) just after sunset.

(c) about midnight.

(d) just before sunset.

(e) None of the above

4. Which of the following planets can be at inferior conjunction?

(a) Jupiter

(b) Mars

(c) Uranus

(d) Venus

(e) All of them

5. When your eye is dark-adapted

(a) your pupils are smallest.

(b) your pupils are biggest.

(c) your color vision is at its most sensitive.

(d) your eye is most sensitive to light.

(e) Both (b) and (d)

## FURTHER EXPLORATIONS

Allen, Richard Hinckley. *Star Names: Their Lore and Meaning.* New York: Dover, 1963.

Arnold H. J. P. *Astrophotography: An Introduction to Film and Digital Imaging.* Willowdale, Ontario: Firefly, 2002.

Aveni, Anthony. *Conversing with the Planets,* rev. ed. Boulder: University Press of Colorado, 2002.

Berman, Bob. *Secrets of the Night Sky: The Most Amazing Things in the Universe You Can See with the Naked Eye.* New York: W. Morrow, 1995.

Covington, Michael. *Astrophotography for the Amateur,* 2nd ed. New York: Cambridge University Press.

Hamilton, Edith. *Mythology.* Boston: Back Bay Books, 1998.

Mechler, Gary. *Galaxies and Other Deep-Sky Objects.* New York: Knopf, 1995. (One in a series of National Audubon Society Pocket Guides.)

Muirden, James, ed. *Sky Watcher's Handbook: The Expert Reference Source for the Amateur Astronomer.* New York: W. H. Freeman, 1993.

Nagler, A. "Choosing Your Telescope's Magnification." *Sky and Telescope* 81 (May 1991): 553.

Newton, Jack. *Guide to Amateur Astronomy.* Cambridge, UK: Cambridge University Press, 1995.

*Observer's Handbook.* Toronto: Royal Astronomical Society of Canada.

Pasachoff, Jay. *Field Guide to the Stars and Planets,* 4th ed. Boston: Houghton Mifflin, 2000.

Ridpath, Ian. "The Origin of Our Constellations." *Mercury* 19 (November/December 1990): 163.

Schaaf, Fred. *Wonders of the Sky: Observing Rainbows, Comets, Eclipses, the Stars, and Other Phenomena.* New York: Dover, 1983.

Schaefer, Bradley E. "The Origin of the Greek Constellations." *Scientific American* 295 (November 2006): 96.

Sesti, Giuseppe M. *The Glorious Constellations: History and Mythology.* New York: Harry N. Abrams, 1991. *This beautiful book contains extensive descriptions of the constellations and their mythology from many ancient cultures.*

Staal, Julius D. W. *The New Patterns in the Sky: Myths and Legends of the Stars.* Blacksburg, Va.: McDonald & Woodward, 1988.

Strom, Karen D. "Photographing the Stars." *Physics Teacher* 34 (September 1996): 340.

Tirion, Wil. *Uranometria 2000.0: Deep sky atlas,* 2nd ed. Richmond, Va.: Willmann-Bell, 2001.

*Two excellent magazines for the amateur astronomer are the following: Astronomy,* published by Kalmbach Publishing Co., 21027 Crossroads Circle, P.O. Box 1612, Waukesha, WI 53187.

*Sky and Telescope,* published by Sky Publishing Corporation, P.O. Box 9111, Belmont, MA 02178-9111.

**Website**

"Backyard Astronomy: Tips on Observing the Universe," by *Sky and Telescope:* **http://skyandtelescope.com/howto/ scopes/article_241_1.asp**

An excellent site with many useful links and good advice about choosing telescopes.

## KEY TERMS

| | |
|---|---|
| altitude, 70 | greatest elongation, 72 |
| asterism, 64 | inferior conjunction, 70 |
| averted vision, 72 | Morning Star, 72 |
| azimuth, 70 | opposition, 71 |
| conjunction, 70 | right ascension, 69 |
| dark adaption, 72 | superior conjunction, 70 |
| declination, 69 | synodic period, 72 |
| Evening Star, 72 | transit, 70 |

# Gravity and Motion

The space shuttle blasts off.

## Concepts and Things to Remember

- Gravity controls the motion of most astronomical objects.
- If no forces act on an object, inertia keeps it moving in a straight line at a constant speed.
- If an object is not moving along a straight line at a constant speed, a force must be acting on it.
- Newton's laws of motion allow us to describe and predict an object's motion if we know the forces acting on it.
- The force of gravity between two objects depends on their mass and the distance between them:
  - The greater their separation, the weaker the force.
  - The greater their mass, the greater the force.
- Orbital motion allows us to determine the mass of ("weigh") astronomical objects.

**?What Is This?**
(See chapter summary for the answer.)

**Concepts and Skills to Review**

- Kepler's laws (1.3)
- Forces (P.12)

**G**ravity gives the Universe its structure. It is a universal force that acts on all the objects in the Universe so that every particle is drawn toward every other particle by its pull. Gravity holds together astronomical bodies of all sizes, from the Earth to the Universe itself. But the role of gravity extends beyond giving structure to astronomical bodies. Gravity also controls their motions, holding the Earth in orbit around the Sun, the Sun in orbit around the Milky Way, and the Milky Way within the Local Group. Thus, gravity and motion are tightly connected in the Universe. This connection is the theme of this chapter.

## 2.1    SOLVING THE PROBLEM OF ASTRONOMICAL MOTION

Astronomers of antiquity did not make the connection between gravity and astronomical motion that we recognize today. They were puzzled about what kept the planets moving and why, if the Earth moved, objects thrown into the air were not left behind.

The solutions to these mysteries began with a series of careful experiments conducted by Galileo in the 1600s. Apart from his famous—and perhaps fictitious—demonstration of weights dropped from the Leaning Tower of Pisa, Galileo experimented with projectiles and with balls rolling down planks. Such experiments led him to propose several laws of motion. More important, perhaps, these experiments demonstrated the power and importance of the experimental method for verifying scientific conjectures.

## 2.2    INERTIA

Central to Galileo's laws of motion is the concept of **inertia.** Inertia is the tendency of a body at rest to remain at rest and a body in motion to keep moving in a straight line at a constant speed. Aristotle noted that bodies at rest resist being moved, but he failed to link this property to the tendency of objects to keep moving once they are set in motion. Kepler also recognized inertia's importance and in fact was first to use that term. However, Galileo not only proposed this property of matter but also demonstrated it by real experiment.

In one such experiment, Galileo rolled a ball down a sloping board over and over again and noticed that it always sped up as it rolled down the slope (fig. 2.1). He next rolled the ball up a sloping board and noticed that it always slowed down as it approached the top. He hypothesized that if a ball rolled on a flat surface and no forces—such as friction—acted on it, its speed would neither increase nor decrease but remain constant. That is, in the absence of forces, inertia keeps an object already in motion moving at a fixed speed. Inertia is familiar to us all even in everyday life. Apply the brakes of your car suddenly, and the inertia of the bag of groceries beside you keeps the bag moving forward at its previous speed until it hits the dashboard or spills onto the floor.

Newton recognized the special importance of inertia. He described it in what is now called **Newton's first law of motion** (sometimes referred to simply as the law of inertia). The law can be stated as follows:

**A body continues in a state of rest or motion in a straight line at a constant speed unless made to change that state by forces acting on it.**

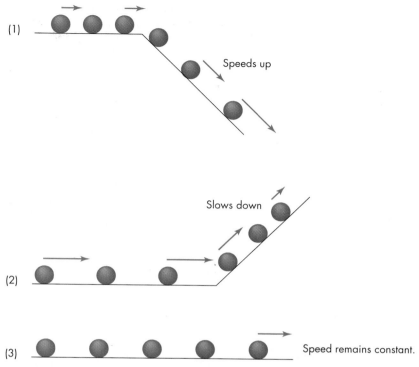

**FIGURE 2.1**
A ball rolling down a slope speeds up. A ball rolling up a slope slows down. A ball rolling on a flat surface rolls at a constant speed if no forces act on it.

Balanced forces = no change in motion

**FIGURE 2.2**
Balanced forces lead to no acceleration.

In applying Newton's first law, we should note two important points. First, we have not defined force yet but have relied on our intuitive feeling that a force is anything that pushes or pulls. Second, we need to note that when we use the term *force*, we are talking about *net* force; that is, the total of all forces acting on a body. For example, if a brick at rest is pushed equally by two opposing forces, the forces are balanced. Therefore, the brick experiences no net force and accordingly does not move (fig. 2.2).

Newton's first law may not sound impressive at first, but it carries the idea that is crucial in astronomy, that if a body is *not* moving in a straight line at a constant speed, some net force *must* be acting on it.

Actually, Newton was preceded in stating this law by the seventeenth-century Dutch scientist Christiaan Huygens. However, Newton went on to develop additional physical laws and—more important for astronomy—showed how to apply them to the Universe. For example, let us look at what happens if we swing a mass tied to a string in a circle. Newton's law tells us that the mass's inertia will carry it in a straight line if no forces act. What force, then, is acting on the circling mass? The force is the one exerted by the string, preventing the mass from moving in a straight line and keeping it in a circle. We can feel that force as a tug on the string, and we can see its importance if we suddenly let go of the string. With the force no longer acting on it, the mass flies off in a straight line, demonstrating the first law, as illustrated in figure 2.3.

We can translate this example to an astronomical setting and apply it to the orbit of the Moon around the Earth, or of the Earth around the Sun, or to the Sun in orbit around the Milky Way. Each of these bodies follows a curved path. Therefore, each must have a force acting on it, the origin and nature of which we will now describe.

**FIGURE 2.3**
For a mass on a string to travel in a circle, a force (green arrow) must act along the string to overcome inertia. Without that force, inertia makes the mass move in a straight line.

Side view

If string is released when ball is here, ball goes straight toward A, *not* toward B, *nor* toward C.

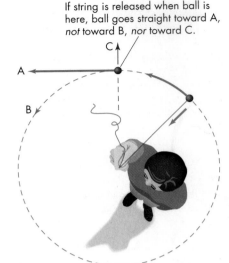

Top view

## 2.3  ORBITAL MOTION AND GRAVITY

Newton was not the first person to attempt to discover and define the force that holds planets in orbit around the Sun. Nearly 100 years before Newton, Kepler recognized that some force must hold the planets in their orbits and proposed that something similar to magnetism might be responsible. Newton was not even the first person to suggest that gravity was responsible. That honor belongs to Robert Hooke, another Englishman, who noted gravity's role in celestial motions several years before Newton published his law of gravity in 1687. Newton's contribution is nevertheless special because he demonstrated the *properties* that gravity must have if it is to control planetary motion. Moreover, Newton went on to derive equations that describe not only gravity but also its effects on motion. The solution of these equations allowed astronomers to predict the position and motion of the planets and other astronomical bodies.

According to legend, Newton realized gravity's role when he saw an apple falling from a tree. The falling apple drawn downward to the Earth's surface presumably made him speculate whether Earth's gravity might extend to the Moon. Influenced by an apple or not, Newton correctly deduced that Earth's gravity, if weakened by distance, could explain the Moon's motion.

Most of Newton's work is highly mathematical, but as part of his discussion of orbital motion, he described a thought experiment to demonstrate how an object can move in orbit. Thought experiments are not actually performed; rather, they serve as a way to think about problems. In Newton's thought experiment, we imagine a cannon on a mountain peak firing a projectile (fig. 2.4A). From our everyday experience, we know that whenever an object is thrown horizontally, gravity pulls it downward so that its path is an arc. Moreover, the faster we throw, the farther the object travels before striking the ground.

We now imagine increasing the projectile speed more and more, allowing it to travel ever farther. However, as the distance traveled becomes very large, we see that the Earth's surface curves away below the projectile (fig. 2.4B). Therefore, if the projectile moves sufficiently fast, the Earth's surface may curve away from it in such a way that the projectile will never hit the ground. Such is the nature of orbital motion and

*Newton's cannon*

**FIGURE 2.4**

(A) A cannon on a mountain peak fires a projectile. If the projectile is fired faster, it travels farther before hitting the ground. (B) At a sufficiently high speed, the projectile travels so far that the Earth's surface curves out from under it, and the projectile is in orbit.

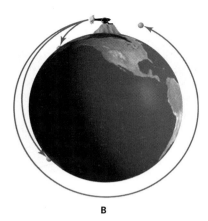

A

B

Q* Why are space vehicles normally launched from regions near the equator? Why are space vehicles normally launched to the east?

A* To take advantage of the Earth's spin in helping the vehicle reach orbital velocity.

how the Moon orbits the Earth. The balance between inertia and the force of gravity maintains the orbit.

We can analyze this thought experiment more specifically with Newton's first law of motion. According to that law, in the absence of forces, the projectile will travel in a straight line at constant speed. But because a force, gravity, is acting on the projectile, its path is not straight but curved. Moreover, the law helps us understand that the projectile does not stop because its inertia carries it forward.

If we apply the same reasoning to the Moon's motion, we conclude that

• the Moon moves along its orbit because its inertia carries it forward, and
• its path is curved (not a straight line) because the force of gravity deflects it.

Notice that in the above discussion we used *no* formulas. All we needed was Newton's first law and the idea that gravity supplies the deflecting force. However, if we are to understand the particulars of orbital motion, we require additional laws. For example, to determine how rapidly the projectile must move to be in orbit, we need laws that have a mathematical formulation.

## 2.4   NEWTON'S SECOND LAW OF MOTION

We stated that an object's inertia causes it to move at a constant speed in a straight line in the absence of forces. However, suppose forces do act on the object. How much deviation from straight-line motion will such forces produce? To answer that question, we need to define more carefully what we mean by motion.

Motion of an object is a change in its position, which we can characterize in two ways: by the direction the object is moving and by its speed. For example, a car is moving east at 40 miles per hour. If the car's speed and direction remain constant, we say it is in uniform motion (fig. 2.5A). If the car changes either its speed or direction, it is no longer moving uniformly, as depicted in figure 2.5B and C. Such nonuniform motion is defined as **acceleration.**

### Acceleration

We are all familiar with acceleration as a change in speed. For example, when we step on the accelerator in a car and it speeds up from 30 to 40 mph, we say the car is accelerating. Its speed has changed, and its motion is therefore nonuniform. Although in everyday usage acceleration implies an increase in speed, scientifically any change in speed is an acceleration. Thus, technically, a car "accelerates" when we apply the brakes, and it slows down.

Uniform motion
(Same speed (V), same direction)

Acceleration
(A change in speed)

Acceleration
(A change in direction)

A

B

C

**FIGURE 2.5**

Views of a car in uniform motion and accelerating. (A) Uniform motion implies no change in speed or direction. The car moves in a straight line at a constant speed. If an object's (B) speed or (C) direction changes, the object undergoes an acceleration.

Q★ Is it possible for an object to travel with constant speed and still accelerate?

A★ Yes—by traveling along a curved path.

In the example above, we produced an acceleration by changing an object's speed.* We can also produce an acceleration by changing an object's direction of motion. For example, suppose we drive a car around a circular track at a steady speed of 30 miles per hour. At each moment, the car's direction of travel is changing, and therefore, it is not in uniform motion. Similarly, a mass swung on a string or a planet orbiting the Sun is experiencing nonuniform motion and is therefore, accelerating. In fact, a body moving in a circular orbit constantly accelerates, even though its speed is not changing.

How do we produce an acceleration? Newton realized that for an object to accelerate, a force must act on it. For example, to accelerate—change the direction of—the mass whirling on a string, we must constantly exert a pull on the string. Similarly, to accelerate a shopping cart, we must exert a force on it. In addition, experiments show that the acceleration we get is proportional to the force we apply. That is, a larger force produces a larger acceleration. For example, if we push a shopping cart gently, its acceleration is slight. If we push harder, its acceleration is greater. But experience shows us that more than just force is at work here. For a given push, the amount of acceleration also depends on how full the cart is. A lightly loaded cart may scoot away under a slight push, but a heavily loaded cart hardly budges, as illustrated in figure 2.6. Thus, the acceleration produced by a given force also depends on the amount of matter being accelerated.

## Mass

The amount of matter an object contains is determined by a quantity that scientists call mass. Technically, **mass** measures an object's inertia. The more inertia, the more mass, and vice versa.

Scientists generally measure the mass of ordinary objects in kilograms (abbreviated kg). One kilogram equals 1000 grams or about 2.2 pounds of mass. For example, under normal conditions, a liter of water (roughly a quart) has a mass of 1 kilogram, but it is important to remember that mass is not the same as weight. Because an object's measured mass describes the amount of matter in it, its mass in kilograms is a fixed quantity. An object's weight, however, measures the force of gravity on it, a point we will explore more later. Thus, although an object's mass is fixed, its weight changes if the local gravity changes. For example, on Earth we have one weight but on

---

*In our discussion, we use the word *speed* to denote the rate of motion, irrespective of its direction. Were we to be more technically correct, we would use the term *velocity,* which means the speed in a *given direction.* Thus, a body's velocity changes if *either* its speed or direction changes. With velocity so defined, acceleration is simply the change in velocity over some interval of time.

**FIGURE 2.6**
A loaded cart will not accelerate as easily as an empty cart.

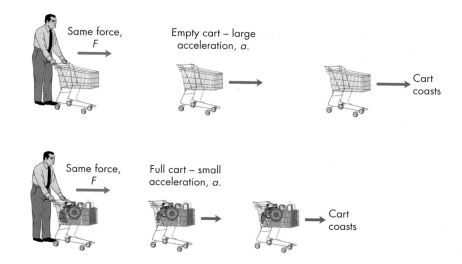

the Moon, where gravity is less, we have a lesser weight, but no matter where we are, we have the same mass.

Mass is the final quantity we need to define before we can understand **Newton's second law of motion** in its full form. Mathematically, the law states that when a force, $F$, acts on an object whose mass is $m$, it produces an acceleration, $a$.

$F$ = Any force
$m$ = Mass of the body being accelerated
$a$ = Amount of acceleration

$$F = ma$$

or, in other words:

> **The amount of acceleration ($a$) that a force ($F$) can produce depends on the mass ($m$) of the object being accelerated.**

This astonishingly simple equation allows scientists to predict virtually all features of a body's motion. With $F = ma$ and with knowledge of the masses and the forces in action, engineers and scientists can, for example, drop a spacecraft safely between Saturn and its rings or use a computer to design an airplane that will fly successfully without being tested.

## **2.5** THE LAW OF GRAVITY

*Gravity produces a force of attraction between bodies.*

Using Newton's second law, we now can determine an object's motion if we know its initial state of motion and the forces acting on it. For astronomical bodies, that force is often limited to gravity, and so to predict their motion, we need to know how to calculate gravity's force. Once again we encounter Newton's work, for it was he who first worked out the **law of gravity.** On the basis of his study of the Moon's motion, Newton concluded the following:

> **Every mass exerts a force of attraction on every other mass. The strength of the force is directly proportional to the product of the masses divided by the square of their separation.**

We can write this extremely important result in a shorthand mathematical manner as follows:

Let $m$ and $M$ be the masses of the two bodies (fig. 2.7) and let the separation between their centers be $r$. Then the strength of the force between them, $F$, is

$F$ = Strength of the force between two bodies
$M$ = Mass of one body
$m$ = Mass of second body
$r$ = Separation between the bodies' centers
$G$ = Constant

$$F = \frac{GMm}{r^2}$$

$$F_{gravity} = \frac{GmM}{r^2}$$

**FIGURE 2.7**
Gravity produces a force of attraction (green arrows) between bodies. The strength of the force depends on the product of their masses, $m$ and $M$, and the square of their separation, $r$. $G$ is the universal gravitational constant.

The factor G is a constant whose value is found by measuring the force between two bodies of known mass and separation. The resulting number for G depends on the units chosen to measure $M$, $m$, $r$, and $F$, but once determined, G is the same as long as the same units are used. For example, if $M$ and $m$ are measured in kilograms, $r$ in meters, and $F$ in SI units,* then $G = 6.67 \times 10^{-11}$ meters³/(kilogram-second²) [$m^3$-$kg^{-1}$-$s^{-2}$].

Writing the law of gravity as an equation helps us see several important points. If either $M$ or $m$ increases and the other factors remain the same, the force increases. If $r$ (the distance between two objects) increases, the force gets weaker. Moreover, the force weakens as the square of the distance. That is, if the distance between two masses is doubled, the gravitational force between them decreases by a factor of four, not two. Finally, although one object's gravitational force on another weakens with increasing distance, the gravitational force never completely disappears. Thus, the gravitational attraction of an object reaches across the entire Universe, so the Earth's gravity not only holds you on to its surface but also extends to the Moon and exerts the force that holds the Moon in orbit around the Earth.

## 2.6    NEWTON'S THIRD LAW OF MOTION

Newton's studies of motion and gravity led him to yet another critical law, which relates the forces that bodies exert on each other. This additional relation, **Newton's third law of motion,** is sometimes called the law of action–reaction. This law states:

**When two objects interact, they create equal and opposite forces on each other.**

Two skateboarders side by side may serve as a simple example of the third law (fig. 2.8). If A pushes on B, *both* move. According to Newton's law, when A exerts a force on B, B exerts a force on A so that both are accelerated.

The gravitational force between the Earth and the Sun affords an astronomical example of Newton's third law and at the same time leads us a step closer to understanding orbital motion. The gravitational force of the Earth on the Sun is exactly equal to the gravitational force of the Sun on the Earth. We can see this perhaps surprising result from Newton's law of gravity where the gravitational force between two objects depends on the product of their masses. We thus get the same force regardless of whether we let the Earth act on the Sun or vice versa. Why, then, does the Earth orbit the Sun and not the other way around?

**FIGURE 2.8**
Skateboarders illustrate Newton's third law of motion. When A pushes on B, an equal push is given to A by B.

*SI units (System International) use the kilogram, meter, and second as the fundamental units.

The second law supplies the answer. If we translate $F = ma$ into $a = F/m$, we see that the acceleration an object feels is inversely proportional to its mass; that is, the more massive it is, the more force is required to accelerate it. Thus, even though the forces acting on the Earth and Sun are precisely equal, the Sun accelerates 300,000 times *less* because it is 300,000 times *more* massive than the Earth. As a result, because the Earth's acceleration is so much larger than the Sun's, the Earth does most of the moving. In fact, however, the Sun does move a little bit as the Earth orbits it, much as you must move if you swing a child around you in play.

## 2.7    MEASURING AN OBJECT'S MASS USING ORBITAL MOTION

**INTERACTIVE**

*Orbital velocity*

Knowledge of orbital motion is important for more than just understanding the paths of astronomical objects. From the orbit's properties (such as size and orbital period), astronomers can deduce physical properties of the orbiting objects, such as their mass.

The basic method for determining an astronomical object's mass uses a modified form of Kepler's Third Law and was first worked out by Newton using his laws of motion and gravity.* The underlying idea is very simple:

• The masses of the orbiting objects determine the gravitational force between them.
• The gravitational force in turn sets the properties of the orbit.

Thus, from knowledge of the orbit, astronomers can work backward to find the masses of the objects. To see how this can be done, we consider a very simple case: orbital motion in a perfect circle with the orbiting object having a mass so small that it can be ignored compared with that of the central body. These restrictions are met to high precision in many astronomical systems, such as the Earth's motion around the Sun and the Sun's motion around the Milky Way. By assuming that the mass of the central body is large compared with the orbiting body, we can ignore the acceleration of the central body (as we just discussed) and assume it is at rest. These assumptions simplify the problem but are not essential for its solution.

To work out the orbital properties of an object moving around another, we use Newton's laws of motion and his law of gravity. From the first law, we know that if an object moves along a circular path, there is a net force (an unbalanced force) acting on it because balanced forces give straight-line motion. This force† must be applied to any object moving in a circle, whether it is a car rounding a curve, a mass swung on a string, or the Earth orbiting the Sun.

Using Newton's second law of motion and some algebra and geometry gives us an equation that shows that if a mass ($m$) moves with a velocity ($V$) around a circle of radius ($r$), the force needed to hold it in a circular orbit is

$F$ = Force needed to hold a body in orbit
$m$ = Mass of the body
$r$ = Radius of the orbit
$V$ = Velocity of the body

$$F = \frac{mV^2}{r}$$

Using the above equation, we can find the orbital velocity of a planet around the Sun. Let the Sun's mass be $M$ and the planet's mass be $m$, the latter of which is

---

*In fact, Newton drew upon Kepler's third law in deriving his law of gravity. Although it is not simple to see, it turns out that the exponents in Kepler's law (the square of P and the cube of a) are set by the power (r squared) in the law of gravity.

†This force is called a "centripetal force," which is the force applied to draw an object *inward* toward the center of the orbit.

assumed to be much smaller than $M$. Assume the planet moves in a circular orbit of radius ($r$) at a velocity ($V$) so that the force required to hold the planet in orbit is $mV^2/r$. That force—the force that deflects the planet from its tendency to move in a straight line—is supplied by the gravitational force between the Sun and the planet. We have already defined that force as $GMm/r^2$.

Because $mV^2/r$ is the force required to hold the planet in orbit, it must equal the force of gravity.

$$\frac{mV^2}{r} = \frac{GMm}{r^2}$$

We can cancel $m$ out, and one of the $r$'s to obtain

$$V^2 = \frac{GM}{r}$$

Finally, we take the square root of both sides to obtain the orbital velocity.

$$V = \sqrt{\frac{GM}{r}}$$

The above equation, giving the orbital velocity in terms of $M$ and $r$, can be used to determine the mass of the central object if the orbiting object's velocity and distance from it are known.

## EXTENDING OUR REACH

### WEIGHING THE SUN

To find the mass ($M$) of the Sun, we go back to our equation for $V^2$, cross-multiply by $r$, and divide by G to obtain

$$M = \frac{V^2 r}{G}$$

Therefore, to evaluate $M$, we need $V$, the Earth's orbital velocity (the speed with which it is moving in its orbit). We find $V$ by dividing the circumference of its orbit, C, by the time it takes the Earth to complete one orbit, a length of time we call the orbital period, $P$. Thus

$$V = \frac{C}{P}$$

The formula for the circumference of a circle is C = $2\pi r$. Thus

$$V = \frac{2\pi r}{P}$$

Putting this expression for $V$ into the equation for $M$, above, gives

$$M = \left(\frac{2\pi r}{P}\right)^2 \frac{r}{G}$$

$$= \frac{4\pi^2 r^3}{GP^2}$$

This is a modified form of Kepler's third law and it is important because we can use it to measure the mass of an object given the orbital period, $P$, and orbital radius, $r$,

of a much smaller object moving around it. For example, suppose we want to find the Sun's mass, $M$.

According to the expression for $M$, we need $P$ and $r$ for the Earth's orbit. If we measure $P$ in seconds and $r$ in meters, then the gravitational constant G = $6.67 \times 10^{-11}$ m³-kg$^{-1}$-s$^{-2}$.

To find $P$ in seconds, we remember that it takes the Earth one year to orbit the Sun. Thus, $P$ is one year. We can express $P$ in seconds by multiplying the number of seconds in a minute (60), times the number of minutes in an hour (60), times the number of hours in a day (24), times the number of days in a year (365.25). The result of that calculation, rounded off to three significant figures, is

$$P = 3.16 \times 10^7 \text{ seconds}$$

Similarly, we need $r$, the Earth–Sun distance, in meters, which we can look up in the appendix and find is $1.50 \times 10^{11}$ m. Putting these values and the value of $\pi$ and G into the expression for $M$ we find

$$M = \frac{4(3.14)^2 \times (1.50 \times 10^{11} \text{ m})^3}{6.67 \times 10^{-11} \text{ m}^3\text{-kg}^{-1}\text{-s}^{-2} \times (3.16 \times 10^7 \text{ s})^2}$$

$$= \frac{4 \times 9.86 \times 3.38 \times 10^{33} \text{ m}^3}{6.67 \times 9.99 \times 10^{14-11} \text{ m}^3\text{-kg}^{-1}\text{s}^{-2}\text{s}^2}$$

$$= 2 \times 10^{30} \text{ kg}$$

The same method can be used to find the mass of any object around which another object orbits. Thus, gravity becomes a tool for determining the mass of astronomical objects, and we shall use this method many times throughout our study of the Universe.

There are other ways we can use the law of gravity. For example, we can use it to find out how much we would weigh on another planet and how fast a spacecraft must move to escape from a planet's surface.

## **2.8**    SURFACE GRAVITY

**Surface gravity** measures the gravitational attraction at a planet's or star's surface. It is the *acceleration* on a mass created by the local gravitational force, not the force itself. This acceleration determines how fast objects fall. To understand the importance of surface gravity, recall that mass measures the amount of material an object contains and is therefore constant. On the other hand, an object's weight depends on its mass and the acceleration of gravity. Thus, surface gravity determines what a mass weighs. In addition to determining weight on a planet, surface gravity also influences a planet's shape and whether it has an atmosphere. For example, small objects such as asteroids are not spherical because their surface gravity is too weak to crush them into round shapes. Likewise, a small surface gravity makes it hard for a planet to keep an atmosphere.

We determine the strength of a planet's surface gravity as follows.

The law of gravity states that a planet of mass *M* exerts a gravitational force *F* on a body of mass *m* at a distance *r* from its *center* given by

$$F = \frac{GMm}{r^2}$$

At the planet's surface, $r = R$, the planet's radius, so

$$F = \frac{GMm}{R^2}$$

Newton's second law ($F = ma$) tells us that for any force, *F*, the acceleration it produces on a body of mass *m* is $a = F/m$. Therefore,

$$ma = \frac{GMm}{R^2}$$

Canceling out the *m*'s then gives

$$a = \frac{GM}{R^2}$$

Thus, a planet's surface gravity depends on its mass and radius. That dependence is such that two planets with the same radius but different masses will have different surface gravities. In particular, if two planets have the same radius but different masses, the planet with the larger mass has the larger surface gravity. Similarly, if two planets have the same mass but different radii, the planet with the larger radius has less surface gravity.

Surface gravity is usually denoted by the letter *g* (the origin of the phrase "Pulling *g*'s," which is used by pilots). We can therefore write that

INTERACTIVE

*Gravity variations*

*g* = Surface gravity
G = Constant
*M* = Mass of the attracting body
*R* = Radius of the attracting body

$$g = \frac{GM}{R^2}$$

Because the surface gravity depends on the mass and radius of the attracting body, the strength of the surface gravity is different from body to body. For example, we can compare the surface gravity of the Moon with that of the Earth to show why you would weigh less on the Moon than on the Earth.

**FIGURE 2.9**
An astronaut can make huge leaps in the Moon's low gravity.

🔭★ Why is the sky black in this photo, when the Sun is obviously shining? Why does the flag appear to be waving in the airless atmosphere of the moon?

🔭★ There is no air present to scatter blue light, and the flag was set up with wires to look that way.

| TABLE 2.1 | | |
|---|---|---|
| **Mass and Radius of the Earth and Moon\*** | | |
| | Mass | Radius |
| Earth | $6.0 \times 10^{24}$ kg | $6.4 \times 10^6$ meters |
| Moon | $7.3 \times 10^{22}$ kg | $1.7 \times 10^6$ meters |
| Ratio Earth/Moon | 81 | 3.8 |

\*Numbers have been rounded slightly.

To make the calculation, we need to know the mass and radius of the Earth and the Moon. Those numbers are given in table 2.1.

Because the Earth has greater mass, we might guess that its surface gravity, $g$, is greater than the Moon's. However, the Moon's radius is smaller, so its surface gravity might be greater. Thus, to determine which body has the larger surface gravity, we need to evaluate $g$ mathematically.

We can make this calculation two ways. The first way is simply to plug into the equation for $g$ the values of $M$ and $R$ appropriate for the Earth and then repeat the process with values appropriate for the Moon. A far easier way is to work with ratios. To do that we write out on one line the expression for $g$ on the Earth and then write out on the line below it the value for $g$ on the Moon. Next we draw a horizontal line between them to indicate division, as shown below. This gives

$$\frac{g_{\text{Earth}}}{g_{\text{Moon}}} = \frac{GM_{\text{Earth}}/(R_{\text{Earth}})^2}{GM_{\text{Moon}}/(R_{\text{Moon}})^2}$$

Notice that by doing this, the value of G, the gravitational constant, cancels out. If we then group the terms in $M$ and the terms in $R$ separately, we find

$$= \frac{M_{\text{Earth}}/M_{\text{Moon}}}{(R_{\text{Earth}}/R_{\text{Moon}})^2}$$

$$= \frac{(81)}{(3.8)^2} = 5.6 \sim 6$$

Thus, the ratio of $g$ on the Earth to $g$ on the Moon is about 6 : 1 so that you weigh about 6 times more on the Earth than you would on the Moon. That fact allowed the astronauts to make such huge leaps on the Moon (fig. 2.9).

## 2.9    ESCAPE VELOCITY

**INTERACTIVE**

*Escape velocity*

To overcome a planet's gravitational force and escape into space, a rocket must achieve a critical speed known as the **escape velocity.** Escape velocity is the speed an object needs to move away from a body and not be drawn back by its gravity. We can understand how such a speed might exist if we think about throwing an object into the air. The faster the object is tossed upward, the higher it goes and the longer it takes to fall back. Escape velocity is the speed an object needs so that it will *never* fall back, as depicted in figure 2.10. Thus, escape velocity is of great importance in space travel if craft are to move away from one object and not be drawn back to it. However, escape velocity is also important in many astronomical phenomena, such as whether a planet has an atmosphere and the nature of black holes.\*

---

\*In chapter 14, we will see that a black hole is an object whose escape velocity equals the speed of light. Thus, light cannot escape from it, thereby making it black.

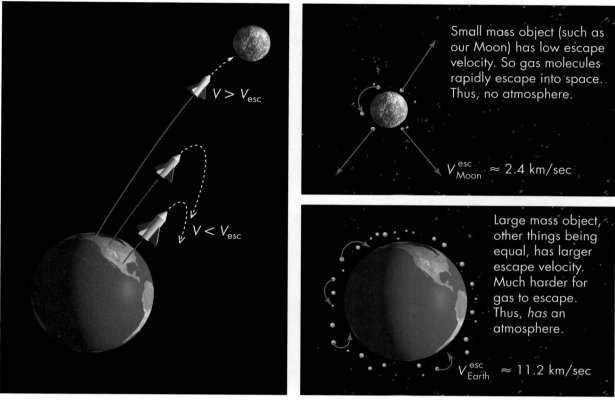

**FIGURE 2.10**
Escape velocity is the speed an object must have to overcome the gravitational force of a planet or star and not fall back. A low escape velocity, in general, leads to the absence of an atmosphere on a planet or satellite.

The escape velocity, $V_{esc}$, for a spherical body such as a planet or star, can be found from the law of gravity and Newton's laws of motion and is given by the following formula:

$V_{esc}$ = Escape velocity
$G$   = Constant
$M$   = Mass of the body to be escaped
     from
$R$   = Radius of the body to be escaped
     from

$$V_{esc} = \sqrt{2GM/R}$$

Here, $M$ stands for the mass of the object from which we are attempting to escape, and $R$ is its radius, as shown in figure 2.11.

Notice in the equation for $V_{esc}$ that if two objects of the same radius are compared, the larger mass will have the larger escape velocity. Likewise, if two objects of the same mass are compared, the one with the smaller radius will have the greater escape velocity. We will see in chapter 14 that the huge escape velocity of a black hole arises from its abnormally small radius.

To illustrate the use of the formula, we calculate the escape velocity from the Moon. From the data in table 2.1, we find the Moon's radius and mass. We insert these values in the formula for escape velocity and find

$$V_{esc} (\text{Moon}) = \sqrt{2GM/R}$$

$$V_{esc} (\text{Moon}) = \sqrt{2 \times 6.7 \times 10^{-11}\ \text{m}^3\text{-s}^{-2}\text{-kg}^{-1} \times 7.3 \times 10^{22}\ \text{kg}/1.7 \times 10^{6}\ \text{m}}$$

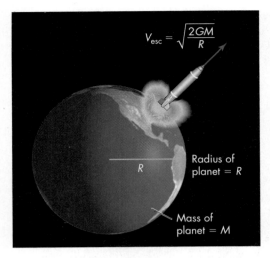

**FIGURE 2.11**
Calculating the escape velocity from a body.

We can simplify the math by collecting the numerical values, powers-of-ten, and units separately to get

$$V_\infty \text{ (Moon)} = \sqrt{2 \times 6.7 \times 7.3/1.7 \times 10^{-11} \times 10^{22}/10^6 \times \text{m}^3\text{-s}^{-2}\text{-kg}^{-1}\,\text{kg/m}}$$
$$= \sqrt{58 \times 10^5 \text{ m}^2\text{-s}^{-2}}$$

Taking the square root* now gives

$$= 2.4 \times 10^3 \text{ m/s} = 2.4 \text{ km/s}$$

A similar calculation shows that the escape velocity from the Earth is 11 kilometers per second. Thus, it is much easier to blast a rocket off the Moon than the Earth.[†] In chapter 6, we will see that this low escape velocity is partly responsible for the Moon's lack of an atmosphere.

## SUMMARY

A gravitational force exists between any two objects in the Universe. The strength of this force depends on the masses of the bodies and their separation. Gravitational forces hold astronomical bodies together and in orbit about one another.

An object's inertia makes it remain at rest or move in a straight line at a constant speed unless the object is acted on by a net force. Thus, for a planet to orbit the Sun, the Sun's force of gravity must act on it. The law of inertia and Newton's other laws of motion, when combined with the law of gravity, allow us to relate the size and speed of orbital motion to the mass of the central object.

The gravitational force exerted by a planet determines its surface gravity and escape velocity. The former determines your weight on a planet. The latter is the speed necessary to leave the surface and escape without falling back.

### **?** What It Is.

Chapter opener: This is a time-lapse photo of the launch of the space shuttle *Atlantis* on July 12, 2001.

## QUESTIONS FOR REVIEW

1. What is meant by inertia?
2. What does Newton's first law of motion tell you about the difference between motion in a straight line and motion along a curve?
3. What is Newton's law of gravity?
4. How does mass differ from weight?
5. If your mass is 70 kilograms on Earth, what is it on the Moon?
6. If you weigh 110 pounds on the Earth, do you weigh 110 pounds on the Moon? Why?
7. What does surface gravity measure?
8. What is meant by escape velocity?

## THOUGHT QUESTIONS

1. Which do you think has more inertia: a small, inflated balloon or a cinder block? If each were moving toward you at 1 meter/second, which would be easier to catch?
2. A cinder block can be weightless in space. Would you want to kick it with your bare foot? Even if it is weightless, does it have mass?

---

*Note: You must take the square root of the units too. This is easy to do. Just divide their exponents by 2. Thus: $\sqrt{\text{m}^2\text{-s}^{-2}} = \text{m-s}^{-1}$.

[†]If an object's mass and radius are given in units of the Earth's, then $V_{esc} \approx 11\sqrt{M/R}$ km/s.

3. In some amusement park rides, you are spun in a cylinder and are pressed against the wall as a result of the spin. People sometimes describe that effect as being due to "centrifugal force." What is really holding you against the wall of the spinning cylinder? Drawing a sketch and using Newton's first law may help you answer the question.

# PROBLEMS

1. How many times greater is the Earth's gravitational force on the Moon than the Moon's gravitational force on the Earth? Think about Newton's third law of motion before answering this.

2. Calculate the escape velocity from the Earth, given that the mass of the Earth is $6 \times 10^{24}$ kilograms and its radius $6 \times 10^6$ meters. In this problem, round off G to $7 \times 10^{-11}$ meters$^3$/(kg-s$^2$).

3. Calculate the escape velocity from the Sun, given that its mass is $2 \times 10^{30}$ kg and its radius is $7 \times 10^8$ meters.

4. Which body has a larger escape velocity, Mars or Saturn?

   $M_{\text{Mars}} = 0.1\,M_{\text{Earth}}$

   $M_{\text{Saturn}} = 95\,M_{\text{Earth}}$

   $R_{\text{Mars}} = 0.5\,R_{\text{Earth}}$

   $R_{\text{Saturn}} = 9.4\,R_{\text{Earth}}$

5. Calculate the ratio of the escape velocities from the Moon and Earth.

6. Calculate your weight on the Moon.

7. Given that Jupiter is about 5 times farther from the Sun than the Earth, calculate its orbital velocity. How many years does it take Jupiter to complete an orbit around the Sun?

8. Given that the mass of the Milky Way galaxy is $10^{11}$ times that of the Sun and that the Sun is $2.6 \times 10^{20}$ meters from its center, what is the Sun's orbital speed around the center of the galaxy? How long does it take the Sun to orbit the Milky Way? (In this problem, we assume that the galaxy can be treated as a single spherical blob of matter. Strictly speaking, this isn't correct, but the far more elaborate math needed to calculate the problem properly ends up giving almost the same answer.)

9. A good baseball pitcher can throw a ball at 100 miles/hr (about 45 meters/s). If the pitcher were on the fictional asteroid Cochise, could the pitcher throw the ball fast enough so it would escape from Cochise? (Note: Assume Cochise is a sphere and that its mass is $M = 9.6 \times 10^{16}$ kg and its radius is 20 km = $2.0 \times 10^4$ meters.)

10. Use the modified form of Kepler's third law to find the mass of the imaginary star 57 Fungaloid, given that a planet is in a circular orbit around it at a distance of $3 \times 10^{11}$ meters with an orbital period of 3 years. Divide your answer by the Sun's mass to see how much more (less) massive the star is than our Sun.

# TEST YOURSELF

1. Which of the following demonstrate the property of inertia?
   (a) A car skidding on a slippery road
   (b) The oil tanker *Exxon Valdez* running aground
   (c) A brick sitting on a table top
   (d) Whipping a table cloth out from under the dishes set on a table
   (e) All of the above

2. If an object moves along a curved path at a constant speed, you can infer that
   (a) a force is acting on it.
   (b) it is accelerating.
   (c) it is in uniform motion.
   (d) both (a) and (b) are true.
   (e) neither (a) nor (b) is true.

3. If the distance between two bodies is quadrupled, the gravitational force between them is
   (a) increased by a factor of 4.
   (b) decreased by a factor of 4.
   (c) decreased by a factor of 8.
   (d) decreased by a factor of 16.
   (e) decreased by a factor of 64.

4. The strength of the gravitational force exerted by the Sun on the Earth is the same as the strength of the gravitational force exerted by the Earth on the Sun.
   (a) True
   (b) False

5. Two planets have identical diameters but differ in mass by a factor of 25. The more massive planet therefore has an escape velocity
   (a) 25 times larger than the other.
   (b) 25 times smaller than the other.
   (c) 5 times larger than the other.
   (d) 625 times larger than the other.
   (e) 50 times larger than the other.

## FURTHER EXPLORATIONS

Duzen, Carl, Jane Nelson, and Jim Nelson. "Classifying Motion." *Physics Teacher* 30 (October 1992): 414.

*The following books discuss the laws of motion in greater detail:*

Casper, Barry M., and Richard J. Noer. *Revolutions in Physics.* New York: W. W. Norton, 1972.

Hecht, Eugene. *Physics in Perspective.* Reading, Mass.: Addison-Wesley, 1980.

Hewitt, Paul G. *Conceptual Physics.* 10th Edition. San Francisco: Pearson Addison-Wesley, 2006.

Kirkpatrick, Larry D., and Gregory E. Francis. *Physics: A World View.* Belmont, CA: Thomson Brooks/Cole, 2007.

### Website

Please visit the *Explorations* website at **http://www.mhhe.com/arny** for additional online resources on these topics.

## KEY TERMS

acceleration, 79

escape velocity, 86

inertia, 76

law of gravity, 81

mass, 80

Newton's first law of motion, 76

Newton's second law of motion, 81

Newton's third law of motion, 82

surface gravity, 85

# Light and Atoms

Newton studies the Sun's spectrum.

**?What Is This?**
(See chapter summary for the answer.)

## Concepts and Things to Remember

- Light can be thought of either as a wave of energy or a stream of particles called photons.
- White light is a mix of all colors.
- The color of light is determined by its wavelength.
- Atoms can emit light when their electrons shift orbits.
- Each kind of atom emits light at a unique set of wavelengths (colors).
- For many kinds of objects, the color and intensity of their light are set by their temperature:
  - The hotter the object, the bluer the light it emits.
  - The hotter the object, the more light it emits.
- These properties allow us to deduce an object's temperature from its color.
- The wavelengths of the radiation emitted by a moving object are shifted.

## Concepts and Skills to Review

- Structure of atoms (P. 12)
- Forces (P. 12)

Our home planet is separated from other astronomical bodies by such vast distances that, with few exceptions, we cannot learn about them by direct measurements of their properties. For example, if we want to know how hot the Sun is, we cannot stick a thermometer into it. Similarly, we cannot directly sample the composition of the atmosphere of Saturn or a distant star. However, we can sample such remote bodies indirectly by analyzing their light. Light from a distant star or planet can tell us what the body is made of, its temperature, and many of its other properties. Light, therefore, is our key to studying the Universe. To use the key, however, we need to understand some of its properties.

In this chapter, we will discover that light is a form of energy that can be thought of either as a wave or as a stream of particles. Furthermore, we will discover that the light we see is just part of the radiation emitted by astronomical objects. We will also learn that light can be produced within an atom by changes in its electrons' energies. These changes imprint the atom's "signature" on the light. However, the light may also bear unwanted messages. For example, when light reaches our atmosphere, gases there alter its properties, blocking some rays, and bending and blurring others. These distortions limit what astronomers can learn from the ground.

The goal of this chapter is to explain the nature of light, how it is produced, and how it interacts with our atmosphere. Our first step toward this goal is to better understand what light is.

## 3.1    PROPERTIES OF LIGHT

**Light** is radiant energy; that is, it is energy that can travel through space from one point to another without the need of a direct physical link. Therefore, light is very different in its basic nature from, for example, sound. Sound can reach us only if it is carried by a medium such as air or water, whereas light can reach us even across empty space. In empty space, we can see the burst of light of an explosion, but we will hear no sound from it at all.

Light's capacity to travel through the vacuum of space is paralleled by another very special property: its high speed. In fact, the speed of light is an upper limit to all motion. In empty space, light travels at the incredible speed of 300,000* kilometers per second. An object traveling that fast could circle the Earth in a mere seventh of a second.

The speed of light in empty space is a constant and is denoted by "c." However, in transparent materials, such as glass, water, and gases, the speed of light is reduced. Furthermore, different colors of light are slowed differently. For example, in nearly all materials, blue light travels slightly more slowly than red light. As we will see in chapter 4, lenses and prisms work because they slow the light as it travels through them.

### The Nature of Light—Waves or Particles?

Observation and experimentation on light throughout the last few centuries have produced two very different models of what it is and how it works. According to one model, light is a wave that is a mix of electric and magnetic energy, changing together, as depicted in figure 3.1. Because light is a mix of electric and magnetic energy, it is often called an **electromagnetic wave** or **electromagnetic radiation.** The ability of such radiation to travel through empty space comes from the interrelatedness of electricity and magnetism.

---

*More precisely, 299,792.458 km-s$^{-1}$.

**FIGURE 3.1**
A wave of electromagnetic energy moves through empty space at the speed of light, 299,792.5 kilometers per second. The wave carries itself along by continually changing its electric energy into magnetic energy and vice versa.

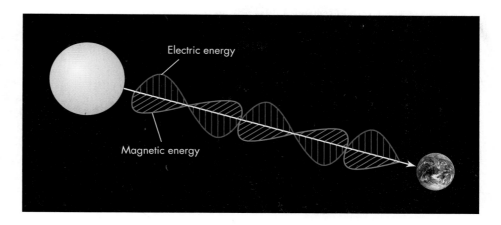

Electric energy

Magnetic energy

You can see this relationship between electricity and magnetism in everyday life. For example, when you start your car, turning the ignition key sends an electric current from the battery to the starter. There, the current generates a magnetic force that turns over the engine. Similarly, when you pull the cord on a lawn mower, you spin a magnet that generates an electric current that creates the spark to start its engine.

This interrelatedness between electricity and magnetism is what allows light to travel through empty space. A small disturbance of an electric field creates a magnetic disturbance in the adjacent space, which in turn creates a new electric disturbance in the space adjacent to it, and so on. Thus, a fluctuation of electric and magnetic field spreads out from its source carried by the fields. In this fashion, light can move through empty space "carrying itself by its own bootstraps."

As the electromagnetic wave travels through matter, it may disturb the atoms, causing them to vibrate the way a water wave makes a boat rock. It is from such disturbances in our eyes, a piece of film, or an electronic sensor that we detect the light.

The model of light as a wave works well to explain many phenomena, but it fails to explain some of light's properties. In those circumstances, it is necessary (and easier!) to use a different model.

In this model, light may be thought of as a stream of particles called **photons** (fig. 3.2). The photons are packets of energy, and when they enter your eye, they produce the sensation of light.

In empty space, photons move in a straight line at the speed of light. Although they are described as particles, photons can behave as waves, but they are not unique in this respect. According to the laws of quantum physics, subatomic particles such as electrons and protons can also behave like waves. For this reason, scientists often speak of light and sub-atomic particles as having a **wave-particle duality** and they use whichever model—wave or particle—best describes a particular phenomenon. For example, reflection of light off a mirror is easily understood if you imagine photons striking the mirror and bouncing back just the way a ball rebounds when thrown at a wall. On the other hand, the focusing of light by a lens is best explained by the wave model.

In most of the rest of this chapter, we will discuss light using the wave model, so that we do not have to constantly refer first to photons and then to waves. But regardless of which model we use, light has two important properties that we need to describe. The first of these properties is its brightness (or intensity). Brightness measures the amount of energy carried by the light. In the wave picture, brightness is related to the height of the wave. In the photon picture, brightness is related to the number of photons traveling in a given direction. The second important property of light is its color.

ANIMATION

*Photons stream away from a light source at the speed of light.*

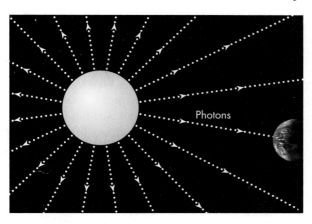

Photons

**FIGURE 3.2**
Photons—particles of energy—stream away from a light source at the speed of light.

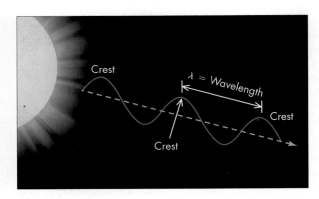

**FIGURE 3.3**
The distance between crests defines the wavelength, λ, for any kind of wave, be it water or electromagnetic.

| TABLE 3.1 | | |
|---|---|---|
| **Colors and Wavelengths*** | | |
| Red | 700 nm | 0.7 micrometers |
| Yellow | 580 | 0.58 |
| Blue | 480 | 0.48 |
| Violet | 400 | 0.40 |

*These color equivalences are only approximate.

## Light and Color

Human beings can see colors ranging from deep red through orange and yellow into green, blue, and violet, and we call these colors the **visible spectrum.** But what property of photons or electromagnetic waves corresponds to light's different colors?

According to the wave theory, the color of light is determined by the light's **wavelength,** which is the spacing between wave crests (fig. 3.3). That is, instead of describing a quality (color), we specify a quantity (wavelength), which we usually denote by the Greek letter lambda, λ. For example, the wavelength of deep red light is about $7 \times 10^{-7}$ meters. The wavelength of violet light is about $4 \times 10^{-7}$ meters. Intermediate colors have intermediate wavelengths. The wave-particle duality model allows us to make a similar connection between wavelength and color for photons. Thus, we can also characterize photons by their wavelength.

The wavelengths of visible light are very small (roughly the size of a bacterium). They are therefore usually measured not in meters but in billionths of a meter, a unit called the **nanometer,** abbreviated nm. Thus, the wavelength of red light is about 700 nanometers and that of violet light is about 400 nanometers.* Table 3.1 lists the wavelengths of the primary colors in nm and some other units sometimes used to measure wavelengths.

The relation between wavelength and color is important, and we will refer to it repeatedly in later chapters. In doing so, we will often mention only the red and blue colors of visible light. The other colors are not missing: they are just not referred to explicitly. Given this simplification, you should be sure to remember that *red* colors refer to *long* wavelengths of visible light and *blue* colors refer to *short* wavelengths.

---

*Scientists sometimes use other units of length to measure wavelengths. For example, astronomers have traditionally used angstrom units and micrometers. One angstrom unit is $10^{-10}$ meters (1 ten-billionth of a meter) and is thus the same as 1/10 of a nanometer. The wavelength of red light is thus about 7000 angstroms. The micrometer (also called a micron and abbreviated μm) is used especially at infrared wavelengths (wavelengths longer than visible light that we perceive as heat). One micrometer is $10^{-6}$ meters. The wavelength of red light is about 0.7 micrometers.

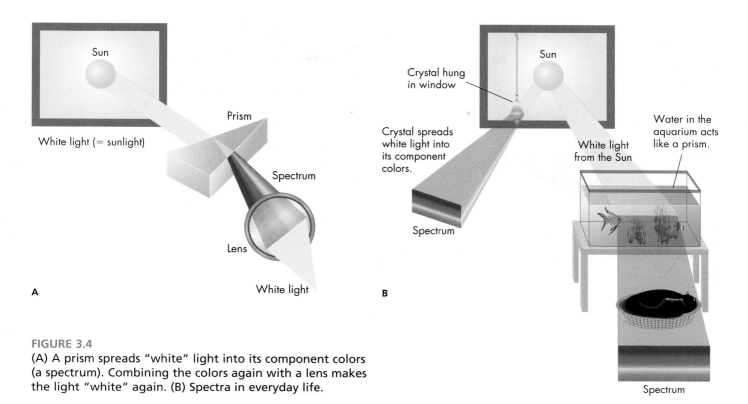

**FIGURE 3.4**
(A) A prism spreads "white" light into its component colors (a spectrum). Combining the colors again with a lens makes the light "white" again. (B) Spectra in everyday life.

## Characterizing Electromagnetic Waves by Their Frequency

Sometimes it is useful to describe electromagnetic waves by their frequency rather than their wavelength. **Frequency** is the number of wave crests that pass a given point in 1 second. It is measured in hertz (abbreviated as Hz) and is usually denoted by the Greek letter nu, $\nu$. You can see an everyday example of this on a radio dial, where you tune in a station by its frequency rather than its wavelength.

For all kinds of waves, the frequency and wavelength are related to the wave speed because in one vibration, a wave must travel a distance equal to one wavelength. This implies that for light, the product of the wavelength ($\lambda$) and the wave frequency ($\nu$) equals the speed of light (c): that is, $\lambda \nu = c^*$. Because all light travels at the same speed (in empty space), we can treat c as a constant. Thus, specifying $\lambda$ determines $\nu$ and vice versa. We will generally use $\lambda$ to characterize electromagnetic waves, but $\nu$ is just as good.

## White Light

Although wavelength is an excellent way to specify most colors of light, some light seems to have no color. For example, the Sun when it is seen high in the sky and an ordinary light-bulb appear to have no dominant color. Light from such sources is called **white light.**

White light is not a special color of light; rather it is a mixture of all colors. That is, the sunlight we see is made up of all the wavelengths of visible light—a blend of red, yellow, green, blue, and so on—and our eyes perceive this blend as white. Newton demonstrated this property of sunlight by a very simple but elegant experiment. He passed sunlight through a prism (fig. 3.4A) so that the light was spread out into the visible spectrum (or rainbow of colors). He then recombined the separated colors with a lens and reformed the beam of white light.

You can see how colors of light mix if you look at a color television screen close up. You will notice that the screen is covered with tiny red, green, and blue dots. In a red object, only the red spots are lit. In a blue one, only the blue spots. In a white object, all three are lit, and the brain mixes these three colors to form white. Other colors are made by appropriate blending of red, green, and blue. Notice this is very different from the way that pigments of paint mix. Red, green, and blue paint when mixed give a brownish color.

---

*In using this equation, if $\nu$ is in hertz, the units of $\lambda$ will be set by the units you choose for c. Thus, if c is in meters/second, $\lambda$ will be in meters.

Why do we see sunlight as white? Presumably because our senses have evolved to make us aware of *changes* in our surroundings. Thus, we ignore the ambient "color" of sunlight just as we come in time to ignore a steady background sound or smell.

But there is more to light than what meets the eye. Just as red is but one part of the visible spectrum, so too the visible spectrum itself is but one part of a much wider spectrum of electromagnetic waves.

## 3.2 THE ELECTROMAGNETIC SPECTRUM: BEYOND VISIBLE LIGHT

Visible light is but one of the many kinds of electromagnetic waves that exist in our Universe. For example, radio waves, X rays, and ultraviolet radiation are also electromagnetic waves. They differ from visible light only in their wavelength, and to indicate this fundamental unity, scientists refer to them in general as the **electromagnetic spectrum.**

The electromagnetic spectrum has been studied over a huge range of wavelengths. The longest electromagnetic waves yet detected have wavelengths thousands of kilometers long.* The shortest have wavelengths of $10^{-18}$ meters or less. Ordinary visible light falls in a very narrow section in about the middle of the known spectral range (see table 3.2 and fig. 3.5).

Although our eyes can detect only a tiny portion of the electromagnetic spectrum—namely, the part we call visible light—various instruments allow us to explore most of the other wavelength regions, too. In fact, new instruments allow astronomers to "see" such astronomically important events as the formation of stars, the remnants left behind when stars die, and, indirectly, black holes. In the following sections, we discuss these different wavelength regions in the order in which they were discovered.

### Infrared Radiation

Although humans cannot see infrared radiation, several kinds of snakes, including the rattlesnake, have special infrared sensors located just below their eyes. These allow the snake to "see" in total darkness, helping it to find warm-blooded prey such as rats.

The exploration of the electromagnetic spectrum began in 1800, when Sir William Herschel (discoverer of the planet Uranus) showed that heat radiation, such as you feel from the Sun or from a warm radiator, though invisible, was related to visible light.

### TABLE 3.2

### Electromagnetic Spectrum

| Wavelength | Kind of Radiation | Astronomical Sources |
|---|---|---|
| 100–500 meters | Radio (AM broadcast) | Pulsars (remnants of exploded stars—also emit X rays) |
| 10–100 meters | Short-wave radio | Active galaxies |
| 1–10 meters | TV, FM radio | Solar radio outbursts, interstellar gas |
| 1–100 centimeters | Radar, cell phones | Planets, active galaxies |
| 1–10 millimeters | Microwaves | Interstellar clouds, cosmic background radiation |
| 1–1000 $\mu$m | Infrared (heat) | Young stars, planets, interstellar dust |
| 400–700 nanometers | Visible light | Stars, Sun |
| 1–300 nanometers | Ultraviolet | Stars |
| 0.01–1 nanometers | X rays | Collapsed stars, hot gas in galaxy clusters |
| $10^{-7}$–0.01 nanometers | Gamma rays | Active galaxies and gamma-ray bursters |

Note: 1 mm = 1000 $\mu$m; 1 $\mu$m = 1000 nm

*Such long waves have not been detected from astronomical sources, however, and cannot pass through our atmosphere.

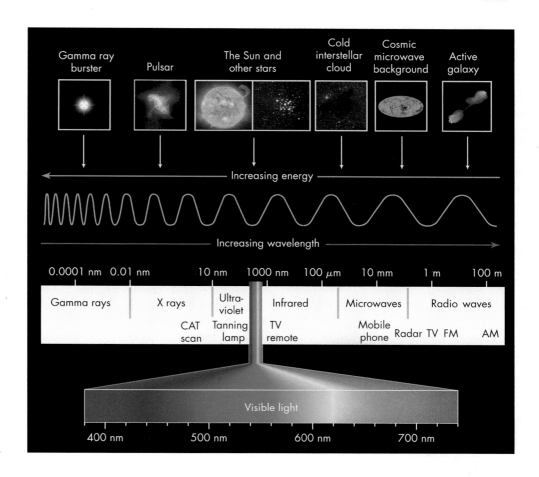

**FIGURE 3.5**
The electromagnetic spectrum.

Herschel was trying to measure heat radiated by astronomical sources. He projected a spectrum of sunlight onto a table top and placed a thermometer in each color to measure its energy. He was surprised that when he put a thermometer just off the red end of the visible spectrum, the thermometer registered an elevated temperature there just as it did in the red part of the spectrum. He concluded that some form of invisible energy perceptible as heat existed beyond the red end of the spectrum and he therefore called it **infrared.** Even though our eyes cannot see infrared light, nerves in our skin can feel it as heat.

## Ultraviolet Light

Another important part of the electromagnetic spectrum, **ultraviolet** radiation, was discovered in 1801 by J. Ritter while he was experimenting with chemicals that might be sensitive to light. Ritter noted that when he shined a spectrum of sunlight on a layer of silver chloride, the chemical blackened most strongly in the region just beyond the violet end of the spectrum, implying the presence there of some invisible radiation.

Infrared and ultraviolet radiation differ in no physical way from visible light except in their wavelength. Infrared has longer wavelengths and ultraviolet shorter wavelengths than visible light (see table 3.2). Exploration of those parts of the electromagnetic spectrum with wavelengths much larger and much smaller than visible light had to await the growth of new technology, as the development of radio astronomy demonstrates.

## Radio Waves

James Clerk Maxwell, a Scottish physicist, predicted the existence of radio waves in the mid-1800s. It was some 20 years later, however, before Heinrich Hertz produced them experimentally in 1888, and another 50 years had to pass before Karl Jansky discovered naturally occurring radio waves coming from cosmic sources. Jansky's discovery in the 1930s that the center of the Milky Way was a strong source of radio emission was the birth of radio astronomy.

Radio waves range in length from about 10 centimeter to hundreds of meters, making them much longer than visible and infrared waves. Today we can generate radio waves and use them in many ways, ranging from communication to radar. Astronomers too use radio waves, detecting them with radio telescopes. Radio signals, generated by natural processes, allow astronomers to obtain radio "views" of forming stars, exploding stars, active galaxies, and interstellar gas clouds. Radio wavelengths are even being searched for signals that might reveal the existence of extraterrestrial civilizations (see essay 3).

## Other Wavelength Regions

Many decades also passed between the discovery of X rays by Wilhelm Roentgen in 1895 and the detection in the late l940s of X rays coming from the Sun. X-ray wavelengths turn out to be far shorter than those of visible light, typically between 0.01 and 10 nanometers, but they too are important. Doctors and dentists use X rays to probe our bones and organs. Astronomers use X-ray telescopes to detect X rays emitted by the hot gas surrounding black holes and the tenuous gas in distant groups of galaxies.

Two parts of the electromagnetic spectrum were relatively unexplored until recently: microwaves, the region between infrared and radio waves, and gamma rays, which are the shortest wavelengths known. Both of these wavelength regions are difficult to study from the ground because, as we will see in section 3.7, they fall in wavelength bands that are strongly blocked by the Earth's atmosphere. However, orbiting telescopes have now given astronomers preliminary views of the sky in these wavelengths.

Despite the enormous variety of electromagnetic radiation, it is all the same physical phenomenon: the vibration of electric and magnetic energy traveling at the speed of light or a stream of photons. (In fact, scientists refer to radio photons, visible photons, ultraviolet photons, etc.) The essential difference between these many kinds of electromagnetic radiation is simply their wavelength (or frequency). This difference alters not only how we perceive them but also how much energy they can carry.

## Energy Carried by Electromagnetic Radiation

The warmth we feel on our face from a beam of sunlight demonstrates that light carries energy, but not all wavelengths carry the same amount of energy. It turns out that the amount of energy, $E$, carried by electromagnetic radiation depends on its wavelength, $\lambda$. Each photon of wavelength $\lambda$ is an energy packet that carries an amount of energy, $E$, given by

$E$ = Energy carried by a photon of
    wavelength = $\lambda$
$h$ = Constant
$c$ = Speed of light (constant)

$$E = \frac{hc}{\lambda}$$

The speed of light c, and the constant, h, are unchanging,* so if the wavelength of the light decreases, the energy it carries increases. Thus

**Short-wavelength radiation carries proportionally more energy than long wavelength radiation.**

Ultraviolet light with its short wavelength therefore carries proportionally more energy than infrared light with its longer wavelength. In fact, ultraviolet light of sufficiently

---

*If $E$ is measured in joules and $\lambda$ in meters, h = $6.63 \times 10^{-34}$ joule-second (known as Planck's constant) and hc = $1.99 \times 10^{-25}$ joule-meters.

short wavelengths can carry so much energy that it can break apart atomic and molecular bonds. As we will see in chapter 13, this can cause intense heating of gas near stars. Nearer to home, it is the reason ultraviolet light gives you a sunburn while an infrared heat lamp does not.

We have now described the main properties of light itself. We now need to discuss how light is produced and what it can tell us about the objects emitting it.

## RADIATION AND TEMPERATURE

Hot objects emit light, as you can easily demonstrate if you turn on a burner on an electric stove. As the stove's burner warms up, it begins to glow. Initially the burner emits only a dim, deep red color of light. However, as the burner grows hotter, the light it emits becomes brighter and its color changes (figure 3.6A), becoming brighter red and eventually yellow. If we could make the burner even hotter, it would begin to glow blue.

Recalling that blue light has a shorter wavelength than red light, you can see from this simple demonstration an important relation between an object's temperature and the color of the light it emits:

**As an object's temperature increases, the object radiates light more strongly at shorter wavelengths.**

That is, as an object heats up, the color of the light it emits shifts from red (long) wavelengths toward yellow (medium) wavelengths and, if hot enough, to blue (short) wavelengths. This connection between an object's temperature and the color of the light it emits applies to more than just electric stove burners. It is a general property of many hot objects, including stars, and it allows astronomers to measure the temperature of stars (and many other astronomical objects) from their light.

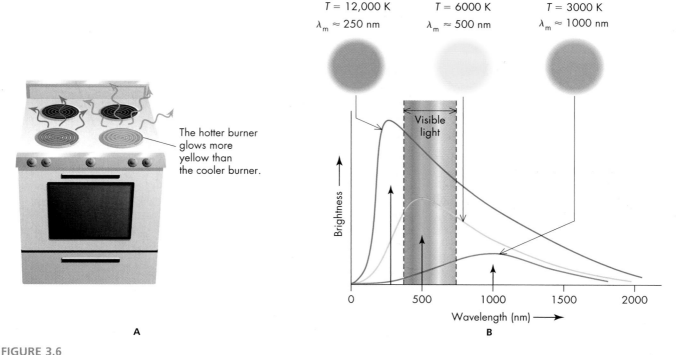

**FIGURE 3.6**
(A) The hotter burner glows more yellow than the cooler burner. (B) As an object is heated, the wavelength at which it radiates most strongly, $\lambda_m$, shifts to shorter wavelengths, a relation known as "Wien's law." Note also that as the object's temperature rises, the amount of energy radiated increases at *all* wavelengths.

In the above discussion we have not given a value to the temperature but merely said "hot" or "hotter." However, we can find a numerical value for the temperature using a relation first worked out by the German physicist Wilhelm Wien (pronounced "Vēn") about 1900. **Wien's law** states that the wavelength (color) at which an object radiates most strongly is inversely proportional to the object's temperature.

In figure 3.6B we illustrate this principle by plotting the amount of energy radiated at each wavelength (color) for three objects of different temperatures. You can see from the curves for these objects that the hottest one is brightest (emits its greatest amount of energy) at 250 nanometers. That is, its curve is highest at that wavelength. On the other hand, the coolest one emits its greatest amount of energy at 1000 nanometers. Finally, the object of intermediate temperature has the peak of its curve at an intermediate wavelength (500 nm). You can therefore see in this figure the relation between temperature and wavelength we mentioned above—namely, that hotter objects emit more strongly at shorter wavelengths. In the Extending Our Reach box, we show how to *use* this relation to take the temperature of the Sun.

You might note that the wavelength at which the Sun radiates most strongly corresponds to a blue-green color, yet the Sun looks yellow-white to us. The reason we see it as whitish is related to how our eyes perceive color. Physiologists have found that the human eye interprets sunlight (and light from all extremely hot bodies) as whitish, with only tints of color. Keep in mind that although hot bodies emit most strongly at a wavelength indicated by Wien's law, they emit at all other wavelengths too. Thus, cool stars look white tinged with red, while very hot stars look white tinged with blue.

Although Wien's law works accurately for most stars and planets, it has some important exceptions. For example, the red color of an apple and the green color of a lime come from the light they *reflect* and have nothing to do with their temperature. The apple does emit some radiation, but if it is at normal room temperature, its radiation will be mostly in the infrared.

Wien's law makes good sense if you think about the relation between energy and temperature. Hotter things carry more energy (other quantities being equal) than cooler things. Also, bluer light carries more energy than red. Thus, it is reasonable to expect that hotter bodies emit bluer light.

## EXTENDING OUR REACH

### TAKING THE TEMPERATURE OF THE SUN

To measure the temperature of an object using Wien's law, we proceed as follows. First we measure the object's brightness at many different wavelengths to find at which particular wavelength it is brightest (that is, its wavelength of maximum emission). Then we use the law to calculate the object's temperature. To see how this is done, however, we need a mathematical expression for the law.

If we let $T$ be the object's temperature, measured in Kelvin* and $\lambda_m$ be the wavelength in nanometers at which it radiates most strongly (fig. 3.6), Wien's law can be written in the form

$$T = \frac{3 \times 10^{6\dagger}}{\lambda_m}$$

The subscript m on $\lambda$ is to remind us that it is the wavelength of maximum emission.

As an example, let's measure the Sun's temperature. The Sun turns out to radiate most strongly at a wavelength of about 500 nanometers. That is, its $\lambda_{max} = 500$ nm. Then, substituting that value for $\lambda_m$ in our expression for $T$, we find

$$T = \frac{3 \times 10^6}{500} = \frac{3 \times 10^6}{5 \times 10^2}$$

$$= 0.6 \times 10^4 = 6000 \text{ Kelvin}$$

This is within a few hundred degrees of the actual value.

---

*See p. 56 for discussion of the Kelvin temperature scale.

†The constant $3 \times 10^6$ K-nm is more accurately $2.898 \times 10^6$. We round it off here to make calculations easier. The error this creates is small.

Our discussion above has been qualified several times by terms such as *usually* and *most.* The reason for these qualifications is that Wien's law applies only to a class of objects known as blackbodies.

## Blackbodies and Wien's Law

A **blackbody** is an object that absorbs all the radiation falling upon it. Because such a body reflects no light, it looks black to us when it is cold; hence its name. Experiments show that when blackbodies are heated, they radiate more efficiently than other kinds of objects. Thus, they are both excellent absorbers and excellent emitters. Moreover, the intensity of their radiation changes smoothly from one wavelength to the next with no gaps or narrow peaks of brightness. Very few objects are perfect blackbodies, but many of the objects we will study are near enough to being blackbodies that we can use Wien's law with little fear of its being in error. For example, the electric stove burner, the Sun, and the Earth all obey Wien's law quite satisfactorily.

On the other hand, gases (unless compressed to a high density) are generally not blackbodies and do not obey Wien's law. Interstellar clouds, for example, can radiate strongly only in narrow wavelength ranges, such as the red part of the visible spectrum or the millimeter wavelength part of the radio spectrum. The clouds' color is determined by composition more than temperature. Ordinary fluorescent lights and neon signs are not blackbodies either.

You can easily demonstrate the importance of composition in determining color with a gas flame on a stove or Bunsen burner. Normally the flame has a blue part and a yellow part. The yellow part is blackbody radiation from very hot specks of carbon soot. However, the blue part is caused by nonblackbody emission from carbon atoms. If you add chemicals to the flame, the flame's color may change dramatically. For example, if you hold some copper sulfate crystals with a pair of pliers in the flame, the flame will take on a greenish-blue color caused by the emission wavelengths of copper (fig. 3.7A). Likewise, the strontium in a highway emergency flare gives its light a strong red color (fig. 3.7B).

So far we have described some of the general properties of electromagnetic radiation, such as its wave nature and the existence of different wavelength regions, but we have not explained the origin of the radiation. Radiation originates in matter, and we must thus look more closely at the structure of matter if we are to understand radiation's origin. That is our next goal, and, as we will see, it will lead us not only to an understanding of the nature of atoms but also to an explanation of how we can detect those atoms in remote astronomical bodies.

**INTERACTIVE**

*Blackbody radiation and stellar luminosity*

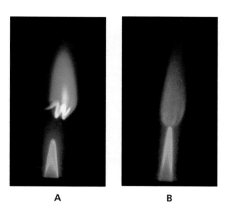

A          B

**FIGURE 3.7**

These photographs show the effect of (A) copper (green) and (B) strontium (red) on the burner's flame.

## 3.3 ATOMS

The structure of atoms determines both their chemical properties and their light-emitting and light-absorbing properties. For example, iron and hydrogen not only have very different atomic structures but also emit very different wavelengths of light. From those differences astronomers can deduce whether an astronomical body—a star or a planet—contains iron, hydrogen, or whatever chemicals happen to be present. Therefore, an understanding of the structure of atoms ultimately leads us to an understanding of the nature of stars.

## Structure of Atoms

We described in the preview section of this text how an atom has a dense core called a *nucleus* around which smaller particles called *electrons* orbit. The nucleus is in turn composed of particles called *protons* and *neutrons;* the protons have a positive charge, the neutrons have no charge, and the electrons have a negative charge. Moreover, it

Level "3"

Level "2"

Ground level = "1"

Hydrogen atom

Electron cloud
orbit n = 3

Electron cloud
orbit n = 2

Electron cloud
orbit n = 1

Proton in
nucleus

$r$ = 0.053 nm

$r$ = 0.21 nm

**FIGURE 3.8**
Just as the painters can only be at levels 1, 2, 3, . . . of the scaffold (and cannot "float in between"),
so too an electron must be in orbit 1, 2, 3, . . . etc.

is the positive electrical charge of the protons that attracts and holds the negatively charged electrons in their orbits (recall our discussion in the preview chapter of how static electricity makes clothes stick together in a dryer).

The orbits of electrons in an atom are generally extremely small. For example, the diameter of the smallest electron orbit in a hydrogen atom is only about $10^{-10}$ (1 ten-billionth) meter. This infinitesimal size leads to effects that operate at an atomic level and have no counterpart in larger systems. The most important of these effects is that the electron orbits may have only certain prescribed sizes. Although a planet may orbit the Sun at any distance, an electron may orbit an atomic nucleus at only certain distances, much as when you climb a set of stairs, you can be only at certain discrete heights. For example, in a hydrogen atom the electron must have an orbital radius of 0.053 $n^2$ nanometers, where $n = 1$, or 2, or 3, . . . etc. That is, the radius can be 0.053, 0.21, 0.48, etc. nanometers, but it cannot have intermediate values. We describe this restriction on the allowable size of orbit by saying that they are **quantized.** Figure 3.8 illustrates this property of electron orbits and compares it to painters being confined to only certain levels when they work on a scaffold.

The above restriction on orbital sizes results from the electron's acting not just as a particle but also as a wave. That is, just as light itself has a wave–particle duality, so too does an electron. The electron's wave nature forces the electron to move only in orbits whose circumference is a whole number of wavelengths. If it were to move in other orbits, the electron's wave nature would "cancel" it out.

The wave nature of the electron has another important effect. It "smears" the electrons. As a result, although we have described the electrons as orbiting like tiny spheres around the nucleus, most scientists prefer to think of them as moving within an electron cloud.

Electrons in orbit have another property totally unlike those of planets in orbit: they routinely shift from one orbit to another. This shifting changes their energy, as can be understood by a simple analogy. The electrical attraction between the nucleus and the electron creates a force between them like a spring. If the electron increases its distance from the nucleus, the spring must stretch. This requires giving energy to the atom. Likewise, if the electron moves closer to the nucleus, the spring contracts and the atom must give up, or emit, energy. We perceive that emitted energy as light or, more generally, electromagnetic radiation. The wavelength of that radiation is not the same for all atoms. But to understand why, we need to say a bit more about atomic structure. In particular, we need to describe what makes one kind of atom different from another—for example, what makes iron different from hydrogen.

## The Chemical Elements

Iron and hydrogen are examples of what are called chemical **elements.** A chemical element is a substance composed only of atoms that all have the same number of protons. For example, hydrogen consists exclusively of atoms that contain 1 proton; helium, of atoms that contain 2 protons; carbon, 6; oxygen, 8; and so forth. Although the identity of an element is determined by the number of protons in its nucleus, the chemical properties of each element are determined by the number of electrons orbiting its nucleus. However, the number of electrons must equal the number of protons. This means that each atom has an equal number of positive and negative electrical charges and is therefore electrically neutral.[†]

Table 3.3 lists some of the more important elements we will discuss during our exploration of the Universe and the number of protons each contains.

In summary then, atoms consist of a nucleus containing protons and neutrons around which electrons orbit. The electrons are bound to the nucleus by the electric attraction between the protons and electrons. Electrons may shift from one orbit to another accompanied by a change in the atom's energy. The identity of the atom—the element—is determined by the number of protons in its nucleus. With this picture of the atom in mind, we can now turn to how light is generated within atoms.

### TABLE 3.3

#### Astronomically Important Elements

| Element | Number of Protons | Number of Neutrons* |
|---------|-------------------|---------------------|
| Hydrogen | 1 | 0 |
| Helium | 2 | 2 |
| Carbon | 6 | 6 |
| Nitrogen | 7 | 7 |
| Oxygen | 8 | 8 |
| Silicon | 14 | 14 |
| Iron | 26 | 30 |

*The number of neutrons listed here is the number found in the most abundant form of the element. Different neutron numbers can occur and lead to what are called *isotopes* of the element.

---

[†]Under some circumstances, an atom may lose or gain one or more electrons. Such atoms are said to be ionized, as we will discuss in chapters 11 and 12.

## 3.4 THE ORIGIN OF LIGHT

We saw above that when an electron moves from one orbit to another, the energy of the atom changes. If the atom's energy is increased, the electron moves outward from an inner orbit. Such an atom is said to be **excited.** On the other hand, if the electron moves inward toward the nucleus, the atom's energy is decreased.

Although the energy of an atom may change, the energy cannot just disappear. One of the fundamental laws of nature is the **conservation of energy.** This law states that energy can never be created or destroyed, it can only be changed in form. According to this principle, if an atom loses energy, that energy *must reappear in some other form.* One important form in which the energy reappears is light, or, more generally, electromagnetic radiation.

How is the electromagnetic radiation created? When the electron drops from one orbit to another, it alters the electric energy of the atom. As we described in section 3.1, such an electrical disturbance generates a magnetic disturbance, which in turn generates a new electrical disturbance. Thus, the energy released when an electron drops from a higher to a lower orbit becomes an electromagnetic wave, a process called **emission** (fig. 3.9).

Emission plays an important role in many astronomical phenomena. The aurora (northern lights) is an example of emission by atoms in the Earth's upper atmosphere, and sunlight and starlight are examples of emission in those bodies.

The reverse process, in which light is stored in an atom as energy, is called **absorption** (fig. 3.10). Absorption lifts an electron from a lower to a higher orbit and excites the atom by increasing the electron's energy. Absorption is important in understanding such diverse phenomena as the temperature of a planet and the identification of star types, as we will discover in later chapters.

Emission and absorption are particularly easy to understand if we use the photon model of light. According to this model, an atom emits a photon when one of its electrons drops from an upper to a lower orbit. Similarly, an atom absorbs light when a photon collides with it and "knocks" one of its electrons into an upper level.

*Energy is released when an electron drops from an upper to a lower orbit.*

*Absorption*

**Emission of light by a hydrogen atom**

**FIGURE 3.9**
Energy is released when an electron drops from an upper to a lower orbit, causing the atom to emit electromagnetic radiation.

Energy of red light matches energy difference between orbits 2 and 3. Energy of light "lifts" electron to upper level and light disappears.

n = 4
n = 3
n = 2
n = 1
+ Proton

Energy of yellow and green light does not match any energy difference between orbits. Thus, they pass by atom with no interaction.

**Absorption of light by a hydrogen atom**

**FIGURE 3.10**
An atom can absorb light if its energy matches the energy difference between two orbits.

You may find it helpful in understanding emission and absorption if you think of an analogy. Absorption is a bit like drawing an arrow back preparatory to shooting it from a bow. Emission is like the arrow being shot. In one case, energy of your muscles is transferred to and stored in the flexed bow. In the other, it is released as the arrow takes flight.

## 3.5 FORMATION OF A SPECTRUM

The key to determining the composition and conditions of an astronomical body is its spectrum. The technique used to capture and analyze such a spectrum is called **spectroscopy.** In spectroscopy, the light (or more generally the electromagnetic radiation) emitted or reflected by the object being studied is collected with a telescope and spread into its component colors to form a spectrum by passing it through a prism or a grating consisting of numerous, tiny, parallel lines. Figure 3.11 shows not only how to form a spectrum but also what the spectrum looks like, in this case a band of rainbow colors. In general, we can show a spectrum as it would look to us or as a plot of the light's brightness at each color. In the case shown in figure 3.11, all colors are present and are more or less equally bright. As we will discover later, not all spectra look like this. Sometimes only a few colors are present.

Because light is emitted from atoms as electrons shift between orbits, we might expect that the light will bear some imprint of the kind of atom that creates it. That is usually the case, and astronomers can search for the atom's "signature" by measuring how much light is present at each wavelength.

Spectroscopy is such an important tool for astronomers that we should look in greater detail at how it works. Specifically, why does an atom produce a unique spectral signature? To understand that, we need to recall how light is produced.

### How a Spectrum Is Formed

We saw earlier that each kind of atom has a different number of electrons. This means that each kind of atom has a different set of electron orbits. We also saw that the orbit

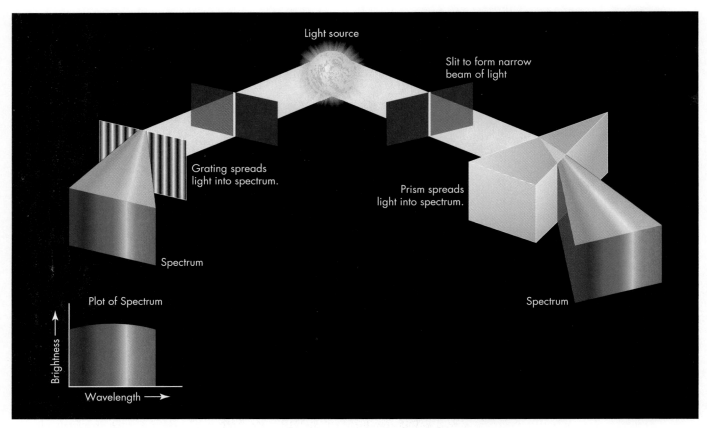

**FIGURE 3.11**
Sketch of a spectroscope and how it forms a spectrum. Either a prism or a grating may be used to spread the light into its component colors.

in which an electron is located at any given moment sets the atom's energy. For this reason scientists sometimes refer to the orbits as **energy levels.**

When an electron moves from one energy level (orbit) to another, the atom's energy changes by an amount equal to the *difference* in the energy between the two levels. As an example, suppose we look at light from heated hydrogen. Heating speeds up the atoms, causing more forceful and frequent collisions, knocking each excited atom's electron to outer orbits. However, the electrical attraction between the nucleus and the electron draws the electron back almost at once. Suppose we look at an electron shifting from orbit 3 to orbit 2, as shown in figure 3.12. As the electron shifts downward, the atom's energy decreases, and the energy lost appears as light.

The wavelength of the emitted light can be calculated from the energy difference of the levels and the relation we mentioned earlier between energy ($E$) and wavelength ($\lambda$) ($E = hc/\lambda$). If we evaluate the wavelength of this light, we find that it is 656 nanometers, a bright red color. An electron dropping from orbit 3 to orbit 2 in a hydrogen atom will always produce light of this wavelength.

If, instead, the electron moves between orbit 4 and orbit 2, there will be a different change in energy because orbit 4 has a different energy from that of orbit 3. That different energy will have a wavelength different from 656 nanometers. A calculation of its energy change leads in this case to a wavelength of 486 nanometers, a turquoise blue color.

Similar calculations show that when the electron jumps from orbit 5 to orbit 2 or from orbit 6 to orbit 2, other spectrum lines are emitted. However, we will see no light at most other colors because hydrogen has no electron orbits corresponding to those

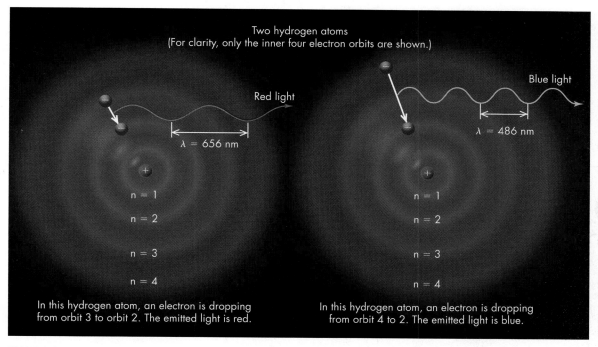

Two hydrogen atoms
(For clarity, only the inner four electron orbits are shown.)

Red light

$\lambda = 656$ nm

Blue light

$\lambda = 486$ nm

n = 1

n = 2

n = 3

n = 4

n = 1

n = 2

n = 3

n = 4

In this hydrogen atom, an electron is dropping from orbit 3 to orbit 2. The emitted light is red.

In this hydrogen atom, an electron is dropping from orbit 4 to 2. The emitted light is blue.

**FIGURE 3.12**
Emission of light from a hydrogen atom. The energy of an electron dropping from an upper to lower orbit is converted to light. The light's color depends on the orbits involved.

energies. Therefore, the hydrogen spectrum shows a set of brightly colored lines separated by wide, dark gaps. This is how an emission-line spectrum is formed.*

You can see these emission lines in figure 3.13, which shows not only what the spectrum of hydrogen looks like but also a plot of how bright the spectrum is at each wavelength.

We might now imagine making the same kind of calculations for a different chemical element, such as helium. If we did, we would discover that its wavelengths are in general different from those of hydrogen (see fig. 3.14). Thus, hydrogen's signature is its red 656-nanometer and blue 486-nanometer lines, and that signature offers astronomers a way to determine what astronomical objects are made of.

## Identifying Atoms by Their Light

In the previous paragraphs we have described how atoms emit light. Moreover, we have seen that each chemical element emits a particular set of spectrum lines and that these emission lines provide a way to identify the presence of that element in a hot gas.

It is also possible to identify atoms in a gas from the way they absorb light. Light is absorbed if the energy of its wavelength corresponds to an energy that matches the difference between two energy levels in the atom. If the wavelength *does not* match, the light will not be absorbed, and it will simply move past the atom, leaving itself and the atom unaffected.

For example, suppose we shine a beam of light that initially contains all the colors of the visible spectrum through a box full of hydrogen atoms. If we examine the spectrum of the light after it has passed through the box, we will find that certain

---

*The model for the atom that we have used here and that so successfully explains its spectrum was developed by Danish physicist Niels Bohr. He won the 1922 Nobel Prize for Physics for his work.

**Hydrogen atoms in tube**
Atom emits at wavelength set by the orbit its electron happens to be in. Thus, if electron jumps from orbit $3 \rightarrow 2$, the atom emits red light. If the electron jumps from $5 \rightarrow 2$, it emits violet, etc. *No* orbit jump corresponds to yellow or green light so those colors do not appear in the hydrogen spectrum.

**FIGURE 3.13**
The spectrum of hydrogen in the visible wavelength range.

wavelengths of the light have been removed and are missing from the spectrum (fig. 3.15). In particular, the spectrum will contain gaps that appear as dark lines at 656 nanometers and 486 nanometers, precisely the wavelengths at which the hydrogen atoms emit. The absorption spectrum is, in effect, the opposite of the emission spectrum and an atom's absorption lines have exactly the same wavelengths as its emission lines.

These gaps are created by the light at 656 nanometers interacting with the hydrogen and lifting the electron in some atoms from orbit 2 to orbit 3 while the light at 486 nanometers lifts the electron in other hydrogen atoms from orbit 2 to orbit 4. Light at other wavelengths in this range has no effect on the atom. Thus, we can tell that hydrogen is present from either its emission or its absorption spectral lines.

In our discussion above, we have considered light emitted and absorbed by individual atoms in a gas. If the atoms are linked to one another to form molecules, such

*Atomic emission and absorption*

http://www.mhhe.com/arny

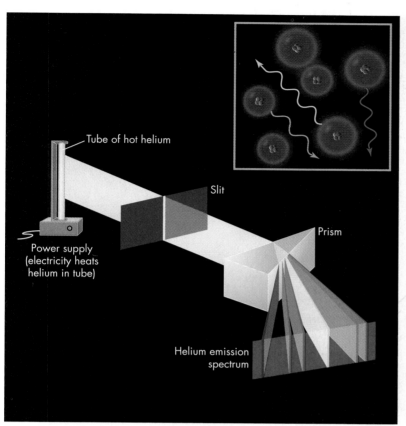

**Helium atoms in tube**
The electron orbits for helium atoms are different from the orbits in hydrogen. The light they emit therefore differs from that of hydrogen.

FIGURE 3.14
The spectrum of helium in the visible wavelength range.

as water or carbon dioxide, they too produce emission and absorption lines. In fact, even solid objects may imprint spectral lines on light that reflects off them.

In the above examples, we have considered light emitted from or absorbed by the body we wish to identify. But even reflected light generally bears some imprint of the surface from which it came. For example, when light from the Sun reflects from an asteroid, spectral features appear that were not present in the original sunlight. This gives astronomers information about the surface composition of bodies too cool to emit significant light of their own.

We conclude that in general we can identify the kind of atoms or molecules that are present by examining either the bright or dark spectrum lines. Gaps in the spectrum at 656 nanometers and 486 nanometers imply that hydrogen is present. Similar gaps at other wavelengths would show that other elements are present. By matching the observed gaps to a directory of absorption lines, we can identify the atoms that are present. *This is the fundamental way astronomers determine the chemical composition of astronomical bodies.*

## Types of Spectra

Although a spectrum may bear the imprint of the matter that emitted the light, it may also have certain general properties. For example, the spectrum of a hot, tenuous gas is usually different from that of a hot, dense solid, regardless of the composition of either the gas or the solid. Therefore, it will be helpful to learn to recognize these general properties of spectra as well.

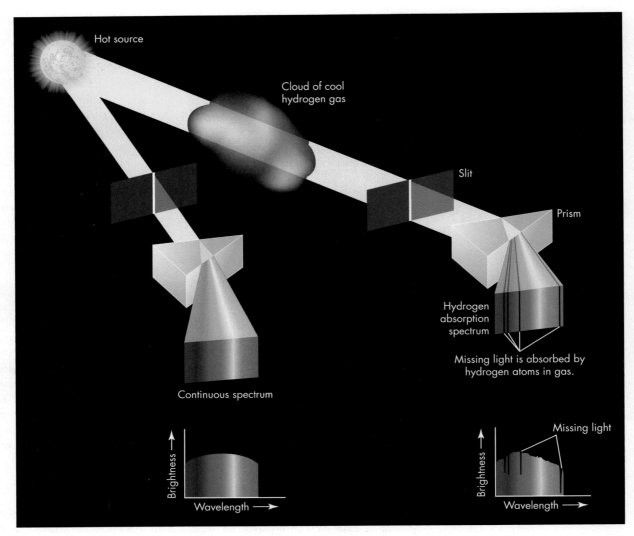

**FIGURE 3.15**
Gas between an observer and a source of light that is hotter than the gas creates an absorption-line spectrum. Atoms in the gas absorb only those wavelengths whose energy equals the energy difference between their electron orbits. The absorbed energy lifts the electrons to upper orbits. The lost light makes the spectrum darker at the wavelengths where it is absorbed.

Spectra have the following three basic forms:

1. For some sources, the brightness of the emitted light changes smoothly with wavelength and all colors are present. We say such a light has a **continuous spectrum** (fig. 3.16A). For a source to emit a continuous spectrum, its atoms must in general be packed so closely that the electron orbits of one atom are distorted by the presence of neighboring atoms. Such conditions are typical of solid or dense objects such as the heated filament of an incandescent lightbulb or a nail heated by a blowtorch.

2. Some heated objects have a spectrum in which light is emitted at only a few particular wavelengths while most of the other wavelengths remain dark (fig. 3.16B). This type of spectrum is called an **emission-line spectrum.** Emission-line spectra are usually produced by *hot, tenuous gas,* such as that in a fluorescent tube, the aurora, and many interstellar gas clouds.

3. A still different type of spectrum arises when light from a hot, dense body passes through cooler gas between it and the observer. In this case, nearly all the colors are present, but light is either missing or much dimmer at some wavelengths, as we

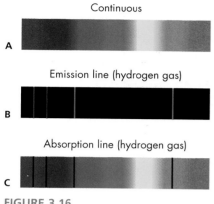

Continuous

A

Emission line (hydrogen gas)

B

Absorption line (hydrogen gas)

C

**FIGURE 3.16**

Types of spectra: (A) continuous, (B) emission-line, and (C) absorption-line.

discussed earlier. This causes the bright background to be crossed with narrow dark lines where the light of some colors is fainter or absent altogether (fig. 3.16C). The resulting spectrum is therefore called a **dark-line** or **absorption-line spectrum.**

Absorption lines were first detected astronomically in 1802, when the English scientist William H. Wollaston viewed sunlight through a prism and a narrow slit. Interested in learning whether each color blended smoothly into its neighbor (red into orange, orange into yellow, and so forth), he noticed dark lines between some of the colors but paid little attention to them. These dark lines in the Sun's spectrum were independently discovered a few years later by the German scientist Joseph Fraunhofer, who catalogued them and discovered similar lines in other stars. In fact, because nearly all stars have absorption-line spectra, this spectrum type is especially important in astronomy. However, if we consider the physical process of spectrum formation, such absorption-line spectra are really just a special case of continuous spectra with light missing at some wavelengths.

## Analyzing the Spectrum

The first step facing an astronomer who wants to analyze a spectrum is to identify the spectrum lines. This is done by measuring the wavelengths of the lines and then consulting a directory of spectrum lines. By matching the wavelength of the line of interest to a line in the table, astronomers can determine what kind of atom or molecule created the line. A look at a typical spectrum will show you that some lines are hard to see, being faint and weak. On the other hand, some lines may be very obvious and strong. The strength or weakness of a given line turns out to depend on the number of atoms or molecules absorbing (or emitting, if we are looking at an emission line) at that wavelength. Unfortunately, the number of atoms or molecules that can absorb or emit depends not just on how many of them are present but also on their temperature, as we will discuss more fully in chapter 12. Nevertheless, astronomers can deduce from the strength of emission and absorption lines the relative quantity of each atom producing a line and thereby deduce the composition of the material in the light source. Table 3.4 shows the result of such an analysis for our Sun, a typical star.

## Astronomical Spectra

Let us now apply what we know about spectra to astronomical bodies. We begin by using a telescope to obtain a spectrum of the object of interest. Next we measure the wavelengths and identify the lines. As an example, consider the spectrum of the Sun in figure 3.17A.

**TABLE 3.4**

### Composition of a Typical Star, Our Sun*

| Element | Relative Number of Atoms | Percent by Mass |
|---|---|---|
| Hydrogen | $10^{12}$ | 70.6% |
| Helium | $9.77 \times 10^{10}$ | 27.4% |
| Carbon | $3.63 \times 10^{8}$ | 0.31% |
| Nitrogen | $1.12 \times 10^{8}$ | 0.11% |
| Oxygen | $8.51 \times 10^{8}$ | 0.96% |
| Neon | $1.23 \times 10^{8}$ | 0.18% |
| Silicon | $3.55 \times 10^{7}$ | 0.07% |
| Iron | $4.68 \times 10^{7}$ | 0.18% |
| Gold | 10.0 | $1.4 \times 10^{-7}\%$ |
| Uranium | less than 0.3 | less than $5.7 \times 10^{-9}\%$ |

Reprinted from *Geochim Cosmochim Acta* 53 (1989): 197, by Anders and Grevesse. Copyright 1989, with permission from Elsevier.

*The table lists some of the most common elements. Gold and uranium are included only to illustrate how extremely rare they are. Notice that they are a million or more times rarer than iron, which is itself thousands of times rarer than hydrogen.

Spectrum of a comet

Solar spectrum

A

B

**FIGURE 3.17**
(A) The spectrum of a comet (an emission-line spectrum) and the Sun (an absorption-line spectrum).
(B) A graphical representation of their spectra.

**FIGURE 3.18**
(A) A radio spectrum of a cold interstellar cloud. (B) An X-ray spectrum of hot gas from an exploding star.

We can see from the spectral lines that the Sun contains hydrogen. In fact, when a detailed calculation is made of the strength of the lines, it turns out that the Sun is about 71% hydrogen.* Similar observations show that the spectrum of a comet consists mainly of emission lines from such substances as the molecules carbon dioxide and cyanogen (CN). Thus, we know that comets contain these substances. Moreover, recalling our earlier discussion of types of spectra (continuous, emission-line, or absorption-line), we can tell that the CN and carbon dioxide must be gaseous because the spectrum consists of emission lines. There may be other gases present too, but without seeing their spectral features, we cannot tell for sure.

Although the examples we have used above involve spectra of visible light, one of the most useful features of spectroscopy is that it may be used in *any* wavelength region. For example, figure 3.18 shows a radio spectrum and an X-ray spectrum.

Regardless of the wavelength region we use, the spectrum allows us to determine what kind of material is present. In addition, it can sometimes reveal to us in what direction and how fast that material is moving.

---

*Expressed in a slightly different way, this means that roughly 9 out of 10 atoms in the Sun are hydrogen. Hydrogen contributes only about 71% of the mass because it is the lightest element.

## 3.6   THE DOPPLER SHIFT: DETECTING MOTION

**INTERACTIVE**

*Doppler shift*

If we observe light from a source that is moving toward or away from us, we will find that the wavelengths we receive from it are altered by the motion. If the source moves toward us, the wavelengths of its light will be shorter. If it moves away from us, the wavelengths will be longer, as illustrated in figure 3.19A. Furthermore, the faster the source moves, the greater those changes in wavelength will be. This change in wavelength caused by motion is called the **Doppler shift,** and it is a powerful tool for measuring the speed and direction of motion of astronomical objects.*

The Doppler shift occurs for all kinds of waves. You have perhaps heard the Doppler shift of sound waves from the siren of an emergency vehicle as it passes you and moves away: the siren's pitch drops as the wavelength of its sound increases (fig. 3.19B). Likewise, the Doppler shift of a radar beam that bounces off your car reveals to a law enforcement officer how fast your car is moving (fig. 3.19C).

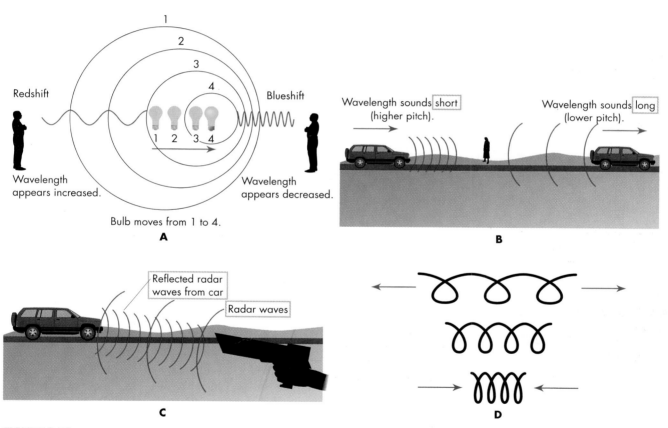

**FIGURE 3.19**

(A) The Doppler shift: waves appear to shorten as a source approaches and to lengthen as it recedes. (B) The Doppler shift of sound waves from a passing car. (C) The Doppler shift of radar waves in a speed trap. (D) A Slinky illustrates the shortening of the space between its coils as its ends move toward each other and a lengthening of the space as the ends move apart.

*These wavelength changes occur irrespective of whether the observer or the source is moving, as long as the distance between them changes. Motion perpendicular to the line of sight creates no observable Doppler shift unless the velocity is nearly the speed of light.

*The Doppler effect*

It is easy to see why the Doppler shift occurs. Imagine that the waves from the moving source are a wiggly line. If the source is moving away from you, the wiggles get stretched so the spacing of the waves increases. If the source is moving toward you, the wiggles are scrunched up so the spacing of the waves decreases (fig. 3.19D).

Mathematically, the Doppler shift arises because the wavelength we observe ($\lambda$) is the original wavelength ($\lambda_0$) plus the distance the source travels during the time a single wave is emitted. That distance depends on the speed, $V$, of the source along the line from the source to the observer, a speed astronomers call the radial velocity. Some mathematics (omitted here) then leads to the Doppler shift formula

$\Delta\lambda$ = Wavelength shift
$\lambda$ = Measured wavelength (what we observe)
$\lambda_0$ = Emitted wavelength
$c$ = Speed of light
$V$ = Velocity of source along the line of sight (radial velocity)

$$V = c(\lambda - \lambda_0)/\lambda_0 = c(\Delta\lambda/\lambda_0)$$

where c is the speed of the wave and $\Delta\lambda = (\lambda - \lambda_0)$ is simply shorthand for the change in wavelength.* We will see applications of this law in later chapters. Here, our goal is simply to indicate that the Doppler shift allows us to find out how fast a source is moving toward or away from us.[†]

Doppler shift measurements can be made at any wavelength of the electromagnetic spectrum. But regardless of the wavelength region observed or whether the waves are of visible light, astronomers generally refer to shifts that increase the measured wavelength as redshifts and those that decrease the measured wavelength as blueshifts. Thus, even though we may be describing radio waves, we will say that an approaching source is blueshifted and a receding one is redshifted.

The previous sections show how important a source of information electromagnetic radiation is. It provides information about the temperature, composition, and motion of astronomical objects. However, we often observe that radiation after it has passed through our atmosphere. Thus, we must ask how the atmosphere affects the radiation passing through it.

## 3.7   ABSORPTION IN THE ATMOSPHERE

Gases in the Earth's atmosphere absorb electromagnetic radiation, affecting the flow of heat and light through it. The amount of this absorption depends strongly on wavelength. For example, the gases affect visible light hardly at all and so our atmosphere is nearly completely transparent at the wavelengths we see with our eyes. On the other hand, some of the gases strongly absorb infrared radiation while others strongly block ultraviolet radiation.

This nearly total blockage of infrared and ultraviolet radiation results from the ability of molecules such as carbon dioxide, water, and ozone to absorb at a wide range of wavelengths. For example, carbon dioxide and water molecules strongly absorb infrared wavelengths. Likewise, ozone ($O_3$) and ordinary oxygen ($O_2$) strongly absorb ultraviolet radiation, while oxygen and nitrogen absorb X rays and gamma radiation.

Molecules in general are excellent absorbers (and emitters) because they can store energy in more ways than atoms can. As we have described earlier, atoms store energy by lifting electrons to upper orbits. Molecules can store energy not only by lifting electrons but also by the spinning and vibrating motions of the molecule as a whole. These added ways to store energy are what make molecules such powerful blockers of radiation at many wavelengths.

---

*The Greek letter $\Delta$, or delta, is widely used to stand for "the change in quantity."

[†]A negative V means the distance between the source and observer is decreasing. A positive V means that distance is increasing. That is, the source and observer are moving apart. This formula must be modified if V is close to c.

As a result of this absorption by molecules, virtually no infrared, ultraviolet, X-ray, or gamma-ray radiation can pass through our atmosphere.

The transparency of the atmosphere to visible light compared to its opacity (nontransparency) to infrared and ultraviolet radiation creates what is called an **atmospheric window.** An atmospheric window is a wavelength region in which energy comes through easily compared to other wavelengths (fig. 3.20).

Without atmospheric windows, it would be impossible for us to study astronomical objects from the ground. As it is, the visible window allows us to study stars and galaxies (which radiate lots of visible energy), but the lack of ultraviolet and the rarity of infrared windows makes it very difficult to observe objects that radiate strongly in those spectral regions. This is one of several reasons why astronomers so badly need telescopes in space, where there is no absorption by our atmosphere.

*Absorption by atmosphere*

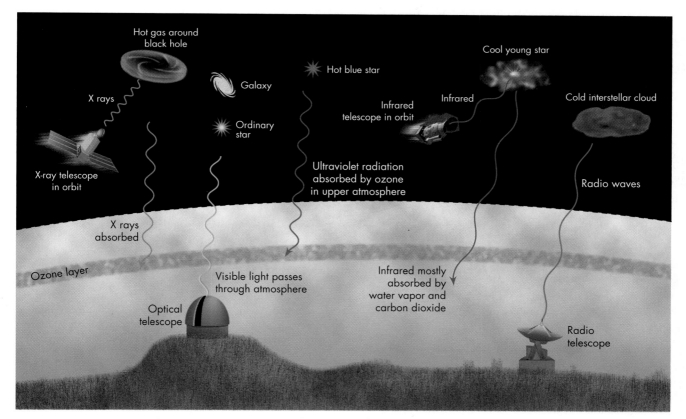

**FIGURE 3.20**

Atmospheric absorption. Wavelength regions where the atmosphere is essentially transparent, such as the visible spectrum, are called "atmospheric windows." Wavelengths and atmosphere are not drawn to scale.

# SUMMARY

Light can be described in two complementary ways: as a stream of particles called photons, or as electromagnetic waves. In the wave picture, the energy increases as the wavelength decreases. The wavelength of light determines its color. Red light has a longer wavelength than blue light. In addition to the electromagnetic radiation that we see as visible light, many electromagnetic waves are invisible to the eye, such as infrared, ultraviolet, radio, X rays, and gamma rays. The entire assemblage of electromagnetic waves is called the electromagnetic spectrum.

Energy can be absorbed by or released from an atom when an electron moves to a higher or lower orbit, respectively. If an electron drops from an upper to a lower orbit, the energy appears as light. If light of the appropriate energy (wavelength) hits an atom, it may lift an electron in the atom from a lower to an upper orbit. The generation of light is called emission. The removal of light from a beam of radiation is called absorption.

Each kind of atom has a unique set of wavelengths at which it emits and absorbs. These create the atom's spectrum and allow it to be identified. Motion of the emitting material alters the wavelengths, creating a Doppler shift from which the material's speed and direction of motion can be deduced.

Atmospheric gases also absorb light, although there is little absorption in the wavelengths of the visible atmospheric window. Carbon dioxide and water absorb infrared radiation, and oxygen and ozone in our upper atmosphere absorb ultraviolet radiation. These gases therefore strongly hinder observations of astronomical objects at infrared and ultraviolet wavelengths from the ground.

## ? What It Is.

The ring of light around the Sun is a halo and is caused by sunlight refracted in tiny atmospheric ice crystals. Haloes are quite common (perhaps one a week) and may be seen around both the Sun and the Moon. They are easier to see around the Sun if you cover the Sun with your hand or block its direct light with a building or tree, as shown here.

# QUESTIONS FOR REVIEW

1. Why is light called electromagnetic radiation?
2. How are color and wavelength related?
3. What is a photon?
4. Name the regions of the electromagnetic spectrum from short to long wavelengths.
5. What is the difference between emission and absorption in terms of what happens to an electron in an atom?
6. What is the Doppler shift?
7. Which gases in the atmosphere absorb infrared radiation? Which gases absorb ultraviolet?
8. What is meant by the *electromagnetic spectrum*?
9. What are some of the things astronomers can learn about astronomical objects from their spectra?

# THOUGHT QUESTIONS

1. Why don't atoms emit a continuous spectrum?
2. How can you tell what sort of gas is emitting light?
3. How would a spectrum help you learn what the atmosphere of Venus is made of?
4. If you added more water or carbon dioxide to our atmosphere, how would it alter the loss of heat from our planet? Would you expect the Earth to get warmer or colder? Why?
5. Given that water absorbs microwaves very strongly, can you explain why a Pop-Tart gets very hot inside while its crust stays cool if you heat it in a microwave oven?
6. Your body temperature is about 300 K. At what wavelength do you radiate most strongly? What region of the electromagnetic spectrum is this? Do you understand now how a rattlesnake can bite you in the dark?
7. Why do night-vision cameras use infrared detectors?
8. If you were to look at the spectrum of the gas flame of a stove or the blue part of a Bunsen burner flame, what sort of spectrum would you expect to see: absorption, emission, or continuous? Why?
9. Can you explain why the atmospheric layer containing ozone is much warmer than the levels above and below it?

## PROBLEMS

1. Given that the Sun is 150 million kilometers from the Earth, how long does it take light to travel from the Sun to the Earth?

2. Suppose you are operating a remote-controlled spacecraft on Mars from a station here on Earth. How long will it take the craft to respond to your command if Mars is at its nearest point to Earth? You will need to look up a few numbers here.

3. Sketch an atom emitting light. Does the electron end up in a higher or lower orbit? Repeat for an atom absorbing light.

4. A lightbulb radiates most strongly at a wavelength of about 3000 nanometers. How hot is its filament?

5. An electric stove burner on "high" radiates most strongly at about 2000 nanometers. What is its temperature?

6. The Earth's temperature averaged over the year is about 300 Kelvin. At what wavelength does it radiate most strongly? What part of the electromagnetic spectrum does this wavelength lie in? Can you see it?

## TEST YOURSELF

1. Which type of electromagnetic radiation has the longest wavelength?
   (a) Ultraviolet
   (b) Visible
   (c) X ray
   (d) Infrared
   (e) Radio

2. Which kind of photon has the highest energy?
   (a) Ultraviolet
   (b) Visible
   (c) X ray
   (d) Infrared
   (e) Radio

3. An astronomer finds that the visible spectrum of a mysterious object shows bright emission lines. What can she conclude about the source?
   (a) It contains cold gas.
   (b) It is an incandescent solid body.
   (c) It is rotating very fast.
   (d) It contains hot, relatively tenuous gas.
   (e) It is moving toward Earth at high speed.

4. A star's radiation is brightest at a wavelength of 400 nanometers. Its temperature is about
   (a) 4000 K.
   (b) 12,000 K.
   (c) 1500 K.
   (d) 750 K.
   (e) 7500 K.

5. If an object's spectral lines are shifted to longer wavelengths, the object is
   (a) moving away from us.
   (b) moving toward us.
   (c) very hot.
   (d) very cold.
   (e) emitting X rays.

## FURTHER EXPLORATIONS

Steffy, Philip C. "The Truth about Star Colors." *Sky and Telescope* 84 (September 1992): 266.

Zajonc, Arthur. *Catching the Light: The Entwined History of Light and Mind.* New York: Bantam Books, 1993.

**Website**

Please visit the *Explorations* website at **http://www.mhhe.com/arny** for additional online resources on these topics.

## KEY TERMS

absorption, 104

atmospheric window, 115

blackbody, 101

conservation of
energy, 104

continuous spectrum, 110

dark-line or absorption-
line spectrum, 111

Doppler shift, 113

electromagnetic
radiation, 92

electromagnetic
spectrum, 96

electromagnetic wave, 92

elements, 103

emission, 104

emission-line
spectrum, 110

energy levels, 106

excited, 104

frequency, 95

infrared, 97

light, 92

nanometer, 94

photons, 93

quantized, 102

spectroscopy, 105

ultraviolet, 97

visible spectrum, 94

wavelength, 94

wave–particle duality, 93

white light, 95

Wien's law, 100

# Telescopes

# 4

One of the twin Keck Telescopes on the summit of Mauna Kea, Hawaii, during a total eclipse of the Sun. These are the two largest individual optical telescopes in the world.

## Chapter Objectives

- Astronomers use telescopes to gather light and thereby make dim objects observable.
- The larger the diameter of the lens or mirror used in a telescope:
  - The greater its ability to gather light
  - The finer the detail that the instrument will show
- The Earth's atmosphere interferes seriously with observing.
  - It absorbs and totally blocks many wavelengths.
  - It blurs details.
  - Astronomers use special-purpose telescopes to observe nonvisible wavelengths. Many of these are in orbit, so they are not affected by blurring or atmospheric absorption.

**?What Is This?**
(See chapter summary for the answer.)

## Concepts and Skills to Review

- The nature of light (wave or photons) (3.1)
- Red light has a longer wavelength than blue (3.1)
- The parts of the electromagnetic spectrum (3.2)
- Absorption in Earth's atmosphere (3.7)

Astronomers, like all scientists, rely heavily on observations to guide them in theorizing and in testing theories already developed. Unlike most scientists, however, astronomers cannot directly probe the objects they study. Rather, they must perform their observations from vast distances and can only passively collect radiation emitted by the bodies they seek to study. Collecting enough radiation to be useful in studying astronomical objects is difficult because most objects are so remote that their radiation is extremely faint by the time it reaches Earth. Moreover, special instruments must be used to extract the information desired from the radiation—instruments that can measure the brightness, the spectrum, and the position of objects to high precision. For example, to collect enough light to detect remote galaxies, astronomers use telescopes with mirrors the size of a small swimming pool. To avoid the blurring and blocking effects of our atmosphere, they use orbiting observatories. To analyze and display the observations, they use a computer. This chapter describes some of the more important devices and how they work. We will see that modern telescopes bear little resemblance to the long tubes depicted in cartoons. Moreover, modern astronomers rarely sit at the eyepiece of a telescope. They are more likely to be sitting at a computer terminal operating a telescope remotely, examining the data collected, or solving equations that describe, for example, the paths of stars when two galaxies collide.

## 4.1    TELESCOPES

A telescope enables the astronomer to observe things not visible to the naked eye. Although our eyes are superb detectors, they cannot see extremely faint objects or fine details on distant sources. For example, we are unable to read standard newspaper print on the opposite wall in a dimly lit room. A telescope overcomes these difficulties, first, by collecting more light than the eye can and, second, by increasing the detail discernible. The first of these properties is called "light-gathering power"; the second is called "resolving power."

### Light-Gathering Power

For our eyes to see an object, photons (light) from it must strike the retina. How bright that object appears to us depends on the number of its photons that enter our eye per second, a number limited by the size of our eyes. Astronomers overcome that limit by "collecting" photons with a telescope, which then "funnels" the photons to our eye. The bigger the telescope's collecting area, be it a lens or mirror, the more photons it collects, as shown in figure 4.1. Thus, a larger-diameter mirror or lens gives a telescope a greater **light-gathering power.** The result is a brighter image, which allows us to see dim objects that are invisible in telescopes with smaller gathering areas. Because the gathering area of a circular lens or mirror of radius $r$ is $\pi r^2$, a small increase in the radius of the gathering area gives a large increase in the number of photons caught. For example, doubling the radius of a lens or mirror increases its light-gathering area by a factor of 4. Because the light-gathering area is so important to a telescope's performance, astronomers often describe a telescope by the diameter of its lens or mirror. Thus, the 10-meter Keck Telescope in Hawaii has mirrors spanning 10 meters (roughly 30 feet) in diameter.

### Focusing the Light

Once light has been gathered, it must be focused to form an image or to concentrate it on a detector. Telescopes in which light is gathered and focused by a lens are called

**FIGURE 4.1**

A large lens collects more light (photons) than a small one, leading to a brighter image. We therefore say that the larger lens has a greater "light-gathering power."

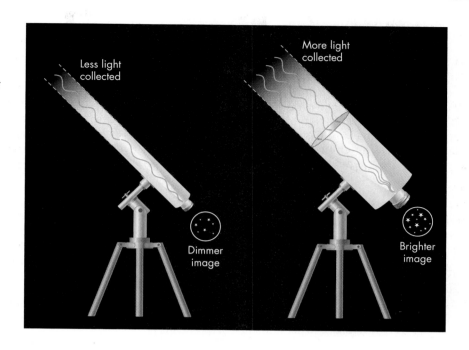

Less light collected

Dimmer image

More light collected

Brighter image

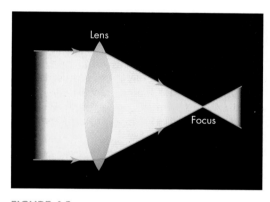

Lens

Focus

**FIGURE 4.2**

How a lens focuses light.

"refracting telescopes," or **refractors** for short. The lens of a refractor focuses the light by bending the rays, as shown in figure 4.2. This bending is called **refraction,** and it happens when light moves from one substance (such as air) into a different substance (such as glass), as discussed in the Extending Our Reach box on refraction. It is this refraction that gives such telescopes their name. Figure 4.3 shows a photograph of the world's largest refractor, the 1-meter (40-inch) diameter Yerkes Telescope of the University of Chicago.

Lenses have many serious disadvantages in large telescopes, however. First, large-diameter lenses are extremely expensive to fabricate. Moreover, a lens must be supported at its edges so as not to block light passing through it. This makes the lens "sag" in the middle (though by only tiny amounts), distorting its images. A third difficulty with lenses is that most transparent materials bring

## EXTENDING OUR REACH

### REFRACTION

When light moves at an angle from one material into another (for example, from air into water), its direction of travel generally bends. This phenomenon is called refraction. Refraction is the principle by which our eyes and eyeglasses focus light. You can easily see its effects by sticking a pencil in a glass of water and noticing that the pencil appears bent, as seen in Box figure 4.1. The pencil in water also illustrates an important property of refraction. If you look along a pencil placed partly in the water and change the pencil's tilt, you will see that the amount

of bending (refraction) changes. Exactly vertical rays are not bent at all, nearly vertical rays are bent only a little, and rays entering at a grazing angle are bent most.

Refraction occurs because the speed of light changes as it enters matter, generally becoming slower in denser material. This decrease in the speed of light arises from its interaction with the atoms through which it moves. To understand how this reduction in light's speed makes it bend, imagine a light wave approaching a slab of material. The part of the wave that enters the material first is

*(continued)*

## EXTENDING OUR REACH (CONTINUED)

slowed while the part remaining outside is unaffected, as depicted in Box figure 4.2A. To see why slowing part of the wave makes it bend, imagine what would happen if the wheels on the right-hand side of your car turned more slowly than those on the left. Your car would swerve to the right, a result that lies behind the reason why cars have a differential. By allowing one wheel to turn faster than the other, the differential "swings" your car smoothly around corners. A similar effect occurs if you walk hand-in-hand with a friend, and your friend walks more slowly than you do. You will soon find yourself traveling in a curve (see Box fig. 4.2B). So, too, if one portion of a light wave moves more slowly than another, the light's path will bend.

**BOX FIGURE 4.1**
Refraction of light in a glass of water. Note how the pencil appears to be bent.

Refraction not only bends light but also generally spreads the light into its component colors, breaking white light into a spectrum, or rainbow. This spreading occurs because different colors of light travel at different speeds in most materials and are therefore bent by different amounts in a process called **dispersion**. Thus, if light consisting of a mix of colors enters a block of glass, each color is slowed to a different speed and is therefore deflected differently. The result is that colors initially traveling together separate into different beams. This is how a prism creates a spectrum.

### Atmospheric Refraction
Refraction is easy to see in the Earth's atmosphere where it distorts the shape of the Sun when it rises and sets and makes the stars twinkle at night.

### Distortion of the Sun's Shape
Refraction distorts the shape of the rising or setting Sun because when sunlight enters the Earth's atmosphere from space, it is refracted and bent slightly toward the ground. Thus, if you are on the ground and look back along the light ray, the light seems to come from slightly higher than it really does (Box fig. 4.3A). That is, refraction makes astronomical objects look higher in the sky than they actually are. This effect is greatest when objects are near the horizon, as shown in Box figure 4.3B. The result is that light from the lower edge of the Sun is refracted more than light from its upper edge, which "lifts" the lower edge more than the upper and makes the Sun look flattened (Box fig. 4.3C).

Refraction also slightly alters the time at which the Sun seems to rise or set. When it is at the horizon, the Sun is "lifted" by about the height of its diameter. Thus, at the

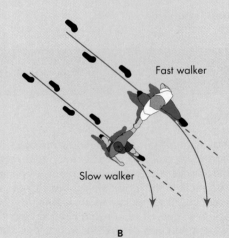

**BOX FIGURE 4.2**
Cause of refraction. (A) Light entering the denser medium is slowed, while the portion still in the less dense medium proceeds at its original speed. (B) A similar effect occurs when you walk hand-in-hand with someone who walks slower than you do.

**BOX FIGURE 4.3**
(A) Atmospheric refraction makes the Sun or a star look higher in the sky than it really is. (B) Refraction is stronger for objects nearer the horizon. (C) The Sun is flattened because refraction "lifts" its lower edge more than its upper edge. The Sun's reflection in a band below it on the water is called the glitter path.

moment we see the Sun touch the horizon, it has actually set. By "lifting" the Sun's image above the horizon, even though it has set, refraction slightly affects the length of the daylight hours. As a result, the day of the year the Sun is above the horizon for exactly 12 hours is not the equinox, but rather a few days before the spring and a few days after the autumnal equinox. It turns out that near latitude 40° N, St. Patrick's Day (March 17) is the day with almost exactly 12 hours between sunrise and sunset in the spring.

### The Moon Illusion
Atmospheric refraction is not, however, the reason why the Moon sometimes appears to be so huge when you see it rising. In fact, if you measure the Moon's apparent diameter carefully, you will find it to be smaller when it is near the horizon than when it is overhead, regardless of how huge it looks. This misperception, known as the **Moon illusion,** is still not well understood but is an

optical illusion caused, at least in part, by the observer's comparing the Moon with objects seen near it on the horizon, such as distant hills and buildings. You know those objects are big even though their distance makes them appear small. Therefore, you unconsciously magnify both them and the Moon, making the Moon seem larger. You can verify this sense of illusory magnification by looking at the Moon through a narrow tube that blocks out objects near it on the sky line. Seen through such a tube, the Moon appears to be its usual size.

Box figure 4.4 shows a similar effect. Because you know that the rails are really parallel, your brain ignores the apparent convergence of the railroad tracks and mentally spreads the rails apart. That is, your brain provides the same kind of enlargement to the circle near the rails' convergence point as it does to the rails, causing you to perceive the middle circle as larger than the lower one, even though they are both the same size.

*(continued)*

**EXTENDING OUR REACH (CONTINUED)**

**BOX FIGURE 4.4**
Circles beside converging rails illustrate how your perception may be fooled. The bottom circle looks smaller than the circle on the horizon but is in fact the same size. Similarly, the circle high in the sky looks smaller than the circle on the horizon.

Lens 1 meter (40 inches) in diameter

19.5 meters (63.5 feet)

Pivots to allow telescope to track stars

Instruments

light of different colors to a focus at slightly different distances from the lens. This creates images fringed with color, a flaw called "chromatic aberration." Finally, many lens materials completely absorb short-wavelength light, making them, for some purposes, as useless in a telescope as a chunk of concrete.

To avoid such difficulties, most modern telescopes use mirrors rather than lenses to gather and focus light, and they are therefore called **reflectors.** The mirrors are made of glass that has been shaped to a smooth curve, polished, and then coated with a thin layer of aluminum or some other highly reflective material. As figure 4.4 shows, such a curved mirror can focus light rays reflected from it, creating an image just as well as a lens can. Moreover, because the light does not pass through the mirror, it focuses all colors equally well and does not absorb short-wavelength light. Furthermore, because the light does not have to pass through the mirror, the mirror can be supported from behind, thereby reducing the sagging problem that affects large lenses. For these and other reasons, astronomers now use reflecting telescopes almost exclusively. Figure 4.5 shows photographs of two large reflecting telescopes—one of the 10-meter Keck reflectors (a multimirror instrument) and one of the twin 8-meter Gemini reflectors.

Figure 4.4 shows that light striking a mirror is focused in front of the mirror. Thus, to see the image, the observer would ordinarily have to stand in front of the mirror, thereby blocking some of the light. To overcome this difficulty, a secondary mirror is often used

**FIGURE 4.3**

A refracting telescope. Completed in 1897 for the University of Chicago's Yerkes Observatory in Williams Bay, Wisconsin, this refractor has a lens approximately 1 meter (40 inches) in diameter, making it the world's largest refracting telescope.

**FIGURE 4.4**
How a curved mirror focuses light.

to deflect the light either off to the side or back toward the mirror and out through a hole in its center (fig. 4.6).

Most telescopes are mounted on huge pivots that allow them to follow astronomical objects as they move across the sky. Swinging the many tons of metal and glass smoothly and with precision requires great care in construction and design. Moreover, as the telescope moves, its lenses or mirrors must keep their same precise shapes and relative positions if the images are to be sharp. This is one of the most technically demanding parts of building a large telescope, because the large pieces of glass used in lenses and mirrors bend slightly when their positions are shifted. One way to counteract this effect in a reflector is to collect and focus the light with several smaller mirrors and then align each one individually. That is, instead of a telescope having a single large mirror, it may have many small ones. Telescopes designed this way are called multimirror instruments. Currently, the largest multimirror instruments are the twin 10-meter Keck Telescopes (fig. 4.7), operated by the California Institute of Technology

A

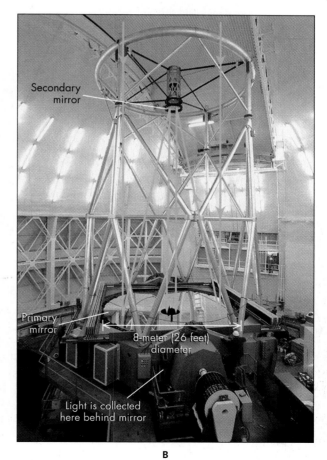

B

**FIGURE 4.5**
(A) Photograph of one of the twin Keck Telescopes. The 36 mirrors cover an area 10 meters (about 33 feet) in diameter, making them the world's largest single optical telescopes. (B) One of the two Gemini telescopes. One of this telescope pair is on Mauna Kea, Hawaii; the other one is in Chile. The mirror of each telescope is about 8 meters (26 feet) in diameter. The yellow lines show the light path through the telescope.

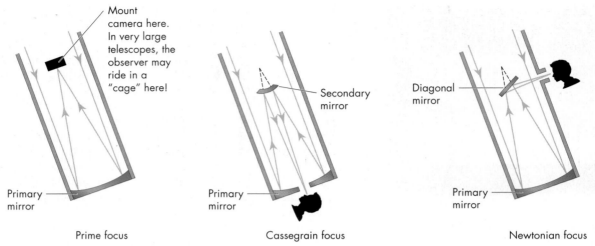

**FIGURE 4.6**
Sketches of different focus arrangements for reflectors.

**FIGURE 4.7**
The twin Keck Telescopes on the summit of Mauna Kea in Hawaii. As the telescopes track the stars or galaxies being observed, all 36 mirrors in each telescope can be measured with a laser and then adjusted to keep them in precise alignment.

and the University of California, and located on the 14,000-foot volcanic peak Mauna Kea in Hawaii. Each Keck Telescope consists of 36 separate mirrors (visible in fig. 4.5) that are kept aligned by lasers that measure precisely the tilt and position of each mirror. If any misalignment is detected, tiny motors shift the offending separate mirror segment to keep the image sharply in focus.

Ironically, the problem of keeping the separate segments of a multimirror telescope aligned has led astronomers back to single mirrors, but mirrors much thinner and lighter than those in most existing telescopes. In the past, astronomers made mirrors thick to make them stiff and reduce sagging of the glass. Large pieces of glass, however, weigh more than smaller pieces and thus sag even worse unless very carefully supported. Just as a smaller gob of whipped cream will keep its shape on a tilted plate whereas a large gob will slump under its own weight, so too a thin piece of glass, if properly supported, keeps its shape better when tilted than a large

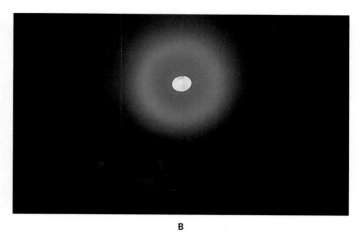

A

B

**FIGURE 4.8**
(A) Water ripples diffracted as they pass through a narrow opening. (B) Photograph showing colored rings created by the diffraction of light waves by the tiny water droplets on a fogged piece of glass.

piece. Thus, astronomers have sought ways to make extremely thin* mirrors that are then kept precisely shaped by alignment systems similar to those used in the Kecks. Such a procedure is used on the Very Large Telescope of the European Southern Observatory in Chile, which, as a result, can achieve several times better resolution than conventional telescopes.

Up to now, we have concentrated on the light-gathering power of telescopes: their ability to make dim objects bright enough to see. Telescopes, however, serve another important function—they increase our eyes' ability to see details.

## Resolving Power

If you mark two tiny black dots close together on a piece of paper and look at them from the other side of the room, your eye may see them as a single dark mark, not as separate spots. Likewise, stars that lie very close together or markings on planets may not be clearly distinguishable. A telescope's ability to discern such detail depends on its **resolving power.**

Resolving power is limited by the wave nature of light. For example, suppose two stars are separated by a very tiny angle. For them to be discernible as separate images, their light waves must not get mixed up. Such mixing, however, always occurs when waves pass through an opening, because as each wave passes the opening, smaller, secondary waves are produced in a phenomenon called **diffraction.** Figure 4.8A shows how water waves are diffracted as they pass through a narrow opening. Light waves are similarly diffracted as they enter a telescope. The result of diffraction is that point sources of light become surrounded by rings of light. You can observe diffraction by looking at a tiny, bright light source, such as a study lamp, through a piece of cloth, such as a T-shirt. The light will be surrounded by colored diffraction rings produced as the light waves pass through the tiny openings in the weave of the fabric. Similarly, if you look at a bright light source through a few strands of hair pulled over your eye, you may see diffraction fringes with rainbowlike colors around

*For example, the 8.2-meter (about 27 feet) diameter mirrors of the new European Southern Observatory's Very Large Telescope are only 17 cm (about 7 inches) thick.

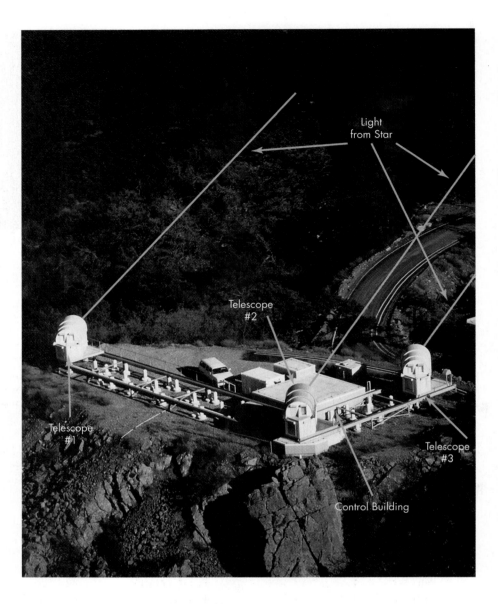

**FIGURE 4.9**
Photograph of an infrared and optical wavelength interferometer (IOTA). Light from the object of interest is collected by the three telescopes and sent to a control room. Computers there combine the light and reconstruct an image of the object.

the hairs. An even better way to see the effects of diffraction is to fog a piece of clear glass with your breath and hold it close to your eye while you look at a small, bright light source. If you wear glasses, just breathe on them, put them back on, and look at a bright light. You will see colored diffraction rings around the light, as figure 4.8B illustrates.

Diffraction seriously limits the detail visible through a telescope. In fact, diffraction theory shows that if two points of light that are separated by an angle $\alpha$ (measured in arc seconds*) are observed at a wavelength $\lambda$ through a telescope whose diameter is $D$, the points cannot be seen as separate sources unless $D$ is greater than ($>$)

$$2.5 \times 10^5 \, \lambda/\alpha$$

If $D$ is expressed in centimeters, $\lambda$ in nanometers, and $\alpha$ in seconds of arc, then

$$D > 0.02 \, \lambda/\alpha$$

_____

*An arc second is a unit of angle and is equal to 1/3600 of a degree.

For example, to resolve two stars separated by 0.1 seconds of arc when observing with visible light ($\lambda = 500$ nanometers), you need a telescope whose lens has a diameter greater than 100 centimeters (about 40 inches). Notice that the diameter needed to resolve two sources increases as the sources get closer together.*

## 4.2 INTERFEROMETERS

Diffraction effects can never be totally eliminated, but they can be reduced by enlarging the opening through which the light passes, so that its waves do not mix as severely. Astronomers sometimes accomplish this with a device called an **interferometer.** With an interferometer, observations are made simultaneously through two or more widely spaced telescopes (fig. 4.9) that direct the light to a common detector that combines the separate light beams.

The interferometer is so named because when it mixes the separate beams, the light waves of one "interfere" with the waves from the other. Where the crests of two waves arrive together, they create a bright region. Where the crest of one wave arrives simultaneously with the trough of another, they cancel and create a dark patch. The result is a complex pattern of alternating light and dark regions, which can be analyzed by a computer to create an image of the object observed. The result of this complicated process is an image in which the resolution is set not by the size of the individual mirrors but rather by their separation.† For example, if the mirrors are 100 meters apart, the interferometer has the same resolving power as a telescope 100 meters in diameter. The resulting high resolution is far beyond what can be obtained in other ways. For example, figure 4.10A shows a view of two closely spaced stars as observed with a small telescope. Their images are severely blended as a result of diffraction. Figure 4.10B shows the same stars observed with an interferometer and after the image has been processed by a computer. The two stars can now be easily distinguished: the two separate mirrors produce the resolving power of a single mirror whose diameter equals the spacing between them.

A

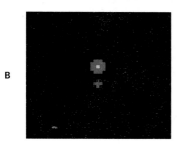

B

**FIGURE 4.10**
(A) A young star observed with an ordinary telescope. (B) The same star observed with an interferometer. The higher resolution of the interferometer reveals that the "star" is actually two stars in orbit around one another.

## 4.3 OBSERVATORIES

The immense telescopes and the associated equipment astronomers use are extremely expensive. Therefore, the largest telescopes are often national or international facilities: for example, the National Optical Astronomical Observatory of the United States and the Anglo-Australian Telescope. Despite the great expense of such facilities, many colleges and universities have their own large research telescopes (in addition to smaller ones near campus for instructional purposes). In addition, some large private groups such as the Carnegie Institution operate observatories. Altogether, several thousand observatories exist around the world on literally every continent. There are even telescopes at the South Pole in Antarctica to take advantage of the extreme dryness of the bitterly cold Antarctic air.

---

*As we will discover later in this chapter, our atmosphere seriously blurs fine details in astronomical objects, degrading the resolving power of large ground-based telescopes to far below their diffraction limits.
†The twin Keck Telescopes are designed to be used this way.

Mirror is
100 meters across

**FIGURE 4.11**
Proposed new OWL
("overwhelmingly large") telescope.

The largest optical telescopes in the United States at this time are the twin 10-meter Keck Telescopes described earlier. Next largest in collecting area in the United States is the Hobby-Eberly Telescope in Texas, which uses 91 mirrors each of 1-meter diameter set in an 11-meter disk. To avoid distortions in this immense assembly of mirrors, the telescope can only be pointed at a limited area of the sky. Moreover, the main mirror assembly can be aimed only approximately toward the star or galaxy of interest. Then, a much smaller "secondary" mirror adjusts the beam of light so it falls on the detectors.

The largest optical telescope in the world is the VLT (for "very large telescope"), a group of four 8-meter telescopes that work as a unit. The VLT is operated by a consortium of European countries and Chile and is located in the extremely dry northern part of Chile. Several other large telescopes have begun operation recently, such as the 8.2-meter Subaru* Telescope operated by Japan and located in Hawaii. Others are the Gemini telescopes—two identical 8.1-meter instruments. One is located in Hawaii; the other is in Chile. These twin telescopes are run by a consortium consisting of the United States, the United Kingdom, Canada, Chile, Australia, Brazil, and Argentina.

To observe the faint trickle of light from forming galaxies billions of light-years away, astronomers are planning even bigger telescopes, such as the "Overwhelmingly Large" Telescope (OWL, for short) (fig. 4.11). Its proposed collecting area would span a football field. However, many technical problems need to be overcome before such a colossus can be built, and no starting date for it has been set.

Other large existing telescopes are the 236-inch Russian telescope located in the Caucasus area east of the Black Sea; the two 4-meter (150-inch) telescopes of the National Optical Astronomy Observatory, one located about 50 miles west of Tucson, Arizona, on Kitt Peak, and its twin, located in Chile; and the Anglo-Australian Telescope operated jointly by Great Britain and Australia and located west of Sydney in the dry Australian outback. Astronomers need these large instruments in the Southern Hemisphere for complete sky coverage: a telescope located in the United States cannot see the sky near the south celestial pole.

*Subaru is the Japanese name for the Pleiades star cluster. Look carefully at the logo on a Subaru car and you will see the stars in the design.

## 4.4   DETECTING THE LIGHT

Once light has been collected, it must be detected and recorded. In olden days, the detector was the eye of an astronomer who sat at the telescope eyepiece and wrote down data or made sketches of the object being observed. The human eye, marvel that it is, has difficulty seeing very faint light. Many astronomical bodies are too distant or too dim for their few photons to create a sensible effect on the eye. For example, if you were to look at any but the nearest galaxies through even one of the Keck 10-meter Telescopes, the galaxies would appear merely as dim smudges. Only by storing up their light, sometimes for hours, can a quality picture of them be taken. Thus, to see very faint objects, astronomers use detectors that can store light in some manner. Such storage can be done chemically with film or electronically with detectors similar to those used in video camcorders.

From the late 1800s until the 1980s, astronomers generally used photographic film to record the light from the bodies they were studying. Film forms an image by absorbing photons that cause a chemical change, making the film dark where light has fallen on it. This process, however, is very inefficient: fewer than 4% of the photons striking the film produce a useful image. The result of such low efficiency is that many hours are needed to accumulate enough light to create an image of faint objects. Moreover, the film must be developed, thereby delaying the observing process even further.

Astronomers today, however, almost always use electronic detectors similar to those in camcorders. In these devices, the incoming light strikes a semiconductor surface, allowing electrons to move within the material. The surface is divided into millions of little squares, or pixels, in which the electrons are temporarily stored. The number of electrons in each pixel is proportional to the number of photons hitting it (that is, proportional to the intensity of the light). An electronic device coupled to a computer then scans the detector, counting the number of electrons in each pixel and generating a picture, in much the same way as a TV screen or printed photo creates a picture made up of many separate tiny dots.*

Such electronic devices are extremely efficient, recording 75% (or more) of the photons striking them, allowing astronomers to record images much faster than with film. Electronic detectors have other advantages as well. For example, they record the signal digitally, essentially counting every photon that falls on each part of the detector. Such digital images can be processed by computers to sharpen them, remove extraneous light, and enhance contrast.

Astronomers use many kinds of electronic detectors, but one of the best types is the **CCD** (charge-coupled device). Modern CCDs can make pictures virtually indistinguishable from photographs in their detail and with a sensitivity to faint light approximately 20 times greater.

## 4.5   OBSERVING AT NONVISIBLE WAVELENGTHS

Visible light, which we can see because its wavelengths are detectable by our eyes, is just one of many wave bands of electromagnetic radiation. Many astronomical objects, however, radiate at wavelengths that our eyes cannot see, and so astronomers have devised ways to observe such objects. For example, cold clouds of gas in interstellar space emit little visible light but large amounts of radio energy. Thus, to observe them, astronomers use radio telescopes. Likewise, dust clouds in space are too cold to emit visible light, but they do radiate infrared energy, which astronomers observe

---

*This is easy to see if you look closely at a color picture printed in a newspaper.

Radio waves
from space

Signals are focused here
and carried by cable to
the control room.

An antenna 10.4 meters
(about 34 feet) in diameter
collects radio waves and
reflects them to focus.

Mounting allows
telescope to track
sources.

Instrument room

**FIGURE 4.12**
Photograph of the Owens Valley Millimeter Wavelength radio telescope, operated by the California Institute of Technology. In the background you can see the Sierras. The "telescope" is an array of six separate dishes which collect the radio waves. The captured radiation is then combined by computer to increase the resolution of the instrument.

with infrared telescopes. Similarly, astronomers use X-ray telescopes orbiting in space to observe the hot gas that accumulates around black holes. Radio, infrared, and X-ray telescopes have the same fundamental purposes as optical telescopes: to collect radiation and to resolve details. Figure 4.12 shows a typical radio telescope with its large "mirror" to collect the radiation and focus it on a detector. However, because our eyes cannot see these other wavelengths, astronomers must devise ways to depict what such instruments record.

One way astronomers depict observations made at nonvisible wavelengths is with false-color pictures, as shown in figure 4.13. In a false-color picture, the colors used represent the intensity of radiation that we cannot see. For example, in figure 4.13A (a radio "picture" of a radio galaxy and the jet of hot gas spurting from its core), astronomers color the regions emitting the most intense radio energy red; they color areas emitting somewhat less energy yellow and the faintest areas blue. Thus, if we could "see" radio waves, the red areas would look brightest and the blue areas dimmest. Similar techniques may be used to depict astronomical objects at other wavelengths. For example, figure 4.13B shows a false-color X-ray "photograph" of the gas shell ejected by an exploding star.

Telescopes operating at infrared, ultraviolet, and X-ray wavelengths face an additional obstacle: most of the radiation they seek to measure cannot penetrate the Earth's atmosphere. Gases in our atmosphere such as ozone, carbon dioxide, and water strongly absorb infrared, ultraviolet, and shorter wavelengths. For example, infrared radiation whose wavelength is 50 micrometers is almost totally absorbed by water and carbon dioxide in our atmosphere. However, some infrared radiation penetrates the

FIGURE 4.13

(A) False-color picture of a radio galaxy. We can't see radio waves, so colors are used to represent their brightness—red brightest, blue dimmest, (B) False-color X-ray picture of Cas A, an exploding star.

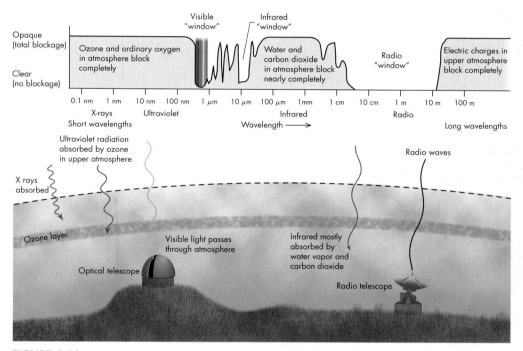

FIGURE 4.14

Diagram showing the transmission of light through the Earth's atmosphere. (See also fig. 3.20.)

Q★ Would 100 μm be a good wavelength to try to observe a cool star with a ground-based telescope? Would 10 μm be better?

A★ 100 μm would not be good because the transmission of our atmosphere is very low at that wavelength. 10 μm would be much better because there is a narrow "window" of transmission there.

atmosphere in narrow-wavelength regions called "atmospheric windows," as shown in figure 4.14. Thus, radiation whose wavelength is between 8 and 15 micrometers is only weakly absorbed. If astronomers want to view an object in a blocked wavelength, they must use a telescope in space, operated remotely from the ground or, more rarely, by a scientist-astronaut in space.

## EXTENDING OUR REACH

### EXPLORING NEW WAVELENGTHS: GAMMA RAYS

Astronomers have made many of their most exciting discoveries when new telescopes allowed them to observe the sky at wavelengths not previously detectable. Gamma-ray wavelengths are among the last of the wavelength regions to be explored, and astronomers are still trying to interpret what they see.

Gamma-ray astronomy began in 1965 when a small and (by modern standards) primitive satellite detected cosmic gamma rays. A few years later, a slightly more advanced satellite detected gamma rays coming from the center of our galaxy, the Milky Way. By the 1970s, astronomers had discovered that many familiar sources, such as the Crab Nebula and the remnants of other exploded stars, emit gamma rays. Ironically, perhaps the most interesting gamma-ray sources were discovered accidentally earlier.

In 1967, the United States placed several military surveillance satellites in orbit to watch for the gamma rays produced when a nuclear bomb explodes. The satellites were designed to monitor the United States–Soviet Union ban on nuclear bomb tests in the atmosphere. Curiously, on a number of occasions, the satellites detected gamma-ray bursts coming not from the Earth but from space. Unfortunately for astronomers, the discovery of the bursts was top secret at the time and was not made public until 1973.

Astronomers' thirst for more information about these high-energy sources was unsatisfied for many years because our atmosphere absorbs gamma rays, and ordinary telescopes cannot focus gamma rays. Nevertheless, with ever more complex instruments in satellites, astronomers discovered that gamma-ray sources—apart from the bursts—coincided with known astronomical objects such as dying stars and some peculiar galaxies. The gamma-ray burst, on the other hand, would appear suddenly in otherwise blank areas of the sky, flare in intensity for a few seconds, and then fade to invisibility. It has taken nearly 30 years of study to even answer the simple question, "Are they near or far?" The breakthrough came in December 1997, when astronomers detected a gamma-ray burst that coincided with a distant galaxy. This solved the mystery of the bursters' distance but leaves unanswered what they are. Theories to explain the bursts abound. According to one hypothesis, the bursts occur when a pair of dead, collapsed objects called neutron stars* collide in a distant galaxy. According to another hypothesis, the bursts are "hypernovas," stellar explosions caused when massive stars run out of fuel and collapse to form black holes. This latter proposal gained support when, in 2002, astronomers obtained spectra of a burst that showed emission lines suggestive of the explosion of a massive star. Today, gamma-ray bursts still are mysterious, but that is what makes doing science exciting.

---

*Neutron stars are the remnants of massive stars that have exploded. They are discussed in more detail in chapter 14.

## EXTENDING OUR REACH

### OBSERVING THE CRAB NEBULA AT MANY WAVELENGTHS

In midsummer A.D. 1054 just after sunset, astronomers in ancient China and other far eastern countries noticed a brilliant star near the crescent moon in a part of the sky where no bright star had previously been seen. They wrote of this event:

"In the last year of the period Chih-ho, ... a guest star appeared.... After more than a year it became invisible."

We know today that these astronomers of long ago witnessed a supernova explosion, the violent event that marks the death of a massive star. Their record—nearly 1000 years old—begins a story that continues today as astronomers try to understand what causes a star explosion. Although the story began with naked-eye observations, it continues with observations made with telescopes on the ground and in space. Moreover, the story illustrates how astronomers have come to rely on observing radiation at many wavelengths, not just visible light.

Despite its initial brilliance, the dying star seen so long ago faded and disappeared from the sky and astronomical records. Then, in 1731, John Bevis, a British physician and amateur astronomer, noticed with his telescope a faint dim patch of light in the constellation Taurus. (You can see Taurus and where the Crab Nebula lies in "Looking Up at Taurus" on p. 136.) Twenty-seven years later, Charles Messier, a French astronomer and comet hunter, rediscovered the glowing cloud and made it the first entry in his catalog of fuzzy patches of light that were not comets. In 1844, Lord Rosse, a British astronomer and telescope builder, noticed that the fuzzy patch contained filaments (Box fig. 4.5A) that to his eye resembled a crab. He therefore named it the Crab Nebula.

In 1921, John Duncan, an American astronomer, compared two photographs of the nebula taken 12 years apart and noticed that it had increased slightly in diameter.

**BOX FIGURE 4.5**
(A) Visible-light photograph of the Crab Nebula. (B) Radio image of the Crab Nebula. (C) X-ray image in false-color of the core of the Crab Nebula.

⟲⋆ What does the flattened round shape of the glowing gas in Box figure 4.5C suggest about the gas's motion?

ⓐ⋆ It is spinning.

He therefore deduced that the nebula was expanding. At the same time, several other astronomers came across the ancient Chinese records and noticed the coincidence in position of the nebula with the report of the exploding star. Then, 7 years later, Edwin Hubble, at Mount Wilson Observatory in California, measured the increase of size more accurately and calculated from the rate of expansion that the nebula was about 900 years old—roughly the same age as the dying star seen nearly a millennium earlier by the Chinese astronomers. Thus, astronomers realized that the Chinese astronomers had witnessed a supernova explosion, marking the death of a star.

Since then, astronomers have examined the Crab Nebula with telescopes using virtually all wavelength bands and, in doing so, have added yet more to their understanding of a star's demise. For example, in 1948, Australian astronomers using a radio telescope discovered that the Crab Nebula is a powerful source of radio waves (Box fig. 4.5B). In 1968, further observations at radio wavelengths revealed that a faint peculiar star near the center of the nebula is spinning about 30 times per second and that it is the core of the star whose explosion created the Crab Nebula. Likewise, with an X-ray telescope above the atmosphere, astronomers discovered that it is a source of X-ray radiation (Box fig. 4.5C).

What have all these observations shown? They have given astronomers their best view yet of the last moments of a star's life. From visible-wavelength observations, astronomers measure that the gas ejected when the star exploded is expanding with a speed of about 1000 km/s. From radio-wavelength observations, they deduce that the nebula contains charged particles moving at nearly the speed of light and that the central star pulses on and off about 30 times per second. The X-ray observations confirm this picture. Thus, by observing the Crab Nebula and its stellar corpse at a variety of wavelengths, astronomers have shown that it is a far richer and more mysterious object than could be deduced from observations at one wavelength alone.

## 4.6    OBSERVATORIES IN SPACE

Figure 4.15 shows several of the many orbiting telescopes astronomers use. Some, like the Hubble Space Telescope, which was launched in 1990, have operated for many years. Others, such as the Spitzer Space Telescope, have a much shorter lifetime. This NASA infrared observatory needs liquid helium to keep its instruments cold for maximum sensitivity and once the helium on board the craft is used up (by 2009 or so),

# LOOKING UP AT... TAURUS

### M1 The Crab Nebula

~7 ly

The Crab Nebula is the remnant of a star that blew up in the year AD 1054 as a supernova. It is about 5000 ly away from us.

### Zeta Tauri

**Taurus — The Bull**
One of the constellations of the zodiac and one of the creatures hunted by Orion. Taurus is visible in the evening sky from November through March. The brightest star in Taurus is Aldebaran, the eye of the bull.

This open star cluster is easy to see with the naked eye and looks like a tiny dipper. It is about 400 ly from Earth.

### M45 The Pleiades

~8 ly

### Aldebaran

Aldebaran is a red giant star. It is about 65 ly from us and its diameter is about 45 times the diameter of the Sun. It lies between us and the Hyades.

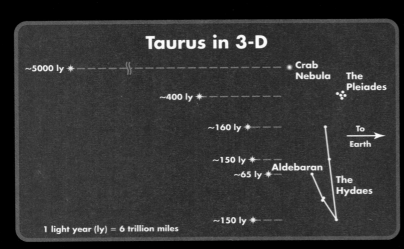

## Taurus in 3-D

~5000 ly ✳ –––––––––⌇⌇––––––––––• Crab Nebula

The Pleiades

~400 ly ✳ ––––––––– ✳

~160 ly ✳ –––––––

To Earth

~150 ly ✳ ––––

~65 ly ✳ – **Aldebaran**

The Hydaes

~150 ly ✳ –––––

1 light year (ly) = 6 trillion miles

### The Hyades

The "V" in Taurus is the nearest star cluster to us, about 137 ly away. It is easy to see its many stars with binoculars.

SOUTH

Cover opening = 2.4 m
(approx. 94 inch)

Hubble Space Telescope (13.6 m long) – HST

Extreme Ultraviolet Explorer – EUVE

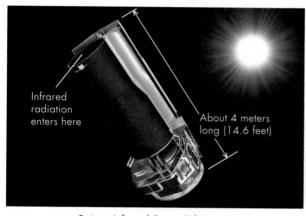

Infrared radiation enters here

About 4 meters long (14.6 feet)

Spitzer Infrared Space Telescope

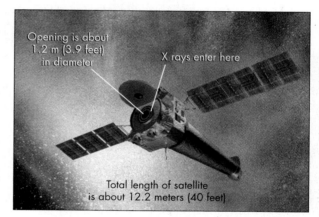

Opening is about 1.2 m (3.9 feet) in diameter

X rays enter here

Total length of satellite is about 12.2 meters (40 feet)

Chandra X-ray Telescope Satellite

**FIGURE 4.15**
Photograph of the Hubble Space Telescope and drawings of some other orbiting observatories: EUVE, Spitzer, and Chandra.

the telescope will no longer operate. In the meantime, other wavelength regions not observable from the ground are being probed, such as the X-ray and gamma-ray parts of the electromagnetic spectrum.

X rays are absorbed in our atmosphere by ozone and oxygen. To observe cosmic X-ray sources, a number of countries have launched X-ray telescope satellites. Currently, the largest such orbiting X-ray observatory is Chandra,* launched in 1998 by NASA.

Of the many orbiting telescopes used by astronomers, the Hubble Space Telescope (HST) is the most ambitious to date. The HST is designed to observe at visible and ultraviolet wavelengths and has a mirror 2.4 meters (about 94 inches) in diameter that directs light to a set of instruments. Its two electronic cameras, one for wide-field views and the other for detailed, large-scale images, have already taken many striking pictures. The HST also has two spectrographs for analyzing light from stars and galaxies.

---

*Chandra is short for Chandrasekhar, the Indian astrophysicist Subrahmanyan Chandrasekhar who won the Nobel Prize in Physics in 1983 for work he did in the 1930s on collapsed stars. Some of his work is discussed in chapter 14.

An international team including scientists from Germany, America, and the United Kingdom designed and built the detectors and other instruments for Chandra.

Sombrero Galaxy

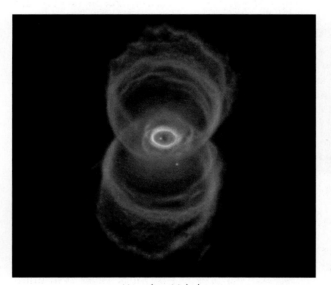

Hourglass Nebula
A dying star

A bubble of hot gas rising from the center of the galaxy NGC 4438 where
a huge black hole is swallowing stars and other matter.

**FIGURE 4.16**
Some images obtained with the Hubble Space Telescope.

Although the HST initially had a number of problems, astronauts have repaired the major defects, and astronomers are now delighted with the clarity of its images (fig. 4.16). These images reveal details never before seen by telescopes on the ground because such telescopes must peer through the blurring effects of our atmosphere.

## Atmospheric Blurring

Anyone who has ever watched the stars on a clear night has seen the blurring that our atmosphere creates. The light we see from the stars flickers, making them "twinkle." Twinkling, more properly called **scintillation,** is caused by atmospheric irregularities refracting the star's light. These irregularities result from slight variations in the air's density caused by small temperature differences. As light moves through these irregularities, its speed changes by a tiny amount, and the light is slightly refracted. If there were no irregularities, a star's light would reach your eye along a single path and be steady.

**FIGURE 4.17**
Twinkling of stars (seeing) is caused by moving atmospheric irregularities that refract light in random directions.

Q* Why do stars twinkle more when they are low in the sky and are close to the horizon than when they are nearly overhead?

A* Their light passes through much more atmosphere and thus is more likely to encounter atmospheric irregularities. (Box figure 4.3A and B show the greater path length through the atmosphere for a star that is low in the sky.)

*Why stars twinkle*

However, the many irregularities create many paths by which the light reaches you. Moreover, these paths change direction rapidly and erratically as the irregularities move, carried by the wind. As a result, the starlight you see at any instant is a blend of light from many slightly different directions, a blend that smears the star's image and makes it dance (fig. 4.17). You can see a similar effect looking at something through water. If the surface of the water has even slight disturbances, a pebble or coin on the bottom seems to dance around. Atmospheric irregularities also slightly disperse the light, making the star's color dance, too. Such refractive twinkling, though very pretty to watch, seriously limits the ability of Earthbound observers to see fine details in astronomical objects. The dancing image of a star or planet distorts its picture when recorded by a camera or other device. Outside the atmosphere (in space) this distortion, technically called **seeing,** disappears. The elimination of seeing problems caused by the turbulence of our atmosphere is one reason why astronomers find space observatories so useful.

Until recently, ground-based astronomers had to simply accept the distortions of seeing, but now they can partially compensate for such seeing in several ways. One technique involves observing a "reference" star simultaneously with the object of interest. By measuring carefully how the atmosphere distorts the known star's image, corrections can be made in the pictures of other objects.

Another method to improve seeing uses a powerful laser beam to create an artificial star. The laser beam is projected on the atmosphere, as shown in figure 4.18. The distortions of the artificial star image are recorded by a computer that then triggers tiny actuators on a correcting mirror placed in the telescope's light beam. The actuators create tiny distortions in the correcting mirror that cancel out those created by the atmosphere. This technique, called **adaptive optics,** has already given astronomers dramatically improved views through the turbulence of our atmosphere.

## Space Observatories Versus Ground-Based Observatories

Despite the freedom from atmospheric blurring and absorption that space observatories enjoy, much astronomical work will be done from the ground for the foreseeable future. For many years, ground-based telescopes will be much larger than orbiting telescopes. Moreover, equipment problems can be corrected easily without the expense, delay, danger, and complexity of a space-shuttle launch.

Because huge telescopes in space or even on the Moon will remain dreams for years to come, astronomers therefore choose with care the location of ground-based observatories. Sites are picked to minimize clouds and the inevitable distortions and absorption of even clear air. Thus, nearly all observatories are in dry, relatively cloud-free regions of the world such as the American southwest, the Chilean desert, Australia, and a few islands such as Hawaii and the Canaries. Moreover, astronomers try to locate observatories on mountain peaks to get them above the haze that often develops close to the ground in such dry locales.

Recently, astronomers have had to contend with another factor that affects the location of observatories: light pollution. Most inhabited areas are peppered with nighttime lighting such as street lights, advertising displays, and automobile headlights (fig. 4.19A and B). Although some such lighting may increase safety, much of it is wasted energy, illuminating unessential areas and spilling light upward into the sky, where it serves no purpose at all. Figure 4.19C shows a satellite view of North America at night and illustrates the waste of energy created by light pollution. Such stray light can seriously interfere with astronomical observations. In fact, some observatories have been essentially shut down because of light pollution. In some places,

**FIGURE 4.18**
A laser beam creates an artificial star whose image serves as a reference to eliminate the atmosphere's distortion of real stars. This photograph was taken at the Starfire Optical Range of the Phillips Laboratory at Kirtland Air Force Base in New Mexico.

astronomers have persuaded regional planning bodies to develop lighting codes to minimize light pollution. Light pollution, however, not only wastes energy and interferes with astronomy, it also destroys a part of our heritage—the ability to see stars at night. The night sky is a beautiful sight, and it is shameful to deprive people of it.

## 4.7 GOING OBSERVING

When astronomers want to observe at a telescope, they do not just run off to the observatory. Generally, they must submit an observing proposal that describes what objects they wish to look at and why such observations are important. Proposals are then screened by a committee that allocates telescope time. If the proposal is accepted, the astronomers must prepare for the observing run, being sure that all necessary equipment will be available. This may include backup equipment and computers as well as charts for finding the objects to be observed. Because many observatories are at remote locations, sometimes travel plans must also be made.

Observing sessions at a large telescope are typically assigned in blocks of several nights so that the astronomer's risk of clouds or equipment failure is minimized. If the astronomer is using an optical telescope and must therefore observe at night, he or she must also get used to switching schedules to be awake all night and sleep during the day. Most observatories have small dorm rooms with special shades to make the room dark and quiet during daylight hours for astronomers trying to catch up on sleep. Many also have small cafeterias where food (and coffee) is available at odd

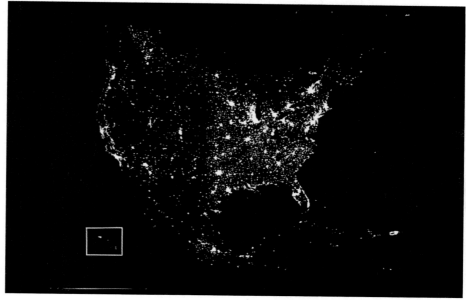

**FIGURE 4.19**
Photographs illustrating light pollution. (A) Los Angeles basin viewed from Mount Wilson Observatory in 1908. (B) Los Angeles at night in 1988. (C) Notice the pattern of the interstate highway system visible in the satellite picture of North America at night.

hours. Sometimes observing runs go smoothly, and the astronomer can return to his or her home institution laden down with data. At other times weather or equipment will not cooperate, and the run is a total loss.

For astronomers wishing to avoid weather problems, a new "observing" procedure using the World Wide Web may help. Over the last decade, astronomers from around the world have compiled and stored immense data archives at numerous websites. These archives are typically based on observations in which the sky is systematically mapped at one or more wavelengths. Instead of gathering data on a few particular objects, such maps record information about *every* astronomical object they can detect. For example, astronomers working on 2MASS (the Two Micron All Sky Survey) have mapped the entire sky at short infrared wavelengths that get through our atmosphere, compiling a database of several *million* images. You can see these images of gas clouds, galaxies, star clusters, and so on at the relevant websites from which they can be downloaded. Similarly, astronomers working on the Sloan Digital Sky Survey are mapping portions of the sky at visible wavelengths. These colossal data files are far too large to analyze all at once. Instead, an astronomer interested in a particular object or class of object goes to a web archive and downloads the data and images of interest.

The use of archives does not make direct observing obsolete, but it can give a first look at a problem so that when the astronomer later goes to a telescope, he or she can use the time there more efficiently.

# 4.8 COMPUTERS

In the last few decades, the computer has become one of the astronomer's most important tools. In fact, for many astronomers today, operating a computer and being able to program is more important than knowing how to use a telescope. Astronomers use computers not only to solve equations but also to move the telescope, feed the information to detectors, and convert the data obtained to a useful form. Moreover, electronic computer networks allow astronomers to communicate and exchange their data and ideas with other astronomers around the world in a matter of minutes. For such purposes, observatories typically have a range of computers for different tasks, ranging from pocket calculators through desktop personal computers to workstations. If a huge computer is needed, for example, for the mathematical simulation of the collision of two galaxies or the explosion of a star, the astronomer may use the computer network to link his or her desktop machine to a supercomputer, perhaps thousands of miles away. The hard calculations are then done on the remote machine, but the output ends up at the desktop.

## SUMMARY

Astronomers use telescopes to collect radiation from astronomical sources. Telescopes generally have large-diameter mirrors or lenses to collect as much radiation as possible and allow faint objects to be seen. This gives them large light-gathering power. A large mirror or lens also increases the telescope's ability to resolve detail, giving sharper images, but such gains are seriously limited for ground-based visible light telescopes by the blurring effects of our atmosphere. Interferometers give the resolving power of a single large area by combining radiation detected with two or more small but widely separated collectors.

Although many astronomical objects radiate visible light, some of the most interesting bodies radiate in radio, infrared, X-ray or gamma-ray regions of the electromagnetic spectrum. Astronomers use special telescopes to observe in these other wavelength regions. Because many of these wavelengths do not penetrate our atmosphere, telescopes have been put in space to observe them.

### **?** What It Is.

This is alleged to be Galileo's telescope. It is in the Florence Museum of Science in Italy.

## QUESTIONS FOR REVIEW

1. What is meant by the term *light-gathering power?* How does it affect the ability to see faint objects?

2. What is resolution of a telescope? What physical process limits it?

3. How is resolution affected by the size of a telescope's mirror or lens?

4. What is the difference between a reflecting and a refracting telescope?

5. What are some reasons for using mirrors rather than lenses in telescopes?

6. What is the purpose of an interferometer?

7. Why do astronomers put X-ray observatories in space rather than just on a high mountain?

8. What is meant by adaptive optics?

## THOUGHT QUESTIONS

1. How, apart from magnification, do binoculars help you see better?

2. Which binoculars should you choose to observe at night: a pair with 50-millimeter lenses or a pair with 35-millimeter lenses? Why?

3. Why does the useful resolving power of a ground-based telescope with a 2-meter diameter mirror not match its theoretical value?

4. If you look with binoculars down a beach on a hot day, you will see that distant objects appear shimmery. How is this related to astronomical "seeing"?

5. Put a pencil straight down into a glass of water. Notice whether the pencil looks bent. Now tilt the pencil and note how its apparent bending changes. How does this illustrate why light from objects near the horizon is refracted more strongly?

## PROBLEMS

1. Compare the light-gathering power of a telescope with a 10-centimeter (about 4-inch) diameter mirror to that of the human eye. (Take the diameter of the pupil of the eye to be about 5 millimeters.)

2. Estimate your eye's resolving power by drawing two lines 1 millimeter apart on a piece of paper. Put the paper on the wall and then step back until the two lines appear as one, measuring that distance. From the distance and the separation of the lines (1 millimeter), estimate their angular separation. How does your result for the eye's resolving power compare with that calculated from the resolving-power formula, using a pupil diameter of 5 millimeters and a wavelength of 500 nanometers?

3. Can the unaided human eye resolve a crater on the Moon whose angular diameter is 1 minute of arc (=60 seconds of arc)? (Take the diameter of the pupil of the eye to be about 5 millimeters and the wavelength of the light to be 500 nanometers.)

## TEST YOURSELF

1. Telescope A's mirror has three times the diameter of telescope B's. How much greater is A's light-gathering power?
   (a) 3 times
   (b) 6 times
   (c) 8 times
   (d) 9 times
   (e) 27 times

2. A telescope's resolving power measures its ability to see
   (a) fainter sources.
   (b) more distant sources.
   (c) finer details in sources.
   (d) larger sources.
   (e) more rapidly moving sources.

3. A ground-based telescope to observe X rays would
   (a) be a powerful tool for studying abnormally cold stars or distant planets.
   (b) give astronomers the chance to study the insides of stars and planets.
   (c) be worthless because no astronomical objects emit X rays.
   (d) be worthless because X rays cannot get through the Earth's atmosphere.
   (e) be worthless because astronomers have not yet devised detectors for X rays.

4. Astronomers use interferometers to
   (a) observe extremely dim sources.
   (b) measure the speed of remote objects.
   (c) detect radiation that otherwise cannot pass through our atmosphere.
   (d) enhance the resolving power (see fine details) in sources.
   (e) measure accurately the composition of sources.

5. One way to increase the resolving power of a telescope is to
   (a) make its mirror bigger.
   (b) make its mirror smaller.
   (c) replace its mirror with a lens of the same diameter.
   (d) use a mirror made of gold.
   (e) observe objects using longer wavelengths.

## FURTHER EXPLORATIONS

Berger, David H., Jason P. Aufdenberg, and Nils H. Turner. "Resolving the Faces of Stars." *Sky and Telescope* 113 (February 2007): 40.

Chen, P. K. "Visions of Today's Giant Eyes." *Sky and Telescope* 100 (August 2000): 34.

Fugate, Robert Q., and Walter J. Wild. "Untwinkling the Stars— Part I." *Sky and Telescope* 87 (May 1994): 24.

Jayawardhana, Ray. "The Age of Behemoths." *Sky and Telescope* 103 (February 2002): 30.

Kellerman, Kenneth I. "Radio Astronomy in the 21st Century." *Sky and Telescope* 93 (February 1997): 26.

Schilling, Govert. "Giant Eyes of the Future." *Sky and Telescope* 100 (August 2000): 52.

Sinnott, Roger W., and David Tytell. "A Galaxy of Telescopes: Coming of Age in the 20th Century." *Sky and Telescope* 100 (August 2000): 42.

Tytell, David. "The Heat is On." *Sky and Telescope* 107 (April 2004): 30.

Waller, William H. "NASA's Space Infrared Telescope Facility: Seeking Warmth in the Cosmos." *Sky and Telescope* 105 (Feb 2003): 43.

Wanjek, Christopher. "Gamma-Ray Astronomy Achieves Stardom." *Sky and Telescope* 106 (Aug 2003): 38.

**Website**

Please visit the *Explorations* website at **http://www.mhhe.com/arny** for additional online resources on these topics.

## PLANETARIUM EXERCISE

Set the location you are viewing from to the nearest large city, or your latitude and longitude. Verify that your local time is correct. Set the field of view to 60°, the magnitude limit to 5.0, and the time step to 30 minutes. Find and center on Polaris. Step the program ahead for 6 hours. How long does it take a star close to Polaris in the sky to complete one rotation? How about a star farther away? What does this say about the relative speeds of stars as they move across the sky? Though they all move through the same angle, which stars move a greater distance on the sky over time?

## KEY TERMS

adaptive optics, 139

CCD, 131

diffraction, 127

dispersion, 122

interferometer, 129

light-gathering power, 120

Moon illusion, 123

reflectors, 124

refraction, 121

refractors, 121

resolving power, 127

scintillation, 138

seeing, 139

## PROJECTS

1. Make a simple spectroscope using aluminum foil, a cardboard tube, and a grating. Plans are sketched below. Put aluminum foil over a small hole at one end of the tube. Cut a very thin slit in the foil with a razor blade or sharp knife. The slit should be straight, about an inch long, and as narrow as you can make it and still have light get through it. At the other end of the tube mount a small square of grating material. Tape it over a small hole cut in the tube's end. If you use a toilet-paper tube, just bunch the foil over one end and tape it on. Fold some thin cardboard over the other end and cut a small square hole about an inch across. Hold the grating material up to light and notice to which side it creates a rainbow. Tape the grating along its edges oriented so that the rainbow runs perpendicular to the slit. Ask your instructor about details of construction and how to get the grating. To use, hold the grating end of tube right up to your eye and look at the light source. Rotate the tube so that you see a rainbow off to the side. Sketch the spectra you see from (a) a fluorescent light, (b) a mercury vapor street light, (c) an ordinary incandescent lightbulb, (d) the blue sky.*

2. Use your spectroscope to look at the spectrum of flames. Use a gas-burning stove and add a pinch of table salt to see sodium emission lines (yellow) or use copper sulfate crystals or a scrap of copper wire to see green copper emission lines.

---

*Do not look directly at the Sun.

Old oatmeal box–or even toilet paper tube. A square box will work as well. Length should be at least 4 inches.

Narrow slit–as thin as possible (approximately 1 inch long).

Grating material taped over hole. Hole should be about 1 inch square.

**PROJECT FIGURE 4.1**

# The Earth

# 5

The Earth.

Volcanoes, such as this one, add gases to our atmosphere.

## Chapter Objectives

- Understanding the Earth helps us better understand other planets.
- Scientists can determine properties of the Earth's interior by analyzing earthquake (seismic) waves.
- The Earth's structure consists of:
  - a solid crust,
  - a mantle of rocky material, and
  - a core of mainly iron, part of which is liquid.
- The Earth's interior is very hot. The heat is probably mainly left over from the time of the Earth's formation.
- Radioactive material in the Earth contributes heat to its interior.
- Analysis of the amount of radioactivity gives the Earth's age—about 4.6 billion years.
- Heat inside the Earth causes motions in the interior, which cause the surface to creep slowly.
- Such surface motions build mountains and alter our planet's surface.
- Earth's atmosphere contains mostly nitrogen and oxygen.
- The atmosphere formed from gases liberated mainly during Earth's formation. Oxygen, however, has been added by green plants.

**?What Is This?**
(See chapter summary for the answer.)

145

## Concepts and Skills to Review

- Newton's law of gravity (2.5)
- The law of inertia (2.2)
- Absorption in the Earth's atmosphere (3.7)

**E**arth is a beautiful planet. Even from space we can see its beauty—blue seas, green jungles, red deserts, and white clouds. Much of our appreciation of the Earth comes from knowing that it is home for us and the billions of other living things that share this special and precious corner of the Universe. But why study Earth in an astronomy course? The reason is simple: we know Earth better than any other astronomical body, and we can learn from it about many of the properties that shape other worlds.

Earth's special characteristics result in large measure from its dynamic nature. The Earth is *not* a dead ball of rock; both its surface and its atmosphere have changed greatly during its vast lifetime. Even today, the ground below our feet sometimes trembles and wrenches in response to dynamic forces, crumpling our planet's crust into mountains, stretching it, and tearing it open to form new ocean basins.

These slow but violent motions within the Earth arise as heat generated deep within flows toward the surface. That heat also drives volcanic eruptions, which vent gases and molten rock. Over billions of years, such gases accumulated, in part creating our atmosphere—an atmosphere that has itself been changed by the presence of abundant water and life.

## 5.1 THE EARTH AS A PLANET

In the simplest terms, Earth is a huge, rocky sphere spinning in space and hurtling around the Sun. In the time it took you to read that sentence, the Earth carried you about 100 miles through the black hostile space around the Sun. But you were protected by a blanket of air, a screen of magnetism, and filters of molecules that blocked most of the hazards of interplanetary space. Other planets share many of these properties but not in the right mix to make it possible for life as we know it to live on them.

### Shape and Size of the Earth

Astronomers in ancient Greece knew* that the Earth is a sphere with a radius of about 6400 km (4000 miles).

Aristotle even argued that the Earth formed by material being pulled together in such a way that the jostling of the pieces led to a spherical shape. He didn't refer to gravity, specifically, but we know today that it is gravity, in fact, that makes the Earth round.

Gravity is the great leveler. Over millions of years, the force of gravity crushes and deforms rock, pulling high points down and rounding large bodies off. However, this shaping process is effective only if the body exceeds a critical size that depends on the object's composition. For bodies made of rock, the critical radius is about 350 kilometers. An object with a radius larger than about 350 kilometers has a strong enough gravity that it can pull itself into a sphere even if it was initially irregular. Smaller objects retain their irregular shape, as you can see by comparing the Earth with the asteroid Gaspra (fig. 5.1), which has a longest dimension of about 19 km (about 12 miles).

Although the Earth is approximately a sphere, it is not a perfect one. It bulges out at the equator, as illustrated in figure 5.2A. The existence of this bulge was first demonstrated in the eighteenth century, when detailed mappings of the Earth showed that its equatorial radius is about 21 kilometers (about 13 miles) greater than its polar radius. This discovery was not a complete surprise because astronomers such as Newton and Hooke had already suggested that the Earth's spinning motion might make its equator bulge into a shape technically known as an oblate spheroid. Moreover, telescopic views of other planets showed that they bulged (fig. 5.2B), making it likely the Earth did too.

You can demonstrate rotationally caused bulges with a water balloon. If you toss a water balloon (gingerly!) into the air, it will take on an almost spherical shape. If you set it spinning as you toss it up, it will become noticeably bulged. This is a great demonstration, but choose an appropriate place.

---

*Recall their reasoning based on the shape of the Earth's shadow during a lunar eclipse (section 1.2).

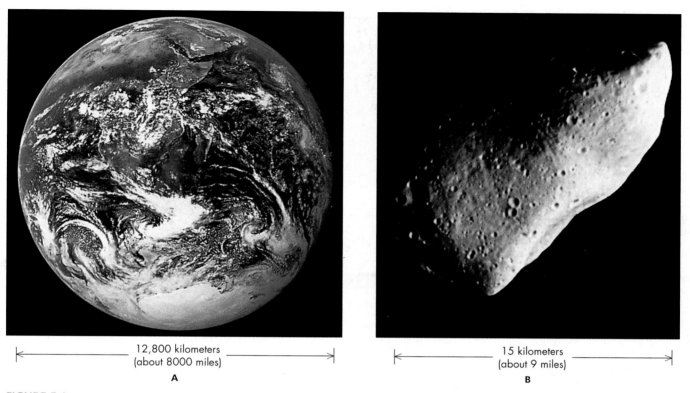

| 12,800 kilometers (about 8000 miles) | 15 kilometers (about 9 miles) |
|---|---|
| A | B |

**FIGURE 5.1**
Photographs show that the Earth is round but the asteroid Gaspra is not. Gaspra is too small for its gravity to make it spherical.

Inertia makes material move away from a spinning object.

To understand why the Earth's equator bulges, think of what happens when you lift a spinning electric beater from a bowl of cake batter. Particles on such a spinning object fly outward as a result of their inertia, the tendency of all moving objects to keep moving in a straight line, as described by Newton's laws of motion. This same tendency moves matter away from the Earth's rotation axis and is strongest at the equator because the Earth rotates fastest there.

All points on the Earth take the same time (one day) to rotate once around its axis, but because points near the equator travel farther than points near the pole, they must travel faster. At the equator, a point on the Earth's surface moves at about 1000 miles per hour, while at middle latitudes, such a point moves at about 700 miles per hour. The greater speed of the equator is harder for the Earth's gravity to overcome, so the equator bulges outward.

## Composition of the Earth

Although we may call the Earth a ball of rock, the statement is not very informative because so many different kinds of rock exist. Rocks are composed of minerals, and minerals in turn are composed of chemical elements. Analysis of the surface rocks of the Earth shows that the most common elements in them are oxygen, silicon, aluminum, magnesium, and iron. Furthermore, silicon and oxygen usually occur together as **silicates.** For example, ordinary sand (particles of the silicate mineral quartz) is nearly pure silicon dioxide ($SiO_2$). Table 5.1 lists a few of the most abundant elements in the Earth.

Other kinds of minerals are more complicated, with atoms of calcium, magnesium, or iron included with the silicates. For example, much of the Earth's interior is

**FIGURE 5.2**
(A) Rotation makes the Earth's equator bulge. (B) Jupiter's rapid rotation creates an equatorial bulge visible in this photograph.

| TABLE 5.1 | | |
|---|---|---|
| **Composition of Earth** | | |
| **Chemical Element (Symbol)** | **% of Element in Crust by Mass** | **Density (g/cm³)** |
| Oxygen (O) | 45.5% | |
| Silicon (Si) | 27.2% | 2.42 |
| Aluminum (Al) | 8.3% | 2.70 |
| Iron (Fe) | 6.2% | 7.9 |
| Calcium (Ca) | 4.66% | 1.55 |
| Magnesium (Mg) | 2.76% | 1.74 |
| Sodium (Na) | 2.27% | 0.97 |
| Potassium (K) | 1.84% | 0.87 |
| Titanium (Ti) | 0.63% | 4.5 |
| Others | 1% | |

**FIGURE 5.3**
Olivine (the greenish crystals) in a rock sample.

composed of the mineral olivine, which is an iron-magnesium silicate. It gets its name from its color: in some solidified lavas, it forms tiny green crystals that look like little pieces of chopped olive (fig. 5.3).

At this point you might ask how we can tell what the interior of the Earth is made of. We can infer what lies deep inside our planet in several ways. One is by studying earthquake waves, a point we will take up in more detail in the next section. Another way is by analyzing the Earth's density.

## Density of the Earth

**Density** is a measure of how much material is packed into a given volume. It is defined as an object's mass divided by its volume and is usually measured in terms of the mass in grams of one cubic centimeter of the substance. In these units, for example, water has a density of 1 gram per cubic centimeter. Ordinary rocks, on the other hand, have a density of about 3 grams per cubic centimeter, and iron has a density of almost 8 grams per cubic centimeter. In other words, a volume of iron has about 8 times as much mass as a similar volume of water; that is, iron is much denser than water.

We see from this that density gives some clue to an object's composition. For example, it would be easy to tell if a closed box contained a block of iron or a block of wood. This is the basis of the famous story of the ancient scientist Archimedes leaping from his bath and shouting "Eureka!" after realizing that he could test whether the king's crown was pure gold by measuring its density. Likewise, we can use the density of the Earth to estimate *its* composition.

We find the Earth's density by dividing its mass by its volume. Its mass is $6.0 \times 10^{21}$ tons (metric), or $6.0 \times 10^{27}$ grams, and its radius is about 6400 kilometers, or $6.4 \times 10^{8}$ centimeters. To actually make the calculation, we divide the mass of the Earth by its volume, $\left(\frac{4}{3}\right)\pi R^{3}$, assuming it is a sphere. Thus, the density is

$M$ = Mass of the Earth
$R$ = Radius of the Earth
$\frac{4}{3}\pi R^{3}$ = Volume of a sphere

$$\frac{M}{\left(\frac{4}{3}\right)\pi R^{3}} = \frac{6.0 \times 10^{27} \text{ g}}{\left(\frac{4}{3}\right)\pi (6.4 \times 10^{8})^{3} \text{ cm}^{3}} = \frac{6.0}{\frac{4}{3}\pi \times 6.4^{3}} \times \frac{10^{27}}{10^{24}} \frac{\text{g}}{\text{cm}^{3}}$$
$$= 5.5 \text{ g/cm}^{3}$$

That is about twice the density of ordinary rock.

The density as defined on page 148 is really an *average density* over the whole planet. Because we can measure directly that the average density of surface rocks is much less than 5.5 grams per cubic centimeter, we can therefore infer that other parts of the Earth must have a much greater density than 5.5 grams per cubic centimeter. That by itself does not tell us what lies inside the Earth, but if we ask ourselves what substances are both dense and abundant in nature, we find that iron is a likely choice, as table 5.1 shows. We therefore deduce that the Earth has an iron core. But we can do better than merely deduce. We can test that hypothesis by taking advantage of one of nature's most violent phenomena: earthquakes.

## 5.2   THE EARTH'S INTERIOR

If we ask how the Earth's interior can be studied, your first reaction might be to say, "Why not drill a very deep hole and take a look or pull out samples?" Unfortunately, the deepest hole yet drilled in the Earth penetrates only 12 kilometers, a mere scratch when compared to the Earth's 6400-kilometer radius. If the Earth were an apple, the deepest holes yet drilled would not have broken the apple's skin. Thus, to study the Earth's interior, we rely on indirect means such as earthquake waves.

### Probing the Interior with Earthquake Waves

When earthquakes shake and shatter rock within the Earth, they generate **seismic waves** that travel outward from the location of the quake through the body of the Earth (fig. 5.4). Seismic waves are disturbances inside the Earth that slightly compress

## EXTENDING OUR REACH

### MEASURING THE EARTH'S MASS

We can measure the Earth's mass by comparing its gravitational attraction to that of a huge lead ball of known mass. In such an experiment, a sensitive laboratory weighing scale is carefully balanced with identical standard masses in its opposite pans. Then a huge lead ball is moved under one of the pans. The lead ball exerts a tiny extra gravitational force on that pan, making it dip slightly lower. A tiny extra mass is then added to the opposite pan to bring the scale back into balance. Using the mass required to balance the pans with the lead ball present and the equation for Newton's law of gravity, we can then calculate the relative mass of the Earth and the lead ball.

FIGURE 5.4
Seismic waves spread out through
the Earth from an earthquake.

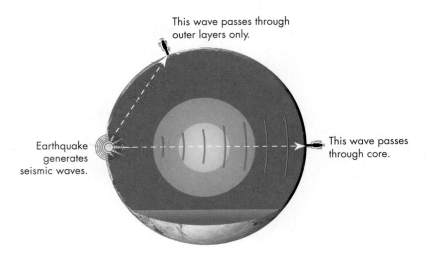

This wave passes through
outer layers only.

Earthquake
generates
seismic waves.

This wave passes
through core.

FIGURE 5.5
A sonogram allows a doctor to
"see" inside a patient. Here, a
developing fetus in the womb is
visible.

*S and P waves generated by earthquakes*

rock or cause it to vibrate up and down. The speed of the waves depends on the properties of the material through which they move. A wave's speed can be determined by
carefully timing its arrival at remote points of the world. From that speed, scientists
can deduce a picture of the Earth's interior along the path of that wave. Thus, seismic
waves allow us to "see" inside the Earth much as doctors use sound waves to "see"
inside our bodies.

To make a picture of your internal organs, sound waves are sent through your body
and are then picked up with a sensitive microphone. Because the sound travels at different speeds in bone, tissue, cartilage, and so forth, a medical technician can analyze
the signals with a computer to make a picture of your anatomy or of an unborn child
(fig. 5.5). You can use a similar, though obviously much cruder, technique to locate
wall studs by thumping areas of a plaster wall with your knuckle.

Seismic waves in the Earth are of two main types: S waves and P waves.* P waves
form as matter in one place pushes against adjacent matter—whether solid or

---

*P and S stand for "primary" and "secondary" and refer to their arrival time at a distant site. That is, the
primary waves arrive first.

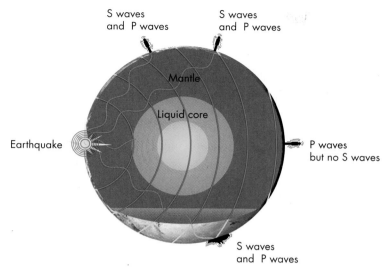

**FIGURE 5.6**
P and S waves move through the Earth, but the S waves cannot travel through the liquid core.

↻* P-type seismic waves have velocity of about 10 km-s$^{-1}$. About how long does it take such a wave to travel straight through the Earth from surface to core to the opposite surface?

ℚ* The time, $t$, for something moving at a speed $v$ to travel a distance D is given by $t = D/v$. The Earth's diameter is about 12,000 km. Therefore $t = 12{,}000$ km/10 km-s$^{-1}$ = 1200 seconds, or about 20 minutes.

*The differentiation of the Earth's core*

liquid—compressing it. They travel easily through both solids and liquids. By contrast, S waves form as matter "jerks" adjacent material up and down or from side to side, like a wriggle in a shaken rope. In a liquid, material easily slips past adjacent matter, preventing S waves from spreading. Thus, S waves can travel only through solids. Therefore, if a laboratory on the far side of the Earth from an earthquake detects P waves but no S waves, the seismic waves must have encountered a region of liquid on their way from the earthquake to the detecting station (fig. 5.6), an indication that the Earth has a liquid interior. Observations show precisely this effect, from which we infer that the Earth has a liquid core. More complicated analyses can then reveal the density of the material and give clues to its composition.

Seismic studies show that the Earth's interior has four distinct regions. The surface layer is a solid, low-density **crust** about 20 to 70 kilometers (12 to 43 miles) thick and composed of rocks that are mainly silicates. Beneath the crust is a region of hot, essentially solid rock called the **mantle.** This region is also composed of silicates, the most common of which is the mineral olivine. The mantle extends roughly halfway to the Earth's center and, despite being basically solid, is capable of slow flow when stressed, much the way a wax candle can be bent by a steady pressure.

Beneath the mantle is a region of dense liquid material, probably a mixture of iron, nickel, and perhaps sulfur, called the **liquid** (or **outer**) **core.** At the very center is a **solid** (or **inner**) **core,** probably also composed of iron and nickel. Figure 5.7 illustrates these different layers and their relative sizes.

You can see from this discussion that the Earth's interior structure is layered so that the dense heavier material (the iron and nickel) is at the center and the lower-density lighter materials (silicates) are near the surface in the crust and mantle. Scientists call this separation of materials by density **differentiation.**

**FIGURE 5.7**
An artist's view of the Earth's interior.

**FIGURE 5.8**

Melting ice cream "differentiates" as the dense chocolate chips sink to the bottom of the box. So too, melting has made much of the Earth's iron sink to its core.

Differentiation can occur if a mixture of high- and low-density material melts, allowing the dense substances to sink and the lighter ones to rise. You have seen differentiation at work if you have ever had the misfortune to have a half gallon of mint chocolate chip ice cream melt. When you open the box, you find all the chips have sunk to the bottom and the air in the ice cream has risen to the top as foam. Because the Earth is differentiated, we can infer that it must have been almost entirely melted at some time in the past (fig. 5.8).

You may be puzzled as to why there is a solid core inside a liquid core at the Earth's center. If it is hot enough to melt part of the interior of the Earth, why is the very center not liquid as well? The solid core is not cooler, but rather it has a higher melting point because it is under greater pressure. At very high pressures, a previously melted material may resolidify. You can understand why this happens in the following way. For a solid to form, the atoms composing it must be able to link up to their neighbors to form rigid bonds. Heating makes the atoms move faster and breaks the bonds between neighboring atoms. With the bonds broken, the solid has nothing to hold it together, and so it becomes liquid. However, if the material is highly compressed, the atoms may be forced so close together that, despite the high temperature, bonds to neighbors may hold and keep the substance solid.

The compression needed to solidify the Earth's inner core comes from the weight of the overlying material. The thousands of miles of rock above the Earth's deep interior generate an enormous pressure there. To help visualize that pressure, imagine what it would feel like to have a pile of cinderblocks a mile high put on your stomach. In the Earth's core, the huge pressure squeezes what would otherwise be molten iron into a solid.

## Heating of the Earth's Core

The Earth's interior is much hotter than its surface (a fact that figures in folklore and theology!). Anyone who has ever seen a volcano erupt can hardly doubt this. In fact, the rise in temperature as you move deeper into the Earth can be measured easily in deep mines where air conditioning must be used to create a tolerable working environment. Near the surface, the temperature rises about 2 K every 100 meters you descend. If this temperature increase continued at this same rate all the way to the center, the Earth's core would be a torrid 120,000 K. However, by measuring the amount of heat escaping from the deep interior and from laboratory studies of the properties of heated rock, geologists estimate that the temperature in the core of the Earth is "only" about 6500 K (hotter than the Sun's surface). What makes our planet's core so hot?

Scientists think that the Earth was born hot. According to this theory, the Earth formed from many smaller bodies drawn together by their mutual gravity. As each body hit the accumulating young Earth, the impact generated heat.* When the bombardment stopped, the Earth's surface cooled, but its interior has remained hot. That is, the Earth has behaved much like a baked potato taken from the oven, cooling on the outside but remaining hot inside because heat leaks only slowly from its interior to its surface.

According to another theory, even if the Earth were cold at its birth, it has heated up since its formation with the heat being supplied by small amounts of natural radio-activity in the Earth's interior. That is, the Earth generates heat much as a nuclear reactor does. All rock contains trace amounts of naturally occurring **radioactive elements** such as uranium. A radioactive element is one that breaks down into another element by ejecting a subatomic particle from its nucleus, a process called **radioactive decay.** Radioactive decay releases energy, generating heat. If that heat is created in a small piece of rock at the Earth's surface, it simply escapes into the surroundings. Thus, the rock's temperature does not increase. However, in the Earth's deep interior, the heat is trapped by the outer rocky layers, slowing its escape. The amount of heat lost depends on the surface area, but the amount of heat contained depends on the volume. Because a smaller body has proportionately a much larger surface area compared to its volume than a larger body, the smaller body cools quickly, but the larger one remains hot (fig. 5.9). Thus, with the heat trapped, the temperature of the Earth's interior gradually rises and the rock eventually melts.

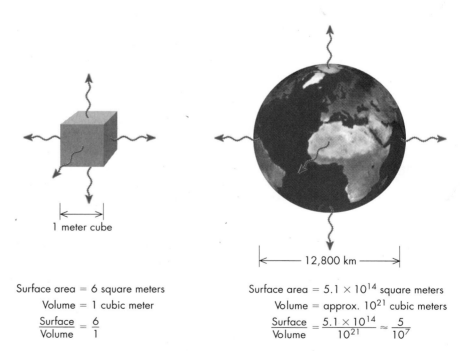

**FIGURE 5.9**

Heat readily escapes from small rocks but is retained in larger bodies.

1 meter cube

Surface area = 6 square meters

Volume = 1 cubic meter

$$\frac{Surface}{Volume} = \frac{6}{1}$$

12,800 km

Surface area = $5.1 \times 10^{14}$ square meters

Volume = approx. $10^{21}$ cubic meters

$$\frac{Surface}{Volume} = \frac{5.1 \times 10^{14}}{10^{21}} \approx \frac{5}{10^7}$$

*You can demonstrate that impact generates heat by hitting a small piece of metal repeatedly with a hammer and then feeling the metal. It will be warmer than it was before you began hitting it. For the forming Earth, the impacting objects act as the hammer, and gravity is the force that drives them onto the young Earth. Technically, therefore, impact heating is release of gravitational potential energy.

Scientists are unsure which of the above heating mechanisms is more important.* In fact, many scientists now think that the most important effect of radioactive heating is that the extra heat it supplies simply slows our planet's cooling. But regardless of its exact role in heating the Earth, radioactivity in rocks is important because it gives scientists a powerful tool for measuring our planet's age.

## 5.3 THE AGE OF THE EARTH

One can find the age of a rock sample by measuring the amount of radioactive material it contains. As the rock ages, its radioactive atoms decay into so-called **daughter atoms.** For example, uranium decays into lead, and radioactive potassium decays into calcium and the gas argon. The more daughter atoms a rock contains relative to the original radioactive atoms, the older the rock is. For instance, suppose we have a rock sample containing both potassium and argon trapped within the rock crystals. To apply the method, we assume that when the rock formed from molten material, any argon—being a gas—in it escaped. Suppose the sample happened to contain a million atoms of radioactive potassium and no argon when it solidified. Laboratory studies of radioactive potassium show that it decays into calcium and argon at a constant rate such that at any time, half the remaining potassium changes into these elements in 1.28 billion years.[†] Only about 10% of the decayed atoms turn into argon. The other 90% turn into calcium. Neither the argon nor the calcium are radioactive. Although argon is thus the rarer of the decay products, its source is easier to identify, and so we will ignore the calcium from now on in this discussion. The argon is trapped within the rock unless the rock melts, and so at the end of 1.28 billion years, the sample will contain half a million potassium atoms. The other 500,000 potassium atoms have decayed into 450,000 calcium atoms plus 50,000 argon atoms (10% of the 500,000). The argon/potassium ratio at this point will be 50,000/500,000 = 0.1, as shown in table 5.2. In another 1.28 billion years, half the surviving potassium will decay, leaving 250,000 potassium atoms and creating another 25,000 argon atoms (again, 10% of the decays). The argon/potassium ratio will now be (25,000 + 50,000)/250,000 = 0.3. The ratio of argon to potassium in the rock therefore changes with time from 0, when it solidified, to 0.1, after 1.28 billion years, and so on. This ratio gives us the age of the rock sample, as table 5.2 and figure 5.10 show.

From such studies, scientists have found that the oldest rocks on the Earth have an age of about 4 billion years. These ancient rocks are found in such diverse places

### TABLE 5.2
#### Radioactive Decay and Age of Rock

| Number of Radioactive Atoms (Potassium) | Number Decayed (Calcium + Argon) | Argon/ Potassium Ratio | Time (billion years) |
|---|---|---|---|
| 1,000,000 | 0 | 0.0 | 0 |
| 500,000 | 500,000 = 450,000 + 50,000 | 0.1 | 1.28 |
| 250,000 | 750,000 = 675,000 + 75,000 | 0.3 | 2.56 |
| 125,000 | 875,000 = 787,500 + 87,500 | 0.7 | 3.84 |
| 62,500 | 937,500 = 843,750 + 93,750 | 1.5 | 5.12 |

*Some recent proposals suggest that chemical changes during the formation of Earth's core may also have been important heat sources.

[†]Technically, the time it takes for half of a radioactive element's atoms to decay is called its half-life. Thus, 1.28 billion years is the half-life of this form of potassium.

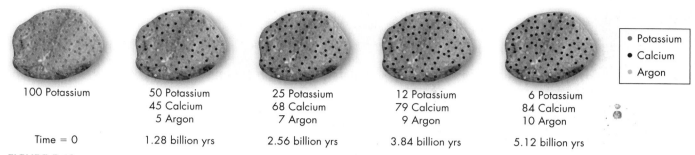

**FIGURE 5.10**

The amount of argon compared to potassium in a sample of rock gives information about the rock's age.

as northern Canada, southern Africa, and Australia. Thus, the Earth must be at least 4 billion years old.

However, scientists believe that the Earth is even older, because rock samples from other Solar System bodies, such as the Moon, have ages of a little more than 4.5 billion years. Moreover, samples of some material from asteroid fragments (fallen to Earth as meteorites) are nearly 4.6 billion years old, and, although we have no samples from it, the Sun also appears to be about this old, as we will see in chapter 13. All of this evidence suggests a common age for the Solar System of 4.6 billion years, an age that we will also take to be the Earth's. Why, then, are there no rocks this old on Earth? They were probably destroyed by processes we will discuss in the next section.

An age of 4.6 billion years is immense. To illustrate, if those billions of years were compressed into a single year, all of human existence would be a mere 3 hours, all human recorded history would have happened during the last minute of the year, and a human life span would be less than 1 second.

The brevity of human life compared to the vast age of the Earth prevents us from seeing how dynamic our planet is. Mountains and seas appear to us permanent and unchanging, but even they change over the vast epochs Earth has existed. Such changes have their ultimate cause in the heat slowly flowing from Earth's interior, a flow that creates motions in the Earth's interior and crust.

## 5.4    MOTIONS IN THE EARTH'S INTERIOR

Heat in the Earth's interior, whether left over from our planet's birth or generated by radioactive decay, creates movement of the material inside the Earth. In fact, heating often causes motion, as you can see by watching a pan of soup on a hot stove. If you look into the pan as it heats, you will see some of the soup, usually right over the burner, slowly rising from the bottom to the top, while some will be sinking again (fig. 5.11B). Such circulating movement of a heated liquid or gas is called **convection.**

### Convection in the Earth's Interior

Convection occurs because heated matter expands and becomes slightly less dense than the cooler material around it. Being less dense, it rises, the principle on which a hot-air balloon operates. As the hotter material flows upward, it carries heat along with it. Thus, convection not only causes motion but also carries heat.

Convective motions in a pot of soup are easy to see—here a lima bean rises; there a noodle sinks. Such motions are less obvious in the Earth. Our planet's crust and mantle are not bubbling and heaving like the soup; rather, they are solid rock. Nevertheless, when rock is heated, it too may develop convective motions, though they are very, very slow and therefore difficult to observe. Despite its slowness, the results of convective

**FIGURE 5.11**
Examples of convection: (A) In our atmosphere, puffy cumulus clouds form when the Sun heats the ground and warms the air so that it rises. (B) You can see rising and sinking motions in a pan of heated soup. (C) An artist's view of convection in the Earth's interior.

**FIGURE 5.12**
(A) Rifting may occur where rising material in the mantle nears a planet's surface. (B) Subduction builds mountains where material sinks back toward the interior of the Earth.

motion are evident around us. They create such diverse phenomena as earthquakes, volcanoes, the Earth's magnetic field, and perhaps even the atmosphere itself.

## Plate Tectonics

Deep within the Earth, hot molten material rises in great, slow plumes. When such a plume nears the crust, it spreads out and flows parallel to the surface below the crust (fig. 5.11C). There, the hot material drags the surface layers, shifting and stretching the crust in a process called **plate tectonics.** In some places, the stretching breaks the crust apart in a process called **rifting** (fig. 5.12A).* Gradually, the hot material loses its heat to the outside and then cools and sinks downward again to be rewarmed and rise once more.

---

*The Mid-Atlantic Ridge (see the Extending Our Reach box) is an excellent example of rifting.

**FIGURE 5.13**
Locations of earthquakes on Earth identify many plate boundaries.

Where cool material sinks, it may drag pieces of continental crust together, buckling them upward where they collide to form mountain ranges. Sometimes one piece of crust may slip beneath another and be pushed under it in a process called **subduction** (fig. 5.12B). Rifting and subduction are the dominant forces that sculpt the landscape of our planet and contribute to Earth's uniqueness in the Solar System. These motions also trigger earthquakes and cause volcanoes. For example, friction may temporarily stop two plates from sliding past each other, causing them to stick. Pressure may then build until the rock breaks, freeing the stuck plates and generating a sudden lurch in the crust, an earthquake. A map of the Earth showing the locations of such quakes (fig. 5.13) helps us see the shapes and locations of the plates.

The shifting of large blocks of the Earth's surface used to be called *continental drift,* but contemporary geologists prefer the term plate tectonics. The term *plate* is used because the pieces of the Earth's crust that move are very thin (perhaps only 50 kilometers) but many thousands of kilometers across. Some of the more important plates are those that form the continents, such as the North American and African plates, as seen in figure 5.14. The ocean bottoms also are composed of plates, such as the Pacific and Caribbean plates.

Plate motion is a little like the movement of the crust on bubbling oatmeal. The crust breaks apart where hot oatmeal bubbles up and collects where the slightly cooler oatmeal sinks.

The development of plate tectonic theory is an interesting example of how science works. As early as 1596, Abraham Ortelius, a geographer from Antwerp, in what is now Belgium, noticed that maps depicting the new discoveries of the period showed that the coastlines of South America and Africa approximately matched like two pieces of a giant jigsaw puzzle. In 1858, a French scientist, Antonio Snider-Pellegrini, also remarked how similar the coastlines were and noted that fossils found at matching locales on both sides of the Atlantic were also very alike. He conjectured that the continents had broken apart, creating the Atlantic Ocean in the opening rift between, but apart from the similarities in fossils, he offered little in the way of supporting evidence

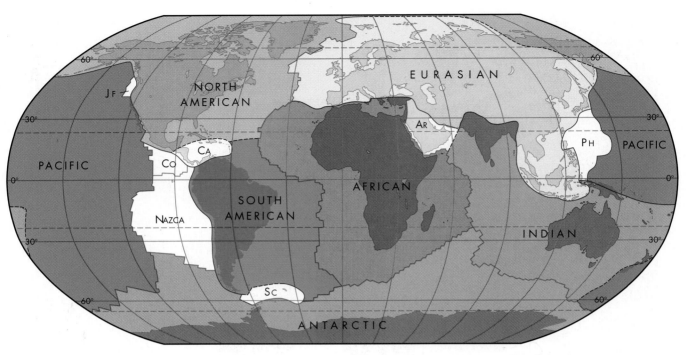

**FIGURE 5.14**
Map of the Earth, showing its plates. Smaller plates include the Cocos (Co), Caribbean (Ca), Juan de Fuca (Jf), Arabia (Ar), Philippines (Ph), and Scotia (Sc).

⟳★ Look carefully at the southern point where the Pacific and the North American plates meet. What famous geologically active area lies there? Given that the plates are sliding past each other at that location, how might you explain that activity?

Ⓐ★ The San Andreas fault in California.

*Plate motion over time*

for his idea. Similarly, in 1910, the American geologist F. B. Taylor published a paper proposing that South America and Africa had once been joined, but he too offered only slight evidence supporting his hypothesis. The first significant evidence came in 1912, when the German meteorologist Alfred Wegener published a paper called *The Origin of the Continents,* in which he developed the modern theory.

Wegener proposed that all the continents were originally assembled in a single supercontinent, which he called Pangea (literally "all-Earth"). For reasons that are still obscure, Pangea began to split into smaller plates that became the familiar continents of today (fig. 5.15), taking about 250 million years for the plates to move into their present locations.

Although he had amassed fossil and geological evidence to support his theory, Wegener's ideas were not well received at first. In fact, there are still problems in understanding what drives the motions and causes a new rift to develop. Nevertheless, over the last several decades, the cumulative effect of many separate pieces of evidence that support the model of plate tectonics have made the theory almost universally accepted.

Thus, our world is literally changing beneath our feet, growing new crust at mid-oceanic ridges and devouring it at subduction zones. But this devoured rock is not lost. As it is carried downward, it is heated, causing it to rise again toward the surface. Such molten, rising rock creates volcanic eruptions. Volcanic eruptions do more than just spew out lava, however. They also eject enormous quantities of gases, which over billions of years in the past may have given birth to some part of our atmosphere.

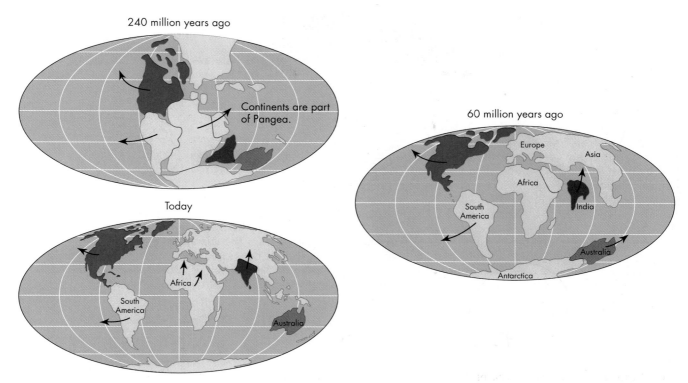

240 million years ago

Continents are part of Pangea.

Today

60 million years ago

Europe   Asia

Africa

South America   India

Australia

Antarctica

South America

Africa

Australia

**FIGURE 5.15**
Breakup of Pangea and the Earth today. Notice the close match of the African and South American coastlines.

## 5.5   THE EARTH'S ATMOSPHERE

Surrounding the solid body of the Earth is a veil of gases that constitutes our atmosphere. Most planets in the Solar System have an atmosphere, but the Earth's has many unique features. The Earth's atmosphere is of interest not just for its unique properties but also for what it can tell us about atmospheres in general.

INTERACTIVE

*Planetary variations*

### Composition of the Atmosphere

One of the most striking differences between the atmosphere of the Earth and that of other planets is its composition. For example, the atmospheres of Mars and Venus are nearly completely carbon dioxide, while the atmospheres of Jupiter and Saturn are mostly hydrogen and helium. On the other hand, the atmosphere of the Earth is primarily a mixture of the two gases nitrogen and oxygen. Nitrogen molecules make up about 78% of the gas and oxygen about 21%. The remaining 1% includes carbon dioxide, ozone, and water, gases crucial for protecting us and making life possible.

It may surprise you to see water listed as a gas. But even at low temperatures liquid water evaporates into individual water molecules that mingle with the other gases in our atmosphere. We call such free water molecules *water vapor*. Table 5.3 lists the most important gases in our atmosphere.

ANIMATION

*The origin of Earth's atmosphere by volcanoes, comet impacts, and planetesimal collision*

### Origin of the Atmosphere

Astronomers have proposed several theories to explain how our atmosphere formed. According to one theory, the gases of our atmosphere were originally trapped inside the

# EXTENDING OUR REACH

## MEASURING THE MOTION OF PLATES ACROSS TIME

Geologists have at least two ways to deduce that the Earth's plates move. Probably the most persuasive evidence comes from measurements of the age of rocks across the bottom of the Atlantic Ocean. Running north–south in a zigzag right down the middle of the Atlantic Ocean is a huge underwater mountain range called the **Mid-Atlantic Ridge**. The ridge originates from hot material rising from the Earth's interior to its surface (Box fig. 5.1).

The age of rock on the ocean floor increases with distance from the ridge on both sides. New crust is created at the ridge and spreads away from it like paper peeling off a roll. One can calculate the speed of the spreading motion by dividing a rock's distance from the ridge by its age. For example, if rocks 100 kilometers from the ridge are 10 million years old, the plates must have shifted at an average speed of 100 kilometers per 10 million years, or 1 centimeter per year, a fairly typical plate velocity.

Geologists can also measure plate motion directly by using radio signals from the orbiting satellites of the global positioning system (GPS). With specialized instruments similar to those used by airline pilots, hikers, and boaters—

but far more precise—they can measure the distance between continents to an accuracy of millimeters and can "see" the separation grow from one year to the next. Moreover, from the change in the separation, geologists can calculate how rapidly the continents are moving. Such calculations agree well with the speed determined when we divide the width of the Atlantic by its age.

"Stripes" represent rocks of different ages.

Older rocks  ←  Young rocks  →  Older rocks

Mid-Atlantic Ridge

10   5   0   5   10
Age in millions of years

0        100 km

**BOX FIGURE 5.1**
Artist's view of the Mid-Atlantic Ridge (from *World Ocean Floor* by Bruce C. Heezen and Marie Tharp, 1977) and the increasing age of rocks away from it.
(Copyright by Marie Tharp 1977/2003. Reproduced by permission of Marie Tharp Oceanographic Cartographer, One Washington Ave., South Nyack, New York 10960.)

solid material that eventually became the Earth. When that material was heated—either by volcanic activity (fig. 5.16A) or by the violent impact of asteroids hitting the surface of the young Earth (fig. 5.16B)—the gases escaped and formed our atmosphere.

Recently, some astronomers have proposed a very different explanation to account for our atmosphere. According to this theory, the gases were not originally part of the Earth but were brought here by comets.

| TABLE 5.3 | |
| --- | --- |
| **Atmospheric Gases** | |
| **Gas** | **% of Molecules by Number in Dry Air** |
| Nitrogen ($N_2$) | 78.08 |
| Oxygen ($O_2$) | 20.95 |
| Argon (Ar) | 0.93 |
| Water ($H_2O$) | Variable—typically between about 0.1 and 3.0 |
| Carbon dioxide ($CO_2$) | 0.03 |
| Trace gases (less than 100 parts per million) Neon, helium, ozone, krypton, hydrogen, methane, carbon monoxide, and many pollutants both natural and human-made. | |

A

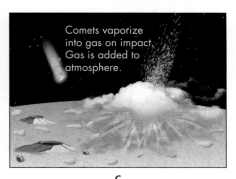

B                                                C

**FIGURE 5.16**

Sources of our atmosphere. (A) Volcanic gas vent today. Gas from ancient eruptions built some of our atmosphere. (B) Planetesimals collide with young Earth and release gas—another source of our atmosphere. (C) Comets striking young Earth and vaporizing. The released gases also contributed to our atmosphere.

As we will see in chapter 10, comets are made mostly of a mixture of frozen water and gases. When a comet strikes the Earth, the impact melts the ices and vaporizes the frozen gas. Given a large enough number of impacts, comets could have delivered enough gas to form the atmosphere (fig. 5.16C). We know from the collision of Comet Shoemaker-Levy 9 with Jupiter in July 1994 that comets collide with planets even today. But such collisions were almost certainly far more common billions of years ago, when the Earth was young, because at that remote time, the Solar System was full of smaller objects—the pieces from which the planets themselves grew.

In both these theories, the early atmosphere had a very different composition than the air we breathe today. For example, our planet's ancient atmosphere probably contained far more methane ($CH_4$) and ammonia ($NH_3$) than it does now. Although these gases are still abundant in the giant planets such as Jupiter and Saturn, they have all but disappeared from our atmosphere, which is fortunate, because both methane and ammonia are poisonous.

What has rid Earth of these noxious gases? Astronomers think that sunlight is responsible. Solar ultraviolet radiation is intense enough at Earth's distance from the Sun to break the hydrogen atoms out of both methane and ammonia, leaving carbon and nitrogen atoms, respectively. The nitrogen and carbon remain behind, supplying at least some of the nitrogen in our atmosphere. The hydrogen, however, gradually escapes into space because Earth's gravity is too weak to hold it. Only huge planets such as Jupiter and Saturn have strong enough gravities to retain their hydrogen and thus preserve the large amounts of methane and ammonia we see there today.

Which of these theories of the origin of our atmosphere—delivered by comet or liberated from Earth's own material—is correct? Scientists have tried hard to test these very different theories. For example, they have studied whether volcanoes erupt enough gas to have supplied our atmosphere and whether the composition of these gases can explain the mix of molecules in the air around us.

It requires a certain nonchalance to walk up to the lip of a bubbling volcano and hold a collecting tube over the edge to sample the foul-smelling exhalation, but when geologists make such dangerous tests, they find that nitrogen, water, and carbon dioxide are added to the atmosphere even now by volcanic eruptions. Thus, the theory can account for several of the gases in our atmosphere. The theory requires, however, a second important test: Can it account for the amount of these gases we see today? Again the answer is yes, if we assume that eruptions have been about as frequent in the past as they are now and that they ejected comparable amounts of gas.

Thus, these tests confirm that the gases of our atmosphere might have been released by heating the material from which our planet formed.

Testing the comet delivery theory is difficult because of our lack of precise knowledge about what frozen gases and ices comets contain. The evidence currently available, however, confirms that comet impacts could account for at least part of our planet's atmospheric gases. Thus, scientists remain divided about which theory is correct. Perhaps, as with so many differences of opinion, each side is partly right and some gases came from comets while others came from volcanoes and the heating of rocks during our planet's birth.

Neither of the above hypotheses for the origin of our atmosphere—volcanic exhalations or comet impacts—can account for the large amount of oxygen in our atmosphere. Where, therefore, did that vital ingredient originate? Chemical analysis of ancient rocks, particularly those rich in iron compounds that react with oxygen, shows that our atmosphere once contained much less oxygen than it does today. In fact, over the past 3 billion years, the amount of oxygen in our atmosphere has steadily increased, a rise paralleled by the spread of plant life across our planet. Most scientists therefore agree that the bulk of the free oxygen, which we breathe, was created from $H_2O$ and $CO_2$ by photosynthesis of ancient plants. This intimate connection between life and the environment of our planet is a fact that we ignore at our peril.

Plants have created most of our oxygen by photosynthesis, but not all. Some has come from water molecules split by solar ultraviolet radiation into hydrogen and oxygen. The lighter hydrogen slowly drifts to the top of the atmosphere and escapes, leaving oxygen behind. This mechanism for adding oxygen to the atmosphere was probably the dominant source of that gas in the early history of the Earth.

## The Ozone Layer

The oxygen in our atmosphere not only is important to us for breathing, but it also forms a vitally protective blanket, shielding us from harsh solar ultraviolet radiation. Some of that shielding is provided by $O_2$ (the normal form of oxygen), but most of it comes from another molecular form of oxygen, $O_3$, or **ozone.**

Most of the ozone in our atmosphere is located in the ozone layer* at an altitude of about 25 kilometers (80,000 feet). Ozone is formed because at these upper levels solar ultraviolet radiation is intense enough to split $O_2$ into individual oxygen atoms.[†] The splitting occurs because the ultraviolet radiation makes the molecule vibrate so energetically that it flies apart. These individual oxygen atoms then combine with other $O_2$ molecules to form $O_3$.

Ozone is important because it is a strong absorber of ultraviolet radiation; without the ozone layer, solar ultraviolet radiation would pour into the lower atmosphere. The short wavelength (and therefore the high energy) of the radiation would damage living organisms. Without the protective ozone layer, you would get a severe sunburn on exposed skin simply by stepping outside. In fact, it is doubtful that life could exist on the Earth's surface without the ozone layer to shield us. For this reason, governments around the world are cooperating to limit the use of chemicals that might mix into the upper atmosphere and chemically combine with the ozone layer and destroy it.

Ozone is not the only gas that absorbs radiation in our atmosphere. Carbon dioxide and water are also strong absorbers, but they absorb infrared rather than ultraviolet radiation. Therefore, they play a crucial role in regulating our planet's temperature through the greenhouse effect.

## The Greenhouse Effect

The transparency of the Earth's atmosphere to visible radiation allows sunlight to enter the atmosphere and reach the surface, where it is converted to heat. The surface radiates infrared energy, but the opacity of the atmosphere (its blocking power) at infrared wavelengths reduces the heat loss and makes the surface warmer than it would be if the infrared energy could escape freely, a phenomenon shown in figure 5.17 and known as the **greenhouse effect.**

You can get some indication of how effectively water vapor traps heat by noticing how the temperature drops dramatically at night in desert regions or on very clear nights. All gardeners know that it is clear nights (with no clouds and little water vapor) that are most likely to have frost. On cloudy nights, heat is retained.

It is important to recognize that the greenhouse effect does not generate heat; rather, it limits the heat loss to space. The greenhouse effect therefore warms the Earth the same way a blanket warms you. The blanket doesn't make you generate more heat; it simply slows down the loss of heat already there. Likewise, the water and carbon dioxide do not create heat of their own; they simply slow down the loss of heat from the ground by absorbing the infrared energy. Eventually they re-emit it, but more is re-emitted toward the ground than is lost into space. That extra infrared energy reradiated to the ground helps keep it warm at night.

---

*Ozone is, in fact, found throughout our atmosphere, but its strongest concentration is at 25 km.
[†]Recall that ultraviolet radiation, having shorter wavelengths, is more energetic than visible light.

**FIGURE 5.17**

The greenhouse effect. Radiation at visible wavelengths passes freely through the atmosphere and is absorbed at the ground. The ground heats up and emits infrared radiation. Atmospheric gases absorb the infrared radiation and warm the atmosphere, which in turn warms the ground.

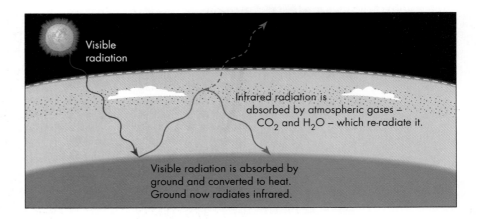

Q★ If you look at the red and blue light waves in this figure, you will notice that the red ones are drawn with longer "wiggles." Why might this have been done?

A★ Red light has a longer wavelength than blue light.

We can see how important our atmosphere is as a heat blanket by comparing the temperature of the Earth and the Moon. Although they are the same distance from the Sun, the average temperature of the Earth is much higher than the average temperature of the Moon. When averaged over seasons and latitude, the Earth's average temperature is 59° F (15° C), whereas the Moon's average temperature is a frigid −4° F (−20° C).

Ironically, the heat trapping of an atmosphere by the blockage of infrared energy is not what makes an actual greenhouse warm. Air inside a greenhouse is warmer than air outside because the greenhouse confines the air within it, preventing it from rising and cooling the way outside air can. Accordingly, some scientists prefer the term *atmosphere effect* rather than greenhouse effect.

## Structure of the Atmosphere

Our atmosphere extends from the ground to an elevation of hundreds of kilometers, but at the highest altitudes, the air is extremely tenuous. In fact, the density of the atmospheric gases decreases steadily with height. Gases near the ground are compressed by the weight of gases above them. Thus, the atmosphere is a little like a tremendous pile of pillows. The pillow at the bottom is squashed by the weight of all those above it. Likewise, a block of air near sea level is more compressed and therefore has a greater density than a block of air near the top of the atmosphere. That is why it is so difficult to breathe at 30,000 feet (~10,000 m) where the air is far less dense.* Because of compression by the overlying layers, roughly half the mass of the atmosphere is within 4 kilometers (13,000 feet) of the Earth's surface and a jet aircraft at a typical cruising altitude is flying above three-fourths of the atmosphere's bulk.

Although traces of atmosphere can be detected by satellites above 100 km, the air there is extremely thin and gradually merges with the near vacuum of interplanetary space. Nevertheless, this very rarified gas has an important cumulative effect on spacecraft in low orbits, slowing them down and gradually causing them to spiral into the denser lower regions where they burn up. But, although these tenuous gases may influence the motion of spacecraft, they may themselves be influenced by magnetic forces of the planet.

---

*One cubic centimeter (a volume roughly the size of the end of your little finger) of the air around you contains about $10^{19}$ molecules. When you take even a tiny sniff, you inhale a number of molecules roughly comparable to the number of grains of sand in a pile the size of the Astrodome.

## 5.6   THE EARTH'S MAGNETIC FIELD

Many astronomical bodies have magnetic properties, and the Earth is no exception. The English natural philosopher William Gilbert (1540–1603) was the first to appreciate that the Earth acted like a magnet, though the ancient Chinese had used the Earth's magnetism in their invention of the compass many hundreds of years earlier.

Magnetic forces are communicated by what is called a **magnetic field.** Although some forces are transmitted directly from one body to another (for example, when two billiard balls collide), other forces, such as gravity, need no such direct physical link.

Magnetic fields are often depicted by a diagram showing **magnetic lines of force.** Each line represents the direction that a tiny compass would point in response to the field. The concentration of lines indicates the field's strength, with more lines implying a stronger field. For example, the field lines of an ordinary toy magnet emanate from one end of the magnet, loop out into the space around it, and return to the other end. The Earth's magnetic field has a similar shape, as represented in figure 5.18.

Magnetic fields have an important property called **polarity.** Polarity gives field lines a direction, and so they always start at a north magnetic pole and end at a south magnetic pole. Thus, all magnets must have both a north pole and a south pole. The existence of north and south poles allows magnets to either attract or repel. Two north poles or two south poles repel each other, but a north and a south pole attract. A compass works on this principle. Its needle is a magnet, and its north pole is attracted to the Earth's magnetic south pole, and its south pole is attracted to the Earth's magnetic north pole.

An important question about any field is, How is it generated? Gravitational fields are generated by masses. Magnetic fields are generated by electric currents, either large-scale currents or currents on the scale of atoms. You can easily demonstrate the former by wrapping a few coils of insulated wire around an iron nail and attaching the wire ends to a battery. The nail will now act like a magnet and be able to pick up small pieces of iron and will deflect a compass needle. Microscopic electric currents on the atomic scale create the magnetism of toy magnets.

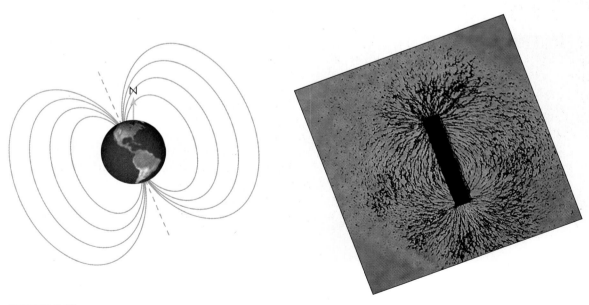

**FIGURE 5.18**
Schematic view of Earth's magnetic field lines and photograph of iron filings sprinkled on a toy magnet, revealing its magnetic field lines. "N" indicates the Earth's north pole.

## Origin of the Earth's Magnetic Field

The magnetic field of the Earth is generated by electric currents flowing in its molten iron core. Scientists are still unsure about how such currents originate but hypothesize that they originate from a combination of rotational motion and convection.* Studies of the magnetic fields of other Solar System bodies support this view. For example, bodies with weak or no magnetic fields, such as the Moon and Venus, are either too small to have a large convecting core or rotate very slowly. On the other hand, bodies with large magnetic fields, such as Jupiter and Saturn, rotate very rapidly and probably have very active cores.

The Earth's magnetic poles do not coincide with the poles determined by its rotation axis (true north and south). Therefore, a compass needle does not in general point true north but, instead, several degrees away to what is called "magnetic north." Both the position and strength of the Earth's magnetic poles change slightly but measurably, from year to year, even reversing their polarity about every 10,000 years. Thus, at some time in the distant future, a compass that now points north will point south. These changes in our planet's magnetic field probably come from irregular motions of the molten iron swirling in the Earth's core.

## Magnetic Effects in the Upper Atmosphere

The Earth's magnetic field does more than make a compass work. For example, it partially screens us from electrically charged particles emitted by the Sun. Such particles are often energetic enough to damage living cells and are therefore potentially harmful to us. The magnetic field protects us because when charged particles encounter it, they are deflected by its magnetic force into a spiraling motion around the field lines (fig. 5.19). This slows the particles streaming from the Sun, causing them to flow around the Earth much as water in a stream is diverted by a rock. Thus, we are spared the full impact of the Sun's charged particles.

As the solar particles stream past the Earth, they generate electric currents in our upper atmosphere. These currents, circulating around the magnetic poles, force electrons along the magnetic field lines. The moving electrons spiral around the field lines, colliding with molecules of nitrogen and oxygen. Such collisions excite atmospheric

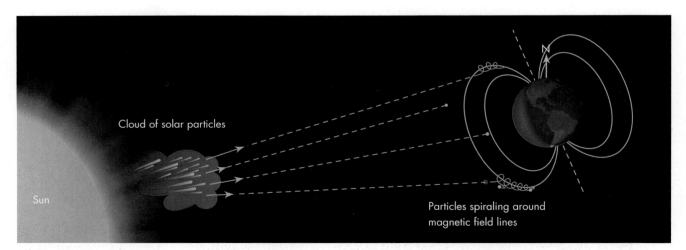

**FIGURE 5.19**
Electrically charged particles from the Sun spiral in the Earth's magnetic field. "N" indicates the Earth's north pole.

---

*Astronomers call this process a magnetic dynamo.

gases, lifting their electrons to upper orbits. As the electrons drop back to lower orbits, they emit the lovely light we see as the **aurora** (fig. 5.20). The exact process by which the aurora forms is still controversial, but there is no doubt that its beautiful streamers are shaped by the Earth's magnetic field, much as sprinkled iron filings outline the field of an ordinary toy magnet (fig. 5.18).

The magnetic field exerts an even stronger influence on particles at the top of the atmosphere in what is called the **magnetosphere.** Here, roughly two or three Earth radii out in space (fig. 5.21), the magnetic field controls the structure of the atmosphere, trapping charged particles in two doughnut-shaped rings called the **Van Allen radiation belts.**

The particles trapped in the Van Allen belts are energetic enough to penetrate spacecraft and be a hazard to space travelers, damaging their genetic material or other

A                                                            B

**FIGURE 5.20**
Photographs of an aurora from (A) the ground and (B) from space.

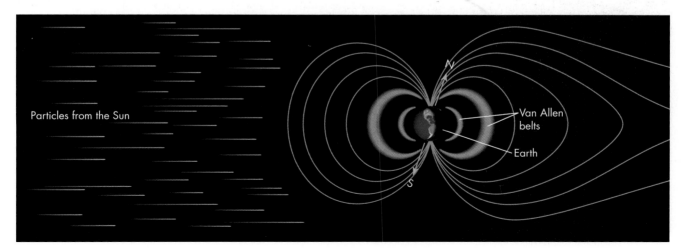

**FIGURE 5.21**
Artist's view of the Van Allen radiation belts (side view). "N" and "S" indicate the Earth's north and south poles.

tissue. Astronauts therefore try to avoid passing through the belts or to go through them as quickly as possible.

We described earlier how the combined motion of the molten material in the Earth's core and our planet's spin create its magnetic field. The Earth undergoes many other motions as well.

## 5.7    MOTIONS OF THE EARTH

Considering what happens to many of us on an amusement park ride, it is just as well that we are unaware of the Earth's many motions. Our planet spins on its axis, orbits the Sun, is dragged along by the Sun around the galaxy, and moves through the Universe with the Milky Way galaxy (fig. 5.22). We have already discussed how the Earth's rotational and orbital motions define the day and year, and cause seasons. But our planet's motions have other effects. For example, its spin strongly influences winds and ocean currents, and, over thousands of years, a slow "wobble" of its rotation axis plays a role in the onset of ice ages.

ANIMATION

*The Coriolis effect*

### Air and Ocean Circulation: The Coriolis Effect

If you sit with a friend on a rotating schoolyard merry-go-round and toss a ball back and forth, you will discover that the ball does not travel in the direction in which you aim but instead curves off to the side. Similarly, ocean and air currents sweeping across a spinning planet like Earth are deflected from their original direction of motion. This phenomenon is called the **Coriolis effect.**

The Coriolis effect, named for the French engineer who first studied it, alters the path of objects moving over a rotating body, such as the Earth, other planets, or stars. To understand why the Coriolis effect occurs, imagine standing at the North Pole and throwing a rock as far as you can toward the equator (fig. 5.23). As the rock arcs through the air, the Earth rotates under it. Thus, if you were aiming at a particular point on the equator, you will miss because the surface has turned beneath the rock's path, making the rock appear to have been pushed to

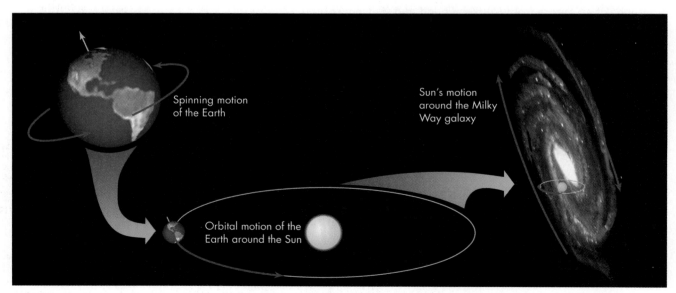

**FIGURE 5.22**
The Earth's many motions in space.

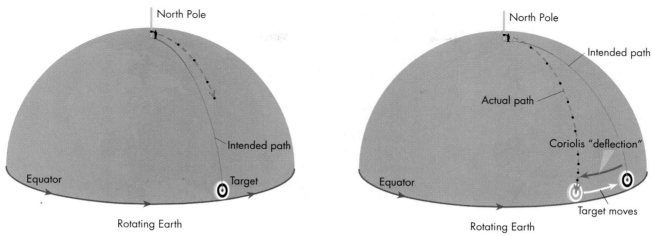

**FIGURE 5.23**
Coriolis effect on a rock thrown toward the equator from the North Pole.

**FIGURE 5.24**
Weather satellite pictures show clearly the spiral pattern of spinning air around a storm that results from the Coriolis effect.

the right.* Air, water, rockets, or anything moving across the rotating Earth in any direction are affected similarly.

You can clearly see the results of the Coriolis effect when you see weather satellite pictures of a large storm system. The Coriolis effect spins the air flowing into the storm, creating the spiral pattern seen in figure 5.24. Likewise, the Coriolis effect

*We must not be guilty of "hemispherism." The Coriolis effect deflects objects to the *left* in the Southern Hemisphere. There is no truth to the story that water spirals down the drain clockwise in the Northern Hemisphere and counterclockwise in the Southern Hemisphere—at least, unless you ensure that the water is initially absolutely motionless or use an ocean-size basin! The Coriolis effect is totally overwhelmed by the persistence of the motion from filling the basin unless you allow the water to come to a complete rest, a state that can take weeks to achieve.

deflects ocean currents, creating flows such as the Gulf Stream. It also creates atmospheric currents such as the trade winds.

The trade winds blow relatively steadily around our planet from east to west in two broad bands, one north and the other south of the equator. They form in response to the Sun's heating of equatorial air, which rises and expands, flowing away from the equator toward the poles. However, before this air can reach the poles, it cools and sinks toward the surface, where it flows back toward the equator. As that air approaches the equator, the Coriolis force deflects it into the pattern we observe in the trade winds.

The Coriolis force also establishes the direction of the **jet streams,** which are narrow bands of rapid, high-altitude winds. Jet streams are an important feature of the Earth's weather and are found on other planets as well. On rapidly rotating Jupiter, Saturn, and Neptune, the Coriolis effect is much stronger than on the Earth, creating extremely fast jet streams. The striking cloud bands we see on these planets are partly caused by this effect (fig. 5.25).

At one time, the Coriolis effect was of special interest because it is an indirect proof that the Earth rotates. Another such indirect proof is the Foucault pendulum, the huge swinging ball on a long wire that you may have seen at a science museum. (fig. 5.26). If you watch the pendulum for half an hour or so, you will notice that its swing changes direction. This can be made more evident by setting pegs in a circle around the pendulum; the pendulum will knock the pegs over, one by one. However, if you could watch the pendulum from space, its swing would not change direction. This is because the ball's inertia keeps its path fixed in space. It is the Earth (and therefore you and the building) that is turning beneath the swinging ball. Today, such indirect proofs are unnecessary: we can *see* the spin of our planet from pictures taken in space.

## Precession

As the Earth moves around the Sun over long periods of time, the direction in which its rotation axis points changes very slowly. This motion, similar to the wobble that occurs when a spinning coin or toy top begins to slow down, is called **precession.** If the Earth were perfectly spherical, precession would not occur. But the Earth's spin makes its equator bulge slightly, so the Sun and Moon exert an unbalanced gravitational attraction on our planet, twisting it slightly. That twisting makes the Earth's rotation axis slowly change direction, completing one swing in about 26,000 years (fig. 5.27). Currently, the North Pole points almost at the star Polaris. In about A.D. 14,000, the North Pole will point instead nearly at the bright star Vega. Thirteen thousand years later, the North Pole will again point nearly

**ANIMATION**

*The Foucault pendulum*

A                    B                                                          C

**FIGURE 5.25**
Cloud bands on Jupiter, Saturn, and Neptune, created in part by the Coriolis effect.

**FIGURE 5.26**
A Foucault pendulum.

Earth's rotation axis
swings to new position.

To Polaris          Toward Vega

North Pole now     North Pole
                   in approx.
                   A.D. 14,000

Earth's
orbit

Spinning and       Spinning top       Spinning and
precessing coin                       precessing top

**FIGURE 5.27**
Precession makes the Earth's rotation axis swing slowly in a circle.

at Polaris. "Looking Up at the Northern Circumpolar Constellations" shows part of this region of the sky, including the star Thuban in the constellation Draco (see page 172). Thuban is of interest because it was the pole star at the time the pyramids were built in ancient Egypt, and the passage into the Great Pyramid pointed at that star.

Precession is of minor importance in day-to-day life, but over long periods of time, it may cause climate changes. As noted earlier, we are currently nearest the Sun when it is winter in the Northern Hemisphere. In 13,000 years, the Earth will be farthest from the Sun in Northern Hemisphere winter. That will make winters slightly colder in the Northern Hemisphere and perhaps trigger a new ice age.

# LOOKING UP AT THE...
## NORTHERN
## CIRCUMPOLAR
## CONSTELLATIONS

There are 5 northern circumpolar constellations: Ursa Major (the Big Dipper), Ursa Minor (the Little Dipper), Cepheus (the King), Cassiopeia (the Queen), and Draco (the Dragon). They are visible year round after dark over most of the northern hemisphere.

**Pointer Stars**

**Big Dipper**

**M97 The Owl Nebula**

This nebula is ~2500 ly distant.

**~1.6 ly**

**Thuban**

This was the pole star when the pyramids were built in ancient Egypt.

**Draco**

**M52**

This is an open star cluster. Its distance is uncertain — perhaps 3000-5000 ly.

**~12 ly**

**Polaris (The North Star)**

**Little Dipper**

**M101**

This galaxy is ~27 million light years from us.

**~170,000 ly**

**NORTH**

**Draco**

**Cassiopeia**

**Cepheus**

**Delta Cephei**

This is a pulsating variable star.

### Cassiopeia in 3-D

To Earth

450 ly ⁕
100 ly ⁕
230 ly ⁕
600 ly ⁕
54 ly ⁕

1 light year (ly) = 6 trillion miles

## SUMMARY

The Earth is roughly spherical, and its radius is about 6400 kilometers (4000 miles). Trace amounts of radioactivity in rocks here and elsewhere in the Solar System reveal that the Earth formed about 4.6 billion years ago. Radioactive material also adds heat to its interior. The flow of heat to the Earth's surface stirs slow convective motions, which shift the Earth's crust (plate tectonics) creating mountains, volcanoes, ocean basins, and earthquakes.

The waves generated by earthquakes (seismic waves) allow us to study the Earth's interior. They show it is stratified into four distinct regions: a very thin crust of ordinary rock, a mantle of hot but essentially solid silicates, an outer core of liquid iron and nickel, and an inner core of solid iron and nickel.

Currents created by motions in the Earth's core generate the Earth's magnetic field. That field in turn affects the motion of charged particles in the upper atmosphere. Such particles may create auroral displays when they collide with oxygen and nitrogen in the upper atmosphere.

The nitrogen, carbon dioxide, and water of our atmosphere may have come from volcanic gases vented over the Earth's history. Alternatively, these atmospheric gases may be the evaporated remains of comets that hit the Earth in its infancy. Plant life has created the atmosphere's oxygen by photosynthesis.

Some atmospheric gases absorb radiation. High-altitude ozone absorbs ultraviolet radiation, thereby protecting us from its biologically harmful effects. Carbon dioxide and water vapor absorb infrared radiation, trapping heat radiated from the Earth's surface. By slowing heat loss from our planet into space, these gases create the greenhouse effect and make Earth slightly warmer than it would be if the infrared radiation could escape freely.

The Earth's spinning motion creates a Coriolis effect that deflects objects moving over its surface. The Coriolis effect makes large storm systems rotate and is essential for driving the circulation of the atmosphere and the oceans.

### **?What It Is.**

This is a photo of Earth and our Moon as seen from Mars by the *Mars Global Surveyor* spacecraft.

## QUESTIONS FOR REVIEW

1. Why is the Earth not perfectly round?
2. What are some of the more common elements composing the Earth?
3. What are two explanations that scientists offer for why the interior of the Earth is hot? How hot is it?
4. What is convection? What are some other examples of convection besides hot soup?
5. What is the relation between rising and sinking material in the Earth's interior and subduction and rifting?
6. What is precession? What are some of its possible consequences?
7. What is the Mid-Atlantic Ridge?
8. On what plate of the crust are you located? Which way is it taking you?
9. What is happening where your plate is smashing into another?
10. How do we know that the Earth has a liquid core?
11. Why is the inner core solid even though it is hotter than the outer liquid core?
12. What factors are thought to be responsible for the Earth's magnetic field?
13. How is the aurora related to the Earth's magnetic field?
14. If the Earth rotated more slowly, would you expect it to have as strong a magnetic field?

## THOUGHT QUESTIONS

1. Thinking about a stone thrown from the pole toward the equator, convince yourself that if the Earth rotated faster, the Coriolis effect would be larger.
2. As seen from space, the Earth rotates counterclockwise as seen from above the North Pole. Using the thrown-stone argument of Question 1, explain why the Coriolis effect deflects objects to the right of their motion in the Northern Hemisphere.
3. When you choose fruit at a supermarket, you may heft the fruit in your hand to test its weight. How does this tell you

whether the fruit is dried out inside? How is that similar to using mean density as an indicator of the composition of the Earth's interior?

4. Flicking your finger against your cheek makes a different sound from flicking it against your forehead. How is that similar to studying the interior of the Earth with seismic waves?

5. How does the eventual acceptance of the plate tectonic theory illustrate some aspects of the scientific method?

6. Astronomers are still uncertain about how the Earth's atmosphere formed. How does this illustrate the workings of the scientific method?

7. According to the *Guiness Book of Mountains and Mountaineering,* the summit of the volcano Chimborazo in Ecuador is the point on the Earth's surface farthest from the center. However, the book also states that the summit of Mount Everest is the highest point above sea level. Is something wrong here? Why?

## PROBLEMS

1. Given that Pluto's mass is $1.3 \times 10^{25}$ grams and its radius is $1.1 \times 10^8$ centimeters, what is its average density? Does this indicate that Pluto has a large iron core like the Earth? Why?

2. Seismic waves are partly reflected when they cross a boundary such as that between the mantle and the liquid core. Suppose that a P wave has a constant velocity of 8.0 km/sec. Suppose

further that 700 seconds (about 12 minutes) after an earthquake near the surface, a seismometer detects a reflected P wave. How far below the surface is the liquid core–mantle boundary? Compare your answer to the distance in figure 5.7. Hint: Remember this is an echo.

## TEST YOURSELF

1. Scientists think the Earth's core is composed mainly of
   (a) silicate rocks.
   (b) uranium.
   (c) lead.
   (d) sulfur.
   (e) iron.

2. Our knowledge of the composition of the Earth's core comes from
   (a) analysis of earthquake waves.
   (b) X-ray pictures taken with a powerful device in Russia.
   (c) samples obtained by drilling deep holes.
   (d) analysis of the material erupted from volcanoes.
   (e) Both (c) and (d).

3. The slow shifts of our planet's crust are believed to arise from
   (a) the gravitational force of the Moon pulling on the crust.
   (b) the gravitational force of the Sun pulling on our planet's crust.
   (c) the Earth's magnetic field drawing iron in crustal rocks toward the poles.
   (d) heat from the interior causing convective motion, which pushes on the crust.

   (e) the great weight of mountain ranges forcing the crust down and outward from their bases.

4. Scientists use radioactivity in rock samples to measure
   (a) the temperature in the Earth's core.
   (b) the depth of the oceans.
   (c) the Earth's age.
   (d) the composition of the mantle.
   (e) the composition of the inner core.

5. What evidence indicates that part of the Earth's interior is liquid?
   (a) With sensitive microphones, sloshing sounds can be heard.
   (b) We know the core is lead, and we know the core's temperature is far above lead's melting point.
   (c) Deep bore holes have brought up liquid from a depth of about 4000 kilometers.
   (d) No S-type seismic waves are detectable at some locations after an earthquake.
   (e) S-type waves are especially pronounced at all locations around the Earth after an earthquake.

## FURTHER EXPLORATIONS

Akasofu, Syun-lchi. "The Dynamic Aurora." *Scientific American* 260 (May 1989): 90.

Allègre, Claude J., and Stephen H. Schneider. "The Evolution of the Earth." *Scientific American* 271 (October 1994): 66.

Davis, Neil. *The Aurora Watcher's Handbook.* Fairbanks: University of Alaska Press, 1992.

Fernie, J. D. "The Shape of the Earth (Parts I–III)." *American Scientist* 79 (March/April 1991): 108; ibid., 79 (September/October 1991): 393; ibid., 80 (March/April 1992): 125.

Murphy, J. Brenden, and R. Damian Nance. "Mountain Belts and the Supercontinent Cycle." *Scientific American* 266 (April 1992): 84.

### Website

Please visit the *Explorations* website at **http://www.mhhe.com/arny** for additional online resources on these topics.

## VIDEO

*Planet Earth,* produced by WQED in association with the National Academy of Sciences and funded by the Annenberg/CPB Project, 1986.

## KEY TERMS

aurora, 167

convection, 155

Coriolis effect, 168

crust, 151

daughter atoms, 154

density, 148

differentiation, 151

greenhouse effect, 163

jet streams, 170

liquid or outer core, 151

magnetic field, 165

magnetic lines of force, 165

magnetosphere, 167

mantle, 151

Mid-Atlantic Ridge, 160

ozone, 163

plate tectonics, 156

polarity, 165

precession, 170

radioactive decay, 153

radioactive elements, 153

rifting, 156

seismic waves, 149

silicates, 147

solid or inner core, 151

subduction, 157

Van Allen radiation belts, 167

## PROJECT

Make a photocopy or tracing of a map of the world (fig. E2.5 showing world time zones will do). Cut out North and South America and place them against the western edge of Europe and Africa. Do they fit? Does this help you see what led early scientists to the idea that the continents shift?

# Keeping Time

From before recorded history, people have used events in the heavens to mark the passage of time. The day, the month, and the year were all originally defined in terms of obvious astronomical phenomena. The day was the time interval from sunrise to sunrise. The month was the interval from new moon to new moon. The year was the time it takes for the Sun to complete one circle of the zodiac. Astronomical events are not perfect time markers, however: even the day and year need to be defined with care if we are to have reliable clocks and calendars. Otherwise, we may end up with snow in "summer" and heat waves in "winter."

## LENGTH OF THE DAYLIGHT HOURS

Although each day lasts 24 hours, the number of hours of daylight, or the amount of time the Sun is above the horizon, changes drastically throughout the year unless you are very close to the equator. For example, in northern middle latitudes, including most of the United States, southern Canada, and Europe, summer has about 15 hours of daylight and only 9 hours of night. In the winter, the reverse is true.

Dalí, Salvador. *The Persistence of Memory [Persistence de la Mémoire]*. 1931. Oil on canvas, $9\frac{1}{2} \times 13$ inches (24.1 × 33.0 cm).

*The change in number of hours of daylight as seasons change*

This variation in the number of daylight hours is caused by the Earth's tilted rotation axis. Remember that as the Earth moves around the Sun, its rotation axis points in roughly a fixed direction. Thus, the Sun shines more directly on the Northern Hemisphere during its summer and less directly during its winter. The result (as you can see in fig. E2.1) is that only a small part of the Northern Hemisphere is unlit in the summer, but a large part is unlit in the winter. Thus, as the Earth's rotation carries us around, only a relatively few hours of a summer day are unlit, but a relatively large number of winter hours are dark. Figure E2.1 also shows that on the first day of spring and autumn (the equinoxes), the hemispheres are equally lit, so that day and night are of equal length everywhere on Earth.

If we change our perspective and look out from the Earth, we see that during the summer, the Sun's path is high in the sky, so that the Sun spends a larger portion of the day above the horizon. This gives us not only more heat (because the sunlight falls more directly on the ground) but also more hours of daylight. On the other hand, in winter the Sun's path across the sky is much shorter, giving us less heat (because the sunlight falls less directly on the ground) and fewer hours of light.

## THE DAY

The length of the day is set by the Earth's rotation speed on its axis. One day is defined as one rotation. However, we must be careful how we measure our planet's rotation. For example, we might use the time from one sunrise to the next to define a day. That, after all, is what sets the day–night cycle around which we structure our activities. However, we would soon discover that the time from sunrise to sunrise changes steadily throughout the year as a result of the seasonal change in the number of daylight hours. A better time marker might be the time it takes the Sun to move from its highest point in the sky on one day (what we technically call apparent noon) to its highest point in the sky on the next day—a time interval that we call the **solar day.**

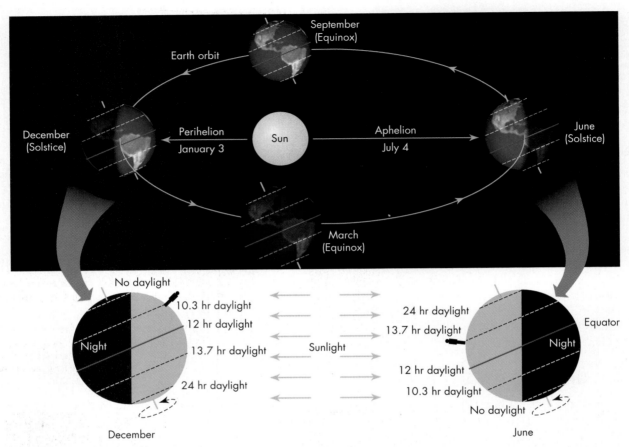

**FIGURE E2.1**
The tilt of the Earth affects the number of daylight hours. Locations near the equator always receive about 12 hours of daylight, but locations toward the poles have more hours of dark in winter than in summer. In fact, above latitudes 66.5°, the Sun never sets for part of the year and never rises for another part of the year (the midnight sun phenomena). At the equinoxes, all parts of the Earth receive the same number of hours of light and dark. Sizes and separation of the Earth and Sun are *not* to scale.

If we measure the length of the solar day, however, we will discover that it is, in general, *not* exactly 24 hours. Moreover, its length changes by several minutes over the course of the year. As we will discover later, this variation arises from the Earth's motion around the Sun. Thus, although the Sun's motion across the sky determines the day–night cycle, the Sun's motion is *not* a good reference for the actual time it takes our planet to complete one spin.

We can avoid most of this variation in the day's length if, instead of using the Sun, we use a star as our reference. For example, if we pick a star that lies exactly overhead at a given moment and measure the time it takes for that same star to return to exactly overhead, we will find that the time interval is not 24 hours but an essentially unchanging 23 hours and 56 minutes. This day length, measured with respect to the stars, we call a **sidereal day.**

Why do the solar and sidereal day differ in length? We can see the reason by looking at figure E2.2, where we measure the interval between successive apparent noons—a solar day. Let us imagine that at the same time we are watching the Sun, we can also watch a star and that we measure the time interval between its passages overhead, a sidereal day.

As we wait for the Sun and star to move back overhead, the Earth moves along its orbit. The distance the Earth moves in one day is so small compared with the star's distance that we see the star in essentially the same direction as on the previous day. However, we see the Sun in a measurably different direction, as figure E2.2 shows. The Earth must therefore rotate a bit more before the Sun is again overhead. That extra rotation, needed to compensate for the Earth's orbital motion, makes the solar day slightly longer than the sidereal day.

It is easy to figure out how much longer on the average the solar day must be. Because it takes us $365\frac{1}{4}$ days to orbit the Sun and because there are by definition 360 degrees in a circle, the Earth moves approximately 1 degree per day in its orbit around the Sun. That means that for the Sun to reach its noon position, the Earth must rotate approximately 1 degree past

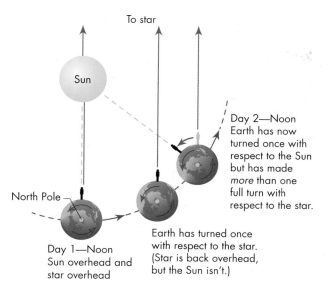

**FIGURE E2.2**
The length of the day measured with respect to the stars is not the same as the length measured with respect to the Sun. The Earth's orbital motion around the Sun makes it necessary for the Earth to rotate a tiny bit more before the Sun will be back overhead. (Motion is exaggerated for clarity.)

its position at the previous noon.* In 24 hours = 24 × 60 = 1440 minutes, the Earth rotates 360 degrees. Therefore, to rotate 1 degree takes 1440/360 minutes, or about 4 minutes. The solar day is therefore about 4 minutes (3 minutes 56 seconds, to be precise) longer than the sidereal day.

As mentioned earlier, the motion of the Earth around the Sun alters the length of the solar day. If you measure carefully the time interval from one apparent noon to the next, in general, it is not 24 hours but may differ slightly either way. This variation arises because the Earth's orbit is not circular, and therefore, our orbital velocity changes according to Kepler's second law.

The Earth moves along its orbit faster when it is near the Sun and slower when it is farther away. This means that it takes a little longer for the Earth to swing you around into the morning Sun (slightly lengthening the interval between successive noons) when the Earth is moving rapidly in its orbit than when it is moving slowly (fig. E2.3). Hence, the solar day is longer when we are near the Sun and shorter

---

*Another way of thinking about this is that the Sun is slowly moving eastward across the sky through the stars at the same time the Earth is rotating. Thus, in a given "day," the Earth must rotate a bit more to keep pace with the Sun than it would to keep pace with the stars.

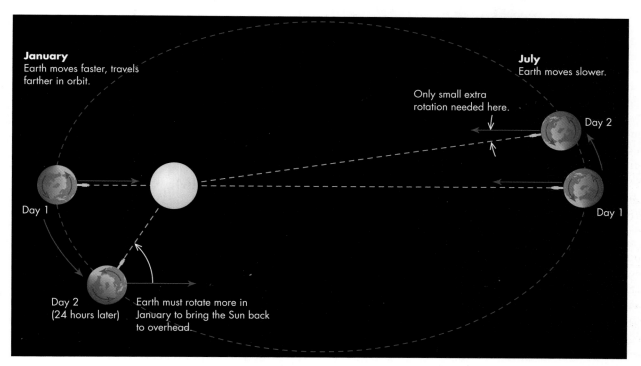

**FIGURE E2.3**
As the Earth moves around the Sun, its orbital speed changes as a result of Kepler's second law of motion. For example, the Earth moves faster in January when it is near the Sun than in July when it is far from the Sun. Thus, in 24 hours the Earth moves farther along its orbit in January than in July. As a result, the Earth must turn slightly more in January to bring the Sun back to overhead. This makes the interval between successive noons longer in January than in July and means they are not exactly 24 hours. For that reason, time is kept using a "mean Sun" that moves across the sky at the real Sun's average rate.

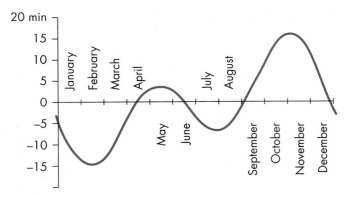

**FIGURE E2.4**
The equation of time is the correction that must be applied to the true Sun to determine mean solar time. It can be shown as a graph (as here) or as a figure-8 shape called an "analemma," often seen on globes of the world.

when we are farther away. The amount by which the length of the solar day varies is small, but it must be accounted for if our clocks are to always read about noon when the Sun is highest in the sky.

We could design clocks so that the hour is of different lengths at different times of the year. That could be done to ensure that our clocks advance to conform with the changing length of the solar day. However, it is much easier to define the length of the day differently, using not the true interval from one apparent noon to the next, but the *average* value of that interval over the year. That average daylength is called the **mean solar day,** and it has, by definition, 24 hours of clock time. We therefore use mean solar time in our daily timekeeping.

Over the course of the year, the difference in length between the mean solar day and the true solar day accumulates and leads to a difference of several minutes between clock time and time based on the position of the Sun. This difference is called the "equation of time" and is shown graphically in figure E2.4. The equation of time gives the correction needed on a sundial if it is to give the same time as your watch.

Although we use solar time in regulating our daily activities, astronomers find sidereal time extremely useful. One reason is that, at a given location, a given star always rises at the same sidereal time. To avoid the nuisance of A.M. and P.M., sidereal time is measured on a 24-hour basis. For example, the bright star Betelgeuse in the constellation Orion rises at about 10 P.M. in November but at about 8 P.M. in December. However, on a clock keeping sidereal time, it always rises at the same time at a given location, about 23 hours 50 minutes.

## TIME ZONES

Because the Sun is our basic timekeeping reference, most people like to measure time so that the Sun is highest in the sky at about noon. This is unnecessary now that we have good

electronic clocks that can keep time independent of the Sun. Nevertheless, it is a tradition that is hard to break, and as a result, clocks in different parts of the world are set to read different times. Because the Earth is round, the Sun can't be "overhead" everywhere at the same time, so it can't be noon everywhere at the same time.

The Earth is therefore divided into 24 major **time zones** in which the time differs by one hour from one zone to the next. There are a few exceptions, but we'll ignore them here.

Across the contiguous 48 United States, the time zones are, from east to west, Eastern, Central, Mountain, and Pacific (see fig. E2.5). The time within each zone is the same and is called **standard time.** Thus, in the eastern zone, the time is denoted Eastern Standard Time (EST), whereas in the central zone, it is denoted Central Standard Time (CST). As you travel across the country, it is therefore necessary to reset your watch if you cross from one time zone to another, adding 1 hour for each time zone as you move from west to east and subtracting 1 hour when you move from east to west.

If you travel a very large number of time zones, you may need to make such a large time correction that you shift your watch past midnight. For example, if you could travel fast enough and far enough, setting your watch back each time you cross a time zone, you could end up at your starting point with your watch set to a time 24 hours before you left!

A high-speed traveler cannot actually "gain" a day, however, because when you cross longitude 180° (roughly down the middle of the Pacific Ocean), you add a day to the calendar if you are traveling west and subtract a day if you are traveling east. The precise location where the day shifts is called the **international date line** (fig. E2.5). It generally follows 180° longitude but bends around extreme eastern Siberia and some island groups to ensure they keep the same calendar time as their neighbors.

## UNIVERSAL TIME

The nuisance of having different times at different locations can be avoided by using **Universal time,** abbreviated as UT. Universal time is the time kept in the time zone containing the longitude zero, which passes through Greenwich, England. By using UT, which is based on a 24-hour system to avoid confusion between A.M. and P.M., two people at remote locations can decide to do something at the same time without worrying about what time zone they are in.

## DAYLIGHT SAVING TIME

In many parts of the world, people set clocks ahead of standard time during the summer months and then back again to standard time during the winter months. This has the effect of shifting sunrise and sunset to later hours during the day, thereby creating more hours of daylight during the time most people are awake. Time kept in this fashion is therefore called

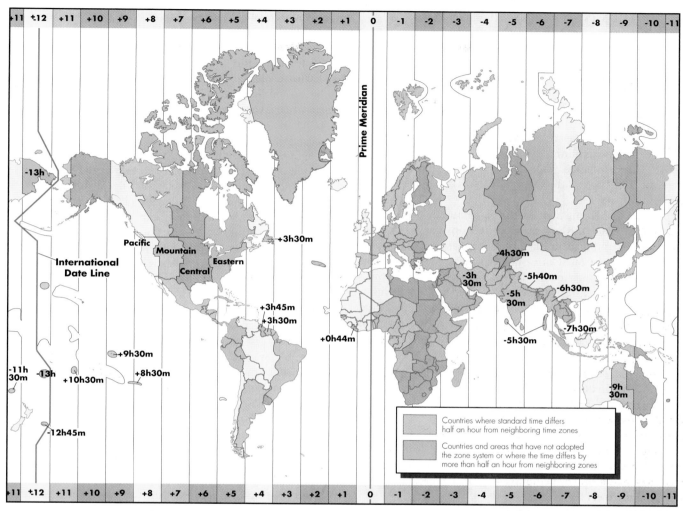

**FIGURE E2.5**
Time zones of the world and the international date line. Local time = Universal time − numbers on top or bottom of chart.

**daylight saving time** in the United States. In other parts of the world, it is called "Summer Time."

Daylight saving time was originally established during World War I as a way to save energy. By setting clocks ahead, less artificial light was needed during work hours late in the day. Nowadays, it allows us more daylight hours for recreation after work during the summer. As of March 2007, daylight saving time in the United States runs from the second Sunday in March to the second Sunday in November.

## THE MONTH

The day with its 24 hours is excellent for keeping track of short periods of time, but for longer intervals, longer time units are helpful. Of these other units, the month is the next largest unit used by nearly all cultures.

As you know from looking at a calendar, the month is usually about 30 days long. This time interval, and its name, derives from the Moon's cycle of phases. The time interval between full moons is about 29.5 days, which for use on the calendar is rounded off to 30. Because the year has about 365 days in it, there are about 12 lunar cycles per year. That is the reason that we have 12 months in the year.

You will notice, however, that 12 lunar cycles of 30 days ends up 5 days short of a full year. For that reason, some of the months are made 31 days long. In fact, if every other month, starting with January, were a 31-day month, the year would total 366 days. To make the days add up to 365, February is trimmed 1 day, to 29.

But, you protest, that is not the way the calendar looks. Although January, March, May, and July have 31 days, the sequence is broken in the later months. You see at work there the politics of ancient Rome.

# THE CALENDAR

Our calendar is based on one developed about 200 B.C. by the Romans. In fact, the word *calendar* is itself of Roman origin. There is some controversy about how the original Roman calendar was organized. It may have had only 10 months, and it probably began on the first day of spring (the vernal equinox) rather than in January.

Certainly the names of our months date from that calendar and its modifications. For example, if the year began in March, then September, October, November, and December were the 7th (Sept.), 8th (Oct.), 9th (Nov.), and 10th (Dec.) months, respectively.

Because it did not contain the right number of months and days to match the astronomical phenomena, this original calendar became a form of political patronage. The priests who regulated the calendar would add days and even months to please one group, and take days off to punish another. Such confusion resulted from these abuses that in 46 B.C., Julius Caesar asked the astronomer Sosigenes to design a calendar that would fit the astronomical events better and give less room for the priests and politicians to tinker with it. The resulting calendar, known as the **Julian calendar,** consisted of 12 months, which with the exception of February, alternated between 31 and 30 days in length.

The Julian calendar barely survived Caesar before the politicians were at it again. First, the name of the seventh month was changed to Julio to honor Julius Caesar—hence our July. Next, on the death of Julius Caesar's successor, Augustus Caesar, a very able and highly respected leader, it was decided to name the eighth month in *his* honor—hence, the name August. However, because it would have been impolitic to have his month a day shorter than Julius's, August became a 31-day month, and all the following months had the number of their days changed to maintain the alternation. Unfortunately, this led to using up one more day than there were days in the year. Thus, poor February, already one day short, was trimmed a second day, leaving it with only 28 days. With only minor modifications, this is the calendar we use today. However, those modifications are important, as we will see.

# LEAP YEAR

The ancient Egyptians knew that the year is not exactly 365 days long. It turns out that it takes about 365 and $\frac{1}{4}$ days for the Earth to complete an orbit around the Sun, which is how we measure a year. Because we can't have fractions of a day in the calendar, a calendar based on a year of 365 days will come up 1 day short every 4 years.

Your first reaction might be, So what? However, the seasons are set by the orientation of the Earth's rotation axis with respect to the Sun, not by how many days have elapsed. We therefore want to make sure that we start each year with the Earth having the proper orientation. Otherwise, the seasons get out of step with the calendar. For example, because

in 4 years you will lose 1 day, in 120 years you will lose a month, and in 360 years, you will lose an entire season. With a 365-day year, in a little over three centuries, April would be coming in what is now January.

This problem is corrected by leap year, a device used by the ancient Romans to keep the calendar in step with the seasons. Leap year corrects by adding a day to the calendar every fourth year. The extra day is traditionally added to February because it is the shortest month.

Unfortunately, the year is actually a little bit shorter than $365\frac{1}{4}$ days. Thus, leap year corrects a tiny bit too much. To deal with this problem, centuries that are not divisible by 400 are not leap years. Thus, 1900 was not a leap year, but 2000 was. This modification of omitting leap year for all century years not divisible by 400 was not part of the Julian calendar but was added in 1582 at the direction of Pope Gregory XIII. The calendar we use today is thus known as the **Gregorian calendar.**

The inauguration of the Gregorian calendar in 1582 was not a peaceful affair. Because it was adopted roughly 1600 years after the Julian calendar, the error in the length of the year had grown to about 10 days. To bring the calendar back into synchrony with the seasons, Pope Gregory simply eliminated 10 days from the year 1582 so that the day after October 4 became October 15.

Although the changeover went smoothly in most places, non-Catholic countries such as Protestant England refused to abide by the Pope's edict. The calendar in England and in a few other northern European countries was not altered. This made commerce between Catholic and non-Catholic countries very difficult because the day and sometimes even the month and year were different from one country to the next.

Eventually, the Gregorian calendar was adopted essentially worldwide, but the change was not made in England until 1752.* In Russia, the change was not made until the revolution in the early part of the 20th century. Other countries (Greece and Turkey, for example) changed in the 1920s.

# RELIGIOUS CALENDARS

Although the Gregorian calendar is used nearly worldwide, many religions use traditional calendars for setting their feast and holy days. Two of the best known examples are the Jewish and Muslim calendars, which are particularly interesting to astronomers because of the role played in them by the lunar cycle.

The Muslim calendar is in fact a purely lunar one consisting of 12 months of either 29 or 30 days. The year thus comes out about 11 days shorter than those in the Gregorian calendar. This means that the Muslim calendar is totally out of synchrony with the seasons. As a result, the holy month of Ramadan can fall at any time during the year, irrespective of the season. This

---

*This elimination of some 10 days from the calendar supposedly led to riots by people fearing they would be charged a full month's rent for only 20 days.

may seem odd to people living in climates with strong seasonal variations, but for people living in the Middle East where seasons are not so extreme, it makes little difference.

Likewise, the Jewish calendar is based on the lunar cycle. To correct for the missing days, from time to time an extra month is added in the middle of the year to keep it in step with the seasons. The Jewish calendar is especially interesting astronomically because it begins near the autumnal equinox and the extra month is added near the vernal equinox. Also, the holy days of Yom Kippur and Passover are located near the equinoxes, a feature shared by the Gregorian calendar, wherein Easter is near the vernal equinox (actually the first Sunday after the first full moon after the equinox) and Christmas is near the winter solstice.

## OTHER CALENDARS

Another calendar of note is the Chinese one, which, like the Jewish calendar, combines lunar and seasonal aspects. The Chinese calendar is based on a 365.2444-day year broken up into 6 cycles of 60 days. The years are themselves grouped into 60-year cycles composed of 5 cycles repeating every 12 years. The years in each 12-year cycle are given names such as the Year of the Rat, the Year of the Dog, and so forth. Days and months are added so that it works out to match the cycle of the seasons.

## NAMES OF THE MONTHS AND DAYS

That there are seven days in the week is probably a result of there being seven visible objects that move across the sky with respect to the stars: the Sun, the Moon, Mercury, Venus, Mars, Jupiter, and Saturn. We can see the names of some of these bodies in our English day names (Sunday, Monday, and Saturday). The influence is even clearer in the romance languages, such as Spanish (*lunes, martes, miércoles, jueves, viernes*).

Some English day names come to us through the names of Germanic gods (many of whom have a direct parallel with the Greco-Roman gods after whom the planets are named). For example, Tuesday is from *Tīw*, god of war, like Mars (matching Spanish *martes*). Wednesday is named for *Wōden*, the chief god of Germanic peoples and identified with the Roman Mercury (matching Spanish *miércoles*). Thursday is named for *Thōr*, the thunder god (matching Spanish *jueves*, "Jove's day"). Friday is named for *Freya*, a love goddess, like Venus (matching *viernes*).

The names of the months have less certain origins. Table E2.1 lists possible origins.

## THE ABBREVIATIONS A.M., P.M., B.C., A.D., B.C.E., AND C.E.

Four abbreviations are used frequently in the measure of time and calendars. They are the familiar letters A.M., P.M., B.C., and A.D. The first two have specific astronomical meaning. The last two have cultural meaning.

| TABLE E2.1 | |
|---|---|
| **Origin of the Names of the Months** | |
| January | Janus (gate), the two-faced god looking to the past and future; hence, beginnings. |
| February | Februa (expiatory offerings). |
| March | The god Mars. |
| April | Etruscan *apru* (April), probably shortened from the Greek Aphrodite, goddess of love and earlier of the underworld. |
| May | Maia's month; Maia ("she who is great"), the eldest of the Pleiades and the mother of Hermes by Zeus. |
| June | Junius, an old Roman noble family (from Juno, wife and sister of Jupiter, equal to Greek Hera). |
| July | Julius Caesar (Julius "descended from Jupiter"; The *Ju* of June and July are the same: Jupiter, "Sky-father"). |
| August | Augustus Caesar (*augustus* means sacred, grand). |
| September–December | "Seventh-month" to "tenth-month." The *-ember* may come from the same root as *month*. |

A.M. and P.M. stand for "ante meridian" and "post meridian," respectively. The meridian is the line passing from due north to due south and passing directly through the point exactly overhead (also called the zenith).

As the Sun moves across the sky, it crosses the meridian at the time called apparent noon. Before noon, it lies before (ante) the meridian. After noon, it lies past (post) the meridian. Hence, A.M. and P.M.

B.C. stands for "before Christ," referring to the year of his birth. Oddly, by convention 1 B.C. refers to the year before his birth, to avoid having a year "0." Most historians believe this chronology was inaccurate and that Jesus would have been born about 5 B.C.

A.D. stands, not for "after death," but for *anno Domini*, meaning "in the year of the Lord." The term A.D. was introduced by the sixth-century monk Dionysius Exiguus, about A.D. 528, in his attempts to trace the chronology of the Bible.

Recently, two different abbreviations have begun to replace A.D. and B.C. They are B.C.E. and C.E., which stand, respectively, for "before common era" and "common era." "Common era" refers to our present calendar, which is used nearly worldwide for most business purposes and thereby avoids reference to a particular religion. Yet another abbreviation—B.P.—is used, especially in anthropological and geological works. B.P. stands for "before present (era)" and is used for dates determined by analyzing the radioactive carbon in the object of interest. It takes 1950 C.E. as its base year.

## SUMMARY

Our system for keeping time is based on the motion of the Earth, Moon, and Sun. The day is determined by the Earth's spin, the month by the Moon's orbital motion around the Earth, and the year by the Earth's orbital motion around the Sun.

The solar day is based on the time interval between one apparent noon and the next. The sidereal day, or the interval between the time of star-rise for a given star and the time of its next rising, is about 4 minutes shorter than the solar day. This difference arises because, as the Earth moves along its orbit, the direction to the Sun shifts slightly. We must therefore wait a little longer to allow the Earth's rotation to carry us into the same position with respect to the Sun.

Time zones divide the Earth into regions such that the time differs by 1 hour (in general) from zone to zone. The resulting time difference allows the Sun to be approximately at its highest point above the horizon at noon in each zone.

The Earth makes approximately 365.25 rotations in the time it takes it to complete one orbit around the Sun. Thus, every 4 years, an extra day accumulates, which in leap years we add to the calendar as February 29.

## QUESTIONS FOR REVIEW

1. How is the solar day defined? How is the sidereal day defined?
2. Why do the sidereal and solar days differ in length?
3. Why do we need a leap year?
4. Why isn't the solar day always exactly 24 hours long?
5. What do A.M., P.M., A.D., and B.C. stand for?
6. Suppose you were asked to revise the calendar. What changes would you make?

## TEST YOURSELF

1. Suppose that the length of the year was 365.2 days instead of 365.25 days. How often would we have leap year? Every
   (a) 2 years.
   (b) 5 years.
   (c) 10 years.
   (d) 20 years.
   (e) 50 years.
2. Suppose the Earth's rotation axis was not tilted with respect to its orbit. How would the number of daylight hours change throughout the year?
   (a) The number would be no different.
   (b) Days would be longer and nights shorter at all times of the year.
   (c) Days and nights would be of equal length throughout the year.
   (d) Days would be shorter and nights longer throughout the year.
   (e) None of the above.
3. Why is February the shortest month?
   (a) The Earth is moving most slowly in its orbit then.
   (b) The Earth is moving fastest in its orbit then.
   (c) The Earth spins faster in February than at other times of the year.
   (d) The Earth spins slower in February than at other times of the year.
   (e) When the calendar was revised, days were taken from February to make other months longer.
4. If on a given date the night is 24 hours long at the North Pole, how many hours long is the night at the South Pole?
   (a) 24 hours
   (b) 12 hours
   (c) 36 hours
   (d) 48 hours
   (e) There is no night then.
5. On what day(s) of the year are nights longest at the equator?
   (a) They are the same length throughout the year there.
   (b) The solstices
   (c) The equinoxes
   (d) Approximately June 21
   (e) Approximately December 21

## FURTHER EXPLORATIONS

Bartky, Ian R., and Elizabeth Harrison. "Standard and Daylight-Saving Time." *Scientific American* 240 (May 1979): 46.

Cleere, G. S. "Eleven Lost Days." *Natural History* 100 (September 1991): 78.

Daniel, Glyn. "Megalithic Monuments." *Scientific American* 263 (July 1990): 78.

Duncan, David E. *Calendar: Humanity's Epic Struggle to Determine a True and Accurate Year.* New York: Avon Press, 1998.

Jespersen, James, and Jane Fitz-Randolph. *From Sundials to Atomic Clocks: Understanding Time and Frequency.* National Bureau of Standards Monograph 155. Washington, D.C.: U.S. Government Printing Office, 1977.

Monson, B. "A Simple Method of Measuring the Length of the Sidereal Day." *Physics Teacher* 30 (December 1992): 558.

Moyer, Gordon. "The Gregorian Calendar." *Scientific American* 246 (May 1982): 144.

## PLANETARIUM EXERCISES

1. *Sunrise and sunset times and number of daylight hours.* Set the location that you are viewing from to the nearest large city or your latitude and longitude. Set the local time to the summer solstice, June 21—the longest day of the year in the Northern hemisphere. Set the view to the East at sunrise. Note the time. Step the day through and note the sunset time. Using sunrise and sunset times, determine the length of the day for your latitude on this, the longest day. Using what you know about the summer and winter solstices, calculate the length of day for the shortest day of the year, December 21 (northern hemisphere). Verify your results by setting the program to that date and once again calculating the length of the day.

2. *How hours of daylight change with latitude.* In the previous planetarium exercise, you found the number of daytime hours at the summer and winter solstice at your location on Earth. In this exercise you will see how the number of daylight hours changes with latitude.

Begin by choosing two locations in your hemisphere, one that lies at least a few hundred miles north of your location on Earth and another that lies a similar distance south of your location. For each location, calculate the number of hours of daylight on the winter and summer solstice, as you did in the previous exercise. Which location has the most daylight hours at the summer solstice? Which location has the least number of daylight hours at the winter solstice? From your findings, what can you conclude about how latitude affects the number of hours of daylight?

## KEY TERMS

daylight saving time, 181

Gregorian calendar, 182

international date line, 180

Julian calendar, 182

mean solar day, 180

sidereal day, 178

solar day, 177

standard time, 180

time zone, 180

Universal time, 180

# The Moon

# 6

The Moon.

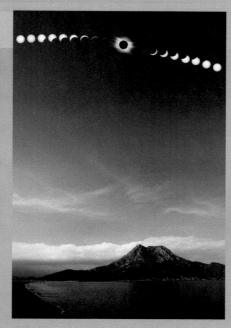

A total eclipse of the Sun.

## Chapter Objectives

- The Moon is much smaller than the Earth.
- The Moon's small size gives it a weak gravity, so that today:
  - It lacks an atmosphere.
  - It has cooled much more since the time of its formation than the Earth has.
- Because its interior is so relatively cool, the Moon is not active and lacks crustal motions and active volcanoes.
- The Moon's surface is covered with craters blasted into its surface by asteroid impacts.
- The Moon was probably formed from debris blasted out of the Earth by the impact of a planet-size object shortly after the Earth's formation.
- The Moon periodically aligns with the Sun and Earth to create eclipses.
- The Moon's gravity creates tides in the Earth's oceans.

**?What Is This?**
(See chapter summary for the answer.)

The Moon is our nearest neighbor in space, a natural satellite orbiting the Earth. It is the frontier of direct human exploration, an outpost that we reached more than a quarter century ago but from which we have since drawn back. But despite our retreat from its surface, the Moon remains of great interest to astronomers. Although originally the Moon was molten, its small mass and radius have made it difficult for it to either generate or retain any appreciable internal heat. It is therefore a dead world, with neither plate tectonic nor volcanic activity. That inactivity when coupled with the Moon's lack of atmosphere means that its surface features are essentially unaltered since its youth. But the Moon has not always been inactive. Shortly after its formation, it was pelted with a hail of rocky fragments whose size ranged up to 200 kilometers (about 100 miles) in diameter. The small fragments made craters, and the big fragments made huge basins. The basins subsequently flooded with lava (long since congealed) to create several dark, nearly circular plains easily visible to the naked eye. The Earth probably once bore such features, but erosion and plate motions have erased them. On the Moon's windless, rainless, airless surface, however, they remain—a record of events in the early Solar System that gives clues not only to the Moon's birth but to that of the Solar System as well.

In this chapter, we will describe the Moon's surface and why astronomers believe so many of its features were carved by impact. We will see that lunar rocks differ significantly from terrestrial ones and how they point to the Moon's having been born in a cataclysmic event early in the Earth's history. We will also discuss how the Moon affects Earth today through tides and eclipses. But we will begin with a short physical description of the Moon to help us visualize this nearest world.

## Concepts and Skills to Review

- Density (5.1)
- Law of gravity (2.5)
- Escape velocity (2.9)

## 6.1    DESCRIPTION OF THE MOON

### General Features

The Moon is about one-fourth the diameter of the Earth, a barren ball of rock, possessing no air, water, or life. In the words of lunar astronaut Buzz Aldrin, the Moon is a place of "magnificent desolation." But you don't have to walk on the Moon to see its desolation. Even to the naked eye, the Moon is a world of grays without the vivid colors of the terrestrial landscape. Yet even gray has its variety, as you can see where the dark roughly circular areas stand out from the lighter background,* as shown in figure 6.1.

### Surface Features

Through a small telescope or even a pair of binoculars, you can see that the dark areas are smooth while the bright areas are covered with numerous large circular pits called **craters,** as illustrated in figure 6.2. Craters usually have a raised rim and range in size from tiny holes less than a centimeter across to gaping scars in the Moon's crust such as Clavius,† about 240 kilometers (150 miles) across. Some of the larger craters have mountain peaks at their center.

---

*These dark features create the face of the "Man in the Moon."

†Most lunar craters are named for famous scientists. For example, Cristoph Clavius (1537–1612) was a German astronomer and mathematician.

**FIGURE 6.1**
The Moon.

✺ If you look very carefully at this picture you will notice something odd about the shadows around craters. How was this picture made?

✺ The shadows on the right half and left half lie in opposite directions. The image is a first quarter and 3rd quarter Moon joined together. This creates strong shadowing and thus shows craters well.

**FIGURE 6.2**
Photograph showing the different appearance of the lunar highlands and maria. The highlands are heavily cratered and rough. The maria are smooth and dark and have few craters.

The large, smooth, dark areas are called **maria** (pronounced MAR-ee-a), from the Latin word for "seas." However, these regions, like the rest of the Moon, are totally devoid of water. This usage comes from early observers who believed the maria looked like oceans and who gave them poetic names such as Mare (pronounced MAR-ay) Serenitatis (Sea of Serenity) or Mare Tranquillitatis (Sea of Tranquility), the site where astronauts first landed on the Moon.

The bright areas that surround the maria are called **highlands.** The highlands and maria differ in brightness because they are composed of different rock types. The maria are basalt, a dark, congealed lava rich in iron, magnesium, and titanium silicates. The highlands, on the other hand, are mainly anorthosite, a rock type rich in calcium and aluminum silicates. This difference has been verified from rock samples obtained by astronauts. Moreover, the samples also show that the highland material is generally less dense than mare rock and considerably older.

The highlands are not only brighter and their rocks less dense than the maria, they are also more rugged, being pitted with craters. In fact, highland craters are so abundant that they often overlap, as shown in figure 6.3A. Contrast this picture with the mare region shown in figure 6.3B, in which only a few, small craters are visible.

From many craters, long, light streaks of pulverized rock called **rays** radiate outward, as shown in figure 6.4. A particularly bright set spreads out from the crater Tycho near the Moon's south pole and can be seen easily with a pair of binoculars when the Moon is full.

A small telescope reveals still other surface features. Lunar canyons known as **rilles,** perhaps carved by ancient lava flows, wind away from some craters, as shown in figure 6.5. Elsewhere, straight rilles gouge the surface, probably the result of crustal cracking. Drying mud and chocolate pudding left too long in the refrigerator show similar cracks.

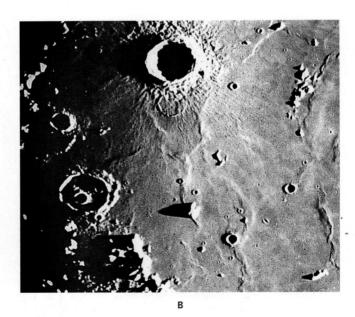

A                                    B

**FIGURE 6.3**
(A) Overlapping craters in the Moon's highlands. (B) Isolated craters in the smooth mare.

☾★ In (A) a small crater lies at the edge of a larger one. Which formed more recently: the small one or the large one?

☾★ The smaller one.

**FIGURE 6.4**
The long, narrow, white streaks radiating away from the crater at the bottom are lunar rays.

☾★ Can you see rays near other craters?

☾★ Some rays cross maria. What does this imply about the relative age of the rays and the maria?

☾★ The rays are younger than the maria.

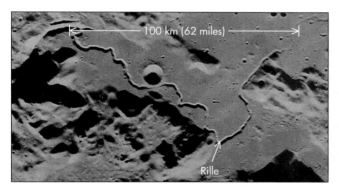

**FIGURE 6.5**
Photographs of some lunar rilles.

## Origin of Lunar Surface Features

Nearly all the surface features we see on the Moon—craters, maria, and lunar rays—were made by the impact of solid bodies on its surface. When such an object hits a solid surface at high speed, it disintegrates in a cloud of vaporized rock and fragments. The resulting explosion blasts a hole whose diameter depends on the mass and velocity of the impacting object. The hole's shape is circular, however, unless the impact is grazing.

   As the vaporized rock expands from the point of impact, it forces surrounding rock outward, piling it into a raised rim. Pulverized rock spatters in all directions, forming rays. Sometimes the impact compresses the rock below the crater sufficiently that it

FIGURE 6.7
A drop of rain falling into water produces an effect similar to the one that creates central peaks in lunar craters. The drop falls into the water but is then pushed up again as the water rebounds, as shown in this high-speed photo.

FIGURE 6.6
Central peak in a crater and slumped inner walls. Apollo astronauts took this photograph of the crater Eratosthenes on the last manned flight to the Moon. This crater is 58 kilometers (approximately 36 miles) in diameter.

FIGURE 6.8
Photograph of mountains along the edge of a mare. These mountains were probably thrown up by the impact that created the mare.

rebounds upward, creating a central peak, as shown in figure 6.6. Figure 6.7 shows a similar process happening in a high-speed picture of a raindrop falling into water. In other cases, material blasted out into the crater's rim may slump back into the crater. As it slides in toward the crater's center, the material pushes up a peak.

Astronomers believe the maria are also impact features, but to understand their formation, we must briefly describe the early history of the Moon. From the great age of the highland rocks (in some cases as old as 4.5 billion years), astronomers deduce that these rugged uplands formed shortly after the Moon's birth. At that time, the Moon was probably molten, allowing heavier (dense), iron-rich material to sink to its interior while lighter (less dense) material floated to the lunar surface. On reaching the surface, the less dense rock cooled and congealed, forming the Moon's crust. A similar process probably formed the Earth's continents. The highlands were then heavily bombarded by solid bodies from space, forming the numerous craters we see there.

As the Moon continued to cool, its crust thickened. But before the crust grew too thick, a small number of exceptionally large bodies—objects over 100 kilometers (about 60 miles)—struck the surface, blasting huge craters and pushing up mountain chains along their edges, as you can see in figure 6.8. Subsequently, molten material from deep within the Moon gradually flooded the vast crater and congealed to form the smooth, dark lava plains that we see now. Because the denser material sunk to the Moon's interior during its molten stage, the erupted lava from those depths was denser than the crustal rock into which it flooded. Moreover, because it was molten more recently, the mare material is therefore younger than the highlands. By the time the maria formed, most of the impacting bodies were gone—collected into the Earth and Moon by earlier collisions. Thus, too few bodies remained to crater the maria, which therefore remain smooth, even to this day.

Our home planet Earth furnishes additional evidence that most lunar craters formed early in the Moon's history. Like the Moon, Earth too was presumably battered by impacts in its youth. Although the vast multitude of these craters have been obliterated by erosion and plate tectonics, a few remain in ancient rock whose measured age is typically hundreds of millions of years. From the scarcity of such craters, astronomers can deduce that the main bombardment must have ended billions of years earlier.

Craters and maria so dominate the lunar landscape that we might not notice the absence of folded mountain ranges and the great rarity of volcanic peaks, land forms common on Earth. Why have such features not formed on the Moon?

## 6.2   STRUCTURE OF THE MOON

The Moon's small size relative to the Earth explains the differences between the two bodies. Because its volume compared to its surface area is small relative to the Earth's, heat escapes far more easily from the Moon. Thus, the Moon has cooled far more than the Earth. (Think of how a small baked potato cools much faster than a big one.) Moreover, because its mass is much smaller than the Earth's, the Moon contains much less radioactive material and so cannot generate as much heat. Thus, having a much cooler interior, the Moon lacks the convection currents that drive plate tectonic activity on the Earth. Confirmation of this comes from studies of the Moon's interior.

### Crust and Interior

The Moon's interior can be studied by seismic waves just as the Earth's can. One of the first instruments set up on the Moon by the Apollo astronauts was a seismic detector. Measurements from that and other seismic detectors placed by later Apollo crews show that the Moon's interior is essentially inactive and has a much simpler structure than the Earth's.

The Moon's surface layer is shattered rock that forms a **regolith**—meaning "blanket of rock"—tens of meters deep. The regolith consists of both rock chunks and fine powder, the result of successive impacts breaking rock into smaller and smaller pieces. This powdery nature is easily seen in the crispness of the astronauts' footprints (fig. 6.9). Samples of the regolith picked up by astronauts show that these surface rocks are typically the same type as the underlying rock. That is, the regolith on maria is generally broken-up basalt, whereas that on the highlands is broken-up highland material. In places the regolith may be several hundred meters thick. Analysis of the regolith shows that over time, its rocks have been broken up by high-velocity impacts, supporting the interpretation that the surface has been bombarded by meteoritic bodies.

Below this surface layer of rocky rubble is the Moon's crust, about 100 kilometers (60 miles) thick, on the average. The crust is much thinner (about 65 kilometers) on the side of the Moon that faces the Earth than on the far side, but the reason for this difference is not clear. It may result from the Earth's gravity shifting the Moon's core slightly toward Earth billions of years ago, when the Moon's interior was hotter. The crust on the near side—being slightly closer to the Moon's core because of that shift—might therefore have become hotter and as a result thinner than that on the far side. Subsequently, the Moon cooled, leaving the crust thinner on one side than the other. The thinner crust on the near side made it much easier for maria to form, as you can see in figure 6.10.

The Moon's crust, like the Earth's, is composed of silicate rocks relatively rich in aluminum and poor in iron. Beneath the crust is a thick mantle of solid rock, extending down a little more than 1000 kilometers (600 miles). The Moon's mantle is probably rich in olivine, the same type of dense, greenish rock that composes most of the Earth's mantle. Unlike the Earth's mantle, however, it appears too cold and rigid to be stirred by the Moon's feeble heat.

**FIGURE 6.9**

Footprint of an astronaut on the Moon.

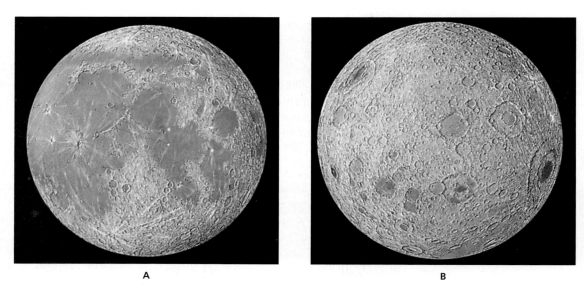

A                                B

**FIGURE 6.10**
(A) The near side of the Moon and (B) the far side. Note how uncommon maria are on the far side.

☾★ Why can't we see the far side of the Moon from Earth?

𝒜★ The Moon rotates on its axis so that it always keeps the same face toward the Earth. That is, it makes exactly one turn for each orbit around the Earth.

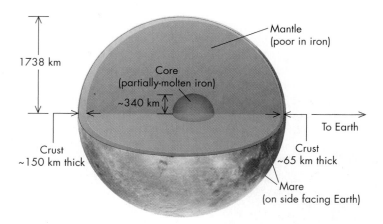

**FIGURE 6.11**
An artist's impression of the Moon's interior. Notice the thinner near-side crust and the displacement (exaggerated for clarity) of the core toward the Earth.

The Moon's low density (3.3 grams per cubic centimeter) tells us its interior contains little iron. Recall how in chapter 5 we saw that the Earth's high density (about 5 grams per cubic centimeter) was an indication that it had a large iron core. Some molten material may lie below the mantle, as illustrated in figure 6.11, but the Moon's core is smaller and contains far less iron and nickel than the Earth's. These factors, plus the fact that the Moon rotates very slowly (as we will see in section 6.3), lead astronomers to think that the Moon's core is unable to generate a magnetic field as the Earth does. Measurements made by the Apollo astronauts confirm that the Moon has essentially no magnetic field. Thus, a compass would be of no use to an astronaut lost on the Moon.

## The Absence of a Lunar Atmosphere

The Moon's surface is never hidden by lunar clouds or haze, nor does the spectrum of sunlight reflected from it show obvious signs of gases. With no atmosphere* to absorb and trap heat, temperatures on the Moon soar during the day and plummet at night. Likewise, no wind blows, and so the lunar surface lies dead and silent under a black sky.

The Moon has no atmosphere for two reasons. First, its interior is too cool to cause volcanic activity, which as we saw in chapter 5 was probably the source of much of the Earth's early atmosphere. Second, and more important, even if volcanos had created an atmosphere in its youth, the Moon's small mass creates too weak a gravitational force for it to retain the erupted gas. In chapter 2, we learned that the Moon's escape velocity is only about a fourth that of the Earth's (2.4 kilometers per second versus 11 kilometers per second), and so atoms in the Moon's atmosphere would have found it easy to escape its gravity. With no atmosphere and no plate tectonics, the Moon has been essentially unchanged for billions of years. The footprints left on the Moon by the astronauts in 1969 will probably still be there a million years from now.

## 6.3   ORBIT AND MOTIONS OF THE MOON

By watching the Moon for a few successive nights, you can see it move against the background stars as it follows its orbit around the Earth. The Moon's orbit is elliptical, with an average distance from Earth of 380,000 kilometers (about 250,000 miles) and a period of 27.3 days.[†] Its distance can be measured by triangulation, radar, or laser beams (fig. 6.12). To triangulate its distance, astronomers observe the Moon from

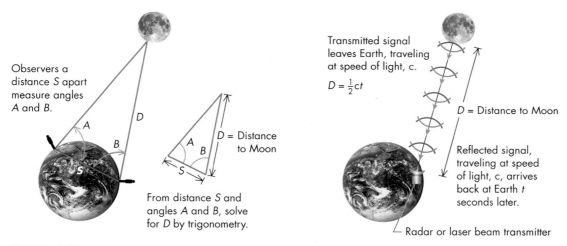

**FIGURE 6.12**
Finding the distance from the Earth to Moon by triangulation and radar and laser ranging.

---

*Lunar scientists have found tiny quantities of gas, mostly helium, above the Moon's surface. This gas is so tenuous (less than one-quadrillionth the density of our atmosphere) that we will ignore it. Perhaps more interesting, however, is the detection of traces of hydrogen near the Moon's poles by a spacecraft orbiting the Moon. Some scientists think the hydrogen comes from the breakdown of frozen water mixed with rock in several craters near the lunar poles. The water itself may have come from comets striking the Moon and vaporizing. The water vapor then condensed in the coldest places on the Moon (the polar craters into which sunlight never shines). You may have seen this tendency for frost to form in cold spots if you have taken something out of a freezer and left it for a while on a table.

[†]This is the time to complete an orbit around the Earth and is shorter than the cycle of the phases, as discussed in chapter 1.

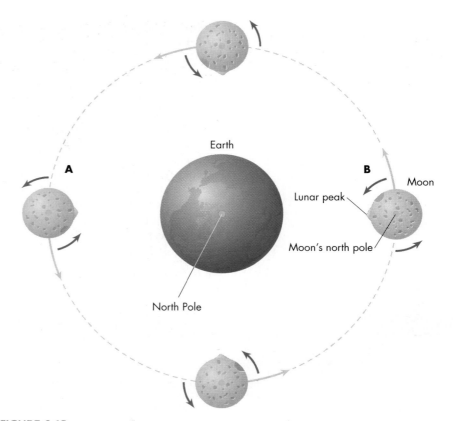

**FIGURE 6.13**

The Moon rotates once each time it orbits the Earth, as can be seen from the changing position of the exaggerated lunar mountain. Notice that at (A) the lunar peak is to the right, while at (B) it is to the left. Thus, from the Earth, we always see the same side of the Moon even though it turns on its axis.

To help see that the Moon rotates even though it keeps the same face toward the Earth, put a coin on the figure of the Moon and move it around the Earth so that the same edge of the coin always faces the Earth.

two different spots on the Earth. The distance between the locations, the angles to the Moon, and a little trigonometry give the Moon's distance.

A more accurate method is to bounce either a radar pulse or a laser beam off the Moon. Half the time interval between the transmission and the return of the reflected radiation multiplied by the speed of light gives the Moon's distance to an accuracy of centimeters.

## The Moon's Rotation

As it orbits, the Moon keeps the same side facing the Earth, as you can see by watching it through a cycle of its phases. You might think from this that the Moon does not rotate. Figure 6.13 shows, however, that the mountain on the side facing the Earth points to the right when the Moon is at A and to the left when the Moon is at B. Thus, the Moon *does* turn on its axis but with a rotation period exactly equal to its orbital period, a condition known as **synchronous rotation.** The Earth's gravity causes this locking of the Moon's spin to its orbital motion, as we will discuss in section 6.6.

## Oddities of the Moon's Orbit

The Moon's orbit is tilted by about 5° with respect to the Earth's orbit around the Sun, as illustrated in figure 6.14. It is also tilted with respect to the Earth's equator and is thus unlike most of the moons of Jupiter, Saturn, and Uranus, which lie nearly exactly

*The rotation of the Moon*

**FIGURE 6.14**

The Moon's orbit is tipped 5° with respect to the Earth's. The angle is exaggerated in this picture for clarity. Sizes and separations of objects are not to scale.

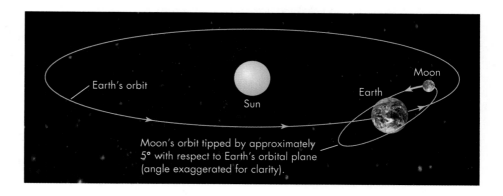

in their planets' equatorial plane. Also, most moons are tiny compared to their planets. Even the largest of the moons of Jupiter and Saturn have masses less than 1/1000 that of their planet. But our Moon's mass is 1/81 that of the Earth. These oddities suggest that our Moon formed differently from the moons of other planets, but how?

## 6.4    ORIGIN AND HISTORY OF THE MOON

Lunar rocks brought back to Earth by the Apollo astronauts caused astronomers to radically revise their ideas of how the Moon formed. Before the Apollo program, lunar scientists had three hypotheses regarding the Moon's origin:

*   The Moon was originally a small planet orbiting the Sun; it approached the Earth and was captured by Earth's gravity (capture theory).
*   The Moon and Earth were "twins," forming side by side from a common cloud of dust and gas (twin formation theory).
*   The Earth initially spun enormously faster than now and formed a bulge that ripped away from the Earth to become the Moon (fission theory).

Each of these hypotheses led to different predictions about the composition of the Moon. For example, had the Moon been a captured planet, its composition might be very unlike the Earth's. If the Earth and Moon had formed as twins, their overall composition should be similar. Finally, if the Moon was once part of the Earth, its composition should be nearly identical to the Earth's crust. When the rock samples were analyzed, astronomers were surprised that for some elements, the composition was the same, but for others, it was very different. For example, the Moon has a relatively high abundance of high-melting-point materials such as titanium and an almost complete lack of low-melting-point materials such as water. It also has much less iron than the Earth, as we pointed out when discussing its interior and low density.

The failure of evidence based on lunar surface samples to confirm any of the three hypotheses led astronomers to consider alternatives, and now, a completely different picture of the Moon's origin has emerged. According to the new hypothesis, the Moon formed from debris blasted out of the Earth by the impact of a Mars-sized body, as shown in figure 6.15A. The great age of lunar rocks and the absence of any impact feature on the Earth indicate that this event must have occurred during the Earth's own formation, at least 4.5 billion years ago. The colliding body melted and vaporized millions of cubic kilometers of the Earth's surface rock and hurled it into space in an incandescent plume. As the debris cooled, its gravity gradually drew it together into what we now see as the Moon.

This violent birth hypothesis explains many of the oddities of the Moon. The impact would vaporize low-melting-point materials and disperse them, leaving, for example, little water to be incorporated into the lunar body. Computer models (fig. 6.15B) of such an event also show that only surface rock would be blasted out of the Earth,

ANIMATION

*The birth of the Moon*

**FIGURE 6.15**
Origin of the Moon. (A) Artist's sketch illustrating main stages in the birth of the Moon. (B) This computer simulation shows how the Moon might have formed when a Mars-size object hit the young Earth and splashed out debris that later assembled into the Moon.

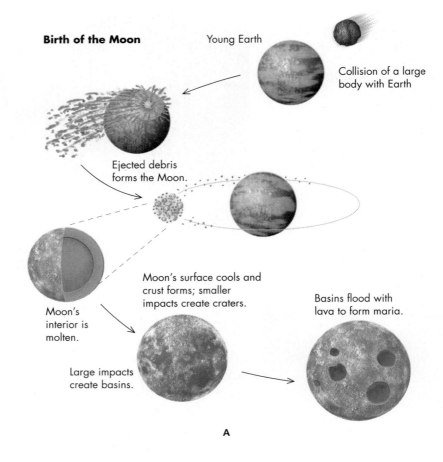

**Birth of the Moon**

Young Earth

Collision of a large body with Earth

Ejected debris forms the Moon.

Moon's surface cools and crust forms; smaller impacts create craters.

Basins flood with lava to form maria.

Moon's interior is molten.

Large impacts create basins.

A

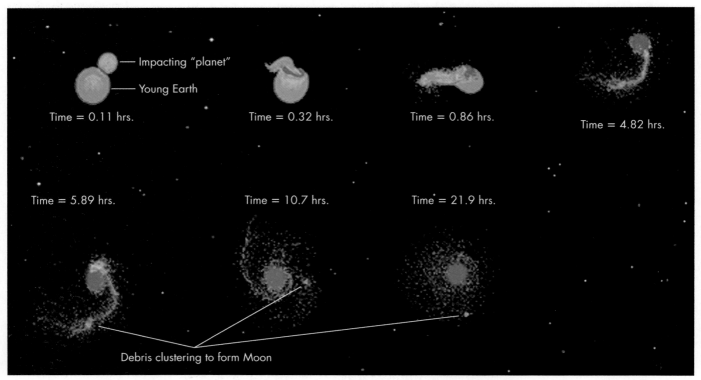

Impacting "planet"

Young Earth

Time = 0.11 hrs.

Time = 0.32 hrs.

Time = 0.86 hrs.

Time = 4.82 hrs.

Time = 5.89 hrs.

Time = 10.7 hrs.

Time = 21.9 hrs.

Debris clustering to form Moon

B

Asteroid

Asteroid impact
blasts huge basin.

Highlands

Mare

Molten rock from
Moon's interior
rises into basin
and floods it.

**FIGURE 6.16**
A rain of debris creates craters on the young Moon. Large impacts late in the process form the maria basins. Lava floods the basins to make the maria.

leaving our planet's iron core intact, thereby also explaining the low iron content of lunar rocks. The splashed-out rock would condense in an orbit whose shape and orientation were determined by the collision rather than by the orientation of the Earth's equator. Furthermore, we would expect both similarities and differences in composition between the Earth and Moon because the Moon was made partly from Earth rock and partly from rock of the impacting object.

After the Moon's birth, stray fragments of the ejected rock pelted its surface, creating the craters that blanket the highlands, as depicted in figure 6.16. A few huge fragments plummeting onto the Moon later in its formation process blasted enormous holes that later flooded with molten interior rock to become the maria. That rock was probably melted in the Moon's interior by radioactive decay, as happened in the Earth. During the time it took the rock to melt, about half a billion years, most of the debris remaining in the Moon's vicinity fell onto its surface. Thus, by the time the maria flooded, little material was left to fall on them, and so they are only lightly cratered. Since that time, the Moon has experienced no major changes. It has been a virtually dead world for all but the earliest times in its history, but it nevertheless creates effects on Earth: eclipses and tides.

## 6.5   ECLIPSES

An **eclipse** occurs when one astronomical body casts its shadow on another. As discussed in chapter 1, for us on the Earth two types of eclipses can occur: lunar and solar. A lunar eclipse happens when the Earth's shadow falls on the Moon. A solar eclipse happens when the Moon's shadow falls on the Earth. The great beauty of eclipses and their rarity make them eagerly awaited, and table 6.1 lists some upcoming ones. The approximate locations of the tabulated solar eclipses are shown in figure 6.17B.

## TABLE 6.1

### Some Upcoming Solar and Lunar Eclipses

**Solar Eclipses**

| | | |
|---|---|---|
| 2008 February 7 | Annular[†] | Antarctica, e. Australia, New Zealand [Annular: Antarctica] |
| 2008 August 1 | Total | n. Canada, Greenland, Siberia, Mongolia, China |
| 2009 January 26 | Annular[†] | S. Africa, Antarctica, se. Asia, Australia, [Annular: S. India, Sumatra, Borneo] |
| 2009 July 22 | Total | India, Nepal, China, c. Pacific |
| 2010 January 15 | Annular[†] | Africa, Asia [Annular: c. Africa, India, Myanmar, China] |
| 2010 July 11 | Total | s. Pacific, Easter Island, Chile, Argentina |

**Lunar Eclipses**

| | | |
|---|---|---|
| 2008 February 21 | Total | c. Pacific, Americas, Europe, Africa |
| 2008 August 16 | Partial* | S. America, Europe, Africa, Asia, Australia |
| 2009 February 9 | Penumbral[‡] | e. Europe, Asia, Australia, Pacific, w. N. Am. |
| 2009 July 7 | Penumbral[‡] | Australia, Pacific, Americas |
| 2009 August 6 | Penumbral[‡] | Americas, Europe, Africa, w. Asia |
| 2009 December 31 | Partial* | Europe, Africa, Asia, Australia |
| 2010 June 26 | Partial* | e. Asia, Australia, Pacific, w. Americas |
| 2010 December 21 | Total | e. Asia, Australia, Pacific, Americas, Europe |

*A partial lunar eclipse occurs when the Earth's shadow falls on only part of the Moon. A partial solar eclipse happens when the Moon covers only part of the Sun. A total lunar eclipse occurs if the Earth's shadow completely covers the Moon. A total solar eclipse occurs if the Moon completely covers the Sun's disk.

[†]A solar eclipse is annular if it occurs when the Moon is near the point in its orbit farthest from Earth. Under that condition, the Moon will not quite cover the Sun, even though it is exactly in line with it. The edge of the Sun, therefore, remains uncovered all around the Moon and appears as a bright ring, or annulus.

[‡]A penumbral eclipse occurs when the Moon passes through the outer part of Earth's shadow. These eclipses may be difficult to see from Earth.

**ANIMATION**

*Eclipses and the Moon's orbital inclination*

### Rarity of Eclipses

Eclipses are rare because the Moon's orbit around the Earth is tilted with respect to the Earth's orbit around the Sun (see fig. 6.14 and figure 1.20). Without that tilt, we would have lunar and solar eclipses every month, but with the tilt, the Moon lies above the Earth's orbit for half the month and below it for the other half. The result is that at full moon, the Earth's shadow generally falls either above or below the Moon, and at new moon, the Moon's shadow falls below or above the Earth, as shown in figure 6.18. How, then, are any eclipses possible?

As the Earth orbits the Sun, the Moon's orbit keeps nearly the same direction of tilt, as shown in figure 6.18. This orbital tilt is kept fixed—like that of the spinning Earth—by a gyroscopic effect or, more technically, by the conservation of angular momentum. The result is that twice each year, the Moon's orbital plane (if extended) passes through the Sun, as shown in figure 6.18. At those times—**eclipse seasons**—eclipses will happen when the Moon crosses the Earth's orbital plane (the ecliptic*). Table 6.2 shows that in 2008, the eclipse seasons are in February and August. Only in those months can eclipses happen: at other times, the shadows of the Earth and Moon always fall on empty space.

---

*As mentioned in chapter 1, this is the reason the ecliptic is so named.

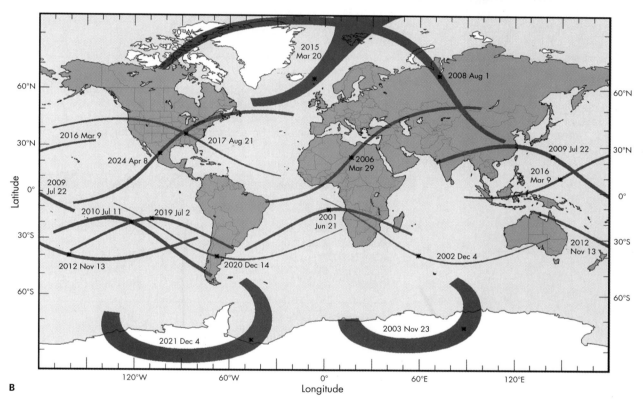

**FIGURE 6.17**
(A) Sketch of how the Moon's shadow travels across the Earth. (B) Location of some recent and upcoming total solar eclipses.

According to the map, when will the next total solar eclipse occur in North America? In Australia?

2017 for North America; 2012 for Australia.

You can also see from figure 6.18 that when a solar eclipse occurs at new moon, conditions are right for a lunar eclipse to happen at either the previous or the following full moon. Thus, eclipses generally occur in pairs, with a solar eclipse followed approximately 14 days later by a lunar eclipse, or vice versa, as shown for the 2008 eclipses listed in table 6.2.

This simple pattern does not always work, because the tilt of the Moon's orbit is not exactly fixed. An imaginary line perpendicular to the orbit (shown in fig. 6.19A) slowly changes direction just as the Earth's rotation axis does. That is, the Moon's orbit precesses, swinging once around about every 18.6 years. This orbital precession makes

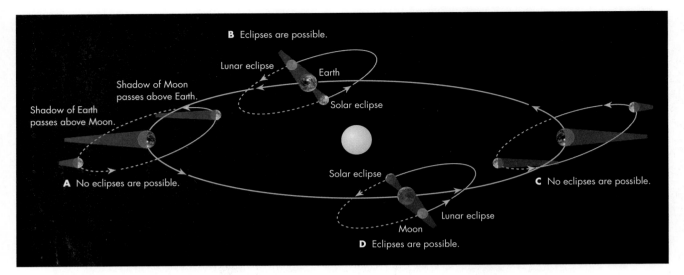

**FIGURE 6.18**

The Moon's orbit keeps approximately the same orientation as the Earth orbits the Sun. Because of its orbital tilt, the Moon generally is either above or below the Earth's orbit. Thus, the Moon's shadow rarely hits the Earth, and the Earth's shadow rarely hits the Moon, as you can see in (A) and (C). Eclipse seasons occur when the Moon's orbital plane, if extended, intersects the Sun. A solar eclipse will then occur at new moon and a lunar eclipse at full moon, as you can see in (B) and (D).

**TABLE 6.2**

**Pairing of Eclipses at Eclipse Seasons**

| | |
|---|---|
| February 7, 2008 Solar (Annular*) | February 21, 2008 Lunar (Total) |
| Antarctica, e. Australia, New Zealand | c. Pacific Americas, Europe, Africa |
| August 1, 2008 Solar (Total) | August 16, 2008 Lunar (Partial) |
| n. Canada, Greenland, Siberia, China | S. America, Europe, Africa, Asia, Australia |

*See footnote to table 6.1.

the dates of the eclipse seasons shift by 1/18.6 year (about 20 days) each year. Thus, in 2008, eclipses occur about 3 weeks earlier than in 2007, as shown in figure 6.19B.

If one of the eclipse seasons occurs in early January with the next in June, a third season may sometimes happen in late December. As a result, as many as five solar and two lunar eclipses or four solar and three lunar eclipses can occur each year. No matter when the eclipse season falls, at least two solar eclipses must happen each year,[†] but that does not mean they will be visible to an observer at a given location, since the eclipse may be visible only from another part of the Earth. Because the Moon is so small compared with the Earth, its shadow is small, and therefore, you can see a solar eclipse only from within a narrow band, as illustrated in figure 6.17A. Lunar eclipses, however, are visible from anywhere the Moon is above the horizon at the time of the eclipse.

## Appearance of Eclipses

Eclipses are beautiful and marvelous events and well worth watching. During a lunar eclipse, the Earth's shadow gradually spreads across the Moon's face, cutting an ever deeper dark semicircle out of it. The shadow takes about an hour to completely cover

Be extremely careful when watching a solar eclipse. Looking at the Sun through improper filters will blind you. A safer way is to *not* look directly at the Sun but to use eyepiece projection to view the Sun. Hold a piece of paper about a foot from the eyepiece of a small telescope (or even binoculars), and a large image of the Sun will be visible on it. This method also allows many people to watch the eclipse simultaneously.

---

[†]At least two lunar eclipses happen each year, although sometimes only the outer part of the Earth's shadow (technically, the penumbra) falls upon the Moon, so that the Moon is only slightly dimmed and the eclipse is not particularly noticeable.

**FIGURE 6.19**

(A) Precession of the Moon's orbit causes eclipses to come at different dates in successive years. Notice its similarity to (B) twisting a tilted book that has one edge resting on a table. Sizes of objects and their separations are not to scale.

the Moon and produce totality. At totality, the Moon generally appears a deep ruddy color, almost as if dipped in blood. Sometimes it even disappears. After totality, the Moon again becomes lit, bit by bit, reverting to its unsullied, silvery light.

A little light falls on the Moon even at totality because the Earth's atmosphere bends some sunlight into the shadow, as shown in figure 6.20. The light reaching the Moon is red because interactions with air molecules remove the blue light as it passes through our atmosphere and is bent, exactly as happens when we see the setting Sun.

A solar eclipse begins with a black "bite" taken out of the Sun's edge as the Moon cuts across its disk. Such partial eclipses are fun to watch but should be observed with care so as not to hurt your eyes.* However, unless a large part of the Sun is covered by the Moon, you may not even notice an eclipse is happening. On the other hand, if you are fortunate enough to be at a location where the eclipse is total, you will see one of the most amazing sights in nature.

As the time when the Moon's disk completely covers the Sun (totality) approaches, the landscape takes on an eerie light. Shadows become incredibly sharp and black: even individual hairs on your head cast crisp shadows. Sunlight filtering through leaves creates tiny bright crescents on the ground. Seconds before totality, pale ripples of light sweep across the ground and to the west the deep purple shadow of the Moon hurtles down on you at more than 1000 miles an hour. In one heartbeat you are plunged into darkness. Overhead, the sky is black, and stars may appear. The corona of the Sun—its outer atmosphere—gleams with a steely light around the Moon's black

*See the observing tip on page 202.

**A**

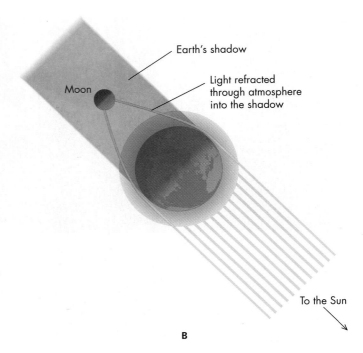

Earth's shadow

Light refracted through atmosphere into the shadow

Moon

To the Sun

**B**

**FIGURE 6.20**
(A) Photograph of a total lunar eclipse. (B) As sunlight falls on the Earth, some passes through the Earth's atmosphere and is slightly bent so that it ends up in the Earth's shadow. In its passages through our atmosphere, most of the blue light is removed, leaving only the red. That red light then falls on the Moon, giving it its ruddy color at totality.

☾★ Sometimes you see clouds after sunset that are lit red. How is this like the red color you see on the totally eclipsed Moon?

☽★ In both cases they are lit by sunlight that has passed through our atmosphere and had most of its blue light removed.

disk. Perhaps a solar prominence—a tiny, glowing, red flamelike cloud in the Sun's atmosphere—may protrude beyond the Moon's black disk (fig. 6.21). Birds call as if it were evening. A deep chill descends because for a few minutes the Sun's warmth is blocked by the Moon. The horizon takes on sunset colors: the deep blue of twilight with perhaps a distant cloud in our atmosphere glowing orange. As the Moon continues in its orbit, it uncovers the Sun, and instantly, it is daylight again. Now the cycle continues in reverse. If you ever have the chance to see a total eclipse, do it. It is worth traveling hundreds of miles to see.

The rarity of eclipses may lead you to think that the astronomical effects of the Moon are uncommon. A day spent by the ocean, however, will reveal a far more common and more powerful lunar influence.

## 6.6 TIDES

Anyone who has spent even a few hours by the sea knows that the ocean's level rises and falls during the day. A blanket set on the sand 10 feet from the water's edge may be inundated an hour later, or a boat pulled ashore may be left high and dry. This regular change in the height of the ocean is called the **tides** and is caused mainly by the Moon.

**A**                                                      **B**

**FIGURE 6.21**
(A) Picture of a total solar eclipse. The bright halo of light is the Sun's corona, its outer atmosphere. The red protrusions are prominences, clouds of glowing gas in the Sun's upper atmosphere. (B) The landscape is eerily lit during a total solar eclipse. The dark color of the sky is the Moon's shadow.

*Tidal forces*

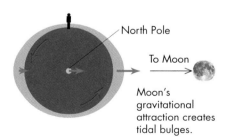

**FIGURE 6.22**
Tides are caused by the Moon's gravity creating tidal bulges.

## Cause of Tides

Just as the Earth exerts a gravitational pull on the Moon, so too the Moon exerts a gravitational attraction on the Earth and its oceans and draws material toward it. The attraction is stronger on the side of the Earth near the Moon and weaker on the far side (see fig. 6.22) because the force of gravity weakens with distance (recall Newton's law of gravity, p. 81). The difference between the strong force on one side and the weaker force on the other is called a **differential gravitational force.**

The differential gravity draws water in the oceans into a **tidal bulge** on the side of the Earth facing the Moon, as shown in figure 6.22.* But curiously, it creates an identical tidal bulge on the Earth's far side. This second tidal bulge can be viewed as a result of the Moon's gravity pulling the Earth "out from under" the water on the far side. A better way to view it, however, is to examine the Moon's gravitational forces on the Earth and its oceans as seen by a person on the Earth, as shown in figure 6.23.

The arrows in figure 6.23 (top) represent the Moon's gravitational pull at several points on the Earth. Points on the side of the Earth near the Moon (B) undergo a stronger pull toward the Moon than points on the far side (D), and so the arrow from point B is longer than the arrow from point D. Likewise, because point C is closer to the Moon than point D, the arrow from point C (which is the Moon's pull on the center of the Earth) is longer than the arrow from point D.

---

*The tidal bulges do not exactly align with the Moon for reasons we will discuss in the section on tidal braking.

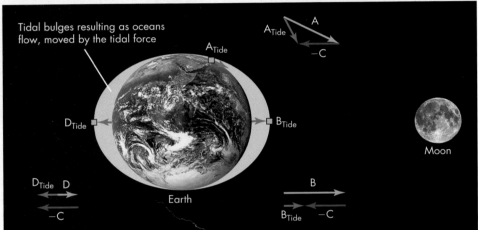

**FIGURE 6.23**
(Top) Arrows schematically show the Moon's gravitational force at different points on the Earth. (Bottom) Tidal forces from the point of view of an observer on the Earth. These arrows represent the difference between the Moon's gravitational force at a given point and its force at the Earth's center (C). Graphically, you can find the tidal force by "adding" the arrows. The figure shows schematically how to do this, but details are omitted.

To see how the tidal bulges form, we need to look at the *difference* between the gravitational force at a given point and at the center of the Earth. For example, at point B, the force is larger than at point C, and so matter at point B will be pulled *away* from point C. This creates one tidal bulge. But matter at point C is in turn pulled away from point D, which creates a second tidal bulge. If we now draw a second set of arrows to represent the *difference* between the force at C and at every other point (the differential gravitational force), we find the forces illustrated in figure 6.23 (bottom). These drive the oceans into the bulges* that we see.

In the previous discussion, we have ignored the Earth's rotation. The tidal bulges are aligned approximately with the Moon, but the Earth spins. Its rotation therefore carries us first into one bulge and then the next. As we enter the bulge, the water level rises, and as we leave it, the level falls. Because there are two bulges, we are carried into high water twice a day, creating two high tides. Between the times of high water, as we move out of the bulge, the water level drops, making two low tides each day (fig. 6.24A).

---

*Tides also occur in the atmosphere and solid ground, but the latter are very small because the ground is rigid and cannot move as easily as water or air.

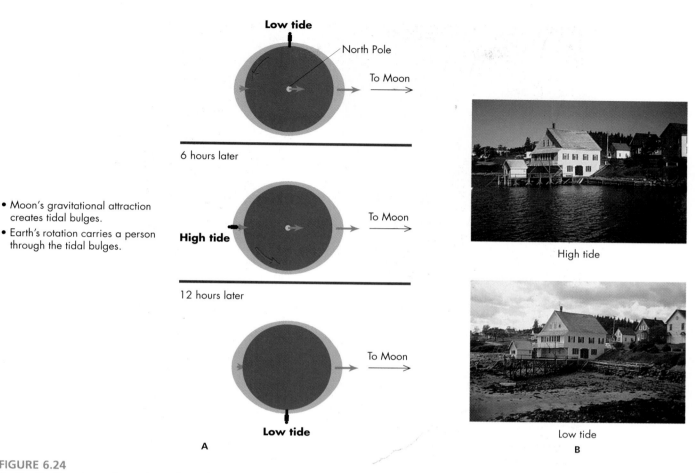

- Moon's gravitational attraction creates tidal bulges.
- Earth's rotation carries a person through the tidal bulges.

High tide

Low tide

**FIGURE 6.24**

As the Earth rotates, it carries points along the coast through the tidal bulges. Because there are two bulges where the water is high and two regions where the water is low, we get two high tides and two low tides each day at most coastal locations.

This simple picture must be altered to account for the inability of the ocean to flow over land areas. Thus, water tends to pile up at coastlines when the tidal bulge reaches shore. In most locations, the tidal bulge has a depth of about 2 meters (6 feet), but it may reach 10 meters (30 feet) or more in some long narrow bays (as you can see in the photographs of high and low tides along the Maine coast in fig. 6.24B) and may even rush upriver as a tidal bore—a cresting wave that flows upstream. On some rivers, surfers ride the bore upstream on the rising tide.

The motion of the Moon in its orbit makes the tidal bulge shift slightly from day to day. Thus, high tides come about 50 minutes later each day, the same delay as in moonrise, discussed in chapter 1.

## Solar Tides

The Sun also creates tides on the Earth, but although the Sun is much more massive than the Moon, it is also much farther away. The result is that the Sun's tidal force on the Earth is only about one-half the Moon's. Nevertheless, it is easy to see the effect of their tidal cooperation in spring tides, which are abnormally large tides that occur at new and full moon. At those times, the lunar and solar tidal forces work together, adding their separate tidal bulges, as illustrated in figure 6.25A. Notice that spring tides have nothing to do with the seasons; rather, they refer to the "springing up" of the water at new and full moon.

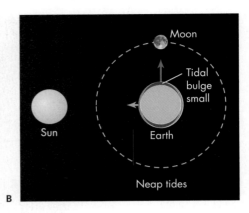

**FIGURE 6.25**

The Sun's gravity creates tides too, though its effect is only about one-half that of the Moon. (A) The Sun and Moon each create tidal bulges on the Earth. When the Sun and Moon are in line, their tidal forces add together to make larger-than-normal tides. (B) When the Sun and Moon are at 90° as seen from Earth, their tidal bulges are at right angles and partially nullify each other, creating smaller-than-normal tidal changes.

It may seem odd that spring tides occur at both new and full moon because the Moon and Sun pull together when the Moon is new but in opposite directions when it is full. However, both the Sun and Moon create two tidal bulges, and the bulges add together regardless of whether the Sun and Moon are on the same or opposite sides of the Earth. On the other hand, at first and third quarters, the Sun and Moon's tidal forces work at cross-purposes, creating tidal bulges at right angles to one another, as shown in figure 6.25B. The so-called neap tides that result are therefore not as extreme as normal high and low tides.

## Tidal Braking

Tides create forces on the Earth and Moon that slow their rotation, a phenomenon known as **tidal braking.** Figure 6.26 shows how the Moon tidally brakes the Earth. As the Earth spins, friction between the ocean and the solid Earth below drags the tidal bulge ahead of the imaginary line joining the Earth and Moon, as depicted in figure 6.26. The Moon's gravity pulls on the bulge, as shown by the long green arrow in the figure, and holds it back. The resulting drag (short green arrows) is transmitted through the ocean to the Earth, slowing its rotation the way a brake shoe on a car or your hand placed on a spinning bicycle wheel slows the wheel.

As the Earth's rotation slows, the Moon accelerates in its orbit, moving farther from the Earth, as required by the need to conserve angular momentum. The Moon accelerates because the tidal bulge it raises on the Earth exerts a gravitational force back on the Moon (as predicted by Newton's third law of motion), which pulls the Moon ahead in its orbit, as shown by the purple arrows in figure 6.26. That acceleration makes the Moon move away from the Earth at about 4 centimeters (roughly $1\frac{1}{2}$ inch) per year, a tiny increase in the Earth-Moon distance, but nevertheless detectable with laser range finders. Thus, the Moon was once much closer to the Earth and the Earth spun much faster, perhaps as rapidly as once every 5 hours several billion years ago. Over that immense period of time, the Moon has receded to its present distance, and the Earth's rotation has slowed to 24 hours. These processes occur even now: tidal braking lengthens the day by about 0.002 seconds each century.

Tidal braking is also the reason the Moon always keeps the same face to the Earth. Just as the Moon raises tides, which slow the Earth, the Earth raises tides on the Moon, which slow it. These lunar tides distort the Moon's crust and have braked the Moon severely, locking it into synchronous rotation. The Moon's braking of the Earth will eventually make the Earth rotate synchronously with the Moon's orbital motion. Billions of years from now, the Earth and Moon will orbit so that each constantly presents the same face

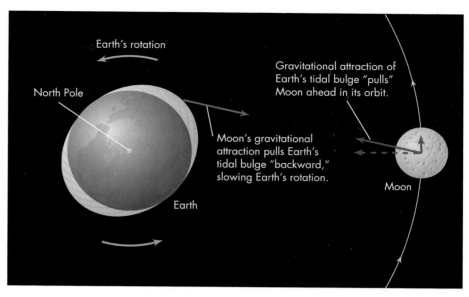

**FIGURE 6.26**
Tidal braking slows the Earth's rotation and speeds up the Moon's motion in its orbit. Friction between the oceans and Earth's solid crust "drags" the bulges of water "ahead" of the Earth—Moon line.

to the other: the Moon will then be visible only from one side of the Earth! Similar tidal effects have locked some of the moons of other planets into synchronous rotation, but the planets themselves have not been noticeably slowed.* On the other hand, tidal braking by the Sun probably slowed the rotation of Mercury and Venus.

The Moon's gravitational pull on the Earth may also stabilize our climate. Astronomers have recently discovered with computer simulations that the tilt of a planet's rotation axis may change erratically by many tens of degrees if the planet has no moon. Because the tilt causes seasons, changes in the tilt will alter the severity of the seasons. Our Moon is large enough that its gravitational attraction on Earth's equatorial bulge helps hold the Earth's tilt relatively fixed, sparing us catastrophically large climate changes.

## 6.7    MOON LORE

The Moon figures prominently in folklore around the world. Most stories concerning its powers are false. For example, people often claim that the full moon triggers antisocial behavior, hence the term *lunatic*. All studies to look for such effects have found nothing. Automobile accidents, murders, admissions to clinics, and so forth show no increase when the Moon is full.

On the other hand, "once in a blue moon," indicating a rare event, is a phrase with a basis in fact because, on rare occasions, the Moon may look blue. This odd coloration comes from particles in the Earth's atmosphere. Normally, our atmosphere filters the blue colors from light better than the red ones. For example, light from the rising or setting Sun passes through so much atmosphere that little blue light remains by the time it reaches us. Therefore, the Sun looks red when it is low in the sky. However, if the atmosphere contains particles whose size falls within a very narrow range, the reverse may occur. Dust from volcanic eruptions or smoke from forest fires may have just the right size to filter out the red light, allowing mainly the blue colors to pass through. Under these unusual circumstances, we may therefore see a "blue Moon."

*The dwarf planet Pluto is an exception. Its moon Charon has locked it into synchronous rotation.

**TABLE 6.3**

## Names Used for Full Moons*

| | | | |
|---|---|---|---|
| January | old moon | July | thunder or hay moon |
| February | hunger moon | August | grain or green corn moon |
| March | sap or crow moon | September | harvest moon |
| April | egg or grass moon | October | hunter's moon |
| May | planting moon | November | frost or beaver moon |
| June | rose or flower moon | December | long night moon |

*Most of these names derive from Native American usage.

A different meaning for "blue moon" has appeared within the last few years. This new meaning applies to months with two full moons. Because the cycle of phases is 29.5 days, unless the Moon is full on the first day of the month, the next full moon will fall in the following month. The odds of the Moon being full on the first of the month are about 1 in 30, and so two full moons in a given month happens about every $2\frac{1}{2}$ years. Why such occasions should be referred to as "blue moons" is controversial.

The harvest moon, the full moon nearest the time of the autumn equinox, is another well-known phrase. As it rises in the east at sunset, the light from the harvest moon helps farmers see to get in the crops. Full moons in other months also have special names, but only the harvest and hunter's moon are widely accepted. Other names occasionally used in American folklore are listed in table 6.3.

## SUMMARY

The Moon is the Earth's satellite. It is much smaller than the Earth: about one-fourth the Earth's radius and about 1/81 its mass. Its small size has allowed its internal heat to escape, keeping its core cool, thereby preventing plate tectonic motions. The Moon has no atmosphere because it is too cool to create one by volcanic outgassing and too small for its low gravity to retain gases even if an atmosphere had formed.

With neither atmosphere nor tectonic activity, the Moon's surface is unaltered except by impact features: craters, rays, and the maria. Maria are enormous lava flows that have flooded into basins made by large impacting bodies late in the Moon's formation.

The Moon may have formed when a Mars-size body collided with the Earth and splashed material from the Earth into orbit. That debris, drawn together by its own gravity, would then have reassembled into the Moon.

The Moon's shadow sometimes falls onto the Earth, causing a solar eclipse. When the Earth's shadow falls onto the Moon, we see a lunar eclipse.

The Moon's gravity creates tides, and as the Earth rotates beneath the tidal bulge of the ocean, our planet's rotation is slowed. Similar tidal braking exerted by the Earth on the Moon has slowed the Moon's spin, making it synchronous with its orbital motion around the Earth.

### ? What It Is.

This picture shows what is sometimes called "The old Moon in the new Moon's arms." You see the crescent Moon (lit by direct sunlight) and the rest of the Moon (dimly lit by sunlight reflected from clouds on the Earth). That is, the dimly lit part of the Moon is illuminated by earthlight. This dim light, sometimes also called "ashen light," is easy to understand if you imagine yourself on the Moon looking at the Earth in the lunar sky and ask yourself what "phase" of the Earth you would see.

## QUESTIONS FOR REVIEW

1. What are the Moon's mass and radius compared with the Earth's?
2. Describe a crater and how it is formed.
3. What are lunar rilles? What are rays?

4. How do the maria differ from the highlands?
5. What formed the maria? Why are they smooth?
6. Why has the Moon no atmosphere?

7. Why is the Moon's surface cratered but the Earth's is not?

8. How do astronomers believe the Moon formed? What supports this theory? How does the theory explain why the Earth and Moon have such different densities?

9. How are tides formed on the Earth?

10. Why does the Moon form two tidal bulges on the Earth?

11. Why aren't there eclipses each month?

12. What is an eclipse season?

13. How does the Moon rotate? Why does it spin in this manner?

## THOUGHT QUESTIONS

1. Why has the Moon's interior cooled more than the Earth's?

2. Bergmann's rule states that individuals of a given species—for example, bears—will be larger in cold climates than in warmer climates. How is an explanation of this rule similar to an explanation of the temperature difference between the Earth's and the Moon's interior?

3. If the day were 12 hours long, what would be the approximate time interval between high and low tide?

4. Highway surfaces develop "potholes" over time. How can you use the number of potholes as an indication of the "age" of the paving? How is this like using craters to estimate the age of the Moon's surface?

5. Why will an astronaut's footprint on the Moon last so long?

## PROBLEMS

1. A laser pulse takes 2.56 seconds to travel from Earth to the Moon and return. Given that the speed of light is 300,000 kilometers per second, how far away is the Moon? How might this time delay affect conversations between an astronaut on the Moon and someone back on Earth?

2. A lunar crater has an angular diameter of 1 minute of arc. What is its diameter in kilometers?

3. A spacecraft orbits the Moon 100 kilometers above its surface and with an orbital period of 114 minutes. What is the Moon's mass?

4. The Moon's orbital period is 27.2122 days. Its synodic period, the period of the phases, is 29.5306 days. Show that 242 orbital

periods very nearly equals 223 synodic periods. How long is this in years? What does this suggest about eclipses and why? (This match of cycles is called the *saros* and was used by ancient astronomers to predict eclipses.)

5. Calculate the Moon's density (see the end of section 5.1 in the Earth chapter for how to calculate density). The Moon's mass is about $7.3 \times 10^{25}$ gm, and its radius is about $1.7 \times 10^8$ cm. On the basis of your value for the density, what can you say about the amount of iron in the Moon? (See table 5.1 for iron's density.)

## TEST YOURSELF

1. The large number of craters on the lunar highlands compared to those on the maria is evidence that
   (a) the surface of the maria is liquid and craters quickly disappear there.
   (b) the material composing the highlands is very soft and easily cratered.
   (c) the bodies that struck the Moon and made the craters were clumped in such a manner that they missed hitting mare areas.
   (d) the maria are much younger than the highlands.
   (e) the maria are much older than the highlands.

2. Figure 6.21 shows an eclipse of the Sun. The black circle in the middle is
   (a) the Earth's shadow on the Sun.
   (b) the Sun's shadow on the Moon.
   (c) the Moon covering the Sun.
   (d) the Earth's shadow on the Moon.
   (e) a dark cloud in our atmosphere.

3. The photographs in figure 6.24 were taken at high tide and the next low tide. About how much time elapsed between the pictures?
   (a) 3 hours
   (b) 24 hours
   (c) 12 hours
   (d) 6 hours
   (e) 1 month

4. What evidence indicates that the Moon lacks a large iron core?
   (a) The Moon has a very strong magnetic field.
   (b) The Moon always keeps the same side facing the Earth.

(c)  The Moon has no atmosphere.

(d)  The Moon's average density is about 3.3 grams per cubic centimeter, similar to that of rock.

(e)  The Moon has so many volcanos that all the iron in its core has been erupted onto its surface.

5.  Eclipses do not occur each month because

(a)  the Moon's orbit is so elliptical.

(b)  the Moon's orbit is tilted with respect to the Earth's orbit around the Sun.

(c)  the Moon takes 6 months to complete its orbit around the Earth.

(d)  the Earth's rotation axis is tilted.

(e)  the statement is false. Eclipses do occur each month, but they are generally visible only in very remote parts of the Earth.

## FURTHER EXPLORATIONS

Brueton, Diana. *Many Moons: The Myth and Magic, Fact and Fantasy of Our Nearest Heavenly Body.* Englewood Cliffs, N.J.: Prentice-Hall, 1991.

Foust, Jeffrey A. "The Moon Rediscovered." *Sky and Telescope* 96 (December 1998): 32.

Goldreich, Peter. "Tides and the Earth-Moon System." *Scientific American* 226 (April 1972): 42.

Hiscock, Philip. "Once in a Blue Moon." *Sky and Telescope* 97 (March 1999): 52.

Nicholson, T. D. "The Moon Illusion." *Natural History* (August 1991): 66.

Paul D. Spudis. "The New Moon." *Scientific American* 289 (December 2003): 86.

### Website

Please visit the Explorations website at **http://www.mhhe.com/arny** for additional online resources on these topics.

## PLANETARIUM EXERCISES

*The period of the Moon's phases.* The synodic period of the Moon is the time it takes our satellite to complete a cycle of its phases, from new to full and back to new again, for example. This is slightly longer than the time it takes the Moon to complete an orbit around us. In this exercise you will track the phases of the Moon to see how long the cycle of phases is.

Use the Starry Night planetarium software to estimate what the Moon's phase and percentage of illumination is for tonight.

Set the time step of the program to be one day. Step the time forward and record the name of the new phase and the new percentage of illumination. Repeat until you return to the original phase and percentage of illumination. How many days has this taken? (This should correspond to the Moon's synodic period.) Repeat the process for a second cycle of phases to verify your result.

## KEY TERMS

craters, 188

differential gravitational force, 205

eclipse, 199

eclipse seasons, 200

highlands, 190

maria, 190

rays, 190

regolith, 193

rilles, 190

synchronous rotation, 196

tidal braking, 208

tidal bulge, 205

tides, 204

## PROJECT

Look at the Moon on an evening when it is nearly full. Make a sketch of the light and dark markings that you see on its surface with the naked eye. Then observe the Moon with binoculars and make an enlarged sketch that shows more detail. Mark a few of the craters you can see. Estimate the diameter of these craters from your knowledge that the Moon's radius is about 1000 miles (1700 km). How big is the largest crater you can see compared to the size of your town? Can you see any lunar rays? If so, sketch them on your drawing. How long are the rays?

# Survey of the Solar System

# 7

Terrestrial
planets

Jovian planets

The planets and Sun to scale.

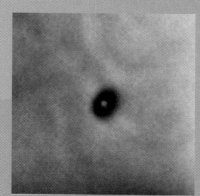

**?What Is This?**
(See chapter summary for the answer)

## Chapter Objectives

- The Solar System consists of the Sun, the eight planets, their moons, and smaller objects (such as dwarf planets, asteroids, and comets).
- The planets orbit the Sun within a flat disk-shaped region, held there by the Sun's gravity.
- The planets form two main families:
  - The four planets near the Sun (Mercury, Venus, Earth, and Mars) are rocky with iron cores and are similar to the Earth in size.
  - The next four planets from the Sun (Jupiter, Saturn, Uranus, and Neptune) are much larger than the Earth and are balls of liquid or gas, rich in hydrogen and its compounds.
  - Pluto, which fits into neither category, is now considered a dwarf planet. It and a few similar small icy objects orbit beyond Neptune.
- The Solar System formed about 4.6 billion years ago from an interstellar gas cloud that collapsed and shrank under the force of its own gravity.
- Rotation of the cloud made it flatten into a disk.
- Within the disk, gravity drew dust and gas into clumps that became the planets.

## Concepts and Skills to Review

- Law of gravity (2.5)
- Density (5.1)
- Modified form of Kepler's third law (2.7)

The Solar System consists of the Sun and the bodies orbiting in its gravitational field: the eight planets, their moons, and swarms of asteroids and comets. In this chapter, we will survey the general properties of this system and describe its birth, reserving for later chapters more detailed descriptions of its components. Although earthlings have not walked on any Solar System objects except the Earth and Moon, we have detailed pictures sent to us from spacecraft of most of the planets and their satellites. Some are naked spheres of rock; others are mostly ice. Some have thin, frigid atmospheres so cold that ordinary gases crystalize as snow above their cratered surfaces; others have thick atmospheres the consistency of wet cement and no solid surface at all. On Mars, a dead volcano over twice the height of Mount Everest looms above a red desert. On Uranus, an ocean of water compressed to the density of rock by the gravity of this massive planet heaves beneath poisonous clouds. Yet despite such diversity, the Solar System possesses an underlying order, an order from which astronomers attempt to read the story of how our Solar System came to be.

As long ago as the eighteenth century, astronomers recognized that the highly ordered pattern in which the planets move is a clue to their origin. All the planets circle the Sun in the same direction, and all lie within a region of space shaped like a flattened disk, as we can directly infer from their motion within the zodiac. That pattern suggested to those early astronomers that the planets formed from a vast, spinning disk of gas and dust surrounding the Sun. As they learned more about the planets from more powerful telescopes and from the spectral analysis of their light, astronomers discovered a second underlying pattern. Despite major surface differences, the planets form two main families: (a) solid, rocky bodies composed mainly of silicates (like the Earth) that orbit near the Sun, and (b) liquid and gaseous bodies rich in hydrogen (like Jupiter) that orbit far from the Sun.

From these and other regularities, astronomers hypothesize that the Sun and planets formed from the collapse of a huge, slowly spinning cloud of gas and dust. Most of the cloud's material fell inward and ended up in the Sun, but in response to rotation, some settled into a swirling disk around it. Then, within that disk, dust particles coagulated—perhaps aided by electrostatic effects such as those that make lint cling to your clothes—to form pebble-sized chunks of material, which in turn collided and sometimes stuck, growing ever larger to become the planets we see today. The type of planet formed—rocky or gaseous—depended strongly on the temperature and hence the place in the disk where it condensed.

Near the Sun's fiery surface, only dust with a high melting point could form from the original gases, and so planets there became masses of rock and metal such as the Earth. But in the bitterly cold regions far from the Sun, ice and frozen gas could condense as well as rock and metals. Thus, planets such as Jupiter are rich in these lighter materials. However, regardless of their composition, the objects that formed in the disk retained the motion of the original gas and dust, and so we see them today, moving in a flattened system, all orbiting the Sun in the same direction. How long ago did all this happen? From the amount of radioactivity in meteorites, astronomers estimate the Solar System formed about 4.6 billion years ago. With this overview in mind, we can now look in more detail at what our Solar System is like and how astronomers think it formed.

## 7.1   COMPONENTS OF THE SOLAR SYSTEM

### The Sun

The Sun is a star, a ball of incandescent gas (fig. 7.1) whose light and heat are generated by nuclear reactions in its core. It is by far the largest body in the **Solar System**—more than 700 times the mass of all the other bodies put together—and its gravitational force holds the planets and other bodies in the system in their orbital patterns about it. This gravitational domination of the planets by the Sun justifies our calling the Sun's family the Solar System.

The Sun is mostly hydrogen (about 71%) and helium (about 27%), but it also contains very small proportions of nearly all the other chemical elements (carbon, iron, uranium, and so forth) in vaporized form, as we can tell from the spectrum of the light it emits.

### The Planets

The planets are much smaller than the Sun and orbit about it. They emit no visible light of their own but shine by reflected sunlight. In order of increasing distance from the Sun, they are Mercury, Venus, Earth, Mars, Jupiter, Saturn, Uranus, and Neptune. The planets

**FIGURE 7.1**
Image of the Sun made with an ultraviolet telescope that reveals high temperature gases in the Sun's atmosphere. Earth and Jupiter are shown to correct relative size for comparison.

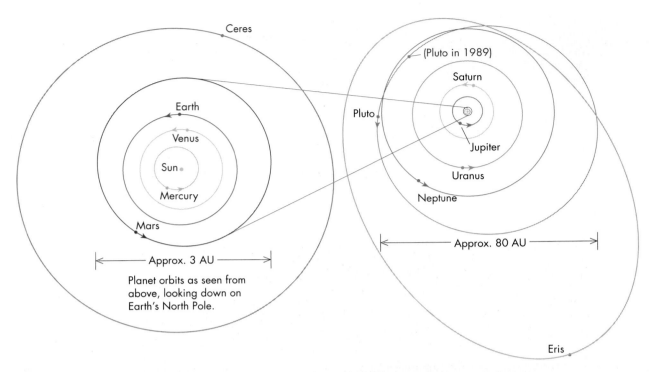

**FIGURE 7.2**
An artist's view of the Solar System from above. The orbits are shown in the correct relative scale in the two drawings.

The Solar System out to Neptune has about the same relative thickness as 3 CDs stacked together.

move around the Sun in approximately circular* orbits all lying in nearly the same plane, as shown in the top view in figure 7.2 and the side view in figure 7.3. Thus, the Solar System is like a spinning pancake, with the planets traveling around the Sun in the same direction: counterclockwise, as seen from above the Earth's North Pole. We can see this flatness from Earth: it makes the planets all lie in an approximate line in the sky (figure 7.4).

As the planets orbit the Sun, each also spins on its rotation axis. The spin is generally in the same direction as each planet's orbital motion around the Sun (again, counterclockwise, as seen from above the Earth's North Pole), and the tilt of the rotation axes relative to the plane of planetary orbits is generally not far from the perpendicular. However, there are two exceptions: Venus and Uranus. Uranus has an extremely large tilt to its rotation axis: it lies nearly in its orbital plane (fig. 7.3). Venus's rotation axis has only a small tilt, but it spins backward, a motion technically called "retrograde rotation." However, despite this backward spin, Venus orbits the Sun in the same direction as the rest of the planets.

The flattened structure (fig. 7.4) and the orderly orbital and spin properties of planets in our Solar System are two of its most fundamental features, and any theory of the Solar System must explain them. But a third and equally important feature is that the planets fall into two families called inner and outer planets based on their size, composition, and location in the Solar System.

## Two Types of Planets

The **inner planets**—Mercury, Venus, Earth, and Mars—are small rocky bodies with relatively thin or no atmospheres. The **outer planets**—Jupiter, Saturn, Uranus, and Neptune—are gaseous and liquid. They are much larger than the inner planets and have deep, hydrogen-rich atmospheres. For example, Jupiter is more than 10 times larger in diameter than the Earth and has 318 times its mass. These differences can be

---

*The circularity of the orbits is only approximate because we know from Kepler's laws that in reality, the orbits are ellipses. However, the amount of ellipticity is, in general, very small.

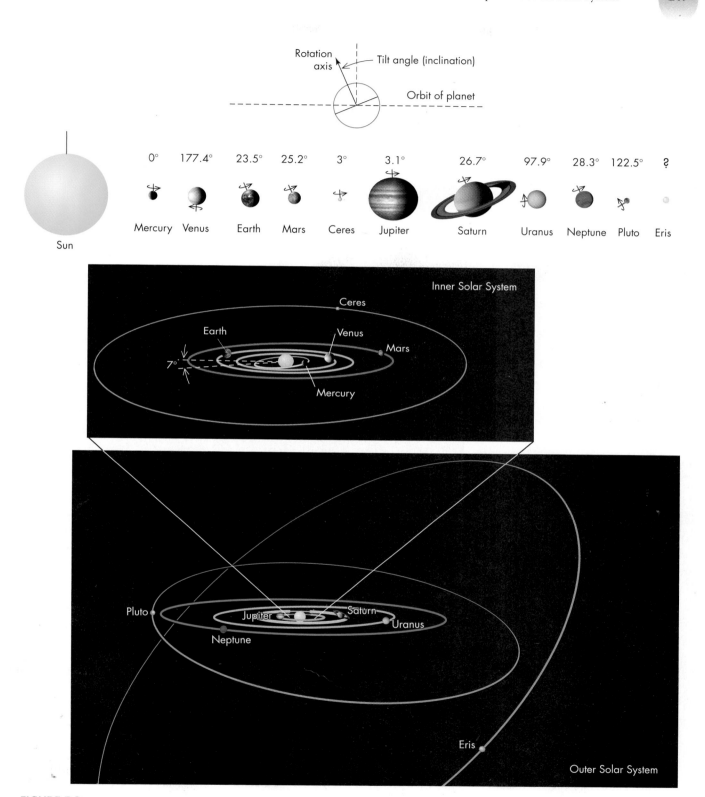

**FIGURE 7.3**

Planets and their orbits from the side. Sketches also show the orientation of the rotation axes of the planets and Sun. Orbits and bodies are not to the same scale.* The dwarf planets Pluto and Eris are also shown, illustrating their highly inclined orbits.

---

*Technically, Venus's tilt angle is about 177°, but it can also be described as 3° (180° − 177°), with a "backward," or retrograde, spin.

FIGURE 7.4
Sunset view of four planets strung along the zodiac on March 1, 1999. Their straight-line arrangement results from the flatness of the Solar System. From top to bottom, you can see Saturn, Venus, Jupiter, and Mercury (nearly lost in the twilight).

seen in figure 7.5, which also shows a small part of the edge of the Sun to illustrate how the Sun dwarfs even the large planets.

In describing the planets, we have used the terms *rock* and *ice*. By rock, we mean material composed of silicon and oxygen ($SiO_2$) with an admixture of other heavy elements such as aluminum (Al), magnesium (Mg), sulfur (S), and iron (Fe). By ice, we mean frozen liquids and gases such as ordinary water ice ($H_2O$), frozen carbon dioxide ($CO_2$), frozen ammonia ($NH_3$), frozen methane ($CH_4$), and so on. If we consider the Solar System as a whole, rock is rare because the silicon atoms that compose it are outnumbered more than 25,000 to 1 by hydrogen. However, in the warmth of the inner Solar System, rock dominates because intrinsically more abundant materials such as hydrogen, water, methane, and ammonia cannot condense to mingle with it. Thus, the inner planets are composed mainly of rock.

The outer planets have no true "surface"; rather their atmospheres thicken with depth and eventually liquefy. There is no distinct boundary between "atmosphere" and "crust" as we have on the Earth. In the deep interior, the liquid may be compressed into a solid, as happens in the Earth's inner core, but the transition from liquid to solid is also probably not sharply defined. Thus, we can never "land" on Jupiter because we would simply sink ever deeper into its interior.

Instead of "inner" and "outer" planets, astronomers sometimes use "terrestrial" and "Jovian" to describe the two types of planets. The **terrestrial planets** (Mercury to Mars) are so named because of their resemblance to the Earth. The **Jovian planets***  (Jupiter to Neptune) are named for their resemblance to Jupiter.

Although two categories of planets neatly describe the larger objects that orbit the Sun, astronomers have found many smaller objects that fit neither category. Pluto has long failed to fit, because of its small size, composition of ice and rock, and odd

---

*Some astronomers go further and divide the Jovian planets into *gas giants* (Jupiter and Saturn) and *ice giants* (Uranus and Neptune).

**FIGURE 7.5**
The planets and Sun to scale.

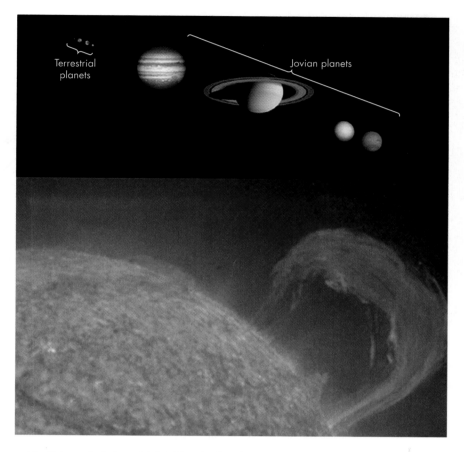

orbit. (Not only is its orbit highly tilted with respect to the other planets, it also crosses Neptune's orbit). Moreover, in the last decade astronomers have discovered some 200 similar (though slightly smaller) objects orbiting at almost the same distance from the Sun as Pluto. Then, in 2005, astronomers announced the discovery of Eris, an icy world slightly *larger* than Pluto that orbits about 68 AU from the Sun, roughly half again Pluto's distance from our star.

In response to the discovery of Eris and half a dozen other objects similar in size but even more distant, the International Astronomical Union introduced in 2006 a new category of Solar System objects called **dwarf planets.**[*] Dwarf planets are objects that orbit the Sun, are massive enough that their gravity compresses them into an approximately spherical shape, but have not swept their orbital region clear of other objects.

## Satellites

As the planets orbit the Sun, most are themselves orbited by satellites. Jupiter, Saturn, and Uranus have large families of 63, 56, and 27 clearly identified moons, respectively.[†] Neptune has 13, Mars has 2, and Earth has just 1. Even some of the dwarf planets have moons. Pluto has 3 and Eris has 1. Only Mercury and Venus are moonless.

The larger moons generally move along approximately circular paths that are roughly in the planet's equatorial plane, their orbits tilted like the planets themselves. Thus, each planet and its moons resemble a miniature Solar System—an important clue to the origin of the satellites.

*Oort cloud and Kuiper belt*

---

[*]The new definition is not without problems of its own, however. See the *Astronomy Education Review* website at aer.noao.edu for many links to this topic.

[†]Astronomers are finding new moons around these planets so rapidly that it is very difficult to keep the numbers up-to-date.

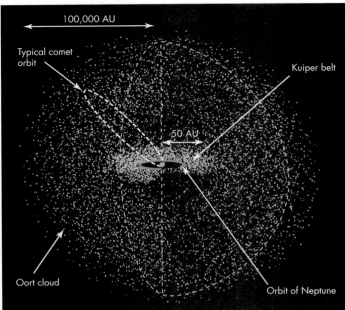

100,000 AU

Typical comet orbit

Kuiper belt

50 AU

Oort cloud

Orbit of Neptune

**FIGURE 7.6**

Sketch of the Oort cloud and the Kuiper belt. The scale shown is only approximate. Orbits and bodies are not to scale.

## Asteroids and Comets

Asteroids and comets are far smaller than planetary bodies. The **asteroids** are rocky or metallic bodies with diameters that range from a few meters up to about 1000 km (about one-tenth the size of the Earth). The **comets,** on the other hand, are icy bodies about 10 km (about 6 miles) or less in diameter that grow huge tails of gas and dust as they near the Sun and are partially vaporized by its heat. Thus, these minor bodies exhibit the same split into two families that we see for the planets; that is, rocky and icy bodies.

Asteroids and comets differ not only in their composition but also in their location within the Solar System. Most asteroids circle the Sun in the large gap between the orbits of Mars and Jupiter, a region called the **asteroid belt.** They are probably material that failed—perhaps as a result of disturbance by Jupiter's gravity—to aggregate into a planet. Most comets, on the other hand, orbit far beyond Pluto in a region of the Solar System called the Oort cloud, and only rarely do they move into the inner Solar System.

The **Oort cloud,** named for the Dutch astronomer who proposed its existence, is thought to be a spherical region that completely surrounds the Solar System and extends from about 40,000 to 100,000 AU from the Sun (fig. 7.6). Although the majority of comets probably originate in the Oort cloud, some come from a disklike swarm of icy objects that lies just beyond the orbit of Neptune and extends to perhaps 50 AU from the Sun, a region called the **Kuiper belt.**\* We will discuss more details of the Oort cloud and Kuiper belt in chapter 10, but for now, we simply note that together they probably contain more than 1 trillion ($10^{12}$) comet nuclei, thousands of larger objects, and probably several dozen additional dwarf planets.

## Composition Differences
## Between the Inner and Outer Planets

We stated earlier that the inner planets are rocky and the outer planets are hydrogen-rich. These composition differences are so important to our understanding of the history of the Solar System that we should look at them more closely and see how we determine them.

Astronomers can deduce a planet's composition in several ways. From its spectrum, they can measure its atmospheric composition and get some information about the nature of its surface rocks. However, spectra give no clue as to what lies deep inside a planet where light cannot penetrate. To learn about the interior, astronomers must therefore use alternative methods.

We saw in chapter 5 how earthquake waves reveal what lies inside the Earth, but this method has not yet been used for other planets. Although quake detectors were landed on Mars, they did not work properly, and such detectors would require very special modification to work on the Jovian planets because they have no surface on which to land! Thus, we must try other means to study the interior of planets. One such technique uses the planet's density.

## Density as a Measure of a Planet's Composition

The average density of a planet is its mass divided by its volume. Both these quantities can be measured relatively easily. For example, we showed in chapter 2 how to

---

\*The dwarf planet Pluto actually lies within the Kuiper belt. Pluto's relation to these other objects is discussed more in chapter 10.

# EXTENDING OUR REACH

## BODE'S LAW: THE SEARCH FOR ORDER

A curious and as yet unexplained feature of the orbits of the planets is their regular spacing. Very roughly, each planet is about twice as far from the Sun as its inner neighbor. This progression of distance from the Sun can be expressed by a simple mathematical relation known as **Bode's law,** which works as follows: write down 0, 3, and then successive numbers by doubling the preceding number until you have nine numbers. That is, 0, 3, 6, 12, 24, and so on. Next, add 4 to each, and divide the result by 10, as shown in table 7.1. The resulting numbers, with two exceptions, are very close to the distances of the planets from the Sun in astronomical units.

Bode's law was worked out before the discovery of Uranus, Neptune, and Pluto, and when Uranus was discovered and found to fit the law, interest was focused on the "gap" at 2.8 AU. Astronomers therefore began to search for a body in the gap, and, as we will see in chapter 10, Giuseppi Piazzi, a Sicilian astronomer, soon discovered the dwarf planet Ceres, which fitted the rule splendidly.

Ironically, the next planet to be found, Neptune, did not fit the rule at all, though the dwarf planet Pluto does, at least approximately! These irregularities make astronomers unsure how to treat Bode's law. Bode's law is not a law like the "law of gravity," and many astronomers prefer to call it "Bode's rule" to emphasize this differ-

ence. It is not based on any (known) physical principles, but computer simulations of planet formation sometimes produce planet spacings that follow similar spacing patterns. Is the "law" just chance, or is it telling us something profound about the Solar System that we have yet to understand? Perhaps it merely shows the human fascination with patterns and our tendency to see order where none may actually exist.

### TABLE 7.1

Bode's Law

| Bode's Law | Number | Object | True Distance |
|---|---|---|---|
| $(0 + 4)/10 =$ | 0.4 | Mercury | 0.39 |
| $(3 + 4)/10 =$ | 0.7 | Venus | 0.72 |
| $(6 + 4)/10 =$ | 1.0 | Earth | 1.00 |
| $(12 + 4)/10 =$ | 1.6 | Mars | 1.52 |
| $(24 + 4)/10 =$ | 2.8 | Ceres | 2.77 |
| $(48 + 4)/10 =$ | 5.2 | Jupiter | 5.2 |
| $(96 + 4)/10 =$ | 10.0 | Saturn | 9.5 |
| $(192 + 4)/10 =$ | 19.6 | Uranus | 19.2 |
| | | Neptune | 30.1 |
| $(384 + 4)/10 =$ | 38.8 | Pluto | 39.5 |

determine a body's mass from its gravitational attraction on a second body orbiting around it by applying Newton's modification of Kepler's third law. Thus, from this law, we can calculate a planet's mass by observing the orbital motion of one of its moons or a passing spacecraft. We can determine a planet's volume ($V$) from the formula $V = 4\pi R^3/3$, where $R$ is the planet's radius. We can measure $R$ in several ways— for example, from its angular size and distance, a technique we used in chapter 1 to measure the radius of the Moon. With the planet's mass, $M$, and volume, $V$, known, we can calculate its average density straightforwardly by dividing $M$ by $V$ (fig. 7.7).

Once the planet's average density is known, we can compare it with the density of abundant, candidate materials to find a likely match. For example, we saw in chapter 5 that the average density of the Earth (5.5 grams per cubic centimeter) was intermediate between silicate rock (about 3 grams per cubic centimeter) and iron (7.8 grams per cubic centimeter). Therefore, we inferred that the Earth has an iron core beneath its rocky crust, a supposition that was verified from studies using earthquake waves.

Although density comparison is a powerful tool for studying planetary composition, it also has drawbacks. First, there may be several different substances that will produce an equally good match to the observed density. Second, the density of a given material can be affected by the planet's gravitational force. For example, a massive planet may crush rock whose normal density is 3 grams per cubic centimeter to a density of 7 or 8 grams per cubic centimeter. Thus, in making a match to determine the composition, we must take into account compression by gravity.

All the terrestrial planets have an average density similar to the Earth's (3.9 to 5.5 grams per cm³). On the other hand, all the Jovian planets have a much smaller average density (0.71 to 1.67 grams per cm³), similar to that of ice. After correcting

**FIGURE 7.7**
Measuring a planet's mass, radius, and average density.

⟳⋆ Suppose you are given a tiny box that has a volume of 10 cubic centimeters and a mass of 30 grams. What is its density? Is it more likely to contain solid iron or rock?

ℚ⋆ The density is the mass (30 grams) divided by the volume (10 cm³). 30 grams/10 cm³ = 3 grams/cm³. Iron's density is about 8 gm/cm³, while a typical silicate rock's density is about 3 gm/cm³. It is thus more likely to be rock.

**Mass**

Observe motion of a satellite orbiting planet. Determine satellite's distance (a) from planet and orbital period, P. Use Newton's form of Kepler's third law.

$$M = \frac{4\pi^2 a^3}{GP^2}$$

Insert measured values of a and P, and value for constant G. Solve for M.

**Average Density**

Average density ($\rho$) is mass (M) divided by volume (V)

$$\rho = \frac{M}{V}$$

**Radius**

Measure angular size of planet and use relation between angular size and distance to solve for R.

**Volume**

$$V = \frac{4\pi R^3}{3}$$

For a spherical planet of radius R

for the above gravitational compression, we conclude that all the inner planets contain large amounts of rock and iron and that the iron has sunk to the core, as shown in figure 7.8. Likewise, the outer planets contain mainly light materials, as borne out by their spectra, which show them to be mostly hydrogen, helium, and hydrogen-rich material such as methane ($CH_4$), ammonia ($NH_3$), and water ($H_2O$).

The outer planets probably have cores of iron and rock about the size of the Earth beneath their deep atmosphere, as illustrated in figure 7.8. Astronomers deduce the existence of these cores in two ways. First, if the outer planets have the same relative amount of heavy elements as the Sun, they should contain several Earth masses of iron and silicates, and because these substances are much denser than hydrogen, they must sink to the planet's core. Secondly, mathematical analysis of these planets' rotation rates shows that the shape of their equatorial bulges can be best explained if they have small, dense cores.

Our discussion of the composition of the planets not only underlines the difference between the two families of planets but also furnishes another clue to their origin: the planets and Sun were all made from the same material. Astronomers come to this conclusion because Jupiter and Saturn have a composition almost identical to that of the Sun, and the inner planets have a similar composition *if we were to remove the Sun's hydrogen and helium*.* Thus, we can explain the compositional difference between the inner and outer planets by proposing a process that would keep the inner planets from collecting and capturing these light gases.

## Age of the Solar System

An important clue to the origin of the Solar System comes from its age. Despite great differences in size, structure, and composition, the planets, asteroids, and comets all seem to have formed at nearly the same time. We can directly measure that date for the Earth, Moon, and some asteroids from the radioactivity of their rocks, and we find that none is more than about 4.6 billion years old. (Rocks from asteroids are the origin of many of the meteors that have reached Earth and are thus available for study, as we will discuss in chapter 10). Likewise, we find a similar age for the Sun based on its current brightness and temperature and its presumed rate of nuclear fuel consumption.

---

*Carbon, nitrogen, neon, and other elements normally in gaseous compounds are also relatively rare in the inner planets.

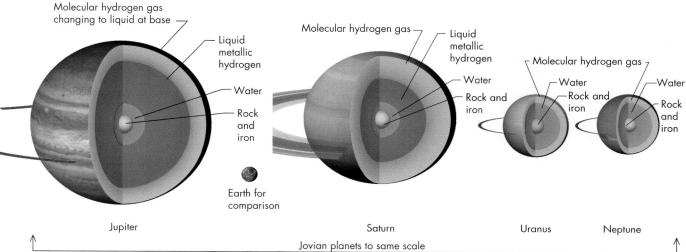

**FIGURE 7.8**

Sketches of the interiors of the planets. Details of sizes and composition of inner regions are uncertain for many of the planets.

# 7.2   ORIGIN OF THE SOLAR SYSTEM

How did the Solar System form? What processes gave it the features we discussed in the previous section, such as its flatness and two main families of planets? Given that we were not around 4.6 billion years ago to witness its birth, our best explanation of its origin must be a reconstruction based on observations that we make now, billions of years after the event. Those observations, discussed above, are summarized below. Each must be explained by whatever theory we devise.

1. The Solar System is flat, with all the planets orbiting in the same direction.
2. There are two types of planets, inner and outer, with the rocky ones near the Sun and the gaseous or liquid ones farther out.
3. The composition of the outer planets is similar to the Sun's, while that of the inner planets is like the Sun's minus the gases that condense only at low temperatures.
4. All the bodies in the Solar System whose ages have so far been determined are less than about 4.6 billion years old.

We have listed only the most important observed features that our theory must explain. There are many additional clues from the structure of asteroids, the number

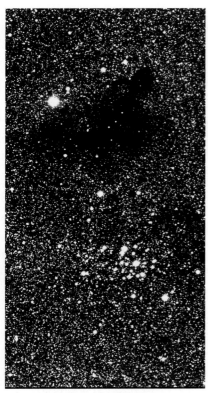

**FIGURE 7.9**
Photograph of an interstellar cloud (the dark region at top) that may be similar to the one from which the Solar System formed. The dark cloud is known as Barnard 86. The star cluster is NGC 6520.

*Solar System builder*

of craters on planetary and satellite surfaces, and the detailed chemical composition of surface rocks and atmospheres.

The currently favored theory for the origin of the Solar System derives from theories proposed in the eighteenth century by Immanuel Kant, the great German philosopher, and Pierre Simon Laplace, a French mathematician. Kant and Laplace independently proposed what is now called the **solar nebula theory** that the Solar System originated from a rotating, flattened disk of gas and dust, with the outer part of the disk becoming the planets and the center becoming the Sun. This theory offers a natural explanation for the flattened shape of the system and the common direction of motion of the planets around the Sun.

The modern form of the theory proposes that the Solar System was born 4.6 billion years ago from an **interstellar cloud,** an enormous rotating aggregate of gas and dust like the one shown in figure 7.9. Such clouds are common between the stars in our galaxy even today, and astronomers now think all stars have formed from them. Thus, although our main concern in this chapter is with the birth of the Solar System, we should bear in mind that our theory applies more broadly and implies that *most* stars could have planets, or at least surrounding disks of dust and gas from which planets might form.

## Interstellar Clouds

Because interstellar clouds are the raw material of the Solar System, we need to describe them more fully. Although such clouds are found in many shapes and sizes, the one that became the Sun and planets was probably a few light years in diameter and contained about twice the present mass of the Sun. If it was like typical clouds we see today, it was made mostly of hydrogen (71%) and helium (27%) gas, with tiny traces of other chemical elements, such as gaseous carbon, oxygen, and silicon. In addition to the gases, interstellar clouds also contain tiny dust particles called **interstellar grains.**

Interstellar grains range in size from large molecules to micrometers or larger and are believed to be made of a mixture of silicates, iron compounds, carbon compounds, and water frozen into ice. Astronomers deduce the presence of these substances from their spectral lines that are seen in starlight that has passed through dense dust clouds. Moreover, a few hardy interstellar dust grains, including tiny diamonds, have been found in ancient meteorites. This direct evidence from grains and the data from spectral lines shows that the elements occur in proportions similar to those we observe in the Sun. This is additional evidence that the Sun and its planets could have formed from an interstellar cloud.

The cloud began its transformation into the Sun and planets when the gravitational attraction between the particles in the densest parts of the cloud caused it to collapse inward, as shown in figure 7.10A. The collapse may have been triggered by a star exploding nearby or by a collision with another cloud. But regardless of its initial cause, the infall was not directly to the center. Instead, because the cloud was rotating, it flattened, as shown in figure 7.10B.

Flattening occurred because rotation retarded the collapse perpendicular to the cloud's rotation axis. A similar effect happens in an old-fashioned pizza parlor where the chef flattens the dough by tossing it into the air with a spin.

## Formation of the Solar Nebula

It took a few million years for the cloud to collapse and become a rotating disk with a bulge in the center. The disk is called the **solar nebula,** and it eventually condensed into the planets while the bulge became the Sun. This explains the first obvious property of the Solar System—its disklike structure—which we mentioned at the beginning of the section.

The solar nebula was probably about 200 AU in diameter and perhaps 10 AU thick. Its inner parts were hot, heated by the young Sun and the impact of gas falling on the disk during its collapse, but the outer parts were cold, far below the freezing point of water. We are fairly certain of these dimensions and temperatures because we can

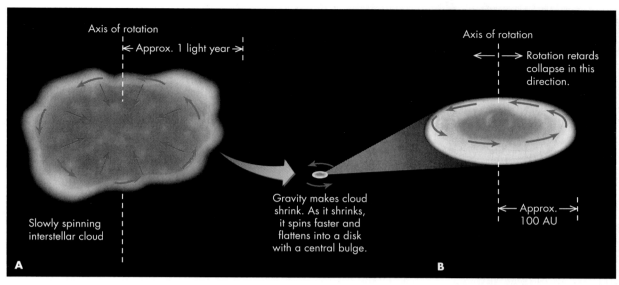

**FIGURE 7.10**
A sketch illustrating the (A) collapse of an interstellar cloud and (B) the cloud's flattening. Note that the final size of the disk is not shown to scale—in actuality it would be about 60,000 times smaller than the cloud from which it formed!

*Flattening and spreading up of a collapsing interstellar cloud*

observe disks around other stars and, in a few cases, can even detect other planets. For example, figure 7.11A shows a picture made with the Hubble Space Telescope of gas and dust disks near the Orion Nebula. The stars at the centers of these disks have not yet become hot enough to emit much visible light. Figure 7.11B, on the other hand, shows a disk (seen nearly edge on) in which the star has reached full brilliance. Although the picture is grainy and in false color in order to emphasize the limited detail, you can see the disk edge on.

## Condensation in the Solar Nebula

**Condensation** occurs when a gas cools and its molecules stick together to form liquid or solid particles.\* For condensation to happen, the gas must cool below a critical temperature (the value of which depends on the substance condensing and the surrounding pressure). For example, suppose we start with a cloud of vaporized iron at a

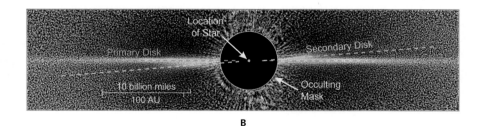

**FIGURE 7.11**
(A) The small blobs in this picture are stars in the process of formation (protostars) and their surrounding disk of dust and gas. These are in the Orion Nebula, a huge gas cloud about 1500 light years from Earth. (B) Picture in false color of a disk of dust around the young star β Pictoris made with the Hubble Space Telescope. A small mask in the telescope blots out the star's direct light, which would otherwise overexpose the image. Note the second faint dust disk.

---

\*Technically, condensation is the change from gas to liquid, and deposition is the change from gas to solid. However, we will not make that distinction here.

**FIGURE 7.12**

Water vapor cools as it leaves the kettle. The cooling makes the vapor condense into tiny liquid water droplets, which we see as the "steam."

Q* How does the process illustrated in Figure 7.12 explain why you can see your breath on a cold morning.

A* When you breathe out, warm moist air from your lungs comes in contact with the cold air outside. The moisture in your breath condenses and makes a tiny "cloud."

temperature of 2000 K. If we now cool the iron vapor to about 1300 K, tiny flakes of iron will condense from it. Likewise, if we cool a gas of silicates to about 1200 K, flakes of rocky material will condense.

At lower temperatures, other substances will condense. For example, water can condense at room temperature, as you can see as steam escapes from a boiling kettle (fig. 7.12). Here, water molecules in the hot steam come into contact with the cooler air of the room. As the vaporized water cools, its molecules move more slowly, so that when they collide, electrical forces can bind them together, first into pairs, then into small clumps, and eventually into the tiny droplets that make up the cloud we see at the spout.

An important feature of condensation is that when a mixture of vaporized materials cools, the materials with the highest vaporization temperatures condense first. Thus, as a mixture of gaseous iron, silicate, and water cools, it will make iron grit when its temperature reaches 1300 K, silicate grit when it reaches 1200 K, and finally water drops when it cools to only a few hundred degrees K. It is a bit like putting a jar of chicken soup in the freezer. First the fat freezes, then the broth, and finally the bits of chicken and celery.

However, the condensation process stops if the temperature never drops sufficiently low. Thus, in the example above, if the temperature never cools below 500 K, water will not condense and the only solid material that forms from the gaseous mixture will be iron and silicates.

This kind of condensation sequence occurred in the solar nebula as it cooled after its collapse to a disk. But because the Sun heated the inner part of the disk, the temperature from the Sun to almost the orbit of Jupiter never dropped low enough for water and other substances with similar condensation temperatures to condense there. On the other hand, iron and silicate, which condense even at relatively high temperatures, could condense everywhere within the disk. Thus, the nebula became divided into two regions: an inner zone of silicate-iron particles, and an outer zone of similar particles on which ices also condensed, as illustrated schematically in figure 7.13. Water, hydrogen, and other easily vaporized substances were present as gases in the inner solar nebula, but they could not form solid particles there. However, some of these substances combined chemically with silicate grains so that the rocky material from which the inner planets formed contained within it small quantities of water and other gases.

## Accretion and Planetesimals

In the next stage of planet formation, the tiny particles that condensed from the nebula must have begun to stick together into bigger pieces in a process called **accretion.**

The process of accretion is a bit like building a snowman. You begin with a handful of loose snowflakes and squeeze them together to make a snowball. Then you add more snow by rolling the ball on the ground. As the ball gets bigger, it is easier for snow to stick to it, and it rapidly grows in size.

Similarly in the solar nebula, tiny grains stuck together and formed bigger grains that grew into clumps, perhaps held together by electrical forces similar to those that make lint stick to your clothes. Subsequent collisions, if not too violent, allowed these smaller particles to grow into objects ranging in size from millimeters to kilometers. These larger objects are called **planetesimals** (that is, small, planetlike bodies) (fig. 7.13).

Because the planetesimals near the Sun formed from silicate and iron particles, while those farther out were cold enough that they could incorporate ice and frozen gases as well, there were two main types of planetesimals: rocky-iron ones near the Sun and icy-rocky-iron ones farther out. This then explains the second observation we described at the beginning of this section—that there are two types of planets—as described below.

## Formation of the Planets

As planetesimals moved within the disk and collided with one another, planets formed. Computer simulations show that some collisions led to the shattering of

*Planet formation from the solar nebula*

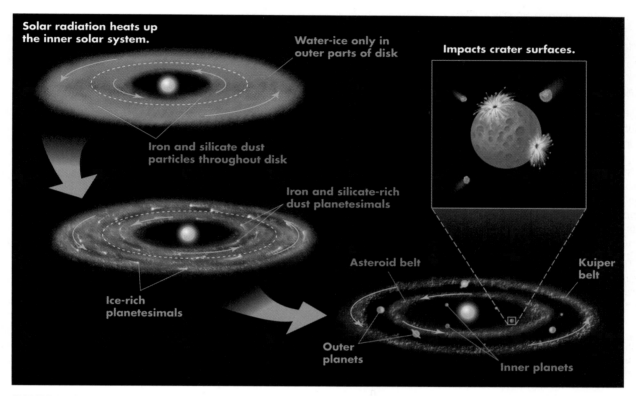

**Solar radiation heats up the inner solar system.**

**Water-ice only in outer parts of disk**

**Impacts crater surfaces.**

**Iron and silicate dust particles throughout disk**

**Iron and silicate-rich dust planetesimals**

**Ice-rich planetesimals**

**Asteroid belt**

**Kuiper belt**

**Outer planets**

**Inner planets**

**FIGURE 7.13**
An artist's depiction of how the planets may have formed in the solar nebula.

both bodies, but gentler collisions led to merging, with the planetary orbits gradually becoming approximately circular. Moreover, in some such simulations, the distance between orbits is similar to that given by Bode's law.

Merging of the planetesimals increased their mass and thus their gravitational attraction. That, in turn, helped them grow even more massive by drawing planetesimals into clumps or rings around the Sun. Within these clumps, growth went even faster, so that over a time lasting about 100,000 years, larger and larger objects formed.

Planet growth was especially rapid in the outer parts of the solar nebula. Planetesimals there had more material from which to grow because ice was about 10 times more abundant than silicate and iron compounds. Thus, planetesimals in the outer solar nebula could in principle become 10 times larger than those in the inner nebula.

Additionally, once a planet grew somewhat larger than the diameter and mass of the Earth, it was able to attract and retain gas by its own gravity. Because hydrogen was overwhelmingly the most abundant material in the solar nebula, planets large enough to tap that reservoir could grow vastly larger than those that formed only from solid material. Thus, Jupiter, Saturn, Uranus, and Neptune may have begun as Earth-size bodies of ice and rock, but their gravitational attraction resulted in their becoming surrounded by the huge envelopes of hydrogen-rich gases that we see today. The smaller and warmer bodies of the inner Solar System could not capture hydrogen and therefore remained small and lack that gas. This explains the third observation we mentioned at the beginning of the section—that the outer planets have a composition similar to the Sun's.

As planetesimals struck the growing planets, their impact released gravitational energy that heated both the planetesimal and the planet. Gravitational energy is liberated whenever something falls. For example, when a cinderblock falls onto a box of tennis balls, the impact scatters the balls in all directions, giving them kinetic

## DIRECT FORMATION OF GIANT PLANETS

Because astronomers have no direct way to observe how the Solar System formed, they rely heavily on computer simulations to study that distant time. Computer simulations try to solve Newton's laws of motion for the complex mix of dust and gas that we believe made up the solar nebula. The solutions then can reveal what might have happened as the dust particles stuck together to form planetesimals and how the planetesimals then drew together under the influence of their gravity to form planets.

One of the more interesting findings of such calculations is that the giant planets may have formed directly from slightly denser regions of gas in the disk. Far from the Sun, where the gas is cold, gravity can more easily overcome the resistance of warmer gas to being squeezed into a smaller region. (Think of how a balloon resists being squeezed.) This may have allowed gravity to pull gas together to make the giant planets without the need to first form cores from planetesimals. Does this make the planetesimal theory wrong? No, just incomplete. Moreover, because this is an area of active research, astronomers still await definitive answers.

energy—energy of motion. In much the same manner, planetesimals falling onto a planet's surface give energy to the atoms in the crustal layers, energy that appears as heating. You can easily demonstrate that motion can generate heat by hitting a steel nail a dozen or so times with a hammer and then touching the nail to your lip: the metal will feel distinctly hot. Imagine now the vastly greater heating created as mountain-size masses of rock plummet onto a planet. The heat so liberated, in combination with the radioactive heating described earlier, melted the planets and allowed matter with high density (such as iron) to sink to their cores, while matter with lower density (such as silicate rock) "floated" to their surfaces. We saw in chapter 5 that the Earth's iron core probably formed by this process, and astronomers believe that the other terrestrial planets formed their iron cores and rocky crusts and mantles the same way. A similar process probably occurred for the outer planets when rock and iron material sank to their cores.

## Formation of Moons

The moons of the outer planets probably were formed from planetesimals orbiting the growing planets. Once a body grew massive enough so that its gravitational force could draw in additional material, it became ringed with debris. Thus, moon formation was a scaled-down version of planet formation, and so the satellites of the outer planets have the same regularities as the planets around the Sun.

All four giant planets have flattened satellite systems in which the larger satellites (with few exceptions) orbit in the same direction. Many of these satellites are about as large as Mercury, and they would be considered full-fledged planets were they orbiting the Sun. A few of these bodies even have atmospheres, but they have too little mass (and thus too weak a gravitational attraction) to have accumulated large quantities of hydrogen and other gases from the solar nebula as their parent planets did. Thus, these moons are composed mainly of rock and ice, giving them solid surfaces—surfaces that are generally cratered and that, in a few cases, show signs of volcanic activity. These distant moons might in the future be ideal bases for studying those planets that have no surface to land on.

## Final Stages of Planet Formation

The last stage of planet formation was a rain of planetesimals that blasted out the huge craters such as those we see on the Moon and on all other bodies in the Solar System with solid surfaces. Figure 7.14 shows some of the planets and moons bearing this vivid testimony to the role of planetesimals in planet building.

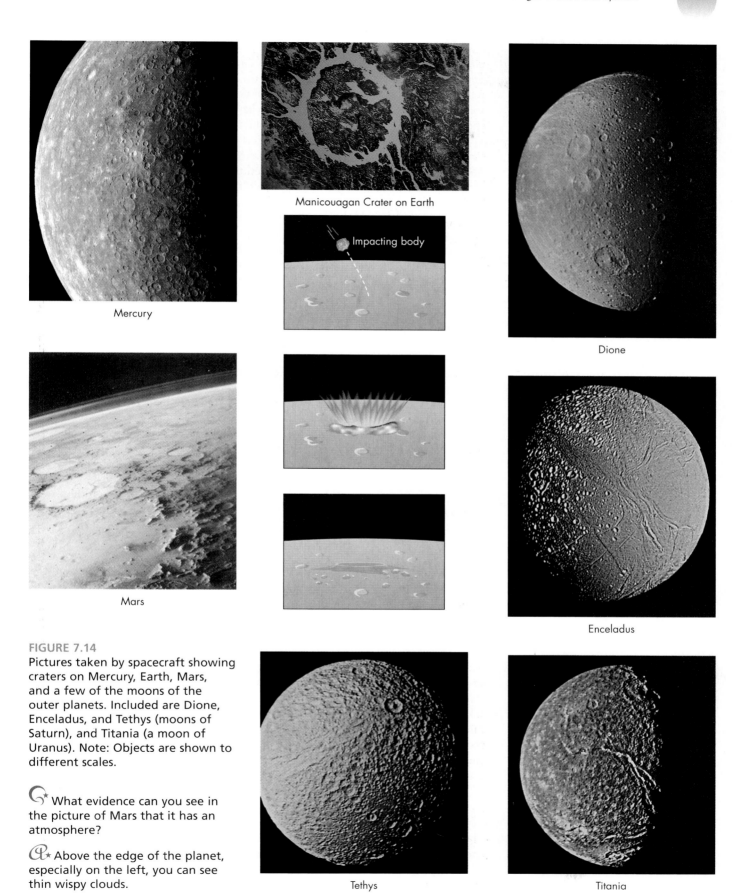

Mercury

Manicouagan Crater on Earth

Impacting body

Dione

Mars

Enceladus

**FIGURE 7.14**
Pictures taken by spacecraft showing craters on Mercury, Earth, Mars, and a few of the moons of the outer planets. Included are Dione, Enceladus, and Tethys (moons of Saturn), and Titania (a moon of Uranus). Note: Objects are shown to different scales.

✩ What evidence can you see in the picture of Mars that it has an atmosphere?

✩ Above the edge of the planet, especially on the left, you can see thin wispy clouds.

Tethys

Titania

Occasionally, an impacting body was so large that it did more than simply leave a crater. For example, we saw in chapter 6 that the Moon may have been created when the Earth was struck by a Mars-size body. Likewise, as we will discuss in chapter 8, Mercury may have suffered a massive impact that blasted away its crust. The extreme tilt of the rotation axes of Uranus and Pluto may also have arisen from planetesimal collisions. In short, planets and satellites were brutally battered by the remaining planetesimals.

Although planet building consumed most of the planetesimals, some survived to form small moons, the asteroids, and comets. Rocky planetesimals and their fragments remained between Mars and Jupiter, where, stirred by Jupiter's gravitational force, they were unable to assemble into a planet. We see them today as the asteroid belt. Jupiter's gravity (and that of the other giant planets) also disturbed the orbits of icy planetesimals, tossing some in toward the Sun and others outward in elongated orbits to form the swarm of comet nuclei of the Oort cloud. The few that remain in the disk from Neptune's orbit out to beyond Pluto's form the Kuiper belt.

## Formation of Atmospheres

Atmospheres were the last part of the planet-forming process. The inner and outer planets are thought to have formed atmospheres differently, a concept that explains their very different atmospheric composition. The outer planets probably captured most of their atmospheres from the solar nebula, as mentioned above; because the nebula was rich in hydrogen, so are their atmospheres.

The inner planets were not massive enough and were too hot to capture gas from the solar nebula (as mentioned earlier) and are therefore deficient in hydrogen and helium. Venus, Earth, and Mars probably created their original atmospheres—by volcanic eruptions and by retaining gases from infalling comets and icy planetesimals that vaporized on impact. In fact, as a general rule, bodies too small to have captured atmospheres directly but that show clear signs of extensive volcanic activity (now or in the past) have atmospheres. More quiescent ones do not. Moreover, small bodies such as Mercury and our Moon keep essentially no atmosphere at all because their weak gravitational force means that their escape velocity is rather small, and atmospheric gases tend to escape easily from them.

## Cleaning Up the Solar System

Only a few hundred thousand or million years were needed to assemble the planets from the solar nebula, though the rain of infalling planetesimals may have lasted longer. Such a time is long in the human time frame but short in the Solar System's. All the objects within the Solar System are about the same age—the fourth property of the Solar System, mentioned at the beginning of this section.

One process still had to occur before the Solar System became what we see today: the residual gas and dust must have been removed. Just as a finished house is swept clean of the debris of construction, so too was the Solar System. In the sweeping process, the Sun was probably the cosmic broom, with its intense heat driving a flow of tenuous gas outward from its atmosphere. As that flow impinges on the remnant gas and dust around the Sun, the debris is pushed away from the Sun to the fringes of the Solar System. Such gas flows are seen in most young stars, and astronomers are confident the Sun was no exception. Even today, some gas flows out from the Sun, but in its youth, the flow was more vigorous.

The above theory for the origin of the Solar System explains many of its features, but astronomers still have many questions about how the Sun and its family of planets and moons formed. Is there any way, therefore, we can confirm the theory? For example, according to the solar nebula theory, planet formation is a normal part of star formation, so, do other planetary systems exist and do they resemble ours? Might we even be able to discover very young stars in the process of forming their planets and see the process at work?

# 7.3   OTHER PLANETARY SYSTEMS

**INTERACTIVE**

*Exoplanets*

Astronomers have long searched for planets orbiting stars other than the Sun. Their interest in such **exoplanets**\* (as these distant worlds are called) is motivated not merely by the wish to detect other planets. Equally important is the hope that study of such systems will help us better understand the formation of the Solar System.

Directly seeing exoplanets is extremely difficult. Planets are, of course, very small, so the light they reflect is drowned by that of their star. However, if we could find a system just beginning to form planets, the young star—according to the solar nebula theory—should be surrounded by a disk of dust and gas tens of astronomical units across. Such a disk should be visible, and in fact, just such disks are seen with newer ground-based instruments and the Hubble Space Telescope (fig. 7.15A). In a few instances, the disks even show "lumps," which may represent the first clustering of dust as planetesimals form (fig. 7.15B). In other cases, the disks show a strong ring (fig. 7.15B), perhaps caused by planets that have already formed there.[†] In short, these images confirm that disks similar to the solar nebula exist. But what about the planets themselves?

Most present evidence for exoplanets comes from their effect on the star they orbit. As a planet orbits its star, the planet exerts a gravitational force back on the star as a result of Newton's third law—the law of action–reaction. That force makes the star's position wobble slightly, just as you wobble a little if you swing a heavy weight around you. The wobble creates a Doppler shift in the star's light that astronomers can measure (fig 7.16). From that shift and its change in time, astronomers can deduce the planet's orbital period, mass, and distance from the star. Using this Doppler method, astronomers have discovered about 200 exoplanets, and figure 7.17 shows sketches of a few of these systems.

Looking at figure 7.17, you can see that none of these exoplanetary systems looks much like our own. First, most of the planets found so far have a mass comparable to

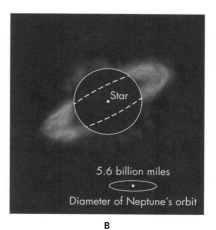

A                                           B

**FIGURE 7.15**

(A) Disks of dust and gas around two forming stars in Orion. (B) Dust disk around the young star HR 4796A. The star itself has been blotted out by a small disk in the telescope so that its glare will not hide the disk.

---

\*A number of astronomers use the term extra-solar planets. However, this is a bit peculiar because, after all, Earth is extra-solar too, in the sense that it is orbiting outside the Sun.

[†]Similar ringlike features occur in the asteroid belt and within the rings of Saturn and Neptune.

**FIGURE 7.16**

Detecting a planet from the motion of the star it orbits.

*The position and Doppler shift of a star orbiting its common center of mass with a planet*

Jupiter.* However, unlike Jupiter, most of these objects orbit extremely close to their star. Does this imply that our model for planet formation is wrong? Perhaps, but more likely, we are seeing what is simply the outcome of the way we search. That is, the technique we use to find planets works best if the planet is massive and near its star. The reason for this selection is that only for such planets is the tug they exert on the star large enough to detect by the Doppler shift method. Given that limitation, is there some way we can search for planets that more closely resemble our Earth? One way is to use a discovery made by Einstein in the early 1900s. He showed, as part of his general theory of relativity, that a mass bends space in its vicinity and that this bending creates the mass's gravity. As a result, if a ray of light passes near a mass, the bent space around the mass deflects the light and can bring it to a focus, as figure 7.18 schematically shows.† By analogy with the focusing ability of an ordinary lens, astronomers call such deflection of light **gravitational lensing.**

Gravitational lensing has proved to be a powerful tool for detecting low-mass planets. The method works approximately as follows. Suppose we look at some distant star and measure its brightness. Suppose further that, by chance, a star at an intermediate distance moves between us and the distant star. Rays of the distant star's light that would have traveled past us in the absence of the intermediate star are now bent so that they reach us (see fig. 7.18). Thus, we observe *more* light from the distant star when an intermediate-distance star is present. Moreover, we receive even more light (although only a very tiny amount more) if a planet is orbiting the intermediate-distance star.

Such bending of light by gravity may sound like science fiction, but as long ago as 1916 astronomers, following Einstein's suggestion, detected the bending of light from a distant star as the star's light traveled past our Sun.

The presence of both an intermediate star and an orbiting planet that is properly positioned is very rare. Thus, to search for planets by this method, astronomers monitor the brightness of millions of stars, and computers scan millions of bits of data for the tiny increase in brightness of a lensing event. In 2005 astronomers were rewarded by the detection of just such an expected brightening caused by a planet orbiting a dim

---

*Astronomers announced in 2007 the discovery of a 5 Earth-mass exoplanet. This planet is so small that it is likely to be rocky. It orbits a little less than 0.1 AU from the dim star Gliese 581, which lies about 20 ly from Earth. The planet's surface temperature is probably above freezing, allowing the possibility of liquid water on its surface

†In extreme cases, this bending of space can bend light so much that it is trapped around a mass, thereby creating a black hole.

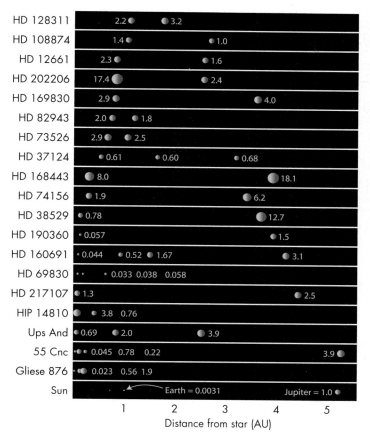

| Star | | | |
|---|---|---|---|
| HD 128311 | 2.2 ● | ● 3.2 | |
| HD 108874 | 1.4 ● | ● 1.0 | |
| HD 12661 | 2.3 ● | ● 1.6 | |
| HD 202206 | 17.4 ● | ● 2.4 | |
| HD 169830 | 2.9 ● | | ● 4.0 |
| HD 82943 | 2.0 ● ● 1.8 | | |
| HD 73526 | 2.9 ● ● 2.5 | | |
| HD 37124 | ● 0.61 | ● 0.60 | ● 0.68 |
| HD 168443 | ● 8.0 | | ● 18.1 |
| HD 74156 | ● 1.9 | ● 6.2 | |
| HD 38529 | ● 0.78 | | ● 12.7 |
| HD 190360 | ● 0.057 | ● 1.5 | |
| HD 160691 | ● 0.044 ● 0.52 ● 1.67 | | ● 3.1 |
| HD 69830 | ● ● ● 0.033 0.038 0.058 | | |
| HD 217107 | ● 1.3 | ● 2.5 | |
| HIP 14810 | ● ● 3.8  0.76 | | |
| Ups And | ● 0.69 ● 2.0 | ● 3.9 | |
| 55 Cnc | ●●● 0.045 0.78  0.22 | | 3.9 ● |
| Gliese 876 | ●● 0.023  0.56  1.9 | | |
| Sun | ● Earth = 0.0031 | | Jupiter = 1.0 ● |

Distance from star (AU)  1  2  3  4  5

**FIGURE 7.17**
Sketches of several newly discovered planetary systems. (The numbers show mass in units of Jupiter's mass.)

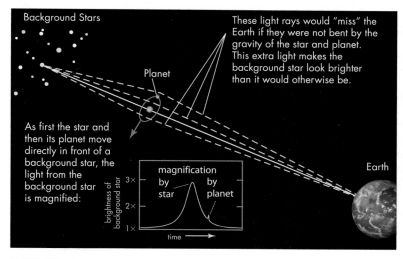

**FIGURE 7.18**
Detecting a planet by the slight bending of light from a background star caused by the planet's gravity.

red star known as OGLE-2005-BLG-390.* Calculations of the light change show that the dim (and cool) star contains a little less than one-quarter as much matter as our Sun and that the planet is only about six times more massive than the Earth and orbits its star at distance of about 2.9 AU (roughly three times the Earth–Sun distance). The planet's small mass implies that it could not have drawn in a significant quantity of hydrogen or helium, so it cannot be a gas giant planet like Jupiter. Moreover, because the star turns out to be so cool and because the planet is farther from its star than the Earth is from the Sun, the planet must also be very cold. Thus, it is probably an icy planet, perhaps like a huge Pluto.

Our inability, so far, to see the majority of exoplanets directly greatly limits what we can learn about them. However, astronomers can partly overcome this limitation for a few planets. For example, one planet of the Sun-like star HD209458 orbits so that it passes between us and the star every 3.5 days. At such times, the planet blocks a tiny amount of the star's light. From the amount of dimming, astronomers can deduce the diameter of the planet, which turns out to be about 1.3 times the diameter of Jupiter. Moreover, a tiny fraction of the star's light leaks through the planet's atmosphere so that gas in the planet's atmosphere imprints very weak absorption lines on the spectrum. The lines are from hydrogen, sodium, carbon, and oxygen, and analysis of the line strengths suggests that the planet is a gas giant planet similar to Jupiter. Notice, however, the extremely short orbital period of 3.5 days. Using Kepler's third law, astronomers deduce that this planet orbits a mere 0.05 AU from its star, roughly one-tenth the distance that Mercury orbits from our Sun—vastly nearer than where we expect a giant planet to orbit.

Having a gas giant planet so near its star raises many questions. How can a gas giant form so close to a star? How can it survive there? Thus, while astronomers are delighted that the discovery of exoplanets confirms the general planet formation hypothesis, they are puzzled by how unlike our own planetary system most of these remote planetary families are. Not only are the giant planets "too close" to their star, they also are often in very elliptical orbits (rather than the essentially circular ones in our own system). This could be additional evidence that our model for planet formation is wrong, but it could also mean that the planets formed farther out from their star and migrated inward toward it (see Science at Work: "Migrating Planets" on p. 234).

Another puzzle to astronomers is that no system of exoplanets yet found looks particularly like our own. The nearest match so far is the system

---

*This name identifies the star as having been found by the *Optical Gravitational Lensing Experiment* against the bulge of stars (= BLG) in the center of our galaxy.

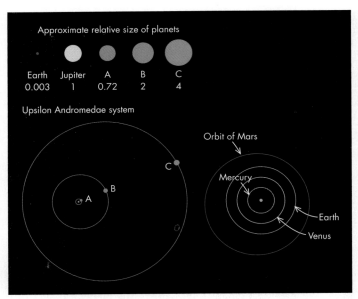

Approximate relative size of planets

| Earth | Jupiter | A | B | C |
|---|---|---|---|---|
| 0.003 | 1 | 0.72 | 2 | 4 |

Upsilon Andromedae system

Orbit of Mars

C

B

A

Mercury

Earth

Venus

**FIGURE 7.19**
Planets in the system Upsilon
Andromedae. Our Solar System's
inner planets are shown at right for
comparison.

of planets orbiting the star Upsilon Andromedae. This Sun-like star has three orbiting planets, but even it differs substantially from our Solar System (see fig. 7.19). One moves in a very elliptical orbit that is a little larger than Mars' orbit around the Sun. Another is extremely close to the star, and all three are roughly the same mass as Jupiter. That so many extra-solar systems have a massive planet on a very elliptical orbit does not bode well for the existence of Earth-like planets in these systems. As a massive planet sweeps into the inner portion of a star system, it will, over time, disturb the orbit of smaller planets, either ejecting them from the system or causing them to fall into their star. Some evidence suggests this fate may have befallen planets in a few of these remote systems. A number of the stars with exoplanets are appreciably richer in iron than our Sun. One suggestion for why these stars are so iron-rich is that they have swallowed Earth-like planets and vaporized them. The iron from the vaporized planet's core then enriches the star, making its spectrum lines of iron stronger. This is not the only interpretation, however. Perhaps it is easier to make planets in the first place if a star has a higher-than-average concentration of iron. Which interpretation is correct? We do not yet know.

## SCIENCE AT WORK

### MIGRATING PLANETS

With the discovery of massive planets orbiting close to other stars, astronomers have been forced to think more critically about how our own Solar System formed.

For example, until very recently, most astronomers assumed that the Solar System's planets move along orbits that lie close to where the planets formed. But the discovery of planets with masses comparable to Jupiter's orbiting close to their star like our terrestrial planets has led astronomers to propose that planets may form at one distance from a star and then "migrate" to a new distance.

Can this proposal be tested? We can't watch real planets shift their orbits, but we can make computer simulations that follow a planet for millions of years under conditions similar to the early stages of our Solar System. These simulations show that interactions between the forming planets and leftover material in the disk of dust and gas can shift the planets' orbits either inward or outward, depending on the circumstances. The amount of such shifting in our Solar System is unknown, but, according to some models, Neptune may have formed about 20 AU from the Sun and drifted out to its present distance of about 30 AU.

Migration of planets has important consequences for planetary systems. For example, if a giant planet migrates inward toward its star, it will probably destroy smaller, Earth-size planets as it passes them. Thus, small planets, suitable for life as we know it, may form but fail to survive in such systems.

Yet another consequence of planet migration is that as a planet changes its orbital distance from the Sun, the planet also changes its distance from the other planets. Because the distance separating planets changes, the gravitational tug they exert on one another also changes. Such gravitational tugs exert a twisting force on a planet and can change the tilt of its rotation axis. Many astronomers now think that such interactions explain why Uranus is so strongly tilted with respect to its orbit (recall fig. 7.3), a point we will discuss further in chapter 9.

## SUMMARY

The Solar System consists of a star (the Sun) and planets, asteroids, and comets, which orbit it in a broad, flat disk. All the planets circle the Sun in the same direction and, with a few exceptions, spin in the same direction. Their moons also form flattened systems, generally orbiting in the same direction. The planets fall into two main categories: small, high-density bodies (the inner, or terrestrial, planets) and large, low-density bodies (the outer, or Jovian, planets). The former are rich in rock and iron; the latter are rich in hydrogen and ice.

These features of the Solar System can be explained by the solar nebula theory. In this theory, the Solar System was born from a cloud of interstellar gas that collapsed to a disk called the solar nebula. The center of the nebula became the Sun, and the disk became the planets. This explains the compositional similarities and the common age of the bodies in the system.

The flat shape of the system and the common direction of motion around the Sun arose because the planets condensed within the nebula's rotating disk. Planet growth occurred in two stages: dust condensed and clumped to form planetesimals; and then later the planetesimals aggregated to form planets and satellites. Two kinds of planets formed because lighter gases and ice could condense easily in the cold outer parts of the nebula, but only rocky and metallic material could condense in the hot inner parts. Impacts of surviving planetesimals late in the formation stages cratered the surfaces and may have tilted the rotation axes of some planets. Some planetesimals (and/or their fragments) survive to this day as the asteroids and comets.

Astronomers have found many planets orbiting other stars. Study of these exoplanets may help us better understand the origin of our own Solar System.

### ?What It Is.

This is a dust and gas disk around a forming star. It is much like the one from which our Solar System formed.

## QUESTIONS FOR REVIEW

1. Name the eight planets in order of increasing distance from the Sun.

2. Make a table listing separately the inner and outer planets.

3. What properties, apart from position, distinguish the inner and outer planets?

4. Make a sketch of the Solar System showing top and side views.

5. What is the Oort cloud? Where is it located, and what kind of objects come from it?

6. What is Bode's law?

7. What is an interstellar cloud? What does it have to do with the Solar System?

8. What is the solar nebula? What is its shape and why?

9. Why are there two main types of planets?

10. How do we know the composition of Jupiter?

11. What are planetesimals?

12. What is the difference between condensation and accretion?

13. Describe the planetesimal theory of planet formation.

14. How does the planetesimal theory of planet formation explain the asteroids?

15. How did the craters we see on many of the planets form?

16. Describe a theory of how planets may have formed their atmospheres.

17. How would you describe the formation of the Solar System to a little brother or sister?

## PROBLEMS

1. Calculate the densities of Venus and Jupiter, given the following data: The mass and radius of Venus are $4.87 \times 10^{27}$ grams and 6051 kilometers, respectively. The mass and radius of Jupiter are about $1.9 \times 10^{30}$ grams and 71,492 kilometers, respectively. How do these numbers compare with the density of rock (about 3 grams per cm$^3$) and water (1 gram per cm$^3$)? (Note: Be sure to convert kilometers to centimeters if you are expressing your answer in grams per cm$^3$.)

2. Look up the mass and radius of Mercury and Jupiter and calculate their escape velocity, using the expression in chapter 2. Does this help you see why the one body has an atmosphere but the other doesn't? (Note: Be sure to convert kilometers to meters or the appropriate unit.)

3. Suppose an exoplanet orbits a star of one solar mass. Suppose the orbital period is 5 days. What is the semimajor axis of the orbit in AU?

## TEST YOURSELF

1. Which of the following objects are primarily rocky with iron cores?
   (a) Venus, Jupiter, and Neptune
   (b) Mercury, Venus, and Pluto
   (c) Mercury, Venus, and Earth
   (d) Jupiter, Uranus, and Neptune
   (e) Mercury, Saturn, and Eris

2. One explanation of why the planets near the Sun are composed mainly of rock and iron is that
   (a) the Sun's magnetic field attracted all the iron in the young Solar System into the region around the Sun.
   (b) the Sun is made mostly of iron. The gas ejected from its surface is therefore iron, so that when it cooled and condensed, it formed iron-rich planets near the Sun.
   (c) the Sun's heat made it difficult for other substances such as ices and gases to condense near it.
   (d) the statement is false. The planets nearest the Sun contain large amounts of hydrogen gas and subsurface water.
   (e) the Sun's gravitational attraction pulled iron and other heavy material inward and allowed the lighter material to float outward.

3. Which of the following features of the Solar System does the solar nebula theory explain?
   (a) All the planets orbit the Sun in the same direction.
   (b) All the planets move in orbits that lie in nearly the same plane.
   (c) The planets nearest the Sun contain only small amounts of substances that condense at low temperatures.
   (d) All the planets and the Sun, to the extent that we know, are the same age.
   (e) All of the above

4. The numerous craters we see on the solid surfaces of so many Solar System bodies are evidence that
   (a) they were so hot in their youth that volcanos were widespread.
   (b) the Sun was so hot that it melted all these bodies and made them boil.
   (c) these bodies were originally a mix of water and rock. As the young Sun heated up, the water boiled, creating hollow pockets in the rock.
   (d) they were bombarded in their youth by many solid objects.
   (e) all the planets were once part of a single, very large and volcanically active mass that subsequently broke into many smaller pieces.

5. In studying planets orbiting other stars than the Sun, which have astronomers *not* found?
   (a) An Earth-mass planet orbiting farther from its star than Pluto orbits the Sun.
   (b) A planet whose mass is several times Jupiter's mass but which orbits nearer its star than Mercury orbits the Sun.
   (c) Planets with orbits considerably more elliptical than those in our Solar System.
   (d) Any star with more than one planet orbiting it.
   (e) No planets with any of these properties have been discovered.

## FURTHER EXPLORATIONS

Beatty, J. Kelly, Carolyn Collins Petersen, and Andrew Chaikin. *The New Solar System.* 4th ed. Cambridge: Sky Publishing Company, 1999.

Gingerich, Owen. "Losing It in Prague: The Inside Story of Pluto's Demotion." *Sky and Telescope* 112 (November 2006): 34.

Malhotra, Renu. "Migrating Planets." *Scientific American* 281 (September 1999): 56.

Seager, Sara. "Unveiling Distant Worlds." *Sky and Telescope* 111 (February 2005): 28.

Soter, Steve. "What is a Planet?" *Scientific American* 296 (January 2007): 34.

## KEY TERMS

accretion, 226
asteroid belt, 220
asteroid, 220
Bode's law, 221
comet, 220

condensation, 225
dwarf planet, 219
exoplanet, 231
gravitational lensing, 232

inner planets, 216
interstellar cloud, 224
interstellar grains, 224
Jovian planet, 218

Kuiper belt, 220
Oort cloud, 220
outer planets, 216
planetesimals, 226

solar nebula, 224
solar nebula theory, 224
Solar System, 215
terrestrial planets, 218

# The Terrestrial Planets

# 8

MERCURY

VENUS

EARTH

MARS

The four inner planets: Mercury, Venus, Earth, and Mars.

## Chapter Objectives

- The terrestrial (or inner) planets are Mercury, Venus, Earth, and Mars.
- These planets are comparable in size to the Earth and like it in composition, having an iron core surrounded by rocky outer layers.
- Heat left over from their formation makes their centers hot.
- The smaller objects (Mercury and Mars) have cooled more than Earth and Venus have.
- The flow of heat drives motions inside Earth and Venus.
- These motions determine the types of surface features found on these planets.
- Their original atmospheres were probably mainly carbon dioxide and nitrogen.
- Liquid water in Earth's early atmosphere gradually removed $CO_2$. Green plants produced Earth's oxygen by photosynthesis.

**?What Is This?**
(See chapter summary for the answer.)

## Concepts and Skills
## to Remember

- Density (5.1)
- Absorption in Earth's atmosphere (3.7)

Terrestrial planets, as their name suggests, have a size and structure similar to Earth's. Within our Solar System, Mercury, Venus, Earth, and Mars fit in this category. Orbiting in the inner part of the Solar System, close to the Sun, these rocky worlds are too small and too warm to have captured massive hydrogen envelopes such as those that cloak the outer planets. Nor have they the array of moons possessed by their cold, giant brethren. In fact, of the terrestrial planets, only Earth has a large moon, although Mars is orbited by two tiny captured asteroids.

## 8.1    PORTRAITS OF THE TERRESTRIAL PLANETS

The terrestrial planets roughly resemble one another in size, as you can see in figure 8.1, which shows them in approximately correct proportion. These four planets have markedly different atmospheres and surfaces, as illustrated in the insets below each one. For example, Mercury, the smallest, looks like the Moon: a gray, bare, cratered body with essentially no atmosphere. Venus, on the other hand, is covered with deep clouds of sulfuric acid droplets suspended in a dense atmosphere of carbon dioxide. This thick atmosphere traps sunlight by the greenhouse effect, heating Venus to a fairly constant 750 K (about 900°F) and making it hotter even than Mercury. From this baking surface, immense volcanic peaks tower above desolate plains. One Venusian peak, Maxwell Montes, rises to 11 kilometers (approximately 6.8 miles) above its surroundings, making it much higher than Earth's highest peak, Mount Everest, which reaches only 8.8 kilometers (approximately 5.5 miles) above sea level. Compared with Earth, with its gleaming blue seas, white clouds and ice caps, red deserts, and green jungles, Mercury looks dead and Venus resembles medieval visions of hell.

Mars, on the other hand, looks more familiar. Caps of ice and frozen carbon dioxide gleam at its poles, surrounded by vast red deserts pocked with craters and furrowed with dunes. Canyons and river beds, now dry but once filled with flowing water, slope toward an immense gash in the Martian crust, the Valles Marineris, which runs for about 5000 kilometers (3000 miles) roughly along the Martian equator. To the northwest of this great rift lies a highland dotted with ancient volcanic peaks. The tallest, Olympus Mons, rises approximately 25 kilometers (about 15.5 miles)—nearly three times the height of Mount Everest—into the cold, thin carbon dioxide atmosphere of Mars.

Our goal in this chapter is to learn more about Mercury, Venus, and Mars so that we can better understand how these neighboring planets came to be so different from Earth. We will discover that size plays a major role. For example, because Mercury is so small, it generates little internal heat to create surface activity, and so its crust is essentially unchanged from its birth. Mars and Venus, on the other hand, are large enough to have hot interiors rather like Earth's, and so they have active surfaces with mountains and volcanic peaks.

Size, coupled with distance from the Sun, creates the great atmospheric differences between these terrestrial worlds. Mercury, for example, is too small and its surface is too hot to retain an atmosphere, while Mars, only slightly larger but farther from the Sun and therefore cooler, has retained one. Venus and Earth are both large enough to have sizeable atmospheres, but Earth's slightly greater distance from the Sun has made it cool enough to have liquid water in its atmosphere. That simple fact has led to the profound difference between the atmospheres of Earth and Venus because liquid water can remove carbon dioxide from air. Moreover, liquid water has allowed life to form and flourish here, and life has not only removed additional carbon dioxide from our air but has also added oxygen.

Mercury
diameter = 4880 km

Venus
diameter = 12,100 km

Earth
diameter = 12,800 km

Mars
diameter = 6800 km

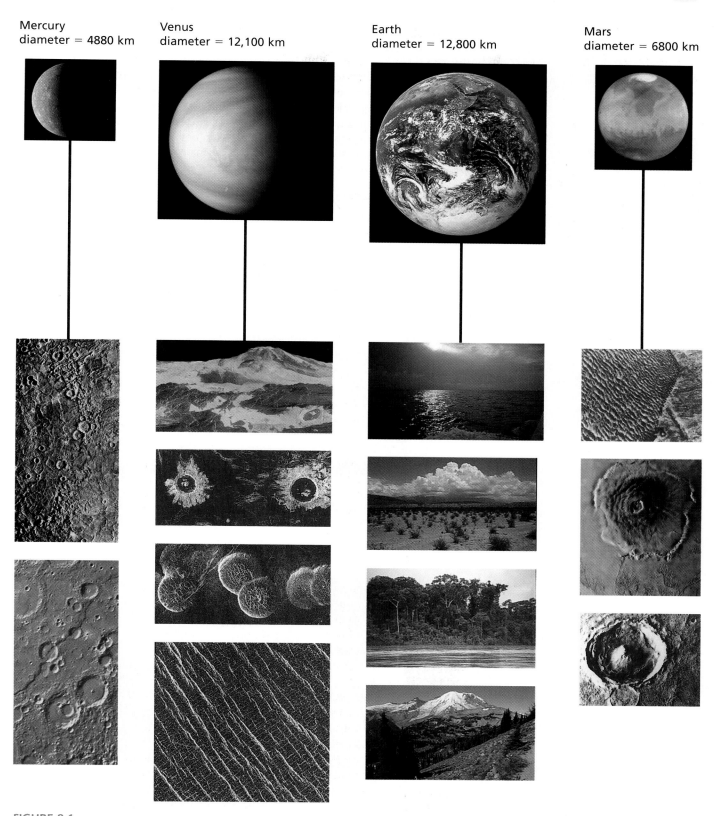

**FIGURE 8.1**
Pictures (left to right) of Mercury, Venus, Earth, and Mars showing representative surface features.

## 8.2 MERCURY

Mercury is the smallest terrestrial planet and is named for the Roman deity who was the messenger of the gods because it changes its position on the sky faster than any other planet. Mercury resembles our Moon in both size and appearance (fig. 8.2), with its radius being about one-third and its mass about one-twentieth that of the Earth. Astronomers can determine Mercury's radius from its angular size and distance, as we discussed in chapter 7 (see fig. 7.7). They can calculate its mass from its gravitational attraction on the *Mariner 10* spacecraft, which flew past the planet in 1974 and 1975, taking pictures of its surface. Those images show that circular craters like those on the Moon cover Mercury's surface, as illustrated in figure 8.3. The largest of these craters by far is the vast Caloris Basin, shown in figure 8.4. With a diameter of 1300 kilometers

**FIGURE 8.2**
Mercury (on the left) and the Moon (on the right) shown to approximately correct relative size. Note that both these airless objects are heavily cratered.

**FIGURE 8.3**
Picture of Mercury's cratered surface (taken by a passing spacecraft). The "patchy" look is caused by inexact matching of the separate images.

(about 800 miles), this mountain-ringed depression is reminiscent of lunar maria. Moreover, its circular shape and surrounding hills indicate that, like the maria, it was formed by impact.

Mercury's surface is not totally Moonlike, however. Congealed lava flows flood many of Mercury's old craters and pave much of its surface. On our Moon, such flows are found almost exclusively within the maria. In addition, enormous scarps—cliffs formed where the crust has shifted—run for hundreds of kilometers across Mercury's surface, as seen in figure 8.5. The scarps may have formed as the planet cooled and shrank, wrinkling like a dried apple. Mercury also possesses the curious landscape illustrated in figure 8.6, which lies on the far side of the planet exactly opposite Caloris Basin. Astronomers call this hummocky, jumbled surface "weird terrain" and think it was churned up by earthquake waves generated by the impact that created Caloris. As the waves traveled around Mercury, they converged on its far side, heaving up the rock, much as dropping cream into coffee creates a tiny splash as the ripples reconverge.

**FIGURE 8.4**
Caloris Basin is the semicircular set of rings here. Only half the basin is shown here. This picture of the basin, a Mercurian mare, was taken by the passing *Mariner* spacecraft.

☾* Does Caloris Basin look like lunar maria? Compare figure 8.4 (above) to figure 6.8. How are they similar? How are they different?

☉* Caloris Basin is round and large, but its interior is much rougher than lunar mare.

Scarp

Notice craters cut by scarp.

Approx. 200 km
(about 120 miles)

**FIGURE 8.5**
Photograph of a scarp, a cliff, running across Mercury's surface.

☾* Which formed first, the scarp or the craters it passes through? On what do you base your answer?

☉* The scarp must be younger because it cuts the craters. If the craters had formed after the scarp, they would have destroyed the scarp within the craters.

**FIGURE 8.6**
Picture of odd terrain that lies on the side of Mercury opposite Caloris Basin.

## Mercury's Temperature and Atmosphere

Mercury's surface is one of the hottest places in the Solar System. At its equator, noon temperatures reach approximately 700 K (about 800°F). On the other hand, nighttime temperatures are among the coldest in the Solar System, dropping to approximately 100 K (about −280°F). These extremes result from Mercury's closeness to the Sun, its slow rotation, and its lack of atmosphere. The slow rotation allows the surface to heat up strongly during the Mercurian day and to cool down greatly at night. The absence of an atmosphere means that nothing moderates the inflow of sunlight during the day or retains heat during the night. Despite a high equatorial temperature, however, Mercury's poles are very cold. Sunlight shines so indirectly on them that they receive little heat, and with no atmosphere to distribute warmth, the poles have grown very cold—so cold, in fact, that radar telescopes have detected features that are thought to be small ice caps.

Mercury lacks an atmosphere for the same reason the Moon does. Although traces of gas, perhaps captured from interplanetary space, have been detected spectroscopically, Mercury's small mass makes its gravitational attraction too small to retain much gas around it. Moreover, its proximity to the Sun makes keeping an atmosphere difficult because the resulting high temperature in the equatorial regions causes molecules to move so fast that they readily escape into space. In fact, Mercury probably never had an extensive atmosphere because it probably was formed from material that, because of its closeness to the Sun, had most of its gases "baked out." Where, then, has the ice at its poles come from? One theory suggests the ice came from comets that have struck its surface. Comets occasionally plunge into the Sun, and so some must occasionally hit Mercury. Such impacts vaporize the comet, creating a wispy and ephemeral atmosphere, most of which quickly escapes into space. Some gas, however, may drift toward the cold polar regions and freeze there, much as frost condenses on automobile windshields on a subfreezing morning. Over time, such frost deposits might build up the caps seen on this otherwise hot planet.

## Mercury's Interior

Mercury probably has an iron core beneath its silicate crust, but astronomers have little proof because no spacecraft has landed there to deploy seismic (earthquake) detectors. Their conclusion is therefore based on Mercury's density and gravitational field. A massive planet's gravity can compress its interior to high density, but Mercury is too small for this effect to be significant. Thus, its high density (5.4 grams per cm$^3$) indicates an iron-rich interior with only a thin rock (silicate) mantle, as depicted in figure 8.7.

Why Mercury is so relatively rich in iron but poor in silicates is unclear. One possibility is that silicates did not condense as easily as iron compounds in the hot, inner solar nebula where Mercury formed. Another possibility is that Mercury once had a thicker rocky crust but that it was blasted off by the impact of an enormous planetesimal, as the computer simulation in figure 8.8 illustrates. (Recall that a similar collision may have happened to Earth and formed our Moon.)

2439 km
(1515 miles)

1800 km
(1100 miles)

Silicate mantle

Iron-nickel core

**FIGURE 8.7**
Artist's depiction of Mercury's interior.

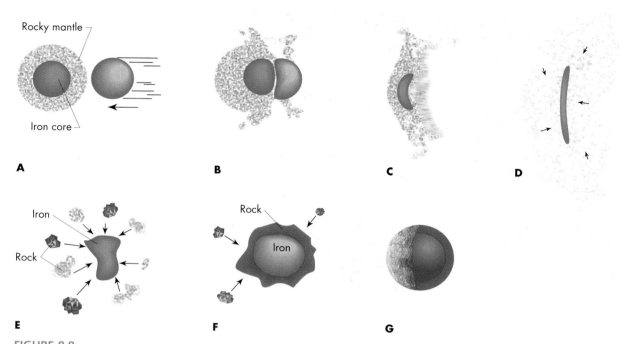

**FIGURE 8.8**
Computer simulation of collision between Mercury and a large planetesimal. The impact strips away most of the outer rocky layers and leaves a highly distorted iron core surrounded by some rocky debris. Gravity eventually reshapes the planet into a sphere.

Recently, astronomers have detected a small "wobble" in Mercury's spin, which they interpret as evidence that its core is at least partly molten.

Whether Mercury has a liquid (molten) inner core like the Earth's is unknown. We infer from its small mass that Mercury has probably been less heated than the Earth by radioactivity, and its small radius would have allowed heat to escape readily. Thus, the core may not be hot enough to be molten, or the molten part may be very small. We discussed in chapter 5 that circulating motions in the Earth's molten iron core when combined with our planet's spin may generate its magnetic field. The small size or solidity of Mercury's core may therefore explain why it has such a weak magnetic field (it is only about 1% that of the Earth's). Moreover, as we shall see below, Mercury spins very slowly. In fact, given that Mercury lacks two of what are thought to be the necessary ingredients for making a planetary magnetic field, astronomers are surprised not so much that Mercury's magnetic field is weak, but that Mercury has any magnetic field at all. Perhaps, however, its magnetism is simply a remnant, preserved from earlier times in its large iron core.

## Mercury's Rotation

Mercury spins very slowly. Its rotation period is 58.646 Earth days, exactly two-thirds its orbital period around the Sun of 87.969 Earth days, as you can verify with a pocket calculator. This means that it spins exactly three times for each two trips it makes around the Sun. How did this proportion occur, or is it just coincidence?

The Sun has affected Mercury's rotation by tidal forces, just as the Earth has affected the Moon's. That is, the Sun's gravity exerts a force on Mercury, which tends to twist the planet and make it rotate with the same period in which it orbits the Sun. Mercury's orbit, however, is very elliptical. Thus, in accordance with Kepler's second law of planetary motion, Mercury's orbital speed changes as it moves around the Sun. Because of that changing speed, the Sun cannot lock Mercury into a purely synchronous spin; the closest to synchrony it can get is three spins for each two orbits (fig. 8.9). Such an integer ratio of periods is called a *resonance*.

**FIGURE 8.9**
Mercury's odd rotation. The planet spins three times for each two orbits it makes around the Sun.

↺★ What resemblance do you see between Mercury's motions and those of our Moon?

ℚ★ Mercury's rotation, like our Moon's, has been slowed by tidal interactions. Mercury tries to rotate so that its spin and orbital motions match, but because its orbit is so elliptical, and its orbital speed therefore changes, it can't make the match occur.

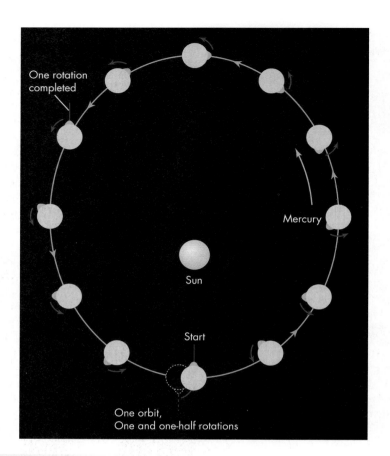

Resonance occurs when a force that acts repeatedly on a body causes its motion to grow ever larger. For example, pushing a child on a swing is a resonance. If you push just as the swing starts to move forward, the child will swing higher and higher, and the pushing force will be in resonance with the motion of the swing. On the other hand, if you push before the backward motion is stopped, the swinging motion will decrease, and no resonance occurs. Likewise, applying power resonantly to a car stuck in a ditch may "rock" it out. A similar resonance exists between the Sun's changing gravitational tug on Mercury as it moves along its elongated orbit and its rotation. The result is the 2:3 relation between its orbital and spin periods.

Mercury's odd rotation gives it an extremely long solar day (the time between successive sunrises) of 176 Earth days. During that time, the Sun sometimes changes its direction of motion across the sky. For example, if sunset occurs when Mercury is at the point of its orbit nearest the Sun, the Sun will set and then briefly rise again before setting a second time!

*The rotation of Mercury during 2 Mercury years*

## 8.3 VENUS

Of all the planets, Venus is most like the Earth in diameter and mass. It is named for the Roman goddess of beauty and love. Because Venus is so similar to the Earth in size, we might therefore expect it to be like the Earth in other ways. However, Venus and the Earth have radically different surfaces and atmospheres. Many of these differences we have only recently discovered because dense clouds perpetually cloak Venus, as shown in figure 8.10. Nevertheless, astronomers know that its surface looks nothing like the Earth's, and its atmosphere is much hotter and denser than ours and has a very different composition.

## The Venusian Atmosphere

The atmosphere of Venus is mainly (96%) carbon dioxide. Astronomers know the atmospheric composition from its spectrum and from measurements with space probes. Gases in its atmosphere absorb some of the sunlight falling on the planet and create absorption lines in addition to those of the Sun itself. These lines then reveal the composition and density of the gas. Moreover, spacecraft have descended through the atmosphere to the surface and have sampled its atmosphere. Thus, we have learned that, in addition to carbon dioxide, Venus's atmosphere contains about 3.5% nitrogen and very small amounts of water vapor and other gases.

Spectra also reveal the nature of the Venusian clouds: they are composed of sulfuric acid droplets that formed when sulfur compounds—perhaps ejected from volcanos—combined with the traces of water in the atmosphere. These clouds permanently cover the planet and are very high and thick, beginning at about 30 kilometers (19 miles) above the surface and extending upward to about 60 kilometers (37 miles). In fact, the clouds are so thick that no surface features can be seen through them with ordinary telescopes. This deep cloud layer strongly reflects sunlight falling on it, making Venus very bright as seen from Earth (fig. 8.11)—so bright that if the air is very clear, you can see Venus in broad daylight. Below the clouds, the Venusian atmosphere is relatively clear, and some sunlight penetrates to the surface. The light is tinged orange, however, because the blue wavelengths are absorbed in the deep cloud layer.

Venus's atmosphere is extremely dense. It exerts a pressure roughly 100 times that of the Earth's, equivalent to the pressure you would feel under 3000 feet of water. We will discuss in section 8.5 why Venus has such a massive atmosphere, but there are other

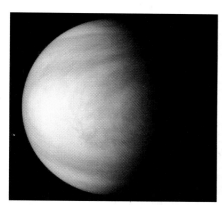

**FIGURE 8.10**
Photograph through ultraviolet filter of the clouds of Venus. The picture is artificially colored and enhanced to show the clouds clearly.

Why does Venus not look round in this photograph? Hint: Think of Galileo's observations of Venus.

Venus does not look round in this image because we are seeing it with the Sun off to the left. Basically, we are seeing here a gibbous phase of Venus.

**FIGURE 8.11**
The Moon and Venus over Ottawa Lake, Wisconsin. Venus is the bright dot below and to the right of the Moon. The left part of the Moon is dimly lit by sunlight reflected from the Earth—earthlight.

features of the planet that we need to understand first. One of the most important of those features is that its lower atmosphere is extremely hot. Observations made with radio telescopes from Earth show that the surface temperature is more than 750 K (about 900°F), a value confirmed by spacecraft landers.* What makes the atmosphere of a planet so similar to the Earth in size, and only slightly nearer the Sun, so very hot?

## The Greenhouse Effect

Venus's carbon dioxide atmosphere creates an extremely strong **greenhouse effect.** We discussed in chapter 5 how gases in a planet's atmosphere may allow sunlight to enter and warm the surface but retard the heat so generated from escaping to space. That is, certain gases may trap a planet's heat by hindering or blocking its infrared radiation from escaping to space, as was illustrated in figure 5.17. In chapter 5, we discussed how carbon dioxide in the Earth's atmosphere is especially effective at trapping heat, creating a weak, but necessary, greenhouse effect here. Venus, however, has about 300,000 times more carbon dioxide than Earth, and so its greenhouse effect is correspondingly much stronger. In fact, it is so effective at trapping heat that the surface of Venus is hotter than Mercury's, even though Venus is farther from the Sun.

## The Surface of Venus

The surface of Venus is hidden beneath its thick clouds, but planetary scientists can map its ground features with radar from both Earth and spacecraft orbiting Venus. Just as radar penetrates terrestrial clouds to show an aircraft pilot a runway even through fog, so too radar penetrates the Venusian clouds, revealing the planet's surface. Figure 8.12 shows a radar map of Venus made by the *Pioneer* Venus orbiter, the first satellite to observe Venus with radar. Such maps show that Venus is less mountainous and rugged than Earth, with most of its surface being low, gently rolling plains. Only two major highland regions, Ishtar and Aphrodite, rise above the lowlands to form land masses similar to terrestrial continents. Ishtar, named for the Babylonian goddess of love, is about the size of Greenland and is studded with volcanic peaks, the highest of which, Maxwell Montes, rises more than 11 kilometers (about 6.8 miles) above the average level of the planet. (Notice that because no oceans exist on Venus, "sea level" has no meaning as a reference height.) The other highland region, Aphrodite, bears the ancient Greek name for Venus and is about the size of South America. Together, Ishtar and Aphrodite comprise only about 8% of Venus's surface, a far smaller fraction than for Earth, where continents and their submerged margins cover about 45% of the planet.

The best pictures of the Venusian surface are the radar maps made by *Magellan*, a U.S. spacecraft previously orbiting the planet. *Magellan* transmitted a radar beam downward to the surface of Venus that was reflected back to the satellite and recorded. The recorded information was stored and transmitted to Earth, where it was processed to create pictures of surface features such as figure 8.13, which shows a view of the entire planet with its numerous volcanic peaks, channels, and wrinkled crust.

Figure 8.14 shows close-ups of some of the more intriguing pictures that *Magellan* has transmitted. Details as small as 100 meters (100 yards) are visible. Thanks to spacecraft, we have better maps of Venus than we had of the Earth itself 50 years ago.

The many odd and unique structures seen in the radar maps have proved puzzling to planetary geologists. Venus is so similar in diameter and mass to the Earth that they

ANIMATION

*Venus*

*Magellan* met a deliberatedly engineered fiery doom in 1994. Its orbit was altered so that it plunged into Venus's atmosphere. Analysis of its final tumblings gave astronomers data on the density of Venus's upper atmosphere.

---

*Such landers do not survive long, however, because Venus is hot enough to melt the lead in their electronic instrumentation.

**FIGURE 8.12**
Global radar map of Venus made by the *Magellan* Venus-orbiting satellite. Colors indicate the relative height of surface features. Lowlands are blue; mountain peaks are red.

**FIGURE 8.13**
Global radar map of Venus made by the spacecraft *Magellan* in orbit around Venus. The light areas are regions of high radar reflectivity and are therefore rougher than the dark areas. The picture is artificially colored to simulate lighting conditions observed by the Russian *Venera* lander.

Congealed lava domes at the eastern edge of
the Alpha Regio Highlands

20 km (about 12 miles)

Craters in the Lavinia region

Approx. 50 km
(about 31 miles)

Maat Mons volcano
(Vertical scale exaggerated)

Fractured plains in the Lakshmi region

Approx. 37 km (about 23 miles)

**FIGURE 8.14**
Gallery of *Magellan* radar pictures. The orange color is artificially added to match the color of the landscape observed by the Russian *Venera* lander. Features are also exaggerated vertically for clarity.

expected to see landforms there similar to those on the Earth. For example, some planetary geologists predicted that there would be evidence of plate tectonics, such as continental blocks, crustal rifts, and trenches at plate boundaries. But few such features are visible. Instead, Venus has a surface almost totally unlike the Earth's. Although Venus has some craters (often weirdly distorted, however) and crumpled mountains, volcanic landforms dominate. These include peaks with immense lava flows, "pancake-shaped domes" of uplifted rock, long narrow faults (cracks), and peculiar lumpy terrain. All these features indicate a young and active surface, a deduction borne out by the scarcity of impact craters. From the small number of craters, scientists have concluded that virtually all of Venus's original surface has been destroyed by volcanic activity. The surface we see is probably at most half a billion years old, much younger than Earth's continental surface, and some regions may be less than

10 million years old. Such estimates of crustal age are difficult to make, however, because the Venusian atmosphere is so dense that all but the largest infalling bodies (bigger than a few hundred meters) are broken up in it.

Are the Venusian volcanos such as Maxwell Montes still active? Eruptions have not been seen directly, but some lava flows appear very fresh. Moreover, electrical discharges, perhaps lightning, have been detected near some of the larger peaks. On Earth, volcanic eruptions frequently generate lightning, and some astronomers think the electrical activity indicates that Venus's volcanos are still erupting. Such eruptions might also explain brief increases in sulfur content detected in the Venusian atmosphere, changes similar to those produced on Earth by eruptions here.

The numerous volcanic peaks, domes, and uplifted surface regions suggest to some scientists that heat flows less uniformly within Venus than within the Earth. Although some locations on Earth (Yellowstone Park and the Hawaiian Islands for example) are heated anomalously by "plumes" of rising hot rock, such plumes seem to dominate on Venus. As hot rock wells upward, it bulges the crust, stretching and cracking it. We may be viewing on our sister planet what Earth looked like as its crust began to form and before smooth heat flows were established. Alternatively, Venus and Earth may differ for deeper reasons.

Planetary geologists have proposed several theories for why Venus and Earth have such different surface features and subsurface activity. Although these theories differ in detail, they tend to agree that the amount of water in the rocks of these two worlds plays a key role. For example, rocks that contain water trapped in their structure melt at a lower temperature than similar rocks that lack water. Moreover, when they become molten, they are "runnier," which makes it easier for the melted rock to flow. As a result, convection can be more vigorous in a planet whose rocks are rich in water compared to a planet with drier rocks. Furthermore, because the water-rich rock melts at a lower temperature, the solid crust of such a planet will be thinner. Other things being equal, we therefore expect wet Earth to have a thinner crust than dry Venus.

Why does the thickness of the crust affect a planet's surface features? A thin crust, such as we have on Earth, allows a steady loss of heat, so our planet's interior cools steadily. Moreover, its thin crust breaks easily into many small plates. This breaking allows crustal motion and activity to occur more or less continuously and at many places on the surface. For Venus, the thicker crust holds heat in, keeping the interior hot, but the convective motions that develop are unable to break up the thick crust.

Ultimately, however, the trapped heat must escape, and it is here that theories differ strongly. According to one theory, at points where the hot, rising material reaches the crust, the surface bulges upward and weakens, and volcanoes may form.[*] Where cooler material sinks, the thick crust crumples into a continentlike region. This creates a planet with few and small continents and whose surface is active but only in isolated spots.

According to another theory, the trapped heat gradually melts the bottom of the thick crust, thinning it and allowing it to break up. This may happen over widespread areas, flooding almost the entire planetary surface with lava in a brief time. The heat then rapidly escapes to space, the interior cools, and the crust again thickens. Surface activity therefore subsides, but heat is trapped once more, and the process may repeat but only at intervals of hundreds of millions of years.

Although most of what astronomers know of the Venusian surface comes from radar maps, several Russian *Venera* spacecraft have landed there and transmitted pictures back to Earth from the Venusian surface itself. The pictures show a barren surface covered with flat, broken rocks and lit by the pale orange glow of sunlight diffusing through the deep clouds (fig. 8.15). These robotic spacecraft have also sampled the rocks, showing them to be volcanic.

---

[*]A similar process occurs on Earth in some regions such as the Hawaiian Islands.

FIGURE 8.15
Picture made on the surface of Venus by the Russian spacecraft *Venera* that landed there. Sunlight filtering through the thick clouds gives the landscape its orange color.

## The Interior of Venus

The deep interior of Venus is probably like the Earth's, an iron core and rock mantle. However, astronomers have no seismic information to confirm this conjecture and, as with Mercury, must rely on deductions from its gravity and density, which are similar to the Earth's.

## Rotation of Venus

Venus spins on its axis more slowly than any other planet in the Solar System, taking 243 days to complete one rotation. Moreover, its spin is retrograde ("backward") compared with the direction of rotation of the other terrestrial planets. This slow and retrograde spin has led some astronomers to hypothesize that Venus was struck shortly after its birth by a huge planetesimal; the impact slowed Venus and set it spinning backward. A less dramatic explanation of the spin is that Venus has been affected by a combination of tidal forces exerted by the Sun—and perhaps the Earth—so that the tilt angle of its rotation axis may have shifted in time.

Venus rotates so slowly that it cannot generate a strong magnetic field as the Earth does. The slow rotation also makes the solar day there very long, approximately 117 Earth days. Furthermore, because the planet's spin is retrograde (backward), the Sun rises in the west and sets in the east.

## 8.4   MARS

Mars is named for the Roman god of war, presumably because of its distinctly reddish color. Compared with Mercury and Venus, Mars seems positively Earthlike. Although its diameter is only about half and its mass about one-tenth the Earth's, its surface and atmosphere are less alien. For example, on a warm day, the temperature at the Martian equator reaches about 50°F (10°C), and although winds sweep dust and patchy clouds of ice crystals through its sky, the Martian atmosphere is generally clear enough for astronomers on Earth to view its surface clearly. Such views from here and from spacecraft orbiting Mars show a world of familiar features. Polar caps of sparkling white contrast with the reddish color of most of the planet and are visible from Earth, as depicted in figure 8.16.

But the pictures sent back to Earth from a long series of spacecraft that have orbited Mars now and in the past—*Mariner, Viking I* and *II, Mars Global Surveyor, Mars Odyssey, Mars Express*—reveal the true marvels of the planet.

Along the equator runs a rift—Valles Marineris—that stretches 5000 kilometers (3000 miles) long, 100 kilometers (60 miles) wide, and 10 kilometers (6 miles) deep, as shown in figure 8.17. This canyon, named for the *Mariner* spacecraft whose pictures led to its discovery, dwarfs the Grand Canyon and would span the continental United States. In figures 8.16 and 8.17, you can clearly see the reddish color of Mars, but what makes the planet so red? Essentially, Mars' color comes from the iron minerals in its surface rocks. We know from everyday experience that a piece of iron will become rust-colored when exposed to air. Here on Earth, such rusting occurs because the iron

**FIGURE 8.16**
Picture of Mars made by the Hubble Space Telescope orbiting Earth.

☞★ How big across is the polar cap in this picture? (Estimate its size from the radius of Mars, which you can look up in the appendix.)

𝒜★ About 1900 km.

**FIGURE 8.17**
Photograph of Valles Marineris, the Grand Canyon of Mars. This enormous gash in Mars's crust may be a rift that began to split apart the Martian crust but failed to open farther. The canyon is about 5000 kilometers (approximately 3000 miles) long. Were it on Earth, it would stretch from California to Florida.

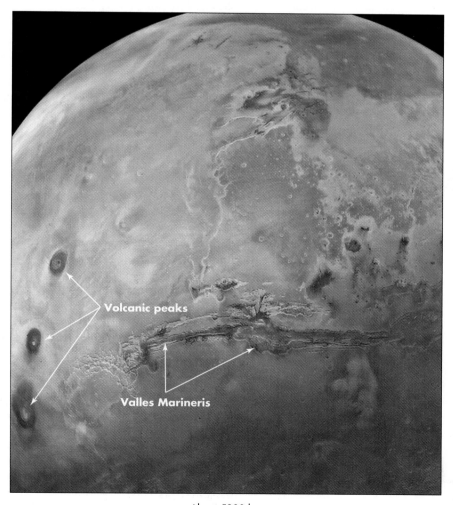

Volcanic peaks

Valles Marineris

About 5000 km
(about 3000 miles)

**A**     **B**     **C**

**FIGURE 8.18**
Pictures of (A) the north Martian polar cap and (B) the south Martian polar cap. (C) Note the layered structure visible in this enlarged view of the north polar cap.

combines with oxygen in our atmosphere to form iron oxides and other compounds. On Mars, even though there is little oxygen in its atmosphere, other chemical reactions with the iron in its surface minerals lead to the same effect.

Figure 8.18 shows the Martian polar caps. These frozen regions change in size during the cycle of the Martian seasons, a cycle resulting from the tilt of Mars' rotation axis in the same way that our cycle of seasons is caused by the tilt of Earth's rotation axis. The Martian seasons are more extreme than terrestrial ones because the Martian atmosphere is much less dense than Earth's, and therefore, it does not retain heat as well. Because Mars' seasonal changes are so extreme, its polar caps vary greatly in size, shrinking during the Martian summer and growing again during the winter. The southern cap is frozen carbon dioxide—dry ice—and in winter, its frost extends in a thin layer across a region some 5900 kilometers (about 3700 miles) in diameter, from the south pole to latitude 40°, much as snow cover extends to middle latitudes such as New York in our winters. But because the frost is very thin over most of this vast cap, it shrinks in the summer to a diameter of about 350 kilometers (approximately 220 miles). The northern cap shrinks to a diameter of about 1000 kilometers (approximately 600 miles), and although it has a surface layer of $CO_2$, the bulk of the frozen material there is ordinary water ice, as deduced from its temperature. Despite the great extent of the Martian polar caps during the Martian winter, the total amount of water in the Martian polar caps is small—far less, for example, than in the ice caps of our own planet.

The northern cap consists of numerous separate layers, as can be seen in figure 8.17C. These strata indicate that the Martian climate changes cyclically. Thus, Mars may have "ice ages" similar to those on Earth.

Why do the Martian polar caps differ so? Altitude measurements made by the Mars *Global Surveyor* spacecraft show that Mars' south pole is considerably higher in elevation and is thus much colder than its north pole. This creates a strong wind pattern that carries water vapor and carbon dioxide away from the south pole toward the north pole. There, it precipitates out, leading to a larger north polar cap.

The Martian poles are bordered by immense deserts with dunes blown into parallel ridges by the Martian winds, as illustrated in figure 8.19. At midlatitudes, a huge uplands called the Tharsis bulge is dotted with volcanic peaks, two of which can be seen in figure 8.20A. Another peak on Tharsis, Olympus Mons, rises about 25 kilometers (about 16 miles) above its surroundings, nearly three times the height of Earth's highest peaks, and is illustrated in figure 8.20B. If ever interplanetary parks are established, Olympus Mons should lead the list.

Planetary geologists believe that the Tharsis region formed as hot material rose from the deep interior of the planet and forced the surface upward as it reached the crust. The hot matter then erupted through the crust to form the volcanos, some of which appear relatively young. For example, the small number of impact craters in its slopes implies that Olympus Mons is no older than 250 million years and that it may in fact have been active much more recently. Some planetary geologists think the Tharsis bulge may also have created the gigantic Valles Marineris, which lies to the southeast. According to this theory, Valles Marineris formed as the Tharsis region swelled, stretching and cracking the crust. Other planetary scientists believe that

**FIGURE 8.19**
Picture of dune fields in the Martian desert.

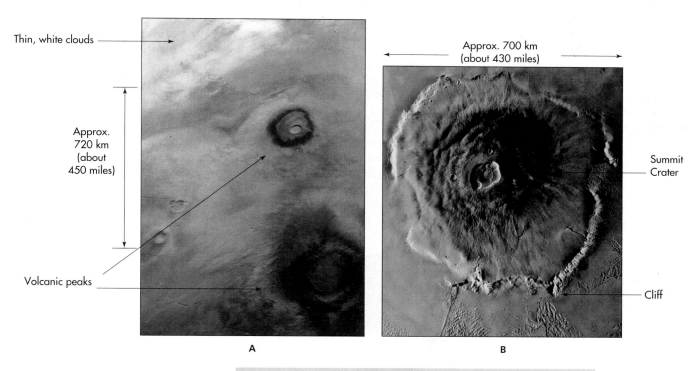

Thin, white clouds

Approx. 720 km (about 450 miles)

Volcanic peaks

A

Approx. 700 km (about 430 miles)

Summit Crater

Cliff

B

Olympus Mons — 15 miles (26 km)

Mount Everest 5.5 miles (8.85 km)

Mauna Kea 6.3 miles (10.2 km) above the ocean bottom

Sea level

C

Height scale is exaggerated by ~ factor of two.

**FIGURE 8.20**
Pictures of (A) two volcanos on the Tharsis bulge and (B) Olympus Mons, the largest known volcano (probably inactive) in the Solar System. (C) Profile of Olympus Mons; Mount Everest, the highest mountain above sea level on Earth; and the Hawaiian volcano, Mauna Kea, rising from the sea floor.

**FIGURE 8.21**
Picture of channels probably carved by running water on Mars.

this vast chasm is evidence for plate tectonic activity, like that of Earth, and that the Martian crust began to split, but the motion ceased as the planet aged and cooled.

Perhaps the most surprising features revealed by the *Viking* spacecraft are the huge channels and dry river beds, such as those seen in figure 8.21. We infer from these features, which wind across the Martian surface and often contain "islands," that liquid water once flowed on Mars, even though no surface liquid is present now. In fact, many astronomers now believe that huge lakes and small oceans once existed on Mars. The strongest evidence for these ancient bodies of water is smooth terraces that look like old beaches around the inner edges of craters and basins, as you can see in the lower left part of figure 8.22. Moreover, some of these features are cut by narrow canyons that breach their rims and appear to drain into lowland areas (fig. 8.22). But where has the water gone? To understand, we need to look at the properties of the Martian atmosphere.

## The Martian Atmosphere

Clouds and wind-blown dust are visible evidence that Mars has an atmosphere. Spectra and direct sampling by spacecraft landers confirm this and show that the atmosphere is mostly (95%) carbon dioxide with small amounts (3%) of nitrogen and traces of oxygen and water. From this, astronomers can determine the density of Mars' atmosphere, which turns out to be very low—only about 1% the density of Earth's.

This density is so low that, although the Martian atmosphere is mostly carbon dioxide, it creates only a very weak greenhouse effect. The consequent lack of heat trapping and Mars' greater distance from the Sun make the planet very cold. Temperatures at noon at the equator may reach a little above the freezing point of water, but at night they plummet to far below zero on the Fahrenheit scale. The resulting average temperature is a frigid 218 K (−67°F). Thus, although some water exists on Mars, most of it is frozen, locked up either below the surface in the form of permafrost or in the polar caps as solid water ice.

Clouds of dry ice (frozen $CO_2$) and water-ice crystals ($H_2O$) drift through the Martian atmosphere carried by the Martian winds. These winds, like the large-scale winds on Earth, arise because air that is warmed near the equator rises and moves toward the poles. This flow from equator to poles, however, is deflected by the Coriolis effect arising from the planet's rotation. The result is winds that blow around the

**FIGURE 8.22**
A Martian crater thought to have once been a "Crater Lake." The crater is roughly 50 km (about 30 miles) across. Note the inflow channel on the upper right and the outflow channel on the left. The smooth floor (apart from a few small craters) suggests that the crater bottom is covered with sediment left behind as the lake dried out.

A

B

**FIGURE 8.23**
(A) Fog in Martian valleys and (B) frost on surface rocks near the *Viking* lander.

planet approximately parallel to its equator. The Martian winds are generally gentle, but seasonally and near the poles, they become gales, which sometimes pick up large amounts of dust from the surface. The resulting vast dust storms occasionally cover the planet completely and turn its sky pink.

No rain falls from the Martian sky, despite its clouds, because the atmosphere is too cold and contains too little water. In fact, there is so little water in the Martian atmosphere that even if all of it were to fall as rain, it would make a layer only about 12 micrometers deep (less than 1/2000 inch). For comparison, Earth's atmosphere holds enough water to make a layer a few centimeters (inches) deep. Despite such dryness, however, fog sometimes forms in some Martian valleys, and frost condenses on the ground on cold nights, as seen in figure 8.23. In addition, during the Martian winter, $CO_2$ "snow" falls on the Martian poles.

Mars has not always been so dry; the numerous channels in its highlands show that water once flowed freely there. But for a planet to have liquid water, it must have a warm atmosphere with a pressure similar to that of Earth's. If the pressure on a liquid is very low, molecules can break free from its surface, evaporating easily because no external force restrains them. On the other hand, if the pressure is high, molecules in a liquid must be heated strongly to turn them into gas. For example, at normal atmospheric pressure, water boils at 100° Celsius. If the pressure is reduced, however, the boiling point drops, an effect used by food producers to make "freeze-dried" foods, such as instant coffee. Coffee is brewed normally, then frozen and placed in a chamber from which the air is pumped out. The reduced pressure makes the liquid "boil" without heating and evaporate, leaving only a powder residue—instant coffee. Similarly, on Mars, any liquid water on its surface today would evaporate.

The existence of channels carved by liquid water on Mars is therefore strong evidence that in its past, Mars was warmer and had a denser atmosphere. However, that milder climate must have ended billions of years ago. How do we know? The large number of craters on Mars shows that its surface has not been significantly eroded by rain or flowing water for about 3 billion years. Why has Mars dried out, and where has its water and atmosphere gone?

**FIGURE 8.24**
Pictures of (A) a channel cut by water released as subsurface ice melted. Note the lumpy terrain at the broad end of the channel where the surface collapsed as the water drained away. (B) Crater with surrounding flow patterns. Heat released by the impact that formed the crater has melted subsurface ice. The thawed material has oozed out.

Some water probably lies buried below the Martian surface as ice as indicated by measurements from the *Mars Odyssey* orbiting spacecraft. If the Martian climate was once warmer and then cooled drastically, water would condense from its atmosphere and freeze, forming sheets of surface ice. Wind might then bury this ice under protective layers of dust, as happens in polar and high mountain regions of Earth. In fact, figure 8.24A shows indirect pictorial evidence of such buried ice. Planetary geologists think that subsurface ice melted, perhaps from volcanic activity, and drained away, causing the ground to collapse and leaving the lumpy, fractured terrain visible at the broad end of the channel. The water from the melted ice then flowed downstream, carving the 20-kilometer (12 mile)-wide canyon. Similarly, figure 8.24B shows a crater from which partially melted matter "squished" out on impact. Thus, Mars' water may now mostly be subsurface ice. But why did the planet cool so?

If Mars had a denser atmosphere in the past, as deduced from the higher pressure needed to allow liquid water to exist, then the greenhouse effect might have made the planet significantly warmer than it is now. The loss of such an atmosphere would then weaken the greenhouse effect and plunge the planet into a permanent ice age. Such a loss could happen in at least two ways. According to one theory, repeated asteroid impacts on Mars when it was young may have blasted its original atmosphere off into space. Such impacts, although rare now, do occur, and our own planet may have been struck about 65 million years ago with results nearly as dire, as we shall learn in chapter 10.

A less dramatic explanation for how Mars lost most of its atmosphere is that Mars's low gravity allowed gas molecules to escape over the first 1 to 2 billion years of the planet's history. Regardless of which explanation is correct, the loss of its atmosphere would have cooled the planet and locked its remaining water up as permafrost. But why have the Martian volcanos not replenished its atmosphere, keeping the planet warm? Astronomers believe that the blame lies with Mars' low level of tectonic activity, a level set by conditions in its interior.

## The Martian Interior

Astronomers believe that the interior of Mars is differentiated like the Earth's into a crust, mantle, and iron core. However, because Mars is so small compared with the Earth, its interior is probably cooler. Mars' smaller mass supplies less heat, and its smaller radius allows the heat to escape more rapidly. Unfortunately, astronomers have no direct confirmation of Mars' interior structure because the seismic detectors landed there by the *Viking* spacecraft failed. Thus, as is the case for Mercury and Venus, astronomers must rely on indirect evidence from its density and gravitational field to learn about the interior of Mars. Having a mass between that of dead Mercury and lively Earth and Venus implies that Mars should be intermediate in its tectonic activity. Such seems to be the case. Using the *Mars Global Surveyor* spacecraft in orbit around the planet, astronomers have measured Mars' magnetic field and internal structure. They concluded that Mars has a metallic core whose radius is between 1200 km and 2400 km, or about 0.4 to 0.7 the overall radius of the planet. (For comparison, Earth's iron core is about 0.5 our planet's radius.) But Mars, unlike Earth, has no planetwide magnetic field, so its core is probably no longer molten.

Although it possesses numerous volcanic peaks and uplifted highlands, implying that it had an active crust, at least in the past, Mars bears no evidence of large-scale crustal motion like the Earth's. For example, it has no folded mountains. Astronomers therefore think that Mars has cooled and its crust thickened to perhaps twice the thickness of the Earth's crust. As a result, the now-weak interior heat flow can no longer drive tectonic motions. A thick Martian crust may also explain why Mars has a small number of very large volcanos, while the Earth has a large number of small ones. Mars' immense volcanos are thus mute testimony to a more active past.

Mars' current low level of tectonic activity is also demonstrated by the many craters that cover its older terrain, far more than are seen on either Earth or Venus. The number of those craters implies that Mars has been geologically quiet for billions of years. Mars is probably not dead, however, because some regions (for example, the slopes of Olympus Mons and other volcanos) are essentially free of craters. Thus, these immense peaks may still occasionally erupt. They do not erupt often enough, however, to replace the gas lost to space because of the planet's low gravity. Thus, Mars may have entered a phase of planetary old age.

## The Martian Moons

Mars has two tiny moons, Phobos and Deimos, which are named for the demigods of Fear and Panic. These bodies are only about 20 kilometers across and are probably captured asteroids. They are far too small for their gravity to have pulled them into spherical shapes. Both moons are cratered, implying bombardment by smaller objects. Phobos (see fig. 8.25) has cracks, suggesting that it may have been struck by a body large enough to split it nearly apart.

Approx. 20 km (about 12 miles)

**FIGURE 8.25**
Picture of Phobos and Deimos, the moons of Mars. These tiny bodies are probably captured asteroids.

⟲⋆ What does the irregular shape of these bodies tell you about the strength of their surface gravity? Is it likely these moons have any atmosphere of their own?

⟳⋆ Their irregular shape implies that they have only a weak surface gravity. If they had a stronger surface gravity, it would crush them into a rounder shape. Their weak surface gravity would make it virtually impossible for them to retain any atmosphere.

Phobos and Deimos were discovered in 1877, but by chance they appeared in literature nearly two centuries earlier in Jonathan Swift's book *Gulliver's Travels*. Gulliver stops at the imaginary country Laputa whose inhabitants include numerous astronomers. Among the accomplishments of these people is the discovery of two tiny moons of Mars. Even earlier, Kepler guessed that Mars might have two moons because the Earth has one moon and Jupiter, at least in Kepler's time, was known to have four. Mars, lying between these two bodies should therefore (according to Kepler's mystic argument) have a number of moons lying between 1 and 4, and he chose 2 as the more likely case.

## Life on Mars?

Scientists have long wondered whether living organisms developed on Mars. Much of that interest grew from a misinterpretation of observations made in 1877 by the Italian astronomer Giovanni Schiaparelli. Schiaparelli saw what he took to be straight-line features on Mars and called them *canali,* by which he meant "channels." In English-speaking countries, the Martian *canali* became canals, with the implication that Mars must be inhabited by intelligent beings who built them. The interest in these canals had become so great by 1894 that the wealthy Bostonian Percival Lowell built an observatory in northern Arizona to study Mars and search for signs of life there.

Most astronomers could see no trace of the alleged canals, but they did note seasonal changes in the shape of dark regions, changes that some interpreted as the spread of plant life in the Martian spring. By the early 1970s, scientists were excited by satellite photographs of water-carved canyons and old river beds because water—at least on Earth—is so important for life. Therefore, to further the search for life on Mars, the United States landed two *Viking* spacecraft on the planet in 1976. These craft carried instruments to search for signs of carbon chemistry in the soil and to look for metabolic activity in soil samples that were put in a nutrient broth carried on the lander. All tests either were negative or ambiguous.

Then, in 1996, a group of American and English scientists reported possible signs of life in rocks from Mars. These were not samples returned to Earth by a spacecraft but of meteorites found on Earth. They arrived here after being blasted off the surface of Mars presumably by the impact of a small asteroid. Such impacts are not uncommon, but most fragments are scattered in space or fall back to Mars. Moreover, of those that are shot into space, only a tiny fraction have just the right combination of speed and direction to reach Earth.

How can astronomers tell if a meteorite has come from Mars? One way is to sample the gas trapped in tiny bubbles in the meteorite and see if it matches the composition of Mars' atmosphere as measured by the *Viking* Mars landers. For the meteorite in question, the match was excellent, assuring that it came from Mars. What was the evidence suggesting life? That turns out to be far more controversial. Microscopic examination of samples from the interior of the meteorite revealed many tiny, rod-shaped structures (fig. 8.26). These look very much like ancient terrestrial bacteria but are much smaller. To some scientists, they look like "fossilized" primitive Martian life. Moreover, the meteorite contains traces of organic chemicals known as polycyclic aromatic hydrocarbons (PAHs, for short). Terrestrial bacteria make such chemicals when they die and decay, but PAHs can also form spontaneously, given the proper mix of chemicals. In fact, they have been found in a number of non-Martian meteorites and have also been detected by their spectrum lines in the radio emission from interstellar gas and dust clouds. Other structures in the meteorite can also be interpreted as having a biological origin. But over the last few years, scientists have shown that ordinary chemical weathering can form very similar structures. As a result, most scientists today are unconvinced that any meteorite yet studied shows evidence of Martian life.

It will take years of further analysis of these and other Martian meteorites or perhaps a robot Mars explorer to tell us for sure whether life exists or once existed on that remote red world.

**FIGURE 8.26**
Fossils of ancient Martian life? The tiny rod-shaped structures look similar to primitive fossils found in ancient rocks on Earth. However, some scientists think these structures formed chemically.

## 8.5   WHY ARE THE TERRESTRIAL PLANETS SO DIFFERENT?

We have seen in sections 8.1–8.4 that the four terrestrial planets have little in common apart from being rocky spheres. They have different surfaces, atmospheres, and interiors. Astronomers think these differences arise from their different masses, radii, and distance from the Sun.

### Role of Mass and Radius

As discussed earlier, a planet's mass and radius affect its interior temperature and thus its level of tectonic activity, with low-mass, small-radius planets being cooler inside than larger bodies. We see, therefore, a progression of activity from small, relatively inert Mercury, to slightly larger and once-active Mars, to the larger and far more active surfaces of Venus and Earth. Mercury's surface still bears the craters made as it was assembled from planetesimals. Mars has some craters, from which we infer that much of its surface is very old, but being larger and more tectonically active, it also has younger surface features such as volcanos, canyons formed by surface cracking as hot material rose inside it, and erosional features such as canyons and river beds carved by running water. In contrast, Earth and Venus retain essentially none of their original crust; their surfaces have been enormously modified by activity in their interiors over the lifetimes of these planets.

### Role of Internal Activity

Internal activity, as we have seen, also affects a planet's atmosphere. In fact, the atmospheres of the terrestrial planets, though now greatly modified, are probably mostly volcanic gases vented as the result of their internal activity. Thus small, inactive Mercury probably never had much atmosphere, and Mars, active once but now geologically quiet, likewise could create and maintain only a thin atmosphere. Moreover, these planets have so little mass and consequently such a small surface gravity relative to Venus and Earth that they have difficulty retaining what little gases they might once have had. As a result, Mercury is virtually without an atmosphere today, and Mars has only a small remnant of its original atmosphere. On the other hand, highly active Venus and Earth have extensive atmospheres.

Astronomers think that the atmospheres that remain on Venus, Earth, and Mars have changed their composition appreciably over time from chemical processes. The atmospheres of all three bodies were probably originally nearly the same composition: primarily $CO_2$ but with small amounts of nitrogen and water. But these original atmospheres have been modified by sunlight, tectonic activity, and, in the case of the Earth, life.

### Role of Sunlight

Sunlight affects a planet's atmosphere in several ways. First, of course, it warms a planet by an amount that depends on the planet's distance from the Sun, and so we expect that Venus will be warmer than Earth and Earth will be warmer than Mars. These expected temperature differences are increased, however, by the atmospheres of these bodies. Moreover, even relatively small differences in temperature can lead to large differences in physical behavior and chemical reactions within an atmosphere. For example, Venus is just enough nearer the Sun than the Earth that even without a strong greenhouse effect, most of its atmosphere would be so warm that water would have difficulty condensing and turning to rain. Moreover, with the Venusian atmosphere being warmer throughout, water vapor can rise to great heights in the Venusian atmosphere, whereas on cooler Earth, water vapor condenses to ice at about 30,000 feet, making our upper atmosphere almost totally devoid of water. You may have noticed this extreme dryness on a long plane trip because after such a flight, your skin may feel itchy and the inside of your nose may feel cracked or stuffy. That dryness is the result of bringing upper-atmospheric air into the cabin and compressing it to make it breathable but not adding moisture.

## Role of Water Content

The great difference in the water content of the upper atmosphere of Earth and Venus has led to a drastic difference between their atmospheres at lower levels. At high altitudes, ultraviolet light from the Sun is intense enough to break apart any water molecules in that region into their component oxygen and hydrogen atoms, a process called **photo dissociation.** Being very light, the hydrogen atoms so liberated escape into space, while the heavier oxygen atoms remain. Because water can rise to great heights in the Venusian atmosphere, over billions of years it has steadily been dissociated there and almost completely lost from our sister planet. In our atmosphere, however, water has survived.

Water makes possible chemical reactions that profoundly alter the composition of our atmosphere. For example, $CO_2$ dissolves in liquid water and creates carbonic acid, which in dilute form we drink as "soda water." In fact, the bubbles in soda water are just carbon dioxide that is coming out of solution. As rain falls through our atmosphere, it picks up $CO_2$, making it slightly acidic, even in unpolluted air. As the rain falls on the ground, it reacts chemically with rocks to form carbonates, locking some $CO_2$ into the rock. The amount of carbon dioxide that can be removed from an atmosphere by this process is impressive. For example, if here on Earth the $CO_2$ that is now bound into carbonates were to be freed again, our atmosphere's pressure would increase by a factor of about 60. This would make our atmosphere, apart from its oxygen, comparable in composition and mass to the existing atmosphere of Venus.

## Role of Biological Processes

Biological processes also remove some $CO_2$ from the atmosphere. For example, plants use it to make the large organic molecules such as cellulose, of which they are composed. This $CO_2$ is usually stored for only short periods of time, however, before decay or burning releases it back into the atmosphere. More permanent removal occurs when rain carrying dissolved $CO_2$ runs off into the oceans, where sea creatures use it to make shells of calcium carbonate. As these creatures die, they sink to the bottom, where their shells form sediment that eventually is changed to rock. Thus, carbon dioxide is swept from our atmosphere and locked up both chemically and biologically in the crust of our planet. With most of the carbon dioxide removed from our atmosphere, mostly nitrogen is left. In fact, our atmosphere contains roughly the same total amount of nitrogen as the atmosphere of Venus.

Our atmosphere is also rich in oxygen, a gas found in such relative abundance nowhere else in the Solar System. Our planet's oxygen is almost certainly the product of green plants breaking down the $H_2O$ molecule during photosynthesis, as we discussed in chapter 5. Thus, much of the cause of the great difference between the atmospheres of Earth and Venus may be that life was able to start on Earth. On our planet, liquid water removed much of the carbon dioxide from our atmosphere (as we discussed in the previous paragraph) and life swept out most of the rest of it, leaving only the tiny amount (0.03%) we have today. In removing the carbon dioxide, plants also produced the oxygen on which all animal life depends. What evidence supports this idea that liquid water and living things removed most of the early Earth's carbon dioxide? It is the carbonate rock we mentioned earlier. If the carbon dioxide locked chemically in that rock were released back into our atmosphere and the oxygen were removed, our planet's atmosphere would closely resemble Venus's.

If water is so effective at removing carbon dioxide from our atmosphere, why does any $CO_2$ remain? We add small amounts of $CO_2$ by burning wood and fossil fuels, but the major contribution is from natural processes. Atmospheric chemists hypothesize that tectonic activity gradually releases $CO_2$ from rock back into our atmosphere. At plate boundaries, sedimentary rock is carried downward into the mantle, where it is melted. Heating breaks down the carbonate rock so that it releases carbon dioxide, which then rises with the heated rock to the surface and reenters the atmosphere. A similar process may once have occurred on Mars to remove carbon dioxide from its atmosphere, locking it up in rock there. Mars' lower level of tectonic activity, however, prevents its $CO_2$ from being recycled. Thus, with so little of its original carbon dioxide left, Mars has grown progressively colder. Our Earth, because it is active, has retained enough $CO_2$ in its atmosphere to maintain a moderate greenhouse effect, making our planet habitable. Thus, poised between one planet that is too hot and another that is too cold, Earth has been blessed with a relatively stable atmosphere, one factor in the complex web of our environment to which we owe our existence.

## 8.6 EXPLORING MARS

Over the last decade, scientists have sent a number of spacecraft to Mars. Three of those have successfully landed there and sent back to Earth pictures and measurements of the Martian landscape and close-up images of rocks. For example, in figure 8.27 you can see the view at the *Spirit* lander site near Bonneville Crater.

These missions have many goals, but perhaps the major one is searching for evidence that liquid water was once present on what now is a cold, dry world. Why is liquid water of such interest? The answer is simple: scientists who study the possibility of life elsewhere in the Universe think that liquid water is a critical ingredient for living organisms of almost any type. Thus the search for water is a first step in the search for life.

The spacecrafts *Opportunity* and *Spirit,* sent by NASA to explore Mars's surface, landed at sites that were chosen because pictures and spectral data taken from orbit suggested that water might have been present there long ago. *Spirit* landed in Gusev Crater, a smooth-floored feature at the end of a narrow Martian valley. *Opportunity* landed on the flat plains along the Martian equator called Meridiani Planum, on the opposite side of Mars from *Spirit*. Both craft then deployed rovers—small, wheeled vehicles that can move away from the landing site and explore interesting features at close range.

Both rovers have been highly successful in their searches. For example, *Opportunity* took the picture shown in figure 8.28, which shows a rock outcropping thought to be material deposited in an ancient, now dried-up small sea or ocean. Examination of the rocks shows that they contain layers that are typical of sediment that sank to the bottom of a body of water and later was transformed into rock. A close-up picture, again made by the *Opportunity* rover, shows details of the layers (fig. 8.29). *Opportunity* has found more than just rock layers, however. It has also found in those layers features that are similar

**FIGURE 8.27**
A panoramic view of the Bonneville Crater, lying in the floor of the much larger Gusev Crater on Mars. This spot is near where the *Spirit* spacecraft landed.

**FIGURE 8.28**
Rock outcropping at *Opportunity* landing site. These rocks show thin layers and contain minerals that suggest that they were formed on the bottom of a salty lake or ocean.

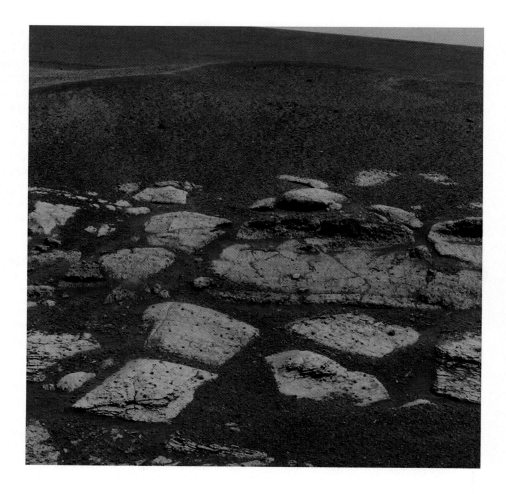

**FIGURE 8.29**
A close-up image of a rock at the *Opportunity* landing site. The lines in the rock suggest that the rock formed from sediment in flowing water.

2 cm

to the ripple marks you see at the beach as water washes back and forth across the sand. Moreover, minerals in the rocks at the *Opportunity* site have a chemical makeup consistent with their having been deposited in a salty lake or small ocean. Although scientists are still analyzing the images and the readings from the many instruments carried on these landers, the evidence has already convinced many scientists that the locations on Mars that *Spirit* and *Opportunity* have explored were indeed once under liquid water.

Meanwhile, the *Mars Global Surveyor* and *Mars Odyssey* orbiters (launched by NASA) and the *Mars Express* orbiter (launched by the European Space Agency) continue to take detailed pictures of the surface of Mars, such as the "mesa" shown in figure 8.30 and the fan of sediment in figure 8.31, which is thought to be the delta of an ancient river. But in addition to charting such surface features, these orbiters also carry instruments that can measure the spectrum of light reflected from Mars. From such spectra, astronomers can deduce what minerals are present at a given point on the Martian surface. Matching the composition of those minerals with data on whether water is needed to produce those minerals gives additional evidence that Mars was once much wetter. For example, the presence of the mineral hematite, a compound of iron that forms by minerals precipitating in water, is an indicator that a region now dry was once covered by a salty lake or sea.

Recent radar observations of Mars' south polar cap reveal huge deposits of buried ice, which are so thick that if melted, the released water would cover Mars to a depth of 10 meters (30 feet).

Other instruments on the *Mars Odyssey* orbiter can detect hydrogen atoms locked up in water in the Martian soil. Data from the *Odyssey* orbiter has astonished astronomers by revealing what appear to be huge amounts of water, presumably ice, buried under a thin layer of Martian dust.

Astronomers are both puzzled and excited by the possibility that water has flowed "recently" on Mars. They are puzzled because, as mentioned earlier, Mars' atmosphere is too cold and has too low a pressure for liquid water to exist there now, so how have the channels been cut? On the other hand, they are excited because of the belief that where there is water, there may be life—even if only microbes.

**FIGURE 8.30**
Image from the *Mars Global Surveyor*. Note the layering visible in several places along the edges of the flat-topped mesa. Such layers often form when sediment is deposited by liquid water.

**FIGURE 8.31**
A view of what appears to be a dried-up river delta. The fan of sediment is about 13 km (approx. 8 miles) wide.

# SUMMARY

The terrestrial planets—Mercury, Venus, Earth, and Mars—are alike in being rocky bodies of comparable size and internal structure, but they nevertheless differ dramatically from one another (table 8.1 and fig. 8.32). Their slight differences in mass and diameter have led to great differences in their internal heat and surface activity. In fact, when ordered by mass, the terrestrial planets are also ordered by activity. Mercury, the least massive, is least active and has the oldest surface. The many impact craters created late in its formation make Mercury look much like our Moon.

Mars, intermediate in mass and radius between Mercury and Earth, shows a surface with an intermediate level of activity. Much of it is cratered, implying a great age. Some regions, however, show volcanic uplift like that seen on Venus, implying that its interior was once much hotter than now. Moreover, its surface shows an amazing variety of landforms: polar caps of ordinary ice and frozen carbon dioxide, immense deserts with dune fields, and canyons created both by crustal cracking and erosion from running water.

Venus and Earth have young surfaces, extensively altered by volcanic activity. On Earth, the flow of heat from its core has led to plate motion. On Venus, perhaps because of its hotter surface, the flow of heat to its surface is less uniform. As a result its surface is marked by isolated regions of intense volcanic uplift. In fact, Venus is probably as active geologically as the Earth.

Mercury, Venus, and Mars have very different atmospheres. Mercury was too small to either make or retain an atmosphere and is now essentially airless. Venus and Mars both have carbon dioxide atmospheres, but Venus's is about 100 times denser than Earth's, while Mars' is about 100 times less dense. The great thickness of the Venusian atmosphere creates a strong greenhouse effect, heating it to about 900°F. High in the Venusian atmosphere, thick clouds of sulfuric acid droplets block our view of the Venusian surface. Astronomers have nevertheless been able to study the Venusian surface with radar telescopes both on Earth and in orbit around Venus.

Mars' atmosphere is too thin now for a strong greenhouse effect, but the old river channels, cut presumably by running water, imply that Mars once had a denser and warmer atmosphere, perhaps even resembling Earth's.

The great differences between the atmospheres of Venus, Earth, and Mars can probably be explained by the differences in their water content and the fact that life evolved on Earth.

## ?What It Is.

These are sand dunes on Mars imaged by the *Mars Global Surveyor*.

## TABLE 8.1

## Comparison of the Terrestrial Planets

|  | Mercury | Venus | Earth | Mars | Moon |
|---|---|---|---|---|---|
| $R_{equator}$ (Earth units) | 0.382 | 0.949 | 1.0 | 0.533 | 0.272 |
| (km) | 2439 | 6052 | 6378 | 3398 | 1738 |
| Mass (Earth units) | 0.055 | 0.815 | 1.0 | 0.107 | 0.012 |
| (Kg) | $3.30 \times 10^{23}$ | $4.87 \times 10^{24}$ | $5.98 \times 10^{24}$ | $6.42 \times 10^{23}$ | $7.35 \times 10^{22}$ |
| Density (grams/cm$^3$) | 5.43 | 5.25 | 5.52 | 3.94 | 3.34 |
| Atmospheric composition | None | $CO_2$ (96%) $N_2$ (3%) | $N_2$ (78%) $O_2$ (21%) | $CO_2$ (95%) $N_2$ (2.7%) | None |
| Pressure (bars) | 0.0 | about 90 | 1.0 | about 0.007 | 0.0 |
| Surface features | Craters, maria, scarps | Some craters, mountains, volcanic peaks, congealed lava plains | Oceans, mountains, volcanic peaks | Deserts, canyons, volcanic peaks | Craters, maria |
| Sidereal day | 58.65 Earth days | 243.02 Earth days | 23.9345 hours | 24.62 hours | 27.32 Earth days |
| Solar day | 176 Earth days | 116.8 Earth days | 24 hours | 24.66 hours | 29.53 Earth days |
| Satellites | None | None | Moon | Phobos and Deimos | — |
| Distance from Sun | 0.387 AU | 0.723 AU | 1 AU | 1.524 AU | 1 AU |
| Orbital period | 87.969 Earth days | 224.70 Earth days | 365.26 Earth days | 686.98 Earth days | — |
| Axial tilt | 0° | 177.4° | 23.45° | 23.98° | 5.15° |

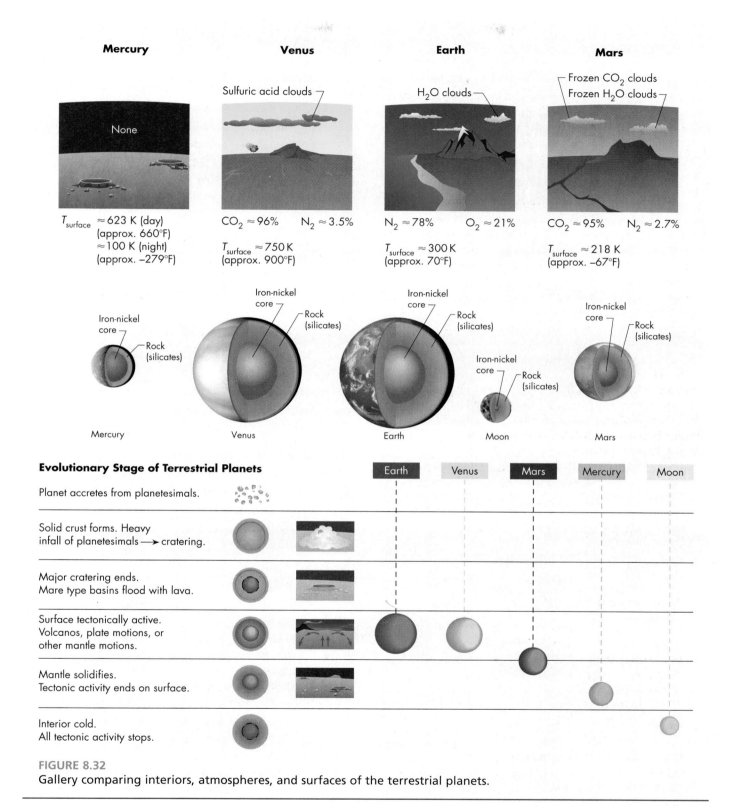

**FIGURE 8.32**

Gallery comparing interiors, atmospheres, and surfaces of the terrestrial planets.

## QUESTIONS FOR REVIEW

1. How do Mercury's radius and mass compare with the Earth's?
2. Does Mercury have an atmosphere? Why or why not?
3. What is the surface of Mercury like?
4. What is peculiar about Mercury's rotation? What causes this oddity?
5. How does Venus compare with the Earth in mass and diameter?
6. Can we see the surface of Venus? Why or why not?
7. How do astronomers know what the surface of Venus is like?
8. What is the dominant gas in Venus's atmosphere? How do astronomers know this?
9. What are the clouds of Venus made of?
10. Why is Venus so hot?
11. What sort of features are seen on Venus's surface? Is the surface young or old? How do astronomers know these things?
12. How does Mars compare with the Earth in mass and diameter?
13. Describe some of the surface features seen on Mars.
14. What is the Martian atmosphere like?
15. What explanation(s) have been offered for why the atmospheres of the terrestrial planets are so different?
16. What are the Martian polar caps composed of?
17. What is the evidence that Mars once had running water on its surface?
18. How do astronomers explain why the Earth's atmosphere ended up with so little $CO_2$, compared with that of Mars and Venus?
19. What is the evidence that leads some scientists to believe life may have existed on Mars?

## THOUGHT QUESTIONS

1. Some scientists have proposed "terraforming" Venus and Mars. That is, they propose introducing organisms that would use carbon dioxide to photosynthesize oxygen and make both planets more suitable for humans. Do you think this is a good idea? Why or why not?
2. What evidence do we see from the terrestrial planets that supports the theory for the origin of the Solar System discussed in chapter 7?
3. What are some examples of resonances in everyday life?
4. What role might planetesimal impacts have played in the history of the terrestrial planets?
5. How does the surface temperature of Venus relate to concern about global warming of the Earth?
6. Suppose scientists discover simple life forms on Mars. Would that alter the way you look at life here on Earth?

## TEST YOURSELF

1. Why does Mercury have so many craters and the Earth so few?
   (a) Mercury is far more volcanically active than the Earth.
   (b) Mercury is much more massive than the Earth and therefore attracts more impacting bodies.
   (c) The Sun has heated Mercury's surface to the boiling point of rock, and the resulting bubbles left craters.
   (d) Erosion and plate tectonic activity have destroyed most of the craters on the Earth.
   (e) Mercury's iron core and its resulting strong magnetic field have attracted impacting bodies.
2. Which body in the inner Solar System has the densest atmosphere?
   (a) Mercury
   (b) Venus
   (c) Earth
   (d) Mars
   (e) Our Moon
3. Why is Venus's surface hotter than Mercury's?
   (a) Venus rotates more slowly, so it "bakes" more in the Sun's heat.
   (b) Clouds in Mercury's atmosphere reflect sunlight back into space and keep its surface cool.
   (c) Carbon dioxide in Venus's atmosphere traps heat radiating from its surface, thereby making it warmer.
   (d) Venus is closer to the Sun.
   (e) Venus's rapid rotation generates strong winds that heat the ground by friction as they blow.
4. What are the Venusian clouds made of?
   (a) Mostly water droplets
   (b) Frozen carbon dioxide
   (c) Sodium chloride
   (d) Sulfuric acid droplets
   (e) Ice crystals

5. Mercury's average density is about 1.5 times greater than the Moon's, even though the two bodies have similar radii. What does this suggest about Mercury's composition?

    (a) Mercury's interior is much richer in iron than the Moon's.

    (b) Mercury contains proportionately far more rock than the Moon.

    (c) Mercury's greater mass has prevented its gravitational attraction from compressing it as much as the Moon is compressed.

    (d) Mercury must have a uranium core.

    (e) Mercury must have a liquid water core.

## FURTHER EXPLORATIONS

Albee, Arden L. "The Unearthly Landscapes of Mars." *Scientific American* 288 (June 2003): 44.

Beatty, J. Kelly, Carolyn Collins Petersen, and Andrew Chaikin. *The New Solar System.* 4th ed. Cambridge: Sky, 1999.

Bell, Jim. "The Red Planet's Watery Past." *Scientific American* 295 (December 2006): 62.

Bullock, Mark A., and David H. Grinspoon. "Global Climate Change on Venus." *Scientific American* 280 (March 1999): 50.

Clifford, Stephen M. "The Iceball Next Door." *Sky and Telescope* 106 (August 2003): 30.

DiGregorio, Barry E. "Life on Mars? 27 Years of Questions." *Sky and Telescope* 107 (February 2004): 40.

Hartmann, William K. "Mysteries of Mars." *Sky and Telescope* 106 (July 2003): 36.

Kargel, Jeffrey S., and Robert G. Strom. "Global Climatic Change on Mars." *Scientific American* 275 (November 1996): 80.

Kaula, W. M. "Venus: A Contrast in Evolution to Earth." *Science* 247 (9 March 1990): 1191.

Musser, George. "The Spirit of Exploration." *Scientific American* 290 (March 2004): 52.

Naeye, Robert. "Europe's Eve on Mars." *Sky and Telescope* 110 (December 2005): 30.

Schubert, G., and C. Covey. "The Atmosphere of Venus." *Scientific American* 245 (July 1981): 66.

Stofan, Ellen R. "The New Face of Venus." *Sky and Telescope* 86 (August 1993): 22.

Strom, R. G. "Mercury: The Forgotten Planet." *Sky and Telescope* 80 (September 1990): 256.

### Website

Please visit the *Explorations* website at **http://www.mhhe.com/arny** for additional online resources on these topics.

## VIDEOS

*Flying by the Planets: The Videos* (available from the Astronomical Society of the Pacific, 390 Ashton Ave., San Francisco, CA 94122).

## PLANETARIUM EXERCISE

1. Follow Mars in its orbit. Set the location that you are viewing from to the nearest large city, or your latitude and longitude. Verify that the local time is correct. Set the time step to 7 days. Use the Selection heading and set the Find option to locate Mars. Set the field of view to 60° using the toolbar. Step by step, advance the time forward and observe the position of Mars with respect to the background stars and constellations. Advance the steps faster and make note of the path of Mars. Note that there are times when it appears to move backward (retrograde). About how often does this happen? Consult chapter 1 to determine why this happens.

## KEY TERMS

greenhouse effect, 246      photo dissociation, 260      resonance, 244

# The Outer Planets

Saturn

## Chapter Objectives

- The outer planets are Jupiter, Saturn, Uranus, and Neptune.
- The outer planets (also called the Jovian planets) are much larger than the Earth.
  - They are rich in hydrogen gas and/or its compounds, such as water, methane, and ammonia.
  - They have deep, thick atmospheres and no solid surface.
  - They have many moons and rings.
- Pluto was formerly considered one of the outer planets. It and newly discovered Eris are now officially called *dwarf planets*. They are composed mainly of rock and ice and are far smaller than even Mercury.

**?What Is This?**
(See chapter summary for the answer.)

## Concepts and Skills to Remember

- Density (5.1)
- Coriolis force (5.7)

**B**eyond Mars, the Solar System is a realm of ice and frozen gas. In this frigid zone, far from the Sun, where solar heat is only a vestige of what we receive on Earth, the giant planets formed. The low temperatures—about −100° Fahrenheit and colder—allowed objects condensing there, within the solar nebula, to capture hydrogen-rich materials such as methane, ammonia, and especially water. Because these substances were far more abundant in the young Solar System than the silicate- and iron-rich material from which the terrestrial planets condensed, planets that formed in this cold environment had more material available for their growth. As a result, these cold planets became vastly larger than those near the Sun, and they developed very different structure and composition. The four largest planets—Jupiter, Saturn, Uranus, and Neptune—are composed mainly of gaseous and liquid hydrogen and its compounds. Although these giant bodies may have cores of molten rocky matter, they lack solid surfaces and, consequently, have no surface features, such as mountains and valleys. Rather, it is their atmospheric features that give them such different appearances.

The dwarf planets Pluto and Eris are also members of this remote part of the Solar System. They are far smaller than the giant planets and are composed of ice and rock. They closely resembles the larger moons of these planets.

The moons of the outer Solar System range in size from very small bodies to those as large as Mercury. As they orbit their parent planets, they form families rather like miniature Solar Systems. Some of the larger moons have brightly colored surfaces, others have numerous craters, and a few have surface features unlike anything seen elsewhere in the Solar System. A few moons even have active volcanoes. In fact, astronomers consider these diverse bodies, virtually unknown before the space age, to be some of the most interesting members of the Sun's family.

## 9.1    JUPITER

To the ancient Romans, Jupiter was the king of the gods. Although they did not know how immense this planet is, they nevertheless chose its name appropriately.

### Jupiter's Appearance and Physical Properties

Jupiter is the largest planet in our Solar System, both in radius and in mass. In fact, its mass is larger than that of all other planets in the Solar System combined. It is slightly more than 10 times the Earth's diameter and more than 300 times its mass. Dense, richly colored parallel bands of clouds cloak the planet, as shown in figure 9.1. Spectra of the sunlight reflected from these clouds show that Jupiter's atmosphere consists mostly of hydrogen, helium, and hydrogen-rich gases such as methane ($CH_4$), ammonia ($NH_3$), and water ($H_2O$). These gases were also directly detected in December 1995 when a probe dropped from the *Galileo* spacecraft parachuted into Jupiter's atmosphere. The clouds themselves are harder to analyze, but theoretical calculations of the chemistry of Jupiter's atmosphere suggest they are particles of water-ice, and ammonia compounds. Their bright colors may come from complex organic molecules or compounds of sulfur or phosphorus. Time-lapse pictures show that Jupiter's clouds move swiftly, sweeping around the planet in jet streams that are far faster than those of Earth. Moreover, Jupiter itself rotates once every 10 hours, spinning so fast that its equator bulges significantly.

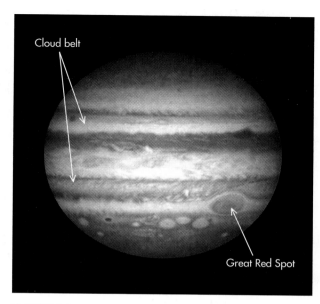

Cloud belt

Great Red Spot

Earth for
comparison

**FIGURE 9.1**
Jupiter and its clouds.

## Jupiter's Interior

Astronomers cannot see through Jupiter's cloud layers to its interior, nor can they probe its interior with seismic detectors. Instead, they must rely on theory to tell them what lies inside this giant planet. For example, despite its great mass, Jupiter is much less dense, on the average, than the Earth. To determine a planet's average density, astronomers divide its mass by its volume. The mass can be found by calculating the planet's gravitational attraction acting on one of its moons, using the method described in chapter 7 (see fig. 7.7). We also described in that figure how to find the planet's radius and, hence, its volume. When such calculations are made for Jupiter, we find that its average density is only slightly greater than that of water—1.3 grams per cubic centimeter—showing that the bulk of the planet, not just its atmosphere, must be composed mainly of very light elements, such as hydrogen. Calculations based on Jupiter's low density, its shape, and measurement of its gravitational attraction on its satellites and passing spacecraft all give astronomers information on what lies below its clouds.

Jupiter's immense mass exerts a tremendous gravitational force on its atmosphere and interior, holding the planet together and compressing its gas. Near the cloud tops, this compression is only slight, so the gas density there is low. Deep in the interior, however, the weight of thousands of kilometers of gas compresses the matter to a high density (approximately 3 times the density of iron). Thus, the density of the gas increases with depth, as illustrated in figure 9.2A. Everywhere in the planet, however, the forces of gravity and pressure are nearly in balance, just as they are in our own atmosphere.

Deep within Jupiter, the compression created by its gravity presses molecules so close together that the gas changes to liquid. Thus, about 10,000 kilometers below the cloud tops—about one-sixth of the way into the planet—Jupiter's interior is a vast sea of liquid hydrogen. Deeper still, the weight of the liquid hydrogen compresses matter below it into a state known as liquid metallic hydrogen, a form of hydrogen that scientists on Earth may have created in tiny high-pressure chambers. Astronomers theorize that if the material from which Jupiter formed has the same overall composition as the Sun's, then Jupiter must also contain heavy elements such as silicon and

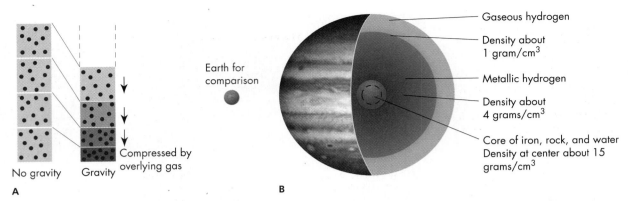

**FIGURE 9.2**
(A) The density of gas increases with depth because the overlying gas compresses the matter. (B) A sketch of what astronomers think Jupiter's interior is like.

From the change in volume shown here in (A), how many times more dense is the bottom, compressed box than the top box?

The bottom box is about twice as dense as the upper box.

iron. Because of their high density, these elements have probably sunk to Jupiter's center, forming a core of iron and rocky material a few times larger than that of the Earth (see fig. 9.2B).

Jupiter's interior is extremely hot, perhaps 30,000 K, which is about five times hotter than the Earth's core. This heat rises slowly to the planet's surface and escapes into space as low-energy infrared radiation. Astronomers have long known from measurements of this infrared radiation that Jupiter emits more energy than it receives from the Sun, but they are still uncertain what supplies that heat. Much of that energy is left over from Jupiter's formation. As we saw in discussing the birth of the Earth, planet building is a hot process. The hail of gas and planetesimals onto the forming world releases a huge amount of gravitational energy. Subsequently, additional heat may be generated by slow but steady shrinkage of the planet as it adjusts to the greater gravity created by the mass added in the last stages of its formation. In fact, giant gas planets such as Jupiter may still be shrinking (and therefore heating) slightly. Yet more heat is released as matter slightly heavier than hydrogen, such as helium, sinks toward Jupiter's core. Whatever its source, the heat generates convection currents similar to those in the Earth's interior. These currents stir both Jupiter's deep interior and its atmosphere.

## Jupiter's Atmosphere

Heat within Jupiter generates convection currents in its outer layers that carry warm gas upward to the top of its atmosphere. Here, the gas radiates heat to space, becoming cooler and sinking again, as illustrated in figure 9.3A. Because of its motion, this rising and sinking gas is subject to a Coriolis effect* created by Jupiter's rapid rotation that deflects the gas into powerful winds called jet streams. We see these winds as the cloud belts, illustrated in figure 9.3B. Winds in these jet streams can have velocities of up to 300 kilometers per hour (nearly 200 miles per hour) with respect to the planet's overall rotation. Yet despite such high speeds, winds in adjacent regions may blow in opposite directions, as shown in figure 9.3C. Such reversals of wind direction

*The rotation of Jupiter*

*The Coriolis effect is a consequence of the conservation of angular momentum. It is not some new force. Recall our discussion in chapter 5, p. 168.

**FIGURE 9.3**

(A) Rising gas from Jupiter's hot interior cools near the top of the atmosphere and sinks. (B) The Coriolis effect, arising from Jupiter's rotation, deflects the gas, creating winds that blow as narrow jet streams. (C) The wind varies widely in speed and direction from region to region.

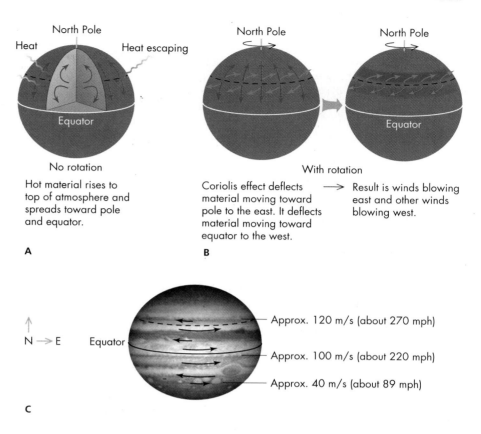

(wind shear) from place to place also occur on Earth, where equatorial winds generally blow from east to west, while midlatitude winds generally blow from west to east.

These reversals occur because the Coriolis effect deflects northern hemisphere winds to the right and southern hemisphere winds to the left of the direction they would travel in the absence of the Coriolis effect. As a result, winds moving toward the equator are deflected to the west, while winds moving toward the poles are deflected to the east, as shown in figure 9.3B. The cumulative effect of these deflections is a set of jet streams blowing in opposite directions with respect to Jupiter's overall rotation.

As the various jet streams circle Jupiter, gas between them rising from below may be spun into huge, whirling atmospheric vortices, much as a pencil between your palms twirls as you rub your hands together. Some of these spinning regions are brightly colored, as figure 9.4 shows. Brown and shades of white dominate, but one exceptionally large vortex—bigger across than the Earth—is nearly brick red. Known as the Great Red Spot, this vortex was discovered in the seventeenth century. Since then, it has changed slightly in appearance and location, but both direct observations of the spot and theoretical models show that it is an essentially permanent feature of Jupiter's atmospheric circulation.

Matter also circulates in Jupiter's deep interior. There, convection in the metallic liquid hydrogen combines with the planet's rapid rotation to generate a magnetic field by a natural dynamo process similar to that which generates the Earth's magnetic field. Jupiter's dynamo process is far more powerful than Earth's, however, and creates the strongest magnetic field of any planet in the Solar System. Measurements taken by passing spacecraft show that the field near the top of Jupiter's cloud layer is about 20 times the strength of the magnetic field on the Earth's surface. When Jupiter's larger radius is taken into account, its field is some 20,000 times as powerful as Earth's, as was deduced even before it was directly measured. For example,

FIGURE 9.4
Vortices form between atmospheric streams of different velocities. Note the Great Red Spot and the white oval below it.

← **Great Red Spot**

A

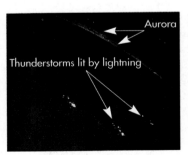

B

FIGURE 9.5
(A) Jupiter's auroral zone as seen by the Hubble Space Telescope. (B) Aurora and lightning on Jupiter, as observed by the *Voyager* spacecraft as it flew over Jupiter's nightside.

astronomers inferred that Jupiter had a strong magnetic field from its auroral activity—just barely detectable from Earth—and from the intense radio emission from the giant planet.

The Earth's auroral activity is linked to its magnetic field, which steers incoming energetic particles from the Sun into our upper atmosphere, where the particles trigger the lovely pale glow of the northern and southern lights. Our planet's magnetic field also traps such particles in the Van Allen radiation belts. Jupiter's magnetic field does much the same: particles trapped magnetically in radiation belts far above the planet emit radio waves, making Jupiter a powerful source of radio emission. Some of those particles descend into Jupiter's upper atmosphere, where they create an aurora, as you can see in figure 9.5A, a picture made with the Hubble Space Telescope. Jupiter's aurora also shows in figure 9.5B as the pale glow along the edge of the planet in a picture taken by the passing *Voyager* spacecraft. This figure also depicts another familiar phenomenon in Jupiter's atmosphere: lightning. Cloud particles carried up and down by the rising and sinking atmospheric motions collide and generate atmospheric electricity—thunderstorms—just as such motions do on Earth. In the picture, you can see many storms lighting up the Jovian night.

## Jupiter's Ring

For centuries, astronomers believed that the only Solar System planet with rings was Saturn. But in 1977, thin rings were detected around Uranus, leading astronomers to wonder if similar rings might surround Jupiter. The opportunity to look for such rings came with the *Voyager I* spacecraft, which flew by Jupiter in 1979. Pictures taken from the craft clearly show that Jupiter has a ring, although only a very thin one, as illustrated in figure 9.6. Jupiter's ring is thought to be made of tiny particles of rock

Jupiter

Neptune

Saturn

Uranus

**FIGURE 9.6**

Rings of the outer planets. The images of Jupiter's, Saturn's, and Neptune's rings were made by the *Voyager* spacecraft. The image of Uranus's rings was made with the Hubble Space Telescope.

dust held in orbit by Jupiter's immense gravitational attraction. These particles are so tiny, however, that radiation from the Sun and collisions with gas trapped in Jupiter's magnetic field exert frictional forces on them, making them gradually drift down into Jupiter's atmosphere. There, they mingle with the swirling gas and are lost from sight. Thus, to maintain the ring, new dust particles must constantly be added to it. Where does this new ring material come from? According to current theories, the tiny satellites of Jupiter's system occasionally collide and fragment, creating new dust to replenish the rings.

## Jupiter's Moons

When Galileo first viewed Jupiter with his telescope, he saw four moons orbiting the planet, bodies now called the "Galilean satellites." Over the past three centuries, astronomers have found additional Jovian moons, with the total now reaching 63. Most of these are too small to be readily seen from Earth and were discovered by examining pictures taken by the *Voyager* spacecraft as it flew past Jupiter.

The Galilean satellites—Io, Europa, Ganymede, and Callisto—are very large: all but Europa are larger than our Moon, and Ganymede has a diameter bigger than Mercury's, making it the largest moon in the Solar System. They orbit along with most of the rest of Jupiter's moons, approximately in Jupiter's equatorial plane, forming a flattened disk, rather like a miniature Solar System. In fact, the Galilean moons most likely formed by a scaled-down version of the process that created the Solar System. That is, they probably aggregated from planetesimals and gas that collected around Jupiter during its formation. Some astronomers believe that Jupiter

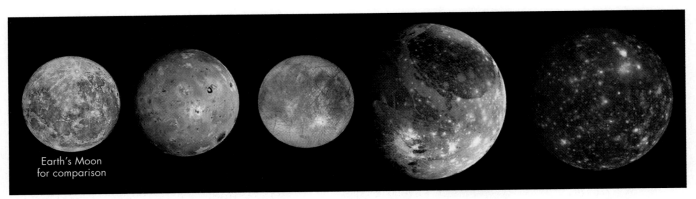

**FIGURE 9.7**
Photographs of the four Galilean satellites of Jupiter and our Moon. From left to right, they are: our Moon, Io, Europa, Ganymede, and Callisto.

⟲★ Why are these moons called the Galilean satellites?

ℚ★ They are called the Galilean satellites because they were discovered by Galileo.

heated this orbiting debris, affecting the composition and density of the moons that formed from it. Such heating would melt ices and partially evaporate them,* so that the moons nearest Jupiter would have less ice and gas and therefore be denser, much as the Sun heated the solar nebula, thereby creating the difference between the terrestrial and outer planets. Evidence that Jupiter heated its moons before or during their formation comes from their density: the densest Galilean satellites are those nearest Jupiter. In order of increasing distance, the densities (in units of grams per cm$^3$) are 3.53 for Io, 2.99 for Europa, 1.94 for Ganymede, and 1.85 for Callisto. Further evidence of heating comes from their strange, often colorful surface features, as illustrated in figure 9.7, and in the case of Io, from its active volcanos.

Io (pronounced[†] *eye-oh*) is named for a mythological maiden with whom Jupiter fell in love and whom he changed into a heifer (a young cow) so that his wife, Hera, would not suspect his infidelity. Io is the nearest to Jupiter of the Galilean moons and therefore is subject to a strong tidal force created by Jupiter's gravity. That tidal force locks Io's spin to its orbital motion the way our Moon's spin is locked to its motion around the Earth. But Io also undergoes a strong gravitational attraction from Europa, the Galilean satellite next closest to Jupiter, and Europa's gravitational tug twists Io from side to side. Moreover, Europa's pull forces Io into an orbit whose shape constantly changes. The outcome of these two effects is that Io is subject to a strong and changing gravitational force from Jupiter that distorts its shape. This deformation heats it by internal friction, much as bending a paperclip until it breaks heats the wire (touch the freshly broken end and it will feel hot). Over billions of years, the heating has melted the rocky matter in Io's interior. As molten matter oozes to the surface, it erupts, creating volcanic plumes and lava flows, as shown in figure 9.8. Sulfur, common in terrestrial volcanoes as well, is now the major component of Io's volcanic outpourings,[‡] and the erupted sulfur compounds give Io its rich red, yellow, and orange colors.

---

*Technically, we should use the word *sublimate* to describe the change of ice to vapor. For simplicity, we will continue to use *evaporate* for this change.

[†]American astronomers tend to say *eye-oh*, but European ones prefer *ee-oh*.

[‡]Although eruptions of sulfur-rich material dominate, astronomers have observed silicate lava (molten rock) in a few eruptions.

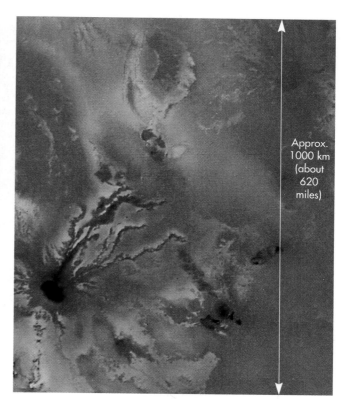

←— Cloud from volcanic eruption

Approx. 1000 km (about 620 miles)

**FIGURE 9.8**
A volcanic eruption and lava flow on Io. The lava is believed to be mostly molten sulfur. These pictures were taken by the *Voyager* spacecraft as it flew past Jupiter.

Ꮐ⋆ What explanation has been offered for the source of heat in Io?

Ꭺ⋆ Io is heated by the gravitational (tidal) force exerted on it by Jupiter, which distorts and thereby heats it.

A recent sulfur lava flow can be seen in figure 9.8 where the molten material has snaked down the slope from the dark volcanic peak. (The peak's darker color indicates that the sulfur was molten there.)

Europa, the smallest of the Galilean moons, is named for another maiden whom Jupiter pursued. According to the legend, Jupiter disguised himself as a bull and carried her on his back across the Hellespont from Asia to Europe, thereby giving the continent its name. Despite this romantic myth, Europa looks rather like a cracked egg. Long, thin lines score its surface, as shown in figure 9.9A. The white material is probably a crust of ice, while the red material is probably mineral-rich water that oozed to the surface through the cracks and then froze. The absence of large craters on its surface suggests that some process has eradicated them, for Europa's surface—like that of Ganymede and Callisto—must surely have been cratered during and following its formation. To explain the lack of craters, astronomers think that Europa may be heated by Jupiter's gravitational forces deforming it, although not as strongly as Io is. That heat, in combination with a small amount from radioactive decay of rocky material in its core, may be sufficient to keep a layer of water melted beneath Europa's crust, forming an ocean (fig. 9.9B). Moreover, the heat may soften the surface ice, allowing it to "flow," glacier-like, and obliterate craters as they form. Close-up images made by the *Galileo* spacecraft show features reminiscent of ice floes (fig. 9.9C). Some astronomers speculate that Europa's ocean might harbor life forms, but no evidence exists to support that view yet.

Ganymede and Callisto look somewhat like our own Moon, being basically grayish brown and covered with craters made during the late stages of their formation, much as our own Moon was pockmarked by infalling debris. But these similarities to our Moon are only superficial, because the surfaces of Ganymede and Callisto are probably mostly ice, the white of the craters being similar to the white you see on ice cubes when they are shattered. Recent observations suggest that Callisto, like Europa, may have a liquid water ocean below its icy crust. Ganymede's surface is

**FIGURE 9.9**
Pictures of Europa. (A) Europa, as imaged by the *Voyager* spacecraft. (B) A model of Europa's interior. Note possible liquid water ocean beneath ice. (C) Close-up picture of ice chunks on Europa's surface.

less heavily cratered than Callisto's, implying that craters in the smoother areas have been destroyed, perhaps much as happened on our own Moon. That is, astronomers think that after Ganymede's surface solidified and was cratered by infalling debris, a few large bodies may have hit it. These large bodies created basins that subsequently flooded with water and then froze into a relatively smooth surface with few craters. However, the smooth regions do show curious parallel ridges, possibly created by tectonic forces as the water in them froze.

From the average density of the Galilean satellites, astronomers deduce that their interiors are composed of a mix of icy and rocky material. Heating of these moons during their youth may have allowed iron in them to sink to their centers and form cores, as happened with our own planet. Indirect evidence for just such a dense core in Ganymede came from the *Galileo* spacecraft, which orbited Jupiter.* As the craft swung past that moon, its trajectory was slightly altered by Ganymede's gravity—in a way that suggests that Ganymede has an ice mantle surrounding a rocky core. Moreover, sensors on *Galileo* detected a magnetic field around Ganymede. The existence of such a field is strong evidence that inside its rocky core Ganymede may have a small iron core.

Jupiter's other moons are much smaller than the Galilean satellites, and are heavily cratered. The orbits of the outermost of these moons are steeply tilted relative to the others, suggesting that these objects may be captured asteroids.

---

*When the *Galileo* probe ran low on the fuel with which to maneuver, NASA deliberately crashed it into Jupiter, where it burned up. This eliminated the chance of a crash into one of Jupiter's moons (especially Europa). That is important because some scientists think that life might have developed in the liquid water below Europa's icy crust, and they don't want to risk accidentally contaminating it or any of Jupiter's moons with terrestrial microorganisms.

## 9.2 SATURN

Saturn is the second largest planet in our Solar System and lies about 10 AU from the Sun. Surrounded by its lovely rings (illustrated in fig. 9.10), Saturn bears the name of an ancient Roman harvest god. In later mythology, Saturn came to be identified with Cronus (also spelled Kronos), whom the ancient Greeks considered the father of the gods.

### Saturn's Appearance and Physical Properties

Saturn's diameter, like Jupiter's, is about 10 times larger than the Earth's diameter. Saturn's mass is about 95 times that of the Earth and its average density is very small—only 0.7 grams per cubic centimeter, which is less than the density of water. Such a low density suggests that Saturn, like Jupiter, is composed mostly of hydrogen and hydrogen-rich compounds. Spectra of the planet bear this out, and astronomers believe these two giant planets have very similar compositions and internal structures, as depicted in figure 9.11.

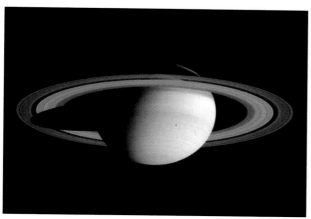

**FIGURE 9.10**
Saturn as pictured by the *Cassini* spacecraft.

Saturn radiates more energy than it gains from the Sun, implying that, like Jupiter, it has an internal heat source. Astronomers think that Saturn's heat does not come from its slow gravitational contraction, however, as most of Jupiter's does. Instead, they hypothesize that deep beneath Saturn's cold clouds, helium droplets condense in its atmosphere, much as water droplets condense in Earth's atmosphere. As the helium droplets fall toward Saturn's core, they release gravitational energy that heats the planet's interior.

If both Jupiter and Saturn have hot interiors and similar compositions, why do they look so different externally? In particular, why does Saturn show only faint cloud belts and markings compared with the striking patterns seen on Jupiter? Saturn's greater distance from the Sun and its consequently lower temperature may provide an answer. Saturn's atmosphere is cold enough for ammonia gas to freeze into cloud particles that veil its atmosphere's deeper layers, making markings below the clouds indistinct.

### Saturn's Rings

Saturn's spectacular rings, illustrated in figure 9.12, were first seen by the astronomer Galileo. Through his small, primitive telescope, however, they looked like "handles" on each side of the planet, and it was not until 1659 that Christiaan Huygens, a Dutch scientist, observed that the rings were detached from Saturn and encircled it.

Molecular hydrogen gas

Helium droplets in hydrogen

Metallic hydrogen

Rock, iron, and water core

Earth for comparison

**FIGURE 9.11**
Internal structure of Saturn.

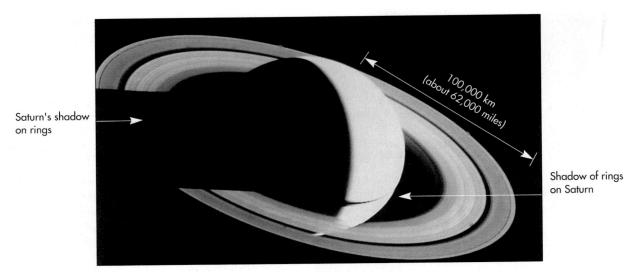

**FIGURE 9.12**
Rings of Saturn.

Saturn's shadow
on rings

100,000 km
(about 62,000 miles)

Shadow of rings
on Saturn

**ANIMATION**

*A 2:1 resonance*

The rings are very wide but very thin. The main band extends from about 30,000 kilometers above the top of Saturn's atmosphere to a little more than twice the planet's radius (136,000 kilometers, or about 84,000 miles), as illustrated in figure 9.12. Some faint inner rings can be seen even closer to Saturn, and faint outer rings extend considerably farther from the planet. Yet despite the rings' immense breadth, they are probably less than a few hundred meters thick—thin enough to allow stars to be seen through them.

The British physicist James Clerk Maxwell, a pioneer in the study of electromagnetism, demonstrated that the rings must be a swarm of particles. He showed mathematically that no material could plausibly be strong enough to hold together in a solid sheet of such vast size. Spectra of the rings support Maxwell's theory: the inner and outer parts orbit Saturn at different velocities, obeying Kepler's third law, as shown by their Doppler shift. Thus, the rings must be a swarm of individual bodies.

Astronomers have since discovered that the ring particles are relatively small, only a few centimeters to a few meters across. Although these particles are far too small to be seen individually with telescopes, they reflect radar signals bounced off them. From the strength of the radar "echo," astronomers can estimate the particle sizes. More precise measurements were made with radio signals from the *Voyager* spacecraft. As it transmitted data to Earth, the signals passed through the rings, scattering slightly from the particles. From the amount of scattering of the *Voyager* signal and the radar waves, astronomers can deduce not only the size of the ring particles but also their composition. Better information about the composition of the rings, however, comes from analyzing the spectrum of sunlight reflected from them. Such analyses show that the rings are composed primarily of water-ice. However, the *Voyager* and *Cassini* spacecraft pictures, such as figure 9.13, show that some parts of the rings are much darker than others, implying that the composition of the rings is not the same everywhere. Particles in the darker ring segments may be rich in carbon compounds similar to those found in some asteroidal material. Figure 9.13 also shows that the rings are not uniformly filled: they consist of numerous

**FIGURE 9.13**
Image of the ringlets, showing the substructure of Saturn's rings.

**FIGURE 9.14**

Two shepherding satellites (the dots inside the small circles) and a portion of the narrow ring around Uranus that they created. The picture's contrast is strongly enhanced to show these faint moons.

separate ringlets. Large gaps in the rings had been seen from Earth, but the many narrow gaps in the rings came as a surprise. What causes these gaps that create the multitude of ringlets?

As long ago as 1866, Daniel Kirkwood, the astronomer who also discovered gaps in the distribution of asteroids, noticed that the largest gap—known as Cassini's division—occurs where ring particles orbit Saturn in exactly one-third the time (one-third the orbital period) of its moon, Enceladus. Thus, any ring particle that attempted to orbit in the gap would undergo a strong and repeated gravitational force from Enceladus every third orbit. Kirkwood concluded that, over long periods of time, the cumulative effect of Enceladus's force would pull particles from the gap. He therefore hypothesized that Enceladus's gravitational attraction creates Cassini's division, just as Jupiter's creates the gaps in the asteroid belt. A few years later, Kirkwood revised his theory of the gap to account for the action of the four largest Saturnian moons. Today, astronomers believe that Saturn's moon Mimas, whose period is twice that of particles in Cassini's division, causes that large gap, but they think the many narrow gaps apparent in figure 9.13 have a different cause.

Narrow gaps in the rings probably arise from a complex interaction between the ring particles and the tiny moons that orbit within the rings. As these moonlets—only tens of kilometers or less in size—orbit Saturn, their gravitational attraction on the ring particles generates waves. These waves spread through the rings much like ripples in a cup of coffee that is lightly tapped. Such ripples are circular in a cup or on a still pond, but in a planetary ring system, they take a different form. Because the inner part of the ring orbits faster than the outer part—a consequence of Kepler's laws—the spreading waves wrap into a tightly wound pattern called "spiral density waves." The crests of these density waves form the narrow rings.

Moonlets may generate gaps within the rings in yet another way. If two moonlets move along orbits that lie very close together, their combined gravitational force may deflect ring particles into a narrow stream between them (fig. 9.14). Such **"shepherding satellites"** occur not only in Saturn's rings but also in Uranus's. But what created planetary rings in the first place?

## Origin of Planetary Rings

Earlier in the 1900s, some astronomers thought that planetary rings were material left over from a planet's formation, perhaps matter that had failed to condense into a satellite. But they now realize that rings are short-lived because they are subject to forces in addition to gravity. For example, gas trapped in a planet's magnetic field may exert a frictional force on the ring particles, gradually causing them to spiral into the planet's atmosphere, as may happen to the material in Jupiter's ring. Thus, new material must be added to the rings from time to time, for without such replenishment, they would disappear in a few million years. One source of new material is the satellites orbiting the planet. A moon in a satellite system as complex as those of Jupiter and Saturn is subject not only to the gravitational forces of the planet but also to that of the other moons. The cumulative effect of such forces alters the satellites' orbits and may lead to collisions between them. Alternatively, a moon may drift so close to its planet that tidal forces break the moon apart, a theory first suggested more than a century ago.

## The Roche Limit

In 1849, the French scientist M. E. Roche (pronounced *rohsh*), while studying the problem of a planet's gravitational effect on its moons, demonstrated mathematically that if a moon gets too close to its planet, the planet's gravity could rip the moon apart. This disruption occurs because the planet pulls harder on one side of the satellite than the other. If the difference in this pull exceeds the moon's own internal gravitational

ANIMATION

*Roche breakup of a moon*

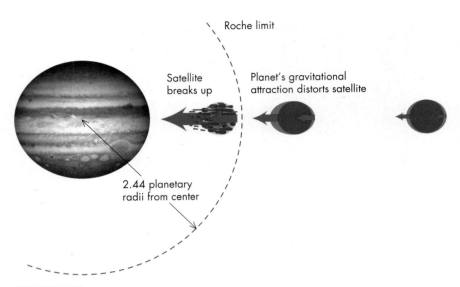

Roche limit

Satellite
breaks up

Planet's gravitational
attraction distorts satellite

2.44 planetary
radii from center

**FIGURE 9.15**
The Roche limit. A planet's gravity pulls more strongly on the near side of a
satellite than on its far side, stretching it and, if strong enough, pulling it apart.

force, the moon will be pulled apart, as shown in figure 9.15. Thus, if a moon—or
any body held together by gravity—approaches a planet too closely, the planet raises
a tide so large it pulls the encroaching object to pieces. Roche calculated the distance
at which the tide becomes fatally large and showed that for a moon and planet of the
same density, breakup occurs if the moon comes nearer to its planet than 2.44 planetary
radii, a distance now called the **Roche limit.** All planetary rings lie near their planet's
Roche limit, suggesting that rings might be caused by satellite disruption. But moons
are not the only bodies that can stray into the danger zone. Asteroids or comets may
occasionally pass too close to a planet, and shattered fragments of such bodies (fig. 9.16)
may help keep rings filled.* The existence side-by-side of ringlets with different compositions (some rich in ice, others rich in carbon) is additional evidence that the rings
formed by the breakup of many different small objects.

The Roche limit applies only to bodies held together by gravity, however. Artificial
satellites or small bodies bonded together by chemical forces can pass safely through
the Roche limit without effect.

## Saturn's Moons

Saturn has several large moons and about four dozen known smaller ones. The larger
of these bodies, like Jupiter's moons, orbit in a flat "mini–Solar System" aligned with
Saturn's equator. Saturn's moons have a smaller average density than the Galilean satellites of Jupiter, from which astronomers deduce that their interiors must be mostly
ice. Moreover, all are about the same density, implying that they were not strongly
heated by Saturn as they formed. (Recall that the Galilean satellites nearest Jupiter
have a higher density than those farther away, presumably a result of Jupiter's heat
driving away much of the icy material from the nearer bodies, leaving them with a
higher proportion of rocky material.)

Titan, the largest Saturnian moon, has a diameter of about 5000 kilometers
(3000 miles), making it slightly bigger in diameter than the planet Mercury and comparable in mass and radius with Jupiter's large moons Ganymede and Callisto. Because
Titan is farther from the Sun than these bodies, it is much colder. Thus, as gas molecules

**FIGURE 9.16**
Comet Shoemaker-Levy 9. This
image shows 20 or so of the
fragments into which it was broken
by Jupiter's tidal force.

---

*In 1992, Comet Shoemaker-Levy 9 passed extremely close to Jupiter and was broken into at least 22 large
fragments. These pieces struck Jupiter in July 1994.

**FIGURE 9.17**
Titan, as seen from the *Cassini* spacecraft. This false color image shows some surface features, glimpsed through the clouds of this mysterious moon.

leak from its interior, they move relatively slowly and are unable to escape Titan's gravitational attraction. This immense moon therefore possesses its own atmosphere, which spectra show to be mostly nitrogen. Clouds in Titan's atmosphere hide its surface, but in 2005, during one of several flybys of Titan by NASA's *Cassini* probe, astronomers began mapping Titan's surface with radar and infrared cameras (fig. 9.17). At the same time, a small robot lander, carried by *Cassini* and built by the European Space Agency, parachuted into Titan's atmosphere. This probe, named *Huygens,* in honor of Christiaan Huygens, descended through Titan's clouds and landed on the surface, from where it sent pictures back to Earth (fig. 9.18A).

Astronomers had shown on the basis of calculations that, because of Titan's extreme cold (94 kelvin, or about −290°F), its clouds would be composed of methane ($CH_4$) and other hydrocarbons rather than water. Such models also predicted that liquid methane or ethane ($C_2H_6$), would "rain" from Titan's clouds and might even form oceans on this frigid moon. No oceans have been found, but radar maps of Titan's polar regions (fig. 9.18B) clearly show lakes, which are presumed to be methane. In addition to the lakes, other images show a landscape with diverse features: channels that appear to have been carved by some kind of liquid (fig. 9.18C); rows of dunes (fig. 9.18D), presumably built and aligned by winds; and at least one "ice" volcano. But these

A
Surface view

~4 cm

~60 km

B
Lakes of liquid methane

C
"River" channels

D
Dunes

**FIGURE 9.18**
The landscape of Titan: (A) "Rocks" of ice on the surface; (B) lakes of liquid methane; (C) channels carved by methane "rain"; (D) dunes.

Titan

~5000 km

400 km (about 250 miles)

Mimas

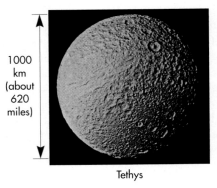

1000 km (about 620 miles)

Tethys

1100 km (about 680 miles)

Dione

500 km (about 310 miles)

Enceladus

**FIGURE 9.19**

Pictures of some of Saturn's moons. Note: The scale is such that Titan is shown at about one-half its correct size relative to the other moons.

Ⓠ★ Given the many craters on these moons, do you think they are tectonically active now? Why?

Ⓐ★ The moons with many craters probably are tectonically inactive or the craters would have been largely covered up or otherwise destroyed by eruptions or surface tectonics.

seemingly familiar surface features differ in composition from similar looking features here on Earth. On Titan, the bitter cold makes ordinary water-ice as hard as rock, so Titan's dunes may be ice crystals (not particles of sand) and its river channels may have been cut by methane rain falling and eroding water-ice (not rock).

Figure 9.19 shows pictures of a few of Saturn's moons taken by spacecraft. Notice that most of these bodies are heavily cratered, implying that they have been extensively bombarded by infalling bodies. The smoother surface seen, for example, on Enceladus suggests that water has erupted from its interior, flooding old craters and drowning them as it freezes. The white markings surrounding many of the craters are probably shattered ice. Thus, these distant and alien moons still have many unsolved mysteries.

## 9.3    URANUS

Uranus, although small compared with Jupiter and Saturn, is much larger than the Earth. Its diameter is about four times that of the Earth, and its mass is about 15 Earth masses. Lying approximately 19 AU from the Sun (more than twice Saturn's distance), Uranus is difficult to study from Earth, visible only as a blue but featureless disk. Even pictures of it taken by the *Voyager* spacecraft show few details (fig. 9.20), although computer processing of the images shows that it has faint cloud bands.

**FIGURE 9.20**

Uranus as pictured from the *Voyager* spacecraft. Because of Uranus's odd tilt (recall fig. 7.3), we are viewing it nearly pole-on. Note the lack of clearly defined cloud belts.

Earth for comparison

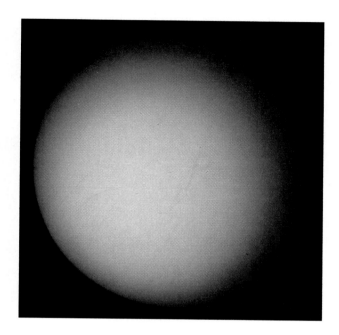

Uranus was unknown to the ancients, even though it is just visible to the naked eye. It was discovered by Sir William Herschel, a German émigré to England. Herschel, a musician, was at the time only an amateur astronomer interested in hunting comets, a task in which he collaborated with his sister Caroline. In 1781, he observed a pale blue object whose position in the sky changed from night to night. Herschel at first thought he had discovered a comet, but observations over several months showed that the body's orbit was nearly circular, and he therefore concluded that he had found a new planet. For this discovery, King George III named Herschel his personal astronomer, and to honor the king, Uranus was briefly known as "Georgium Sidus," or "George's Star." In ancient Greek mythology, Uranus was the father of Cronus and was identified as the God of the Heavens.

## Uranus's Structure

Uranus, like Jupiter and Saturn, is rich in hydrogen and its compounds water and methane. We know this from spectra of its atmosphere, which show very strong absorption lines of methane. In fact, it is methane that gives the planet its deep blue color. When sunlight falls on Uranus's atmosphere, the methane gas strongly absorbs the red light. The remaining light, now blue, scatters from cloud particles in the Uranian atmosphere and is reflected into space, as depicted in figure 9.21. The cloud particles that cause the scattering are thought to be primarily crystals of frozen methane. Such crystals can form in Uranus's atmosphere because, being so far from the Sun, it is extremely cold.

Astronomers rely on indirect methods to study the interior of Uranus, using, for example, its density and shape. From its mass and radius, astronomers can calculate that Uranus has an average density of about 1.2 grams per cubic centimeter. This density is nearly twice that of Saturn and almost as large as Jupiter's. But a planet's density depends on both its composition and the amount by which its gravity compresses it. Given that some part of the density of Jupiter and Saturn is simply a result of their greater mass, astronomers deduce that Uranus must contain proportionally fewer light elements, such as hydrogen, than those more massive worlds. On the other hand, the density is too low for Uranus to contain much rock or iron material. Astronomers therefore believe it must be composed of material that is light and abundant, such as ordinary water mixed with methane and ammonia. This mix satisfactorily explains both the density and the spectrum of Uranus. Confirmation of the abundance of water also

**FIGURE 9.21**

Sketch illustrating why Uranus is blue. Methane absorbs red light, removing the red wavelengths from the sunlight that falls on the planet. The surviving light—now missing its red colors—is therefore predominantly blue. As that light scatters off cloud particles in the Uranian atmosphere and returns to space, it gives the planet its blue color.

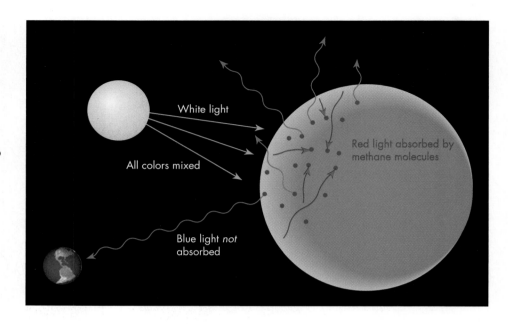

White light

All colors mixed

Red light absorbed by methane molecules

Blue light *not* absorbed

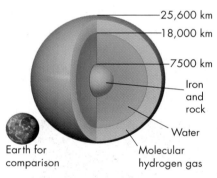

25,600 km
18,000 km
7500 km
Iron and rock
Water
Earth for comparison
Molecular hydrogen gas

**FIGURE 9.22**

Artist's view of a recently suggested model for the interior of Uranus. Note how different it is from Jupiter and Saturn, which both contain large regions of liquid and metallic hydrogen.

The five large Uranian moons are named for characters in English literature. Titania and Oberon are the queen and king of the fairies in Shakespeare's *A Midsummer Night's Dream*. Ariel and Miranda are characters in Shakespeare's *The Tempest*. Umbriel is a sprite in Alexander Pope's poem "The Rape of the Lock," in which Ariel also appears. The smaller Uranian moons are also named for characters from Shakespeare, such as Puck and Cordelia.

*Miranda*

http://www.mhhe.com/arny

comes from studies of the planet's shape. Uranus, like Jupiter and Saturn, rotates moderately fast, with its equator rotating faster than its poles. At its equator, Uranus spins once every 17 hours, bulging the planet's equator. The size of such a bulge depends in part on the planet's gravitational attraction and therefore on how the mass generating that attraction is distributed inside the planet. Thus, astronomers can deduce the density and composition deep inside Uranus from the size of its bulge. Such studies are consistent with the hypothesis that Uranus is composed of mostly water and hydrogen-rich gases and that it may have a core of rock and iron-rich material, as illustrated in figure 9.22. It is not known whether the core formed first and then attracted the lighter gases and ices that condensed around it, or whether the core formed by heavy material sinking to the center after the planet formed. In fact, some astronomers think Uranus's core may be simply highly compressed ices with little rocky material.

## Uranus's Rings and Moons

Uranus is encircled by a set of narrow rings, as illustrated in figure 9.6. The rings, like those of Saturn, are composed of a myriad of small particles, perhaps a meter or so in diameter, moving in individual orbits. The Uranian rings are very dark, however, implying that they are not made of, or coated with, ice like the bright rings of Saturn; instead, they may be rich in carbon particles or organic molecules. Uranus's rings are also extremely narrow compared with those of Saturn. According to one theory, the ring particles are held in such narrow zones by "shepherding satellites," as discussed for Saturn's rings.

Uranus has five large moons and about 20 smaller ones. Like the moons of Jupiter and Saturn, they form a regular system and are probably composed mainly of ice and rock. Many of the Uranian moons are heavily cratered, but Miranda, the smallest of the five large moons, has a surface totally unlike that of any other Solar System body (fig. 9.23). The surface is broken into distinct areas that seem to bear no relation to one another. One region is wrinkled, while an adjacent one has small hills and craters, rather like our Moon. Miranda's patchwork appearance leads some astronomers to think it may have been shattered by impact with another large body. The pieces were subsequently drawn back together by their mutual gravity, giving this peculiar moon its jumbled appearance. Alternatively, the curious surface might have been caused by rising and sinking motions driven by heat in Miranda's interior. Regardless of the cause, Miranda has some extremely curious and unexplained surface features such as a set of cliffs, visible in the top corner of figure 9.23, that are twice the height of Mount Everest.

Cliffs
approx. 20 km—more than 12 miles—high

Enlargement
of cliff region

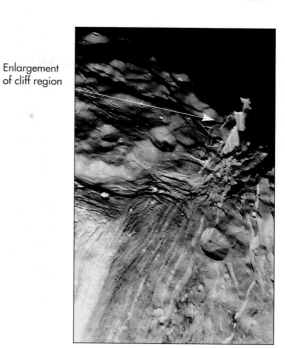

**FIGURE 9.23**
Miranda, an extremely puzzling moon, as observed by the *Voyager* spacecraft. Note the enormous cliffs glinting in the sunlight at the top of the picture.

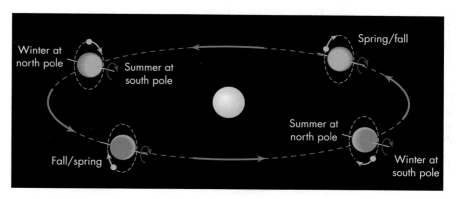

**FIGURE 9.24**
Sketch of the odd tilt of Uranus, perhaps caused by a huge impact early in the planet's history. Because of this tilt, when Uranus's north pole points toward the Sun (its northern summer), the Sun will be above the horizon for many Earth years. Likewise, the other pole will be in night for a corresponding period. On the other hand, during the Uranian spring and fall, the Sun rises and sets approximately every 17 hours. Size of bodies and orbits are not to scale.

## Uranus's Odd Tilt

Uranus's rotation axis is tipped so that its equator is nearly perpendicular to its orbit. That is, it spins nearly on its side, as illustrated in figure 9.24. Moreover, the orbits of Uranus's moons are similarly tilted. They orbit Uranus in its equatorial plane, and, as a result, their orbits are also tilted at approximately 90° with respect to the planet's orbit. Some astronomers therefore hypothesize that during its formation, Uranus was struck by an enormous planetesimal whose impact tilted the planet and splashed out material to create its family of moons. Other astronomers think Uranus was tilted by gravitational tugs exerted on it by neighboring planets (especially Saturn).

Regardless of what caused it, the strong tilt of its axis gives Uranus an odd pattern of day and night. For part of its orbit, one pole is in "perpetual" day and the other pole is in "perpetual" night. Thus, sunlight heats the planet very unevenly, perhaps explaining why Uranus lacks the cloud bands seen on the other giant, gaseous planets. The lack of cloud bands may be temporary, however. Recent observations with an infrared telescope on Earth reveal an odd dark marking in Uranus's atmosphere, perhaps a feature that fades and reappears.

## 9.4 NEPTUNE

**ANIMATION**

*The rotation of Neptune*

Neptune is the outermost of the large planets and is very similar to Uranus in size, with a diameter about 3.9 times that of the Earth and a mass about 17 times the Earth's. Through a telescope on the Earth, it looks like a chip of sapphire, a lovely blue color, but because of its great distance from the Sun—about 30 AU—it is difficult to study. Pictures taken of it by passing spacecraft show it to be a deep blue world, with markings reminiscent of Jupiter's, as shown in figure 9.25. Cloud bands encircle it, and it even has briefly shown a Great Dark Spot, a huge dark blue atmospheric vortex that looks a little like Jupiter's Great Red Spot but is blue instead of red.

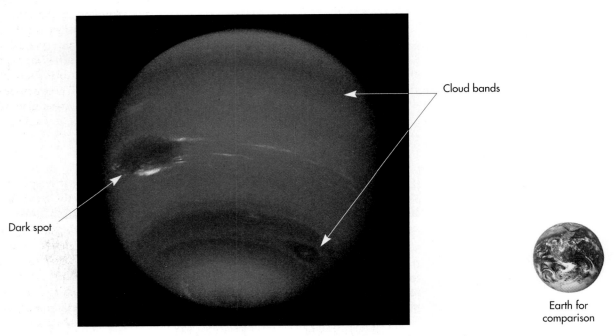

Cloud bands

Dark spot

Earth for comparison

**FIGURE 9.25**
Image of Neptune taken by the *Voyager* spacecraft. Notice the dark blue oval, an atmospheric vortex whose origin is probably similar to that of Jupiter's Great Red Spot.

✦ Approximately how big is this spot compared to the Earth?

✦ Measure the spot and Neptune's disk, and divide the spot's size by Neptune's. The spot is not round, but is roughly 0.2 the size of Neptune. Neptune is about 3.8 times the size of the Earth, so the spot is 0.2 × 3.8 = 0.76 times the size of the Earth (about three-fourths Earth's size).

Neptune, named for the Roman god of the sea, was discovered in the 1840s from predictions made independently by a young English astronomer, John Couch Adams, and a French astronomer, Urbain Leverrier. Adams and Leverrier both noticed that Uranus was not precisely following its predicted orbit, and they therefore inferred that its motion was being disturbed by the gravitational force of an as yet unknown planet. From the size of these orbital disturbances, Adams and Leverrier predicted where the unseen body must lie.

Adams completed his calculations in 1845, but when he reported his results, the astronomer royal, Sir George Airy, was unconvinced and gave a low priority to the search for the unseen planet. In 1846, however, Airy was startled to read a paper by Leverrier detailing calculations nearly identical to those made by Adams. This spurred Airy to begin a search in earnest, but by then it was too late: Leverrier had given his predicted positions to Johann Galle, a German astronomer who that same night pointed his telescope to the predicted location and saw Neptune. Assignment of credit for the discovery of the new planet led to a rancorous dispute tinged with national pride that lasted decades. The discovery is now credited equally to Adams and Leverrier. Ironically, Galileo had seen Neptune in 1613 while observing Jupiter's moons. His observation notes record a dim object whose position changed with respect to the stars, as would be expected for a planet. Galileo failed, however, to appreciate the significance of that motion, so Neptune eluded discovery for another two centuries.

## Neptune's Structure

Neptune's structure is probably similar to Uranus's. That is, the planet is composed mostly of ordinary water surrounded by a thin atmosphere rich in hydrogen and hydrogen compounds, such as methane. Our understanding of the atmosphere's composition is derived from the planet's spectrum. The composition of the interior is inferred from the planet's low average density, which is 1.67 grams per cubic centimeter. As we have seen for the other giant planets, such a low density implies that the planet must be composed mostly of light atoms. But during its formation, Neptune must also have collected heavy elements such as silicon and iron. These denser materials probably sank to the center of Neptune (just as they did in the other giant planets), where they now form a core of rock and iron a little smaller than the Earth's.

## Neptune's Atmosphere

Neptune's blue color is caused, like Uranus's, by methane in its atmosphere. Unlike Uranus, however, Neptune has distinctive cloud belts. Why do these two so similar bodies have such different cloud formations? Infrared observations offer a partial explanation. Neptune, like Jupiter and Saturn, radiates more energy than it gains from the Sun. That excess energy (also like Jupiter's and Saturn's) is probably left over from the planet's birth or is supplied by heavier material sinking even now to the planet's core. Whatever its source, the heat deep within Neptune generates convection currents that rise to its outer atmosphere. There, the rising gas is deflected into a system of winds by the Coriolis effect caused by Neptune's spin—one rotation every 16 hours. The resulting winds create cloud bands similar to those seen on Jupiter and Saturn but tinted deep blue by Neptune's methane-rich atmosphere.

Neptune's winds are extremely fast. For example, near its equator, strong easterly winds blow opposite to the planet's direction of rotation, much like the trade winds on Earth. But these Neptunian winds reach speeds of nearly 2200 kilometers per hour (about 1300 miles per hour). As these gales sweep around Neptune, gas sandwiched between adjacent streams spins, as described in our discussion of Jupiter's atmosphere. Such localized spinning motion creates the Great Dark Spot, as the large, dark blue spot illustrated in figure 9.25 is called. Curiously, pictures of Neptune taken by the Hubble Space Telescope in the mid-1990s showed that the spot is gone. Perhaps such spots grow and dissipate like terrestrial storms. Support for that possibility comes from recent Hubble Space Telescope observations of Uranus, which show that its previously bland appearing atmosphere now has a small dark spot.

## Neptune's Rings and Moons

Neptune, like the other giant planets, has rings, but they are very narrow, more like those of Uranus than those of Saturn, as shown in figure 9.6. They are probably composed of debris from small satellites or comets that have collided and broken up, and they contain proportionally more dust than the rings of either Saturn or Uranus. Moreover, photographs of the rings show that in some places, the ring particles are not distributed uniformly around the ring but are gathered into arcs. This clumping—probably the result of "shepherding satellites"—leads astronomers to think that Neptune's rings may be relatively short-lived and that they may dissipate in a million years or perhaps even faster unless new ring material is added from the breakup of other small bodies.

Neptune has six small moons orbiting close to the planet in roughly circular orbits and seven other moons at much greater distances. One of these, Triton, is immense and orbits "backward" (counter to Neptune's rotation) compared with the motion of most other satellites. Moreover, its orbit is highly tilted with respect to Neptune's equator (most large moons follow orbits approximately parallel to their planet's equator). These orbital peculiarities lead many astronomers to think that Triton is a surviving icy planetesimal from the inner Kuiper belt. When Neptune captured it, the encounter destroyed or expelled any midsize moons that Neptune originally possessed. In fact, Neptune's outermost moon, Nereid, follows such a highly elliptical path far beyond the other moons that it just barely missed escaping from its parent world.

Triton intrigues astronomers for more than its orbital oddity. It is massive enough that its gravity, in combination with its low temperature, allows it to retain gases around it. Triton is thus one of the few moons in the Solar System with an atmosphere.

Triton's surface also has many unusual features, as shown in figure 9.26. Wrinkles give much of the surface a texture that looks like a cantaloupe. Craters pock the surface elsewhere, and dark streaks extend from some of them. At least one of these streaks originates from a geyser caught in eruption by the passing *Voyager II* spacecraft. Astronomers think that the matter ejected by the geysers is a mixture of nitrogen, ice, and carbon compounds. Sunlight, pale though it is at Triton's immense distance from the Sun, may warm such matter trapped below the surface and make it expand and burst through surface cracks. The erupted material cools and condenses in Triton's cold, thin atmosphere, where winds carry and deposit it as a black "soot."

Neptune's moons are named for mythological sea deities, in keeping with the planet's own name.

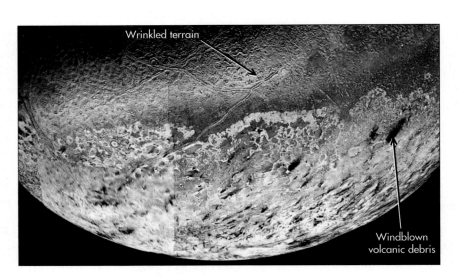

**FIGURE 9.26**
The surface of Triton.

## 9.5   PLUTO AND BEYOND

Pluto, named for the Greek and Roman god of the underworld, was discovered in 1930 by Clyde Tombaugh, an astronomer at Lowell Observatory in Flagstaff, Arizona. Tombaugh painstakingly examined pairs of photographs of the sky for over a year—scanning millions of star images—searching for objects whose position changed between the exposures, the telltale motion that distinguishes a planet from a star. For more than 75 years Pluto was considered a planet, although almost from the time of its discovery, it was clear that it differed in many ways from the previously known planets. But Pluto's great distance from us and the Sun combine with its small size to make it very dim and thus difficult to study. Even in the largest telescopes on the ground, Pluto looks like a dim star, as illustrated in figure 9.27. For years after its discovery, astronomers knew only that Pluto was smaller than the Earth. Then, in 1978, James Christy of the U.S. Naval Observatory discovered a moon orbiting Pluto, which made it possible for astronomers to measure Pluto's radius and mass.

Pluto's moon, Charon, is named for the boatman who, in mythology, ferries dead souls across the river Styx to the underworld. Circling Pluto in a tiny orbit whose average radius is 19,600 kilometers, Charon takes only 6.4 days to complete a trip around the planet. From this orbital data, Pluto's mass* can be calculated using Newton's modified form of Kepler's third law of planetary motion (see fig. 7.7). Such a calculation shows that Pluto's mass is about 0.002 times the Earth's, making Pluto less than $\frac{1}{20}$th the mass of Mercury, the smallest planet.

Observations of Charon's motion around Pluto show that its orbit is tilted steeply with respect to Pluto's orbital plane and that Pluto's rotation axis is similarly tipped, reminiscent of Uranus's odd tilt.

As Charon orbits Pluto, it occasionally[†] eclipses the planet from our view. From these eclipses, the diameters of both Pluto and Charon can be measured. Pluto's diameter turns out to be a little less than one-fifth the Earth's, and Charon, though smaller, turns out to be surprisingly large. Satellites are generally dwarfed by their planets, but Charon's diameter is slightly more than half Pluto's. Moreover, its orbit is locked in synchrony with Pluto's rotation, suggesting a tight gravitational coupling between the bodies. Such eclipse studies are not the only way to study Pluto and Charon, however. Observations of them made with interferometers (see chapter 4) and from the Hubble Space Telescope show Pluto's and Charon's tiny disks, nicely confirming the eclipse data.

From Pluto's mass and radius, astronomers can deduce its density to be about 2.1 grams per cubic centimeter, a value suggesting that Pluto must be mainly a mix of rock, ordinary water-ice, and frozen nitrogen and methane. From spectra of its surface, they can directly determine that Pluto is covered with a mixture of water-ice and ices of methane, nitrogen, and carbon monoxide. Despite this knowledge of the dwarf planet's composition, however, astronomers still lack information about its surface features. Although astronomers have made pictures of Pluto and its moon, Charon, with the Hubble Space Telescope, the best "images" of this distant system come from computer analysis of its changing brightness during eclipses by Charon. These images (fig. 9.28) show that Pluto's south pole is much brighter than its equator, implying the presence of a polar cap there. Spectra suggest that the cap is frozen methane.

Astronomers have also detected a very tenuous atmosphere on Pluto. From measurements of the temperature and surface composition, they deduce its atmosphere is mostly nitrogen and carbon monoxide, with traces of methane. In the bitter cold (40 K, or about −387°F) of this remote world, molecules move so slowly that, despite its tiny mass, Pluto's gravity is strong enough to retain a thin atmosphere.

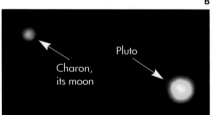

**FIGURE 9.27**
(A) Pluto looks like merely another dim star in this photograph taken at Lick Observatory in California. Only its change in position from night to night shows it to be a planet. (B) A picture of Pluto and its moon Charon obtained with the Hubble Space Telescope and therefore free of the blurring effects of our atmosphere.

☞ Why does the position of Pluto change but not that of the stars?

𝒬 Pluto's shift across the background stars is simply its orbital motion around the Sun.

---

*Technically, this method gives the *combined* mass of Pluto and Charon.

[†]Because of Charon's orbital tilt, Pluto-Charon eclipses occur about every six days for a few years but then do not reoccur for half a Plutonian year, or about 124 Earth years!

Orbit of Pluto

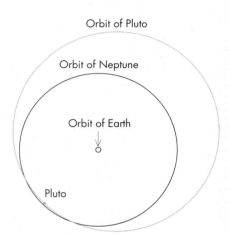

Orbit of Neptune

Orbit of Earth

Pluto

**FIGURE 9.29**

Pluto's odd orbit is highly tilted and highly eccentric (elongated)—so much so that it crosses Neptune's orbit.

↻⋆ If Pluto's orbit crosses Neptune's, why don't they collide? Referring to figure 7.3 may help you answer the question.

ɑ⋆ Pluto's orbit is relatively steeply tilted with respect to Neptune's. As a result, when Pluto "crosses" Neptune's orbit, Pluto is actually well above or below Neptune's path.

Recently astronomers have discovered that Pluto has two other moons in addition to Charon. These moons, called Nix and Hydra, are both very small (40 to 160 km in diameter) and little is yet known of them except that they orbit about two to three times farther from Pluto than Charon orbits.

**FIGURE 9.28**

(Top) Pictures of Pluto made with the Hubble Space Telescope.
(Bottom) Computer-generated maps of Pluto based on such pictures.

Pluto's small size and peculiar orbit (it crosses Neptune's orbit, as shown in fig. 9.29) once led some astronomers to hypothesize that it was originally a satellite of Neptune that escaped and now orbits the Sun independently.

Today, however, astronomers think almost the reverse—that Neptune has "captured" Pluto. Pluto's orbital period is 247.7 years, very close to one and one-half times Neptune's. Thus, Pluto makes two orbits around the Sun for every three made by Neptune. This match of orbital periods has created a cumulative gravitational attraction on Pluto that "tugged" it into its current orbit. In fact, several hundred other presumably icy, objects a few hundred kilometers in diameter orbit at nearly the same distance from the Sun as Pluto. But even larger icy objects orbit farther out.

In 2005 astronomers reported the discovery of a dim object orbiting at about 97 AU from the Sun. Now officially named Eris⋆ (for the Greek Goddess of Discord), this remote world turns out to have a diameter of about 2400 km (about 1500 miles), making it slightly larger than Pluto. Its orbit is both highly elliptical (ranging from 38 to 97 AU from the Sun) and highly inclined (about 44°) to the main disk of the Solar System. Because of its size, some astronomers initially described it as a tenth planet, despite its orbit being so different from that of the other planets.

Eris is like Pluto in many ways besides its size. Its surface is covered with frozen methane, and it has a moon, Dysnomia,† approximately the same size as Pluto's moon, Charon. Moreover, it is not the only large icy object in the outer parts of the Solar System. Some half dozen other large (~1000 km diameter) icy worlds, such as Orcus and Sedna (see fig. 9.30), orbit there as well. Sedna is of particular interest to astronomers because its orbit extends well beyond the standard Kuiper Belt.

What should these objects—presumably surviving icy planetesimals—be called? In August 2006, as mentioned in chapter 7, the International Astronomical Union created a new category of Solar System objects called *dwarf planets*. This dealt with Pluto and Eris, but what of the other ones such as Orcus and Sedna? Many astronomers now call these and similarly distant bodies **Trans-Neptunian Objects (TNOs).** But terminology

---

⋆Eris has had several names. It was formally called 2003 UB313, but it has been informally called Xena.
†In Greek mythology, the goddess of lawlessness.

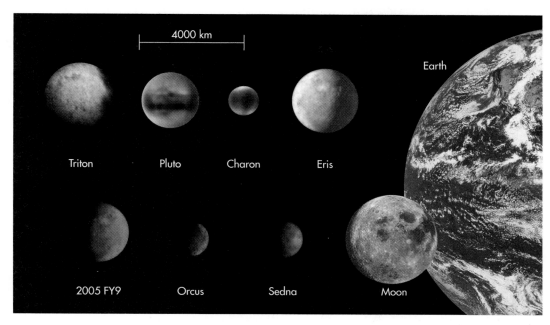

**FIGURE 9.30**

Some of the icy objects of the outer Solar System shown to correct relative scale, along with the Moon and Earth. Most of these images are artist's representations because these icy worlds appear as little more than specks in even in the largest telescopes.

is not the only issue posed by the TNOs. Many of them are too large to have formed in the low-density regions of the outer Solar System where we see them now. Thus, astronomers hypothesize that the TNOs formed nearer the Sun, perhaps near Neptune, and have been ejected into their present orbits by successive gravitational tugs exerted by the massive outer planets.

## SUMMARY

The outer planets formed in the solar nebula far from the Sun, where it was cold enough for them to accumulate hydrogen-rich material. Because of hydrogen's abundance relative to other elements, Jupiter, Saturn, Uranus, and Neptune grew much larger than the terrestrial planets in both diameter and mass (see table 9.1) and are composed mostly of hydrogen and its compounds, such as methane ($CH_4$), ammonia ($NH_3$), and water ($H_2O$). These gases form a deep atmosphere, often richly colored, that grows denser with depth and eventually becomes liquid. The interiors of Jupiter and Saturn are composed mainly of hydrogen—liquid just below the atmosphere and metallic deeper down. Each is believed to have a rocky core whose diameter—within a factor of 2 or 3—is comparable to the Earth's. Uranus and Neptune may have some liquid hydrogen below their hydrogen- and methane-rich atmospheres, but they are probably mostly water, ammonia, and methane surrounding a rocky core. Because they are smaller bodies, their weaker gravity has not allowed them to capture or retain as much hydrogen and helium, but they have retained the heavier gases.

The giant planets are probably heated by continued gravitational contraction and settling of heavier matter toward their cores. As this heat flows outward, it generates convective motions that stir them. The planets' rotation creates a Coriolis effect on the rising gas, drawing it into the cloud belts seen on Jupiter, Saturn, and Neptune.

Rotation also creates an obvious equatorial bulge on Jupiter and Saturn, which spin in about 10 hours. The spinning motion, when combined with convection in their metallic hydrogen cores, generates powerful magnetic fields that in turn contribute to the formation of auroras and energetic, charged particle belts like the Van Allen belts of our Earth.

Each of the four giant planets has a ring system composed of small orbiting particles. The rings are probably debris from small objects that have broken up as the result of either collisions or tidal forces exerted by the planet. Collisions continue to break up small particles, replacing ring material that falls into the planets' atmospheres.

Many satellites orbit each of the four giant planets. Some of these moons are comparable in diameter to Mercury. They are thought to be made mostly of water, ice, and rock, although Io, an exceptionally active moon of Jupiter, may be mostly sulfur and rock. Many of the moons are heavily cratered, but a few are smooth, indicating surface activity (Io's volcanos of sulfur, for example) that has filled in old craters. Saturn's moon, Titan, has an atmosphere of mainly nitrogen, but because of its great distance from the Sun, it is bitterly cold. "Rain" falling from Titan's clouds is not water, but liquid methane, which flows across Titan's surface, carving rivers and collecting in lakes.

The dwarf planets Pluto and Eris are like some of the moons of the outer planets in their size and mass. They are probably mostly ice and rock, with thin methane atmospheres. Astronomers think that these distant worlds are surviving icy planetesimals.

## ?What It Is.

This shows where two pieces of Comet Shoemaker-Levy 9 struck Jupiter in 1994. The lower impact generated a blast in Jupiter's atmosphere that traveled outward to form the dark surrounding ring and arc. These dark markings could be seen even with a small backyard telescope here on Earth.

## TABLE 9.1

### Comparison of the Four Giant Planets

|  | Jupiter | Saturn | Uranus | Neptune |
|---|---|---|---|---|
| $R_{equator}$ (Earth units) | 11.19 | 9.46 | 3.98 | 3.81 |
| (km) | (71,492) | (60,268) | (25,559) | (24,764) |
| M (Earth units) | 317.9 | 95.18 | 14.54 | 17.13 |
| (kg) | $(1.9 \times 10^{27})$ | $(5.68 \times 10^{26})$ | $(8.68 \times 10^{25})$ | $(1.02 \times 10^{26})$ |
| Density (g/cm³) | 1.33 | 0.69 | 1.32 | 1.64 |
| Sidereal day (hours) | 9.924 | 10.656 | 17.24 | 16.11 |
| Distance from Sun (AU) | 5.203 | 9.539 | 19.19 | 30.06 |
| Orbital period (years) | 11.8622 | 29.4577 | 84.014 | 164.793 |
| Axial tilt (degrees) (with respect to orbit) | 3.12 | 26.73 | 97.86 | 29.56 |

## QUESTIONS FOR REVIEW

1. How do Jupiter's mass and radius compare with the Earth's? How do they compare with those of the other outer planets?

2. What are the major gaseous substances that make up Jupiter and Saturn?

3. Why are the outer planets so large?

4. What is the interior structure of Jupiter and Saturn thought to be?

5. What does Jupiter look like?

6. What is the Great Red Spot?

7. Do Jupiter and Saturn have solid surfaces?

8. What sorts of atmospheric motion and activity are observed in Jupiter?

9. What sort of activity has been seen on Io? What is Io's heat source thought to be?

10. How do astronomers know what lies inside the outer planets?

11. What are the rings of Saturn made of? How do astronomers know this?

12. What creates the gaps between the rings?

13. How might the rings have formed?

14. What is the Roche limit? Why does such a limit exist?

15. How do Uranus and Neptune differ inside from Jupiter?

16. Why are Uranus and Neptune so blue?

17. What are the satellites of the outer planets thought to be composed of?

18. What evidence makes astronomers believe that Pluto is strongly influenced by Neptune?

19. What is unusual about Uranus's rotation axis? What might explain this peculiarity?

20. How did the discovery of a moon orbiting Pluto help astronomers better understand this object?

## THOUGHT QUESTIONS

1. Why do some astronomers think that Pluto should be considered a planet? Do you agree?

2. If Jupiter were moved closer to the Sun, what do you think would happen to it?

3. Why might Uranus and Neptune have proportionally less hydrogen than their more massive companions, Jupiter and Saturn?

4. Why do the relatively uncratered surfaces of Europa and Enceladus imply that these moons may have been "active" recently?

## PROBLEMS

1. Use the modified form of Kepler's third law, illustrated in figure 7.7 (and discussed in detail in chapter 2), to calculate Pluto's mass from the orbital data for Charon that is given in the text. Be sure to convert the orbital period to seconds and the orbital radius to meters before putting those numbers into the formula. (Assume that Charon's mass is negligible compared to Pluto's.)

2. How long does it take sunlight to reach Pluto?

3. Calculate the density of Charon, given that its radius is approximately 593 kilometers and its mass is about $1.1 \times 10^{24}$ grams. (Be sure to convert kilometers to centimeters or meters.) Is it likely that Charon has a large iron core? Why?

4. Show that Pluto's orbital period is very close to one and one-half times Neptune's. Use the data in table 4 of the appendix.

## TEST YOURSELF

1. The low average densities of Jupiter and Saturn compared with the Earth's suggest
   (a) they are hollow.
   (b) their gravitational attraction has compressed their cores into a rare form of iron.
   (c) they contain large quantities of light elements, such as hydrogen and helium.
   (d) they are very hot.
   (e) volcanic eruptions have ejected all the iron that was originally in their cores.

2. What makes some astronomers think that Uranus was hit by a large body early in its history?
   (a) It goes around the Sun in a direction opposite to the other planets.
   (b) Its rotation axis has such a large tilt.
   (c) Its composition is so different from that of Neptune, Jupiter, and Saturn.
   (d) It has no moons.
   (e) All of the above

3. Why would it be foolish to send astronauts to land on Jupiter?
   (a) It has no solid surface for them to land on.
   (b) Its clouds are so hot that any spacecraft getting near it would burn up.
   (c) Its gravitational attraction is so weak that they would float off.
   (d) All of the above
   (e) The idea is perfectly reasonable.

4. What is the Roche limit?
   (a) The mass a planet must exceed to have satellites
   (b) The smallest mass a planet can have and still be composed mainly of hydrogen
   (c) The greatest distance from a planet that its satellites can orbit without falling into the Sun
   (d) The distance at which a moon held together by gravity will be broken apart by the planet's gravitational attraction
   (e) The distance astronomers can see into a planet's clouds

5. Astronomers think that the inner core of Jupiter is composed mainly of
   (a) hydrogen.
   (b) helium.
   (c) uranium.
   (d) rock and iron.
   (e) water.

# FURTHER EXPLORATIONS

Beatty, J. Kelly, Carolyn Collins Petersen, and Andrew Chaikin. *The New Solar System.* 4th ed. Cambridge: Sky, 1999. (This fine book covers the Solar System in great detail.)

Beatty, J. Kelly, and Stuart J. Goldman. "The Great Crash of 1994: A First Report." *Sky and Telescope* 88 (October 1994): 18.

Binzel, Richard. "Pluto." *Scientific American* 262 (June 1990): 50.

Burns, Joseph A., Douglas P. Hamilton, and Mark R. Showalter. "Bejeweled Worlds." *Scientific American* 286 (February 2002): 64.

Gingerich, Owen. "Losing It in Prague: The Inside Story of Pluto's Demotion." *Sky and Telescope* 112 (November 2006): 34.

Johnson, Torrence V. "The Galileo Mission to Jupiter and Its Moons." *Scientific American* 282 (February 2000): 40.

Lunine, Jonathan I. "Neptune at 150." *Sky and Telescope* 92 (September 1996): 38.

Pappalardo, Robert T., James W. Head, and Ronald Greeley. "The Hidden Ocean of Europa." *Scientific American* 281 (October 1999): 54.

Stern, S. Alan. "Journey to the Farthest Planet." *Scientific American* 286 (May 2002): 56.

———. "The 3rd Zone: Exploring the Kuiper Belt." *Sky and Telescope* 106 (November 2003): 30.

Tytell, David. "Titan: A Whole New World." *Sky and Telescope* 109 (April 2005): 34.

## Website

Please visit the *Explorations* website at **http://www.mhhe.com/arny** for additional online resources on these topics.

# PLANETARIUM EXERCISE

1. Following the same procedure used in the planetarium exercise in chapter 8, follow the paths of Jupiter and Saturn over a year or more. Describe your results in comparison to Mars. How often does retrograde motion take place for them? Discuss your results in relation to their orbital periods and that of the Earth.

# KEY TERMS

Roche limit, 282      shepherding satellites, 281      trans-Neptunian object (TNO), 292

# Meteors, Asteroids, and Comets

# 10

## Chapter Objectives

- Numerous objects much smaller than the planets orbit the Sun.
  - Rocky objects—the asteroids
  - Icy objects—the comets
- These objects are left over from the formation of the Solar System.
- Fragments of these objects reach Earth from time to time as meteoroids.
- When such a fragment enters our atmosphere, it is heated by friction with the air and glows, creating a "shooting star."
- Earth has been struck in the past by very large objects, which may have caused mass extinctions.

Comet Hale-Bopp.

The asteroid Eros.

**?What Is This?**
(See chapter summary for the answer.)

## Concepts and Skills to Remember

- Density (5.1)

Orbiting the Sun and scattered throughout the Solar System are numerous bodies much smaller than the planets—the asteroids and comets. The asteroids are generally rocky objects in the inner Solar System. The comets are icy bodies and spend most of their time in the outer Solar System. These small members of the Sun's family, remnants from the formation of the Solar System, are of great interest to astronomers because they are our best source of information about how long ago and under what conditions the planets formed. In fact, some asteroids and comets may be planetesimals—the solid bodies from which the planets were assembled—that have survived nearly unchanged from the birth of the Solar System.

Apart from their scientific value, asteroids and comets merit study because they can be both beautiful and deadly. A comet in the dawn sky with its tail a shining plume is a sight not to miss. But "miss" is precisely what we hope will happen if a large comet or asteroid is on a collision course with Earth. Such a collision in the past may have exterminated much of Earth's ancient life, and such an event in the future could well have equally disastrous effects on today's living things.

In this chapter, we will see why astronomers think asteroids and comets are related to planetesimals. We will also discover why meteorites (which are simply fragments of asteroids and comets that by chance fall into our atmosphere) are such important clues to the time of birth and the structure of the ancient Solar System. Likewise, we will study how a comet changes from a 10-kilometer diameter ball of ice into a beautiful banner of light in the night sky and why an asteroid may be the reason you have hair rather than scales.

## 10.1    METEORS, METEOROIDS, AND METEORITES

If you have spent even an hour looking at the night sky, you have probably seen a "shooting star," a streak of light that appears in a fraction of a second and as quickly fades (fig. 10.1). Astronomers call this brief but lovely phenomenon a **meteor.**

A meteor is the glowing trail of hot gas and vaporized debris left by a solid object heated by friction as it moves through the Earth's atmosphere. Most of the heating occurs between about 100 and 50 kilometers in the outer fringes of the atmosphere. The solid body, while in space and before it reaches the atmosphere, is called a **meteoroid.**

### Heating of Meteoroids

Meteoroids heat up on entering the atmosphere for the same reason a reentering spacecraft does. When an object plunges from space into the upper layers of our atmosphere, it collides with atmospheric molecules and atoms. These collisions convert some of the body's energy of motion (kinetic energy) into heat, as shown in figure 10.2. In a matter of seconds, the outer layer of the meteor reaches thousands of degrees Kelvin and glows. Given that reentry speeds are typically at least 10 kilometers per second and often 30 to 40 kilometers per second, the collisions with air molecules are extremely violent and tear atoms off the body, vaporizing the surface layers. The trail of hot evaporated matter and atmospheric gas emits light, making the glow that we see.

If the meteoroid is larger than a few centimeters, it creates a ball of incandescent gas around it and may leave a luminous or smoky trail. Such exceptional meteors, sometimes visible in daylight, are called "fireballs."

Meteoroids bombard the Earth continually: a hail of solid particles that astronomers estimate amounts to hundreds of *tons* of material each day. Slightly more strike between midnight and dawn than in the evening hours, and so it is best to watch for meteors in

**FIGURE 10.1**
A time-exposure photograph captures a "shooting star" (meteor) flashing overhead. The curved streaks are star trails.

☞⋆ Why is the meteor's track straight, whereas the star images are curved?

ℚ⋆ The star trail shows that this was a fairly long time exposure, during which the Earth rotated, smearing the stars' images into long arcs. The meteor passed by very quickly; the Earth didn't rotate very far during its passage, and as a result, the meteor's track is straight.

Some stony meteorites have a smooth structure with no sign of chondrules. Astronomers think that these so-called achondritic meteorites were melted and resolidified, probably in a large asteroid.

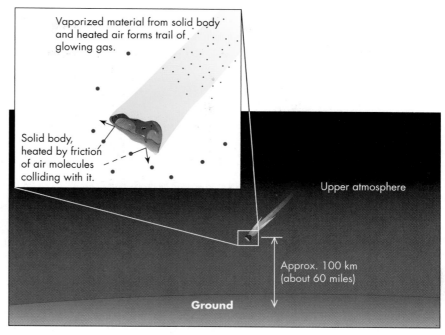

**FIGURE 10.2**
Sketch depicting how air friction heats an object entering our atmosphere, creating the glow we see as a meteor.

the early morning. This difference arises for the same reason that if you run through rain, your front will get wetter than your back. That is, the dawn side of our planet advances into the meteoritic debris near us in space, while the night side moves away from it.

Most meteors that we see last only a few seconds and are made by meteoroids the size of a raisin or smaller. These tiny objects are heated so strongly that they completely vaporize. Larger pieces, though heated and partially vaporized, are so drastically slowed by air resistance that they may survive the ordeal and reach the ground.* We call these fragments found on the Earth **meteorites.**

## Meteorites

Astronomers classify meteorites into three broad categories based on their composition: iron, stony (that is, composed mainly of silicate compounds), and stony-iron. Most stony meteorites are composed of smaller rounded chunks of rocky material stuck together. The grains are called **chondrules,**† and meteorites that have this lumpy structure are called **chondritic meteorites.** Chondrules appear to have been rapidly melted and cooled in the solar nebula. The cause of the heating is not known. Figure 10.3 shows a meteorite of uniform composition and a chondritic meteorite.

Chondrules contain traces of radioactive material, which can be used to measure their age, as described in chapter 5. They are extremely old, 4.5 billion years, and are believed to be the first solid material that condensed within the solar nebula. Many chondrules contain even older material: dust grains that have survived from before the birth of the Solar System. Thus, chondritic meteorites offer us valuable information about the early history of the Solar System.

---

*People are sometimes hit by meteorites but only rarely. There is one well-documented case of a person being killed, and another case is known where a meteor punched a hole through a roof and hit a woman, badly bruising her. See section 10.4 for other reports.

†The *ch* is pronounced as a *k* sound. Thus, *chondritic* is pronounced *kon-DRIT-ic* and means "coarse grained" in Greek.

**FIGURE 10.3**

Meteorites. (A) A fragment of an approximately 100 ton iron meteorite that struck the Sikhote-Alin mountains, in extreme western Russia in 1947. The meteor was seen in broad daylight as a brilliant ball of fire, brighter than the Sun that left a dark trail of smoke. The event was painted on the spot by an artist who witnessed it (reproduced on a Russian stamp shown in inset). (B) Slices showing the chondrules inside two chondritic meteorites, one an ordinary (rocky) chondrite, the other having a matrix of carbon-rich material.

A

Iron

0  1  2  3  4 cm

B

Chondrules

Ordinary chondrite

Carbonaceous chondrite

0   1   2 cm

Amino acids occur in two molecular forms—right-handed and left-handed. All living things on Earth contain only left-handed amino acids. The amino acids found in meteorites contain both forms, however. Their presence indicates that they are not contamination by terrestrial organisms but were formed in space by nonbiological processes.

In some chondritic meteorites, the chondrules are embedded in a black, carbon-rich, coal-like substance and are therefore called **carbonaceous chondrites.** This carbonaceous matter contains organic compounds, including amino acids, the same complex molecules used by living things for the construction of the proteins on which they depend. Thus, the presence of amino acids in meteoritic matter indicates that the raw material of life can form in space and that it might therefore have been available right from the start within the Solar System. Irrespective of the existence of amino acids in meteorites, we can still ask how and where these bodies formed, and what brought them to Earth. Astronomers think that most of them are fragments of asteroids* and comets.

---

*Although most meteorites are fragments of asteroids or comets, some are chunks of rock from the surface of the Moon and Mars, blasted into space by the collision of an asteroid with these bodies.

## **10.2**   ASTEROIDS

**Asteroids** are small, generally rocky bodies that orbit the Sun. They are found throughout the Solar System, but most lie in the **asteroid belt,** a region between the orbits of Mars and Jupiter, stretching from about 2 to 4 AU from the Sun, as shown in figure 10.4. Giuseppe Piazzi discovered the first of this swarm of bodies in 1801 during his search for the "missing planet," which, according to Bode's law, should have been at 2.8 AU from the Sun. He named the asteroid that he found Ceres in honor of the patron goddess of Sicily, his home. More than 100,000 other asteroids have subsequently been found, but Ceres remains the largest one we know of. Despite the huge number of asteroids, their combined mass is very small, amounting to probably less than $\frac{1}{1000}$ the mass of the Earth.

### Size and Shape

The diameter of an asteroid is difficult to measure because nearly all are so small that they appear in ground-based telescopes merely as points of light. Moreover, the amount of light reflected from an asteroid is not a good clue to its size because a large,

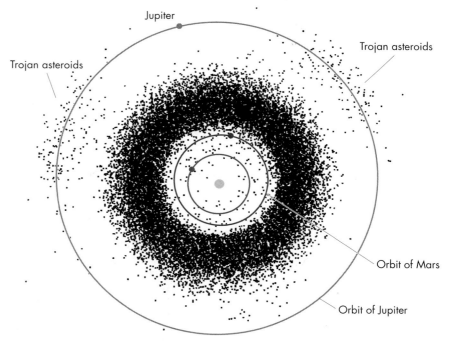

**FIGURE 10.4**

Diagram showing the distribution of 21,785 known asteroids. Notice that most lie between Mars and Jupiter, but a small number form two loose clumps—the Trojan asteroids—located on Jupiter's orbit. Making the plotted points large enough to see causes them to appear far more closely packed than they really are.

With a ruler, estimate how far (in AU) the center of the Trojan swarms are from the Sun and Jupiter.
   How does this compare with Jupiter's distance from the Sun? Does this help you understand why the Trojan asteroids are located where they are?

The Trojan asteroids are about equidistant from the Sun and Jupiter. Thus, the Sun pulls on them about as strongly as it does on Jupiter.

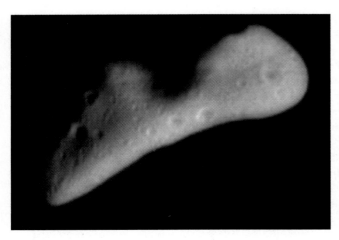

**FIGURE 10.5**
The asteroid Eros, as observed by the *NEAR* spacecraft.

poorly reflective object will look as bright as a small, highly reflective one. For this reason, the emitted infrared radiation is a better measure of diameter; bigger bodies emit more than smaller ones of the same temperature. From such measurements, astronomers have found that asteroids range tremendously in diameter, from Ceres—about 1000 kilometers (less than one-tenth the Earth's size) across—down to kilometer-sized bodies and even smaller. For example, the diameter of the tiny asteroid 1991 BA, which passed about 170,000 kilometers (less than half the distance to the Moon) from Earth in January 1991 is probably less than 9 meters (approximately 30 feet).

The smallest asteroids are far too tiny to observe unless they come extremely close to Earth or strike it. In that case, we see them as meteors. That is, meteorites are in a real sense just the very smallest (we hope!) asteroids that happen to reach our planet.

Most asteroids—such as Eros (fig. 10.5), whose picture was taken by the *NEAR* spacecraft, and Ida (fig. 10.6A)—are irregularly shaped. Only Ceres (fig. 10.6B) and a few other large asteroids are approximately spherical.* Their gravitational force is strong enough to crush their material into a sphere. Small bodies with weak gravities remain irregular and are made more so by collisions blasting away pieces. Collisions leave the parent body pitted and lumpy, and the fragments become smaller asteroids in their own right.

Although we have described asteroids as rocky or iron-rich, that does not mean they are all solid chunks of such material. For example, the density of the asteroid Mathilde is very low (1.4 gm/cm³), suggesting that it is a loose pile of rubble, just barely held together by its gravity. Such a loose structure might also explain the extremely irregular shape of such asteroids as Itokawa, whose picture (fig. 10.7) was taken by the Japanese space probe *Hayabusa*.[†] This probe landed on Itokawa in November 2005, and from measurement of the asteroid's gravitational attraction on the probe, astronomers calculated the asteroid's mass. From the mass and with knowledge of the asteroid's dimensions, astronomers calculated its density, which turned out to be 1.9 gm/cm³. This too is so low a density that the asteroid must be mainly loose rock lumps, only weakly held together by its gravity. Figure 10.7 clearly shows this lumpy structure.

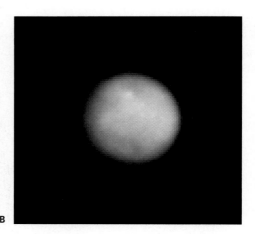

**FIGURE 10.6**
(A) The asteroid Ida and its "Moon" Dactyl. (B) The asteroid Ceres, as viewed by the Hubble Space Telescope. Note that Ceres looks approximately spherical.

---

*Because of its approximately spherical shape, Ceres is now officially also a dwarf planet.

[†]*Hayabusa* means "Falcon" in Japanese.

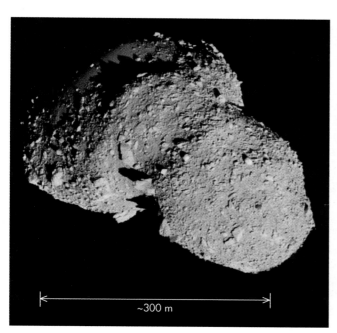

**FIGURE 10.7**
Asteroid Itokawa. Note its rubbly appearance.

*Differentiation in asteroids and their subsequent breakup by collisions to form iron and stony bodies*

Following its landing, the probe attempted to scoop up a sample from the asteroid and then lifted off to return to Earth. It is scheduled to arrive here in 2010, but scientists will not know until then if it succeeded in picking up a sample.

## Composition

When sunlight falls on an asteroid, the minerals in its surface create absorption features in the spectrum of the reflected light, from which we can determine the asteroid's composition. Such spectra show that asteroids belong to three main compositional groups similar to those of meteorites: carbonaceous bodies, silicate bodies, and metallic iron-nickel bodies. The groups are not mixed randomly throughout the asteroid belt: inner-belt asteroids tend to be silicate-rich, and outer-belt ones tend to be carbon-rich.

## Origin of Asteroids

The properties of asteroids that we have just discussed (composition, size, and their location between Mars and Jupiter) give us clues to their origin and support the solar nebula hypothesis for the origin of the Solar System. As we described in chapter 7, the asteroids are probably fragments of planetesimals, the bodies from which the planets were built.

According to the solar nebula hypothesis, bodies that condensed in the inner asteroid belt have a different composition from those that condensed farther out. The inner belt, being warmer, is richer in the easy-to-condense silicate and iron materials and contains less of the hard-to-condense water and carbon-rich materials, as is indeed observed.

The existence of stony and iron asteroids might at first seem to be evidence *against* the solar nebula theory. How could a swirling mass of gas and dust have separated so as to form some bodies of rock and some bodies of iron? Such separation would be a bit like shaking a piece of cake and having it disintegrate into eggs, flour, sugar, and milk.

We saw in chapter 5, however, that chemical elements *can* be separated by differentiation. That is, astronomers think that the Earth's rocky crust and iron core formed as a result of melting, with the iron then sinking to the core and the lighter rock floating. At least some asteroids have also differentiated (fig. 10.8A), and at least one (Vesta) has had volcanic eruptions—this activity, now long dormant, is deduced from its spectrum, which shows the presence of basalt, a volcanic rock.

After differentiation, collisions with neighboring asteroids broke up most of the large bodies (fig. 10.8B). The fragments are what we see today: pieces of crust became stony asteroids, while pieces of core became iron asteroids. But for a body to differentiate,

Asteroid not differentiated    Asteroid differentiated

Differentiated asteroid
broken up by collision

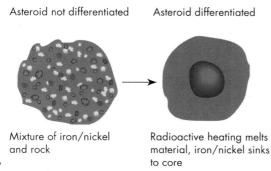

Mixture of iron/nickel
and rock

Radioactive heating melts
material, iron/nickel sinks
to core

**A**

**B**

Some fragments are iron/nickel;
others are rock

**FIGURE 10.8**
Sketch depicting (A) differentiation in asteroids and (B) their subsequent breakup by collision to form iron and stony bodies.

**FIGURE 10.9**

The number of asteroids at each distance from the Sun within the asteroid belt. Notice the conspicuous gaps where there are few, if any, objects. These empty zones are called Kirkwood gaps.

☞★ What feature in Saturn's rings is similar to these gaps?

α★ The gaps in the distribution of the asteroids are very like the gaps in Saturn's rings, such as Cassini's division.

it must be large enough to be able to melt from the heat liberated within it by radioactivity. Thus, the existence of stony and iron asteroids is strong evidence that the early Solar System contained intermediate-sized bodies, that is, planetesimals.

The solar nebula hypothesis also offers an explanation for the reason the asteroids are concentrated between Mars and Jupiter. Any small planet there would have to compete for material with Jupiter, whose immense gravity would disturb the accretion process and prevent the planet's growth.*

## Unusual Asteroids

Even today, Jupiter affects the asteroid belt. Figure 10.9 shows a partial census of the number of asteroids found at each distance from the Sun within the belt. A clear gap can be seen at 2.5 AU, and slight concentrations occur at 3.4 and 4.0 AU.

The seemingly empty regions in the asteroid belt are called **Kirkwood gaps,** and they are caused by the same process that creates the gaps in Saturn's rings: gravitational forces of an outlying body. Saturn's moons, particularly Mimas, create the gaps in the rings. Jupiter creates the Kirkwood gaps. As it orbits the Sun, Jupiter exerts a gravitational force on asteroids in the belt that slightly alters their orbits. If an asteroid has an orbital period that when multiplied by an integer equals Jupiter's period, the asteroid may be subject to a cumulative force that makes it drift to a new orbit. The gap at 2.5 AU arises because, according to Kepler's third law, an asteroid at that distance would have an orbital period exactly one-third that of Jupiter. Every third trip

An asteroid 1 kilometer in diameter striking the Earth will release on impact about the same amount of energy as a 40,000-megaton nuclear explosion. Some astronomers have therefore urged the construction of an international monitoring system to warn of the approach of small asteroids. For example, LINEAR (the Lincoln Near Earth Asteroid Network, operated by MIT's Lincoln Laboratory and the U.S. Air Force) and the British NEO (Near Earth Object) office are now in operation. The hope is that if an asteroid on a collision course with Earth is detected early enough, it might be deflected away from Earth by a nuclear bomb launched toward it and exploded on its surface.

*This is a less dramatic view than the one that was common 50 years ago, when asteroids were believed to be fragments of a terrestrial planet destroyed by collision. This no longer seems likely, given the tiny amount of material in the asteroid belt.

around the Sun, the asteroid would undergo exactly the same tug from Jupiter. Over time, it will be shifted to a new orbit, leaving a gap.

Not all asteroids are found in the main belt. A few, the so-called Trojan asteroids, travel along Jupiter's orbit in two loose swarms, 60° ahead and 60° behind it, as illustrated in figure 10.4.

Nearer the Sun from the main belt are the **Apollo asteroids,** whose orbits carry them into the inner Solar System across the Earth's orbit. Fortunately for us, there are only about 700 such bodies, and so the chance of collision is slim. Nevertheless, on the average, one such body hits the Earth about every 10,000 years. The Apollo asteroids have diameters of about 1 kilometer or less. They may be related to comets, but "dead" ones, shifted to their peculiar orbit by Jupiter's gravitational force and stripped of ice and gas by their repeated passage around the Sun. We will describe how this happens in more detail later in this chapter.

Chiron is another odd "asteroid." Its orbit stretches from just inside that of Saturn almost to Uranus, putting it far outside the main asteroid belt. In addition to this odd orbit, Chiron changes brightness oddly; sometimes it flares up and ejects gas. Such behavior is more like that of a comet than a "normal" asteroid. But if Chiron is a comet, it is not a "normal" one. From its brightness, astronomers deduce that Chiron has a diameter of about 180 kilometers (approximately 112 miles), making it much larger than most comets that enter the inner Solar System.

## 10.3   COMETS

A bright comet is a stunning sight, as you can see in figure 10.10. But such sights are now sadly rare because light pollution from our cities drowns the view for most people. Comets have long been held in fear and reverence, and their sudden appearance and equally sudden disappearance after a few days—or, in some cases, weeks—have added to their mystery.

### Structure of Comets

Comets consist of three main parts—the tail, the coma, and the nucleus—as illustrated in figure 10.11. The largest part is the long **tail,** a narrow column of dust and gas

Light pollution not only spoils our view of the night sky, but it also wastes money. Groups such as the International Dark-Sky Association have shown that better designed street and other outdoor lights make the ground brighter, the sky darker, and use less electricity.

**FIGURE 10.10**
Picture of Comet McNaught as seen from Australia in January 2007.

**FIGURE 10.11**
Artist's depiction of the structure of a comet, showing the tiny nucleus, surrounding coma, and the long tail.

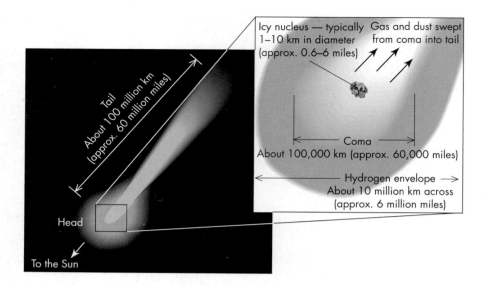

that may stretch across the inner Solar System for as much as 100 million kilometers (nearly an AU!).

The tail emerges from a cloud of gas called the **coma,** which may be some 100,000 kilometers in diameter (10 times or so the size of the Earth). However, despite the great volume of the coma and the tail, these parts of the comet contain very little mass. The gas and dust are extremely tenuous, and so a cubic centimeter of the gas contains only a few thousand atoms and molecules. By terrestrial standards, this would be considered a superb vacuum. This extremely rarified gas is matter that the Sun's heat has boiled off the heart of the comet, its nucleus.

The comet **nucleus** is a block of ice and gases that have frozen in the extreme cold of the outer Solar System into an irregular mass whose approximate diameter is 10 kilometers. The nucleus of a comet has been described as a giant "iceberg" or "dirty snowball," and it contains most of the comet's mass.

Astronomers' first close look at a comet nucleus came from *Giotto,** a spacecraft launched by the European Space Agency as part of an international study of Comet Halley.[†] *Giotto* approached to within 600 kilometers of Halley's nucleus and sent pictures back to Earth of the nucleus. From these its size could be measured. From the size and estimates of the mass of the nucleus, its density can be calculated. This turns out to be about 0.2 grams per cubic centimeter, a value implying that the icy material of the nucleus is "fluffy," like snow, not hard and compacted like pure ice. Unfortunately, the mass estimates are not very accurate, and so the density we infer from them is uncertain.

Despite its icy composition, the nucleus is extremely dark, as you can see in figure 10.12A, which is one of the pictures made by the *Giotto* spacecraft. Astronomers think that the dark color comes from dust and carbon-rich material (similar to that of the carbonaceous chondritic meteorites) coating the surface of the nucleus. Other visible features of the nucleus are its irregular shape and the jets of gas erupting from

---

*The name *Giotto* was given to the spacecraft in honor of the Italian artist who painted a portrait of a comet as part of a Christmas scene for a church altarpiece. Some astronomers believe his painting depicts Halley's comet as it appeared in 1301. Other astronomers believe that Giotto based it on a different comet.

[†]Comets are generally named for their discoverer or for the year they were first seen. Halley's gets its name because Sir Edmund Halley was the first to propose that some comets move around the Sun like planets, and he predicted that the great comet he saw in 1682 would reappear in 1759. It did, but Halley did not live to see his prediction verified. There is controversy about how to pronounce *Halley*. Most astronomers say it rhymes with *Sally*, although there is some evidence that Halley himself pronounced his name as *haw-lee*.

A

B

**FIGURE 10.12**

(A) Picture made by the *Giotto* spacecraft of the nucleus of Halley's comet. The spacecraft was approximately 1000 kilometers (about 620 miles) from the nucleus—deep inside Halley's coma of gas—during the time this picture was made. (B) The nucleus of Comet Wild 2 (*Wild* is said "vilt"), which is about 5 kilometers (about 3 miles) in diameter. Note the many craters. The image was made by NASA's *Stardust* spacecraft, which looped back past Earth in 2006 and parachuted into our atmosphere a canister containing samples of dust collected from near the comet.

the frozen surface. The jets form when sunlight heats and vaporizes the icy material. The irregular shape is probably the outcome of uneven melting of the nucleus during passage by the Sun on previous orbits. You can see similar features in the picture of Comet Wild 2 in figure 10.12B.

Astronomers have recently gotten even closer looks at comets. The craft that took pictures of Comet Wild 2 collected samples of dust from near the comet and successfully returned them to Earth by parachute in early 2006. A more intrusive comet sampling was made by NASA's *Deep Impact* probe, which in 2005 smashed a 370 kilogram (~800 lb) mass into Comet Temple 1. The impact was designed to break through the comet's icy crust and stir up and release dust and gas. The data from these two comet probes are still being analyzed.

## Composition of Comets

The escaped gas from the comet offers astronomers a way to probe the comet's composition. Spectra of gas in the coma and tail show that comets are rich in water, $CO_2$, $CO$, and small amounts of other gases that condensed from the primordial solar nebula. Evaporating water is broken up by solar ultraviolet radiation to create oxygen and hydrogen gas, and most comets are surrounded by a vast cloud of hydrogen created in this way.

If it passes by the Sun too often, the escape of gas from the comet eventually erodes it away. Also, some comets literally fall into the Sun. For example, the SOHO satellite (which observes the Sun's outer atmosphere) has taken pictures of dozens per year falling into the Sun. Since new comets show up frequently, there must be a source to replace those devoured by the Sun, and it is to their origin that we now turn.

## Origin of Comets

Astronomers think that most comets come from the **Oort cloud,** the swarm of trillions of icy bodies* believed to lie far beyond the orbit of Pluto, as we discussed in chapter 7. You may recall that astronomers think the Oort cloud formed from planetesimals that originally orbited near the giant planets and were tossed into the outer parts of the Solar System by the gravitational force of those planets. There, they form a spherical shell that completely surrounds the Solar System and extends to perhaps as much as 150,000 AU from the Sun, as shown in figure 10.13. Astronomers deduce this shape for the Oort cloud from the many comet orbits that are highly tilted with respect to the main plane of the Solar System. However, as we will discuss later, some comets also seem to come from a flatter, less remote region—the so-called **Kuiper belt,** also shown schematically in figure 10.13. The Kuiper belt begins about the orbit of Neptune and extends to approximately 50 AU.

Each comet nucleus moves along its own path, and those in the Oort cloud take millions of years to complete an orbit. With orbits so far from the Sun, these icy bodies receive essentially no heat from the Sun, and calculations indicate that their temperature is a mere 3 K, or about −454°F. Thus, the gases and ices remain deeply frozen.

Such cold and distant objects are invisible to us on Earth; so, if we are to see a comet, its orbit must somehow be altered to carry it closer to us and the Sun. Astronomers think that such orbital changes may arise from the chance passage of a star far beyond the outskirts of the Solar System or from tidal forces exerted on the Oort cloud by the Milky Way. Such gravitational effects disturb the orbits of the comet nuclei in the Oort cloud, altering their paths and making some drop in toward the inner Solar System, as shown in figure 10.14. A single disturbance may shift enough orbits to supply comets to the inner Solar System for tens of thousands of years.

**ANIMATION**

*Oort cloud and Kuiper belt*

**FIGURE 10.13**
Schematic drawing of the Oort cloud, a swarm of icy comet nuclei orbiting the Sun at 40,000 to 100,000 AU. Also shown is the Kuiper belt, another source of comet nuclei. The scale is exaggerated for clarity.

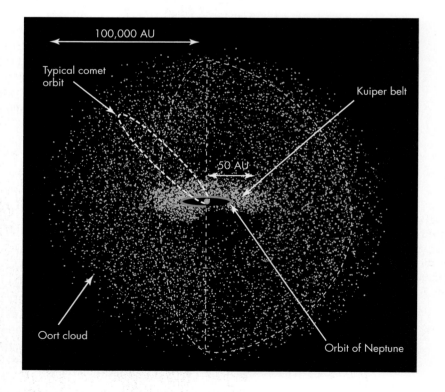

*Astronomers call these cold and inert bodies "comets," even though they do not have tails.

**FIGURE 10.14**
Sketch of how a passing star alters the orbit of a comet nucleus. On its new orbit, the comet will pass by the Sun and be visible from Earth. Although only one comet's orbit is shown, such encounters typically will affect many comets. Note: The distance and size scales are greatly exaggerated. No star presently gets anywhere near this close to the Oort cloud.

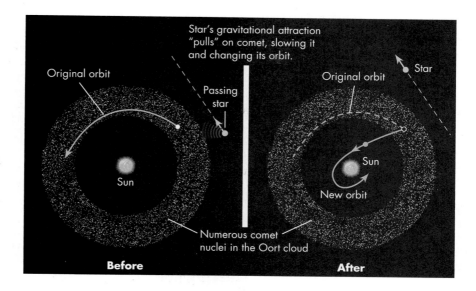

As the comet falls inward toward the inner Solar System, the Sun's radiation heats it and begins to melt the ices. At a distance of about 5 AU from the Sun (Jupiter's orbit), the heat is enough to vaporize the ices, forming gas that escapes to make a coma around the comet nucleus. The escaping gas carries tiny dust grains that were frozen into the nucleus with it. The comet then appears through a telescope as a dim, fuzzy ball. As the comet falls ever nearer the Sun, its gas boils off even faster, but now the Sun begins to exert additional forces on the cometary gas and dust.

## Formation of the Comet's Tail

Sunlight striking dust grains imparts a tiny force to them, a process known as **radiation pressure.** We don't feel radiation pressure when sunlight falls on us because the force is tiny and the human body is far too massive to be shoved around by solar photons. However, the microscopic dust grains in the coma do respond to radiation pressure and are pushed away from the Sun, as shown in figure 10.15. Because all the grains move in the same direction, away from the Sun, a tail begins to form.

The tail pushed out by radiation pressure is made of dust particles, but figure 10.16 shows that comets often have a second tail. That tail is created by an outflow of gas that streams from the Sun into space, a flow called the **solar wind.**

**FIGURE 10.15**
Sketch of how radiation pressure pushes on dust particles. Photons hit the dust, and their impact drives the dust away from the Sun, forming a dust tail. Sizes and distances are not to scale.

**FIGURE 10.16**
Photograph of Comet Hale-Bopp in the evening sky, clearly illustrating the two tails, one of dust, one of gas. These two tails point in slightly different directions because the dust particles feel the effect of the Sun's gravity as well as its radiation pressure. Responding to the gravity, the dust particles follow orbits around the Sun, but according to Kepler's laws, particles farther from the Sun move more slowly, so the dust tail arcs behind the comet. The gas tail, on the other hand, is effected almost exclusively by the solar wind, so it points nearly directly away from the Sun.

*The orientation of comet tails*

Some laundry detergents have "whiteners" in them that fluoresce. They convert ultraviolet radiation into visible light and thereby make a white shirt or blouse look brighter. You can see this effect strongly if you shine a black light on a freshly washed shirt in a darkened room.

The solar wind blows away from the Sun at about 400 kilometers per second. It is very tenuous, containing only a few atoms per cubic centimeter. But the material in the comet's coma is tenuous too, and the solar wind is dense enough by comparison to blow it into a long plume.

Magnetic fields carried along by the solar wind enhance its effect on the comet's tail, helping to drag matter out of the coma and channel its flow, just as magnetic fields in the Earth's atmosphere channel particles to form the aurora. Thus, two forces, radiation pressure and the solar wind, act on the comet to drive out a tail. Because those forces are directed away from the Sun, the comet's tail always points away from the Sun, and the tail may even point out ahead of the comet as it moves away from the Sun (fig. 10.17).

It might help you understand this seemingly odd phenomenon if you think of a runner carrying a torch. If the air is still, the smoke from the torch will, of course, trail behind the runner. However, if a strong wind is blowing at, say, 40 miles per hour, the smoke will be carried along in the direction of the wind regardless of which way the runner moves. Likewise, the high velocity of the solar wind (400 kilometers per second versus about 40 kilometers per second for the comet) carries the tail of a comet outward from the Sun regardless of the comet's motion. Therefore the tails always point away from the Sun.

## Light from the Comet's Tail

The gases and dust swept out into the tail are lit up by the Sun. The dust particles reflect sunlight, and the gases emit light of their own by a process called **fluorescence.** Fluorescence is produced when light at one wavelength is converted to light at another wavelength. A familiar example is the so-called black light that you may have seen for illuminating posters.

Black light is really ultraviolet radiation that we have difficulty seeing because of its short wavelength. When such ultraviolet radiation falls on certain paints or dyes, the chemicals in the pigment absorb the ultraviolet radiation and convert it into visible light.

A major part of a comet's light is created by fluorescence. A photon of ultraviolet—and thus energetic—radiation from the Sun lifts electrons in the atoms of the comet's gas molecules to an upper, excited level in a single leap. The electron then returns to its original level in two or more steps, emitting a photon each time it drops. The combined energy of these photons must equal that of the absorbed ultraviolet photon to conserve energy. Thus, the energy of each emitted photon must be less than that of the original ultraviolet one. That smaller energy then gives them a longer wavelength, which we

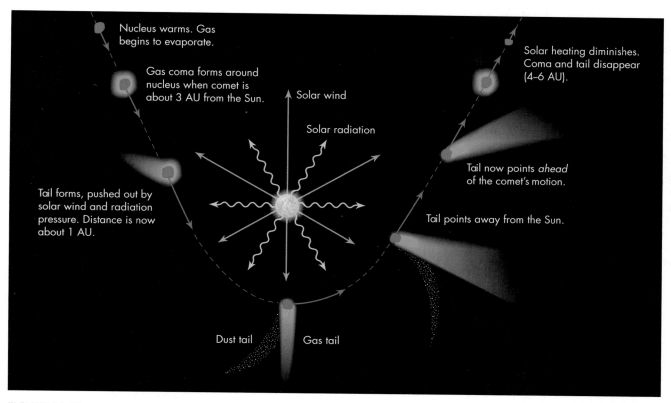

**FIGURE 10.17**
Sketch illustrating how radiation pressure and the solar wind make a comet's tail always point approximately away from the Sun. Sizes and distances are not to scale. Comets may orbit in any direction around the Sun.

can see with our eyes. Thus, fluorescence creates the soft glow of the comet's light. In addition, the spectrum of the fluorescing gas tells us of what the comet is made.

## Short-Period Comets and the Kuiper Belt

Although most comets that we see from Earth swing by the Sun on orbits that will bring them back to the inner Solar System only after millions of years, a small number of comets reappear at time intervals less than 200 years. These **short-period comets** include Halley's, which has a period of 76 years.

The origin of short-period comets is still under study. At one time, it was thought that they came from the Oort cloud, but as they moved through the region of the Solar System containing the giant planets, their orbits were shifted by a close encounter with one of the planets into smaller orbits with periods of centuries rather than millenniums. However, astronomers now think that short-period comets come from the icy nuclei, orbiting beyond Neptune, in the Kuiper belt. Support for this origin comes from the detection of hundreds of small, presumably icy, bodies orbiting near and somewhat beyond Pluto. Astronomers estimate that the Kuiper belt contains well over 30,000 icy objects bigger than 100 kilometers in diameter, and its total mass is hundreds of times larger than the asteroid belt between Mars and Jupiter. These frozen objects are probably survivors of the Solar System's birth—icy planetesimals still orbiting in the disk—but they are too far apart to form additional planets.

Astronomers at one time thought the outer edge of the Kuiper belt might lie as much as 1000 AU from the Sun. However, searches for extremely dim objects have failed to find significant numbers more than about 50 AU from the Sun. On the other hand, in such searches they have discovered a small number of much larger icy objects, such as Sedna, Orcus, Quaoar, and the dwarf planet Eris, which, as we mentioned in chapter 9, is larger than Pluto. Thus, much still remains to be learned about this remote part of our Solar System.

## Fate of Short-Period Comets

Repeated orbits of a short-period comet past the Sun gradually whittle it away: all the ices and gases evaporate, and only the small amount of solid matter, dust and grit, remains. This fate is like that of a snowball made from snow scooped up alongside the road, where small amounts of gravel have been packed into it. If such a snowball is brought inside, it melts and evaporates, leaving behind only the grit accidentally incorporated in it. So too, the evaporated comet leaves behind in its orbit grit that continues to circle the Sun. The material left by the comet produces a delightful benefit: it is a source of meteors.

## Meteor Showers

If you go outside on a clear night and have an unobscured view of the sky, you will see on the average one meteor every 15 or so minutes.* Most of these meteors are stray fragments of asteroids that arrive at the Earth randomly.

At some times of year, instead of one meteor per quarter hour or so, you may observe one every few minutes. Furthermore, if you watch such meteors carefully, you will see that they appear to come from the same general direction in the sky. Meteors of this type are called "shower meteors," and the events that produce them are called **meteor showers.**

The most famous meteor shower occurs each year in mid-August. From August 11 to 13, meteoroids rain into our atmosphere from a direction that lies toward the constellation Perseus. The meteors themselves have no association with Perseus. Rather, they are following an orbit around the Sun that happens to lie roughly in that direction (fig. 10.18A), and the Earth happens to cross their orbits in mid-August. Thus, at that time we encounter far more meteoroids than usual.

This encounter creates an effect similar to what you observe when you drive at night through falling snow: the flakes seem to radiate from a point in front of you, the location of which depends on a combination of the direction and speed of both the wind and your car. Thus, during the time that the Earth crosses the path followed by the meteoroids, they seem to diverge from a common point (fig. 10.18B) called the **radiant.** Meteor showers are generally named for the constellation from which they appear to diverge, and appendix table 7 lists several of the brighter and more impressive showers and their dates. Each shower therefore marks when the Earth crosses the path of a still-active comet or even a "dead" one.

As a comet orbits the Sun and its icy and gaseous material evaporates, it leaves in its path a trail of dust and small bits of solid material ejected from its nucleus. When we cross through or closely approach such a trail, our planet is blasted by this microscopic debris that rains into our atmosphere, burning up, and creating a meteor shower.

On rare occasions, the Earth will pass through a particularly dense clump of material left by the comet. If that happens, thousands of meteors per hour may spangle the sky. Such a display happened in November 1966, when dawn observers on the west coast of the United States and Canada saw literally dozens of meteors per second! The sky looked as if someone had lit a celestial Fourth of July sparkler. Likewise, in November 2001, observers were treated to an amazing display.

Spectacles of this kind are one of the delights of astronomy. However, on even rarer occasions, far more sinister meteoritic events may occur.

A comet is a bit like the "Peanuts" cartoon character Pigpen, who trails dirt wherever he goes.

A meteor shower

---

*Returning from work or late parties, I often wait outside looking at the sky until I see a meteor. I seldom have to wait more than 20 minutes.

**FIGURE 10.18**
Sketch showing how (A) in mid-August, at the time of the Perseid meteor shower, the Earth is moving along its orbit. (B) When the Earth crosses the debris strewn along a comet's orbit, the scattered material plunges into our atmosphere, producing the diverging pattern of meteors characteristic of a meteor shower. (Bodies and orbits are not to scale.)

A

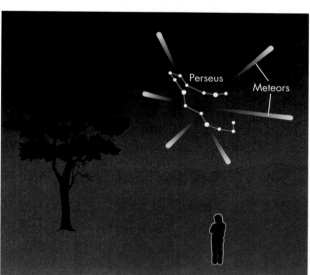

B

# 10.4 GIANT IMPACTS

Every few thousand years, the Earth is hit by a huge meteoroid, a body tens of meters or more in size. Such bodies will produce not only a spectacular glare as they pass through the atmosphere but also an enormous blast on impact.

As we discussed earlier in this chapter, meteoroids may have a very large kinetic energy, that is, energy of motion. If the meteoroid does not burn up on passage through the atmosphere, its remaining kinetic energy is released when it hits the ground or when it breaks up in the atmosphere.

The energy so released can be huge, as we can easily show from the formula for a body's kinetic energy, $E$, given by the expression

$E$ = Kinetic energy of an object
$v$ = Velocity of the object
$m$ = Mass of the object.

$$E = \frac{mv^2}{2}$$

where $m$ is its mass and $v$ is its velocity. For a meteoroid weighing 100 kilograms (about 200 pounds) and traveling at 30 kilometers per second ($= 3 \times 10^4$ meters per second), the kinetic energy of impact will be $100 \times (3 \times 10^4)^2/2 = 4.5 \times 10^{10}$ joules, which is about the same as that released when a hundred tons of dynamite are exploded. Such an object hitting the Moon would make a crater roughly 30 meters (100 feet) in diameter. Fortunately, the Earth's atmosphere significantly slows down meteoroids this small. Even so, they can cause damage if they hit an object such a house or car. Our atmosphere is less able to slow meteoroids that are much larger than this, and so although rare, they could potentially cause much more damage. For example, a body 10 meters in diameter, the size of a small house, would have on impact the explosive power of a thermonuclear bomb and make a crater nearly a kilometer in diameter. Were such a body to hit a heavily populated area, it would have catastrophic results.

Fortunately, we have been spared such disasters recently, but there have been some close calls and some truly horrific impacts in the distant past.

## Giant Meteor Craters

Many giant meteor craters scar our planet. One of the most famous is in northern Arizona. About 50,000 years ago, a meteoroid estimated to have been some 50 meters in diameter hit the Earth about 40 miles east of what is now Flagstaff. Its impact vaporized tons of rock, which expanded and peeled back the ground, creating a crater about 1.2 kilometers across and 200 meters deep (fig. 10.19).

A

B

**FIGURE 10.19**
(A) Photograph of the Arizona meteor crater. The inset shows an iron meteoritic fragment found at the site. (B) A computer simulation of a crater's formation.

Approx. 70 km
(about 43 miles)

**FIGURE 10.20**
Picture (from Earth orbit) of the Manicouagan Crater in Quebec. This winter view shows the lake that fills the crater frozen and covered with snow.

What might explain why this crater is so much less sharply defined than the Arizona meteor crater?

Manicouagan Crater is far enough north that it was buried by glaciers during the last ice age. Moreover, being in a wet climate, it erodes faster and more extensively than the Arizona Meteor Crater, which is in a dry climate.

More recently, in 1908, an asteroid broke up in our atmosphere over a largely uninhabited part of north-central Siberia. This so-called Tunguska event, named for the region where it hit, leveled trees radially outward from the blast point to a distance of some 30 kilometers. The blast was preceded by a brilliant fireball in the sky and was followed by clouds of dust that rose to the upper atmosphere. Sunlight reflected off this dust gave an eerie glow to the night sky for several days. According to some accounts, the blast killed two people. Casualties were few because the area was so remote.

Unfortunately, scientists didn't visit the site until about two decades later because of the political turmoil in Russia at that time. When they did reach the site, they found no crater, just the felled trees. Interestingly, trees at the center of the damaged area were left standing vertically but with their branches stripped off. Thus, the explosion must have occurred in the air.

The Tunguska event is just the most recent giant impact. More ancient impact scars occur in many places on our planet. The huge, ring-shaped Manicouagan Lake, about 70 kilometers (approximately 43 miles) in diameter and pictured in figure 10.20, is a meteor crater, as is Wolf Creek Crater in northwestern Australia.

Astronomers have found even larger craters but are not sure they are impact features. Two such craters are the vast arc (nearly 500 kilometers across) on the east edge

## SCIENCE AT WORK

### GHOST CRATERS OR NO TELL-TALE FRAGMENTS

Hypotheses to explain the Tunguska event offer a good example of how science tests and retests ideas. The lack of a big crater and the absence of meteorites at the site at first seemed to be evidence against the idea that it was caused by a small asteroid. This led some astronomers to propose that a comet might have caused the blast. Being icy, a comet would break up more easily—perhaps even while still in the atmosphere—so there would be no crater. Moreover, any fragments that survived would rapidly disappear, explaining the absence of large rock or iron meteorites.* However, computer simulations of how an asteroid travels at high speed through our atmosphere showed what at first seemed a surprising result: stony or iron asteroids can also create devastation without leaving a crater or tell-tale fragments. A typical asteroid approaches Earth at a speed of 10 to 30 kilometers per second. When it hits our atmosphere, it compresses and heats the air ahead of it. The hot compressed air obeys

Newton's law of action–reaction and exerts a tremendous force back on the asteroid that may shatter it. The resulting fragments plunge deeper into our atmosphere, where air resistance heats them further until they vaporize, creating a fireball at a height of 20 kilometers (12 miles) or so above the ground. No trace of the asteroid survives to form a crater or fragments at the ground. All that remains is the 6000 K incandescent air of the fireball that blasts downward and out, crushing and igniting whatever lies below it, exactly the way a nuclear bomb burst would.

Such calculations do not prove that an asteroid caused the Tunguska event, but they do show that the lack of meteorites does not imply that a comet was to blame.

---

*Stories of unusual radioactivity at the site are totally false but persistent, especially in the kinds of supermarket tabloids that feature articles about 165-pound babies and "How Aliens Abducted My Wife."

of Hudson Bay and a basin (about 300 kilometers across) in central Europe. Still other craters may lie hidden under sediment on land or beneath the oceans.

## Mass Extinction and Asteroid/Comet Impacts

At the end of the Cretaceous period, about 65 million years ago, an asteroid or comet hit the Earth. Its impact and the subsequent disruption of the atmosphere are blamed for exterminating the dinosaurs and many less conspicuous but widespread creatures and plants. In fact, the sudden disappearance of large numbers of life forms at the end of the Cretaceous period defines the end of the Mesozoic era.

The evidence that an extraterrestrial body caused this devastation comes from the relatively high abundance of the otherwise rare element iridium found in sediments from that time. Iridium, a heavy element similar to platinum, is very rare in terrestrial surface rocks because it dissolves easily in molten iron, so most of it sank to the Earth's deep interior at the time our planet formed its core. On the other hand, samples of meteoritic material contain moderate amounts of iridium because most of these bodies have not formed iron cores. Thus, the presence of so much iridium in a layer of clay laid down 65 million years ago is suggestive of a link to meteoritic material. The amount of iridium in the earth sediments at that layer is the quantity that would be dispersed from a piece of meteoritic material 10 kilometers in diameter. Therefore, many astronomers and paleontologists believe the Earth was hit by an asteroid of that size.

A 10-kilometer asteroid hitting the Earth would produce an explosion on impact equivalent to that of several billion nuclear weapons. The impact would not only make an immense crater, but it would also blast huge amounts of dust and molten rock into the air. The molten rock raining down would raise the surface temperature as high as that under an electric broiler and ignite global wildfires. The hot fragments and blast would also create nitrogen oxides, which would combine with water to form a rain of highly concentrated nitric acid. This devastating combination of heat, acid rain, and blast would then be followed by months of darkness and intense cold caused by the dust shroud blotting out the Sun. It seems likely that the biosphere would be devastated, leading to mass extinctions, just as the fossil record shows.

This frightening picture is supported not only by the iridium layer and the sudden disappearance of dinosaurs but also by a layer of soot, as well as a layer of tiny quartz pellets believed to have been created by the melting and blast of a violent impact. Moreover, scientists believe that a faint, circular depression about 65 million years old near Chicxulub* (*cheek-shoo-loob*) in the Yucatán region of Mexico may be the crater formed by this impact.

The Cretaceous mass extinction may have played an especially important role in our own evolution. Before that event, reptiles were the largest animals on Earth. Subsequently, mammals have assumed that niche. Small mammals may have escaped the fury of the heat and acid rain by remaining in burrows, and they may have survived the subsequent cold by virtue of their fur. You may be running your fingers through your hair, rather than your claws across your scales, because of that impact.

Other mass extinctions have occurred earlier and later than the Cretaceous event, but many scientists believe that massive volcanic eruptions or drastic changes in sea level might have the same effect on life forms as a giant impact. Thus, like so many of the most interesting issues in science, no single accepted explanation has yet emerged.

One well-documented event that proved harmless only because of its remoteness was a brilliant flash detected in February 1994 over the South Pacific ocean by a spy satellite. Analysis of the flash suggests that it marked the breakup of a 7-meter (20-foot)-diameter meteoroid. The resulting blast had an energy between 10 and 20 kilotons of TNT, comparable to the bomb that destroyed Hiroshima. Had the object burst over a city, the casualties might have been in the hundreds of thousands.

---

*"Flea of the devil" in Mayan.

# SCIENCE AT WORK

## METEORITES CAN BE DEADLY

As astronomers have studied the details of the impact believed responsible for the Cretaceous extinction, they have come to realize that meteor impacts may be more common (and deadly!) than once believed. Although we have no recent, well-documented cases of people being killed by meteorites such deaths surely do occur. Careful searches through newspapers have also turned up many very near misses. Moreover, reports of meteorites hitting buildings and even cars are surprisingly common. Searching old records from Europe and China, astronomers have found many instances of fatalities attributed to meteorites. One of the grimmer of these reports is from a fifteenth-century Chinese chronicle that describes "stones that fell like rain" that killed 10,000 people. Even allowing for exaggeration in such old records, we can conclude that meteor falls can be far from harmless. In fact, astronomers conclude that although meteoritic disasters are rare, when they occur, they have the potential to kill enormous numbers of people.

# SUMMARY

Our Solar System contains numerous small bodies: asteroids, meteors, and comets. They are important astronomically because they give us information about the time of formation, composition, and physical conditions in the solar nebula.

Asteroids are rocky, metallic, or carbon-rich objects found mainly in the asteroid belt between the orbits of Jupiter and Mars. Comets are icy bodies found mainly in the Oort cloud, far beyond the orbit of Pluto.

A comet becomes visible if a passing star or some other event disturbs the comet's orbit so that it drops in toward the inner Solar System. There, solar heating thaws the frozen nucleus and evaporates gases. Radiation pressure and the solar wind then sweep the liberated gas and dust into a tail.

Some comets are caught in short-period orbits where they may be melted away to a collection of dust and grit. If the Earth crosses or nears the path of such a skeletal comet, the debris falling into the atmosphere causes a meteor shower.

The Earth is hit now and then by large asteroidal or cometary bodies, producing craters or, in very rare instances, mass extinctions.

## ? What It Is.

This image was made by combining a number of separate images of the Leonid meteor shower taken just before dawn November 19, 2002. You can clearly see that the meteors appear to radiate from a small area on the sky. That area, it turns out, lies in the constellation Leo, hence the name of the shower. The bright spot near the center of the image is the planet Jupiter.

# QUESTIONS FOR REVIEW

1. What is the difference between a meteor, a meteoroid, and a meteorite?

2. How do we know what asteroids are made of?

3. Where are most asteroids found?

4. What are Apollo asteroids?

5. What shape are typical asteroids and how do we know? Why is Ceres not so shaped?

6. What do asteroids tell us about the formation of the Solar System?

7. What makes a "shooting star"?

8. How is a meteor heated?

9. What creates meteor showers? When do some occur?

10. What is the Oort cloud? What is the Kuiper belt?

11. What is the life history of a comet from the Oort cloud to an object we see?

12. What creates a comet's tail?

13. Why are there two tails to some comets?

14. How do we know asteroids have a composition similar to that of some meteorites?

15. What evidence is there that the Earth has been hit by asteroids or comets?

16. What was the Tunguska event?

17. Why do some scientists believe that asteroids and comets play a role in mass extinctions?

## PROBLEMS

1. Use Kepler's third law to find the semimajor axis of Halley's comet, given that its orbital period is 76 years.

2. With the result of problem 1 and the fact that for a very elliptical orbit the distance farthest from the Sun is roughly twice the semimajor axis, estimate how far from the Sun Halley's comet gets. What planet is about that same distance from the Sun? How does this distance compare with the distance to the Kuiper belt?

3. Use the formula for the kinetic energy of a moving body to estimate the energy of impact of a 1000-kilogram (roughly 1-ton) object hitting the Earth at 30 kilometers per second. Express your answer in kilotons of TNT, using the conversion that 1 kiloton is about $4 \times 10^{12}$ joules. Note: Be sure to convert kilometers/second to meters/second.

4. Use Kepler's third law to determine the period of a comet whose orbit extends to 50,000 AU, within the inner Oort cloud.

5. Given that the temperature of a body decreases as the square root of its distance from the Sun increases, estimate the temperature of a comet nucleus in the Oort cloud. Take the temperature at 1 AU to be 300 Kelvin.

## TEST YOURSELF

1. The tail of a comet
   (a) is gas and dust pulled off the comet by the Sun's gravity.
   (b) always points away from the Sun.
   (c) trails behind the comet, pointing away from the Sun as the comet approaches it and toward the Sun as the comet moves out of the inner Solar System.
   (d) is gas and dust expelled from the comet's nucleus by the Sun's heat and radiation pressure.
   (e) Both (b) and (d)

2. The bright streak of light we see as a meteoroid enters our atmosphere is caused by
   (a) sunlight reflected from the solid body of the meteoroid.
   (b) radioactive decay of material in the meteoroid.
   (c) a process similar to the aurora that is triggered by the meteoroid's disturbing the Earth's magnetic field.
   (d) frictional heating as the meteoroid speeds through the gases of our atmosphere.
   (e) the meteoroid's disturbing the atmosphere so that sunlight is refracted in unusual directions.

3. Meteor showers such as the Perseids in August are caused by
   (a) the breakup of asteroids that hit our atmosphere at predictable times.
   (b) the Earth passing through the debris left behind by a comet as it moves through the inner Solar System.
   (c) passing asteroids triggering auroral displays.
   (d) nuclear reactions in the upper atmosphere triggered by an abnormally large meteoritic particle entering the upper atmosphere.
   (e) None of the above

4. Astronomers think that most comets come from
   (a) interstellar space.
   (b) material ejected by volcanic eruptions on the moons of the outer planets.
   (c) condensation of gas in the Sun's hot outer atmosphere.
   (d) small icy bodies in the extreme outer parts of the Solar System that are disturbed into orbits that bring them closer to the Sun.
   (e) luminous clouds in the Earth's upper atmosphere created when a small asteroid is captured by the Earth's gravitational force.

5. The asteroid belt lies between the orbits of
   (a) Earth and Mars.
   (b) Saturn and Jupiter.
   (c) Venus and Earth.
   (d) Mars and Jupiter.
   (e) Pluto and the Oort cloud.

## FURTHER EXPLORATIONS

Asphaug, Erik. "The Small Planets." *Scientific American* 282 (May 2000): 46.

Becker, Luan. "Repeated Blows." *Scientific American* 286 (March 2002): 76.

Binzel, Richard P., M. Antonietta Barucci, and Marcello Fulchignoni. "The Origin of the Asteroids." *Scientific American* 265 (October 1991): 88.

Cowen, Ron. "The Day the Dinosaurs Died." *Astronomy* 24 (April 1996): 34.

"Debate: What Caused the Mass Extinction?" *Scientific American* 263 (October 1990): 76.

Alvarez, Walter, and Frank Asaro. "An Extraterrestrial Impact."

Courtillot, Vincent E. "A Volcanic Eruption."

Delsemme, A. H. "Whence Come Comets?" *Sky and Telescope* 77 (March 1989): 260.

Dodd, R. T. *Thunderstones and Shooting Stars.* Cambridge, Mass.: Harvard University Press, 1986.

Gallant, Roy A. "Journey to Tunguska." *Sky and Telescope* 87 (June 1994): 38.

Gehrels, Tom. "Collision with Comets and Asteroids." *Scientific American* 274 (March 1996): 54.

Grieve, Richard A. "Impact Cratering on the Earth." *Scientific American* 262 (April 1990): 60.

Lewis, John S. *Rain of Iron and Ice.* Reading, Mass.: Addison-Wesley, 1996.

Luu, Jane X., and David C. Jewitt. "The Kuiper Belt." *Scientific American* 274 (May 1996): 46.

Schaffer, Bradley E. "Comets That Changed the World." *Sky and Telescope* 93 (May 1997): 46.

Semeniuk, Ivan. "Asteroid Impact." *Mercury* 31 (November/December 2002): 24.

Stern, Alan. "The Sun's Fab Four [Ceres, Pallas, Juno, and Vesta]." *Astronomy* 23 (June 1995): 30.

———. "The 3rd Zone: Exploring the Kuiper Belt." *Sky and Telescope* 106 (Nov. 2003): 30.

Warner, Elizabeth M., and Greg Redfern. "Deep Impact: Our First Look Inside a Comet." *Sky and Telescope* 109 (June 2005): 40.

Weissman, Paul R. "The Oort Cloud." *Scientific American* 279 (September 1998): 84.

Wynn, Jeffrey C., and Eugene M. Shoemaker. "The Day the Sands Caught Fire." *Scientific American* 279 (November 1998): 64.

**Website**

Please visit the *Explorations* website at **http://www.mhhe.com/arny** for additional online resources on these topics.

# KEY TERMS

Apollo asteroids, 305

asteroid belt, 301

asteroids, 301

carbonaceous chondrites, 300

chondritic meteorites, 299

chondrules, 299

coma, 306

fluorescence, 310

Kirkwood gaps, 304

Kuiper belt, 308

meteor, 298

meteor showers, 312

meteorites, 299

meteoroid, 298

nucleus, 306

Oort cloud, 308

radiant, 312

radiation pressure, 309

short-period comets, 311

solar wind, 309

tail, 305

# PROJECTS

1. Pick a clear night and watch the night sky with a friend for half an hour. Count the number of meteors you see. Note the direction they come from. From your knowledge of these bodies, were the ones you saw (if any!) cometary or asteroidal?

2. Watch a meteor shower and count the number you see. Note the position the meteors seem to radiate from and mark it on a star map. If you are in a summer course, try for the Perseids in mid-August (August 10–14), but check with your instructor, a website, or some popular astronomy magazine for more precise dates. For autumn courses, try for the Orionids (October 18–23) or the Leonids (early morning November 14–19). Later in the season, you might try for the Geminids (as early as 10 P.M. December 12–14). For spring courses, try for the Lyrids (April 20–22) or the $\eta$ Aquarids (May 2–7). What was the average time interval between meteors? Does your estimate of the radiant match the predicted position?

# The Sun, Our Star

# 11

## Chapter Objectives

- The Sun is an immense ball of gas.
- Gravity holds it together.
- Gas pressure prevents it from collapsing under the force of its gravity.
- Nuclear reactions in its core supply its power.
- Motions in the Sun's interior generate magnetic effects.
- The Sun's magnetism changes over time.
- Solar magnetism affects Earth:
  - It creates the aurora.
  - It appears to affect our climate.

Magnetic loops in the Sun's lower atmosphere. Gas trapped along the loops is heated by magnetic activity and in turn heats the Sun's corona.

**?What Is This?**
(See chapter summary for the answer.)

## Concepts and Skills to Remember

- Relation between temperature of a hot object and the color of the light it emits (Wien's law) (3.2)
- Doppler shift (3.6)
- Structure of atoms (3.3)

The Sun is a star, a dazzling, luminous ball of gas more than 100 times bigger in diameter than the Earth. Although it gleams peacefully in our daytime sky, a telescope reveals the Sun's surface to be violently agitated, with rising fountains of incandescent gas and a twisted magnetic field. Even greater violence wracks its core. There, a nuclear "furnace" every second burns 600 million tons of hydrogen into helium, producing in one heartbeat the energy of 100 billion nuclear bombs. How that energy is released and how the Sun manages to hold itself together is the main theme of this chapter. We will begin by describing the Sun: its radius, mass, and so on. Then we will discuss how the crushing force of its gravity balances the explosive power in its core. Finally, we will see how energy escaping from the core stirs its atmosphere into the inferno we see.

## 11.1   SIZE AND STRUCTURE

The Sun is immense, dwarfing the Earth and even Jupiter, and its immensity is what makes it shine. With a radius over 100 times that of the Earth and a mass 300,000 times the Earth's, the Sun has an enormous gravity that crushes the material in its interior. To offset that crushing force and prevent its own collapse, the Sun must be extremely hot. But hot objects always lose energy, and the Sun is no exception. We see that lost energy as sunshine and welcome it as the source of our life. But sunshine is a death warrant for the Sun because the energy it carries off must be replenished or the Sun will collapse. Fortunately for us, the Sun does replace its lost energy, but only at the cost of consuming itself—a dilemma that is not unique to the Sun but is shared by most stars.

Before we discuss how the Sun replaces its lost energy, we need to understand better some of its overall properties. How much mass does it contain? How rapidly does it lose energy? What resources are available to supply its energy needs? Table 11.1 lists some of these vital statistics, and figure 11.1 shows the Sun's portrait, illustrating its main features. It will be helpful many times in this chapter to refer to this picture.

Astronomers can measure some of the Sun's properties, such as its diameter and surface temperature. However, most of its other properties, for example, its internal temperature and density, they can only deduce. Such deductions are based on computer models that use the laws of physics to calculate the Sun's properties. The correctness of the model's predictions is then judged by whether the predictions agree with observable properties. For example, if a model predicted the Sun's surface temperature to be 10,000 K, it would be rejected as incorrect. Thus, our understanding of the Sun's properties comes from a combination of theory and measurement.

### TABLE 11.1

#### Properties of the Sun

| | |
|---|---|
| Radius ($R_\odot$*) | $7 \times 10^8$ m = $7 \times 10^5$ km, or about 109 $R_{Earth}$ |
| Mass ($M_\odot$*) | $2 \times 10^{30}$ kg, or about 333,000 $M_{Earth}$ |
| Distance from Earth | $1.5 \times 10^8$ km, or 1 AU |
| Temperature of surface | 5780 Kelvin (approx 9900 °F) |
| Temperature of core | 15 million Kelvin (approx 27 million °F) |
| Composition | 71% hydrogen, 27% helium, 2% heavier elements |
| Power output | $4 \times 10^{26}$ watts |

*Astronomers use the symbol $\odot$ to stand for the Sun. Thus, $R_\odot$ is the Sun's radius and $M_\odot$ is its mass. The symbol is the ancient Egyptian hieroglyph for "Sun."

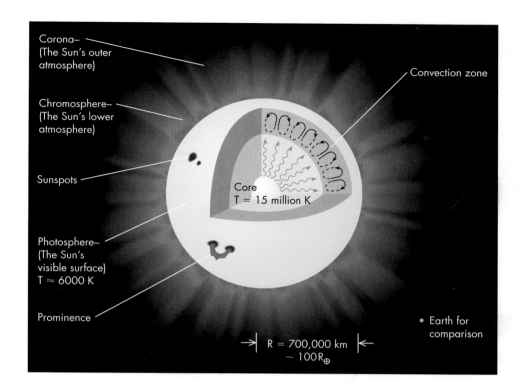

**FIGURE 11.1**
A cutaway sketch of the Sun.

## Measuring the Sun's Properties

The Sun is about 150 million kilometers (1 AU, or 93 million miles) from Earth. Astronomers originally measured its distance by triangulation, but they now use radar, either bounced directly from the Sun or from other bodies whose distance is known in AU from Kepler's third law.

Once the Sun's distance is known, we can find its radius from its angular size, as we showed in chapter 1. We also need to know the Sun's distance from Earth if we are to measure its mass with the help of Kepler's third law, as we showed in chapter 2. From its mass and radius, we can calculate that its surface gravity is about 30 times the Earth's.

The Sun's internal heat balances this crushing force of gravity. Its surface temperature can be found from its color and Wien's law using the methods discussed in chapter 3. Astronomers cannot measure its interior temperature directly but rather from calculations and indirect measurements (discussed in section 11.3), they deduce that its core temperature is about 15 million K (approximately 27 million °F).

From measurements of the amount of solar energy that reaches the Earth, astronomers calculate that the Sun radiates a total of $4 \times 10^{26}$ watts of power into space from its surface, energy that must be replenished by the fusion of hydrogen in its core, as we will see in section 11.2. Fortunately, the Sun has a plentiful supply of hydrogen: its spectrum shows it is about 71% hydrogen, 27% helium, and 2% vaporized heavier elements such as carbon and iron, a composition similar to Jupiter's and Saturn's. But unlike these mostly liquid bodies, the Sun is gaseous throughout because its high temperature breaks most molecular bonds, vaporizing even iron, and allowing the atoms to move freely as a gas.

## The Solar Interior

When we observe the Sun, we see through the low-density, tenuous gases of its outer atmosphere. Our vision is ultimately blocked, however, as we peer deeper into the Sun. There, the material is compressed to high density by the weight of

**FIGURE 11.2**

Plots of how density and temperature change through the Sun.

**FIGURE 11.3**

Sketch of how energy flows from the Sun's core to its surface. In the deep interior, radiation carries the energy. Near the surface, convection carries the energy.

the gas above it. In this dense material, the atoms are sufficiently close together that they strongly absorb the light from deeper layers, blocking our view of them much like frosted glass obscures what lies behind it. Above these layers, however, the absorption of light is weaker and the light emitted by hot material there escapes freely into space as the sunlight we see. These layers where the Sun's gases change from transparent to opaque form what is called the **photosphere,** the visible surface of the Sun. Although the blocking of radiation by denser gas below the photosphere limits our view into the Sun, it helps the Sun retain heat and, like a well-insulated house, thereby reduces the amount of fuel it must consume.

If we could see into the Sun, we would find that its density and temperature rise steadily inward. In the photosphere, the density is comparable to that of the air around us; deeper, the material above pushes down on that below, compressing the gas like a pile of pillows. A similar compression occurs in the atmosphere of the Earth and other planets, as we discussed in chapters 5 and 9. But the greater mass of the Sun leads to a vastly greater compression of its gas, and so near the core, the density is more than 100 times that of water. Despite this great density, the Sun is gaseous throughout because its high temperature gives the atoms so much energy of motion that they are unable to bond with one another to form a liquid or solid substance.

The temperature also rises as we plunge into the Sun's interior. The photosphere is almost 6000 K, but below it, heat is partially trapped and the temperature soars to about 15 million K at the core. Figure 11.2, based upon theoretical calculations, illustrates how the temperature and density change through the Sun. No spacecraft has or is ever likely to make the measurements directly, but we are confident that they are correct because the Sun needs such high temperatures and densities to keep it from collapsing under its own gravity.

## Energy Flow in the Sun

Your own experience and experiments in the laboratory show that heat always flows from hot to cold. Applying this principle to the Sun, we can therefore infer that because its core is hotter than its surface, heat will flow outward from its center, as illustrated in figure 11.3. Near the core, the energy moves by radiation carried by photons through what is called the **radiative zone.** Because the gas there is so dense, a photon travels less than an inch before it is absorbed by an atom and

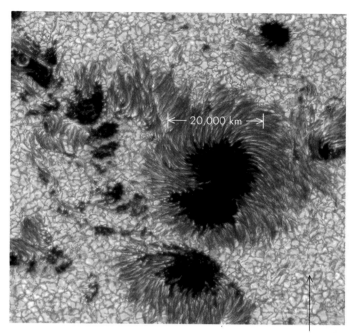

Granulation

**FIGURE 11.4**
Photograph of solar granulation near a sunspot. It has been colorized. The bright patches are immense bubbles of hot gas rising from deep within the Sun. The darker areas around them are cooler gas sinking back into the Sun's interior. The sunspot is about 18,000 kilometers in diameter.

How big across is one of the rising bubbles that forms the granulation? (You can use the sunspot to estimate the size of the granulation.)

Approx 1000 km (600 miles).

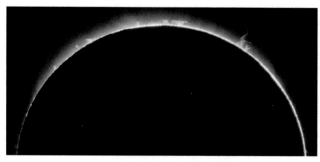

**FIGURE 11.5**
Photograph of a portion of the solar chromosphere at a total solar eclipse.

stopped. The photon is eventually re-emitted, but it will be almost immediately reabsorbed. The constant absorption and re-emission slows photons like cars caught in stop-and-go traffic. Even though photons travel at the speed of light between absorptions, it takes them about 16 million years to move from the core to the surface.\* Thus, today's sunshine was born in the Sun's core before we existed as a species!

The flow of energy from the core toward the Sun's surface is slowed in the region just below the photosphere where the gas is cooler and less transparent. Here, photons are even less effective in moving energy, and convection currents like those in the Earth and giant planets carry the energy to the surface. The rising and sinking gas occupies the **convection zone,** and we can infer the gas's motion there from the numerous tiny, bright regions surrounded by narrow darker zones, called **granulation** (fig. 11.4). The bright areas are bubbles of hot gas many hundreds of kilometers across, rising from deep within the Sun. They are brighter because they are hotter than the gas around them. On reaching the surface, these hot bubbles radiate their heat to space and cool. That cooler matter then sinks back toward the hotter interior where it is reheated and rises again to radiate away more heat. Astronomers can measure the speed of these up-and-down motions using the Doppler effect and find that the bubbles rise at about 1 kilometer per second.

## The Solar Atmosphere

Astronomers refer to the low-density gases that lie above the photosphere as the Sun's atmosphere. This region marks a gradual change from the dense gas of the photosphere to the extremely low-density gas of interplanetary space. A similar transition occurs in our own atmosphere, where the gas density decreases steadily with altitude and eventually merges with the near-vacuum of space.

Although the gas density in the Sun's atmosphere decreases above the photosphere, the gas temperature behaves very differently. Immediately above the photosphere the temperature decreases, but at higher altitudes the gas grows hotter, reaching temperatures of several million Kelvin. Why the Sun's atmosphere is so hot remains a mystery, though astronomers believe that the Sun's magnetic field plays a role in somehow heating these low-density gases, as we will discuss in section 11.4.

The Sun's atmosphere consists of two main regions. Immediately above the photosphere lies the **chromosphere,** the Sun's lower atmosphere. It is usually invisible against the glare of the photosphere but can be seen at a total eclipse of the Sun as a thin red zone around the Sun (fig. 11.5). With

---

\*The question of how long it takes a photon to travel from the Sun's center to its surface turns out to be tricky. One reason is that a single photon never makes such a journey. A typical photon leaving the core is a gamma ray, which will be converted to many lower-energy photons before its energy reaches the surface. Therefore, it is perhaps better to ask how long it takes energy to travel from the Sun's core to the Sun's surface. That time turns out to be about 16 million years, according to one calculation.

**FIGURE 11.6**
Photograph of spicules in the chromosphere. The spicules are the thin, stringy features that look like tufts of grass.

**FIGURE 11.7**
Photograph of the corona at a total eclipse of the Sun.

a telescope equipped with an appropriate filter, you can see that the chromosphere contains millions of thin columns called **spicules,** each a jet of hot gas thousands of kilometers long, as shown in figure 11.6.

The chromosphere's color* comes from the strong red emission-line of hydrogen, H$\alpha$. We saw in chapter 3 that emission lines arise in hot, low-density gas. From these lines astronomers can infer the gas's temperature. Just above the photosphere, the temperature is about 4500 K, but 2000 kilometers higher, it reaches 50,000 K. Here, the chromosphere ends and the temperature shoots up to about 1 million degrees as we enter the **corona,** the Sun's outer atmosphere.

The corona's extremely hot gas has such low density that under most conditions, we look right through it. But like the chromosphere, we can see it at a total solar eclipse when the Moon covers the Sun's brilliant disk. Then the pale glow of the corona can be seen to extend beyond the Sun's edge to several solar radii (fig. 11.7). Pictures of the Sun made at X-ray wavelengths show that the corona is not uniform but has streamers created by the Sun's magnetic field. In addition, the corona contains huge regions of cooler gas called **coronal holes** through which gas escapes from the Sun into space, as we will discuss in section 11.4.

Because the corona is so tenuous, it contains very little energy despite its high temperature. It is like the sparks from a Fourth of July sparkler: despite their high temperature, you hardly feel them if they land on your hand because they are so tiny and thus carry very little total heat.

---

*The prefix *chromo-* means "colored."

**FIGURE 11.8**
A sketch illustrating the condition of hydrostatic equilibrium, the balance of pressure (blue arrows) and gravitational force (purple arrows) in the Sun.

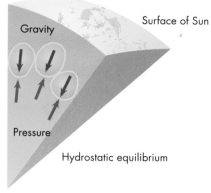

The outward pressure force balances the inward gravitational force everywhere inside the Sun.

Surface of Sun

Gravity

Pressure

Hydrostatic equilibrium

Center of Sun

---

## 11.2   HOW THE SUN WORKS

Our discussion so far has centered on the structure of the Sun. Our task now is to understand *why* it has that structure and how it works. For example, why is the Sun hot? What makes it shine? How does it generate energy?

### Internal Balance (Hydrostatic Equilibrium)

There are many examples of hydrostatic equilibrium in everyday life. Our atmosphere is in hydrostatic equilibrium, its gases pulled downward by the Earth's gravitational force but supported by collisions of the air molecules, creating pressure. Likewise, pressure in an automobile tire supports the weight of the car.

The structure of the Sun depends on a balance between its internal forces. One force holds the Sun together. A second force prevents the Sun from collapsing. This balance is technically called **hydrostatic equilibrium.**

The Sun's inward force arises from its own gravity. The outward force arises from the rapid motion of its atoms, a motion that gives rise to a **pressure.** Thus, in the Sun, as in virtually all stars and planets, the balance of hydrostatic equilibrium requires that the *outward* force created by pressure exactly balance the *inward* force of the Sun's gravity (fig. 11.8). Without such a balance, the Sun would rapidly change. For example, if its pressure were too weak, the Sun's own gravity would rapidly crush it. Therefore, to understand the Sun, we need to discuss in more detail how its pressure arises.

### Pressure in the Sun

Pressure in a gas comes from collisions among its atoms and molecules. If the gas is squeezed, atoms are pushed toward each other. As they collide, they rebound, resisting the compression, as you can feel by squeezing a balloon. The strength of the pressure depends on how often and how hard the collisions occur. Raising the density increases collisions by moving atoms closer together. Raising the temperature speeds atoms up, making them collide harder and more often (fig. 11.9). Thus, the strength of the pressure is proportional to the density times the temperature of the gas; that is:

$$\text{Pressure} = \text{Density} \times \text{Temperature} \times \text{a constant}^*$$

a result known as the **perfect,** or **ideal, gas law.**

---

*The value of the constant depends on the units used to measure the pressure, density, and temperature.

**FIGURE 11.9**

Sketch illustrating the perfect, or ideal, gas law. Gas atoms move faster at the higher temperature so they collide both more forcefully and more often than atoms in a cooler gas. These effects combine to create a higher pressure. Thus, other things being equal, a hotter gas exerts a greater pressure.

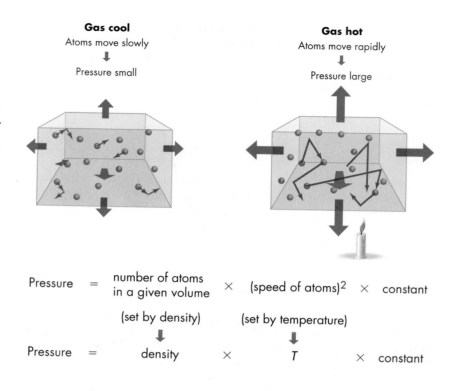

$$\text{Pressure} = \frac{\text{number of atoms}}{\text{in a given volume}} \times (\text{speed of atoms})^2 \times \text{constant}$$

$$\text{(set by density)} \qquad \text{(set by temperature)}$$

$$\text{Pressure} = \text{density} \times T \times \text{constant}$$

*How heating raises pressure*

Although the gas law shows us that the pressure inside the Sun depends on temperature, it does not by itself show us how hot that temperature needs to be. We might reasonably guess that because the Sun has a huge mass (and therefore a huge gravity crushing it), it needs a huge temperature to offset that crushing force.

Indeed, the Sun's core is very hot. To determine exactly how hot, astronomers solve a computer model of the Sun, and the answer turns out to be about 15 million K for the temperature and about 150 grams per cubic centimeter for the density, as we claimed earlier and showed in figure 11.2. But because heat flows out of the hot core, the Sun needs a mechanism to keep replacing its core heat.

## Powering the Sun

Energy that leaves the core eventually escapes into space as sunshine: heat and light. That heat loss must be replaced, or the Sun's internal pressure would drop and the Sun would begin to shrink under the force of its own gravity. The Sun is therefore like an inflatable chair with tiny leaks through which the air escapes. If you sit in the chair, it will gradually collapse under your weight unless you pump air in to replace that which escapes. What acts as the energy pump for the Sun?

Early astronomers believed that the Sun might burn ordinary fuel such as coal. But even if the Sun were pure coal, it could burn only a few thousand years. At the end of the 19th century, the English physicist Lord Kelvin and the German scientist Hermann Helmholtz independently proposed that the Sun is not in hydrostatic equilibrium but that gravity slowly compresses it, making it shrink. In their theory, compression (increasing density) heats the gas and makes the Sun shine. However, gravity could power the Sun by this mechanism only for 10 million years, and we know the Sun has been shining for billions of years because we have fossils of life on Earth that old. Furthermore, if gravity were the power source, the Sun would be steadily shrinking, and it isn't. Therefore, something else must supply energy.

In 1899, T. C. Chamberlin suggested that subatomic energy—energy from the reactions of atomic nuclei—might power stars, but he could offer no explanation of how the energy was liberated. In 1905, Einstein proposed that energy might come from a body's mass. His formula,

$$E = mc^2$$

c = The speed of light: $3 \times 10^8$ m/s
m = An amount of mass
E = An amount of energy

states that a mass, $m$, can become an amount of energy, $E$, equal to the mass multiplied by the square of the speed of light, c, where c = $3 \times 10^8$ meters per second. The amount of energy available from reactions involving atomic nuclei is vastly larger than that from chemical processes such as burning. For example, if 1 gram of mass—less than the amount in an aspirin tablet—is converted to energy, it releases $9 \times 10^{13}$ joules, the equivalent of about 20 kilotons of TNT* or a small nuclear weapon. If the Sun could convert even a tiny amount of its mass into energy, it would have an enormous source of power.

In 1919, the English astrophysicist A. S. Eddington, a pioneer in the study of the physics of stars, showed that the conversion of hydrogen into helium would release enough energy to power the Sun, but even his theory lacked the necessary details, which were supplied independently in the late 1930s by the physicists Hans A. Bethe and Carl F. von Weizsäcker. Their work showed that the Sun could generate its energy by converting hydrogen into helium by a process called **nuclear fusion,** a process that bonds two or more nuclei into a single, heavier one.

**Nuclear Fusion.** Fusion is possible in the Sun because its interior is so hot. Under normal conditions, hydrogen nuclei repel each other, pushed apart by the electrical charge of the protons. However, at high temperatures, nuclei move so fast that when they collide, they are driven extremely close together. At such a collision, the nuclei may get close enough to each other that the electrical repulsion between their protons is overwhelmed by the **nuclear** or **strong force,** the force that holds the protons together in a normal atom.† Thus, the two separate nuclei can merge, or fuse, into a single new nucleus. Because this fusing process requires such a high temperature, the only place in the Sun hot enough for fusion to occur is its core. The core therefore is where the Sun makes its energy. However, before we can understand how fusion creates energy, we need to look at the structure of hydrogen and helium.

**The Structure of Hydrogen and Helium.** Hydrogen consists of one proton and an orbiting electron, and helium consists of two protons, two neutrons, and two orbiting electrons (fig. 11.10). In the Sun's hot interior, however, atoms collide so violently that the electrons are generally removed. Atoms that are missing one or more electrons are said to be ionized. A gas in which the atoms are ionized is technically called a "plasma." Because the electrons are removed from most atoms and move independently in the gas, we can therefore ignore them in much of what follows.

Hydrogen and helium always have one and two protons respectively, but they have other forms (called "isotopes") with different numbers of neutrons. To identify the isotopes, we write their chemical symbol with a superscript that shows the total number of protons and neutrons. The usual form of hydrogen with one proton and no neutrons is $^1$H, while the form of hydrogen containing a proton and a neutron is $^2$H. Likewise, the normal form of helium with two protons and two neutrons is $^4$He, while helium with two protons but only one neutron is $^3$He. As we will see, these isotopes play a role in the energy supply of the Sun.

**The Proton–Proton Chain.** Hydrogen fusion in the Sun occurs in three steps called the **proton–proton chain.** In the first step, two $^1$H nuclei collide and fuse to

Key

Neutron     Proton

$^1$H          $^2$H

$^3$He         $^4$He

**FIGURE 11.10**
Schematic diagrams of hydrogen and helium atoms and isotopes. (Electrons are omitted for clarity.)

---

*One kiloton is about $4.2 \times 10^{12}$ joules of energy.

†Although the bonding of protons and neutrons into nuclei results from the strong force, the first step in the proton–proton chain requires the operation of the weak force to convert one of the protons into a neutron.

*The reactions of the proton–proton chain*

Gamma rays are the highest energy form of electromagnetic radiation known. In some countries, they are used to sterilize food, allowing such products as milk and meat to be stored, if covered, at room temperature without spoiling.

A

B

C

**FIGURE 11.11**
Diagram of the proton–proton chain. Although the Sun's power comes from the conversion of its mass to energy, nuclear fusion creates such a large amount of energy from even a tiny amount of mass that the decrease in the Sun's mass is extremely small and unmeasurable.

form the isotope of hydrogen, $^2$H. In the collision, one proton becomes a neutron by ejecting two particles: a positron (denoted as $e^+$) and a **neutrino** (denoted by $\nu$). The neutrinos so generated play no further role in the Sun's energy generation, but we will encounter them again because they can be detected when they leave the Sun and may help astronomers learn more about conditions in the Sun's interior.

This first step in converting the mass of an atom into the Sun's energy is depicted in figure 11.11A and can be written symbolically as

$$^1H + {}^1H \rightarrow {}^2H + e^+ + \nu$$

The terms to the left of the arrow are the normal hydrogen nuclei that start the process. The terms to the right are the heavy isotope of hydrogen that results ($^2$H), plus the positron and the neutrino. Energy is released by this reaction because the mass of $^2$H is slightly less than the mass of the two $^1$H's that were used to make it. The missing mass is converted to energy, as described in Einstein's formula $E = mc^2$, and ultimately that energy becomes the electromagnetic radiation of sunlight.

In the second step, the $^2$H nucleus collides with a third $^1$H to make the isotope of helium containing a single neutron, $^3$He. This process releases a high-energy photon (gamma ray), which we denote by $\gamma$. Figure 11.11B shows this step, which can be written as:

$$^1H + {}^2H \rightarrow {}^3He + \gamma$$

Here again, the resulting particle, $^3$He, has a smaller mass than the particles from which it was made, and again energy is released.

The third and final step is the collision and fusion of two $^3$He nuclei. Here, the fusion results not in a single particle but rather one $^4$He and two $^1$H nuclei. You can think about this reaction as the attempt to form a nucleus with 4 protons and 2 neutrons, except that 2 protons are ejected by their electric repulsion, as shown in figure 11.11C. Symbolically, we can write it as:

$$^3He + {}^3He \rightarrow {}^4He + {}^1H + {}^1H$$

where again, the final mass is less than the initial mass.

We can find the quantity of energy released by comparing the initial and final masses and using the mass-energy relationship $E = mc^2$. Steps 1 and 2 use three $^1$H, but the first two steps must occur twice to make the two $^3$He's for the last step. Therefore, six $^1$H's are used, but two are returned in step 3, and so a total of four $^1$H's are needed to make each $^4$He. If we now add up their masses, we find the following:

The mass of a single hydrogen nucleus is $1.673 \times 10^{-27}$ kilograms. The mass of a helium nucleus is $6.645 \times 10^{-27}$ kilograms. Thus,

$$4 \times 1 \text{ hydrogen} = 6.693 \times 10^{-27} \text{ kg}$$
$$-1 \text{ helium} = \underline{6.645 \times 10^{-27} \text{ kg}}$$
$$\text{Mass lost} = 0.048 \times 10^{-27} \text{ kg}$$

Multiplying this by $c^2$ gives the energy yield per helium atom made, $E$. That is,

$$E = 0.048 \times 10^{-27} \times (3 \times 10^8)^2 = 0.048 \times 10^{-27} \times 9 \times 10^{16} = 4.3 \times 10^{-12} \text{ joules}$$

This tiny number by itself is insignificant, but when multiplied by the vast number of hydrogen atoms undergoing fusion in the Sun, the total energy released is equivalent to exploding 100 billion-megaton H-bombs per second! Our sunshine has a violent birth.

## 11.3 PROBING THE SUN'S CORE

Although the immense bulk of the Sun hides its core from view, two types of experiments allow us an indirect look. Solar neutrinos escaping from the interior carry information about conditions there, and waves on the Sun's surface reveal the structure of the gas far below as waves in a stream reveal the presence of submerged rocks.

## Solar Neutrinos

We saw in section 11.2 that the Sun makes neutrinos as it converts hydrogen into helium. The number of released neutrinos therefore tells us how rapidly hydrogen is being converted into helium, and from that we can deduce the temperature and density in the Sun's core.

But counting neutrinos is extremely difficult. Neutrinos have no electric charge and only a tiny mass,* which gives them phenomenal penetrating power. They escape from the Sun's core through its outer 700,000 kilometers and into space like bullets through wet Kleenex. They pass straight through the Earth and anything on the Earth, such as you, and keep going. In fact, roughly a trillion neutrinos from the Sun pass harmlessly through your body every second.

The elusiveness that allows neutrinos to slip so easily through the Sun makes them slip with equal ease through detectors on Earth. Nevertheless, because so many neutrinos are produced, we need trap only a tiny fraction of them to get useful information. This trapping is done with neutrino detectors.

Currently, the largest neutrino "telescopes" are the Super-Kamiokande detector located deep in a zinc mine west of Tokyo and the Sudbury Neutrino Observatory (fig. 11.12) more than a mile underground in a nickel mine in northern Ontario, Canada. The Super-Kamiokande detector contains 50,000 tons of water. The Sudbury detector contains about 1000 tons of heavy water, water consisting of molecules in which one of the ordinary hydrogen atoms ($^1H$) is replaced by "heavy hydrogen," $^2H$. This heavy hydrogen gets its name because its nucleus contains a neutron as well as a proton.

**FIGURE 11.12**

An inside view of the neutrino detector located in Sudbury, Canada. The sphere is 12 meters in diameter (about 40 feet). It is filled with heavy water, a form of water in which each hydrogen atom contains a neutron as well as the usual proton. When a neutrino strikes one of the neutrons, the neutron may break down into a proton and electron. As the electron streaks off, it produces a tiny flash of light. Ten thousand detectors (which form the grid visible around the sphere) record the emitted light, thereby allowing scientists to detect the neutrino's passage.

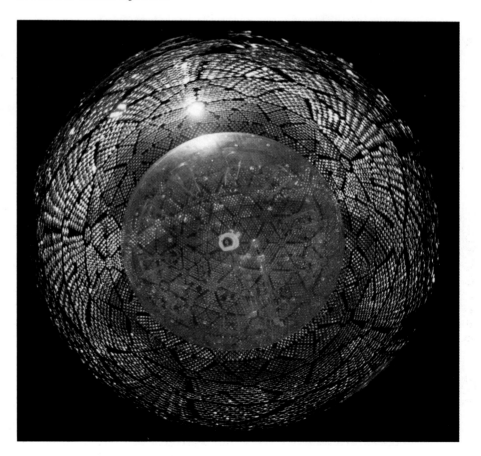

*Exactly how tiny a mass is still under study.

The penetrating power of neutrinos is the theme of a poem called "Cosmic Gall" by novelist and poet John Updike:

Neutrinos, they are very small.
They have no charge and have no mass
And do not interact at all.
The earth is just a silly ball
To them, through which they simply pass,
Like dustmaids down a drafty hall
Or photons through a pane of glass.
They snub the most exquisite gas,
Ignore the most substantial wall,
Cold-shoulder steel and sounding brass,
Insult the stallion in his stall,
And scorning barriers of class,
Infiltrate you and me! Like tall
And painless guillotines, they fall
Down through our heads into the grass.
At night they enter at Nepal
And pierce the lover and his lass
From underneath the bed—you call
It wonderful: I call it crass.

In 2002, the Nobel Prize in Physics was shared by the American scientist Raymond Davis and the Japanese scientist Masatoshi Koshiba for their work on neutrinos from astronomical sources. Davis led the first experiment to detect solar neutrinos and revealed the solar neutrino discrepancy. Koshiba headed the Kamiokande group whose work led to a resolution of the discrepancy. The American scientist Riccardo Giacconi, who made pioneering discoveries in X-ray astronomy, also shared the 2002 prize.

Neutrino detectors are buried to shield them from the many other kinds of particles besides neutrinos that constantly bombard the Earth. For example, protons, electrons, and a variety of subatomic particles constantly shower our planet. These particles, traveling at nearly the speed of light, are called **cosmic rays** and are thought to be particles blasted across space when a massive star explodes.

Cosmic rays can penetrate only a short distance into the Earth; so if a detector is located deep underground, nearly all the cosmic rays are filtered out. Neutrinos, however, are unfazed by a mere mile of solid ground: on the average they could travel through a light year of lead!

Given their virtual unstoppability, how are neutrinos detected? Scientists have several ways. In the Sudbury detector, occasionally a neutrino collides with a neutron in the heavy hydrogen, breaking the neutron into a proton and an electron. As the electron shoots off, it emits a tiny flash of light, which is recorded by photodetectors. But when astronomers first tried to detect solar neutrinos, they found only one-third the expected number. What was wrong? Did the Sun not fuse hydrogen into helium, as predicted? Did neutrinos somehow escape detection? The solution to the puzzle offers us another chance to see the scientific process at work, as explained in the Science at Work box "Solving the Neutrino Puzzle."

## Solar Seismology

Solar seismology is the study of the Sun's interior by analyzing waves in the Sun's atmosphere. We learned in chapter 5 that scientists can study the Earth's interior by analyzing earthquake waves. Astronomers can also learn about the Sun's interior by analyzing waves in its gases.

Waves similar to those of earthquakes travel through the Sun and make its surface heave like the ocean or a bubbling pan of hot oatmeal. The rising and falling surface gas makes a regular pattern (fig. 11.13), which can be detected as a Doppler shift of the moving material. Astronomers next use computer models of the Sun to predict

## SCIENCE AT WORK

### SOLVING THE NEUTRINO PUZZLE

As a first step toward resolving the solar neutrino discrepancy, astronomers checked that their calculations for the predicted number of neutrinos were correct. All such checks led to roughly the same result, implying that there was no obvious flaw in their understanding of how the Sun works. Was it possible, then, that neutrinos had undiscovered properties that affected their detectability on Earth or their production in the Sun?

As scientists looked more closely at the properties of neutrinos, they found that one theory suggested the existence of three kinds of neutrinos, not just one. This seemed promising because the first neutrino detectors could only detect one of the three kinds, thus perhaps explaining why only one-third the expected number were observed. Moreover, the revised theory of neutrinos implied that these elusive particles were not massless and that one kind could change into another. With this new hypothesis about the nature of neutrinos, scientists built a second generation of neutrino detectors that could detect all three varieties. The results from these new detectors agreed perfectly with the predictions of the solar theory. Thus, astronomers have a new and greater confidence in their understanding of the Sun. But in addition, they now know more about neutrinos, including the fact that they are not massless.

FIGURE 11.13
Computer diagram of solar surface waves.

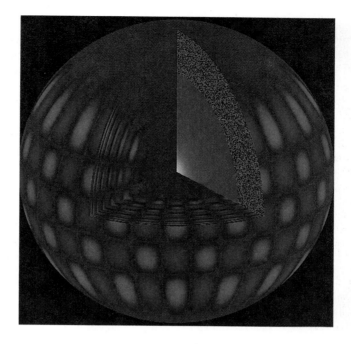

how the observed surface waves are affected by conditions in the Sun's deep interior. With this technique, astronomers can measure the density and temperature deep within the Sun from the pattern and speed of the waves in its atmosphere. The results they find agree well with those of the current models, indicating that we have a good understanding of the Sun's structure.

## 11.4   SOLAR MAGNETIC ACTIVITY

The surface waves described above are but one of many disturbances in the Sun's outer regions. A wide class of dramatic and lovely phenomena on the Sun are caused by its magnetic field. This magnetic activity is also of interest because it affects the Earth, where it triggers auroral displays and appears to alter the climate.

### Sunspots

**Sunspots** are the most common type of solar magnetic activity. They are large, dark-appearing regions (fig. 11.14) ranging in size from a few hundred to many thousand kilometers across. Spots last from a few days to over a month. They are darker than the surrounding gas because they are cooler* (4500 K as opposed to 6000 K of the normal photosphere), and they are cooler because they contain strong magnetic fields. Why do magnetic fields create cool regions? To understand, we must look at how such fields interact with the Sun's hot gas.

### Solar Magnetic Fields

The magnetic field of sunspots is more than a thousand times stronger than the Earth's or the "normal" field of the Sun. In such intense fields, electrons and other

---

*Spattering a few drops of water onto a hot electric stove burner will show a similar effect. Each drop momentarily cools the burner, making a dark spot.

**FIGURE 11.14**
White-light photograph of photosphere with sunspots.

Sunspot groups

Earth for comparison

Magnetic field

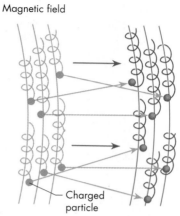

Particles spiral around the field lines.

Field moves. Because the particles must spiral around the field lines, they are dragged along as the field moves.

Charged particle

**FIGURE 11.15**
Charged particles are "frozen" to a magnetic field—so that as the field moves, they move with it.

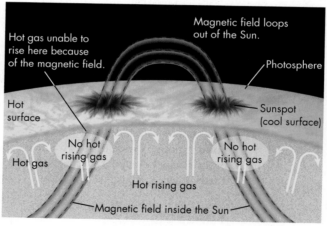

Magnetic field loops out of the Sun.

Hot gas unable to rise here because of the magnetic field.

Photosphere

Hot surface

Sunspot (cool surface)

No hot rising gas

No hot rising gas

Hot gas

Hot rising gas

Magnetic field inside the Sun

**FIGURE 11.16**
Sketch of how magnetic fields cause sunspots.

charged particles spiral around the field, "frozen" to it, as shown in figure 11.15. They are thereby forced to follow the magnetic field as they spiral in it. On the Earth, the field deflects particles toward the poles, where they create auroras. In the Sun, the field slows the ascent of hot gas in the convection zone. With heat rising more slowly where the magnetic field is strong, the surface cools and becomes darker, making a sunspot, as shown in figure 11.16.

## Prominences and Flares

**Prominences** and **solar flares** are magnetic disturbances in the low-density, virtually transparent, hot gases above the Sun's visible surface. Prominences are huge plumes of glowing gas that jut from the lower chromosphere into the corona. Figure 11.17 shows some of their variety. You can get some sense of their immensity from the white dot, which shows the size of the Earth.

ANIMATION

*Prominences on the Sun's limb*

**FIGURE 11.17**
Photographs of some solar prominences.

**FIGURE 11.18**
(A) Sketch illustrating how magnetic fields support a prominence. (B) Photograph of a prominence.

⌖* Approximately how big is the prominence in (B)? Use the fact that the Sun's diameter is about 100 times the Earth's. Hint: Use the curve of the Sun's edge to estimate its size.

ⓐ* The prominence in (B) is ~160,000 km high.

Prominences form where the Sun's magnetic field reduces heat flow to a region. They are *cooler* than the gas around them, which means, according to the perfect gas law, that the pressure inside is less than outside. Thus, the hot external gas "bottles up" the cooler gas of the prominence (fig. 11.18). Under favorable conditions, this cooling gas, trapped in its magnetic prison, may glow for weeks.

Time-lapse movies show that gas streams through prominences, sometimes rising into the corona, sometimes raining down onto the photosphere. The flow is channeled by and supported by the magnetic field, which often arcs between sunspots. Thus, prominences also are related to sunspots.

Sunspots also give birth to solar flares, brief but bright eruptions of hot gas in the chromosphere (fig. 11.19). Over a few minutes or hours, gas near a sunspot may dramatically brighten. Such eruptions, though violent, are so localized they hardly affect the total light output of the Sun at all. Generally, you need a telescope to see their visible light, though they can increase the Sun's radio and X-ray emission by factors of a thousand in a few seconds.

**FIGURE 11.19**
A solar flare—the bright spot and line at the lower right of the image. The green color is to show that this picture has been taken through a special ultraviolet filter.

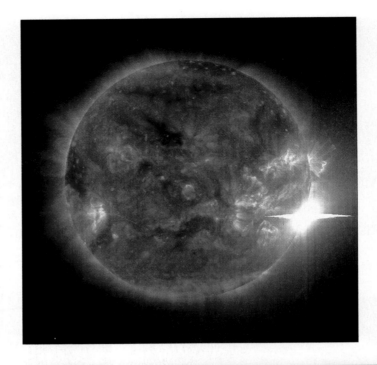

**FIGURE 11.20**
Photograph of the great aurora of March 1989.

↻★ Look up the origin of the word *aurora* in a dictionary. Does the appearance of the aurora indicate why that origin is appropriate?

Ⓐ★ Aurora is the name of the goddess of the dawn in ancient Greek mythology. The aurora's pale glow and rosy tints resemble the pale colors of dawn.

Astronomers today think that solar flares are only part of the cause of auroral displays here on Earth. A more important trigger appears to be *coronal mass ejections.* These ejections are enormous bubbles of hot gas and trapped magnetic fields that burst from the corona out into space. Our most spectacular auroral events occur when one of these ejections strikes Earth.

Flares are poorly understood, but magnetic fields probably play an important role. One theory suggests that the field near a spot gets twisted by gas motions, a bit like winding up a rubber band-powered toy. But such twisting can go only so far before the rubber band breaks. So, too, the magnetic field can be twisted only so far before it suddenly readjusts, whipping the gas in its vicinity into a new configuration. The sudden motion heats the gas, and it expands explosively. Some gas escapes from the Sun and shoots across the inner Solar System to stream down on the Earth. Such a burst created the spectacular auroral displays seen in March 1989 and illustrated in figure 11.20.

## Heating of the Chromosphere and Corona

Although the Sun's magnetic field cools sunspots and prominences, it heats the chromosphere and corona. Other stars also have chromospheres and coronas, and so it is

# EXTENDING OUR REACH

## DETECTING MAGNETIC FIELDS: THE ZEEMAN EFFECT

Astronomers can detect magnetic fields in sunspots and other astronomical bodies by the **Zeeman effect,** a physical process in which the magnetic field splits some of the spectrum lines of the gas into two, three, or more components. The splitting occurs because the magnetic field alters the atom's electron orbits, which in turn alter the wavelength of its emitted light.

Box figure 11.1A shows the Zeeman effect splitting spectrum lines in a sunspot. The line is single outside the spot but triple within, and by mapping the splitting across the Sun's face, astronomers can map the Sun's magnetic field, creating a magnetogram, as seen in box figure 11.1B. The colors in a magnetogram show the strength and polarity of the magnetic field (recall that magnetic fields have a north or south polarity indicating their direction). Notice that the field is strong around spots and weak elsewhere.

Region in sunspot

Region outside of sunspot

A

B

**BOX FIGURE 11.1**
(A) Photograph of the Zeeman effect in a sunspot. Notice that the line is split over the spot where the magnetic field is strong but that the line is unsplit outside the spot where the field is weak or absent. (B) Magnetogram of the Sun. Yellow indicates regions with north polarity, and dark blue indicates regions with south polarity. Notice that the polarity pattern of spot pairs is reversed between the top and bottom hemispheres of the Sun. That is, in the upper hemisphere, blue tends to be on the left and yellow on the right. In the lower hemisphere, blue tends to be on the right and yellow on the left.

additionally important to understand the Sun's. To begin, we need first to recall that the temperature of a gas is a measure of how fast its particles are moving. Anything that speeds atoms up increases their temperature.

An analogy may help you understand how magnetic waves can heat a gas. When you crack a whip, a slow motion of its handle travels as a wave along the whip. As the whip tapers, the wave's energy of motion is given to an ever smaller piece of material. With the same amount of energy and less mass to move, the tip accelerates and eventually breaks the sound barrier. The whip's "crack" is a tiny sonic boom.

**FIGURE 11.21**
Diagram illustrating how magnetic waves (blue) heat the Sun's upper atmosphere. As the waves move outward through the Sun's atmosphere, they grow larger, imparting ever more energy to the gas (green dots) through which they move, accelerating and thereby heating it.

A similar speedup occurs in the Sun's atmosphere when magnetic waves formed in the photosphere move into the corona along the Sun's field lines (fig. 11.21). As the atmospheric gas thins, the wave energy is imparted to an ever smaller number of atoms, making them move faster in random directions. But "faster" in this case means hotter. Thus, the upper atmosphere heats up as the waves travel into it.

What generates the magnetic waves? They probably start in the convection zone where rising bubbles of gas shake the magnetic field and create magnetic waves just as shaking the loose end of a rope makes it wiggle. Thus, the high temperature of the chromosphere and corona is another example of the importance of the Sun's magnetic field and its convection zone. This theory is supported by observations of stars other than the Sun with active convection zones, which also have active chromospheres and coronas.

## The Solar Wind

In addition to the mass it loses in the outbursts of flares, the Sun undergoes a steady, less dramatic loss of mass. The corona's high temperature gives its atoms enough energy to escape the Sun's gravity. As these atoms stream into space, they form the **solar wind,** a tenuous flow of mainly hydrogen and helium that sweeps across the Solar System. The amount of material lost from the Sun is small: less than 1 ten-trillionth of its mass each year. Nevertheless, the solar wind creates comet tails, as we saw in chapter 10, and auroras, as we saw in chapter 5.

The solar wind arises because, unlike the rest of the Sun, the corona is *not* in hydrostatic equilibrium. Recall that the temperature in the Sun's atmosphere increases with altitude, making the corona much hotter than the photosphere. The corona's high temperature, according to the perfect gas law, creates a pressure within it larger than we might otherwise expect for its distance above the photosphere. The pressure is in fact sufficient to overcome the Sun's gravitational force on gas in its upper atmosphere. As a result, it pushes that material outward into space. The expanding gas has a very low density (only a few hundred atoms in a cubic centimeter—the volume of a thimble*). The gas atoms begin their outward motion slowly but accelerate with increasing distance as the Sun's gravitational attraction on them weakens. On the average, the wind speed is about 500 kilometers per second at the Earth's orbit, but it speeds up and slows down in response to changes in the Sun's magnetic field. From the Earth outward, the wind coasts at a relatively steady speed that carries it at least to the orbit of Neptune. At some point, it impinges on the interstellar gas surrounding the Solar System, but where that happens has not yet been found.

## 11.5 THE SOLAR CYCLE

Sunspot and flare activity change from year to year in what is called the **solar cycle.** This variability can be seen in figure 11.22, which shows the number of sunspots detected over the last 140 years or so. The numbers clearly rise and fall approximately every 11 years. For example, the cycle had peaks in 1958, 1969, 1980, 1990, and 2001.

Samuel Heinrich Schwabe, a pharmacist and amateur astronomer, discovered the solar cycle. His work prevented him from observing at night; so he studied the Sun during the day.

Flares and prominences also follow the solar cycle, and climate patterns on Earth may, too. For this reason, astronomers have sought to understand not only the cause of the solar cycle but also how it influences terrestrial climate.

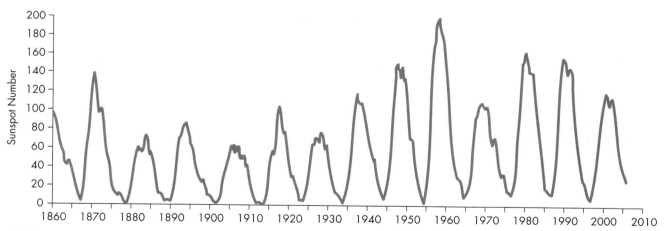

**FIGURE 11.22**
Plot of sunspot numbers showing solar cycle.

Is the interval between the peaks always exactly 11 years? By how much does it change?

The intervals between the peaks are not equal. They vary by about 10% in the time period shown in the figure. That amounts to a difference of about a year.

Are we currently near a high or low of solar activity?

---

*For comparison, a thimbleful of the air we breathe contains about $10^{19}$ molecules!

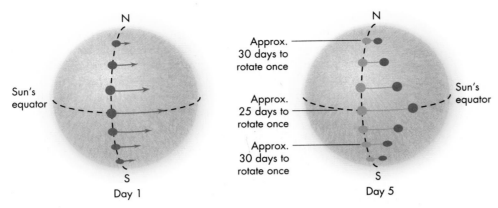

**FIGURE 11.23**
Sketch showing solar differential rotation. Points near the Sun's equator rotate faster than points near the poles.

## Cause of the Solar Cycle

As the Sun rotates, gas near its equator circles the Sun faster than gas near its poles; that is, it spins differentially, a property common in gaseous bodies (recall from chapter 9 that Jupiter and the other giant planets rotate differentially). The Sun's differential rotation is such that its equator rotates in about 25 days and its poles in about 35 days. Thus, a set of points arranged from pole to pole in a straight line would move over the course of time into a curve, as shown in figure 11.23.

Differential rotation should similarly distort the Sun's magnetic field, "winding up" the field below the Sun's surface.* Astronomers think such winding of the Sun's magnetic field may cause the solar cycle, though the exact mechanism is still not well understood. According to one hypothesis, the Sun's rotation wraps the solar magnetic field into coils below the surface, making the field stronger and increasing solar activity: spots, prominences, and flares. The wrapping occurs because the Sun's magnetic field is "frozen" into the gas, as discussed in section 11.4. Thus, if the gas moves, so does the field, and vice versa.

Because the field and gas are tightly connected, differential rotation causes gas at the equator, which is moving faster than the gas at the poles, to drag the magnetic field with it so that a field, initially straight north to south, is wound into two subsurface loops, as shown in figure 11.24A. As the loops are wound tighter, they develop kinks, as when you twist a rubber band too tight. The cycle ends when the field twists too "tightly" and collapses, and the process repeats.

Sunspots form when kinks in the magnetic field rise to the Sun's surface and break through the photosphere (fig. 11.24B). Here, the field slows the outward flow of heat, making the surface cooler and darker than in surrounding areas and thereby creating sunspots, as we discussed in section 11.4. Each kink breaks the photosphere in *two* places—one where it leaves and one where it enters. We therefore expect that spots will occur in pairs or paired groupings, as sketched in figure 11.24C.

In such pairs, one grouping has a north polarity and the other a south polarity.[†] That is, in one the field emerges from the surface, while in the other it descends so that the field direction is reversed. The false-color map of the Sun's field in box figure 11.1B shows this effect clearly. Yellow areas have a north polarity and dark blue areas a south polarity. Notice that the pattern of the polarity differs between the two hemispheres of the Sun. In one, the yellow areas are on the right, while in the other they are on the left. This reversal arises because the subsurface field is coiled in opposite directions in the

*Wrapping of the Sun's subsurface magnetic field*

---

*Spots and cycles of magnetic activity occur on other stars as well.

[†]Polarity of magnetic fields was described briefly in chapter 5, when we discussed the Earth's magnetic field.

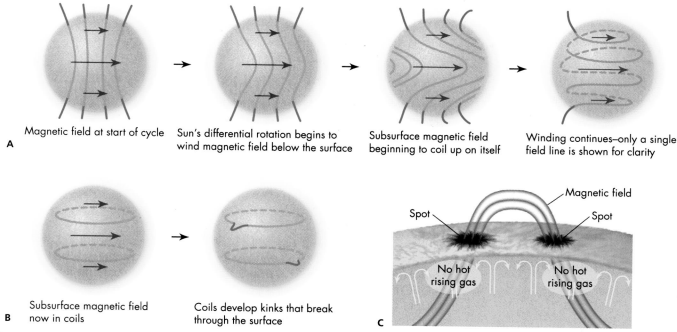

Magnetic field at start of cycle

Sun's differential rotation begins to wind magnetic field below the surface

Subsurface magnetic field beginning to coil up on itself

Winding continues—only a single field line is shown for clarity

**A**

Subsurface magnetic field now in coils

**B**

Coils develop kinks that break through the surface

Spot

Magnetic field

Spot

No hot rising gas

No hot rising gas

**C**

**FIGURE 11.24**

Sketch showing (A) the possible winding up of the subsurface magnetic field; (B) fields penetrating the Sun's surface; and (C) formation of a spot pair.

**FIGURE 11.25**

Diagram illustrating the approximate 22-year periodicity of solar magnetic activity.

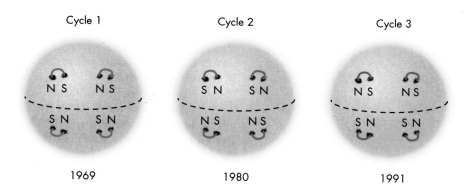

Cycle 1

Cycle 2

Cycle 3

1969

1980

1991

two hemispheres, additional evidence that sunspots and the solar cycle are caused by winding of the Sun's magnetic field.

## Changes in the Solar Cycle

The solar cycle is not always 11 years: it may be as short as 7 or as long as 16 years. Moreover, if we consider the polarity of the spot groups, the cycle averages 22 years, rather than 11, because the polarity of the Sun's field reverses at the end of each 11-year cycle. It therefore takes *two* 11-year cycles for the field to return to its original configuration.

Figure 11.25 illustrates this effect. In the first frame, we see spot pairs as the cycle begins. A right-hand spot in the top hemisphere (technically, the "leading spot" because it leads in the direction the Sun rotates) has a south polarity, while a left-hand spot (technically called a "trailing spot") has a north polarity.

You can also see an additional feature of the cycle: all leading spots in one hemisphere have the same polarity. In the other hemisphere, the leading spots have the opposite polarity. Eleven years later, the polarity of the pairs will be reversed, as you can see by looking at the second frame in figure 11.25. Only in the next cycle, another 11 years later, will the spot fields have returned to their original directions. Thus, the full cycle of magnetic activity takes 22 years on the average.

## Links Between the Solar Cycle and Terrestrial Climate

Climatologists find this 22-year period interesting because of a cycle in which droughts recur approximately every 22 years in the midwestern United States and Canada. Does the Sun's cycle affect the Earth's climate cycle? If so, how? One possibility is as follows.

The Sun's magnetic field heats the corona. The corona drives the solar wind. The solar wind alters the Earth's upper atmosphere; in particular, it changes the way the temperature varies with altitude. This in turn alters the atmosphere's circulation and may shift the jet stream to a new location. The jet stream steers storms and hence rainfall.

Although this hypothesis cannot be verified yet, many scientists think that solar activity affects our climate. The evidence to support this hypothesis is based in part on the work of E. W. Maunder, a British astronomer who studied sunspots. Maunder noted in 1893 that, according to old solar records, very few sunspots were seen between 1645 and 1715 (fig. 11.26). He concluded that the solar cycle turned off during that period. The period is now called the **Maunder minimum** in honor of his discovery.

The Maunder minimum coincides with an approximately 70-year spell of abnormally cold winters in Europe and North America. Glaciers in the Alps advanced; rivers froze early and remained frozen late; the North Sea froze. The cold was so abnormal that meteorologists call the epoch part of the "little ice age." If only one such episode were known, we might dismiss the sunspot-climate connection as a coincidence, but three other cold periods have also occurred during times of low solar activity. This strengthens our belief that somehow the Sun's magnetic activity affects our climate.

Although scientists are still unsure about what creates the link between solar magnetic activity and our climate, they no longer doubt that such a link exists. Figure 11.27 shows how the ocean temperature (expressed as deviations from the normal average) changed from 1860 to 2000. The figure also shows how the number of sunspots changed over the same time span. Notice that from 1860 to 1980, when the number of spots is high, the ocean is warmer than average and when the number of spots is low, the ocean is colder than average. But from 1990 on, the curves no longer are in step. Scientists do not understand why the curves no longer behave the same way.

**FIGURE 11.26**

Plot illustrating that the number of sunspots changes with time, showing the Maunder minimum and the solar cycle.

✪ Can you see a second, slower cycle in the sunspot numbers? (Compare the peak numbers near 1810 and 1900 with the numbers near 1850 and 1950.)

**FIGURE 11.27**

Curves showing the change in ocean temperatures on Earth and the change in sunspot numbers over more than a century. Notice that the curves were approximately in step until the 1980s. Over the last two decades the sea surface temperature has continued to rise while sunspot counts have declined. Astronomers deduce that solar activity affects our climate, but other factors must explain the recent temperature rise.

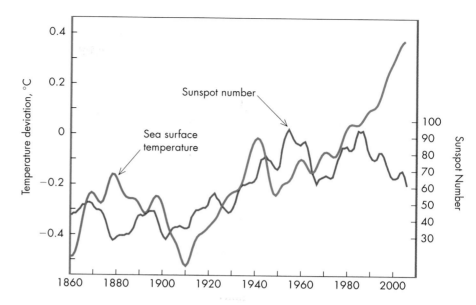

# SUMMARY

The Sun is composed mostly of hydrogen and helium atoms. It is held together by gravity and supported against gravitational collapse by the pressure created by its high internal temperature. This temperature is high enough in the core that hydrogen can be converted into helium by nuclear fusion, thereby supplying the Sun with energy. That energy then flows to the surface by radiation and convection.

Above the Sun's visible surface (the photosphere) lie the chromosphere and corona—low-density, hot atmospheric layers. They are heated by magnetic waves traveling outward from the photosphere. In the corona, the temperature reaches a million degrees and drives the corona gases into space, creating the solar wind.

The Sun's magnetic field changes in structure and strength over time. As the subsurface field strengthens, sunspots form where the field breaks through the surface. Near spots, prominences and flares may occur. Spot numbers rise and fall with a roughly 11-year period. For reasons that are not yet well understood, this variation in solar magnetic activity also creates climatic variation on the Earth.

## **?** What It Is.

This image shows Venus (the round dark circle on the right) passing directly between us and the Sun on the June 8, 2004, transit of Venus. You can also see red columns of hot gas in the Sun's chromosphere and corona projecting beyond the edge of the Sun at right.

# QUESTIONS FOR REVIEW

1. How big is the Sun compared to the Earth?
2. How can we measure the Sun's size, mass, and temperature?
3. What is the Sun made of? How do we know this?
4. What holds the Sun together?
5. Why doesn't the Sun collapse?
6. Why must the interior of the Sun be so hot?
7. How is solar energy generated?
8. How does energy get to the Sun's surface from its core?

9. What visible evidence do we have that the Sun has a convection zone?
10. What are the photosphere, chromosphere, and corona?
11. Which of these layers is hottest? How do we know this?
12. What is meant by solar activity?
13. Why do sunspots appear dark?
14. How do a prominence and a flare differ?
15. What role does magnetic activity play in solar activity?

16. How do we know there are magnetic fields in the Sun?

17. What is meant by solar neutrinos? Why do astronomers say there is a discrepancy? Why is the discrepancy of interest?

18. What is solar seismology? What does it tell us about the Sun?

19. What is the solar cycle?

20. What is the period between maximum sunspot numbers? How does this differ from the period of the full solar cycle?

21. What is the Maunder minimum? Why is it of interest?

## THOUGHT QUESTIONS

1. Why can the Sun not be powered by a chemical process such as the combustion of hydrogen and oxygen to form water?

2. Why is hydroelectric power generation an indirect form of solar power?

3. The Sun's corona has a temperature of 1 million K. Why does it not incinerate us?

4. Why is sailing an indirect form of solar power?

## PROBLEMS

1. Given that the angular diameter of the Sun is 1/2 degree and that its distance is $1.5 \times 10^8$ kilometers, go through the math to determine the Sun's diameter.

2. Suppose you were an astronomy student on Jupiter. Use the orbital data for Jupiter (distance from Sun = 5.2 AU; period = 11.8 years) to measure the Sun's mass using the modified form of Kepler's third law.

3. In this problem, you will calculate approximately the temperature of the Sun's core. You can do it either step by step or by writing out all the algebra to obtain a final result. You will need the following ideas: The pressure, $P$, in a gas is given by $P = $ constant $\times \rho T$. If we measure length in meters, mass in kilograms, and $T$ in Kelvin, then the constant has the value of about 8300 m$^2$-sec$^{-2}$-K$^{-1}$; the density of a body, $\rho$, is its mass per volume, which for a sphere is $M/(4\pi R^3/3)$;

the pressure force from the interior is $P \times A$, where $A$ is the area over which the pressure acts. You can take that as the Sun's cross section, $\pi R^2$. Thus, the pressure force is $\pi R^2 P$. Finally, you need to invoke hydrostatic equilibrium: pressure forces must balance gravitational forces. Approximate the gravitational force holding the Sun together by assuming it is split into two equal halves and apply Newton's law of gravity to calculate the force between the halves. Assume they are separated by 1 solar radius.

4. Calculate the escape velocity from the Sun.

5. Calculate the Sun's density in grams per cubic centimeter. The Sun's mass is approximately $2 \times 10^{33}$ grams. Its radius is approximately $7 \times 10^{10}$ cm. How does the density you find compare with the density of Jupiter?

## TEST YOURSELF

1. The diameter of the Sun is about how large compared with the Earth's?
   (a) Twice as big
   (b) One-half as big
   (c) 10 times as big
   (d) 100 times as big
   (e) 10,000 times as big

2. About how many years elapse between times of maximum solar activity?
   (a) 3
   (b) 5
   (c) 11
   (d) 33
   (e) 105

3. Sunspots are dark because
   (a) they are cool relative to the gas around them.
   (b) they contain 10 times as much iron as surrounding regions.
   (c) nuclear reactions occur in them more slowly than in the surrounding gas.
   (d) clouds in the cool corona block our view of the hot photosphere.
   (e) the gas within them is too hot to emit any light.

4. The Sun is supported against the crushing force of its own gravity by
   (a) magnetic forces.
   (b) its rapid rotation.
   (c) the force exerted by escaping neutrinos.

(d) gas pressure.

(e) the antigravity of its positrons.

5. The Sun produces its energy from

(a) fusion of neutrinos into helium.

(b) fusion of positrons into hydrogen.

(c) disintegration of helium into hydrogen.

(d) fusion of hydrogen into helium.

(e) electric currents generated in its core.

# FURTHER EXPLORATIONS

Foukal, Peter V. "The Variable Sun." *Scientific American* 262 (February 1990): 34.

Hathaway, David H. "Journey to the Heart of the Sun." *Astronomy* 23 (January 1995): 38.

Holman, Gordon D. "The Mysterious Origins of Solar Flares." *Scientific American* 294 (April 2006): 38.

Kennedy, James R. "GONG: Probing the Sun's Hidden Heart." *Sky and Telescope* 92 (October 1996): 20.

Kippenhahn, Rudolph. *Discovering the Secrets of the Sun.* New York: Wiley, 1994.

Lang, Kenneth R. *Sun, Earth, and Sky.* New York: Springer-Verlag, 1995.

———. "Unsolved Mysteries of the Sun [Parts 1 and 2]." *Sky and Telescope* 92 (August 1996): 38; no. 92 (September 1996): 24.

McDonald, Arthur B., Joshua R. Klein, and David L. Wark. "Solving the Solar Neutrino Problem." *Scientific American* 288 (April 2003): 40.

Mims, Forrest M. "Sunspots and How to Observe Them Safely." *Scientific American* 262 (June 1990): 130.

Nesme-Ribes, Elizabeth, Sallie L. Baliunas, and Dmitry Sokoloff. "The Stellar Dynamo." *Scientific American* 275 (August 1996): 45.

Schrijver, Carolus J. The Science Behind the Sun's Corona. *Sky and Telescope* 111 (April 2006): 28.

**Website**

Please visit the *Explorations* website at **http://www.mhhe.com/arny** for additional online resources on these topics.

# KEY TERMS

chromosphere, 325

convection zone, 325

corona, 326

coronal holes, 326

cosmic rays, 332

granulation, 325

hydrostatic equilibrium, 327

Maunder minimum, 342

neutrino, 330

nuclear force, or strong force, 329

nuclear fusion, 329

perfect gas law, or ideal gas law, 327

photosphere, 324

pressure, 327

prominences, 334

proton–proton chain, 329

radiative zone, 324

solar cycle, 339

solar flares, 334

solar wind, 338

spicules, 326

sunspots, 333

Zeeman effect, 337

# PROJECTS

1. **Measure the diameter of the Sun.** Take a piece of thin, dark cardboard and put a small hole in it. Hold it in sunlight about 1 meter (about 3 feet) from a piece of white paper so that a small image of the Sun appears on the paper (see project fig. 11.1A). Carefully measure the distance (d) between the cardboard and the piece of paper and the size of the Sun's image (s) on the paper. On a separate piece of paper, draw two straight lines that cross with a small angle between them (see project fig. 11.1B). Draw two small circles between the lines as shown in project figure 11.1B. Convince yourself that if D is the distance to the Sun (1 AU) and S is its diameter, then $S/D = s/d$, where s is the size of the Sun's image on the paper and d is the distance from the paper to the cardboard. Look up the value of D in table 11.1. Then solve for S, the Sun's diameter. Does your value agree with the value you calculate from the value of the Sun's radius in table 11.1?

# PROJECTS (CONTINUED)

2. **Observing sunspots.** Mount a pair of binoculars on a tripod or some other fixed object. You can use duct tape if you cannot find a binocular clamp and tripod. Hold a piece of white paper about 60 centimeters (roughly 2 feet) from the eyepieces of the binoculars and adjust them so that the Sun's image falls on the paper (project fig. 11.2). You may need to adjust the focus wheel of the binoculars to make the Sun's image sharp and clear on the paper. DO NOT LOOK AT THE SUN THROUGH THE BINOCULARS. THE SUN'S LIGHT CAN QUICKLY AND PERMANENTLY BLIND YOU IF YOU DO. Also, shade the binoculars or telescope from time to time so the eyepiece does not become too hot. Can you see any sunspots on the Sun's image

on the paper? (They will appear as small dark blotches.) Draw a circle on the paper to match the size of the Sun's image. Mark the date on the paper and sketch the location of the spots that you see on the circle. Repeat your observations a few days later. Are the spots in the same location? If they have shifted their position, can you think of a reason why?

3. **Playing photon.** Try the following "game" to get a sense of why it takes energy so long to leak from the center of the Sun to its surface. Recall that in the Sun, photons are constantly absorbed and re-emitted, but in random directions. A given photon thus is as likely to be re-emitted back toward the Sun's core as toward it surface. This is much like what

happens if you try to climb a flight of stairs using the flip of a coin to decide whether you go up a step or down.

Stand at the bottom of a short flight of stairs—a flight of four or five steps is about right. If you can't find such a short flight, then put a book on the fourth step of a longer flight and pretend it's the top step. Flip a coin and go up a step if you get "heads" and down a step (or stay where you are if you are already at the bottom) if you get "tails." Record how many flips you need to reach the top step. Repeat half a dozen or so times. How many flips, on average, does it take you to reach the top step? What factors determine how *long* it takes you to climb the flight?

**PROJECT FIGURE 11.1**
A simple way to measure the Sun's diameter.

**PROJECT FIGURE 11.2**
Observing sunspots with a pair of binoculars.

# Measuring the Properties of Stars

# 12

A region where stars are forming from interstellar dust and gas clouds.

## Chapter Objectives

- Astronomers can learn much about stars, such as their temperature and composition, from the spectra of their light.
- Many stars are members of binary star systems in which one star orbits another.
- Study of binary star motions gives information about the stars' mass.
- Classification of stars by their temperature and intrinsic brightness shows that stars can be divided into three main groups:
  - Main-sequence stars.
  - Red giants.
  - White dwarfs.
- Main-sequence stars obey the mass–luminosity law:
  - More-massive stars are more luminous.
- Some stars pulsate, so that their size and brightness change.

**?What Is This?**
(See chapter summary for the answer.)

## Concepts and Skills to Remember

- Relation between temperature of a hot object and the color of the light it emits (Wien's law) (3.2)
- Doppler shift (3.6)
- Modified form of Kepler's third law (2.7)

It is hard to believe that the stars we see in the night sky as tiny glints of light are in reality huge, dazzling balls of gas and that many are vastly larger and brighter than our Sun. They look dim to us only because they are so far away—several light-years (trillions of miles) to even the nearest. Such remoteness creates tremendous difficulties for astronomers trying to understand the nature of stars. We cannot physically travel to the stars, but in this chapter we will see how astronomers overcome the distance barrier that separates us from stars and how they learn many of the physical properties of these distant objects. How far away are they? How big? What are they made of?

The answers to these questions show us that most stars are remarkably like the Sun. For example, like the Sun, they are composed mostly of hydrogen and helium and have about the same mass. A small percentage, however, are as massive as 30 times the Sun's mass (30 $M_\odot$) and are much hotter than the Sun and blue in color. Others are much less massive than the Sun, only one-tenth its mass, and are cool, red, and dim. Moreover, even stars similar to the Sun in composition and mass may differ enormously from it in their radii and density. For example, some giant stars have a radius hundreds of times larger than the Sun's—so big that were the Sun their size, it would extend beyond the Earth's orbit. On the other hand, some stars are white dwarfs, with as much material as the Sun packed into a volume the size of the Earth.

Astronomers can learn all these properties of stars by using physical laws and theories to interpret measurements made from the Earth. For example, theories of light yield the surface temperature, distance, and motion of a star; theories of atoms yield the composition of a star; and a modified form of Kepler's third law yields the mass of a star. In using such laws, astronomers may sometimes employ more than one method to determine a desired property of a star. For example, a star's temperature may be measured from either its color or its spectrum. Such alternative methods serve as checks on the correctness of the procedures astronomers use to determine the properties of stars.

## 12.1 MEASURING A STAR'S DISTANCE

One of the most difficult problems astronomers face is measuring the distance to stars and galaxies. Yet knowing the distances to such bodies is vital if we are to understand their size and structure. For example, without knowledge of a star's distance, astronomers find it difficult to learn many of the star's other properties, such as its mass, radius, or energy output.

### Measuring Distance by Triangulation and Parallax

Using parallax to find distance may seem unfamiliar, but parallax creates our stereovision, the ability to see things three dimensionally. When we look at something, each eye sends a slightly different image to the brain, which then processes the pictures to determine the object's distance. You can demonstrate the importance of the two images by covering one eye and trying to pick up a pencil quickly. You will probably not succeed in grasping it on the first attempt.

Astronomers have several methods for measuring a star's distance, but for nearby stars the fundamental technique is **triangulation**, the same method we described for finding the distance to the Moon. In triangulation, we construct a triangle in which one side is the distance we seek but cannot measure directly and another side is a distance we can measure—a baseline, as shown in figure 12.1A. For example, to measure the distance across a deep gorge, we construct an imaginary triangle with one side spanning the gorge and another side at right angles to it and running along the edge we are on, as shown in figure 12.1B. By measuring the length of the side along the gorge edge and the angle A, we can determine the distance across the gorge either by a trigonometric calculation or from a scale drawing of the triangle.

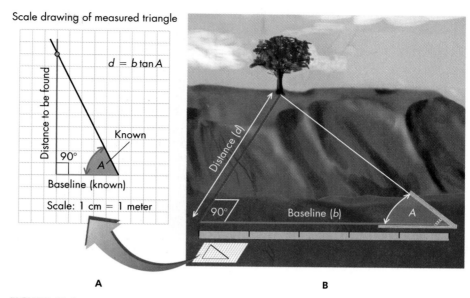

Scale drawing of measured triangle

$d = b \tan A$

Distance to be found

Known

90°

$A$

Baseline (known)

Scale: 1 cm = 1 meter

Distance (d)

90°

Baseline (b)

A

**A**                                    **B**

**FIGURE 12.1**
Sketch illustrating the principle of triangulation.

**ANIMATION**

*Parallax*

Astronomers use a method of triangulation called **parallax** to measure the distance to stars. Parallax is a change in an object's apparent position caused by a change in the observer's position. An example—easy to demonstrate in your room—is to hold your hand motionless at arm's length and shift your head from side to side. Your hand seems to move against the background even though in reality it is your head that has changed position, not your hand.

This simple demonstration illustrates how parallax gives a clue to an object's distance. If you hold your hand at different distances from your face, you will notice that the apparent shift in your hand's position—its parallax—is larger if it is close to your face than if it is at arm's length. That is, *nearby objects exhibit more parallax than more remote ones,* for a given motion of the observer, a result true for your hand and for stars.

To observe stellar parallax, astronomers take advantage of the Earth's motion around the Sun, as shown in figure 12.2A. They observe a star and carefully measure its position against background stars. They then wait 6 months until the Earth has moved to the other side of its orbit, a known distance of 2 AU (about 300 million kilometers), and make a second measurement. As figure 12.2B shows, the star will have a slightly different position compared to the background of stars as seen from the two points. The amount by which the star's apparent position changes depends on its distance from us. The change is larger for nearby stars than for remote stars, but for all stars it is extremely small—so small that it is measured not in degrees but in fractions of a degree called "arc seconds."*

For convenience, astronomers define a star's parallax, $p$, not by the angle its position appears to shift, but by half that angle (see fig. 12.2C). With that definition for parallax, the star's distance, $d$, is simply $1/p$ if we measure $p$ in arc seconds and $d$, not in kilometers or light-years, but in a new unit called "parsecs" (abbreviated pc). That is,

$$d_{pc} = \frac{1}{p_{\text{arc seconds}}}$$

With this choice of units, one **parsec** equals 3.26 light-years ($3.09 \times 10^{13}$ kilometers). The word *parsec* comes from a combination of *par*allax and "arc *sec*ond."

Keep in mind that the change in angle we are talking about here is *very* small. A shift of one second of arc is equivalent to looking at one edge of a U.S. penny 4 km (2.5 miles) away and then looking at its opposite edge.

*One arc second is $\frac{1}{3600}$ of a degree because an arc second is $\frac{1}{60}$ of an arc minute, which is $\frac{1}{60}$ of a degree.

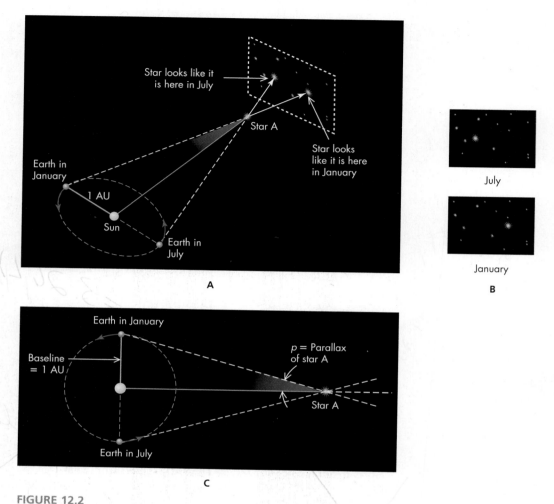

**FIGURE 12.2**

(A) Triangulation to measure a star's distance. The radius of the Earth's orbit is the baseline. (B) As the Earth moves around the Sun, the star's position changes as seen against background stars. (C) Parallax is defined as one-half the angle by which the star's position shifts. Sizes of bodies and their separation are exaggerated for clarity.

To determine a star's distance, we measure its parallax, $p$, and use $d_{pc} = 1/p_{\text{arc seconds}}$. For example, suppose from the shift in position of a nearby star we find its parallax is 0.25 arc seconds. Its distance is then $d = \frac{1}{0.25} = 4$ parsecs. Similarly, a star whose parallax is 0.1 arc second is $\frac{1}{0.1} = 10$ parsecs from the Sun. From this technique, astronomers have discovered that at present the nearest star is Proxima Centauri (See "Looking Up at Centaurus and Crux, The Southern Cross."), which lies 1.3 parsecs (4.3 light-years) from the Sun.* This spacing (about 1 parsec) is typical for stars near the Sun.

Although the parallax–distance relation is mathematically a very simple formula, measuring a star's parallax to use in the formula is very difficult because the angle by which the star shifts is extremely small. It was not until the 1830s that the first parallax was measured by the German astronomer Friedrich Bessel at Königsberg Observatory (now in Kaliningrad). Even now, the method fails for most stars farther away than about 100 parsecs because the Earth's atmosphere blurs the tiny angle of their shift, making it unmeasurable.† Astronomers can avoid such blurring effects by

---

*Proxima Centauri has two companions only slightly more distant. The bigger companion, Alpha Centauri A, is a bright star visible in the southern sky.

†With electronic techniques, astronomers can measure, from the ground, distances to some bright stars as far away as 250 parsecs.

# LOOKING UP AT...
# CENTAURUS AND CRUX,
# THE SOUTHERN CROSS

These constellations are best observed from the southern hemisphere. Northern hemisphere viewers can see Centaurus low in the southern sky in May – July. Crux may be seen just above the southern horizon in May and June from the extreme southern US (Key West and South Texas).

## Centaurus A

This unusual galaxy, ~11 million ly distant, is one of the brightest radio sources in the sky

~50,000 ly

## Alpha Centauri

## Proxima Centauri

This dim star is the nearest star to the Sun, 4.22 ly distant

## Crux
## The Southern Cross

~50 ly

## The Jewel Box

NGC 4755, an open star cluster (Unit 69) ~500 ly from us.

## The Coal Sack

An interstellar dust cloud

## Eta Carinae

A very high-mass star doomed to die young ~8000 ly distant

### Southern Cross in 3-D

352 ly
321 ly
228 ly
363 ly
88 ly

To Earth

1 light year (ly) ≈ 10 trillion km ≈ 6 trillion miles

~200 ly

## Omega Centauri

The largest globular cluster in the Milky Way, ~17,000 ly distant and containing millions of stars

# EXTENDING OUR REACH

## MEASURING THE DISTANCE TO SIRIUS

We can measure the distance to Sirius (or any nearby star) using a geometric construction similar to the one we used in chapter 1 to relate an object's true diameter to its angular diameter and distance. We begin by measuring the star's apparent shift in the sky resulting from our planet's motion around the Sun, as illustrated in figure 12.2A. Let $p$ be half that angle, as shown in figure 12.2C and box figure 12.1, below. To determine the star's distance from the parallax angle, $p$, we construct the triangle $ACS$. The line $AC$ is the radius of the Earth's orbit, with point $A$ chosen so that an imaginary line from Sirius ($S$) to the Sun, $C$, will be perpendicular to $AC$. This makes $ACS$ a right triangle, with angle $C$ equal to 90° and the side $AC$—the distance from the Earth to the Sun—1 astronomical unit. To calculate Sirius's distance from the Sun, $d$, we draw a circle centered on Sirius and passing through the Earth and the Sun. We call the radius of the circle $d$, to stand for Sirius's distance from the Sun. We next form the following proportion: $AC$, the radius of the Earth's orbit, is to the circumference of the circle as $p$ is to the total number of degrees around the circle, which we know is 360. That is,

$$\frac{AC}{2\pi d} = \frac{p}{360}.$$

If we measure $d$ and $AC$ in astronomical units, we can set $AC = 1$. Then, solving for $d$, we obtain

$$d = \frac{360}{2\pi p} \text{ AU}$$

This equation takes on a much simpler form if we express $p$ in arc seconds rather than degrees and $d$ in parsecs (as defined above) rather than in AU. With these new units, the factors 360 and $2\pi$ disappear, leaving $d = 1/p$ as we claimed earlier. (In the above, we assume that Sirius itself has not changed its location significantly compared to its distance in so brief a time as 6 months—a very good assumption.)

Measurements show that for Sirius, the angle $p$ is 0.377 arc seconds. Thus, its distance is $1/0.377 = 2.65$ parsecs. To express this in light-years, we multiply by 3.26 (the number of light-years in a parsec) to get Sirius's distance as about 8.6 light-years.

$$\frac{AC}{\text{Circumference}} = \frac{p}{360}$$

$$\text{Circumference} = 2\pi \times \text{Radius} = 2\pi d$$

$$\frac{AC}{2\pi d} = \frac{p}{360}$$

Cross multiply to get

$$\frac{360\,AC}{2\pi p} = d$$

or

$$d = \frac{360\,AC}{2\pi p}$$

**BOX FIGURE 12.1**
How to determine the relation between a star's distance and its parallax.

observing from above the atmosphere, and an orbiting satellite, *Hipparcos,* has done just that: making parallax measurements from space. With its data, astronomers are able to accurately measure distances to stars as far away as 250 parsecs. For more remote stars, they must use a different method, one based on how bright objects look to us.

## Measuring Distance by the Standard-Candles Method

If you look at an object of known brightness, you can estimate its distance from how bright it appears. For example, if you look at two 100-watt light bulbs, one close and one far away, you can tell fairly accurately how much farther away the dim one is. In fact, if you drive at night, your life depends on making such distance estimates when you see traffic lights or oncoming cars. Astronomers call such distance measurements the **method of standard candles** and use a similar but more refined version of it to find the distance to stars and galaxies. But to use this method to measure an object's distance, astronomers must first determine its true brightness. So until we learn ways to make such brightness determinations, we must set this method aside, but we will return to it later in the chapter.

## 12.2   MEASURING THE PROPERTIES OF STARS FROM THEIR LIGHT

If we were studying flowers or butterflies, we would want to know something about their appearance, size, shape, colors, and structure. So, too, astronomers want to know the sizes, colors, and structure of stars. Such knowledge not only helps us better understand the nature of stars but also is vital in unraveling their life story. Determining a star's physical properties is not easy, however, because we cannot directly probe it. But by analyzing a star's light, astronomers can deduce many of its properties, such as its temperature, composition, radius, mass, and motions.

### Temperature

Stars are extremely hot by earthly standards. The surface temperature of even cool stars is far above the temperatures at which most substances vaporize, and so using a physical probe to take a star's temperature would not succeed, even if we had the technology to send the probe to the star. So, if astronomers want to know how hot a star is, they must once more rely on indirect methods. Yet the method used is familiar. You use it yourself in judging the temperature of an electric stove burner.

An object's temperature can often be deduced from the color of its emitted light. As we saw in chapter 3, hotter objects emit more blue light than red. Thus, hot objects tend to glow blue and cooler ones red. You can see such color differences if you look carefully at stars in the night sky. Some, such as Rigel in Orion, have a blue tint. Others, such as Antares in Scorpius, are obviously reddish. Thus, even our naked eye can tell us that stars differ in temperature.

We can use color in a more precise way to measure a star's temperature with Wien's law, which we discussed in chapter 3. It states that an object's temperature, $T$, in Kelvin is given by the following:

$T$ = Temperature of star
$\lambda_m$ = Strongest emitted wavelength of
    starlight

$$T = \frac{3 \times 10^6}{\lambda_m}$$

**FIGURE 12.3**
Measuring a star's surface temperature using Wien's law.

where $\lambda_m$ is the wavelength in nanometers (nm) at which it radiates most strongly.* Thus, the longer the wavelength of the maximum emitted energy, the lower is the temperature of the radiating object.

For an illustration of how we can use Wien's law to measure a star's temperature, suppose we pick Sirius, a bright nearby star. To determine the wavelength at which it radiates most strongly, we need to measure its brightness at many different wavelengths, as illustrated in figure 12.3. We find that the strongest emission is at 300 nanometers, in the ultraviolet part of the spectrum. That is, $\lambda_m = 300$ nanometers. Inserting this value in Wien's law, we find

$$T = \frac{3 \times 10^6}{300} = 10,000 \text{ K}$$

For another example, we pick the red star Betelgeuse, which radiates most strongly at about 1000 nanometers. Its temperature is therefore about

$$\frac{3 \times 10^6}{1000} = \frac{3 \times 10^6}{10^3} = 3 \times 10^{6-3} = 3 \times 10^3 = 3000 \text{ K}$$

A star's temperature is only one of several properties we can deduce from its light. For example, in some cases we can also measure the star's radius, but before we discuss how, we need to discuss briefly another general property of stars and other hot objects—the amount of energy they radiate.

## Luminosity

Astronomers call the amount of energy an object radiates each second its **luminosity** (abbreviated as $L$). An everyday example of luminosity is the wattage of a light bulb: a typical table lamp has a luminosity of 100 watts, whereas a bulb for an outdoor parking lot light may have a luminosity of 1500 watts. Stars, of course, are enormously more luminous. For example, the Sun has a luminosity of about $4 \times 10^{26}$ watts, which it obtains by "burning" its hydrogen into helium. Thus, a star's luminosity measures how fast it consumes its fuel, a vital quantity for determining its lifetime. Knowing a star's luminosity is also important because from it astronomers can measure a star's radius and distance. But to understand how such measurements are made, we must first discuss a relation between how bright an object appears, how bright it really is (its luminosity), and its distance—a relation known as the *inverse-square law*.

*The subscript $m$ on $\lambda$ is to remind us that it is the wavelength at which the star's emitted energy is at a *maximum*.

## The Inverse-Square Law and Measuring a Star's Luminosity

The **inverse-square law** relates an object's luminosity to its distance and its apparent brightness, that is, how bright it looks to us. We all know that a light looks brighter when we are close to it than when we are farther from it. The inverse-square law puts that everyday experience into a mathematical form, describing how light energy spreads out from a source such as a star. As the light travels outward, it moves in straight lines, spreading its energy uniformly in all directions, as shown in figure 12.4A. Near the source, the light will have spread only a little, and so more enters an observer's eye, making it look brighter than if the observer were far away. That is, the farther away from us a source of light is, the less of its light enters our eyes. As a result, more-distant objects look dimmer.

This decrease in brightness with increasing distance can also be understood if you think of light as photons. Photons leaving a light source such as a star spread evenly in all directions. Now imagine a series of progressively larger spheres drawn around the source. If nothing absorbs the light, the same number of photons pass through each sphere. But because more-distant spheres are larger, the number of photons passing through any given *area* on any one sphere grows smaller as the spheres become more distant and larger, as shown in figure 12.4B. Similarly, as you move away from a light source, the photons reaching you are spread more widely—therefore fewer enter your eyes, and the source appears dimmer. Thus, a more distant source is dimmer because more of its light has spread along lines that never reach you.

*The inverse-square law*

ANIMATION

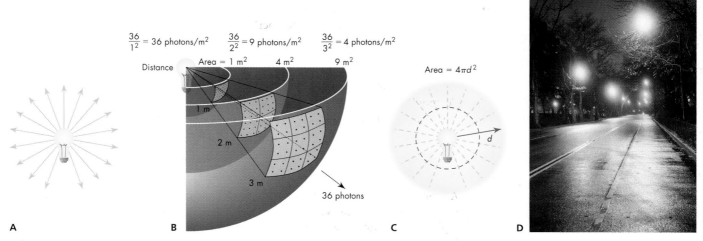

$\frac{36}{1^2} = 36$ photons/m$^2$    $\frac{36}{2^2} = 9$ photons/m$^2$    $\frac{36}{3^2} = 4$ photons/m$^2$

Area = 1 m$^2$    4 m$^2$    9 m$^2$

Distance

Area = $4\pi d^2$

1 m

2 m

3 m

36 photons

A    B    C    D

**FIGURE 12.4**
The inverse-square law. (A) Light spreads out from a point source in all directions. (B) As photons move out from a source, they are spread over a progressively larger area as the distance from the source increases. Thus, a *given* area intercepts fewer photons the farther it is from the source. (C) The area over which light a distance *d* from the source is spread is $4\pi d^2$. (D) You can see the inverse-square law at work in this picture of street lights.

 Count the "photons" (dots) in a small square at each distance to verify the inverse-square law.

 In the inner sphere, there are 36 dots in a small square (although this is hard to see). In the next sphere there are 9 dots, and in the outermost sphere there are 4 dots. Nine is one-fourth of 36, and 4 is one-ninth of 36. Thus, the number of spots drops as the square of the distance.

We can use this argument to show the following: at a distance $d$ from a light source, the source's luminosity, $L$, has spread over a sphere whose radius is $d$, as shown in figure 12.4C. (We use $d$ to stand for radius here to emphasize that we are finding a distance.) Any sphere has a surface area given by $4\pi d^2$, and so the brightness we observe, $B$, is just

$$B = \frac{L}{4\pi d^2}$$

**This relationship is called the inverse-square law because distance appears in the denominator as a square.**

The inverse-square law is one of the most useful mathematical tools available to astronomers not only for measuring a star's luminosity but also its distance. To find a star's luminosity, astronomers measure its distance, $d$, by parallax, as described in section 12.1. Next, with a photometer, a device similar to the electric exposure meter in a camera, they measure how bright the star appears from Earth, $B$. Finally, with $B$ and $d$ known, they calculate the star's luminosity, $L$, using the inverse-square law. Such measurements reveal that the average star has a luminosity similar to that of our Sun. However, astronomers also find some stars that are millions of times more luminous than the Sun and other stars that are thousands of times less luminous than the Sun. Much of this range in luminosity comes from the great range in stellar radii. That is, some stars are vastly larger than others, as we will now show.

## Radius

Common sense tells us that if we have two objects of the same temperature but of different sizes, the larger one will emit more energy than the smaller one. For example, three glowing charcoal briquettes in a barbecue emit more energy than just one briquette at the same temperature. Similarly,

*if two stars have the same temperature but one is more luminous than the other, the brighter star must have a larger surface area, therefore a larger radius than the dimmer star.*

Thus, if we know a star's temperature, we can infer its size from the amount of energy it radiates. To calculate the star's radius, however, we need a mathematical relation between luminosity, temperature, and radius—a relation known as the "Stefan-Boltzmann law."

## The Stefan-Boltzmann Law

Imagine watching an electric stove burner heat up. When the burner is on low and is relatively cool, it glows dimly red and gives off only a slight amount of heat. When the burner is on high and is very hot, it glows bright yellow-orange and gives off far more heat. Thus, you can both see and feel that raising a body's temperature increases the amount of radiation it emits per second—that is, its luminosity. This familiar situation is an example of a law deduced in the late 1800s by two German scientists, Josef Stefan and Ludwig Boltzmann, who showed that the luminosity of a hot object depends on its temperature.

The Stefan-Boltzmann law, as their discovery is now called, states that an object of temperature $T$ radiates an amount of energy each second equal to $\sigma T^4$ per square meter, as shown in figure 12.5A. The quantity $\sigma$ is called the "Stefan-Boltzmann constant," and its value is $5.67 \times 10^{-8}$ watts m$^{-2}$ K$^{-4}$. The Stefan-Boltzmann law affords a mathematical explanation of what we noticed for the electric stove burner. That is, a hotter burner has a larger $T$, and therefore, according to the law, the burner radiates

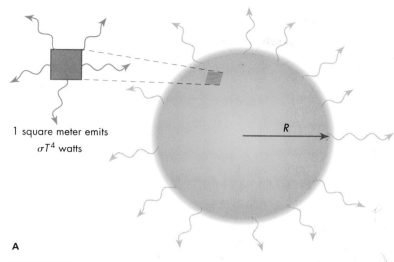

Total energy radiated per second
by the star is its

Luminosity $= L$

$L =$ Energy emitted by 1 square meter
      $\times$ Number of square meters of its surface

$= \sigma T^4 \times$ star's surface area

For a spherical star of radius $R$,
the surface area is $4\pi R^2$

Thus, $L = \sigma T^4 \times 4\pi R^2$

or

$L = 4\pi R^2 \sigma T^4$

1 square meter emits
$\sigma T^4$ watts

$R$

**A**

**B**

**FIGURE 12.5**

The Stefan-Boltzmann law can be used to find a star's radius. (A) Each part of the star's surface radiates $\sigma T^4$. (B) Multiplying $\sigma T^4$ by the star's surface area ($4\pi R^2$) gives its total power output—its luminosity, $L = 4\pi R^2 \sigma T^4$. To find the star's radius, we solve this equation to get $R = \sqrt{L/(4\pi\sigma T^4)}$. Finally, we can use this equation and measured values of the star's $L$ and $T$ to determine $R$.

more strongly. Similarly, the Stefan-Boltzmann law is the basis for dimmer switches on lights: changing the amount of electricity going to the lamp changes its temperature and thereby its brightness. However, the Stefan-Boltzmann law does not apply to all hot objects. For example, it does not accurately describe the radiation from hot, low-density gas, such as that in fluorescent light bulbs or interstellar clouds.

We can apply the Stefan-Boltzmann law to determine a star's luminosity as follows. According to the law, if a star has a temperature $T$, each square meter of its surface radiates an amount of energy per second given by $\sigma T^4$. We can find the total energy the star radiates per second—its luminosity, $L$—by multiplying the energy radiated from 1 square meter ($\sigma T^4$) by the number of square meters of its surface area (fig. 12.5B). If we assume that the star is a sphere, its area is $4\pi R^2$, where $R$ is its radius. Its luminosity, $L$, is therefore

$$L = 4\pi R^2\ \sigma T^4$$

**That is, a star's luminosity equals its surface area times $\sigma T^4$. This relation between $L$, $R$, and $T$, may at first appear complex, but its content is very simple: increasing either the temperature or the radius of a star makes it more luminous. Making $T$ larger makes each square meter of the star brighter. Making $R$ larger increases the number of square meters.**

In the above expression, $L$ is measured in watts, but the "wattage" of stars is so enormous that it is more convenient to use the Sun's luminosity as a standard unit. Likewise, it is easier to use the Sun's radius as a standard size unit rather than meters or even kilometers. If we need to convert to watts or meters, we simply remember that one solar luminosity, $L_\odot$, is $4 \times 10^{26}$ watts and one solar radius, $R_\odot$, is $7 \times 10^5$ kilometers.

Because a star's luminosity depends on its radius and temperature, if we know its luminosity and temperature, we can find its radius, $R$. The method works because if we have a mathematical relation between three quantities (in this case $L$, $T$, and $R$) we can find any one of these given the other two. Thus, if we know a star's luminosity and its temperature, we can use the Stefan-Boltzmann law to solve for its radius. Extending Our Reach: "Measuring the Radius of the Star Sirius" explains how.

# EXTENDING OUR REACH

## MEASURING THE RADIUS OF THE STAR SIRIUS

From its color (the wavelength at which it radiates most strongly), we found that the temperature ($T_s$) of Sirius is about 10,000 K. From the amount of energy we receive from Sirius and its distance, which can be found by parallax, we can find that its luminosity ($L_s$) is about 25 $L_\odot$. We solve for the radius of Sirius ($R_s$) as follows. First write down the Stefan-Boltzmann law for Sirius to get $L_s = 4\pi R_s^2\, \sigma T_s^4$. Then write down the same relation for the Sun: $L_\odot = 4\pi R_\odot^2\, \sigma T_\odot^4$.

Next, we divide the expression for Sirius by the expression for the Sun to get

$$\frac{L_s}{L_\odot} = \frac{4\pi R_s^2\, \sigma T_s^4}{4\pi R_\odot^2\, \sigma T_\odot^4}$$

We can simplify this expression by canceling the identical $4\pi$ and $\sigma$ factors to get

$$\frac{L_s}{L_\odot} = \frac{R_s^2\, T_s^4}{R_\odot^2\, T_\odot^4}$$

We next collect the R's and T's as separate factors, giving us

$$L_s/L_\odot = (R_s/R_\odot)^2\, (T_s/T_\odot)^4$$

We now solve this expression for $(R_s/R_\odot)^2$ to get

$$(R_s/R_\odot)^2 = (L_s/L_\odot)(T_\odot/T_s)^4$$

Finally, we take the square root of both sides to get

$$R_s/R_\odot = (L_s/L_\odot)^{1/2}\, (T_\odot/T_s)^2$$

We can now evaluate $R_s/R_\odot$ by inserting the values for the luminosity and temperature of Sirius and the Sun. Notice that $T_\odot = 6000$ K, and by definition of our units for luminosity, $L_\odot = 1$. This gives us

$$\frac{R_s}{R_\odot} = \left(\frac{25}{1}\right)^{1/2} \left(\frac{6000\ \text{K}}{10,000\ \text{K}}\right)^2 = 5(0.6)^2 = 1.8$$

That is, the radius of Sirius* is 1.8 $R_\odot$, a little less than twice the Sun's radius.

---

*In classical times, the Persians, Greeks, and Romans sometimes called Sirius the Dog Star. It is the brightest star in the constellation Canis Major, the Big Dog, and was associated by many early people with misfortune and fevers. The ancient Greeks and Romans also blamed Sirius for the extreme heat of July and August, because at that time of year, it rose at about the same time as the Sun. They therefore believed it added its brilliance to the Sun's, making the season extra warm. The heat we receive from Sirius is, of course, negligible, but, like the ancient Greeks and Romans, we still refer to the hot days of late summer when the Dog Star shines with the Sun as the "dog days."

Astronomers can also measure a star's radius from its angular size, the technique we used in chapter 11 to measure the radius of the Sun. Unfortunately, the angular size of all stars except the Sun is extremely tiny because they are so far from the Earth, and so even in powerful telescopes stars generally look like a smeary spot of light. That smearing is caused by the blurring effects of our atmosphere and by a physical limitation of telescopes called "diffraction," which we discussed briefly in chapter 4, "Telescopes."

Diffraction limits a telescope's ability to measure tiny angular size, hopelessly blurring the light by an amount that depends on the diameter of the telescope's lens or mirror. Although such diffraction effects are less severe on bigger diameter telescopes, to measure the angular size of most stars, a truly immense diameter telescope is needed. For example, to measure the angular size of a star like the Sun if it were 50 light-years away would require a telescope 300 meters in diameter, about three times the size of a football field. To avoid the need for such enormous (and expensive) telescopes, astronomers have therefore devised an alternative way to measure the angular size of stars by using not one huge telescope but two (or more) smaller ones separated by a large distance (for example, several hundred meters). Such a device is called an "interferometer" (discussed more fully in chapter 4), and its ability to measure angular sizes is equivalent to that of a single telescope whose diameter is equal to the distance that separates the two smaller ones. (Of course, a single enormous telescope would gather more light and enable astronomers to measure the sizes of very faint stars, but it may never be possible to build such a huge instrument.) A computer then combines the information from the two telescopes to give a crude picture of the star. Such interferometric observations are still hampered by blurring caused by our atmosphere. Those effects can be partially offset by a technique called "speckle

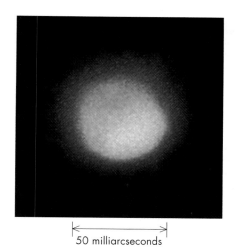

50 milliarcseconds

**FIGURE 12.6**

Image of the star Betelgeuse made by interferometry. The lack of detail in this image reflects the current difficulty in observing the disk of any star but our Sun. One milliarcsecond (mas) is approximately $2.8 \times 10^{-7}$ degrees.

⟲* Use this image of Betelgeuse to measure its diameter. You will need the formula developed in the box in chapter 1, section 1.2, describing how to determine the diameter of a body from its angular size and distance. Betelgeuse is about $6 \times 10^{15}$ kilometers away (200 parsecs).

⟲* An angle of 1 arc second spans 1 AU at 1 parsec (by definition of the parsec). If Betelegeuse is 200 parsecs away, 1 arc second will span 200 AU. (An object's linear size is proportional to its angular size times its distance). An angle of 50 mas (milliarcseconds) is 50/1000 = 1/20th of an arc second. Thus, Betelgeuse's diameter is 1/20 of 200 AU = 10 AU.

interferometry" in which a high-speed camera takes many very short exposures of the star's light. A computer then combines the separate short exposures into a "deblurred" image, such as that of Betelgeuse, illustrated in figure 12.6. With interferometers, astronomers can measure the radius of several dozen nearby stars and a hundred or so more-distant giant stars.

The Stefan-Boltzmann law and interferometer observations show that stars differ enormously in radius. Although most stars have approximately the same radius as the Sun has, some are hundreds of times larger, and so astronomers call them **giants.** Smaller stars (our Sun included) are called **dwarfs.**

In using the Stefan-Boltzmann law to measure a star's radius, we have seen that $L$ can be measured either in watts or in solar units for $L$. Astronomers sometimes use a different set of units, however, to measure stellar brightness.

## The Magnitude System

About 150 B.C., the ancient Greek astronomer Hipparchus measured the apparent brightness of stars in the night sky using units he called **magnitudes.** He designated the stars that looked brightest as magnitude 1 and the dimmest ones (the ones he could just barely see) as magnitude 6. For example, Betelgeuse, a bright red star in the constellation Orion, is magnitude 1, while the somewhat dimmer stars in the Big Dipper's handle are approximately magnitude 2. Astronomers still use this scheme to measure the brightness of astronomical objects, but they now use the term *apparent magnitude* to emphasize that they are measuring how bright a star *looks* to an observer. A star's apparent magnitude depends on its luminosity and its distance; it may look dim either because it has a small luminosity—it does not emit much energy—or because it is very far away.

Astronomers use the magnitude system for many purposes (for example, to indicate the brightness of stars on star charts), but it has several confusing properties. First, the scale is "backward" in the sense that bright stars have small magnitudes, while dim stars have large magnitudes. Moreover, modern measurements show that Hipparchus underestimated the magnitudes of the brightest stars, and so the magnitudes now assigned them are negative numbers.

Second, magnitude *differences* correspond to brightness *ratios*. That is, if we measure the brightness of a first-magnitude star and a sixth-magnitude star, the former is 100 times brighter than the latter. Thus, a *difference* of 5 magnitudes corresponds to a brightness *ratio* of 100; so, when we say a star is 5 magnitudes brighter than another, we mean it is a *factor* of 100 brighter. Each magnitude difference corresponds to a factor of 2.512 . . .—the fifth root of 100—in brightness. Thus, a first-magnitude star is 2.512 times brighter than a second-magnitude star and is $2.512 \times 2.512$, or 6.310, times brighter than a third-magnitude star. Table 12.1 lists the ratios that correspond to various differences in magnitude. For example, let us compare the apparent

| TABLE 12.1 | |
|---|---|
| **Relating Magnitudes to Brightness Ratios** | |
| **Magnitude Difference** | **Ratio of Brightness** |
| 1 | 2.512:1 |
| 2 | $2.512^2 = 6.31$:1 |
| 3 | $2.512^3 = 15.85$:1 |
| 4 | $2.512^4 = 39.8$:1 |
| 5 | $2.512^5 = 100$:1 |
| $\vdots$ | $\vdots$ |
| 10 | $2.512^{10} = 10^4$:1 |

**TABLE 12.2**

**Relating Absolute Magnitude to Luminosity**

| Absolute Magnitude | Approximate Luminosity in Solar Units |
|---|---|
| −5 | 10,000 |
| 0 | 100 |
| 5 | 1 |
| 10 | 0.01 |

brightness of the planet Venus with the star Aldebaran. At its brightest, Venus has an apparent magnitude of −4.2; Aldebaran's apparent magnitude is 0.8, and so the difference in their magnitudes is 0.8 − (−4.2) = 5.0. Therefore we see from table 12.1 that Venus is 100 times brighter to our eye than Aldebaran.

Third, astronomers often use a quantity called "absolute magnitude" to measure a star's luminosity. Recall that a star's apparent magnitude is how bright it *looks* to an observer. But the apparent magnitude of a star depends on the distance from which we observe it. The same star will have one apparent magnitude if it is near to us but a different apparent magnitude if it is far from us. Astronomers, therefore find it useful to have a way to describe a star's brightness that does not depend on its distance. One way to do that is to imagine how bright a star would look if we were to observe it from some *standard* distance. Astronomers have chosen 10 parsecs as that standard distance, and they call the apparent magnitude of a star seen from 10 pc its *absolute magnitude*. Because the absolute magnitude does not depend on distance, it is a measure of a star's true brightness. In other words, absolute magnitude is a measure of a star's luminosity. Table 12.2 illustrates how absolute magnitude is related to a star's luminosity.

## 12.3 SPECTRA OF STARS

A star's spectrum depicts the energy it emits at each wavelength and is perhaps the single most important thing we can know about the star. From the spectrum we can find the star's composition, temperature, luminosity, velocity in space, rotation speed, and some other properties as well. For example, under some circumstances, we may also deduce the star's mass and radius.

Figure 12.7 shows the spectra of three stars. You can easily see differences between them. Spectrum B is from a star similar to the Sun; spectrum A is from a star hotter than the Sun; and spectrum C is from a star cooler than the Sun. Spectrum A has only a few lines, but they are very strong and their spacing follows a regular pattern. Spectrum C, however, shows a welter of lines with no apparent regularity. Understanding such differences and what they tell us about a star is one of the goals of studying stellar spectra.

INTERACTIVE

*Stellar spectroscopy*

**FIGURE 12.7**

Spectra of three stars. A is hotter than the Sun. B is similar to the Sun. C is cooler than the Sun.

**FIGURE 12.8**

Formation of stellar absorption lines. Atoms in the cooler atmospheric gas absorb radiation at wavelengths corresponding to jumps between electron orbits.

## Measuring a Star's Composition

As light moves from a star's core through the gas in its surface layers, atoms there absorb the radiation at some wavelengths, creating dark absorption lines in the star's spectrum, as shown in figure 12.8. Each type of atom—hydrogen, helium, calcium, and so on—absorbs at a unique set of wavelengths. For example, hydrogen absorbs at 656, 486, and 434 nanometers, in the red, blue, and violet part of the spectrum, respectively. Gaseous calcium, on the other hand, absorbs strongly at 393.3 and 396.8 nanometers, producing a strong double line in the violet. Because each atom absorbs a unique combination of wavelengths of light, each has a unique set of absorption lines. From such absorption lines, we can determine what a star is made of.

To measure a star's composition, we make a spectrum of its light and then compare the absorption lines we see with tables that list the lines made by each atom. When we find a match between an absorption line in the star and a line in the table, we can deduce that the element listed exists in the star. To find the quantity of each atom in the star—each element's abundance—we use the darkness of the absorption line, as discussed in chapter 3. Such determinations of the identity and quantity of elements present are difficult, however, because even though an element may be present in the star, the element's atoms may, because of temperature effects, be unable to absorb light and make a spectral line.

## How Temperature Affects a Star's Spectrum

To see why temperature affects a star's spectrum, recall that light is absorbed when its energy matches the energy difference between two electron orbits, as we discussed in chapter 3. For an atom to absorb light, its electrons must therefore be in the proper orbit, or, more technically, energy level. An atom may be abundant in a star's atmosphere and create only very weak lines at a particular wavelength simply because the gas is either too hot or too cold, so that its electrons are in the "wrong" level to absorb light at that wavelength. Hydrogen illustrates this situation dramatically.

The absorption lines of hydrogen that we see at visible wavelengths are made by electrons orbiting in its second level. These lines are sometimes called the "Balmer lines"* to distinguish them from other hydrogen lines in the ultraviolet and infrared wavelengths. Balmer lines occur at wavelengths where light has exactly the amount

---

*Balmer lines are named for Johann Balmer, the scientist who first studied their pattern.

of energy needed to lift an electron from hydrogen's second energy level to the third level, or a higher level. These lines are especially important because their wavelengths are in the visible spectrum and are therefore easily observable. If the hydrogen atoms in a star have no electrons orbiting in level 2, no Balmer absorption lines will appear even though hydrogen may be the most abundant element in the star.

Hydrogen Balmer lines appear weakly or not at all if a star is either very cold or very hot. In a cool star, most hydrogen atoms have their electrons in level 1 and so they cannot absorb Balmer line radiation, which can be absorbed only if the electron is in level 2. In a hot star the atoms move faster, and when they collide, electrons may be excited ("knocked") into higher orbits: the hotter the star, the higher the orbit. Moreover, in very hot stars, the radiation is so energetic that the photons can knock electrons out of the atom, in which case the atom is said to be ionized. As a result of this excitation, proportionally more of the electrons in a very hot star will be in level 3 or higher. Only if the hydrogen has a temperature between about 8000 K and 15,000 K will enough atoms have their electron in level 2 to make strong hydrogen Balmer lines. If we are therefore to deduce correctly the abundance of elements in a star, we must correct for such temperature effects. With these corrections, we discover that virtually all stars are composed mainly of hydrogen: their composition is similar to that of our Sun—about 71% H, 27% He, and a 2% mix of the remaining elements. But despite their uniform composition, stars exhibit a wide range in the appearance of their spectra, as astronomers noted when they began to study stellar spectra.

## Classification of Stellar Spectra

Stellar spectroscopy, the study and classification of spectra, was born early in the 19th century when the German scientist Joseph Fraunhofer discovered dark lines in the spectrum of the Sun. He later observed similar lines in the spectra of stars and noted that different stars had different patterns of lines. In 1866, Pietro Angelo Secchi, an Italian priest and scientist, noticed that the line patterns depended on the star's color. He assigned stars to four color classes—white, yellow, red, and deep red—and considered these classes as evidence for stellar evolution—hot stars cooling from white heat to deep red as they aged.

Technological improvements have allowed astronomers to refine Secchi's system further. The first such improvement came in 1872, when Henry Draper,* a physician and amateur astronomer, recorded spectra on photographs. On Draper's death, his widow endowed a project at Harvard to create a compilation of stellar spectra, known as the Henry Draper Catalog.

The use of photography in compiling the Draper Catalog made it possible to obtain many more spectra and with better details than could be observed visually. In fact, details were so good that E. C. Pickering, the astronomer in charge of the project, began to use letters to subdivide the spectral classes, assigning the letters *A* to *D* to Secchi's white class, *E* to *L* to the yellow class, and *M* and *N* to the red classes.

About 1901, Annie Jump Cannon (fig. 12.9), the astronomer actually doing the classification for the Draper Catalog, discovered that the classes fell in more orderly sequence of appearance if rearranged by temperature. She therefore reordered Secchi's classes to obtain the sequence *O, B, A, F, G, K,* and *M,* with *O* stars being the hottest and *M* stars being the coolest. Her work is the basis for the stellar **spectral classes** we use today.

Stellar spectral classes are based on the appearance of the spectrum. For example, *A*-type stars show extremely strong hydrogen lines, while *B*-type stars show helium and weak hydrogen lines. This scheme, however, despite its wide adoption, lacked a physical

**FIGURE 12.9**
Annie Jump Cannon.

---

*Draper's father was a professor of chemistry and physiology, and the first person we know of to photograph an astronomical object—the Moon.

## SCIENCE AT WORK

### NEW SPECTRUM CLASSES

Since the establishment of the spectrum classes *O* through *M*, astronomers, aided by new technology, have made further refinements in classifying stellar spectra. For example, much classification is now done automatically by computers that scan a star's spectrum and match it against standard spectra stored in memory. Other improvements come from more sensitive infrared detectors that allow astronomers to detect dim stars far cooler than any previously known. Some of these stars are so cool that not only do molecules form in their atmospheres but solid dust particles condense there as well. With atmospheres so different from those of ordinary stars, these extremely cool objects have spectra that are also radically different. Therefore, astronomers have devised new spectrum classes—*L* and *T*—to describe these cool, dim objects. For example, *L* stars show strong molecular lines of iron hydride and chromium hydride, while *T* stars show strong absorption lines of methane.*

---

*Can you think up a mnemonic for the spectral classes that includes these new classes?

**FIGURE 12.10**
Cecilia Payne.

basis in that it employed the appearance of the spectra rather than the physical properties of the stars that produced them. For example, astronomers at that time did not know what made the spectra of *A* and *F* stars differ. That understanding came in the 1920s from the work of the American astronomer Cecilia Payne (later Payne-Gaposhkin) (fig. 12.10), who explained why the strength of the hydrogen lines depended on the star's temperature, as we described above. Her discovery was based on work by the Indian astronomer M. Saha, who showed how to calculate the level in which an atom's electrons were most likely to be found. Saha showed that the levels occupied by electrons—a quantity crucial for interpreting the strength of spectral lines—depend on the star's temperature and density. With Saha's equation, Payne then showed mathematically the correctness and reasonableness of Cannon's order for the spectral classes. She thereby demonstrated that a star's spectral class is determined mainly by its temperature.*

Payne's theory unfortunately left astronomers with Cannon's odd nonalphabetical progression for the spectral classes, which—from hot to cold—ran *O*, *B*, *A*, *F*, *G*, *K*, and *M*. So much effort, however, had been invested in classifying stars using this system that it was easier to keep the classes as assigned with their odd order than to reclassify them (Cannon, in her life, classified some quarter million stars). As a help to remember the peculiar order, astronomers have devised mnemonics. One of the first is "Oh, be a fine girl/guy. Kiss me." A more violent version is "Oh, big and furry gorilla, kill my roommate."† A third version is "Only brilliant, artistic females generate killer mnemonics." Choose whichever appeals most, or make up your own version, to help you learn this important sequence.

### Definition of the Spectral Classes

A star's spectral class is determined by the lines in its spectrum. Figure 12.11 illustrates the seven main types—*O* through *M*—and the differences in the line patterns are easy to see even without knowing the identity of the lines. However, knowing which element makes

---

*Payne also made clear that stars are composed mainly of hydrogen, as mentioned earlier.

†Devised by an astronomy class at Harvard, the *R* for "roommate" is one of four rare spectral classes: *R*, *N*, *S*, and *W*.

**FIGURE 12.11**

The stellar spectral classes. Notice how the hydrogen line at right is weak in the top spectrum and grows stronger down to *A* stars, but then essentially disappears in *K* and *M* stars. The separate curves to the right of the spectra show how the strength of the hydrogen absorption lines and those of other chemical elements change with the spectrum class.

which line can increase our understanding of the different line patterns. As mentioned earlier, almost all stars are made of the same mix of elements, but differing conditions, especially their surface temperature, result in differing spectra. For example, *O* stars have weak absorption lines of hydrogen but strong absorption lines of helium, the second most abundant element, because they are so hot. At their high temperature, an *O* star's hydrogen atoms collide so violently and are excited so much by the star's intense radiation that the electrons are stripped from most of the hydrogen, ionizing it. With its electron missing, a hydrogen atom cannot absorb light. Because most of the *O* star's hydrogen is ionized, such stars have extremely weak hydrogen absorption lines. Helium atoms are more tightly bound, however, and most retain at least one of their electrons, allowing them to absorb light. Thus, *O* stars have absorption lines of helium but only weak lines of hydrogen.

Stars of spectral class *A* have very strong hydrogen lines. Their temperature is just right to put lots of electrons into orbit 2 of hydrogen, which makes for strong Balmer lines. Balmer lines also appear in *F* stars but are weaker. *F* stars are distinguished by the multitude of lines from metals such as calcium and iron, elements that also appear strongly in *G* and *K* stars. Such elements are present in hotter stars but are ionized and generally create only very weak spectral lines under those conditions. In cooler stars, however, metal lines, particularly ionized calcium, are moderately strong, and hydrogen lines become weak.

In the very cool *K* and *M* stars, hydrogen is almost invisible because its electrons are mostly in level 1 and therefore cannot make Balmer lines. These stars have such cool atmospheres that molecules form, and as we saw in chapter 3, they produce very complex spectra. As a result, *K* and *M* stars have numerous lines from such substances as the carbon compound cyanogen (CN) and the carbon radical methylidyne (CH), as well as gaseous titanium oxide (TiO). The small graph at the right-hand side of figure 12.11 shows how the strength of the absorption lines of hydrogen, calcium, and so on, change with spectrum class. The changing strengths of these lines and others are summarized in table 12.3.

We have seen that *O* stars are hot and *M* stars are cool, but what, in fact, are their temperatures? Application of Wien's law and theoretical calculations based on Saha's law of how electrons are distributed in atomic orbits show that temperatures range from more than 25,000 K for *O* stars to less than 3500 K for *M* stars, with *A* stars being about 10,000 K and *G* stars, such as our Sun, being about 6000 K.

Because a star's spectral class is set by its temperature, its class also indicates its color. We know that hot objects are blue and cool objects are red (recall Wien's law), and so we find that *O* and *B* stars (hot classes) are blue, while *K* and *M* stars (cool classes) are red.

To distinguish finer gradations in temperature, astronomers subdivide each class by adding a numerical suffix—for example, *B*1, *B*2, . . . *B*9—with the smaller numbers indicating higher temperatures. With this system, the temperature of a *B*1 star is about 20,000 K, whereas that of a *B*5 star is about 13,500 K. Similarly, our Sun, rather than being just a *G* star, is a *G*2 star.

**TABLE 12.3**

## Summary of Spectral Classes

| Spectral Class | Temperature Range (K) | Features | Representative Star |
|---|---|---|---|
| O | Hotter than 25,000 | Ionized helium, weak hydrogen | |
| B | 11,000–25,000 | Neutral helium, hydrogen stronger | Rigel |
| A | 7500–11,000 | Hydrogen very strong | Sirius |
| F | 6000–7500 | Hydrogen weaker, metals—especially ionized Ca—moderate | Canopus |
| G | 5000–6000 | Ionized Ca strong, hydrogen even weaker | The Sun |
| K | 3500–5000 | Metals strong, CH and CN molecules appearing | Aldebaran |
| M | 2200–3500 | Molecules strong, especially TiO and water | Betelgeuse |
| L | 1300–2200 | TiO disappears. Strong lines of metal hydrides, water, and reactive metals such as potassium and cesium | |
| T | 900?–1300? | Strong lines of water and methane | |

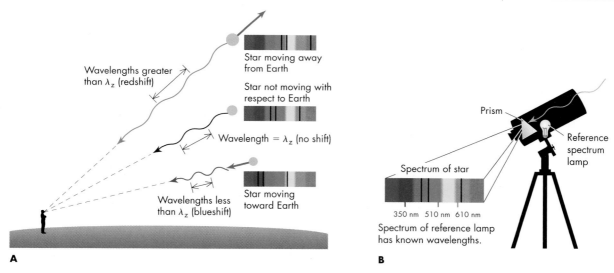

**A**                    **B**

**FIGURE 12.12**

Measuring a star's radial velocity from its Doppler shift. (A) Spectrum lines from a star moving away from Earth are shifted to longer wavelengths (a redshift). Spectrum lines from a star approaching Earth are shifted to shorter wavelengths (a blueshift). (B) A standard lamp attached inside the telescope serves as a comparison spectrum to allow the wavelength shift of the star's light to be found.

## Measuring a Star's Motion

All stars move through space and probably, like our Sun, also spin on their rotation axes. Such motions turn out to be important clues to a star's history and are vital in determining star masses and the structure of galaxies.

We find a star's motion from its spectrum using the Doppler shift, which, as we saw in chapter 3, is the alteration of the wavelength of light from a moving source. If a source moves toward us, its wavelengths are shortened, while if it moves away, they are lengthened, as illustrated in figure 12.12A. The amount of the wavelength shift depends on the source's speed along the line of sight, what astronomers call **radial velocity.** We find the shift by subtracting the wavelength the star would absorb (or emit) if it were not moving relative to the Earth, $\lambda_z$, from the new wavelength, $\lambda$. We will call that difference $\Delta\lambda$. With these definitions, the Doppler shift law states that

$$\frac{\Delta\lambda}{\lambda_z} = \frac{V}{c}$$

**where V is the source's radial velocity, and c is the speed of light. That is, the change in wavelength created by motion of the star (or observer) divided by the wavelength in the absence of motion is just the radial velocity of the star divided by the speed of light. Thus, if we can measure $\Delta\lambda$, we can solve for V. The quantity $\Delta\lambda$ may turn out to be either positive or negative depending on whether the wavelengths of the source are increased or decreased. If the wavelengths are increased, the distance between the source and observer is increasing, meaning that from the observer's point of view the source is moving away. If the wavelengths are decreased, the distance between the source and observer is decreasing, and so from the observer's point of view, the source is approaching. Thus, from the observer's point of view, a positive $\Delta\lambda$ implies the source is receding, while a negative $\Delta\lambda$ implies it is approaching.**

Astronomers measure the Doppler shift of a star by recording its spectrum directly beside the spectrum of a standard, nonmoving source of light attached to the telescope, as shown in figure 12.12B. They then measure the wavelength shift by comparing the spectrum of the star to the standard. From that shift, they calculate the star's radial velocity with the Doppler shift formula. Notice that a star will have a radial velocity only if its motion has some component toward or away from us. Stars moving across our line of sight, maintaining a constant distance from Earth, have no Doppler shift.

We illustrate how to measure a star's motion with the following example. Suppose we observe a star and see a strong absorption line in the red part of the spectrum with a wavelength of 656.380 nanometers. From laboratory measurements on nonmoving sources, we know that hydrogen has a strong red spectral line at a wavelength of 656.280 nanometers. Assuming these are the same lines, we infer that the wavelength is shifted by a $\Delta\lambda$ of 0.100 nanometers. The Doppler shift law tells us that

$\Delta\lambda$ = Change in wavelength due to motion of light source
$\lambda_z$ = Wavelength of unmoving light source
$V$ = Radial velocity of light source*
$c$ = Speed of light (300,000 km/s)

$$\frac{\Delta\lambda}{\lambda_z} = \frac{V}{c}$$

Therefore,

$$\frac{V}{c} = \frac{0.100}{656.28}$$

To find the radial velocity, V, we multiply both sides by c and obtain

$$V = \frac{0.100c}{656.28} = 45.7 \text{ kilometers per second}$$

Because $\Delta\lambda$ is positive, the star is moving *away* from us.

Such measurements reveal that all stars are moving and that those near the Sun share approximately its direction and speed of revolution (about 200 kilometers per second) around the center of our galaxy, the Milky Way. Superimposed on this orbital motion, however, are small random motions of about 20 kilometers per second. Stars therefore move rather like cars on a freeway—some move a little faster and overtake slower-moving ones, which gradually fall behind. Stars are so far away, however, that their motion is imperceptible to the eye, much as a distant airplane seems to hang motionless in the sky. In fact, stellar motions across the sky are difficult to detect even in pictures of the sky taken years apart.

---

*We assume here that V ≪ C.

*Stages in the evolution of a low-mass binary*

**FIGURE 12.13**
Two stars orbiting in a binary system, held together as a pair by their mutual gravitational attraction.

*Binary stars*

*Orbital motion is reflected in the shifting spectral lines of a spectroscopic binary.*

## 12.4 BINARY STARS

Many stars have a special motion: they orbit around each other, held together as stellar companions by their mutual gravitational attraction, as illustrated in figure 12.13. Astronomers call such stellar pairs **binary stars** and value them greatly because they offer one of the few ways to measure stellar masses. To understand how, recall that the gravitational force between two bodies depends on their masses. The gravitational force, in turn, determines the stars' orbital motion. If that motion can be measured, we can work backwards to find the mass.

A star's mass, as we will discover in chapter 13, critically controls its existence, determining both how long it lasts and how its structure changes as it ages. Thus, if we are to understand the nature of stars, we must know their mass. Fortunately, this is not difficult to determine for stars belonging to a binary system, and such stars are relatively common. Among the hottest stars (*O* and *B* type), roughly 80% have orbiting companions. For Sun-like stars, a little over half have companions. But for the coolest stars (*M* type, in particular), the majority lack companions. Because such cool, dim stars vastly outnumber more luminous stars, single stars are the rule, although binary stars predominate among the brighter stars. Although we have described binaries as star pairs, in many cases more than two stars are involved. Some stars are triples, others quadruples, and in at least one case a six-member system is known.

Most binary stars are only a few astronomical units apart. A few, however, are so close that they "touch," orbiting in contact with each other in a common envelope of gas. These contact binaries and other close pairs probably formed at the same time from a common parent gas cloud; that is, rather than becoming a single star surrounded by planets as our Sun did, the cloud formed a close star pair. Widely spaced binaries, on the other hand, perhaps formed when one star captured another that happened to pass nearby, but astronomers still have many unanswered questions about how binary stars form.

A few binary stars are easy to see with even a small telescope. Mizar, the middle star in the handle of the Big Dipper, is a good example. When you look at this star with the naked eye, you will first see a dim star, Alcor, which is not a true binary companion but simply a star lying in the same direction as Mizar. With a telescope, however, Mizar can be seen to be two very close stars: a true binary. You can see where Alcor and Mizar are in the sky in "Looking Up at Ursa Major" on page 368.

### Visual and Spectroscopic Binaries

For some stars (such as Mizar) we can directly see orbital motion of one star about the other* by comparing images made several years apart. Such star pairs are called **visual binaries** because we can see two separate stars and their individual motion.

Some binary stars may be so close together, however, that their light blends into a single blob that defies separation with even the most powerful telescopes. In such cases, the orbital motion cannot be seen directly, but it may nevertheless be inferred from their combined spectra. Figure 12.14A shows a typical **spectroscopic binary,** as such stars are called. As the stars orbit each other about their shared center of mass, each star alternately moves toward and then away from Earth. This motion creates a Doppler shift, and so the spectrum of the star pair shows lines that shift—figure 12.14B—first to the red as the star swings away from us and then toward the blue as it approaches. From a series of spectra, astronomers can measure the orbital speed from the Doppler shift of the star, and by observing a full cycle of the motion, they can find the orbital period. From the orbital period and speed, they can find the size of the orbit, and with the size of the orbit and the period, they can find the stars' masses using a modified form of Kepler's third law, as we will now show.

---

*More technically, the two stars orbit a common center of mass located on the imaginary line joining them. The exact location of the center of mass depends on the star masses—if the masses are the same, the center of mass lies exactly halfway between them.

# LOOKING UP AT...
# URSA MAJOR

Circling in the northern sky is the Big Dipper, part of the well-known constellation Ursa Major, the Big Bear. The Big Dipper is technically not a constellation, but just an asterism — a star grouping. It is easy to see in the early evening looking north from mid-March through mid-September. If you look closely at it, you may notice that the middle star in the "handle" is actually two stars — Alcor and Mizar. With a telescope, look on a dark, clear night for the relatively bright galaxies near the Big Dipper: M51, the "Whirlpool Galaxy," and M81 and M82.

**Big Dipper**

**Location of the Hubble Deep Field**

Gravitational interactions between M81 and M82 have triggered star formation in both galaxies. M82 is one of the most active galaxies known.

**M81 and M82**

**Polaris (The North Star)**

**Little Dipper**

**Mizar and Alcor**

Mizar and Alcor, despite appearing close together in the sky, are probably not in orbit around one another. However, with a small telescope or binoculars, you can see that Mizar (the brighter of the star pair) has a companion star. This companion does in fact orbit Mizar. Moreover, each of Mizar's stars is itself a binary star, making Mizar a quadruple system.

## The Big Dipper in 3-D

79 ly
84 ly
124 ly
81 ly
81 ly
78 ly
100 ly

To Earth

1 light year (ly) = 6 trillion miles

**M51**

170,000 ly

The Whirlpool Galaxy can be seen as a dim patch of light with a small telescope. M51 is about 37 million ly away from Earth.

NORTH-NORTHEAST

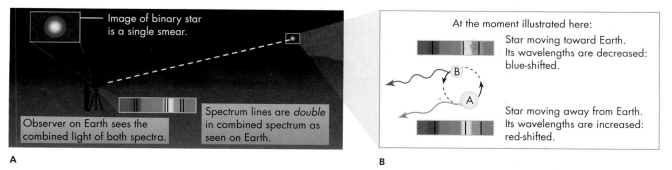

**FIGURE 12.14**

Spectroscopic binary star. (A) The two stars are generally too close to be separated by even powerful telescopes. (B) Their orbital motion creates a different Doppler shift for the light from each star. Thus, the spectrum of the stellar pair contains two sets of lines, one from each star.

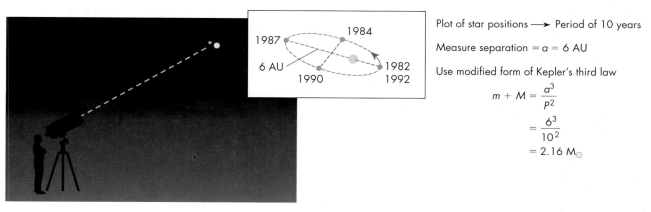

**FIGURE 12.15**

Measuring the combined mass of two stars in a binary system using the modified form of Kepler's third law.

## Measuring Stellar Masses with Binary Stars

In the early seventeenth century, Kepler showed that the time required for a planet to orbit the Sun is related to its distance from the Sun. If $P$ is the orbital period in years and $a$ is the semimajor axis (half the long dimension) of the planet's orbit in AU, then $P^2 = a^3$, a relation called Kepler's third law, as we discussed in chapter 1.

Newton discovered that Kepler's third law in a generalized form applies to any two bodies in orbit around each other. If their masses are $m$ and $M$ and they follow an elliptical path of semimajor axis, $a$, relative to each other with an orbital period, $P$, then

$$(m + M)P^2 = a^3$$

**where $P$ is expressed in years, $a$ in astronomical units, and $m$ and $M$ in solar masses. This relation is our basic tool for measuring stellar masses, as we will now describe.**

To find the mass of the stars in a visual binary, astronomers first plot their orbital motion, as depicted in figure 12.15. It may take many years to observe the entire orbit, but eventually the time required for the stars to complete an orbit, $P$, can be determined. From the plot of the orbit and with knowledge of the star's distance from the Sun, astronomers next measure the semimajor axis, $a$, of the orbit of one star about the other. Suppose $P = 10.0$ years and $a$ turns out to be 6.00 AU. We can then find their combined mass, $m + M$, by solving the modified form of Kepler's law, and we get (after dividing both sides by $P^2$)

$$m + M = \frac{a^3}{P^2}$$

Inserting the measured values for $P$ and $a$, we see that

$$m + M = \frac{6.00^3}{10.0^2}$$

$$= \frac{216}{100}$$

$$= 2.16 \ M_\odot$$

That is, the combined mass of the stars is 2.16 times the Sun's mass.

Additional observations of the stars' orbits allow us to find their individual masses, but for simplicity we will omit the details here. From analyzing many such star pairs, astronomers have discovered that star masses fall within a fairly narrow range from about 30 to 0.1 $M_\odot$. We will discover in the next chapter why the range is so narrow, but once again, we find our Sun is about midway between the extremes.

In the above discussion, we assumed that we can actually see the two stars as separate objects, that is, that they are a visual binary. If the stars are spectroscopic binaries and cannot be distinguished as separate stars, their period and separation can be determined from a series of spectra.

## Eclipsing Binary Stars

On rare occasions, the orbit of a binary star will be almost exactly edge on as seen from the Earth. Then, as the stars orbit, one will eclipse the other as it passes between its companion and the Earth. Such systems are called **eclipsing binary** stars, and if we watch such a system, its light will periodically dim. During most of the orbit, we see the combined light of both stars, but at the times of eclipse, the brightness of the system decreases as one star covers the other, producing a cycle of variation in light intensity called a "light curve." Figure 12.16A is a graph of such change in brightness over time.

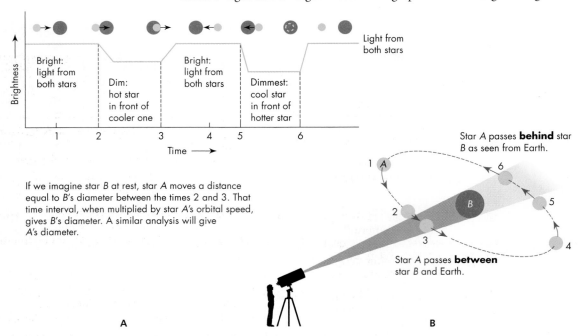

If we imagine star B at rest, star A moves a distance equal to B's diameter between the times 2 and 3. That time interval, when multiplied by star A's orbital speed, gives B's diameter. A similar analysis will give A's diameter.

Star A passes **behind** star B as seen from Earth.

Star A passes **between** star B and Earth.

**A**

**B**

**FIGURE 12.16**
An eclipsing binary and its light curve. From the duration of the eclipse, the diameter of each star may be found. Note: In this illustration, the yellow star is hotter than the red star.

Eclipsing binary stars are useful to astronomers because the duration of the eclipses depends on the stars' diameter. Figure 12.16 shows why. The eclipse begins as the edge of one star first lines up with the edge of the other, as shown in the figure. The eclipse ends when the opposite edges of each star line up. During the eclipse, the covering star has moved relative to the other a distance equal to its diameter plus that of the star it has covered. From such observations and knowledge of the orbital velocities of the eclipsing stars, astronomers can calculate the diameter of each star.

Light curves of eclipsing stars can also give information about the stars' shape and the distribution of brightness across their disks. For example, "star spots"—analogous to sunspots—can be detected on some stars by this method. If, in addition, astronomers can obtain spectra of the eclipsing system, the stars' masses may be found by use of the method we described above for spectroscopic binaries. Thus, astronomers can obtain especially detailed information about eclipsing binary stars.

*An eclipsing binary and its light curve*

## 12.5 SUMMARY OF STELLAR PROPERTIES

Measuring the properties of stars is not easy and is possible for only a tiny fraction of the stars known. Nevertheless, using the methods we have just described, astronomers can determine a star's distance, temperature, composition, radius, mass, and radial velocity. Some of the methods used are summarized in table 12.4. We find from such measurements that all stars have nearly the same composition of about 71% hydrogen and 27% helium, with a trace of the heavier elements. Most have surface temperatures between about 30,000 and 3000 K and masses between about 30 and 0.1 $M_\odot$. We will see in the next chapter that the need to balance their internal gravity is what gives stars their properties, but before we do, we will discuss how astronomers illustrate diagrammatically many of the features we have described above.

### TABLE 12.4

**Methods for Determining Stellar Quantities**

| Quantity | Method |
|---|---|
| Distance | 1. Parallax (triangulation)—for nearby stars (distance less than 250 pc)<br>2. Standard-candle method for more distant stars |
| Temperature | 1. Wien's law (color-temperature relation)<br>2. Spectral class (*O* hot; *M* cool) |
| Luminosity | 1. Measure star's apparent brightness and distance and then calculate with inverse square law.<br>2. Luminosity class of spectrum (to be discussed later) |
| Composition | Spectral lines observed in star |
| Radius | 1. Stefan-Boltzmann law (measure *L* and *T*, solve for *R*)<br>2. Interferometer (gives angular size of star; from distance and angular size, calculate radius)<br>3. Eclipsing binary light curve (duration of eclipse phases) |
| Mass | Modified form of Kepler's third law applied to binary stars |
| Radial velocity | Doppler shift of spectrum lines |

## 12.6    THE H-R DIAGRAM

By the early 1900s, astronomers had learned to *measure* stellar temperatures, masses, radii, and motions but had *understood* little of how stars worked. What made an *A* star different from a *K* star? Why were some *M* stars more luminous and others much less luminous? What supplied the power to make stars shine? We have now reached a similar point in our study of stars. We know that there are many varieties of stars. But what creates that variety and what does it mean? A fundamental step in the discovery of the nature of stars remains: the Hertzsprung-Russell diagram.

In 1912, the Danish astronomer Ejnar Hertzsprung and the American astronomer Henry Norris Russell working independently found that if stars are plotted on a diagram according to their luminosity and their temperature (or, equivalently, spectral class), then most of them lie along a smooth curve. The nearly simultaneous discovery of this relation by astronomers working on opposite sides of the Atlantic Ocean is an interesting example of how conditions can be "ripe" for scientific advances.

A typical Hertzsprung-Russell diagram, now generally called an **H-R diagram** for short, is shown in figure 12.17. Stars fall in the diagram along a diagonal line, with hot luminous stars at the upper left and cool dim stars at the lower right. The Sun lies almost in the middle. The H-R diagram does *not* depict the position of stars at some location in space. It shows merely a correlation between stellar properties much as a height-weight table does for people.

*Stellar evolution and the H-R diagram*

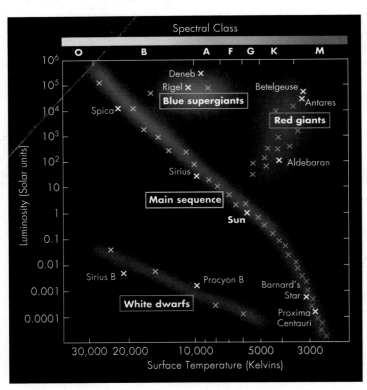

**FIGURE 12.17**
The H-R diagram. The long diagonal line from upper left to lower right is the main sequence. Giant stars lie above the main sequence. White dwarf stars lie below it. Notice that bright stars are at the top of the diagram and dim stars at the bottom. Also notice that hot (blue) stars are on the left and cool (red) stars are on the right.

## Constructing the H-R Diagram

To construct an H-R diagram of a group of stars, astronomers plot each according to its temperature (or spectral class) and luminosity, as shown in figure 12.18. By tradition, they put bright stars at the top and dim stars at the bottom—hot stars on the left and cool stars on the right. Equivalently, blue stars lie to the left and red stars to the right. Notice that temperature therefore increases to the left, rather than to the right, as is conventional in graphs. The approximately straight line along which the majority of stars lies is called the **main sequence.**

When a group of stars is plotted on an H-R diagram, generally about 90% will lie along the main sequence, but some will lie off it. Of these, a few will be in the upper right, where stars are cool but very luminous, and some will be below the main sequence to the lower center, where stars are hot but dim. What makes these stars different?

## Analyzing the H-R Diagram

The Stefan-Boltzmann law, described in section 12.2, gives us the answer. A star's luminosity depends on its surface area and its temperature. If two stars have the same temperature, any difference in luminosity must reflect a difference in area. But a star's area depends on its radius—a large area means a large radius. Stars that lie in the upper right of the H-R diagram have the same temperature as those on the lower main sequence, but because they are more luminous, they must be larger. In fact, they must be immensely larger because some of these bright cool stars emit thousands of times more energy than main-sequence stars of

**FIGURE 12.18**

Constructing an H-R diagram from the spectra and luminosities of stars in a cluster.

Measure luminosity (L) and spectrum of each star in the group. From spectrum, deduce the temperature, T.

To make H-R diagram of star group, plot L and T of each star.

the same temperature, implying that their surface areas are thousands of times larger. Astronomers therefore call these bright, cool stars "giants." The region in the H-R diagram populated by giants is sometimes called the "giant branch." Because many of the stars are cool and therefore red (recall that color is related to temperature), these huge cool stars are called **red giants.** For example, the bright star Aldebaran in the constellation Taurus is a red giant. Its temperature is about 4000 Kelvin and its radius is about 30 times larger than the Sun's.

A similar analysis shows that the stars lying below the main sequence must have very tiny radii if they are both hot and dim. In fact, the radius of a typical hot, dim star is roughly 100 times smaller than the Sun's, making it about the same size as the Earth. Because these stars are so hot, they glow with a white heat and are therefore called **white dwarfs.** Sirius B, a dim companion of the bright star Sirius, is a white dwarf. Its radius is about 0.008 the Sun's—a little smaller than the Earth's radius—and its surface temperature is about 27,000 Kelvin.

We can illustrate such extremes in size by drawing lines in the H-R diagram that represent a star's radius. That is, we use the Stefan-Boltzmann law, which relates the star's luminosity $L$, radius $R$, and temperature $T$. From this law, $L = 4\pi R^2 \sigma T^4$, we plot the values of $L$ that result from a fixed choice of $R$ as we vary $T$. Any stars lying along the plotted line must have equal radii. Figure 12.19 shows that such lines run diagonally from upper left to lower right. Notice that as we move to the upper right, we cross lines of progressively larger radius, implying that stars there are very large. That is, not only are they giants in luminosity, as we mentioned earlier, but they are also giants in radius. Comparing now the radii of stars across the H-R diagram, we see three main types: main-sequence stars, red giants, and white dwarfs.

## Giants and Dwarfs

Giants, white dwarfs, and main-sequence stars differ in more than just diameter. They also differ dramatically in density.* Recall that density is a body's mass divided by its volume. For a given mass, a larger body will therefore have a lower density, and so a giant star is much less dense than a main-sequence star if their masses are similar. For example, the

*Technically, we should use the term *average density* here and in the following.

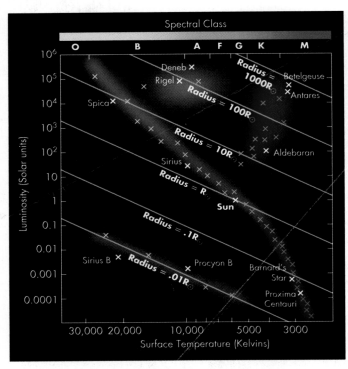

**FIGURE 12.19**
Lines showing where in the H-R diagram stars of a given radius will lie.

⊙⋆ On the basis of these lines, what is the approximate radius of Betelgeuse? Does that agree with the number you found in figure 12.6?

⊙⋆ Betelgeuse lies just a little above the line for a radius of 1000 $R_\odot$, or about $7 \times 10^8$ km. We found in figure 12.6 that Betelgeuse's *diameter* is about 10 AU = $1.5 \times 10^9$ km. Its radius will be half that number, or about $7.5 \times 10^8$ km, pretty close to what we find from reading it off the graph.

$\mathscr{L}$ = Luminosity of stars in solar units
$\mathscr{M}$ = Mass of stars in solar units

density of the Sun, a main-sequence star, is about 1 gram per cubic centimeter, while the density of a typical giant star is about $10^{-6}$ gram per cubic centimeter—1 million times less.

Why do stars have such disparate densities? Russell suggested about 1918 that as a star evolves, its density changes. He believed that stars began their lives as diffuse, low-density red giants. Gravity then drew the stars' atoms toward one another, compressing their matter and making stars smaller, denser, and hotter so that they turned into hot, blue, main-sequence stars. Russell, in fact, believed that gravity compressed these gaseous bodies so much that by the time they were on the main sequence, they had become liquid. Because liquids are difficult to compress, he believed that once a star reached the main sequence, it could no longer shrink. Unable to be compressed further, the star would have no source of heat and would therefore simply cool off and gradually evolve down the main sequence, growing dimmer and dimmer.

As we will discuss in the next chapter, today we know that stars evolve in the opposite direction. They begin life as main-sequence stars and turn into giants as they age. But although Russell was wrong about how stars evolve, his ideas led to many other important discoveries about stars, such as the mass-luminosity relation.

## The Mass–Luminosity Relation

In 1924, the English astrophysicist A. S. Eddington discovered that the luminosity of a main-sequence star is determined by its mass. That is, main-sequence stars obey a **mass–luminosity relation** such that the larger a star's mass, the larger its luminosity is, as illustrated in figure 12.20. A consequence of this relation is that stars near the top of the main sequence, brighter stars, are more massive than stars lower down.

Eddington made his discovery by measuring the masses of stars in binary systems and determining their luminosity from their distance and apparent brightness. Moreover, he discovered that the mass–luminosity relation could be expressed by a simple formula. If $\mathscr{M}$ and $\mathscr{L}$ are given in solar units, then for many stars

$$\mathscr{L} = \mathscr{M}^3$$

approximately. If we apply the relation to the Sun, its mass is 1 and its luminosity is 1 in these units, and the relation is clearly obeyed because $1 = 1^3$. If we consider instead a more massive star, for example, a 2 $M_\odot$ star $\mathscr{L}$ (according to the relation) should be

$$2^3 \ (= 8) \ L_\odot$$

Eddington went on to show from the physics of stellar structure that just such a mass–luminosity relationship is required if a star is to avoid collapsing under its own gravitational forces and if its energy flows through its interior by radiation. In chapter 13, we will see how this relation arises and its importance in determining how long a star can live.

## Luminosity Classes

Because astronomers use a star's luminosity to find its distance, radius, and life span, they have sought other ways to measure it. In the late 1800s, Antonia Maury, an early

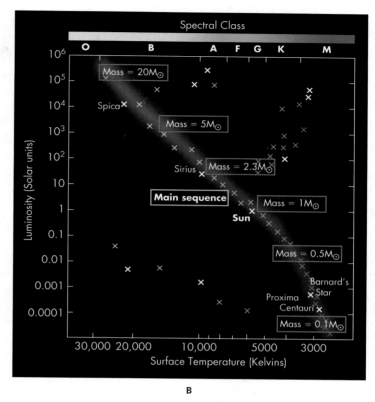

A

B

**FIGURE 12.20**

(A) The mass-luminosity relation shows that along the main sequence more massive stars are more luminous. The law does not work for red giants or white dwarfs. (B) On the H-R diagram, high-mass stars lie higher on the main sequence than low-mass stars.

**FIGURE 12.21**

Spectral lines are narrow in giant (luminous) stars (A and B) and wide in main-sequence (dim) stars (C).

Q★ Suppose that the spectrum of the giant star illustrated here is of Deneb and the spectrum of the main-sequence star is of Sirius. Is that consistent with their position in the H-R diagram (see fig. 12.17)?

Q★ Deneb is actually a supergiant, but the point is that the narrowness of its spectrum lines shows that it must be much brighter than Sirius.

stellar spectroscopist at Harvard, noticed that absorption lines were extremely narrow in some stars compared with other stars of the same temperature. In the early 1900s, Hertzsprung, too, noticed that narrowness and, even before developing the H-R diagram, recognized that luminous stars had narrower lines than less luminous stars, as illustrated in figure 12.21.

The width of the absorption lines in a star's spectrum turns out to depend on the star's density: the lines are wide in high-density stars and narrow in low-density ones. The density of a star's gas is related in turn to its luminosity because a large-diameter star has—other things being equal—a large surface area, causing it to emit more light, and a large volume, giving it a lower density. A small-diameter star, on the other hand, generally has a high density because its gas is compressed into a small volume. Such a star also tends to be less luminous because its surface area is small and thus emits less light.

Using this relationship between spectral line width and luminosity, astronomers divide stars into five luminosity classes (fig. 12.22), denoted by the Roman numerals I to V. Class V stars are the dimmest, and class I stars are the brightest. In fact, class I stars are split into two classes, Ia and Ib. Table 12.5 shows the correspondence between class and luminosity, and although the scheme is not very precise, it allows astronomers to get an indication of a star's luminosity from its spectrum.

A star's luminosity class is often added to its spectral class to give a more complete description of its light. For example, our Sun is a *G*2V star, while the blue giant Rigel is a *B*8Ia star. The luminosity class is especially useful for indicating the difference in luminosity between main-sequence and giant stars of the same spectral class, such as the low-luminosity, nearby, main-sequence star 40 Eridani (*K*1V) and the high-luminosity giant star Arcturus (*K*1III).

**FIGURE 12.22**
Stellar luminosity classes.

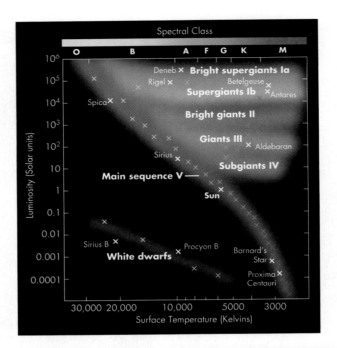

**TABLE 12.5**

**Stellar Luminosity Classes**

| Class | Description | Example |
|-------|-------------|---------|
| Ia | Supergiants | Betelgeuse, Rigel |
| Ib | Dimmer supergiants | Polaris (the North Star) |
| II | Bright giants | Mintaka (a star in Orion's belt) |
| III | Ordinary giants | Arcturus |
| IV | Subgiants | Achernar (a bright southern star) |
| V | Main sequence | The Sun, Sirius |

## Summary of the H-R Diagram

The H-R diagram offers a simple, pictorial way to summarize stellar properties. Most stars lie along the main sequence, with hotter stars being more luminous. Of these, the hottest are blue, and the coolest are red. Yellow stars, such as the Sun, have an intermediate temperature.

Along the main sequence, a star's mass determines its position, as described by the mass-luminosity relation. Massive stars lie near the top of the main sequence; they are hotter and more luminous than low-mass stars. Thus, from top to bottom along the main sequence, the star masses decrease but nevertheless lie in a relatively narrow range, from about 30 to 0.1 $M_\odot$.

Our study of the H-R diagram shows us that most stars fall nicely into one of three regions in the diagram: main-sequence stars, giants, or white dwarfs. But the H-R diagram also clearly shows that some stars fall into yet a fourth region—that of variable stars.

## 12.7    VARIABLE STARS

Not all stars have a constant luminosity. Stars that change in brightness are called **variable stars.** Probably all stars vary slightly in brightness—even the Sun does in the course of its magnetic activity cycle—but relatively few change enough to be

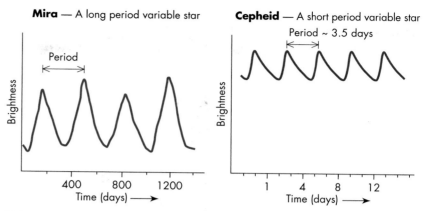

**FIGURE 12.23**

Two variable star light curves. Left: Sketch of the light variation of a long period Mira type variable. Right: Light variation of a Cepheid variable.

☞⋆ Approximately how long is the Mira-type pulsation period?

ℚ⋆ Measure the distance between the peaks and compare with the scale on the diagram. The time between peaks is about 11 months (340 days, or so).

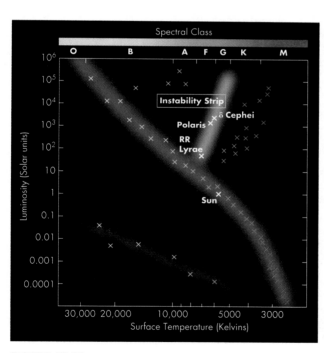

**FIGURE 12.24**

The instability strip in the H-R diagram. Stars that fall in this band generally pulsate.

noticeable to the naked eye. Among those that do vary, however, are several varieties of stars that are very valuable to astronomers as distance indicators, as we will discover in chapter 13. Especially valuable are the pulsating variables—stars whose radius changes, rhythmically swelling and shrinking. As the radius changes, the star's luminosity must also change (recall that a star's luminosity depends on its radius and temperature). The luminosity change, however, is complicated because, as the star pulsates, its temperature also changes, sometimes by as much as a factor of about 2. The result of both changes is that, as the star pulsates, its luminosity changes by a factor of 2 or 3. Although such large variations in luminosity are typical for many pulsating variables, ordinary stars also pulsate but much less dramatically. For them, the pulsations are more like the vibrations of a ringing bell. For example, our Sun's surface vibrates up and down in a complex pattern, as shown in figure 11.13.

The variation in luminosity and temperature is periodic for the majority of pulsating stars; that is, they repeat over some definite time interval called the **period,** as shown in figure 12.23. Astronomers call such a depiction of a star's brightness change its "light curve," and they use the shape of the curve and the time for the star to complete a cycle of brightness—the period—to classify variable stars. On the basis of such light curves, astronomers identify more than a dozen types of regularly pulsating variable stars. For example, some stars pulsate with a period of about half a day, while others take more than a year to complete a pulsation. If variable stars are plotted on the H-R diagram according to their average brightness, most lie in a narrow region called the "instability strip," illustrated in figure 12.24. In chapter 13, we will learn why pulsating stars are confined to this region.

Although many variable stars pulsate regularly, some variable stars change their brightness erratically. Such stars are called "irregular variables," and many very young and very old stars are of this category.

## 12.8    FINDING A STAR'S DISTANCE BY THE METHOD OF STANDARD CANDLES

In our discussion at the start of this chapter, we claimed that astronomers can determine a star's distance from knowledge of its luminosity. As an example of the method, we mentioned using the brightness of an oncoming car's headlights to estimate its distance. We saw that this method works, however, only if we know the luminosity of the light source with some assurance. That is, the luminosity of the light source is some "standard" value. For that reason, astronomers sometimes call this distance-finding scheme the *method of standard candles*, and it is perhaps their single most powerful tool for finding distances.

To measure a star's distance, an astronomer first measures its apparent brightness, $B$, with a photometer, as mentioned in section 12.1. Next, the star's luminosity, $L$, must be determined by any of the number of ways astronomers have to deduce it. For example, they may use the star's spectrum, its position in the H-R diagram, or, for a pulsating variable star, its period (as we will discuss in chapter 13) to reveal the star's luminosity. With $B$ and $L$ known, the astronomers can then calculate the distance to the star, $d$, using the inverse-square law. Unfortunately, all of the above methods to determine the luminosity have uncertainties, and so the distance calculated is uncertain, too.

In using the inverse-square law, however, it is generally much easier to find a star's distance by comparing it to the distance of a known star. For example, suppose we have two stars whose luminosity we know to be the same because they have identical spectra. Suppose one star, the nearer one, is close enough that we can find its distance by parallax to be 10 parsecs. Furthermore, suppose the star whose distance we seek looks 25 times dimmer. We find its distance by writing out the inverse-square law for both stars:

$L$ = Luminosity of star
$B$ = Brightness of star
$d$ = Distance to star

$$B_{\text{near}} = \frac{L_{\text{near}}}{4\pi d_{\text{near}}^2}$$

$$B_{\text{far}} = \frac{L_{\text{far}}}{4\pi d_{\text{far}}^2}$$

Next, we divide the bottom expression into the top one to obtain after cancellation

$$\frac{B_{\text{near}}}{B_{\text{far}}} = \frac{4\pi L_{\text{near}} d_{\text{far}}^2}{4\pi L_{\text{far}} d_{\text{near}}^2} = \frac{L_{\text{near}} d_{\text{far}}^2}{L_{\text{far}} d_{\text{near}}^2}$$

If the two stars have the same luminosity, we can also cancel the $L$s, leaving us with

$$\frac{B_{\text{near}}}{B_{\text{far}}} = \left(\frac{d_{\text{far}}}{d_{\text{near}}}\right)^2$$

That is, for two stars of equal luminosity, the brightness ratio is the inverse of the distance ratio squared. We can therefore find the distance ratio by taking the square root of both sides to get $d_{\text{far}}/d_{\text{near}} = \sqrt{B_{\text{near}}/B_{\text{far}}}$. For the problem we are solving, the brightness ratio is 25. Its square root is 5. Therefore, the farther star is 5 times more distant than the nearer one. Because we know that the nearer star is 10 parsecs away, the further star is 50 parsecs from us.

This method may seem to be rather roundabout, but astronomers have few other options for finding star distances. Moreover, with a little practice, such calculations can be done quickly. **To find a star's distance compared to another star of the same luminosity, find the apparent brightness ratio and take its square root.** This is then the distance ratio. If the stars have different luminosities, the calculation is slightly more difficult. In that case, we must use the full equation above.

# SUMMARY

Astronomers use many techniques to measure the properties of a star (fig. 12.25). Parallax—triangulation of the star from opposite sides of the Earth's orbit—gives a distance for nearby stars, with the nearest being about 1.3 parsecs (about 4 light-years) away from the Sun. The parallax method fails beyond about 250 parsecs because the angle becomes too small to measure. Astronomers must therefore use the inverse-square law, comparing apparent brightness and luminosity to find the distance of more remote stars.

Stars have a wide range of surface temperature as measured by their color. According to Wien's law, hot stars will be blue and cool stars red. Once the temperature and luminosity of a star are known, we can calculate its radius with the Stefan-Boltzmann law.

We can find a star's composition by looking at the lines in its spectrum. The spectrum can also be used to assign stars to the spectral classes: O, B, A, F, G, K, M, where O stars are the hottest and M stars are the coolest.

We can find a star's mass if it is in a binary system—two stars bound together gravitationally. For such stars, the orbital period and size when combined in the modified form of Kepler's third law give the masses of the orbiting stars.

Once a star's luminosity and temperature are known, it may be plotted on an H-R diagram. In this plot, most stars lie along a diagonal line called the main sequence, which runs from hot, blue, luminous stars to cool, red, dim ones. Main-sequence stars obey a mass-luminosity relation such that high-mass stars are more luminous than low-mass ones.

A small percentage of stars fall above the main sequence, being very luminous but cool. Their high luminosity implies they must have a large diameter, and so they are called red giants. Similarly, a few stars fall below the main sequence and are hot but dim. Their low luminosity implies a very small diameter, comparable to that of the Earth. These hot, small stars are called white dwarfs.

Some stars vary in brightness and are called variable stars. Many of these variables lie in a narrow region of the H-R diagram called the "instability strip."

## ? What It Is.

This is a photograph of the Hyades star cluster made through a thin prism. The prism spreads light from each star into a spectrum. The brightest spectrum toward the top is that of Aldebaran, a red giant.

**FIGURE 12.25**

Summary of how astronomers measure the distance, temperature, mass, composition, and radius of stars.

## QUESTIONS FOR REVIEW

1. Describe the method of distance measurement by triangulation.

2. How do astronomers triangulate a star's distance?

3. How is the parsec defined? How big is a parsec compared with a light-year?

4. How do astronomers measure a star's temperature?

5. Why do stars have dark lines in their spectra?

6. What are the stellar spectral classes? Which are hot and which cool?

7. What distinguishes the spectral classes of stars?

8. What is a binary star?

9. How do visual and spectroscopic binaries differ?

10. Why are binary stars useful to astronomers?

11. What is an eclipsing binary? What can be learned from eclipsing binaries?

12. What is the H-R diagram? What are its axes?

13. What is the main sequence?

14. How do we know that giant stars are big and dwarfs small?

15. How does mass vary along the main sequence?

16. What is the mass–luminosity relation?

17. What is a variable star?

18. What is meant by the period of a variable star?

19. Where in the H-R diagram are variable stars found?

## THOUGHT QUESTIONS

1. Measure approximately the distance from where you live to the neighboring building by triangulation. Make a scale drawing, perhaps letting 1/4 inch be 1 foot, and choose a base line. Construct a scale triangle on your drawing and find how far apart the buildings are.

2. Would it be easier to measure a star's parallax from Pluto? Why?

3. Suppose a binary star's orbit is in a plane perpendicular to our line of sight. Can we measure its mass using the methods described in this chapter? Why?

4. Use the data in table 12.6 to plot an H-R diagram. Which star is a red giant? Which is a white dwarf? Note: Plotting will be much easier if you plot the logarithm of the luminosity; that is, express it in powers of 10 and use the power. For example, if the luminosity is 100, plot it as 2 for $10^2$. Alternatively, use a pocket calculator as follows. Enter the luminosity in solar units and hit the "log" key. If the luminosity is 300,000, the answer you get should be 5.477 . . . .

5. Why are the hydrogen Balmer lines weaker in $O$ and $B$ stars than they are in $A$ stars?

### TABLE 12.6

### Data for Plotting an H-R Diagram*

| Star Name | Temperature | Spectrum | Luminosity (Solar Units) |
|---|---|---|---|
| Sun | 6000 | G2 | 1 |
| Sirius | 10,000 | A1 | 25 |
| Rigel | 12,000 | B8 | 60,000[†] |
| Betelgeuse | 3600 | M2 | 60,000[†] |
| 40 Eridani B | 15,000 | DA[‡] | 0.01 |
| Barnard's Star B | 3000 | M5 | 0.001 |
| Spica | 20,000 | B1 | 2000 |

*These data have been rounded off to make plotting easier.
[†]The luminosities of Rigel and Betelgeuse are not accurately known in part because the distance to them is uncertain.
[‡]Given the low luminosity of this star, what do you think the D in its spectral class stands for?

# PROBLEMS

1. The star Rigel radiates most strongly at about 200 nanometers. How hot is it?

2. The bright southern star Alpha Centauri radiates most strongly at about 500 nanometers. What is its temperature? How does this compare to the Sun's?

3. Arcturus is about half as hot as the Sun but is about 100 times more luminous. What is its radius compared to the Sun's?

4. A stellar companion of Sirius has a temperature of about 27,000 K and a luminosity of about $10^{-2}$ L$_\odot$. What is its radius compared to the Sun's? What is its radius compared to the Earth's?

5. Sirius has a parallax of 0.377 arc seconds. How far away is it?

6. The parallax of Proxima Centauri is about 0.763 arc seconds. How far away is it?

7. The parallax of the red giant Betelgeuse is just barely measurable and has a value of about 0.005 arc seconds. What

is its distance? Suppose the measurement is in error by + or − 0.003 arc seconds. What limits can you set on its distance?

8. Two stars in a binary system have an orbital period, $P$, of 5 years and an orbital separation, $a$, of 10 AU. What is their combined mass?

9. Two stars in a binary system have an orbital period, $P$, of 2 years and an orbital separation, $a$, of 4 AU. What is their combined mass?

10. Two stars in a binary system are determined from their position on the H-R diagram and the mass-luminosity relation to have a combined mass of 8 M$_\odot$. Their orbital period, $P$, is 1 year. What is their orbital separation, $a$?

11. A line in a star's spectrum lies at 400.0 nanometers. In the laboratory, that same line lies at 400.2 nanometers. How fast is the star moving along the line of sight; that is, what is its radial velocity? Is it moving toward or away from us?

# TEST YOURSELF

1. A star has a parallax of 0.04 arc seconds. What is its distance?
   (a) 4 light years
   (b) 4 parsecs
   (c) 40 parsecs
   (d) 25 parsecs
   (e) 250 parsecs

2. A star radiates most strongly at 400 nanometers. What is its surface temperature?
   (a) 400 K
   (b) 4000 K
   (c) 40,000 K
   (d) 75,000 K
   (e) 7500 K

3. Which of the following stars is hottest?
   (a) An $M$ star
   (b) An $F$ star

   (c) A $G$ star
   (d) A $B$ star
   (e) An $O$ star

4. A star that is cool and very luminous must have
   (a) a very large radius.
   (b) a very small radius.
   (c) a very small mass.
   (d) a very great distance.
   (e) a very low velocity.

5. In what part of the H-R diagram do white dwarfs lie?
   (a) Upper left
   (b) Lower center
   (c) Upper right
   (d) Lower right
   (e) Just above the Sun on the main sequence

# FURTHER EXPLORATIONS

Berger, David H., Jason P. Aufdenberg, and Nils H. Turner. "Resolving the Faces of Stars." *Sky and Telescope* 113 (February 2007): 40.

Cannizzo, John K., and Ronald H. Kaitchuck. "Accretion Disks in Interacting Binary Stars." *Scientific American* 266 (January 1992): 92.

Davis, John. "Measuring the Stars." *Sky and Telescope* 82 (October 1991): 361.

Hearnshaw, John B. "Origin of the Stellar Magnitude Scale." *Sky and Telescope* 84 (November 1992): 494.

Kaler, James B. *Stars and Their Spectra: An Introduction to the Stellar Spectral Sequence.* New York: Cambridge University Press, 1989. (Much of this material is also covered in a series of articles that appeared in *Sky and Telescope* 71 [February 1986]: 129 and continued at 3-month intervals, culminating with "Journeys on the H-R Diagram," *Sky and Telescope* 75 [May 1988]: 482.)

———. "Stars in the Cellar: Classes Lost and Found." *Sky and Telescope* 100 (September 2000): 38.

Kidwell, P. A. "Three Women of American Astronomy." *American Scientist* 78 (May/June 1990): 244.

MacRobert, Alan M. "The Spectral Types of Stars." *Sky and Telescope* 92 (October 1996): 48.

———. "The Stellar Magnitude System." *Sky and Telescope* 91 (January 1996): 42.

Steffey, Philip C. "The Truth about Star Colors." *Sky and Telescope* 84 (September 1992): 266.

Terrell, Dirk. "Demon Variables (the Algol Family)." *Astronomy* 20 (October 1992): 34.

Trimble, Virginia. "White Dwarfs: The Once and Future Suns." *Sky and Telescope* 72 (October 1986): 348.

**Website**

Please visit the *Explorations* website at **http://www.mhhe.com/arny** for additional online resources on these topics.

## PLANETARIUM EXERCISE

1. Go outside and, with the help of a sketch pad and the star charts located at the back of your text, locate a constellation. Sketch that constellation and estimate the brightness of each star. Rank the stars in order of decreasing brightness, and use the magnitude scale to label your estimates for their brightness.

Load the Starry Sky planetarium software and center on the constellation you chose. Place the cursor on each of the stars and compare your estimate of magnitude and order of brightness to the real numbers in the information boxes. Are there any discrepancies in your estimates? Any surprises?

## KEY TERMS

binary stars, 367
dwarfs, 359
eclipsing binary, 370
giants, 359
H-R diagram, 372
inverse-square law, 355

luminosity, 354
magnitudes, 359
main sequence, 372
mass–luminosity relation, 374

method of standard candles, 353
parallax, 349
parsec, 349
period, 377

radial velocity, 366
red giants, 373
spectral classes, 362
spectroscopic binary, 367

triangulation, 348
variable stars, 376
visual binaries, 367
white dwarfs, 373

## PROJECT

With the help of the star charts, find five of the following stars: Aldebaran, Altair, Antares, Betelgeuse, Capella, Castor, Deneb, Pollux, Procyon, Regulus, Rigel, Spica, Vega. What color does each star look to you? (Try not to let its twinkling affect the color you decide on.) Do your colors match those that you infer from the star's spectrum class as listed in appendix table 8? Which of these stars are giants? How many times bigger a radius does each have compared to the Sun's? Using the information in the appendix, about how long ago did the light from each of the stars start on its journey to Earth? Where were you when the light started its trip? Were you even born yet?

# Stellar Evolution

# 13

Young stars forming from interstellar gas.

## Chapter Objectives

- Stars form from interstellar gas clouds.
- Once a star has formed, nuclear reactions in its core supply it with energy.
- For most of a star's life, its energy is supplied by the fusion of hydrogen atoms into helium atoms.
- When a star uses up the hydrogen in its core, the core shrinks and the outer layers of the star expand and cool. This makes the star turn into a red giant.
- Eventually a star consumes all its available fuel and collapses.
- Stars much more massive than the Sun end their lives by exploding.
- Stars like the Sun may lose material from their outer layers; their core then becomes a white dwarf.

### ?What Is This?

Hint: The right and left pictures were taken about 6 months apart.

(See chapter summary for the answer.)

To us, stars appear permanent and unchanging, but like many other things, they are born, grow old, and die. Stars are not alive, of course. They are merely immense balls of gas, but astronomers nevertheless refer to the changes that stars undergo as stellar evolution. These changes are many, and they are driven by the physical laws that govern a star's structure. For example, gravity holds a star together, and the pressure of its gas supports it against gravity's inward pull. A star generates its supporting pressure from heat energy created by gravity compressing the material in its core. That heat energy steadily escapes from the star as the starlight we see, and it must be replenished, or the star will collapse. A star replenishes its energy by nuclear reactions in its core. In the early stages of its evolution, the nuclear reactions in a star fuse hydrogen into helium. But eventually the star consumes the hydrogen in its core and changes its structure, swelling into a red giant. Stars like the Sun consume their core's hydrogen in about 10 billion years, but more massive stars use up their hydrogen much faster. They therefore evolve much faster than the Sun and burn out in only a few million years. To understand why massive stars evolve faster, recall that according to the mass–luminosity relation, massive stars are far more luminous than stars like the Sun. Thus, to supply their greater luminosity, massive stars must burn their fuel faster than less-massive stars.

When a star runs out of fuel, its evolution is nearly finished. A star like the Sun dies slowly and relatively quietly and becomes a white dwarf. A more massive star explodes and leaves a neutron star or black hole as a remnant. Such stellar explosions, though marking the death of one star, may trigger the birth of another. Moreover, the explosion of a dying star blasts into space elements vital to human life such as carbon and iron, elements that eventually become incorporated into new stars, planets, and, ultimately, people. We therefore owe our very existence to stellar evolution. The goal of this chapter is to illustrate how such evolution occurs.

## Concepts and Skills to Review

- Structure of atoms (3.3)
- Relation between the temperature of a hot object and the color of the light it emits (Wien's law) (3.3)

## 13.1    THE EVOLUTION OF A STAR

The evolutionary changes that affect most stars happen incredibly slowly by human standards and require millions to billions of years. However, we need not wait such a vast span of time to see a star aging. We can "see" a star's aging with computer calculations that solve the equations that govern the star's physics. Stellar models, as such calculations are called, allow us to trace a star's life, revealing on a single page a billion years of a star's history. Figure 13.1 summarizes a few of the most important principles that govern a star's structure and evolution and may serve as a road map in our subsequent discussion.

Stellar models are only one tool in the study of stellar evolution. Astronomers can also deduce the basic features of how a star evolves from observations. For example, the existence of main-sequence stars, red giants, and white dwarfs suggests to astronomers a picture of how stars age, much the way a snapshot of a baby with its parents and grandparents suggests to a viewer a picture of human aging.

With these ideas in mind, let us follow the life story of a star from its birth to its death. In telling that story, we will discover that its ending depends critically on how much material the star contains—its mass. We will therefore divide stars into two groups—low-mass stars, such as the Sun, and high-mass stars. Here, we will consider high-mass stars to be those whose mass is ten or more times larger than that of the Sun. Those whose mass is less than about ten times the Sun's we will consider

**FIGURE 13.1**
Outline of the processes that govern the structure of stars.

## Principles Governing the Structure of a Star

Gravity holds star together; pressure supports star against gravity.
Gravity and pressure forces must balance (hydrostatic equilibrium).

Pressure is created by high temperature in star's core.

High temperature causes heat to flow from core to surface, where it escapes into space as the star's luminosity (starlight).

Escaping heat is replenished by nuclear fusion in core. (Hydrogen fuses into helium, initially.)

Star eventually runs out of fuel.

Low-mass stars turn into **white dwarfs.**   High-mass stars explode, leaving **neutron star** or **black hole.**

Note:  High-mass stars require more pressure to support their greater mass.
Greater pressure is produced by higher temperature.
Higher temperature produces higher luminosity.
Higher luminosity leads to faster fuel usage.
Faster fuel usage means a high-mass star burns out sooner than a low-mass star.

low-mass stars,* and we will begin with the life history of one such low-mass star, our Sun.

## The Sun's Life Story

The Sun began its life as an **interstellar cloud,** a tenuous, cold, dark mass of gas drifting around the Milky Way galaxy. The cloud—perhaps because of a collision with a neighboring cloud or the explosion of a nearby star—began to collapse. As it shrank under the influence of its increasing gravity, it grew smaller and hotter. Within a few million years—an instant, to the Sun—hydrogen began to fuse into helium in its core. The energy released by the fusion raised the pressure inside the Sun, stopping its collapse and leaving it looking as it does today: a small, yellow star. If we plotted it on the H-R diagram, it would lie on the main sequence.

The Sun will remain a main-sequence star until it consumes about 90% of the hydrogen in its core, a process that will take a total of 10 billion years, about half of which time the Sun has already used up. When the hydrogen in the Sun's core is nearly spent, its core will shrink and grow hotter. The rising temperature in the core will make the remaining hydrogen there burn faster, generating more energy. As that energy flows outward through the Sun's outer layers, it will lift them and they will cool as they expand farther from the source of heat. As the surface of this bloated Sun cools, it will turn red, and its light will reveal to our distant descendants a Sun swollen into a red giant. Their view will be brief, however, because the expanding Sun will nearly engulf them and the Earth.

The Sun will shine as a red giant for perhaps another billion years but will then shrink and grow hotter, transforming—when its core finally becomes hot enough to fuse helium—into a yellow giant. During this stage, it will pulsate as if taking slow, deep breaths, but these are dying gasps. Eventually, the Sun will consume most of the helium in its core and will once again change into a red giant, but a giant even larger

*An H-R diagram showing the evolutionary track of a 1 solar-mass star*

---

*Astronomers do not agree on what value of mass to use in this division. In fact, many astronomers feel that a division into three classes would be more appropriate. For simplicity, we will use only two classes.

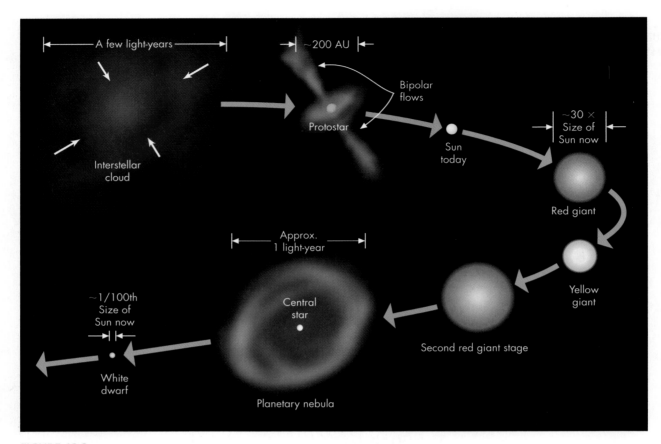

**FIGURE 13.2**
Evolution of a 1 solar-mass star such as the Sun, based on computer calculations.

and more luminous than before. That high luminosity is its death warrant: radiation streaming outward through its atmosphere will drive its gas into space, stripping the Sun to its bare core. The ejected gas will form a shell that will gradually disperse. The tiny core, fiercely hot but with no further energy supply, will cool and dwindle to a white dwarf. These changes, summarized in figure 13.2, are the fate of the Sun.

## The Life Story of a High-Mass Star

A very different fate awaits stars significantly more massive than the Sun: they explode when they run out of fuel. Astronomers are uncertain precisely what mass is needed to make a star explode but believe it is about 10 $M_\odot$. For that reason, we take 10 $M_\odot$* as the dividing line between high-mass and low-mass stars.

We chart the history of a high-mass star in figure 13.3, where we see that its early life is similar to the Sun's: it originates from the collapse of an interstellar cloud but from a larger and more massive clump than the one that formed the Sun. The clump's greater mass "squeezes" it more, thereby heating it more than the young Sun was heated. Thus, when the clump becomes a main-sequence star, it is much hotter, bluer, and more luminous than the Sun. Its greater luminosity implies that it burns its fuel faster, and so rather than taking 10 billion years to expend its hydrogen, it squanders its fuel in 100 million years or less, aging far more rapidly than the Sun. As it runs out of hydrogen, a massive star behaves much like a low-mass star, swelling steadily in size and growing cooler. However, a massive star passes through a pulsating yellow-giant stage before it turns into a red giant.

---

*To be more precise, this dividing line is ~8 $M_\odot$.

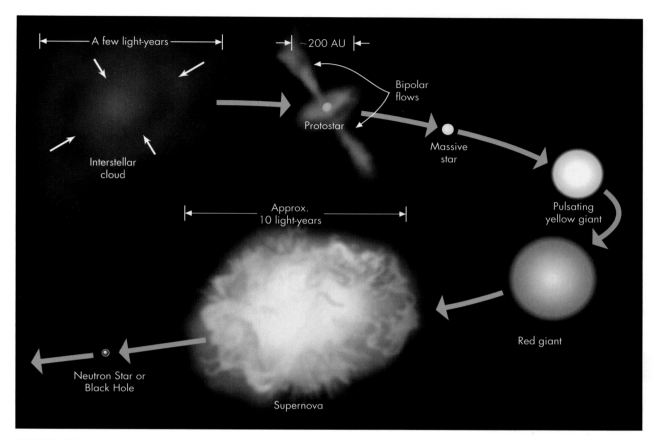

**FIGURE 13.3**
Evolution of a high-mass star, based on computer calculations.

The greater mass of such a star now becomes a primary influence on its evolution, creating an intense gravitational compression of its core. As the core is compressed, its temperature rises, and it burns fuel ever more furiously, thereby maintaining the pressure that supports the star. The higher temperature also permits the star to fuse progressively heavier elements to supply its energy—helium is fused into carbon, carbon into oxygen, and, ultimately, silicon into iron. Iron, however, does *not* release energy when it is fused, and thus a star with a massive iron core cannot support itself against gravity's inward pull. Lacking support, its core collapses in less than a second, triggering a cataclysmic explosion that showers the heavy elements it has made into space. Such a star does not leave a white dwarf; rather its core becomes either a tiny, incredibly compressed ball of neutrons (a neutron star) or an even denser body—a black hole, gravity's ultimate triumph.

## The Importance of Gravity

Gravity drives stellar evolution. From the moment of its birth, when an interstellar gas cloud collapses to become a star, to the moment of its death, gravity constantly works to crush the star. That crushing force makes a star evolve, because physical laws dictate that the star's core must be hot enough to create a pressure sufficient to support the star against gravity's unrelenting pull.

To keep its core hot, a star must generate energy to replace the energy it loses into space as it shines. A star generates that energy by the fusion of light chemical elements into heavier ones in its core. As we saw in chapter 11, released energy arises from the decrease in mass, *m*, that accompanies fusion, with the amount of energy given by

$E = mc^2$. The mass decrease results from the bonding between nuclei caused by the strong force. Thus, the formation of nuclear bonds releases nuclear energy just as the formation of chemical bonds (such as in the burning of some fuel and oxygen) releases chemical energy. However, the bonds created by the strong force are far stronger than chemical bonds, and so the energy released by fusion vastly exceeds that of any chemical reaction. Astronomers, nevertheless, often refer to fusion in stars as "burning," even though fusion is *not* burning in the sense that paper or gasoline burns.

Fusion is critical to a star because the energy it releases maintains the star's internal pressure and thereby supports it against gravity's inward pull. Any change in the energy supply, however, generally changes the star's structure. For example, during the late stages of its formation, the Sun was heated by gravity's compression of its gas. Now, as a main sequence star, the Sun fuses hydrogen into helium in its core for its energy source. When that hydrogen is consumed (billions of years from now), the Sun will change its structure to become a red giant. In the yet more distant future, the Sun will fuse helium into carbon and oxygen and become a yellow giant. Eventually, the Sun will consume its helium and be left with no energy source. It will shine then by leftover heat and become a white dwarf.

All stars need fuel to keep their cores hot, and the radically different evolution of high- and low-mass stars results from the former's ability to burn nuclear fuels heavier than helium. Thus, a major emphasis in this chapter will be on how stars supply their energy needs, but we must first understand how they are born.

## 13.2    STAR FORMATION

### Interstellar Gas Clouds

Stars form from interstellar matter—huge gas and dust clouds that orbit with the stars inside our galaxy. The gas is mostly hydrogen (71%) and helium (27%); the dust is composed of solid, microscopic particles of silicates, carbon, and iron compounds. Interstellar gas is generally cold, perhaps only 10 K (hundreds of degrees below zero on the Fahrenheit scale). At such low temperatures, atoms and molecules in a gas move too slowly to generate much pressure. As a result, the pressure in such cold gas may not be large enough to support the cloud against its own gravity; if the cloud is disturbed, it may then collapse. The collapse may be triggered by collision with a neighboring cloud, the explosion of a nearby star, or other processes.

Maps made with radio telescopes show that gas clouds are not uniform: they contain smaller clumps of gas where the density is higher than the average. When a cloud with such clumps collapses, each clump is also compressed, growing denser. Thus, a single cloud may break into many smaller pieces, each of which may form a star. The result is that stars generally form in groups, not in isolation, and all the stars within a group form at approximately the same time.

Astronomers picture the transformation from gas cloud to star proceeding in several distinct stages. In the first stage, a dense clump within a cloud begins to collapse, its gas drawn inward by gravity, which compresses and heats it. In the second stage, any rotation of the gas clump makes it flatten into a disk, as we described in chapter 7 on the origin of the Solar System. In about a million years, the disk forms a small, hot, dense core at its center called a **protostar,** which marks the third stage. These stages are illustrated in figure 13.4.

### Protostars

A protostar is hotter than the gas from which it condensed—perhaps 1500 K—but it is still much cooler than ordinary stars. Its relatively low temperature makes a protostar "shine" mainly at infrared and radio wavelengths rather than at visible ones. Thus, to observe a protostar, astronomers use telescopes that detect infrared or radio

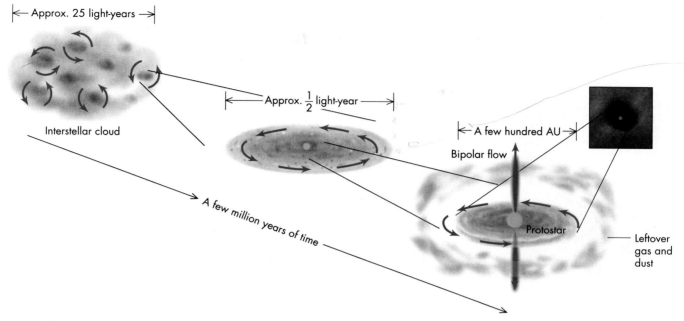

**FIGURE 13.4**

An artist's sketches depicting the birth of a star. (Insert, far right) A Hubble Space Telescope picture of the disk around a star forming in Orion.

**FIGURE 13.5**

False-color infrared and radio maps of protostars.

wavelengths. With such telescopes, astronomers can make infrared and radio maps (such as those in fig. 13.5) of regions in which stars are forming. The small, bright blobs are protostars. These objects are virtually undetectable at visible wavelengths because they are cool and because the dust around them absorbs what little visible light they emit. But protostars do not remain cool for long. As their gravity continues to draw surrounding material inward, they grow hotter from the gravitational energy released by the infalling material.

Gravitational energy is released whenever something falls. For example, when a cinderblock falls onto a box of tennis balls, the impact scatters the balls in all directions, giving them kinetic energy—energy of motion. In a gas, energy of motion appears as heat, raising the temperature; as material falls onto a protostar, its gas atoms speed up and it grows hotter.

When the temperature in the protostar's core reaches a few million or so Kelvins, nuclear reactions begin there. These reactions fuse hydrogen atoms into helium and release energy. With this new and powerful energy supply, the temperature and pressure rise still more in the star's core. Eventually, the outward pressure grows strong enough to stop further collapse in the core, but surrounding material, especially from the disk, continues to fall onto the star for millions of years. The infalling matter releases gravitational energy, which supplies additional heat to the protostar. In fact, for many young stars, the gravitational energy of infalling matter greatly exceeds that released by nuclear reactions in the core.

A protostar passes through these stages in a time that is brief by astronomical standards: a few million years for a star like the Sun and even less time for a more massive star. But for all types of protostars, infall creates violent changes in brightness and, more strangely, ultimately creates a strong *outflow* of gas. This outflow does not take place in all directions. Rather, it generally is "focused" into narrow jets of gas. Figure 13.6 shows a typical jet, but what powers the jet and what makes it so narrow are not yet understood. Computer simulations suggest that the jet's power somehow comes from energy released as matter falls from the inner parts of the disk onto the star. The focusing of the jet probably results from magnetic fields in the star and disk.

**FIGURE 13.6**
(Left) False-color image of a protostar and jet. Notice the Herbig-Haro objects (the blobs) in nearby gas.
(Right) Hubble Space Telescope image of the jet.

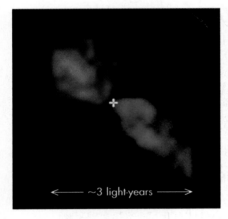

**FIGURE 13.7**
Bipolar flow of gas ejected from a young star. This is a false-color picture of the radio emissions from the gas. The gas receding from us is shown in red, and the gas approaching us is shown in blue. The cross marks the location of the young star.

## Bipolar Flows from Young Stars

Jets from young stars play an important role in "cleaning up" after a star has formed. They may also affect the formation of other stars in their vicinity. To better understand these outflows, let us look in more detail at figure 13.6. You can see there a long, thin jet of gas squirting out from a dark cloud in which a young star lies. The jet has carved a cavity in the gas that remains around the star. Farther out along the jet's direction lie several small, bright blobs called "Herbig-Haro objects." Herbig-Haro objects, named for the astronomers who first discovered them, appear to be splashes similar to those made when water from a hose hits a window and spreads out into blobs. The jet of gas from the star spreads when it hits surrounding gas, and the impact heats both the jet and the gas around it, making them glow, as shown in the pictures. The jet also pushes gas outward from around the protostar to create **bipolar flows.** Bipolar flows (which get their name because there are two jets of gas ejected parallel to the star's polar axes) can be seen in figure 13.7, which shows a radio map of the area around a protostar. Here, a computer has been used to color red the gas moving away from us (that is, red-shifted material) and to color blue the gas moving toward us.

Bipolar flows are important because they clear away gas and dust from around protostars and thereby allow astronomers to see them directly with visible light. However, even powerful bipolar flows leave some of the surrounding material behind. Thus, many young stars are partially immersed in interstellar matter, as illustrated in figure 13.8. Such stars often vary erratically in brightness and have gas streaming away from their surfaces, as shown by the Doppler shift of their spectral lines. This flow occurs in the star's upper atmosphere and is distinct from the bipolar flow we described above. Rather, it is similar to the solar wind but far more intense. Stars that exhibit these features (variable light and outflowing gas) are called **T Tauri stars,** and some of their variability is probably caused by magnetic activity much like that observed on our Sun. For example, T Tauri stars show evidence of giant "starspots" and magnetic flares thousands of times more intense than the activity we observe on our Sun. Astronomers cannot see the spots directly with optical telescopes, but by combining data on variations of the star's brightness obtained with X-ray, radio, and infrared telescopes, they can deduce the presence of spots on these peculiar young stars.

**FIGURE 13.8**
Photograph of two young star clusters still surrounded by the interstellar gas from which they formed. NGC 6559/IC 1274-75 is on the left, and NGC 2264 is on the right.

☾★ What creates the red glow of the gas near these stars? What might account for the dark cone in the bottom part of the cluster on the right? What do you think the squiggly dark blobs silhouetted against the pink glow in the picture on the left might be?

ᘓ★ The red glow is caused by hot hydrogen. The dark cone is a shadow cast by the clump at the top of the cone. The squiggly little things are globules . . . perhaps protostars.

If T Tauri stars are plotted on an H-R diagram, they lie a little above the lower main sequence because they are still shrinking and are thus larger and brighter than main-sequence stars. Figure 13.9 shows the H-R diagram for a very young star group and a photograph of the group itself. Such H-R diagrams help verify our ideas about star formation because older clusters never contain T Tauri stars.

You can see that surrounding the cluster is gas left over from its formation, and if you look closely in the central regions, you can see small, dark blobs that may be protostars that have not yet become hot enough to emit visible light. These tiny, dark globs are called Bok globules in honor of Bart Bok, the astronomer who first studied them in detail. Globules exhibit a diversity of sizes, with diameters ranging from about 0.1 to 2 light-years and with masses ranging from a few up to about 200 times the Sun's mass.

## Stellar Mass Limits

We saw in the previous chapter that the mass of most stars is between about 0.1 and 30 $M_\odot$. These limits are not absolute—higher- and lower-mass stars exist—but the

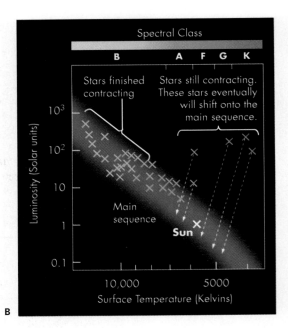

**FIGURE 13.9**
(A) The young star cluster NGC 2264 and (B) its H-R diagram. Notice the large amounts of leftover gas heated by the brilliant young stars.

theory of star formation described above can help us understand why very high- or very low-mass stars are uncommon.

Stars much smaller than about 0.1 $M_\odot$ are rarely seen because their mass is too small to compress them enough to begin nuclear burning. Thus, such low-mass stars are very dim. Astronomers have also searched for objects less massive than this limit, the "brown dwarfs." These objects, midway between stars and planets in mass, might be very numerous, but their dimness makes them extremely hard to detect. Searches with sensitive detectors are just now beginning to detect these tiny, dim, failed stars.

Very high-mass stars are rare for a different reason. As their great gravity compresses them, they rapidly become extremely hot and luminous, emitting such intense radiation that they heat the gas around them, raising its pressure and preventing additional material from falling onto them. Thus, their high temperature limits the amount of material they can accumulate. Furthermore, once formed, high-mass stars are so luminous (recall the mass-luminosity relation) that their radiation strips gas from their outer layers, driving it into space. Thus, a 200 $M_\odot$ star would quickly "shine" itself apart.

We have followed the stages of a protostar's early development. Now we will proceed to the next stage of its life.

## 13.3 MAIN-SEQUENCE STARS

A star stops being a protostar and becomes a main-sequence star when nuclear fuel in its core can supply energy to raise its pressure and stop its collapse. At this stage of its life, the star's interior structure is like the Sun's: it contains a core, where hydrogen is burnt into helium, and an envelope of gas around the core that transports energy to the star's surface. The properties of the core and envelope, however, depend greatly on the star's mass.

## Why a Star's Mass Determines Its Core Temperature

We saw in chapter 12 that high-mass stars are both hotter and more luminous than low-mass stars. This relation between mass, temperature, and luminosity results from the condition that the gravity and pressure inside a star must be in balance, a condition we called "hydrostatic equilibrium." To see how this relation arises, we need to recall that a star forms by gravity drawing interstellar material together. As that material is pulled inward on itself, it is compressed, and that compression heats the material. The amount of heating is set by the forming star's mass, such that the larger the star's mass, the greater its heating.* Thus, high-mass stars end up hotter (and thereby more luminous) than low-mass stars.

## Structure of High-Mass and Low-Mass Stars

The high temperature at the core of a massive star allows it to fuse its hydrogen into helium by a different mechanism from that of the Sun. In the Sun, hydrogen is converted to helium by the proton-proton chain: two protons fuse to form heavy hydrogen, to which a third proton is added to make $^3$He. The $^3$He then fuses to form $^4$He. In a massive star, the conversion of hydrogen to helium takes place by means of the CNO cycle, whereby carbon, nitrogen, and oxygen atoms already present in the star's core act as catalysts to aid the reaction. The CNO cycle, however, requires a very high temperature to operate and thus can only generate energy at the very center of the star. Photons are unable to carry this huge energy flow from so small a volume, and so the gas in the core begins to rise in irregular clumps and carries the heat outward by convection. As we saw in chapter 11, similar convective motions occur in approximately the outer third of the Sun just below its surface, where the radiant energy flow is impeded by the strong absorption of the cool subsurface gas. In a high-mass star, however, the convection currents are confined to the core. They do not extend close enough to the star's surface to allow hydrogen-rich gas from there to mix into its core and replenish the core's depleted fuel. Neither high-mass stars nor low-mass stars such as the Sun can use the fuel in their outer layers. Only very low-mass stars—less than about 0.4 M$_\odot$—mix completely. Thus, a star can be starved for fuel even though its outer layers are fuel-rich. This strongly affects a star's lifetime.

## Main-Sequence Lifetime of a Star

The time a star stays on the main sequence is called its **main-sequence lifetime.** A star's main-sequence lifetime depends on its mass and luminosity. To see why, we can use a simple analogy. Suppose we want to know how long a car can run on a tank of gasoline. That time clearly depends on how much fuel it has and how rapidly the fuel is consumed. For example, if the car has a 16-gallon tank of gasoline and burns fuel at the rate of 2 gallons per hour, it can cruise for 8 hours ($\frac{16}{2}$) before running out of gas. When the same formula is applied to a star, its mass determines how much fuel it has, and its luminosity determines how rapidly the fuel is burned. Thus, to calculate the star's main-sequence lifetime, we divide the amount of fuel by the rate at which it is consumed.

The rate at which a star burns its fuel depends on its luminosity: More luminous stars burn their fuel faster. We can show, using the information in chapter 11, that the Sun consumes its fuel at the rate of about $2 \times 10^{19}$ kilograms of hydrogen per year. For a star whose luminosity is $\mathscr{L}$ solar units, the rate of fuel consumption in kilograms per year is therefore $2 \times 10^{19}\,\mathscr{L}$. Using the car analogy, the lifetime of the star, $t$, is then its mass divided by $2 \times 10^{19}\,\mathscr{L}$, or

$$t = \frac{\text{Mass (kg)}}{2 \times 10^{19}\,\mathscr{L}} \text{ Years}$$

*Technically, the larger mass releases a larger amount of gravitational energy.

A star cannot use all its hydrogen for fuel, however. Only its core is hot enough for nuclear reactions to occur, and the core is generally only about one-tenth of its total mass. Thus, the value of the mass we use above should be about 0.1 times the star's mass, and so for a star whose mass is $M$, the mass of available fuel is $0.1M$ kg. For the Sun, whose mass is $2 \times 10^{30}$ kilograms, the mass of available fuel is $2 \times 10^{30} \times 0.1 = 2 \times 10^{29}$ kilograms. For a star whose mass is $\mathcal{M}$ times the Sun's, the mass of available fuel is therefore $2 \times 10^{29} \mathcal{M}$ kilograms. Using this expression for the amount of fuel available in our expression for a star's lifetime, $t$, we obtain

$$t = \frac{2 \times 10^{29} \mathcal{M}}{2 \times 10^{19} \mathcal{L}} \text{ Years}$$

$$= 10^{10} \frac{\mathcal{M}}{\mathcal{L}} \text{ Years}$$

where $\mathcal{M}$ is in solar masses and $\mathcal{L}$ is in solar luminosities.

For the Sun ($\mathcal{M} = 1$ and $\mathcal{L} = 1$), this is $10^{10}$ (or 10 billion) years. For a star such as Sirius, whose mass is about 2 $M_\odot$ and whose luminosity is about 20 $L_\odot$, its main sequence lifetime is $10^{10}(\frac{2}{20}) = 10^9$ years. For a star whose mass is 30 $M_\odot$ and whose luminosity is $10^5$ $L_\odot$, the main-sequence lifetime is a mere $3 \times 10^6$ years. Notice that *massive stars have much shorter lifetimes* than the Sun, despite having more fuel. This is because they burn their fuel so much faster to supply their greater luminosity. They are like gas-guzzler cars that despite having 20-gallon fuel tanks, run out of fuel sooner than economy-size cars with only 10-gallon tanks.

The shortness of massive stars' lives implies that those we see must be relatively young. Because massive stars are blue, we can conclude that, in general, *blue stars have formed recently.** Their youth is also indicated by the presence of gas, left over from their formation, still surrounding them, as shown in figures 13.8 and 13.9.

## 13.4    GIANT STARS

### Leaving the Main Sequence

**ANIMATION**

*The structure of a star when hydrogen burning in the shell begins*

When a main-sequence star has consumed much of the hydrogen in its core, its structure begins to change. With less fuel available, less energy can be generated. That in turn allows the weight of the star's outer layers to compress the core, which heats it and makes it denser, as shown in figure 13.10A. But at this stage in its life, the star has too little hydrogen left in its core to supply its energy needs. As a result, the hydrogen just *outside* the core begins to fuse, creating what is called a **shell source,** as illustrated in figure 13.10B.

Heat from the shell source and contracting core raises the pressure around the core. That stronger pressure pushes the surrounding gas outward, making the star expand and its radius grow by a factor of anywhere from five to several hundred, depending on the star's mass. But this expansion cools the outer layers, and so the star grows red (recall that cooler bodies radiate more strongly at long [red] wavelengths). Because of its size and color, we call such a star a **red giant.** Thus, a main-sequence star becomes a red giant when it uses up the hydrogen in its core.

A giant star has a very different structure from that of a main-sequence star. As its name implies, its radius is much larger: hundreds of times larger than the Sun's, in some cases. But this enormous size is deceptive. Although most of the star's *volume* is taken up by its very low-density, tenuous atmosphere, most of its *mass* lies in a tiny, hot, compressed core not much larger than the Earth. The core, or its surrounding shell source, supplies the star's energy. The huge envelope is relatively cool and the gas

*Strictly speaking, we are considering only main-sequence stars here.

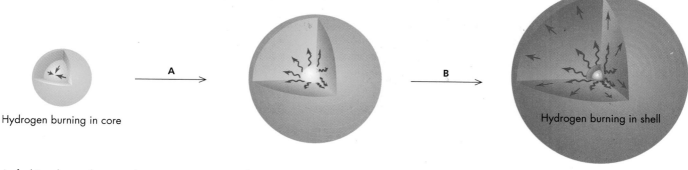

As fuel is exhausted, outward pressure in core drops, and gravity compresses. Core contracts.

Rising heat in contracting core creates pressure that causes outer layers to expand. Shell of H outside core ignites.

Outer layers expand yet more and cool. A red giant forms.

**FIGURE 13.10**

(A) The core of a star begins to shrink as a star uses up the hydrogen in its core. This compresses and heats the core. (B) The heated core ignites the surrounding gas to make a shell source, and the outer layers of the star expand, turning it into a red giant.

**FIGURE 13.11**

How a high temperature acts to overcome the electrical repulsion of the nuclei and brings them close enough for the nuclear force to fuse them.

in it readily absorbs the photons flowing from the core toward the star's surface. This slows the photons, and so convection rather than radiation carries energy to the star's surface. Some of that energy continues to come from burning hydrogen, but many stars switch to a new energy supply: helium burning.

## Nuclear Fuels Heavier Than Hydrogen

Helium burning begins very differently in high-mass and low-mass stars and creates significant differences in their structures. To see why, we need to look in more detail at how nuclear fusion occurs and why it requires a very high temperature.

Two nuclei fuse when they are brought close enough together for the nuclear force to bind them to one another. That force, which is also what holds the protons and neutrons together in a nucleus, operates only over very short distances. If nuclei are more than a few diameters apart, the nuclear force is too weak to bond them. In fact, they will be repelled by the electrical force between the similarly charged protons, as illustrated in figure 13.11. That electric repulsion grows as the number of protons increases. It is weak for hydrogen, with its single proton, but stronger for heavier elements, with more protons. It is enormously more difficult, therefore, for two carbon atoms, with their six protons, to fuse than for two hydrogen atoms to do so.

Nuclei can overcome this electrical repulsion and fuse if they collide at enormously high speed. That speed is achieved in hot gases: the hotter a gas, the faster its nuclei move, and so the more easily it can fuse heavy nuclei. Hydrogen can fuse at about 7 million Kelvin, but helium, with its two protons, must be heated to about 100 million Kelvin. Carbon, with its six protons, must be hotter still to fuse.

If a star's core is as hot as 100 million Kelvin, helium nuclei there will fuse into carbon. The nuclear reaction combines 3 $^4$He into a $^{12}$C. Helium nuclei are sometimes called "alpha particles," and so this reaction is sometimes called the **triple alpha process,** or simply "helium burning." The reaction releases energy but only about one-tenth the amount per kilogram of fuel that hydrogen burning yields. Nevertheless, once a star has consumed all its core hydrogen, it must use its helium or collapse. How does a star make the switch from hydrogen to helium fusion? It must heat its core to 100 million K, and any star more massive than about 0.5 $M_\odot$ can do this by compressing its core.

A high-mass star needs to compress its core only a little before helium begins to fuse because its core is already extremely hot. Thus, helium burning begins easily for high-mass stars. A low-mass star such as the Sun, however, must compress its core enormously to make it hot enough for helium fusion. The compression of such a star packs its gas atoms so closely that they no longer behave like an ordinary gas—in technical terms, the gas becomes a **degenerate gas.**

## Degeneracy in Low-Mass Stars

In a degenerate gas, the matter is so densely packed that the electrons interact not as ordinary charged particles but according to laws of subatomic physics. One consequence of these laws is that no more than two electrons of the same energy can be packed into the same submicroscopic volume.* This packing limit makes a degenerate gas "stiff," so it behaves almost like a solid. As a result, a degenerate gas does not expand on being heated any more than a heated brick swells up. This lack of response to heating strongly affects nuclear burning in a degenerate gas.

When a nuclear fuel begins to fuse in a normal gas, the energy released heats the gas, raising its pressure so that the gas expands. The expansion cools the gas and reduces the rate of nuclear burning, acting like a thermostat that switches off the furnace when the house gets too hot. The reason cooling slows nuclear burning is that only at high temperatures do nuclei collide violently enough to fuse. Thus, as a gas cools, the rate at which its atoms fuse decreases, so that less energy is released. In normal stars, this mechanism keeps them from collapsing or exploding. In degenerate matter, this thermostating mechanism fails.

For a low-mass star with a degenerate core, when a fuel ignites, the energy released does not raise the pressure because the pressure does not depend on the temperature. Thus, the gas does not expand and cool. Instead, it simply gets hotter and releases more energy, which makes it hotter still. Under these circumstances, the release of energy accelerates explosively in what is called the **helium flash.** In only a few minutes, the star's energy production increases several thousand times, yet this outburst is totally hidden from our view by the star's outer layers, just as a firecracker set off under a mattress creates little visible disturbance. However, the energy released heats the core enough to change it back to a normal gas, destroying its degeneracy. With its degeneracy gone, the gas can now expand. That expansion in turn makes the star's outer layers adjust—shrinking, compressing, and heating. The star's reheated surface changes color from red to yellow, and the star becomes a yellow giant. The helium flash therefore ends the red-giant stage of low-mass stars.

High-mass stars do not have a helium flash. As we said earlier, they are so hot that their cores ignite helium with only a little compression, and so they do not become degenerate at this stage of their lives. They continue to expand and change gradually from yellow to red giants, but for both high-mass and low-mass stars, the onset of helium burning may trigger instabilities in the star and make it pulsate.

## 13.5 YELLOW GIANTS AND PULSATING STARS

When stars are plotted on the H-R diagram, some lie between the hot, luminous, blue, main-sequence stars and the red giants. Such stars are called "yellow giants." The most luminous yellow giants are aging high-mass stars. Less luminous yellow giants are old low-mass stars that have completed their first red-giant stage and are burning helium in their cores. Regardless of their mass, many yellow giants have the

---

*Technically, the size of the volume is given by the laws of quantum physics.

An RR Lyrae Variable

**A**

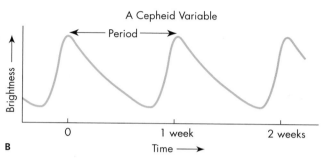

A Cepheid Variable

**B**

**FIGURE 13.12**
Schematic light curves of two pulsating variable stars: (A) an RR Lyrae star; (B) a Cepheid variable.

unusual property of swelling and shrinking rhythmically: they pulsate. As they change size, their luminosity changes, and so astronomers refer to them as "variables," or "pulsating stars."

Two particularly important types of pulsating stars are the RR Lyrae (pronounced *lie-ree*) and Cepheid (pronounced *sef-ee-id*) variables. RR Lyrae stars have a mass comparable to the Sun's and are yellow-white giants with about 40 times the Sun's luminosity. The time it takes them to complete a pulsation cycle from bright to dim and back to bright is called their "period," as illustrated in figure 13.12. RR Lyrae variables have periods of about half a day and are named for RR Lyrae, a star in the constellation Lyra, the harp, which was the first star of this type to be identified.

Cepheid variables are yellow supergiants that are more massive than the Sun and range in luminosity from several hundred times the Sun's luminosity to several tens of thousands times its luminosity. They are named for the star Delta Cephei, and their periods can be as short as about 1 day or as long as about 70 days.

## Why Do Stars Pulsate?

Giant stars pulsate because their atmospheres trap some of their radiated energy. This heats their outer layers, raising the pressure and making the layers expand. The expanded gas cools, and the pressure drops, so gravity pulls the layers downward and recompresses them. The recompressed gas begins once more to absorb energy, leading to a new expansion. These stars continue alternately to trap and release the energy, and so they continue to swell and shrink, as shown in figure 13.13.

A covered pan of water boiling on a stove behaves similarly. The lid will trap the steam so that pressure inside rises. Eventually, the pressure becomes strong enough to tip the lid, and steam escapes. The pressure decreases, and the lid falls back. It again traps the steam, the pressure again builds up, and the cycle is repeated.

A similar process occurs in pulsating stars, with the role of steam played by the star's radiation and the role of the lid played by the star's atmosphere. For a star to trap radiation this way, its atmosphere must have special absorbing properties—technically called "opacity"—that occur only if its surface temperature and radius fall in a narrow

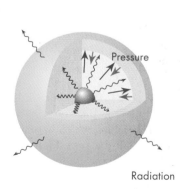

Radiation (purple wiggles) partially trapped. Pressure (blue arrows) increases, overcomes star's gravity (red arrows), and makes it "inflate" and expand.

Expansion allows trapped radiation to escape. Star cools and pressure decreases. Gravity now compresses star back to original size.

Cycle begins again.

**FIGURE 13.13**
Schematic view of a pulsating star.

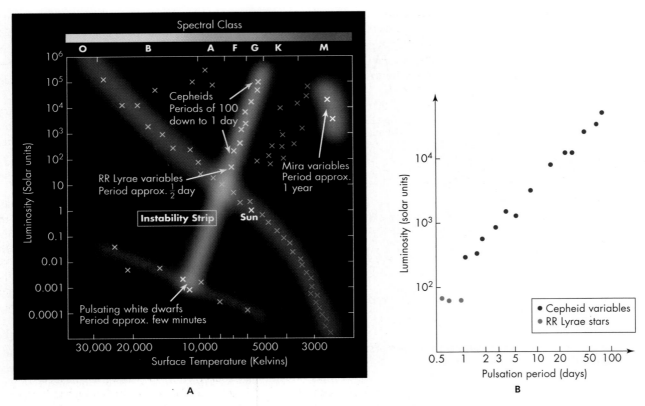

**FIGURE 13.14**

Properties of variable stars. (A) The instability strip in the H-R diagram. Stars in this narrow region pulsate. The locations of a few other types of variable stars are also indicated. (B) The period–luminosity law. More-luminous stars tend to pulsate more slowly.

range. That range, called the **instability strip,** is shown in the H-R diagram illustrated in figure 13.14A. A star that assumes these characteristics as it evolves will begin to pulsate and will continue to pulsate until its temperature or radius changes enough to remove it from the instability strip. For example, when some low-mass stars cross the region of the instability strip, they become RR Lyrae variable stars. When some high-mass stars cross, they, being more luminous, lie above the RR Lyrae variables in the instability strip and instead become Cepheid variables. The amount of time a given star spends in the instability strip depends on its mass. Massive stars such as the highly luminous Cepheids evolve across the strip in less than 1 million years, but some cross the region several times as their interior structures alter. Low-mass stars such as the RR Lyrae variables spend more time in the strip, perhaps a few million years, but cross it less often. In either case, stars pulsate for only a brief portion of their lives.

Astronomers have identified many other types of pulsating variables besides Cepheids and RR Lyrae stars. For example, Mira variables, which have pulsation periods of about 1 year, lie in the upper right of the H-R diagram. Many other kinds of variable stars, such as ZZ Ceti stars, a kind of pulsating white dwarf with periods as short as a few minutes, lie along a rough extension of the instability strip across the main sequence and downward toward the lower left of the H-R diagram.

## The Period–Luminosity Law

Many pulsating variable stars obey a law that relates their luminosity to their period—the time it takes them to complete a pulsation. Observations (and theoretical calculations) show that the more slowly a star pulsates, the more luminous it is (fig. 13.14B). This so-called **period–luminosity law** arises because, other things being equal, more

The period–luminosity law gives astronomers a powerful tool for measuring distances. By measuring the star's period, astronomers can find its luminosity. From its luminosity and apparent brightness, they find the star's distance, as discussed in chapter 12. We will see in chapters 15 and 16 some specific applications of this method to measuring the distance to galaxies.

luminous stars have a larger radius than less luminous ones. This follows simply from the fact that a bigger star has more surface so it can emit more light (see section 12.2). Why, though, does the larger radius make the larger star pulsate more slowly? Again, the answer is simple. Because an object's gravity weakens with increasing distance, other things being equal, a large-radius star has a weaker surface gravity than a small-radius star. Hence, gravity pulls more feebly inward for a large-radius star than for a small-radius one, so its pulsation takes longer. As a result, big (and therefore bright) stars pulsate more slowly than small (and therefore dim) stars.

## 13.6 DEATH OF STARS LIKE THE SUN

Once the Sun has become a yellow giant, it is nearing the end of its life, and as death approaches, its evolution speeds up. The Sun will spend about 10 billion years consuming its hydrogen and becoming a red giant. It will spend less than 100 million years (roughly 1% of its main-sequence lifetime) consuming its helium. The evolution will be faster because helium yields less energy than hydrogen when it is burned, and the Sun will be more luminous, so it will burn it faster. During the time it consumes the helium in its core, the star will lie between the main sequence and the giant region in the H-R diagram, a part of the diagram known as the horizontal branch.

### Ejection of a Low-Mass Star's Outer Layers

As the core burns its helium into carbon, its radius will once again shrink, compressing and heating it. That compression is unable to make the core hot enough to burn carbon, but it does make it hot enough to increase significantly the rate at which the star burns its fuel. The more rapid fuel consumption makes the star more luminous, inflating it to an even larger radius than before, making it a supergiant. As the star swells in size, its outer layers cool to about 2500 K—so cool that carbon and silicon atoms condense and form "flakes," just as water in our atmosphere forms snowflakes when it cools. These carbon and "rock" flakes (technically, what are called grains) do not fall into the star, however. Rather, they are pushed *outward* by the flood of photons pouring from the star's luminous core, just as dust in a comet is blown into a tail by the Sun's radiation. The rising grains drag gas with them and drive it into space, forming a huge shell around the star (fig. 13.15). The carbon grains (as well as other solid particles that condense in the star's cool atmosphere) survive their trip out of the star and drift through space. Some have been found in meteoritic material within our solar system, "fossils" from a bygone star.

**FIGURE 13.15**

Artist's conception of how radiation pressure in a luminous red giant pushes on carbon flakes in its atmosphere and drives off the gas. Magnetic forces probably play a role too. The ejected gas forms a shell around the star—a planetary nebula.

The gas shell expands, thins, and becomes transparent, allowing us to see through to the star's furiously hot core. Because it is so hot, the core's radiation is rich in ultraviolet light, which heats and ionizes the shell around it, making it glow. Astronomers have observed many such glowing shells around dying stars (fig. 13.16) and call them **planetary nebulas.** This term is unfortunate because planetary nebulas have nothing to do with planets. The usage survives from times when observers had only poor telescopes through which planetary nebulas looked like small disks, similar to planets.

## Planetary Nebulas

A planetary nebula shell contains about one-fourth solar mass of glowing gas but may have as much as several solar masses of cooler, nonluminous gas around it. Because of this uncertainty about how much matter is expelled, astronomers are not sure whether the Sun will form such a luminous shell or not. Shells are typically about one-fourth of a light-year in diameter and expand at about 20 kilometers per second. We can see this expansion by comparing photographs taken several decades apart. The shells eventually grow so big and diffuse that they mingle with the interstellar medium. Thus, the cycle of a star's life is complete: gas to star and back to gas. But the core of the star remains behind as a tiny, glowing ball that astronomers call a "white dwarf." Figure 13.17 shows these changes in an H-R diagram, affording a shorthand view of the life of a star like the Sun. A very different fate, however, awaits a more massive star.

## SCIENCE AT WORK

### PLANETARY NEBULAS

For many years, astronomers thought that the typical planetary nebula was ejected uniformly in all directions so that it formed a huge "bubble" around the star's core, as you might deduce from figure 13.16A. Astronomers knew of oddly shaped planetary nebulas but thought them to be the exception. This view has changed markedly over the last decade, especially now that more detailed pictures (such as fig. 13.16C, made with the Hubble Space Telescope) are available. Many such pictures show that the shell is *not* spherical and that the gas has been ejected mainly along two directions, presumably the star's rotation axis. Such ejection creates not a bubble but two oppositely directed cones, as sketched in figure 13.16D. Moreover, many planetary nebulas have extremely narrow jets of hot gas that are directed more or less along the cones, as you can see in figure 13.16E.

Why this happens is unclear, but it has forced astronomers to rethink their hypotheses about how planetary nebula form.

For example, some astronomers think that magnetic fields create the narrow jets and that the fields also create a pressure that helps expel the gas. Such magnetic effects occur on our Sun and help expel gas from it, although on a much smaller scale. Another hypothesis is that a ring of dust and gas in orbit around the star's equator blocks the ejection in those directions. Such a ring might form if a companion star's gravity draws some of the ejected material into a disk around both stars.

Whether such disks or rings give shape to all planetary nebulas is still unclear, but some of those that look circular merely have a rotation axis that happens to point toward us so we see their cones end-on, not from the side. Thus, we see how better data may require new ideas for its interpretation.

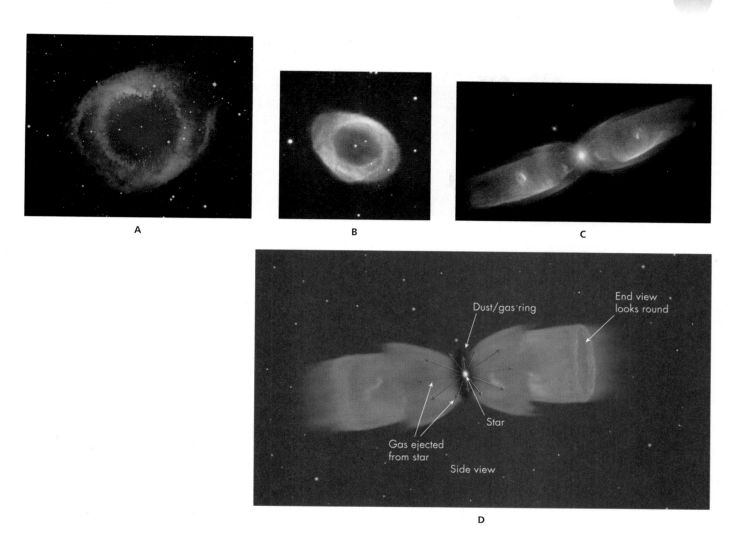

A

B

C

D

**FIGURE 13.16**
Pictures of several planetary nebulas. (A) The Helix Nebula. (B) The Ring Nebula. (C) The Butterfly Nebula. Notice the central star in each. Other stars that look as if they are inside the shell are foreground or background stars. (D) Sketch of how such a curious object as (C) might form. (E) Note the long, thin jets in this planetary nebula.

Q★ What will eventually happen to the central star? What will eventually happen to the nebula itself?

A★ The central star will eventually cool off and become a white dwarf. The nebula will expand, becoming ever larger, and eventually will mingle with the surrounding interstellar gas.

E

**FIGURE 13.17**
The evolution of a low-mass star such as the Sun plotted in an H-R diagram.

## 13.7 OLD AGE OF MASSIVE STARS

Massive stars do not become planetary nebulas or white dwarfs. Their great mass compresses and heats their cores enough to ignite carbon and allows them to keep burning when their helium is gone. A variety of other nuclear reactions may also occur that create heavy elements and at the same time supply energy for the star. Astronomers call the formation of heavy elements by such nuclear burning processes **nucleosynthesis.** Nucleosynthesis in stars was discussed in detail by E. Margaret Burbidge, Geoffrey R. Burbidge, William A. Fowler, and Fred Hoyle, who proposed in 1957 that *all* the chemical elements in our Universe heavier than helium were made in this way. Their idea is one of the triumphs of stellar evolution theory, and an enormous amount of theory and observation support it.

### Formation of Heavy Elements: Nucleosynthesis

Heavy elements form when two or more light nuclei combine and fuse into a single heavier one. The new nucleus must have a number of protons equal to the sum of

**TABLE 13.1**

## Some Nuclear Reactions That Form Heavy Elements

$4 \, ^1H \rightarrow \, ^4He$                    $^{12}C + \, ^{12}C \rightarrow \, ^{20}Ne + \, ^4He$

$3 \, ^4He \rightarrow \, ^{12}C$                  $^{16}O + \, ^{16}O \rightarrow \, ^{28}Si + \, ^4He$

$^4He + \, ^{12}C \rightarrow \, ^{16}O + \gamma$

KEY: H, hydrogen; He, helium; C, carbon; O, oxygen; Ne, neon; Si, silicon; $\gamma$, a gamma ray (high-energy photon).

NOTE: Helium is not considered a heavy element, but its production is shown in the table because it is the first step of nucleosynthesis in stars.

← Sun—for comparison

Red giant
(Approx. 30 times diameter of Sun)

— Hydrogen

— Helium
— Carbon
— Neon
— Oxygen
— Silicon
— Iron

Core of red giant
approx.
diameter of Earth

**FIGURE 13.18**
The layered structure of a massive star as it burns progressively heavier fuels in its core. (These inner zones are enormously magnified for clarity.)

the numbers found in its parent nuclei.* Likewise, it must contain as many neutrons as they possessed. For example, suppose $^{12}C$ and $^{16}O$ fuse. Carbon has 6 protons and 6 neutrons, and oxygen has 8 protons and 8 neutrons. Their fusion makes a nucleus with $6 + 8 = 14$ protons and $6 + 8 = 14$ neutrons (that is, an atom with 14 protons and 28 particles total in the nucleus). Reference to a table of the elements shows that such an atom is silicon, $^{28}Si$. Table 13.1 shows some of the reactions that make heavy elements in high-mass stars. The symbol $\gamma$ is used for a high-energy photon (a gamma ray), which is produced in many of the reactions.

In massive stars, these reactions provide enough energy to support the stellar mass against gravity's unceasing pull. As each fuel—hydrogen, helium, carbon—is exhausted, the star's core contracts and heats by compression. The higher temperature then allows the star to burn still heavier elements, the "ashes" of one set of nuclear reactions becoming the fuel for the next set. Oxygen, neon, magnesium, and, eventually, silicon are formed, but because progressively higher temperatures are needed for each new burning process, each new fuel is confined to a smaller and hotter region around the star's core. Thus, the star develops a layered structure, as illustrated in figure 13.18. The star's surface, where the temperature is too low for nuclear reactions to occur, remains hydrogen, but beneath the surface lies a series of nested shells, each made of a heavier element than the one surrounding it. Helium surrounds a shell of carbon, which in turn surrounds a shell of neon, and so on to the core. By the time the star is burning silicon into iron, its core has shrunk to a diameter smaller than the Earth's, and the core's temperature is about 2 billion K. A massive star (30 M$_\odot$ or more) may take less than 10 million years from its birth to form an iron core. This brief lifetime results because (1) the star burns its fuel rapidly to offset the energy lost because of its high luminosity, and (2) the structure of atomic nuclei is such that fusion of elements heavier than hydrogen yields much less energy than the fusion of a similar amount of hydrogen, and so more fuel must be burned to supply the same amount of energy.

## Core Collapse of Massive Stars

The formation of an iron core signals the end of a massive star's life. Iron cannot burn and release energy: the iron nucleus turns out to be the most tightly bound of all nuclei. As a result, attempting to fuse additional protons or neutrons to it *weakens* the bonds and *absorbs* energy, rather than releasing it. Thus, nuclear fusion stops with iron, and a star with an iron core is out of fuel.

Fuel exhaustion causes a star's core to shrink and heat. But in high-mass stars, the shrinkage presses the iron nuclei so tightly together that a new reaction can occur: protons and electrons may themselves "merge," neutralizing their charge and becoming neutrons. The shrinking core is thus transformed from a sphere of iron into a sphere

---

*In some reactions, such as the proton–proton chain, a proton may turn into a neutron and a positron.

of neutrons, with catastrophic results for the star. Most of the pressure that supported the core was supplied by electrons, but they have been absorbed by protons. Thus, the star's core pressure suddenly drops. Nothing remains to support the star, and so its interior begins to collapse. However, because the matter is so dense, its force of gravity is immense, and it crushes the core. In less than a second, the core shrivels from an iron ball the size of the Earth to a ball of neutrons about 10 kilometers (about 6 miles) in radius. The star's outer layers, with nothing to support them, plummet inward like a tall building whose first floor is suddenly blown away. The plunging outer layers of the star strike the neutron core, crushing it still more, while the impact heats the infalling gas to billions of degrees. The pressure surges and lifts the outer layers away from the star in a titanic explosion—a **supernova.**

Some recent computer simulations of the collapse of a massive star show that its core may be set violently vibrating. These vibrations then heat the gas around the core and thereby help power the star's explosion.

## Supernova Explosions

The explosion of a massive star* mixes the elements synthesized by nuclear burning during its evolution with the star's outer layers and blasts them into space. This incandescent spray expands away from the star's collapsed core at more than 10,000 kilometers per second. Depending on the star's mass, 10 or so solar masses of matter may be flung outward, ultimately to mix with surrounding material and, in time, to form new generations of stars. Interstellar gas is thereby enriched in heavy elements, the atoms needed to build the rock of planets and the bones of living creatures. Indeed, the supernova outburst itself generates even more heavy elements. The explosion creates free neutrons that rapidly combine with atoms in the star to build up heavy and rare elements such as gold, platinum, and uranium.

For the star, the supernova explosion is a quick and glorious death. In a few minutes, it releases more energy than it has generated by nuclear burning during its entire existence. In a matter of hours, the dying star brightens to several billion times the luminosity of the Sun, radiating energy in its death throes at a greater rate than all the other stars in the galaxy combined. Figure 13.19 is a picture of a galaxy with a supernova: the exploding star gleams like a beacon.

A supernova emits more than just visible light: most of the energy of its blast is carried by a burst of neutrinos, the same tiny, highly penetrating particles generated in the Sun as it converts hydrogen into helium. Just as the Sun's neutrinos escape freely into space, so the supernova's neutrinos escape as well. A pulse of such neutrinos was detected in February 1987 when a supernova—SN 1987A—blew up in the Large Magellanic Cloud. You may have "seen" one of those neutrinos because more than 1 trillion of them passed through you, and if one happened to collide with an electron in your eyeball, it would have made a tiny flash of light.

## Supernova Remnants

Gas ejected by the supernova blast plows through the surrounding interstellar space, sweeping up and compressing other gas that it encounters. The huge, glowing cloud of stellar debris—a **supernova remnant**—steadily expands. At the end of one year, it is 300 billion kilometers (about 0.03 light-years) across. At the end of one century, it is several light-years in diameter, but its expansion slows as it runs into more surrounding gas. Figure 13.20 shows pictures of several supernova remnants of different ages. Notice how ragged the remnants in figures 13.20A and 13.20B look compared with the smoothly ejected bubbles of planetary nebulas—evidence that massive stars die more violently than low-mass stars.

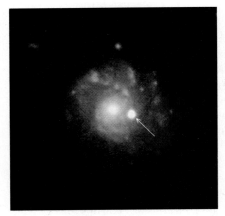

**FIGURE 13.19**
A supernova explosion in a distant galaxy observed in 2002 shines brightly compared to the billions of stars in the galaxy.

---

*We will see in chapter 14 that low-mass stars can also explode as supernova if they are members of a binary star system (see section 14.1).

**A**

**B**

**C**

**FIGURE 13.20**

A few supernova remnants: (A) Color photograph of the Vela supernova remnant. (B) Color picture of the Crab Nebula. (C) X-ray image of the Cassiopeia supernova remnant. In this image, the red on the left is radiation from hot iron atoms and the bright green is radiation from silicon and sulfur atoms.

One such violent outburst was seen about 1000 years ago by astronomers in China and elsewhere in the Far East. On a date that according to our calendar was July 4, 1054, their records describe a "guest star" that appeared in the evening sky and that was visible even in broad daylight for several weeks. Today, with even a small telescope, you can still see the glowing gases ejected from that dying star. Figure 13.20B is a picture of this perhaps most famous of all supernova remnants, now known as the **Crab Nebula.** In chapter 14, we will learn that the star's core survived the blast and lies in the expanding debris.

A supernova explosion marks the death of a massive star, and we chart its complete life in the H-R diagram shown in figure 13.21. We can see that its path through time is far simpler than that of a low-mass star. The massive star begins its life on the upper main sequence as a blue star. As its core hydrogen is consumed, it leaves the main sequence, swelling and cooling to become a yellow supergiant while it begins to burn helium in its core. When the helium there is consumed, the star switches to other fuels, all the while contracting and heating its core and expanding and cooling its outer layers to become a red supergiant. Overall, it keeps its luminosity approximately constant as it ages, but some massive stars, depending on how many heavy elements are incorporated in them at their birth, may return partway to the main sequence as they switch to new fuels. The resultant heating may turn them into blue supergiants before they explode as supernovas. Such was the behavior of Supernova 1987A, whose explosion is documented in figure 13.22.

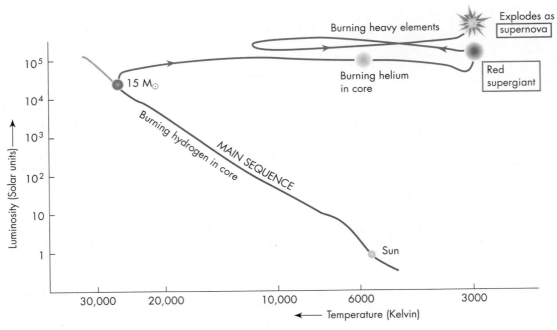

**FIGURE 13.21**
H-R diagram of a massive star's evolution.

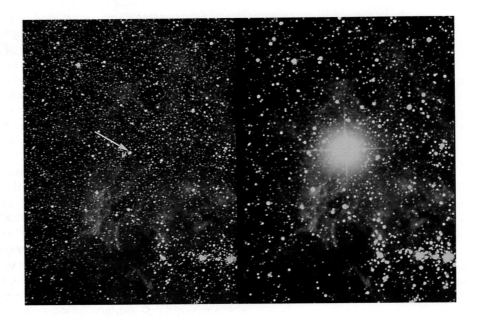

**FIGURE 13.22**
Photograph of a portion of the Large Magellanic Cloud (a small, nearby galaxy) before Supernova 1987A exploded, and a second photograph taken during the event.

Supernova remnants eventually slow down, cool, and mingle with the interstellar clouds around them. But the remnant's gas is rich in heavy elements that were formed in the core of the star that blew up as a supernova. Thus, when the remnant mixes with an interstellar cloud, the cloud is enriched in heavy elements. Eventually the enriched cloud may collapse and turn into a new generation of stars. These new stars will, therefore, contain more heavy elements than earlier generations of stars. In this way, the amount of heavy elements in stars increases from generation to generation.

What remains when the supernova remnant dissipates? The core of the star survives either as a ball of neutrons (a neutron star) or perhaps as an even more compressed body—a black hole—as we will discover in chapter 14.

## 13.8 HISTORY OF STELLAR EVOLUTION THEORIES

The theories we have discussed above about how stars work and evolve are relatively new. For example, Aristotle wrote more than 2000 years ago that stars are heated by their passage through the heavens. He never considered the possibility that they might evolve. In the eighteenth century, the German philosopher Immanuel Kant described stars slightly more scientifically, envisioning the Sun as a fiery sphere, formed when gases gravitated to the center of the solar nebula.

The English physicist Lord Kelvin and the German physicist Hermann von Helmholtz were the first scientists to use the equations of basic physics to arrive at a mathematical description of what we would today call "model stars." In the 1850s and 1860s, they were searching for a mechanism that could explain the Sun's power supply, as discussed in chapter 11. Kelvin and Helmholtz independently solved the mathematical equations for a gas sphere in which pressure is balanced by gravity and showed that a star's core is compressed by the weight of the overlying gas and is much denser than the outer parts of the star. They further showed that as gravity compresses a star, that compression converts gravitational energy into heat, raising the star's temperature. However, the amount of energy available from gravitational energy is far too small to power it for billions of years.

Astronomers had to await discoveries about the nature of atoms in the early years of the twentieth century before they could solve the mystery of stellar energy. The British astrophysicist Sir Arthur Stanley Eddington, who discovered the mass–luminosity relation, was one of the first scientists to recognize the immense power available in the form of nuclear energy, but at the time, too little was known about how atoms interacted to make models of stars powered by hydrogen burning. Moreover, Eddington discovered that to make a model star, more was required than simply balancing pressure and gravitational forces. He demonstrated that a correct model must also account for the outward leakage of energy from the star's core to its surface, a process that made solving the equations much more difficult.

The equations that govern a star's structure express a complicated set of relations between the star's pressure, density, temperature, gravitational force, energy generation, composition, and luminosity. The equations must balance at each point in the star and yield answers that agree with observations. Solving such complex equations is extremely difficult but became easier in the 1940s, when astronomers began using mechanical calculators. These machines—incredibly primitive compared with a modern computer—were operated by turning a hand crank. Nevertheless, they were faster than pencil-and-paper solutions, and the models so produced revealed more clearly the physical processes in stars. Several months might be needed to make the calculations for a half-dozen models that showed how a main-sequence star could change into a red giant. Then, in the 1950s, when astronomers began to use electronic computers to calculate model stars, only a few minutes were needed to solve the equations of a stellar model. By combining the results of dozens of such models, each of which represented a star with successively more of its precious fuel consumed, astronomers could see how stars evolved. Such work continues to this day, with astronomers still trying to understand the late stages of a star's life, when it is losing mass and burning elements heavier than helium in its core.

## 13.9 TESTING STELLAR EVOLUTION THEORY

We have seen how astronomers think stars evolve, but what evidence supports these theories? For example, although we now understand that main-sequence stars turn into red giants, as recently as the 1940s some astronomers thought the reverse. The best demonstration that the modern theory is correct comes from comparing the H-R diagrams of real groups of stars with theoretically determined H-R diagrams.

We learned in section 13.2 that stars probably form in groups—star clusters. The stars in a cluster are bound together by their mutual gravitational attraction, although each star moves along its own orbit within the cluster. All the stars within a given cluster form at approximately the same time and are therefore nearly the same age. If we look at the stars in a cluster shortly after its birth, when they have all completed their contraction stage from protostars, they will all lie on the main sequence. If we look at some later time, some of the stars will have consumed their core hydrogen and evolved off the main sequence. To check our theory of stellar evolution, we calculate evolutionary tracks, such as those in figures 13.17 and 13.21, for every star on the main sequence to show where each star will be, at say, 10 million years, 100 million years, and so on. The resulting curves show us what the H-R diagram of the entire cluster should look like 10 million years, 100 million years, and so on, after its birth. When we compare such curves with H-R diagrams of actual star clusters, the match is excellent. If our theory of stellar evolution were wrong, the shapes would be unlikely to agree so well. In addition, the models help astronomers understand why so many kinds of stars exist. For example, the models not only offer a natural explanation of main-sequence, red-giant, and white-dwarf stars but also help us see how such totally different objects, such as pulsating variables, planetary nebulas, and supernovas, fit into the scheme of stellar evolution. This success in interpreting such stellar diversity is evidence that the theory is essentially correct. Moreover, that success offers astronomers a way to measure the age of a star cluster.

# EXTENDING OUR REACH

## MEASURING THE AGE OF A STAR CLUSTER

Astronomers can deduce the age of a star cluster from the shape of its H-R diagram. To understand why its H-R diagram reveals the cluster's age, recall that in a star cluster, all the stars form at approximately the same time. Moreover, at their birth, all the stars are burning hydrogen in their core. Because the main sequence is determined by the location of stars burning hydrogen in their core, all the stars in a newly formed cluster lie on or near the main sequence. Not all the cluster's stars burn their fuel at the same rate, however. Massive stars burn their fuel more rapidly (to maintain their high luminosity) than low-mass stars do. With their hydrogen used up, high-mass stars leave the main sequence and turn into red giants. The low-mass stars, on the other hand, still have hydrogen to burn, and so they remain on the main sequence longer.

Because the high-mass stars in an old cluster turn into red giants (and thus lie off the main sequence), a line in the H-R diagram connecting the position of the cluster's

These more massive stars have used up their core hydrogen and have left the main sequence.

Less massive stars of the cluster still have unburned fuel. They remain on the main sequence.

Turnoff point. This star is just about to run out of fuel.

Luminosity

← Temperature

**BOX FIGURE 13.1**
The turnoff point in the H-R diagram of a star cluster indicates its age.

stars bends away to the right, as shown in box figure 13.1. The point where that line bends away from the main sequence is called the **turnoff point**. At the turnoff point, a star is not quite old enough to have used up the hydrogen in its core. That is, its age is just a tiny bit less than its main-sequence lifetime. But *all* stars in the cluster are the same age, namely, the age of the cluster. To get the cluster's age, we therefore measure the age of the star at the turnoff point by calculating its main-sequence

lifetime from its mass and luminosity, as described earlier. The answer we get is the age of the star cluster.

Box figure 13.2 shows a series of H-R diagrams for clusters ranging in age from very young to very old (a few million years for the youngest to more than 10 billion years for the oldest). Notice that old clusters have few, if any, stars on the upper part of the main sequence. Young clusters, on the other hand, do have stars on the upper main sequence.

**BOX FIGURE 13.2**
Schematic H-R diagrams of clusters of different ages illustrating the turnoff point and how it shifts down and to the right for older clusters.

# SUMMARY

A star forms from interstellar gas drawn together by gravity, which compresses and heats the gas to form a protostar. Further heating causes the core of the protostar to fuse hydrogen into helium. The energy released keeps the core hot and thereby maintains an outward pressure that stops the protostar's collapse and changes it into a main-sequence star. When the hydrogen is exhausted in a star's core, the core shrinks and heats, making the star swell into a red giant.

A low-mass star such as the Sun continues to burn hydrogen in a shell during its first red-giant stage but eventually compresses the helium in its core enough to ignite the gas in a helium flash. The energy released expands the core, and the outer layers shrink, turning the star into a yellow giant. As the star burns helium, its core again shrinks, and it once more becomes a red giant. It is so luminous now that its radiation drives off its outer layers to form a planetary nebula shell. The core remains as a small, dense star: a white dwarf.

**TABLE 13.2**

## Evolution of Low-Mass and High-Mass Stars

| Low-Mass Star (Mass Less Than 10 M$_\odot$) | High-Mass Star (Mass More Than 10 M$_\odot$) |
|---|---|
| *Protostar* (gravity supplies energy) | *Protostar* (gravity supplies energy) |
| *Main sequence* (hydrogen burns in core)<br>  Core hydrogen is consumed. | *Main sequence* (hydrogen burns in core)<br>  Core hydrogen is consumed. |
| *Red giant* (hydrogen burns in shell)<br>  Core contracts, heats, and ignites helium (helium flash). | *Yellow supergiant* (helium burns in core)<br>  Star pulsates. |
| *Yellow giant* (star pulsates) | *Red supergiant* (heavier elements burn in a series of shells<br>  in the star's core)<br>  Buildup of iron core and its collapse |
| *Red giant* (helium burns in shell)<br>  Outer layers ejected to form: | *Supernova* explosion when core collapses;<br>  remnant may be a neutron star or black hole |
| *Planetary nebula*<br>  Core cools to become: | |
| *White dwarf* (no energy generation) | |

Because of the tremendous heating from gravitational compression of their cores, high-mass stars more easily ignite fuels heavier than hydrogen and undergo less dramatic changes initially than low-mass stars do. They build in their cores layers of heavy elements: carbon, neon, silicon, and, eventually, iron. The iron core cannot burn and will collapse, triggering a supernova explosion. The heavy elements produced during the star's life and in the supernova that destroys it are blasted into space, enriching the interstellar medium. The core remains as a ball of neutrons—a neutron star—or may collapse entirely and become a black hole.

The lifetime of a star is dependent on its mass. High-mass stars burn their fuel quickly to supply the energy they need to support their great weight. These stars are blue because they are so hot. Thus, blue stars burn out quickly and must be young.

Both high- and low-mass stars pass through a stage late in their lives when they pulsate as yellow giants. Low-mass stars become RR Lyrae variables, whereas high-mass stars become Cepheids.

Table 13.2 summarizes these stages.

## ?What It Is.

These two pictures show the red supergiant Mira, a long-period variable star, at its brightest and at its dimmest. The star's brightness changes with a period of about 332 days, as you can easily see by watching it once each month over a year. In diameter it is one of the largest known stars. Its size varies as it pulsates, but it is typically about 10 AU in diameter.

## QUESTIONS FOR REVIEW

1. Sketch an H-R diagram of the evolutionary path of the Sun.
2. Sketch an H-R diagram of the evolutionary path of a 15 M$_\odot$ star.
3. Describe how a protostar is believed to form.
4. What heats a protostar?
5. What is a bipolar flow?
6. How can we observe protostars? Why are they surrounded by dust and gas?
7. What is a T Tauri star?
8. What is a Bok globule?
9. What determines when a star becomes a main-sequence star?
10. What determines how long a star stays on the main sequence?
11. What makes a star move off the main sequence?
12. Where do main-sequence stars end up as they evolve?
13. Why do high- and low-mass stars evolve differently as they become red giants?
14. Why is it easier for a high-mass star than for a low-mass star to burn helium?
15. What is meant by a pulsating star?
16. Why do stars pulsate?
17. What happens to a solar-mass star when it starts to burn helium in its core? What does it turn into?
18. What is a planetary nebula?
19. What is one explanation for how a low-mass star expels its outer layers to make a planetary nebula?

20. What is left when a planetary nebula dissipates?

21. What makes a high-mass star's core collapse?

22. How do the neutrons form in a massive star's remnant core? What other kind of particle is made along with the neutrons?

23. What is a supernova explosion?

24. What kind of subatomic particles have been observed when a supernova explodes?

## THOUGHT QUESTIONS

1. Suppose that stars could mix unburned fuel from their outer layers into their cores. Would that alter the way they evolve? Can you suggest some possible differences?

2. Hold a small rubber ball on top of a basketball, and drop them together toward the floor. What happens to the small ball? Does that help you understand what happens to the outer layers of a supernova as they collapse on the core?

3. Inflate a balloon and carefully measure its size. Put it in the freezer for a few hours. Does it look the same when it is cold? How does this relate to how stars form in cold regions of space?

4. Take a plastic bottle and put a little soapy water in it. Run your finger across the mouth to make a soap film. Now, without breaking the film, run hot water over the bottle. What happens to the soap film? How does this relate to what happens to a star when it is heated?

5. In some very old, dense star clusters, a few blue stars, known as "blue stragglers," are seen. Why are blue stars unexpected in such clusters? If stars that stick together mix their material thoroughly when they collide, how might you explain such blue stars?

## PROBLEMS

1. Calculate the main-sequence lifetime of the Sun by first determining the rate at which it burns hydrogen (refer to chapter 11) and then dividing that rate into its core mass. If your answer is in seconds, convert to years, given that there are about $3.2 \times 10^7$ seconds in a year.

2. Calculate the escape velocity from a red giant's atmosphere (use the formula for escape velocity from chapter 2). Assume that the star's mass is 1 $M_\odot$ and its radius is 100 $R_\odot$ How does this compare with the speed at which a planetary nebula shell is ejected?

3. How long will it take a planetary nebula shell moving at 20 kilometers per second to expand to a radius of one-fourth of a light year?

4. Figure 13.23 shows the H-R diagram of a star cluster. How old is it? (The dashed line is the position of the stars. The solid line is the main sequence.)

5. Calculate the average density of a red giant star in grams per cubic centimeter. Take the star's mass to be $2 \times 10^{33}$ gm and its radius to be $3 \times 10^{12}$ cm. How does this density compare with that of the air we breathe (about $10^{-3}$ gm/cm$^3$)?

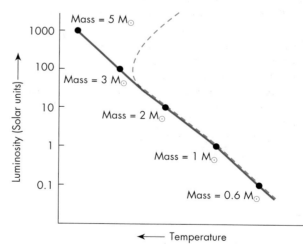

**FIGURE 13.23**
H-R diagram of a star cluster.

## TEST YOURSELF

1. Which of the following sequences correctly describes the evolution of the Sun from young to old?

   (a) White dwarf, red giant, main-sequence, protostar

   (b) Red giant, main-sequence, white dwarf, protostar

   (c) Protostar, red giant, main-sequence, white dwarf

   (d) Protostar, main-sequence, white dwarf, red giant

   (e) Protostar, main-sequence, red giant, white dwarf

2. A planetary nebula is

   (a) another term for the disk of gas around a young star.

   (b) the cloud from which protostars form.

   (c) a shell of gas ejected from a star late in its life.

   (d) what is left when a white dwarf star explodes as a supernova.

   (e) the remnants of the explosion created by the collapse of the iron core in a massive star.

3. Stars like the Sun probably do not form iron cores during their evolution because

   (a) all the iron is ejected when they become planetary nebulas.

   (b) their cores never get hot enough for them to make iron by nucleosynthesis.

   (c) the iron they make by nucleosynthesis is all fused into uranium.

   (d) their strong magnetic fields keep their iron in their atmospheres.

   (e) None of the above

4. As a star like the Sun evolves into a red-giant, its core

   (a) expands and cools.

   (b) contracts and heats.

   (c) expands and heats.

   (d) turns into iron.

   (e) turns into uranium.

5. A star whose mass is 5 times larger than the Sun's and whose luminosity is 100 times larger than the Sun's has a main-sequence lifetime about

   (a) 5 times longer than the Sun's.

   (b) 500 times longer than the Sun's.

   (c) 5 times shorter than the Sun's.

   (d) 100 times shorter than the Sun's.

   (e) 20 times shorter than the Sun's.

## FURTHER EXPLORATIONS

Balick, Bruce, and Adam Frank. "The Extraordinary Deaths of Ordinary Stars." *Scientific American* 291 (July 2004): 51.

Bethe, Hans A., and Gerald Brown. "How a Supernova Explodes." *Scientific American* 252 (May 1985): 60.

de Grasse Tyson, Neil. "Forged in the Stars." *Natural History* 105 (August 1996): 72.

Frank, Adam. "Blowing Cosmic Bubbles." *Astronomy* 25 (February 1997): 36.

Herbst, William, and George C. Assousa. "Supernovas and Star Formation." *Scientific American* 241 (August 1979): 138.

Hillebrandt, Wolfgang, Hans-Thomas Janka, and Ewald Muller. "How to Blow Up a Star." *Scientific American* 295 (October 2006): 42.

Iben, Icko, Jr., and Alexander Tutukov. "The Lives of Stars: From Birth to Death and Beyond (Part I)." *Sky and Telescope* 94 (December 1997): 36.

Jastrow, Robert. *Red Giants and White Dwarfs*. New York: Norton, 1990.

Kawaler, Steven D., and Donald E. Winget. "White Dwarfs: Fossil Stars." *Sky and Telescope* 74 (August 1987): 132.

Kirshner, Robert P. "The Earth's Elements." *Scientific American* 271 (October 1994): 58. (How the elements formed.)

Kwok, Sun. "Butterflies and Crabs of the Southern Sky." *Sky and Telescope* 103 (January 2002): 48.

———. *Cosmic Butterflies: The Colorful Mysteries of Planetary Nebulae*. New York: Cambridge University Press, 2001.

———. "What Is the Real Shape of the Ring Nebula?" *Sky and Telescope* 100 (July 2000): 32.

Ray, Thomas P. "Fountains of Youth: Early Days in the Life of a Star." *Scientific American* 283 (August 2000): 42.

Stahler, Steve W. "The Early Life of Stars." *Scientific American* 265 (July 1991): 48.

Stephens, Sally. "Needles in the Cosmic Haystack." *Astronomy* 23 (September 1995): 50. (About brown dwarfs.)

Tayler, R. "The Birth of Elements." *New Scientist* 124 (December 16, 1989): 25.

Woosley, Stan, and Tom Weaver. "The Great Supernova of 1987." *Scientific American* 261 (August 1989): 32.

**Website**

Please visit the *Explorations* website at **http://www.mhhe.com/arny** for additional online resources on these topics.

## KEY TERMS

bipolar flows, 390

Crab Nebula, 405

degenerate gas, 396

helium flash, 396

instability strip, 398

interstellar cloud, 385

main-sequence lifetime, 393

nucleosynthesis, 402

period–luminosity law, 398

planetary nebulas, 400

protostar, 388

red giant, 394

shell source, 394

supernova, 404

supernova remnant, 404

triple alpha process, 395

T Tauri stars, 390

turnoff point, 409

# Stellar Remnants: White Dwarfs, Neutron Stars, and Black Holes

# 14

Debris from a supernova explosion.

## Chapter Objectives

- When stars use up all their available nuclear fuel, they collapse.
- Low-mass stars like the Sun turn into white dwarfs.
- High-mass stars (ten or so times the Sun's mass) may explode as supernova.
- The core of a high-mass star may survive as either a neutron star or a black hole.
- White dwarfs are about the size of the Earth but have a mass about that of the Sun.
- Neutron stars have a mass slightly more than that of the Sun but are a mere 10 kilometers or so in diameter.
- A black hole forms if the core of a massive star becomes so dense that the escape velocity at its surface reaches the speed of light.
- Black holes emit no electromagnetic radiation but may be detected by observing gas that is heated as it falls into the hole.

**?What Is This?**
(See chapter summary for the answer.)

413

During its life, every star supports itself against gravity by burning nuclear fuel in its core. When its fuel is spent, the star collapses. This fate awaits the Sun 5 billion years from now. Gravity will crush it into a white dwarf—a star about 100 times smaller than the present Sun and roughly the size of the Earth. More massive stars surrender sooner and are more dramatically squeezed to even smaller dimensions, becoming either neutron stars or black holes.

**Compact stars,** as these three kinds of stellar remnants are known, are the end points of stellar evolution. Because nearly all stars must eventually reach this stage, the galaxy is littered with their shriveled bodies. But unlike stars at earlier stages of evolution, no nuclear fuel makes them shine. They shine—if at all—with heat inherited from their previous state. Their matter, too, is unusual. In crushing compact stars to their tiny dimensions, gravity squeezes them into exotic materials. For example, a piece of white dwarf material the size of an ice cube would weigh about 16 tons. Matter in a neutron star is so compressed that electrons have merged with protons, making the star resemble a giant atomic nucleus. So unmercifully has gravity squeezed the most massive of compact stars that they have collapsed completely, their immense gravity warping the space around them so that no light escapes, making them black holes in space.

Compact, however, does not mean inconspicuous: some of these crushed stars radiate intensely despite having no fuel of their own. Rather, a compact star may "parasitize" a companion star, capturing matter from it that may burn explosively and allow the crushed star to "rise from the dead" as a nova—a "new" star. Moreover, because of a compact star's intense gravity, any material that falls on it releases such immense amounts of gravitational energy that this dead star may be thousands of times brighter than a "live" star like the Sun. But this brightness is bought at a dreadful price. The star's mass may increase to the point that gravity causes it to collapse even further or, for some stars, to explode as a type I supernova, blasting the star to atoms.

## 14.1    WHITE DWARFS

### General Properties, Origin, and Fate

**White dwarfs** are hot, compact stars whose mass is comparable to the Sun's but whose diameter is about the same as the Earth's (about 0.01 that of the Sun). Their tiny size gives them so small a surface area that they are very dim despite being hot. Even the brightest white dwarfs are about 100 times fainter than the Sun. Their pale light is not supplied by burning fuel, for they have exhausted all fuel supplies. Rather, it comes from residual heat that leaks from the star's interior. Because they shine by residual heat and are therefore steadily cooling, white dwarfs show a wide range of surface temperatures. Some are as hot as 25,000 K; others are as cool as about 4500 K. A typical surface temperature might be about 10,000 K.

A white dwarf forms when a low-mass star such as the Sun nears the end of its life. When the star swells into a red giant, its intense radiation strips off its outer layers to form a planetary nebula shell. This process exposes the star's core, which becomes the white dwarf, as illustrated in figure 14.1. The composition of the core (and thus the white dwarf) is mainly carbon and oxygen, the end products of the parent star's nuclear burning. The white dwarf's surface is composed of a thin layer of hydrogen and helium, but its core has too little mass and therefore insufficient gravity to contract and heat itself to the ignition temperature of carbon. So, unable to burn further, the dying core simply cools and grows dim.

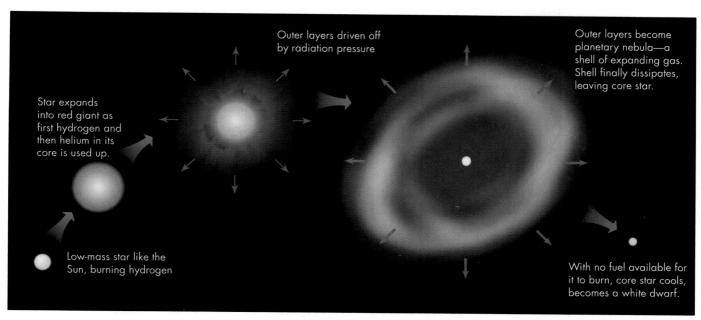

Star expands
into red giant as
first hydrogen and
then helium in its
core is used up.

Low-mass star like the
Sun, burning hydrogen

Outer layers driven off
by radiation pressure

Outer layers become
planetary nebula—a
shell of expanding gas.
Shell finally dissipates,
leaving core star.

With no fuel available for
it to burn, core star cools,
becomes a white dwarf.

**FIGURE 14.1**

Origin of a white dwarf. White dwarfs form when a low-mass star such as the Sun reaches the end of its life. Having expanded into a red giant for the second time (see fig. 13.2), the star loses its outer atmosphere and becomes a planetary nebula. Its core is exposed, and eventually, it turns into a white dwarf, a star with no source of nuclear fuel. (Note: not all stages in a star's evolution are shown.)

Although many white dwarfs form by a roughly solar-mass star ejecting a planetary nebula shell,* in theory, some could form from very low-mass stars (less than about 0.4 or 0.5 $M_\odot$) without such ejection. These stars are relatively cool throughout, and so their gas, like the relatively cool matter just below the Sun's surface, must carry energy by convection. The convection currents in these small stars extend throughout their spheres, mixing their gas and preventing a core from forming. Moreover, their weak gravity cannot compress the spent fuel to make them red giants. Because low-mass stars evolve so slowly, however, they require longer than the age of the Universe to reach this point in their evolution, and therefore none of these very low-mass stars has turned into white dwarfs yet.

What happens to a white dwarf as time passes? Although it is born very hot (recall that it was the core of a star), it has no fuel to burn to keep it hot. Like a dying ember, therefore, it rapidly (by stellar standards) cools. Astronomers calculate that it takes only about 10 million years[†] for the temperature of a white dwarf to drop to 20,000 K and more than the age of the Universe to grow so cool that it emits no visible light. Astronomers call such dead, dark stars **black dwarfs.** Cooling white dwarfs are probably very abundant in our galaxy. How abundant is still controversial.

## Structure of White Dwarfs

Because white dwarfs are very compact and have no fuel supply, their structure differs significantly from that of ordinary stars. Although they are in hydrostatic equilibrium, with pressure balancing gravity, their pressure arises from a peculiar interaction among

---

*"Roughly solar mass" in this case means up to about 8 $M_\odot$. Astronomers are not sure whether all such stars eject a planetary nebula shell, but both theory and observation suggest stars as massive as 8 $M_\odot$ can become white dwarfs if they somehow lose most of their mass, perhaps through a powerful stellar wind.

[†]The equivalent of about one month in a human lifetime.

the electrons that limits the number that can occupy a given volume. This gives white dwarfs an odd property: increasing their mass makes them shrink! Even more crucial, however, white dwarfs must have a mass below a critical limit, or they collapse. To see why, we must consider conditions in a white dwarf's interior.

White dwarfs are very dense, having formed from the shrunken cores of their parent stars. If we divide the mass of a white dwarf by its volume, we find that its density is about $10^5$ grams per cubic centimeter (roughly 1 ton per cubic inch). This high density packs the star's atoms so closely that they are separated by less than the normal radius of an electron orbit. If electrons could remain attached to nuclei under these conditions, their orbits would overlap those of neighboring atoms. But electrons can easily escape from their nuclei at such high density and move freely between them, allowing the electrons to interact in a peculiar way.

## Degeneracy and the Chandrasekhar Limit

A law of subatomic physics called the **exclusion principle** limits the number of electrons that can be put into a given volume. This, in turn, limits how closely electrons can be squeezed, thereby creating a pressure that, to distinguish it from that of an ordinary gas, is called **degeneracy pressure.** Degeneracy pressure depends only on the gas density, not on its temperature. In an ordinary gas, pressure depends on temperature and density. When such a gas is compressed, both the heat and the higher density contribute to raising its pressure. When a degenerate gas is compressed, it heats up, but the heat does not raise its pressure as it does for an ordinary gas. This makes degenerate gas less "springy" than ordinary gas.

If mass is added to a white dwarf, its radius *shrinks*. The extra mass increases the star's gravity, compressing it. Its pressure rises, but because the star is degenerate, the rise is small and is unable to prevent the star from being squashed into a smaller volume. Squashing the star decreases the distance between particles and makes its gravity larger still, squashing the star even more. In fact, gravity may ultimately overwhelm the star's pressure forces and crush the star completely. Thus, too much mass makes a white dwarf collapse, as was demonstrated by theoretical calculations made in 1931 by the Indian astrophysicist Subrahmanyan Chandrasekhar (pronounced *chan-dra-say-car*). Chandrasekhar won the Nobel Prize for Physics in 1983 for these calculations and related work. The limiting mass of a white dwarf is now called the **Chandrasekhar limit** in his honor. For a typical white dwarf, the limiting mass is about 1.4 $M_\odot$, but this depends somewhat on the white dwarf's composition.

Measurement of white dwarf masses made after Chandrasekhar's work was completed show that all white dwarfs are subject to this limit. For those in binary star systems, the masses can be measured using the modified form of Kepler's third law. For isolated white dwarfs, calculations are based on the shift of their spectral lines caused by their immense gravity.

When light escapes from a body, it must work against the force of gravity. The escaping light acts like a ball rolled up a slope: as the ball moves higher, it slows down as it loses energy. Light cannot slow down—it must always travel at the constant speed, c—but it too can lose energy, and it does so. Light's energy, as you have learned, determines its wavelength: lower energy corresponds to longer wavelengths. Longer wavelengths, in turn, correspond to redder colors; thus, as light "climbs" away from a body, the body's gravity stretches the wavelengths and creates a **gravitational redshift,** as shown in figure 14.2. The amount of redshift depends on the star's mass and radius, and from the shift, the white dwarf's mass can be estimated if its radius is known* (from

Ball rolling up a hill loses energy and slows down.

**A**

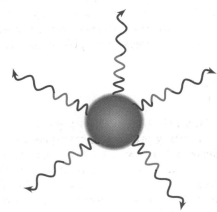

**B**    Light moving away from a star loses energy.

**FIGURE 14.2**

The gravitational redshift. (A) Just as a ball rolling up a hill loses energy, so too does light as it moves away from a star. (B) Light does not slow down, however. Instead, because its energy, $E$, determines its wavelength, $\lambda$, according to the relation $E = $ Constant$/\lambda$, as the light's energy decreases, its wavelength increases. Longer wavelengths correspond to redder colors, and so we refer to the lengthening of the light's wavelength caused by the star's gravity as the gravitational redshift.

---

*To use this method successfully, astronomers need to know the original wavelengths of the spectrum lines. These can be found from the Doppler-shift law if the white dwarf's velocity is known from the motion of a companion star and if the kind of atoms creating the lines can be identified.

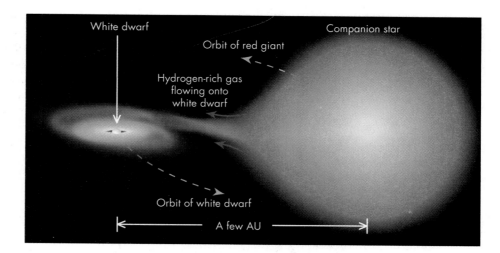

**FIGURE 14.3**

A white dwarf accreting (pulling in) mass from a binary star companion, which in this case is shown as a red giant. The typical separation of the stars is a few AU.

its luminosity and temperature, for example). The greater the redshift, the greater is the mass that causes it.

Spectral lines also permit astronomers to measure the magnetic fields of white dwarfs. We discussed in chapter 11 how the Sun's magnetic field can be measured from the splitting of its spectral lines caused by the Zeeman effect. Similar observations show that some white dwarfs have magnetic fields 10,000 times stronger than sunspots—$10^8$ times that of the Earth. These immense field strengths occur because as the white dwarf shrinks, it draws its magnetic field lines closer together, amplifying the field.

## White Dwarfs in Binary Systems: Novas and Supernovas of Type I

Isolated white dwarfs cool off and disappear from sight, but in binary systems, they may heat up and explode. If a white dwarf has a nearby companion, gas expelled from the companion may fall onto the dwarf, as illustrated in figure 14.3. Coming from the companion's outer layers, such gas is generally rich in hydrogen, and it may briefly replenish the white dwarf's fuel supply. The new fuel forms a layer on the dwarf's surface, where gravity compresses and heats it. The gas layer eventually reaches the ignition temperature for hydrogen, but as we saw in chapter 13, nuclear burning in a degenerate gas can be explosive. The detonating hydrogen is blasted into space and forms an expanding shell of hot gas, as figure 14.4 shows, that radiates far more energy than the white dwarf itself. Sometimes these stellar explosions are visible to the naked eye. When early astronomers saw such an event, they called it a *nova stella,* from the Latin for "new star," because the explosion would make a bright point of light appear in the sky where no star was previously visible. Today, we shorten *nova stella* to **nova.**

A nova may erupt repeatedly if it does not accumulate too much mass. If enough gas is added to push it over the Chandrasekhar limit, however, the white dwarf will collapse and may explode. Carbon and oxygen in the collapsing star will be compressed and heated and, although initially too cold to burn, will eventually reach their ignition temperature and begin nuclear fusion. Carbon and oxygen fuse to form silicon ($^{12}$C + $^{16}$O becomes $^{28}$Si), which in turn fuses into nickel (2 $^{28}$Si becomes $^{56}$Ni). The energy released in this burning may be enough to blow the entire star apart, resulting in a **type I supernova**\* (fig. 14.5A). The nickel, sprayed into space by the explosion, is highly radioactive and rapidly decays into cobalt ($^{56}$Co), which finally decays to iron ($^{56}$Fe), each decay adding additional energy to the outburst.

---

\*Astronomers have recently identified a different form of type I supernova whose origin is uncertain. For this and other reasons, some astronomers prefer nuclear detonation (type I) and core-collapse (type II) supernova.

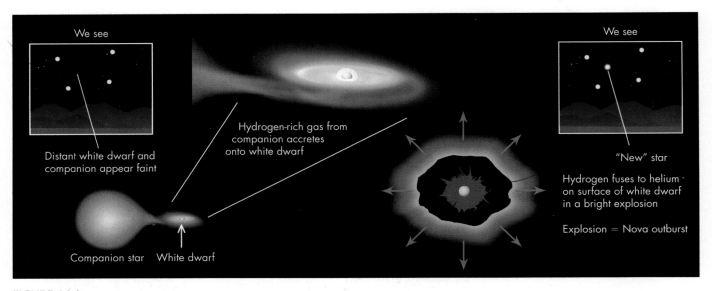

**FIGURE 14.4**

A white dwarf exploding as a nova. The hydrogen that falls onto the surface of a white dwarf from its companion may suddenly fuse into helium, creating an explosion that makes the star brighten.

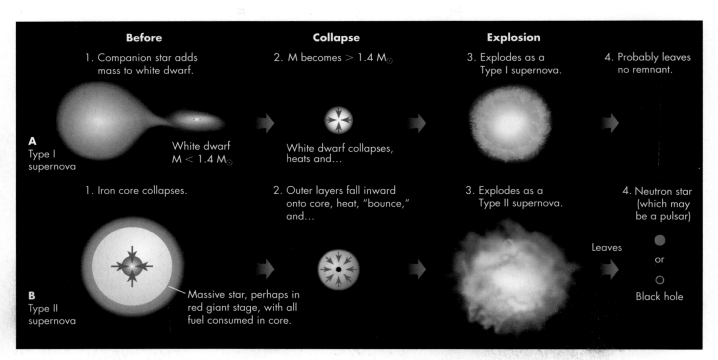

**FIGURE 14.5**

(A) A white dwarf exploding as a type I supernova. If too much hydrogen from a companion accumulates on the white dwarf, it may raise its mass to a value above the Chandrasekhar limit (about 1.4 $M_\odot$). A white dwarf more massive than this limit collapses, heats from the resulting compression, and explodes. (B) Formation of a neutron star or black hole by the collapse of the iron core of a massive star.

A type I supernova leaves no remnant star, unlike a type II supernova, which results from the collapse of a massive star's iron core and which leaves a neutron star or a black hole. Type I supernovas leave only a cloud of rapidly expanding gas rich in the leftover carbon and oxygen, plus the silicon, iron, and other elements made during the nuclear burning. If the iron in your blood was not made by the collapse of a massive star's core, it might have been made when a white dwarf met its doom.

## 14.2 NEUTRON STARS

### General Properties and Origin

**INTERACTIVE**

*Neutron stars*

In 1934, Walter Baade (pronounced *Bah-deh*) and Fritz Zwicky, astrophysicists at Mount Wilson Observatory and the neighboring California Institute of Technology, respectively, suggested that when a massive star reaches the end of its life, its gravity will crush its core and make the star collapse. The collapse of the core will trigger a titanic explosion, which they called a "supernova"* (see fig. 14.5B). Almost as an afterthought, they speculated that the collapsed core might be so dense that the star's protons and electrons would be driven together and merge into neutrons, forming a **neutron star.**

Astronomers readily accepted Baade and Zwicky's idea that a massive star dies as a supernova—such violent explosions had been observed—but they paid little attention to their suggestion that a neutron star might be born in the blast. Such stars would be incredibly tiny compared even with white dwarfs and were thought to be unobservable. Nevertheless, several physicists calculated the properties of neutron stars and found that their radii should be about 10 kilometers (about 6 miles) and their masses should typically be between one and several times that of the Sun. The calculations also showed that, like white dwarfs, neutron stars have a maximum possible mass, which for them is about 2 to 3 solar masses.

### Pulsars and the Discovery of Neutron Stars

Despite this theoretical model of neutron stars, there was no evidence that they existed, and so Baade and Zwicky's concept of tiny, collapsed stars lay dormant for over 30 years. In 1967, a group of English astronomers led by Anthony Hewish was observing fluctuating radio signals from distant, peculiar galaxies. Jocelyn Bell, a graduate student working with the group, noticed an odd radio signal with a rapid and astonishingly precise pulse rate of one burst every 1.33 seconds. The precision of the pulses led some astronomers to wonder whimsically if they had perhaps stumbled onto signals from another civilization, and informally, the signal was known as LGM-1, for "little green men #1." Over the next few months, Hewish's group found several more pulsating radio sources, which came to be called **pulsars** for their rapid and precisely spaced bursts of radiation (fig. 14.6).

The discovery of other pulsars convinced astronomers that they were observing a natural phenomenon, and so their explanations focused on peculiar varieties of pulsating stars. Ordinary pulsating stars—for example, Cepheid variables—obey a period-density relation such that each star's pulsation period depends inversely on the square root of its density. That law applies in a generalized form to all pulsating stars. Thus, the extreme shortness of pulsar periods indicates that they must be extremely dense objects, and the first pulsars discovered had periods that were consistent with their being white dwarfs. But soon other pulsars were discovered whose periods were much shorter, implying densities far larger than those of white dwarfs, and so astronomers turned back to Baade and Zwicky's ideas about neutron stars.

---

*Now referred to as a type II supernova.

Pulsar number PSR 0833−45

Pulsar number PSR 0329+54

**FIGURE 14.6**
Pulsar signals recorded from a radio telescope.

G* Approximately how rapidly do these pulsars spin?

Q* Count the peaks in a time interval of, say, 1 or 2 seconds. Then divide the number of peaks by the time interval. The left one spins 9 times per second. The right one spins 3 times in 2 seconds for a spin rate of 1.5 times per second.

**FIGURE 14.7**
A pulsar's pulses are like the flashes of a lighthouse as its lamp rotates.

Neutron stars, according to their theory, would not only be extremely dense (and thus have very short periods) but would also be associated with supernova explosions. Thus, when a very short-period pulsar was found in the Crab Nebula supernova remnant, it seemed to confirm Baade and Zwicky's model. But calculations of the pulsation period of neutron stars indicated that they should pulsate much faster than any of the pulsars so far observed. The solution to this puzzle came from the independent work of Franco Pacini and Thomas Gold, astrophysicists at Cornell University.

Even before the discovery of pulsars, Pacini had been studying the link between neutron stars and supernova remnants that had been proposed by Baade and Zwicky some 40 years earlier. He found that a neutron star remaining in the debris of a supernova explosion could supply large amounts of energy to the gas in its vicinity if it were rapidly spinning and if it had a very strong magnetic field. Rapid rotation and a strong magnetic field were also the keys to Gold's completely unrelated work, which was an attempt to explain not the energizing of supernova debris but the rapidity of pulsar pulses. Gold suggested that a pulsar is a rapidly *spinning*, not pulsing, neutron star. Furthermore, its magnetic field causes it to emit radiation in two narrow beams so that, as the star spins, its beams sweep across space. We see a burst of radiation when the beam points at Earth. Thus, a pulsar shines like a cosmic lighthouse whose beams swing around as its lamp rotates, as shown in figure 14.7. But the model worried some

*The oblique rotator model of a pulsar*

astronomers because at least one of the pulsars then known—that in the Crab Nebula supernova remnant—rotates 30 times per second. What could make a star spin that fast? For comparison, consider that the Sun spins once a month.

If pulsars are neutron stars, their rapid spin has a natural explanation. When a massive star collapses, forming a neutron star and exploding as a supernova, its core shrinks to a tiny radius—perhaps only 10 kilometers or so—and any rotation it has will be vastly increased. This acceleration results from the "ice-skater effect," which we discussed in describing the formation of the solar system. The ice skater spins faster by drawing his or her arms inward and thereby turning a slow pirouette into a blur of rotation.

*Any* rotating body spins faster when it shrinks—a principle called the **conservation of angular momentum,** illustrated in figure 14.8. Angular momentum measures rotational inertia, which is usually denoted by $L$. For a spherical mass, $L$ is approximately given by the body's mass, $M$, times its equatorial rotation velocity, $V$, times its radius, $R$; or

$L$ = Angular momentum (rotational inertia)
$M$ = Mass of an object
$V$ = Rotation velocity at the object's equator
$R$ = Radius of the object

$$L = MVR$$

The principle of the conservation of angular momentum states that this product must be constant unless some force acts to brake or accelerate the spin. Therefore, for a given mass, $M$, *if we make R smaller, V must increase* to keep the angular momentum constant. Thus, shrinking a spinning body (decreasing its radius) makes it spin faster, and a simple calculation using this principle shows that if even a slowly spinning star such as the Sun were to shrink to the size of a neutron star, it would rotate several thousand times per second! Such a shrinkage is exactly what happens when the iron core of a massive star collapses. Thus, the rapid spin of neutron stars becomes entirely explainable.

Collapse to such a small radius has another effect: it squeezes the star's magnetic field into a denser state, amplifying the field's strength to about 1 trillion times that of the Earth's. Pulsars are thus extremely powerful magnets, and the combination of such a strong magnetic field and the rapid rotation generates the pulsar's radiation beams.

## Emission from Neutron Stars

A varying magnetic field creates an electric field—a principle we use here on Earth to generate electricity by spinning magnets in dynamos. Similarly, the spin of a neutron star and its magnetic field generates powerful electric fields. These fields rip positively and negatively charged particles off the star's surface and accelerate them to nearly the speed of light. No metal wires confine the charges; rather, they are channeled by the pulsar's intense magnetic field. We saw in chapter 5 how the Earth's magnetic field directs solar particles toward the poles to create the aurora. The magnetic field of a spinning pulsar, in a like manner, generates two narrow beams of charged particles at the star's magnetic poles.

To explain how the pulsar's particle beams generate radiation, we invoke the physical law that accelerating charges radiate electromagnetic energy. A radio transmitter works according to this principle: electric currents are pulsed through the broadcast antenna, accelerating electrons in it; the accelerated electrons in turn produce the radio waves we detect. In a pulsar, too, the charges radiate as they accelerate, in this case along the magnetic

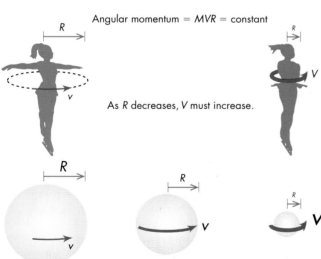

Angular momentum = $MVR$ = constant

As $R$ decreases, $V$ must increase.

**FIGURE 14.8**

Conservation of angular momentum spins up a skater and a spinning star. Angular momentum is proportional to a body's mass, $M$, times its rotation speed, $V$, times its radius, $R$. Conservation of angular momentum requires that mathematically $MVR$ = Constant. Thus, if a given mass changes its $R$, its $V$ must also change. In particular, if $R$ decreases, $V$ must increase, as the ice skater demonstrates by spinning faster as she draws her arms in. Similarly, a rotating star spins faster if its radius shrinks.

NP 0532

**FIGURE 14.9**

The strong magnetic field of a spinning pulsar generates a powerful electric field, which strips charges from its surface. (A) The rapid rotation of the pulsar in the Crab Nebula makes the star appear to turn on and off 30 times per second as its beam of radiation sweeps across the Earth. (B) The charges spiral in the star's magnetic field and emit radiation along their direction of motion. Because the charges stream in a narrow beam confined by the magnetic field, their radiation is also in a narrow beam at each magnetic pole of the star.

field, spiraling as they go (fig. 14.9B). The spiraling charges radiate electromagnetic energy over a range of wavelengths that depends on their energy. Higher-energy charges generate more energetic—and therefore shorter-wavelength—photons. The radiation is beamed because the charges are traveling along the field lines that, near the star's surface, form a tight cone. Because of this beaming, we see only those neutron stars whose sweeping beams point to Earth; those radiating in other directions are invisible to us. Thus, many more neutron stars must exist than the few hundred we have detected so far.

The emission created by the accelerating charges is called **nonthermal radiation** because its properties depend on the charges' acceleration and on the strength of the magnetic field, rather than on the temperature of a heated gas. It is also sometimes called **synchrotron radiation,** after a type of atomic particle accelerator in which such radiation was first observed on Earth. For most pulsars, the charges generate mostly low-energy radiation at radio wavelengths. However, in some very young pulsars, the charges are so energetic that they produce radiation across the entire electromagnetic spectrum, including visible light and gamma rays. Figure 14.9A shows the very young pulsar in the Crab Nebula, which emits visible light flashes 30 times per second in addition to its radio waves. A high-speed television camera has been used to capture the flashes caused by its rapid spin. Pulsars presumably emit some thermal radiation—ordinary visible light and X rays—from their hot surfaces, but because they are so tiny, that radiation is very dim and difficult to detect.

In everyday life, spinning objects slow down. Pulsars are no exception. As a pulsar spins, it drags its magnetic field through the particles that boil off its surface into the surrounding space. The magnetic field exerts a force on those particles and speeds them up. But as the particles speed up, they exert a reaction force back on the magnetic field that slows it and the pulsar down. This process is a little like what happens if you lower the beaters of a rapidly spinning electric mixer into a bowl of thick dough: the dough begins to spin, but the beater blades almost stop spinning. Astronomers measure this "spindown" of a pulsar by precisely timing the interval between pulses. Such measurements indicate that the time interval—the period of the spinning pulsar—is lengthening. The slower rotation also limits the energy of the radiation the pulsar emits. Young, rapidly spinning ones emit visible light and radio waves; old, slowly spinning ones generate only radio waves.

As the pulsar spins ever more slowly, its emission of electromagnetic radiation weakens, and ultimately, the star becomes undetectable. Thus, a pulsar "dies" by becoming invisible.

## Structure of Neutron Stars

Although in general, pulsars are slowing down, some occasionally speed up suddenly. Such jumps in their rotation speed are called **glitches,** and they can tell us about the interior structure of the pulsar or neutron star. Astronomers study the structure of neutron stars with the same methods used to study ordinary stars; that is, they construct mathematical models of the neutron star. Such models are based on physical laws—for example, the requirement that, in a stable star, gravitational forces must balance pressure forces. From these models, astronomers can deduce the structure of the neutron star's interior—its density and pressure. Such calculations show that neutron stars contain at least three distinct regions: a thin, outer gaseous atmosphere about 1 millimeter thick; a thin, solid (perhaps iron) crust a few hundred meters thick, below the atmosphere; and the bulk of the star, a liquid "sea" of neutrons lying below the crust (fig. 14.10).

**FIGURE 14.10**

Schematic structure of the interior of a neutron star.

## Neutron Stars in Binary Systems

Neutron stars, like other types of stars, sometimes belong to a binary star system. But such binary pairs are rare for several reasons. First, the supernova explosion that creates them might expel so much mass that gravity is no longer strong enough to hold the pair of stars together—they rapidly escape from the debris of the explosion that spawned them. This helps explain why so few are seen in supernova remnants. Second, astronomers have calculated that the supernova blast or other processes "kick up" the neutron star to speeds of hundreds of kilometers per second, again making it difficult for its companion to hold on to it. In fact, a number of such speeding pulsars have been detected. Despite such difficulties, some neutron stars do remain bound to a companion, and to such odd couples we now turn.

## X-Ray Binary Stars

Neutron stars in binary systems sometimes emit intense X-ray radiation. Such stars are called **X-ray binaries.** Astronomers have identified several types of X-ray binaries. In one type, the X-ray emission occurs in intense bursts at irregular time intervals, and so these stars are called X-ray bursters. According to one hypothesis, the X rays come from gas that falls onto the neutron star's surface, where the star's intense gravity compresses and heats it to millions of degrees. This heating triggers a thermonuclear explosion like those that cause nova outbursts. The exploding gas is so hot that most of its radiation is at X-ray wavelengths, in accordance with Wien's law (the hotter a gas, the shorter the wavelength at which it radiates). Such explosions on the surface of a

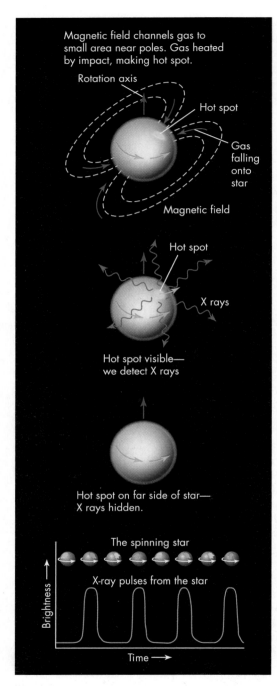

**FIGURE 14.11**
Gas falling onto a neutron star follows the magnetic field lines and makes a hot spot on the star's surface, creating X rays. As the star rotates, we observe X-ray pulses.

neutron star emit thousands of times more energy than the Sun, an impressive achievement for a "dead" star.

In another type of X-ray binary, the X-ray emission comes in rapid, regular pulses every few seconds or less. The X-ray pulses from these stars are reminiscent of the radio pulses from ordinary pulsars, and so astronomers refer to such objects as "X-ray pulsars." An X-ray pulsar generates its emission from small "hot spots" near its magnetic poles. Gas from a companion star flows in along the field lines, much as gas particles from the Sun stream in toward Earth's magnetic poles, creating the aurora. As the gas falls toward the neutron star, the star's intense gravity accelerates the gas. As it crashes onto the neutron star at the magnetic poles (fig. 14.11), the gas is compressed and heated intensely, causing it to emit X rays. As the neutron star spins, the hot regions near each pole move into and then out of our field of view, creating the X-ray pulses that we observe.

Because pulsars generally slow down as they age, astronomers were initially puzzled when they discovered some extremely rapidly spinning—and therefore presumably very young—pulsars in several very old star clusters. Some of these seemingly young pulsars rotate approximately 1000 times per second, so that the pulses we receive from them are separated by little more than a thousandth of a second. Astronomers accordingly call them "millisecond pulsars." Millisecond pulsars are distinguished not only by their rapid rotation but also by having companion stars in most cases. In fact, astronomers think they probably all had companions at one time, for the reason outlined below.

If a neutron star has a companion that ejects mass, the neutron star may capture some of the mass by its gravitational attraction. However, the captured mass generally does not fall directly onto the neutron star. Instead, it temporarily goes into orbit around the neutron star, accumulating in a flat, spinning disk that astronomers call an **accretion disk.** The matter in the accretion disk eventually falls onto the neutron star, and when it does, the material carries some of its spinning motion—technically, angular momentum—with it. That spinning motion is imparted to the neutron star when the gas strikes its surface, making the neutron star spin faster. Thus, the usual slowdown of an aging pulsar may be reversed if it has a companion star. But what of those millisecond pulsars without companions? What made their rotations speed up? One possibility is that such pulsars originally had companions that have since disappeared. In some cases, the companion may have been lost if the pulsar passed too closely to a neighboring star. Gravitational forces from that neighbor may have pulled the pulsar's companion away. A more sinister possibility is that the pulsar destroyed its companion, a process that astronomers believe they may be witnessing in the case of the "black-widow pulsar." This peculiar pulsar has a low-mass companion star that is evaporating under the intense heating created by the pulsar. Given enough time, the companion star may totally evaporate, and future astronomers may see the millisecond pulsar without its original companion.

## 14.3    BLACK HOLES

When a star that is initially more massive than about 10 M$_\odot$ reaches the end of its life and collapses, it may create a compact star whose properties differ dramatically from those of white dwarfs or neutron stars. Its greater mass can compress its core so much that pressure is unable to support it, and it totally collapses to form what astronomers call a **black hole** (see fig. 14.5).

Black holes have very strange properties, but to understand them, we need to review the concept of **escape velocity,** which we discussed in chapter 2. Escape velocity is the speed a mass needs to avoid being drawn back by another object's gravity. For a body of mass $M$ and radius $R$, the escape velocity, $V$, for an object moving away from that body is

$V$ = Velocity necessary to escape an object
$M$ = Object's mass
$R$ = Object's radius

$$V = \sqrt{\frac{2GM}{R}}$$

where G is the gravitational constant and has a value of $6.67 \times 10^{-11}$ if $V$ is in meters per second, $R$ is in meters, and $M$ is in kilograms. For the Earth, the escape velocity is about 11 kilometers per second; for the Sun, it is about 600 kilometers per second. You can see from the formula that the escape velocity for a body of a given mass will be larger if the body has a small radius. The same mass in a smaller radius creates a larger force of gravity, making that body more difficult to escape. Thus, a white dwarf whose mass is the same as the Sun's has an escape velocity much larger than the Sun's because its radius is so small. In fact, because its radius is about 100 times less than the Sun's and because the escape velocity depends on the square root of the radius, its escape velocity is larger by a factor of $\sqrt{100}$ (=10) and equals about 6000 kilometers per second. An object leaving the surface of a white dwarf at a velocity of 6000 kilometers per second would therefore overcome the star's gravity and escape into space. The escape velocity for a neutron star, whose radius is about $10^5$ times smaller than the Sun's but whose mass is comparable, is about 180,000 kilometers per second, or approximately half the speed of light. Thus, if a neutron star could be compressed to about one-fourth of its radius, its escape velocity would exceed the speed of light. Such an object would become a black hole.

As long ago as 1783, the English cleric John Michell discussed the possibility that objects with escape velocities exceeding the speed of light might exist. A little more than a decade later, the French mathematician and physicist Pierre Simon Laplace entertained the same idea. Following their logic, we can calculate the radius at which a body of a given mass would become a black hole by equating its escape velocity to the speed of light. That is, $V = c = (2GM/R)^{1/2}$. Notice that only two quantities of the body enter: its mass and its radius. Thus, we can solve for the radius at which a body of mass $M$ becomes a black hole and find that

$R$ = Radius at which an object of a particular mass becomes a black hole
$M$ = Object's mass
$c$ = Speed of light

$$R = \frac{2GM}{c^2}$$

For a body of 1 solar mass, this turns out to be about 3 kilometers, or 1.9 miles. In other words, if the Sun could be compressed that small, it would become a black hole.

Not until 1916, with the advent of Einstein's general theory of relativity, however, did astronomers obtain the mathematical and physical understanding to properly study black holes. Relativity theory showed that gravity is related to the **curvature of space** and that a black hole is a place where the curvature becomes so extreme that a hole forms. An analogy may help illustrate how gravity and curvature of space are related. Imagine a waterbed on which you have placed a baseball (fig. 14.12). The baseball makes a small depression in the otherwise flat surface of the bed. If a marble is now placed near the baseball, it will roll along the curved surface into the depression. The bending of its environment made by the baseball therefore creates an "attraction" between the baseball and the marble. Now suppose we replace the baseball with a

Baseball:
Marble rolls into depression

Bowling ball:
Marble rolls in faster

Big rock:
Marble disappears into hole

**FIGURE 14.12**

Objects on a waterbed make depressions analogous to the curvature of space created by a mass. According to the general theory of relativity, that curvature produces the effect of gravity. We can see the similarity by placing a marble at the edge of the depression and watching it roll inward, as if attracted to the body that makes the depression. Bigger bodies make bigger depressions, and so a marble rolls in faster. However, a very big body may tear the waterbed, creating an analog of a black hole.

bowling ball. It will make a bigger depression and the marble will roll in farther and be moving faster as it hits the bottom. We therefore infer from the analogy that the strength of the attraction between the bodies depends on the amount by which the surface is curved. Gravity also behaves this way, according to the general theory of relativity. According to that theory, mass creates a curvature of space, and gravitational motion occurs as bodies move along the curvature.

We now extend our analogy to black holes by supposing that we remove the bowling ball and put on the waterbed a steel safe. This will immediately rip the plastic and fall through. A marble placed on the torn surface will follow the safe into the hole and disappear. So too a black hole is a "rip" in space where the curvature has become so strong that the structure of space is disrupted.

Needless to say, this simple analogy is only approximate. The mathematically correct treatment of a black hole's structure is very difficult. The German astrophysicist Karl Schwarzschild pioneered such calculations, and the size of a black hole is now called the **Schwarzschild radius,** $R_S$, in his honor. It turns out, however, that relativity theory gives the same answer for the radius of a black hole that we deduced earlier: that is, the Schwarzschild radius, $R_S = 2GM/c^2 =$ approximately $3\mathcal{M}$ kilometers, where $\mathcal{M}$ is the mass of the body in solar units.

All masses, including the Sun, bend space to some degree, and this curvature can be observed. Far from the Sun, space is "flat," unaffected by the Sun's gravity. As we approach the Sun, however, the curvature increases. If we look at light or radio waves whose paths traverse that curved region, as shown in figure 14.13, the waves' bending can be detected and is found to match precisely the prediction of relativity theory. Thus, astronomers *know* that space is curved.

The extreme curvature of a black hole prevents its light from escaping, and so its boundary is sometimes called the **event horizon.** Just as the Earth's horizon blocks our view of what lies beyond it, so too the Schwarzschild radius prevents our seeing into a black hole. All that happens within it is hidden forever from our view. No radiation of any sort, nor any material body—rocket, spacecraft, and so on—can break free of its gravity. Because we cannot observe the interior of a black hole even in principle, there is only a limited number of physical properties we can ascribe to it. For example, it is meaningless to ask what a black hole is made of. When it warps space inside its event horizon, we can no longer observe its interior—it could be made of neutrons, corn flakes, or gerbils. Only the amount of mass is important, not what composes it.

We can measure the mass of a black hole because the mass generates a gravitational field. In fact, the gravitational field generated by a black hole is no different from that generated by any other body of the same mass. For example, if the Sun were suddenly to become a black hole with the same mass it has now, the Earth would continue to orbit it just as it does now: it would not be pulled in. Nor would Mercury. For something to fall into a black hole, it must approach the black hole to nearly its Schwarzschild radius

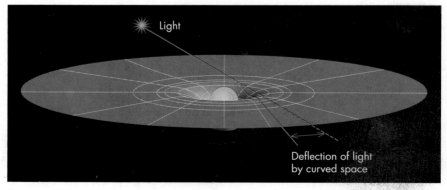

Light

Deflection of light
by curved space

**FIGURE 14.13**

Curved space around the Sun bends a ray of light passing near the Sun.

(recall that this is about 3 kilometers for a solar-mass black hole). Because astronomical objects are generally so immensely far apart, the likelihood that any object will approach another to within a few kilometers is extremely small. Moreover, the black hole's gravity does not give it special attractive powers. In fact, at a given distance, a black hole exerts exactly the same gravitational force as *any* object of the same mass. That force is given (essentially) by Newton's law of gravity, just as it is for the Sun. So, we are no more likely to be pulled into a black hole than we are to be pulled into the Sun.

Black holes can make themselves felt beyond their Schwarzschild radius by more than just their gravity. They may possess electric charge and angular momentum. For example, a black hole may become charged if an excess of one kind of charge, say, positive, falls into it.

Angular momentum is much more likely to be a significant property of a black hole. Material nearly always has some angular momentum, and if it becomes a black hole, its angular momentum is retained. As a result, a spinning black hole will not be a sphere. The distance from its center to the event horizon is smaller along the rotation axis than it is perpendicular to it.

## The Formation and Observation of Black Holes

Astronomers have excellent evidence that some supernova explosions can form a neutron star, and as we showed above, it is only a small step in compression from a neutron star to a black hole. It seems a safe extrapolation to suppose that a black hole would form if a massive star had a core whose mass exceeded 5 to 10 $M_\odot$. It remains, however, to show by *observation* that black holes exist. Furthermore, how could we observe such objects, given that they emit no radiation?

An object that emits no light or other electromagnetic radiation is not easy to observe! But just as you may "see" the wind by its effect on leaves and dust, so too astronomers may be able to see black holes by their effects on their surroundings. Suppose a massive star in a binary system explodes, and its core mass is 10 $M_\odot$, rather than 2 $M_\odot$, so that the core collapses to become a black hole rather than a neutron star. Gas from the companion star may be drawn toward the hole by its gravity, as we know happens for neutron stars that are X-ray bursters. The infalling matter swirls around the black hole and forms an accretion disk whose inner edge lies just outside the hole's Schwarzschild radius, as depicted in figure 14.14. Here, where the disk orbits at nearly the speed of light, turbulence and friction heat the swirling gas to a furious 10 million K, making it emit X rays and gamma rays.

*X rays from a black hole binary*

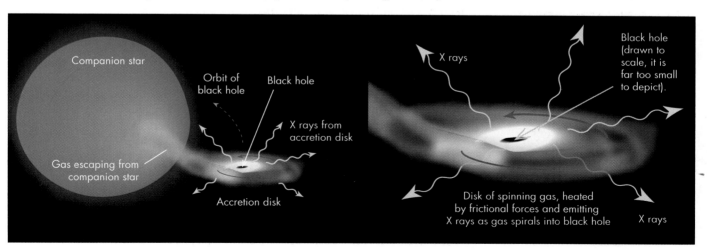

**FIGURE 14.14**
Black holes may reveal themselves by the X rays emitted by gas orbiting them in an accretion disk. Sizes and separations are not to scale.

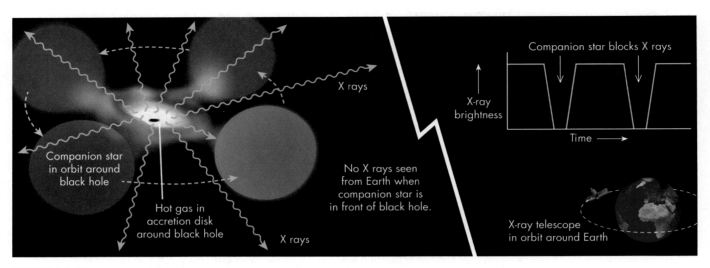

**FIGURE 14.15**
X-ray emission from gas around a black hole. The signal drops as the hole is eclipsed by its companion star. Sizes and separations are not to scale.

As the black hole orbits its neighbor, the X-ray–emitting gas may disappear from our view as it is eclipsed by the companion star. An X-ray telescope trained on such a star system will show a steady X-ray signal that disappears at each eclipse, as shown in figure 14.15. Such a signal might be the sign of a black hole, and at least three cosmic X-ray sources bear that signature. But how do we know an X-ray source is not just a neutron star?

Because these X-ray sources are binary stars, we can measure their masses. If X-ray–emitting gas surrounds a body we cannot see but whose mass exceeds 5 $M_\odot$, we can be reasonably confident that the invisible object is a black hole. Neutron stars cannot be that massive.* For example, Cygnus X-1, the first X-ray source detected in the constellation Cygnus, the swan, consists of a *B* supergiant star and an invisible companion whose mass—based on an application of the modified form of Kepler's third law—is at least 15 $M_\odot$. Nothing we know of but a black hole can be so massive and yet invisible. An even better candidate is A0620-00, an X-ray–emitting object in the constellation Monoceros, the Unicorn. For A0620-00, the invisible star's mass is probably 16 $M_\odot$.

Recently, astronomers have seen in some binary star systems indirect evidence of the event horizon itself. The X rays from those systems thought to contain a black hole are dimmer than the X rays from systems thought to contain a neutron star, presumably because matter disappears into the black hole before it has a chance to radiate as much.

We have just described the evidence for the existence of black holes with mass 10 or so times that of the Sun. Curiously, astronomers have recently found convincing evidence that enormously more massive black holes—with mass millions of times that of the Sun—exist in the centers of many galaxies, including our own. We will discuss the evidence for these supermassive black holes in chapters 15 and 16.

## Gravitational Waves from Double Compact Stars

In our discussion of black holes, we saw that space near them is curved. If a black hole or other compact star orbits a companion, its motion generates **gravitational waves.** Just as ripples spread away from a stone tossed into a pond, so too gravitational waves

---

*In choosing 5 $M_\odot$ as the limit, we have allowed some margin above the theoretical maximum mass of a neutron star of about 3 $M_\odot$.

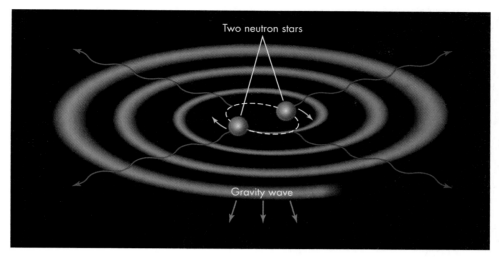

**FIGURE 14.16**

Gravitational waves generated by a rapidly orbiting pair of neutron stars (purple). Such waves are "ripples" in space—and, like ripples in water, they make objects bob (move up and down) slightly as they pass.

spread across space, bending it up and down (fig. 14.16). Scientists are currently starting up a large gravitational wave detector called LIGO, but astronomical "detectors" offer a shortcut. Early this century, Einstein predicted that rapidly orbiting stars should generate gravitational waves, but his proposal could not then be verified. Even now, astronomers cannot detect gravitational waves directly, but they can observe their effect on the bodies that emit the waves.

Gravitational waves carry energy away from the orbiting stars, making them gradually move together. In 1974, Joseph Taylor and Russell Hulse, then at the University of Massachusetts, discovered a binary pulsar: two neutron stars in orbit around each other. From observations of the pulses, Taylor and Hulse deduced the orbital speed and separation of the neutron stars, and to their delight, the speed and separation changed exactly in the manner predicted by Einstein's general theory of relativity. For this discovery, which has persuaded most astronomers that Einstein was right and that gravitational waves do exist, Taylor and Hulse were awarded the 1993 Nobel Prize in Physics. But collapsed stars may emit even more mysterious radiation than gravitational waves.

## Hawking Radiation

In 1974, the English mathematical physicist and cosmologist Stephen Hawking predicted that black holes can radiate! As so often happens in science, others were working on similar ideas. For example, the previous year, the physicist Jacob Bekenstein, who was studying the relation between gravity and thermodynamics, noted that black holes can be assigned a temperature. But Hawking went further and showed that—astonishingly—a black hole emits blackbody radiation, such as we discussed in chapter 3. In fact, we can use Wien's law, which relates a body's temperature to the wavelength at which it radiates most strongly, to estimate the temperature of a black hole. Hawking calculated that the wavelength of maximum emission is approximately 16 times the black hole's Schwarzschild radius ($R_S$), a result also found by Bekenstein. Wien's law states that an object's temperature is inversely proportional to the wavelength of the radiation it emits, and so the temperature of a black hole is equal to a constant divided by $R_S$. For a solar-mass black hole, this turns out to be about $6 \times 10^{-8}$ K. This is very cold, but it is not absolute zero. Accordingly, a black hole, like any other body whose temperature is not absolute zero, emits energy in the form of electromagnetic waves, energy now known as **Hawking radiation.** However, for black holes of stellar mass, the amount of Hawking radiation emitted is far too small to be detected at this time. Nevertheless, the existence of such radiation means that black holes are not truly black.

Hawking radiation is created by quantum physical processes that allow energy to escape from black holes despite their intense gravity. Because the only source of energy available to a black hole is its mass, as a black hole "shines" (albeit very dimly), its mass must decrease. In other words, black holes must eventually "evaporate." However, the time it takes for a solar-mass black hole to disappear by "shining itself away" is *very* long: approximately $10^{67}$ years! This is immensely long—vastly larger than the age of the Universe—but the implications are important: even black holes evolve and "die."

## SUMMARY

White dwarfs, neutron stars, and black holes are the remnants of dying stars. A white dwarf forms when a low-mass star expels its outer layers to form a planetary nebula shell and leaves its hot core exposed. The radius of a white dwarf is about the same as that of the Earth. Its matter is degenerate, and its mass must be less than the Chandrasekhar limit, or it collapses. Having no fuel supply, a white dwarf cools and grows dim. If a white dwarf is in a binary system, it may accrete mass from its neighbor and explode either as a nova or as a type I supernova.

A neutron star forms when a massive star's iron core collapses and triggers a supernova explosion. The collapse compresses the core's protons and electrons together to make neutrons, forming a ball of neutrons with a radius of a mere 10 kilometers but which may contain up to about 2 to 3 solar masses. Conservation of angular momentum during its collapse accelerates its rotation rate to about 1000 times per second. This spin and the star's magnetic field generate beams of radiation that sweep across space, making the spinning neutron star a pulsar.

If the massive star's iron core contains more than about 5 $M_\odot$, a black hole, rather than a neutron star, is born. Gravity overwhelms the pressure forces of the core, crushing it and bending space around it so that light cannot escape. The black hole lies invisibly inside its Schwarzschild radius. If it has a companion star, however, we may be able to see X rays from an accretion disk as gas from the neighbor falls toward the hole.

### ? What It Is.

This image shows two pictures (made with the Hubble Space Telescope) of a nova, an exploding star. The pictures were taken 7½ months apart. You can clearly see that the shell of gas ejected by the nova has expanded.

## QUESTIONS FOR REVIEW

1. What are the approximate mass and radius of a white dwarf compared with those of the Sun?
2. How does a white dwarf form?
3. Can a white dwarf have a mass of 10 solar masses? Why?
4. What is meant by degeneracy pressure?
5. What is a neutron star?
6. What are the mass and radius of a typical neutron star compared with those of the Sun?
7. How does a neutron star form?
8. How do we observe neutron stars?
9. What is a pulsar? Does it pulsate?
10. Are all neutron stars pulsars? Are all pulsars neutron stars?
11. What creates the beams of radiation seen in pulsars?
12. What is nonthermal radiation?
13. What is a black hole?
14. What is the Schwarzschild radius?
15. What is Hawking radiation?

# THOUGHT QUESTIONS

1. Some astronomers have suggested that cooled white dwarfs are made of diamond. Why might it be impractical to mine them?

2. Suppose you jumped into a black hole feet first. What would happen to you as your feet approached its Schwarzschild radius? Hint: Think about tides on the Earth created by the Moon.

3. How would you explain to an 8-year-old child that black holes may exist?

4. Is it surprising that a pulsar is not seen in every supernova remnant? Why?

# PROBLEMS

1. Calculate the density of a white dwarf star of 1 solar mass that has a radius of $10^4$ kilometers.

2. Calculate the escape velocity from a white dwarf and a neutron star. Assume that each is 1 solar mass. Let the white dwarf's radius be $10^4$ kilometers and the neutron star's radius be 10 kilometers.

3. Calculate the Schwarzschild radius of the Sun.

4. Calculate your Schwarzschild radius. How does that compare to the size of an atom? How does it compare to the size of a proton?

5. The mass of a neutron is about $1.7 \times 10^{-27}$ kg. Suppose the mass of a neutron star is about $3.4 \times 10^{30}$ kg. How many neutrons does such a star contain?

6. The volume of a neutron is about $10^{-45}$ cubic meters. Suppose you packed the number of neutrons you found for problem 5 (above) into a cube so that the neutrons touched edge to edge. How big would the volume of the cube be? How big across would the cube be? Hint: The volume of a cube of side L is $L^3$. How does the cube's size compare to the size of a neutron star? What can you conclude about the spacing of neutrons in a neutron star?

# TEST YOURSELF

1. Which of the following has a radius (linear size) closest to that of a neutron star?

   (a) The Sun
   (b) The Earth
   (c) A basketball
   (d) A small city
   (e) A gymnasium

2. What evidence leads astronomers to believe that they have detected black holes?

   (a) They have seen tiny dark spots drift across the face of some distant stars.
   (b) They have detected pulses of ultraviolet radiation coming from within black holes.
   (c) They have seen X rays, perhaps from gas around a black hole, suddenly disappear as a companion star eclipses the hole.
   (d) They have seen a star suddenly disappear as it was swallowed by a black hole.
   (e) They have looked into a black hole with X-ray radar telescopes.

3. What causes the radio pulses of a pulsar?

   (a) The star vibrates.
   (b) As the star spins, beams of radio radiation from it sweep through space. If one of these beams points toward the Earth, we observe a pulse.

   (c) The star undergoes nuclear explosions that generate radio emission.
   (d) The star's dark orbiting companion periodically eclipses the radio waves emitted by the main star.
   (e) A black hole near the star absorbs energy from it and re-emits it as radio pulses.

4. If mass is added to a white dwarf,

   (a) its radius increases.
   (b) its radius decreases.
   (c) its density increases.
   (d) it may exceed the Chandrasekhar limit and collapse.
   (e) All of the above except (a)

5. The Schwarzschild radius of a body is

   (a) the distance from its center at which nuclear fusion ceases.
   (b) the distance from its surface at which an orbiting companion will be broken apart.
   (c) the maximum radius a white dwarf can have before it collapses.
   (d) the maximum radius a neutron star can have before it collapses.
   (e) the radius of a body at which its escape velocity equals the speed of light.

## FURTHER EXPLORATIONS

Bethe, Hans A., and Gerald Brown. "How a Supernova Explodes." *Scientific American* 252 (May 1985): 60.

Bernstein, Jeremy. "The Reluctant Father of Black Holes." *Scientific American* 274 (June 1996): 80. (About Einstein.)

Charles, Philip A., and R. Mark Wagner. "Black Holes in Binary Stars: Weighing the Evidence." *Sky and Telescope* 91 (May 1996): 38.

Drieger, Walter C. "The Care and Feeding of Black Holes." *Astronomy* 23 (May 1995: 19). (A satire.)

Greenstein, George. *Frozen Star*. New York: Freundlich Books, 1984.

Herbst, William, and George C. Assousa. "Supernovas and Star Formation." *Scientific American* 241 (August 1979): 138.

Hillebrandt, Wolfgang, Hans-Thomas Janka, and Ewald Muller. "How to Blow Up a Star." *Scientific American* 295 (October 2006): 42.

Jastrow, Robert. *Red Giants and White Dwarfs*. New York: W. W. Norton, 1990.

Kawaler, Steven D., and Donald E. Winget. "White Dwarfs: Fossil Stars." *Sky and Telescope* 74 (August 1987): 132.

Kirshner, Robert P. "The Earth's Elements." *Scientific American* 271 (October 1994): 58.

Lewin, Walter H. G. "The Sources of Celestial X-ray Bursts." *Scientific American* 244 (May 1981): 72.

Murdin, Paul, and Lesley Murdin. *Supernovae*. New York: Cambridge University Press, 1985.

Piran, Tsvi. "Binary Neutron Stars." *Scientific American* 272 (May 1995): 52.

Ruthen, Russell. "Catching the Wave." *Scientific American* 266 (March 1992): 90.

Starrfield, Sumner, and Steven N. Shore. "The Birth and Death of Nova V1974 Cygni." *Scientific American* 272 (January 1995): 76.

Trimble, Virginia. "Neutron Stars and Black Holes in Binary Systems." *Contemporary Physics* 32 (March/April 1991): 103.

———. "White Dwarfs: The Once and Future Suns." *Sky and Telescope* 72 (October 1986): 348.

Weisberg, Joel M., Joseph H. Taylor, and Lee A. Fowler. "Gravitational Waves from Orbiting Pulsars." *Scientific American* 245 (October 1981): 74.

Williams, Robert E. "The Shells of Novas." *Scientific American* 244 (April 1981): 120.

**Website**

Please visit the *Explorations* website at **http://www.mhhe.com/arny** for additional online resources on these topics.

## PLANETARIUM EXERCISE

Stellar remnants are found throughout the sky. Load the Starry Night planetarium software and set it for your location and time. Locate two stellar remnants, a supernova remnant like M1 (the Crab Nebula) and a planetary nebula like M57 (the Ring Nebula) by centering on them and transferring the information to your star chart. Observe and sketch each object with a moderate-size telescope and compare what you see to the images and information in the text.

## KEY TERMS

accretion disk, 424

black dwarfs, 415

black hole, 425

Chandrasekhar limit, 416

compact stars, 414

conservation of angular momentum, 421

curvature of space, 425

degeneracy pressure, 416

escape velocity, 425

event horizon, 426

exclusion principle, 416

glitches, 423

gravitational redshift, 416

gravitational waves, 428

Hawking radiation, 430

neutron star, 419

nonthermal radiation, 422

nova, 417

pulsars, 419

Schwarzschild radius, 426

synchrotron radiation, 422

type I supernova, 417

white dwarfs, 414

X-ray binaries, 423

# The Milky Way Galaxy

# 15

An all sky view of our galaxy, as we see it from Earth. The image was made by plotting on a map of the whole sky the millions of stars observed in the infrared by the 2-Micron All Sky Survey (2MASS). It is therefore like a map of the Earth, flattened out so as to show all continents. The map clearly reveals the flatness of our galaxy and the bulge of its central regions.

## Chapter Objectives

- We live in the Milky Way galaxy.
- It contains about 100 billion stars (in addition to the Sun), which orbit within it.
- Two main classes of stars occupy the Milky Way:
  - Old red stars
  - Young, often blue, stars
- The Milky Way is about 100,000 light-years across.
- The Milky Way contains gas and dust clouds in the immense spaces between the stars.
- The Milky Way contains dark matter whose nature is unknown.
- Dark matter exerts a gravitational force on the stars and gas in our galaxy, but it emits no as-yet detectable radiation.
- A black hole of more than a million times the Sun's mass lies at the center of the Milky Way.

⊢ ──── 1 light-year ──── ⊣

**?What Is This?**
(See chapter summary for the answer.)

## Concepts and Skills
## to Review

- Inverse square law (12.2)
- Standard candle method (12.1 and 12.8)
- Newton's law of gravity (2.5)
- Modified form of Kepler's third law (2.7)
- Red light has a longer wavelength than blue (3.1)
- Red stars are cooler than blue stars (3.2 and/or 12.2)

O n a clear, moonless night, far from city lights, you can see a pale band of light spangled with stars stretching across the sky (fig. 15.1). The ancient Hindus thought this shimmering river of light in the heavens was the source of the sacred river Ganges. To the ancient Greeks, this dim celestial glow looked like milk spilt across the night sky, and so they called it the **Milky Way.** In the seventeenth century, Galileo showed that the Milky Way is millions of stars too dim to be seen as individual points of light. Now, in the twenty-first century, we know that these stars, along with our Sun, form a huge, slowly spinning disk—our galaxy. The word *galaxy* itself comes from the ancient Greeks and their word for "milk"—*galactos*. Thus, "Milky Way" is both the name of the band of light across the night sky and the name of our galaxy.

A view of the Milky Way on a clear, dark night is one of nature's finest spectacles. Superposed on the dim background glow are most of the bright stars and star clusters that we can see, which all belong to our galaxy. Here and there, however, dark blotches interrupt the glowing backdrop of stars, as you can see in figure 15.2. Ancient Incan

**FIGURE 15.1**
Wide-angle photo of a portion of the Milky Way taken from Mount Graham in Arizona. The dark ring around the edge of the picture is the horizon.

�★ What do you think the whitish blobs along the horizon at the bottom are?

★ The glows are distant city lights.

**FIGURE 15.2**
Dark cloud of dust silhouetted against background stars of Milky Way.

astronomers observing the Milky Way from their temple observatories in the Andes gave these dark areas names, just as peoples of the classical world named the star groups. Today, we know the dark regions are clouds of dust and gas that give birth to new stars, as we discussed in chapter 13.

# 15.1   DISCOVERING THE MILKY WAY

## Shape of the Milky Way

Our understanding of the Milky Way as a star system dates to the eighteenth century, when Thomas Wright, an English astronomer, and Immanuel Kant, the great German philosopher, independently suggested that the Milky Way is a flattened swarm of stars. They argued that if the Solar System were near the center of a spherical cloud of stars, we would see roughly the same number of stars in all directions. Instead, we see vastly more stars in the direction of the Milky Way than in directions perpendicular to it, exactly what we would expect if the Milky Way is a disk rather than a ball of stars, as shown in figure 15.3.

A simple analogy may help you better understand why we see the Milky Way as a band on the sky. Imagine making a lemon Jell-O fruit salad in a big, round flat dish. Imagine that you have spread blueberries throughout the dish. If an ant fell into the Jell-O and was clinging to a blueberry in it, it would see few berries if it looked up through the top of the salad or down through the bottom. However, it would see lots of blueberries if it looked in directions that lay within the dish. So, too, it is with us and stars in the disk of the Milky Way.

These arguments were refined toward the end of the eighteenth century by the English astronomer Sir William Herschel, who, by counting the number of stars visible in different directions, deduced that the disk was about five times wider than it was thick. Moreover, because he saw roughly the same number of stars in all directions around the disk, he concluded that the Sun must lie near its middle, as shown in figure 15.4. However, he had no way to measure the distances to the stars (parallax was not discovered until the next century) and so could not determine the Milky Way's size.

**ANIMATION**

*The Milky Way*

If the Milky Way were spherical, stars would lie in more or less every direction.

We would see stars scattered uniformly.

Earth

Because the Milky Way is a disk, we see lots of stars when we look in directions lying in the disk, and we see very few stars when we look in directions that lie out of the disk.

We see stars in a concentrated band.

**FIGURE 15.3**
If the Milky Way had a spherical shape, we would see about the same number of stars in every direction. However, because it is a disk, we see stars concentrated into a band around the sky.

**FIGURE 15.4**

A copy of Herschel's sketch of the Milky Way made by him in 1784. He based this cross-sectional view on the number of stars he could see in different directions. From this picture, he correctly deduced that the Milky Way was wider than it was thick. However, Herschel *incorrectly* concluded that the Sun (the orange dot) is near the Milky Way's center. He was led astray by his lack of knowledge that dust clouds blot out distant stars and prevent us from seeing our galaxy's true extent.

**FIGURE 15.5**

A typical globular star cluster, so named because of its round shape. This image of the cluster M80 was made with the Hubble Space Telescope.

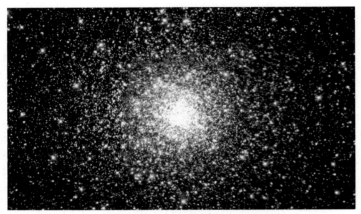

|←—— Approx. 80 light-years ——→|

## Size of the Milky Way

The size of the Milky Way was first measured in the early 1900s by the Dutch astronomer Jacobus C. Kapteyn (pronounced *cap-tine*) and the American astronomer Harlow Shapley. Kapteyn deduced that the Milky Way is a disklike system about 20,000* parsecs in diameter, with the Sun located near the center.

Almost immediately after Kapteyn's publication of his model of the Milky Way, Shapley published a model in strong disagreement with it. Shapley argued that the Milky Way was larger than Kapteyn believed and that the Sun was not near the center but about two-thirds of the way out in the disk. Shapley drew his conclusions from a study of globular star clusters—dense groupings of up to a million stars, such as the one shown in figure 15.5. Because these clusters contain so many stars, they are very

---

*In this and the remaining chapters, we will be dealing with huge objects at immense distances. Perhaps a quick review of some units and powers-of-ten notation is in order.

Astronomers measure large distances in either light-years (ly) or parsecs (pc). One parsec is 3.26 light-years and is about $3 \times 10^{13}$ (30 trillion) kilometers. Recall that $10^{13}$ stands for 1 followed by 13 zeros.

Galaxy dimensions are so huge that even parsecs are inappropriate for measuring their size, and so astronomers often use kiloparsecs (kpc) for that purpose. One kiloparsec = 1000 parsecs. Later, we will find that an even larger unit, the megaparsec (Mpc) is a more appropriate unit for measuring the distances to galaxies.

**FIGURE 15.6**

Schematic version of Shapley's plot of globular clusters from which he inferred the size of the Milky Way and the Sun's location in it. Notice that the clusters fill a roughly elliptical region and that the Sun is *not* at the center of their distribution.

Kapteyn measured the size of the Milky Way by first finding distances to nearby stars from their parallax. He then estimated the distance to more remote stars from their motion, based on the following idea: As stars move in their orbits around the galaxy, their positions in the sky change very slowly. Although such motion is not noticeable to the naked eye, it is detectable when pictures taken many years apart are compared. Kapteyn realized that nearby stars change their position faster than more distant stars, just as a ball tossed past your head moves across your field of view more rapidly than one thrown at the far end of a playing field. Using such motions for remote stars and comparing them with the motion of nearby stars whose distance he knew from parallax, he estimated the distance to the remote stars and thus the size of the Milky Way.

luminous and can be seen at large distances—across the galaxy and beyond. Moreover, many of them lie above the disk, and so we have a clear view of them.

Shapley noticed that the globular clusters are not scattered uniformly across the whole sky but are concentrated in the same place that the stars are concentrated, that is, toward the constellation Sagittarius (see "Looking Up at Sagittarius" on p. 438 for a picture of this region). He hoped, therefore, that a detailed map of the distribution of the globular clusters would show him the shape of the Milky Way.

To make his map of the clusters, Shapley needed to know their distance from the Sun. He found the distances to the clusters by observing the variable stars in them. These stars obey a law* that relates their luminosity to how long it takes them to brighten, dim, and brighten again (their period). From the luminosity of the stars, their apparent brightness, and the inverse-square law, Shapley calculated their distance and thus the distance to the clusters they were in. Then, having found the distance to each cluster, Shapley plotted where they lay in the Milky Way. Figure 15.6 shows his results.

The clusters fill a roughly elliptical region about 100,000 parsecs (roughly 300,000 light-years) across. Moreover, the Sun lies *not* at the middle of the region, but about two-thirds of the way from its center. Shapley therefore concluded that our galaxy is roughly 100,000 parsecs in diameter and that the Sun is nearer to its edge than to its center (fig. 15.6).

The great difference between Shapley's value for the diameter of the Milky Way and the Sun's location and Kapteyn's finding created a major controversy among astronomers. Like so many other controversies, both sides were partially right and partially wrong. For example, Kapteyn deduced a size for the Milky Way that was far too small because he assumed that space was transparent: he did not recognize the dimming effect of interstellar dust. On the other hand, Shapley was correct about the Sun not being at the center, but he greatly overestimated the diameter of the Milky Way, finding

*At that time, this law had been just recently discovered by Henrietta Leavitt.

# LOOKING UP AT...
# SAGITTARIUS

Sagittarius marks the direction to the center of the Milky Way. It is best located by its "teapot" shape, with the Milky Way seeming to rise like steam from the spout. For northern latitude viewers, the constellation is best seen in the evening, July to September. For such viewers, it is low on the southern horizon.

~1 ly

## M16
## The Eagle Nebula

This young star cluster and the hot gas around it lie about 7000 ly from Earth.

## M17

~70 ly

## M20 The Trifid Nebula

The distance to the Trifid Nebula (so named because of the black streaks that divide it into thirds) is very uncertain. It lies between 2000 and 8000 ly away. This makes its size very uncertain, too.

~40–50 ly

## Center of the Milky Way

## The Lagoon Nebula

## M22

~100 ly

M22 is one of the many globular clusters that are concentrated toward the core of our galaxy. Easy to see with binoculars, it is just barely visible to the naked eye. It is about 11,000 ly away from us.

The "teapot" of Sagittarius

### Sagittarius in 3-D

Distance unknown but ~2200 to 8000 ly

Trifid Nebula

77 ly

230 ly
230 ly

120 ly

300 ly

96 ly
89 ly

To Earth

150 ly

1 light year (ly) = 6 trillion miles

SOUTH

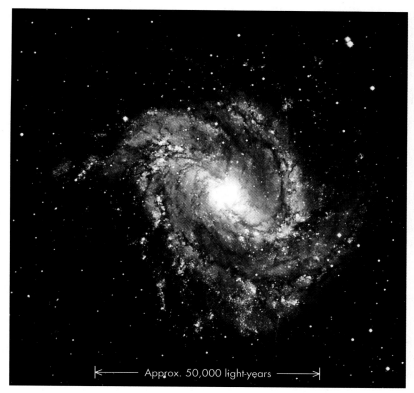

←——— Approx. 50,000 light-years ———→

**FIGURE 15.7**
Photograph of the spiral galaxy NGC 5236 (also known as M83), illustrating its spiral arms.

a value about three times larger than the one currently accepted. Part of the reason for his error was that Shapley, like Kapteyn, did not recognize the dimming effect of interstellar dust. Thus, Shapley mistook the dimness of the globular clusters to mean that they were very far away. Moreover, he did not realize that he was dealing with two types of pulsating variables: RR Lyrae stars and Cepheids. Leavitt had determined the period–luminosity law for Cepheids, but Shapley was measuring the galaxy's size with RR Lyrae stars. Because RR Lyrae stars are much dimmer than Cepheids, Shapley thought they were much farther away than they really are. Thus, he overestimated the size of the Milky Way. Kapteyn came closer to the presently accepted size of our galaxy, but he was wrong about where the Sun lay in it.

Today, astronomers have revised and improved Shapley's model of the Milky Way. Much of that improvement comes from studies of other galaxies. For example, many external galaxies are flat disks with conspicuous spiral arms, as illustrated in figure 15.7. Such arms are extremely hard to detect in our own galaxy because our location in the Milky Way is such that we cannot get an overview of its disk. Nevertheless, given that the Milky Way is a flat disk and that other disk galaxies have spiral arms, it is reasonable to deduce that ours does too. Thus, by combining observations of our own galaxy with the general features known to occur in other galaxies, astronomers can assemble a detailed model for the Milky Way.

## 15.2  OVERVIEW OF THE MILKY WAY

### Structure and Contents

The Milky Way consists of three main parts, as illustrated in figure 15.8: a **disk** about 30 kiloparsecs (100,000 light-years) in diameter, a surrounding **halo,** and a flattened, somewhat elongated **bulge** of stars at its center. Within the disk, numerous bright young stars cluster into **spiral arms** that wind outward from near the center. Our Solar System lies between two arms about 8.5 kiloparsecs (roughly 28,000 light-years) from the center, with the orbits of the planets tilted by about 60° with respect to the galaxy's disk. This tilt is the reason the band of the Milky Way on the sky is tipped with respect to the ecliptic. You may find the following analogy helpful in visualizing the scale of our galaxy: if the Milky Way were the size of the Earth, the Solar System would be the size of a large cookie.

Mingled with the stars of the Milky Way are huge clouds of gas and dust that amount to nearly 15% of the disk's mass. We can see some of these clouds by the visible light they emit, whereas others reveal themselves by blocking the light of background stars, as can be seen in figure 15.2. In fact, dust scattered throughout the galaxy prevents us from seeing farther than about 3000 parsecs within the disk. Thus, we cannot see our **galaxy's nucleus**—its core—at visible wavelengths. However, radio, infrared, and X-ray telescopes can "see" through the dust and reveal that the core of our galaxy contains a dense swarm of stars and gas as well as a massive black hole.

Recently, astronomers have found evidence that the central bulge of our galaxy is not circular. Instead, it has a slightly elongated shape, like a flattened football. Therefore the Milky Way should be described as a "barred" spiral (see chapter 16).

FIGURE 15.8
Artist's sketch of the Milky Way showing top and side views, illustrating the disk, bulge, and halo. Notice how thin the disk is and how the halo surrounds the disk much as a bun surrounds a hamburger. Because the Milky Way's stars gradually thin out and do not just stop at some distance, its size is only labeled approximately. The size of the dark matter halo in which the Milky Way is embedded is very uncertain, so it is shown only schematically. It is omitted from the top view for clarity.

Top view

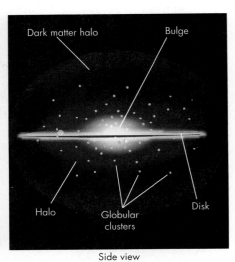

Side view

The gravity of all these components holds our galaxy together—but that gravity would draw all the stars and gas inward and collapse the Milky Way, were it not for the orbital motion of its stars. Astronomers have measured that motion for our Sun and neighboring stars: the Sun and its neighbors move around the center of the Milky Way at a speed of about 220 kilometers per second. That motion around the center is what originally gave the disk its flattened shape, much as the rotation of the presolar nebula made our Solar System flat. Moreover, just as planets near the Sun complete their orbits faster than planets farther out, the stars near the center of the galaxy complete their orbits faster than stars at the edge of the galaxy. Thus, the inner parts of the galaxy spin faster than the outer parts, a phenomenon called "differential rotation." In section 15.5 we will see that this plays an important role in making the galaxy's spiral arms.

The Sun's large orbital speed (roughly 10 times the speed of the Earth around the Sun) might make you think the Milky Way spins rapidly, but because our galaxy is so huge, the Sun takes approximately 240 million years to complete one trip around it. Since the death of the dinosaurs about 65 million years ago, our galaxy has made only about one-quarter of a revolution.

Within this swirling cloud of stars, the Sun is nearly lost like a speck of sand on a beach. But unlike the close packing of grains of sand, stars are widely separated. For example, near the Sun and throughout most of the disk, stars are typically several light-years apart: the Sun's nearest neighbor is 4.2 light-years away, a separation like that of grains of sand 15 miles apart. Near the core of the Milky Way, stars are packed far more densely, with a separation roughly 1500 times less than that near the Sun, a separation like grains of sand at opposite ends of a football field. The density of stars, as measured by the number of stars per cubic light-year, is probably more than 10 million stars per cubic light-year near the core but only about 0.003 stars per cubic light-year near the Sun. Near the edge of the disk, stars are spread even more thinly, much as our atmosphere thins out as it merges into space. Thus, the Milky Way has no sharply defined outer edge.

## Mass of the Milky Way and the Number of Stars

Astronomers can calculate the Milky Way's mass from the gravitational attraction needed to hold its stars in orbit. Such calculations show that the Milky Way must contain at least $10^{11}$ M$_\odot$ of matter. However, it must contain even more matter, because its gravitational effect on nearby galaxies implies a mass as large as $2 \times 10^{12}$ M$_\odot$. This large mass is puzzling, because when astronomers add up the mass of the Milky

Way's *visible* stars and gas, they find only about one-tenth the amount of mass needed to hold our galaxy together. The other nine-tenths is undetectable except by its gravitational effect. Thus, astronomers conclude that the major part of our galaxy is composed of some unseen material, which they therefore call **dark matter.**

Because dark matter emits no radiation of any sort, it is very difficult to study. Its composition is one of the major mysteries in astronomy today, and astronomers know few of its properties. They do know that dark matter is spread throughout our galaxy and that it extends far beyond the visible stars and gas in a huge halo. They also know that it does not interact with ordinary stars and gas except through its gravitational pull. Finally, as we will see in chapter 16, astronomers know that the Milky Way is not alone in having a massive dark-matter halo: other galaxies are also embedded in dark-matter haloes of their own.

The above uncertainty about the nature of dark matter and the value of our galaxy's mass makes it difficult to determine the number of stars our galaxy contains. However, if we assume that the average star in our galaxy has a mass comparable to the Sun's (an assumption based on censusing stars, as we will see in section 15.3), then from the best estimates of the galaxy's total mass, we deduce that it contains about $10^{11}$ (100 billion) stars, each moving along its own orbit, held in its path by the collective gravity of the other stars, gas, and dark matter.

## Age of the Milky Way

The immensity of the Milky Way's size and number of stars is matched by the Milky Way's great age. Astronomers measure this age from the age of our galaxy's oldest stars, using the techniques for measuring stellar ages that we discussed in chapter 13. Such measurements (and others) show that the oldest stars in our galaxy have an age of approximately 13 billion years, an age we will take as that of the Milky Way itself.

Our galaxy's stars were not all created at a single time in that distant past. As we discussed in chapter 13, stars are forming even today and astronomers can estimate the average number of stars born each year by dividing the total number of stars in the galaxy by the galaxy's age. This gives a birthrate of approximately 7 to 8 stars born in the Milky Way each year, although stars probably were born at a much higher rate when our galaxy first formed. Given this rate and the amount of interstellar gas remaining in our galaxy, we can be confident that our galaxy will continue to shine for many billions of years into the future.

## 15.3   STARS OF THE MILKY WAY

### Stellar Censuses

The Milky Way contains many types of stars: giants and dwarfs, hot and cold, young and old, stable and exploding. Counts that list all known stars of a given type in a given region of space—stellar censuses—show that all the types described in chapters 12 to 14 occur in our galaxy. By analyzing the types of stars and their relative numbers, astronomers can deduce some of the ancient history of our galaxy, much as archeologists can learn about life in an ancient city from the kinds of buildings it contained. Such studies of the stellar contents of the Milky Way show that stars similar to or smaller than the Sun vastly outnumber more massive stars. The most numerous stars in our galaxy turn out to be dim, cool, red dwarfs—stars that lie on the main sequence but whose mass is 0.1 to about 0.5 $M_\odot$. Astronomers do not understand yet what determines the distribution of stellar masses, though it probably depends on the conditions in the cloud where the stars form. The average mass for Milky Way stars is a little smaller than 1 $M_\odot$, though some are significantly more massive (up to 100 or so $M_\odot$), whereas others are much less massive (perhaps as small as 0.08 $M_\odot$). Stars much more massive than about 30 $M_\odot$ are

Selection effects can occur whenever one samples a population. For example, when the U.S. Census Bureau counts our population, it must make corrections for people who are homeless or who otherwise might be overlooked. Similarly, astronomers must guard against drawing false conclusions from data that are biased by selection effects.

*The Pop I and II orbits*

very rare because they are nearly unstable and may break up under the force of their own radiation, as we discussed in chapter 13. Objects less massive than about 0.08 $M_{\odot}$ were once also believed to be rare, but astronomers now think they may be common and were simply missed in previous stellar surveys. How could astronomers miss these objects if they are so abundant? Recall that low-mass stars tend to be cool.* In fact, if a star is less massive than about 0.08 $M_{\odot}$, it is so cool that it is unable to fuse hydrogen into helium. Thus, these "brown dwarfs" (as we called them) are more like huge planets than tiny stars, and their extreme dimness and small size make them very difficult to detect. However, searching the sky at the wavelengths at which these objects are brightest greatly increases the chance of finding them. What wavelengths are best? Because they are so cool, brown dwarfs are brightest in the infrared, so it is with infrared surveys such as 2MASS (2 Micron All Sky Survey) that astronomers are now finding large numbers of these "failed" stars. In fact, astronomers now think brown dwarfs may outnumber ordinary stars, making them one of the most common astronomical objects in our galaxy.

Despite their great numbers, low-mass stars are hard to see because they are so intrinsically dim. Thus, when we look out at the night sky, the stars we see tend to be giants. Such stars are rare, but because we can see them even at great distance, we see greater numbers of them. Astronomers refer to this difference between what is seen and what is truly present as a "selection effect."

## Two Stellar Populations: Population I and Population II

Hidden in the great diversity of star types described above is an underlying simplicity astronomers first noted in the 1940s. At that time, the 100-inch reflector at Mount Wilson Observatory, located outside Los Angeles, California, was the largest telescope in the world and the best for observing galaxies. The glow from city lights, however, made it hard to see faint galaxies. But blackouts during World War II darkened the night sky, and Walter Baade, an astronomer at Mount Wilson, took advantage of the darkness to make a series of photographs of neighboring galaxies. He noticed that stars in these nearby galaxies were segregated by color. Red stars were concentrated in the bulges and halos of the galaxies, whereas blue stars were concentrated in their disks and especially in the spiral arms. To distinguish these groups, Baade called the blue stars in the disk **population I** and the red stars of the bulge and halo **population II** (Pop I and II for short). As he pursued his studies of the stellar populations of other galaxies, he saw that Milky Way stars showed the same division.

Following on Baade's work, astronomers today distinguish population I and population II stars by five main properties: location in galaxy, motion, composition, color, and age. Population I stars lie in the disk and follow approximately circular orbits, as shown in figure 15.9. Their composition is like the Sun's, mostly hydrogen and helium, but about 3% of their mass is heavy elements such as carbon and iron. They are generally young ($10^6$ to a few times $10^9$, years old) compared to the galaxy, and many of them are blue.

Population II stars, on the other hand, lie in the bulge and halo of the galaxy and move along highly elliptical orbits that may be tilted strongly with respect to the disk, as shown in figure 15.9. They too are mostly hydrogen and helium, but only a few hundredths percent of their mass is composed of heavy elements, roughly a hundred times less than stars like the Sun. Unlike population I stars, population II stars are very old (about $10^{10}$ years old) and generally red. Table 15.1 summarizes some of these properties.

Division of all stars into two broad categories is an oversimplification. For example, the Sun fits in neither category precisely, but it is considered a Pop I star because of its relatively high heavy-element content and approximately circular orbit in the disk. To avoid forcing a star into a category in which it does not fit, astronomers

---

*"Cool" here means only in comparison to stars like the Sun.

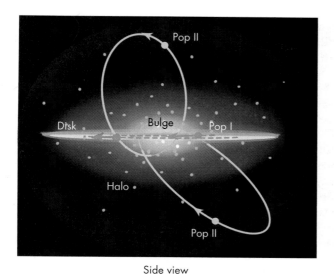

Side view

**A**

**FIGURE 15.9**
Stellar orbits in the Milky Way. Pop I stars orbit in the disk (blue lines). Pop II stars orbit in the halo (yellow lines). Notice the elongation and inclination of their orbits.

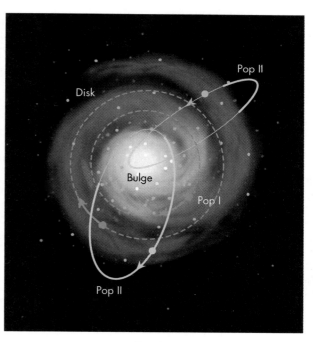

Top view

**B**

**TABLE 15.1**

## Properties of Population I and II Stars

| | Pop I | Pop II |
|---|---|---|
| Age | Young to old ($10^6$ to a few times $10^9$ years) | Old (about $10^{10}$ years) |
| Color | Blue (generally) | Red |
| Location | Disk and concentrated in arms | Halo and bulge |
| Orbit | Approximately circular in disk | Plunging through disk |
| Heavy-element content | High (similar to Sun) | Low ($10^{-2}$ to $10^{-3}$ Sun's) |

In fact, some astronomers now prefer the simple labels of *disk, bulge,* and *halo* stars to describe the stellar populations of the Milky Way and other galaxies.

sometimes subdivide Pop I and Pop II into *extreme* and *intermediate* Pop I and Pop II. Also, some astronomers use the term *old-disk population* to describe stars like the Sun that are located in the disk but that are not extremely young. In short, not all stars fit within the precise division of these two populations. Nevertheless, they are still useful for describing both the structure and history of the Milky Way. For example, many astronomers think that the two different populations show that star formation has not occurred continuously in the Milky Way. Population II stars probably formed in a major burst at the time of the galaxy's birth during its initial collapse (see section 15.8), whereas Population I stars formed much later and continue forming even today. This hypothesis explains the differences between the two populations, as we will discuss in greater detail in section 15.8.

One of the first uses astronomers made of stellar population differences was to map the Milky Way's spiral arms. As Baade noted, the spiral arms of other galaxies are demarcated by their Pop I stars, especially the hot blue ones of spectral classes *O* and *B*. It is therefore logical to assume that the same is true for the Milky Way. By measuring the location of *O* and *B* stars near the Sun, astronomers can make a crude picture of the spiral arms near us, even though they cannot view the arms from a distance. This method is limited, however, because dust in space prevents us from seeing even the brilliant *O* and *B* stars if they are farther from the Sun than about 10,000 light-years. Nevertheless, such maps were the first evidence that we live in a spiral galaxy.

## Star Clusters

Some of the Milky Way's stars are bound together gravitationally into groups called **star clusters.** Within a cluster, each star moves along its own orbit about the center of mass of the cluster, held to it by the group's gravity. You can see several such clusters with your naked eye. The most obvious one is the Pleiades, or the Seven Sisters (shown in figures 15.10 and 15.16), named for the daughters of the giant Atlas who, in Greek mythology, carried the world on his shoulders. The Pleiades (pronounced *plee-a-dees*) are visible from late August through March as a tiny group of stars north of the V in the constellation of Taurus. With binoculars, you can see that the Pleiades contain many more than the half dozen stars visible to the naked eye. Likewise, by scanning the Milky Way with binoculars, you will see many additional clusters dimmer than the Pleiades.

The Milky Way has two main types of star clusters: open clusters and globular clusters. **Open clusters**—the Pleiades, for example—contain up to a few hundred members in a volume with a radius of typically 7 to 20 light-years. They are called "open" because their stars are scattered loosely, as you can see in figure 15.10. Sometimes they are also called "galactic clusters" because most of them lie in the disk of the galaxy.

Astronomers think that open clusters form when giant, cold, interstellar gas clouds move into the galaxy's spiral arms. The clouds are compressed by the stronger gravitational field within the arm and collapse, breaking up into hundreds of stars whose mutual gravity binds them into the cluster. Once formed, an open cluster continues to orbit the galaxy in the disk, but over hundreds of millions of years, stars gradually escape from it and so the cluster eventually dissolves. Assuming that open clusters are as abundant elsewhere in the galaxy as within the regions visible from Earth, astronomers estimate that our galaxy contains about 20,000 open clusters, though dust hides most of them from our view. Our own Sun probably once was a member of such a star group, its companion stars having long since scattered across space.

Within our galaxy's spiral arms extremely young stars sometimes occur in loose groups called **associations** that are a few hundred light-years across. Associations typically spread out from a single large open cluster near their center and may contain other, smaller star groupings. Moreover, the stars in associations are usually still mingled with the massive clouds of dust and gas from which they formed.

**Globular clusters** contain far more stars than open clusters, from a few hundred thousand to several million per cluster. We described them briefly in our discussion of Shapley's study of the shape of the Milky Way. They have larger radii than open clusters have—40 to 160 light-years—but the larger number of stars in them creates a stronger gravity than that found in open clusters. This stronger gravity pulls their stars into a denser ball, as figure 15.10 shows. Astronomers estimate that about 150 to 200 globular clusters orbit the Milky Way, of which about 145 have actually been observed. Table 15.2 summarizes some properties of open clusters and globular clusters and associations.

### TABLE 15.2

**Properties of Clusters and Associations**

| Type | Number of Stars | Radius* | Location |
|------|-----------------|---------|----------|
| Open cluster | Tens to a few thousand | 7 to 20 ly (2 to 6 pc) | Along spiral arms and in disk |
| Globular cluster | $10^5$ to $10^6$ | 40 to 160 ly (12 to 50 pc) | Halo and bulge of galaxy |
| Associations[†] | 5 to 70 *O* or *B* stars | 65 to 325 ly (20 to 100 pc) | Spread along a spiral arm |

*Because star clusters do not have sharp edges but instead gradually thin out, quoted dimensions differ substantially.
[†]Astronomers identify several other types of associations as well. For example, T associations are regions with above-average numbers of T Tauri stars.

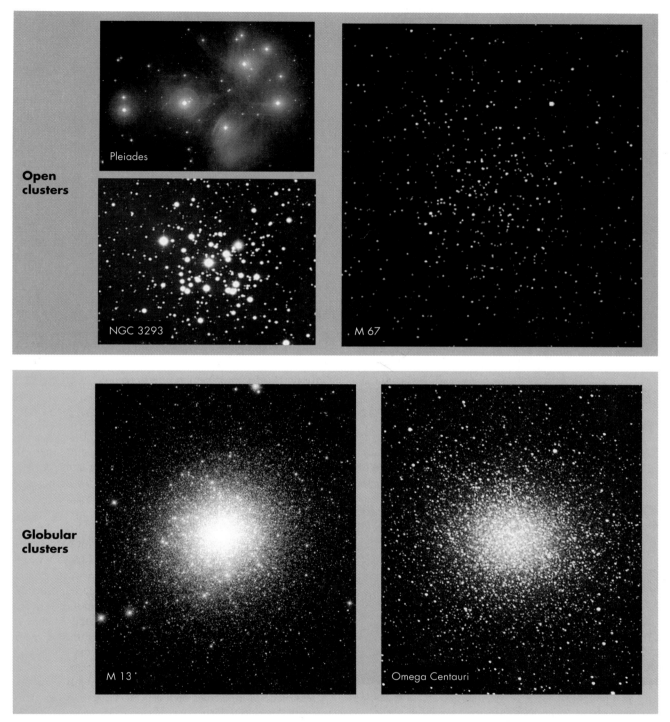

**Open clusters**

Pleiades

NGC 3293

M 67

**Globular clusters**

M 13

Omega Centauri

**FIGURE 15.10**
Star clusters: both open and globular types.

Open clusters and globular clusters reflect the properties of Pop I and II stars that we discussed earlier. Open clusters are generally Pop I, whereas globular clusters are always Pop II. This difference in their stellar population makes clusters especially useful to astronomers for studying the structure of the Milky Way. For example, globular clusters (among the oldest stars in the galaxy, many of them roughly 12 billion years old) outline the halo and bulge. On the other hand, the much younger open clusters (some a mere million years old) trace the galaxy's arms.

## 15.4 GAS AND DUST IN THE MILKY WAY

The space between stars is not empty: it contains **interstellar matter** composed of gas and dust particles. By terrestrial standards, this space is almost a perfect vacuum. On the average, each cubic centimeter contains only a few atoms of gas. For comparison, the air we breathe contains about $10^{19}$ atoms per cubic centimeter. The density of gas in interstellar space compared to air is like one marble in a box 5 miles on a side compared to the same box filled completely with marbles.

Despite interstellar matter's very low density, it plays several important roles in the Milky Way. First, it is the material from which new stars form, as we discussed in chapter 13. With no interstellar matter, star birth would stop—though star deaths would, for a while, replenish the material with gas and dust blown out of stars by stellar winds or explosions. Thus, interstellar matter is a kind of cosmic compost heap with matter from dying stars being recycled back into new stars.

A second reason interstellar matter is important is that the dust mixed in with it strongly limits what we can observe within the Milky Way. Because interstellar dust dims and blocks our view of distant parts of our galaxy, it seriously affects our ability to measure distances to stars and even to detect stars in the more distant regions of the Milky Way. But before we turn to how astronomers overcome the difficulties created by interstellar dust, it will be helpful to look more closely at some general properties of interstellar matter.

### Distribution and Composition of Interstellar Matter

Interstellar matter is scattered throughout the Milky Way, but gravity has pulled most of it into a thin layer in the disk, as can be seen by the dark band of dust in other galaxies (fig. 15.11). Within this layer, gravity and gas pressure have further clumped the dust and gas into clouds whose sizes range from a few light-years to a few hundred light-years. We can see such clouds directly (see fig. 15.2), but we can also detect them by their effect on light from distant stars.

As light from a background star passes through an interstellar cloud, the cloud's gas and dust imprint narrow absorption lines in the star's spectrum (fig. 15.12A). By analyzing these imprinted lines, astronomers can deduce the cloud's composition, temperature, density, and speed. Such observations show that the gas in interstellar clouds has a composition similar to that of the Sun and other stars, namely about 71% hydrogen and 27% helium, with trace amounts of heavier elements such as carbon and neon. Analysis of the light absorbed by the dust particles shows that they are made of silicates and carbon compounds, coated with a thin layer of ices made of molecules such as water, carbon monoxide, and methyl alcohol. These particles are extremely tiny, ranging from about a micrometer to a nanometer or so in diameter (roughly the size of a single large molecule).

**FIGURE 15.11**
Interstellar matter, outlined by dark dust clouds, is concentrated in the plane of a galaxy, forming a dark band across it. This galaxy is sometimes called the "Sombrero galaxy." Technically, it is called either NGC 4594 or M104.

$\mathbb{Q}^\star$ Look carefully above and below the disk of the galaxy. Do you see some small, fuzzy, round spots? On the basis of their location, what do you think they are?

$\mathbb{Q}^\star$ They are globular clusters.

### Interstellar Dust: Dimming and Reddening

We know there is dust in interstellar space because it dims the brightness and alters the color of the stars we see through it, much as haze in our atmosphere dims and reddens the

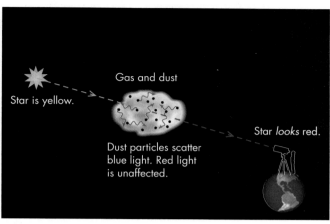

**A**        **B**

**FIGURE 15.12**

(A) Gas atoms in an interstellar cloud between us and a distant star absorb some of the star's light, adding their own absorption lines to those in the star's spectrum. (B) Dust in an interstellar cloud scatters the blue light from a distant star, removing it from the radiation reaching Earth and making the star look redder than it really is.

How is this similar to the red appearance of the Sun when it sets?

The Sun looks red when it sets because molecules and dust in our atmosphere interact more strongly with the blue light than with the red light and scatter the blue light. As a result, less blue light reaches us, so the Sun looks redder. Interstellar dust does the same thing to starlight—it preferentially removes the blue light, leaving the star looking redder than it really is.

setting Sun. Dimming and reddening result when light strikes a tiny particle such as a dust grain. The light is reflected from it in random directions in a process called **scattering.** A particle's ability to scatter light depends on its size compared to the light's wavelength. Light is scattered strongly if its wavelength is smaller than the particles, which for interstellar dust and molecules in our atmosphere means that short-wavelength blue and ultraviolet light is scattered extremely well. Scattering of sunlight creates the blue color of our sky. Scattering of starlight in space also creates visible effects.

Scattering dims a star's light because radiation that would have reached us in the absence of dust is sent off in random directions and is lost to our sight. Because red light scatters less than blue, more of the star's red light reaches us than its blue. As a result the star looks redder than it really is, a phenomenon called **reddening** (fig. 15.12B). Precisely the same effect occurs in our atmosphere and makes the setting Sun look red.

If a dust cloud is extremely thick or dense, it may dim the light of background stars to the point where they become invisible, an effect called "obscuration." This creates a dark region, seemingly empty of stars, which astronomers call a **dark nebula.** Figure 15.13 shows a region of the

Open star cluster        Dark nebula

**FIGURE 15.13**

A dark nebula in the Milky Way. Dust blocks our view of the background stars. The cluster is NGC 6520. The dark nebula is Barnard 86. A few stars lie between us and the nebula.

Milky Way in which a dark nebula is visible, silhouetted against a swarm of faint background stars and blotting out their light.

Only about 1% of the interstellar matter is dust, but this dust scatters visible light so well that it strongly limits our view of the Milky Way. We can see no more than a few thousand light-years away from the Sun into the disk because the dust is concentrated there. Only by looking above or below the disk can we see more distant objects. Dust in the disk also hides galaxies that lie behind it, creating a band around the sky called the **zone of avoidance,** in which essentially no galaxies can be seen. Figure 15.14 shows a chart of over 34,000 galaxies illustrating this effect. The narrow horizontal region where no galaxies appear is the zone of avoidance and coincides with the plane of the Milky Way. We know today that galaxies do occur in this region because they can be seen with radio and infrared telescopes sensitive to radiation that can penetrate the dust.

Infrared and radio waves pass more readily through dust than visible light does because their wavelengths are much larger than the dust particles. This allows infrared

**FIGURE 15.14**

Dust limits our view in the disk of the galaxy and creates the zone of avoidance. We see no galaxies in the plane of our own galaxy because dust in the disk blocks our view. (A) A plot showing how galaxies are distributed on the sky. Notice how few lie along the central line (the galactic equator). This blank region is the zone of avoidance. (B) A sketch illustrating why we see so few galaxies along the plane of the Milky Way. The gray blobs represent dust clouds that block our view toward directions that lie in the Milky Way's plane. The galaxies are in reality about the same size as the Milky Way and are, of course, much farther away than can properly be shown in the figure.

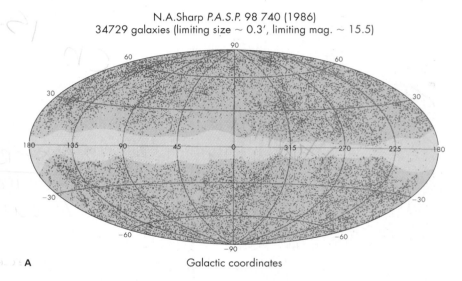

N.A.Sharp *P.A.S.P.* 98 740 (1986)
34729 galaxies (limiting size ~ 0.3′, limiting mag. ~ 15.5)

**A** Galactic coordinates

**B**

**FIGURE 15.15**
An all sky view of our galaxy, as we see it from Earth. The image was made by plotting on a map of the whole sky the millions of stars observed in the infrared by the 2-Micron All Sky Survey (2MASS). It is therefore like a map of the Earth, flattened out so as to show all continents. The map clearly reveals the flatness of our galaxy and the bulge of its central regions.

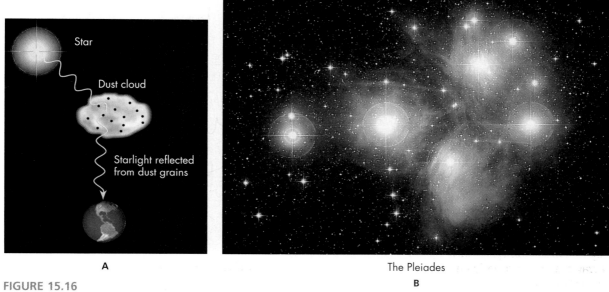

**A**

The Pleiades

**B**

**FIGURE 15.16**
Reflection nebula. (A) Light from a star near a dust cloud is reflected from the dust making it visible. (B) The Pleiades star cluster. Note the bluish reflection nebula around many of the brighter stars.

Q★ Why is red light not scattered?

A★ Red light has a longer wavelength than blue and so it does not interact as strongly with the dust. It is therefore scattered less.

pictures to reveal the structure of the Milky Way far better than visible light pictures can, as illustrated in figure 15.15. The disk and central bulge of our galaxy are easy to see in this wide-angle picture made by plotting the positions of millions of stars observed as part of the 2MASS infrared survey of the sky.

Dust is not all bad;* it may scatter enough light to become visible, thereby creating beautiful **reflection nebulas** (fig. 15.16). Scattered starlight makes the interstellar dust shine just as scattered sunlight makes dust in the air visible. Reflection nebulas are not only lovely but additional proof that dust exists in space.

---

*We should also keep in mind that the atoms that compose the Earth and our bodies came from such dust grains.

**FIGURE 15.17**

(A), (B), (C) Emission nebulas. Notice the bright (and hot) stars near the glowing gas. (D) These stars emit ultraviolet radiation, which heats and ionizes the nearby hydrogen gas. Electrons dropping from orbit 3 to orbit 2 generate the red light.

## Interstellar Gas

Interstellar gas plays several crucial roles in our galaxy. It is the material from which stars form, and it is the repository of matter blown off dying stars. Without interstellar matter to make new stars to replace those that die, the Milky Way would be far dimmer and filled only with stellar corpses and low-mass stars. Interstellar gas is also a help to astronomers seeking to map the Milky Way. Dust blocks our view of the Milky Way and makes it difficult for us to observe its structure, but interstellar gas emits radiation that penetrates the dust and allows us to "see" regions of the galaxy that would otherwise be invisible to us. Finally, interstellar gas creates some of the most spectacular structures in our galaxy: huge glowing clouds heated by stars embedded in them, as shown in figure 15.17. An example of such a gas cloud is the Great Nebula in Orion, which is easy to see with a small telescope or even a pair of binoculars. If you look at the middle star in Orion's sword you will see a pale, fuzzy glow created by luminous interstellar gas surrounding a dense star cluster.

**Visible Emission from Interstellar Gas.** Interstellar gas clouds that emit visible light are called **emission nebulas,** and figure 15.17 illustrates several of the many found within the Milky Way. Emission nebulas give off their own light in visible and other wavelengths, unlike the dust in reflection nebulas, which simply reflects and scatters light from neighboring stars. However, a cloud must be hot to emit visible light, and so such clouds need a source of heat (generally a nearby star), as shown in figure 15.17D. Hot, blue stars are especially effective at heating because they emit large amounts of ultraviolet radiation whose short wavelength is much more energetic than that of visible light. Ultraviolet light whose wavelength is less than 91.2 nm is so energetic that when it is absorbed by hydrogen, it tears the electron free of the nucleus, ionizing the

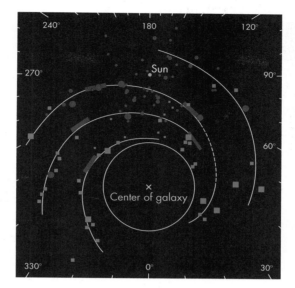

**FIGURE 15.18**

Map of the Milky Way made by combining radio observations (since radio can penetrate dust and thereby allow one to see all the way across the Milky Way's disk) and optical observations of the position of HII regions. Such regions outline the spiral arms of other galaxies. In our galaxy, too, they form a spiral pattern.

gas. In an ionized gas (one in which atoms are missing one or more electrons), the free electrons collide with atoms of oxygen, nitrogen, and other elements and excite them, making them emit light of their own. Eventually, the electrons recombine with the ionized hydrogen atoms and emit more radiation, especially the red light of hydrogen alpha, which gives these nebulas their lovely reddish color. Because hydrogen in these hot gas clouds is ionized, they are sometimes called **HII regions,** the *H* denoting the hydrogen and the *II* indicating that it is ionized.

HII (pronounced *H-two*) regions have a temperature of approximately 10,000 K and are heated by the electrons freed by the ionizing ultraviolet radiation. The liberated electrons collide with gas atoms around them and speed them up, increasing their energy and thus their temperature. The freed electrons are themselves a powerful source of radio emission, emission that helps astronomers see HII regions that might otherwise be hidden from us by dust. Because these HII regions are generally located in spiral arms (fig. 15.7 shows them glowing like rubies interspersed with diamonds on a necklace), radio maps of their location can reveal the spiral structure of our galaxy. In fact, maps made by combining radio observations and optical observations of HII regions are the best evidence we have that the Milky Way has spiral arms (fig. 15.18).

**Radio Waves from Cold Interstellar Gas.** Although we can easily see the glow from gas clouds that happen to be near hot stars, much interstellar gas is too cold to emit visible light, and we can detect it only with radio telescopes. Clouds more than a few tens of light-years from hot stars may be only a few Kelvin (more than 400° below zero on the Fahrenheit scale). Such cold material emits no visible light because it has too little energy to generate visible photons. However, it can emit radio waves of very low energy (recall that radio waves carry much less energy than visible light does); as described in the Extending Our Reach box, Mapping the Milky Way with Radio Waves, page 452.

# 15.5 MOTION OF STARS AND GAS IN THE MILKY WAY

The motions of stars and gas give important clues to the structure of the Milky Way. We saw earlier that in accordance with Newton's first law (the law of inertia), if no force acted on the stars, each would move in a straight line. But a force does act on each star: the gravitational force created collectively by all the other stars and gas. That force acts to draw each star inward toward the center of the galaxy. The result is that each star follows its own orbit around the center of the galaxy, just as each planet moves in orbit about the Sun.

# EXTENDING OUR REACH

## MAPPING THE MILKY WAY WITH RADIO WAVES

One of the most important sources of radio emission is cold hydrogen atoms,* which radiate at a wavelength of 21 centimeters. This radiation, called **21-centimeter radiation,** arises because subatomic particles such as protons and electrons possess a property, known as "spin," that makes them behave like tiny magnets. Magnets have north and south poles with the property that like poles repel and unlike poles attract. If an electron has the same alignment as the proton it is orbiting—both with north poles "up," for example—the electron will flip over, as shown in box figure 15.1. As it flips, the changed magnetic interaction between the electron and the nucleus makes the electron shift to a slightly smaller orbit with a slightly lower energy. However, energy must be conserved, and so when the electron shifts to the new orbit, it emits the excess energy as electromagnetic radiation. We discussed in chapter 3 that this process can produce visible light. In cold hydrogen, however, the flip of the electron liberates too little energy to make visible radiation and can create only radio emission at a wavelength of 21 centimeters.

The 21-centimeter radiation has proved extremely valuable for studying the Milky Way and other galaxies. Hydrogen is abundant in space, and so the signal of the 21-centimeter radiation is strong. From the strength of the signal, astronomers can deduce the amount of hydrogen in an area. If the gas is moving, the wavelength will be Doppler-shifted, from which the gas's speed along our line of sight can be found. Radio observations thereby allow astronomers to map not only where the gas is concentrated in the galaxy but also how it is moving. Although details in such 21-centimeter maps are uncertain because of assumptions that must be made about how the gas moves, they nevertheless are extremely useful because the radiation is not absorbed by interstellar dust, thereby allowing astronomers to "see" the entire Milky Way in its radiation. Such maps show that the gas is confined to a thin disk and its distribution is suggestive of a spiral pattern, confirming the picture deduced by other means.

The 21-centimeter radiation is not a perfect tool, however. It is emitted only by hydrogen if its electron is attached. If the hydrogen is heated so that its electron is torn free—that is, the hydrogen is ionized—it produces no 21-centimeter waves. Likewise, at extremely low temperatures (15 K or so), hydrogen atoms bond to one another to make hydrogen molecules and, in its molecular form, hydrogen cannot emit 21-centimeter

---

*Cold hydrogen atoms with their electrons attached are called "HI" (pronounced *H-one*) to distinguish them from ionized hydrogen atoms, HII.

**BOX FIGURE 15.1**
Radio radiation at 21 centimeters is emitted by cold hydrogen as the electrons "flip."

### TABLE 15.3

#### Some Interstellar Molecules

| | |
|---|---|
| $H_2$ | Molecular hydrogen |
| OH | Hydroxyl radical |
| CN | Cyanogen radical |
| CO | Carbon monoxide |
| $H_2O$ | Water |
| HCN | Hydrogen cyanide |
| $NH_3$ | Ammonia |
| HCOH | Formaldehyde |
| HCCH | Acetylene |
| HCOOH | Formic acid |
| $CH_3OH$ | Methyl alcohol (methanol) |
| $CH_3CH_2OH$ | Ethyl alcohol (ethanol) |

radiation. However, in the cold, dense regions where hydrogen molecules form, other molecules that do emit radio waves form as well.

More than 100 interstellar molecules have been identified so far, including common compounds such as carbon monoxide (CO), formaldehyde (HCHO), and ethyl alcohol ($CH_3CH_2OH$), and also more exotic substances. Table 15.3 lists a few of the many interstellar molecules now known. Collisions among neighboring molecules make them spin and give them energy, as shown in box figure 15.2, but they quickly radiate that energy away and reduce their spin by emitting radio waves. Each molecule has a distinctive set of wavelengths at which it can emit its rotational energy, just as each atom has a distinctive set of wavelengths (its spectral lines) at which it can radiate energy from electron jumps. For example, carbon monoxide emits at wavelengths of 2.6 and 1.3 millimeters, and ethyl alcohol emits at 3.5 millimeters. Both these molecules emit at many other wavelengths as well.

The strength of the molecular radio emission from a gas cloud depends on the number of molecules it contains, giving astronomers information on its density. If the gas is moving, the wavelength of the emission will be slightly altered by the Doppler shift, displacing and broadening the spectrum lines. From such shifts and broadening, astronomers can deduce the speed of the cloud and its rotation. With that information, they can then map out where in the Milky Way the molecular gas is located and thereby obtain additional information about the galaxy's structure.

**BOX FIGURE 15.2**
Emission of radio waves by spinning molecules in cold interstellar clouds. Collisions set the molecules spinning. When their rotation speed changes, they emit radio waves. The wavelength emitted depends on the kind of molecule and by how much its rotation speed changes.

Although all stars within the Milky Way move around its center, the paths followed by stars in the disk and halo are very different, as we discussed in section 15.3. We described there that all the stars in the disk orbit in the same direction and in nearly the same plane, giving the disk its flat shape, just as the planets' motion in approximately the same plane gives the Solar System its flattened shape. On the other hand, stars in the halo and bulge follow orbits that are steeply inclined to the disk's plane (see fig. 15.9).

Motion within the disk also creates the Milky Way's spiral arms. We know that gravity holds a galaxy together, and so we might infer that gravity is also what makes a spiral arm. Gravity does in fact help hold a spiral arm together, but a spiral arm is not a fixed mass of material. Rather, the stars and gas in an arm at one time slowly drift out and are replaced by new stars and gas, according to the density-wave theory of spiral structure. In the **density-wave model,** waves of stars and gas sweep around the galactic disk. The wave is not primarily an up-and-down motion like an ocean wave; rather, it is a place where the density of stars and gas is large compared to the surroundings. This denser region is the wave crest, and we see it as a spiral arm. Spiral waves are not unusual in rotating matter. Batter stirred in a bowl with an electric mixer forms spiral waves, but to better understand how arms form in the Milky Way, consider the following analogy.

As cars move along a freeway, most move at nearly the same speed and keep approximately the same separation. If one car moves slightly slower than the others, however, traffic will begin to bunch up behind it. Cars can pass the slower vehicle but must change lanes to do so. The result is a clump of cars behind the more slowly moving one. But that clump is not composed permanently of the same cars. Cars join the clump from behind, pass the slower moving car, and then leave the clump at its front.

**FIGURE 15.19**

Sketch of a density wave in a galaxy, showing the progression from dust and gas to stars across the arm.

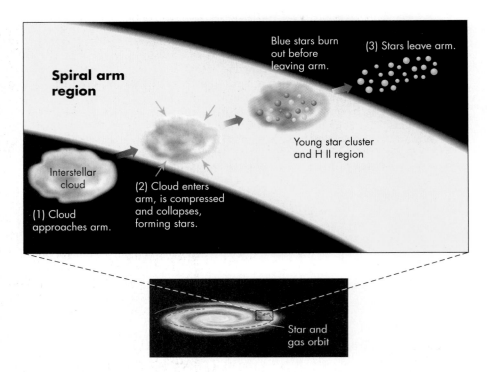

A similar phenomenon happens to stars orbiting in the Milky Way. As stars circle the galaxy, a few closely spaced stars at a given location will make the local gravity slightly stronger than elsewhere in the disk. That excess gravity draws stars into the region so that the clump grows a little. But stars do not remain in the clump. They continue along their orbits and eventually leave the clump. Likewise, gas clouds drift through the clump, initially accelerated inward by the clump's gravity and then slightly slowed as they move out the other side. The result is a wave in the galaxy's disk where stars and gas accumulate before moving through. The accumulation of gas compresses clouds and makes them collide with one another, which in turn makes some of the clouds collapse and form stars. This is the first stage of the star formation process described in chapter 13. The brilliant young O and B stars so born illuminate the arms and make them stand out against the more ordinary stars of the disk. In fact, most of a galaxy's stars form in its spiral arms, and as we will discover in the next chapter, galaxies without arms generally contain extremely few young stars.

Although the density-wave theory has difficulty explaining the longevity of spiral arms, many observations support the theory. For example, by combining radio and visual pictures of a galaxy, astronomers can see a progression through the spiral arm as gas enters on one side, is compressed, and forms stars, taking a few million years to sweep through the arm. On one side of the arm, you can see gas and young stars; on the other side, older, more evolved stars, as indicated in figure 15.19. Doppler-shift studies of the gas motion also clearly show the gas moving into the arm on one side and leaving on the other.

Although most spiral galaxies probably make their arms this way, another theory has been proposed to explain the very ragged-appearing arms in some galaxies, such as the one illustrated in figure 15.20. This second theory is called **self-propagating star formation,** or SSF, for short.

In the SSF theory, star formation starts at some random point in the disk of a galaxy when a gas cloud collapses and turns into stars. As the stars heat the gas around them and explode as supernovas, they generate a disturbance that makes the surrounding gas clouds collapse and turn into stars. The original stars burn out, but the new stars—formed as the old ones evolve and explode—trigger more gas clouds to collapse and form additional stars. In this fashion, the region of star formation spreads across the galaxy's disk much as a forest fire burns outward in a circle through a forest.

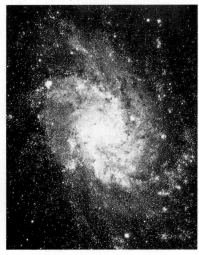

**FIGURE 15.20**

Ragged arms in the spiral galaxy M33, perhaps the result of a self-propagating star formation.

**FIGURE 15.21**
Winding of arms in the SSF model.

But unlike a forest where the trees stand still, stars orbit in a galaxy, and those near the center orbit in less time than those farther out. Thus, the zone in which star formation occurs is drawn out into a spiral by the difference in rotation rate between the inner and outer parts of the disk. The result is spiral arms, as illustrated in figure 15.21.

Astronomers are not certain which process—density wave or SSF—is most important in forming spiral arms. Possibly in some galaxies one of the processes is at work, whereas in other galaxies the other is.

## 15.6  MEASURING THE MILKY WAY

We have described earlier in this chapter how astronomers measure the basic properties of the Milky Way, such as its radius and mass. In this section, we look in greater detail at those methods.

### Diameter of the Milky Way

Astronomers have several ways to measure the Milky Way's diameter, but all such methods depend on first knowing how far from the center the Sun is. We described in section 15.1 how Shapley used globular clusters to map out the structure of the galaxy and thereby find its size and where the Sun lies in it. This method remains one of the best techniques for measuring our galaxy's scale. Astronomers measure the distance and direction from the Sun to a large number of globular clusters or some type of star that can be seen for great distances and then plot their positions, as shown in figure 15.22. They next locate the point around which the clusters seem to be most symmetrically located and assume this point is the center of the galaxy. Finally, they measure the distance from that point to the Sun, as shown in the figure. Modern values for this distance turn out to be about 8.5 kiloparsecs (approximately 28,000 light-years) on the average.

A newer method that may eventually give a more accurate distance uses radio observations of cool giant stars. Some such stars emit powerful beams of energy at radio wavelengths analogous to the light emitted by lasers. These stars are called "maser sources." The word *maser* is an acronym for *m*icrowave-*a*mplification by *s*timulated *e*mission of *r*adiation. The radiation comes from molecules in a cool shell around the star. Maser sources are especially common in the inner bulge of the Milky Way because of the many red giants there, and with the masers, astronomers can measure the distance from the Sun to the Milky Way's center.

The maser sources, like other stars near the galactic center, move along randomly oriented orbits. Some move along our line of sight and therefore show a Doppler shift, which gives their velocity. To obtain their distance, astronomers assume that, on the average, maser sources moving along the line of sight have the same speed as sources moving across the line of sight. Sources moving across the line of sight show a tiny change in position from one year to the next, and from that change and their speed,

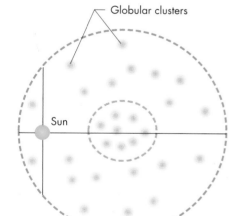

**FIGURE 15.22**
Finding the distance to the center of the Milky Way using globular clusters. Make a scale plot of where the clusters lie with respect to the Sun. Outline the region filled by the clusters and note where they are most concentrated. The center of their concentration is presumably the center of the galaxy. Measure the distance from the center to the Sun and from the known scale of the plot determine Sun–center distance.

astronomers can deduce their distance and therefore the distance to the center of the Milky Way (the point about which they orbit). This method for finding the Sun's distance from the center of the galaxy gives a somewhat smaller value (about 7 kiloparsecs) than is found by other methods, and so astronomers are uncertain which method to believe. Nevertheless, once the Sun's distance from the center is known, the overall size of the Milky Way follows when the distance to objects lying near the galaxy's edge is measured and the two distances are added. From such measures, astronomers believe the Milky Way has an overall diameter spanning 40 kiloparsecs or more, but much of the vast outer part beyond the Sun's orbit is extremely sparsely populated.

## Mass of the Milky Way

Astronomers can deduce the Milky Way's mass from its gravitational attraction on matter in and near it. The procedure they use is very similar to the one described in chapter 12 to determine the mass of stars. You may recall that if one star is in orbit around another, we can calculate their combined mass using the modified form of Kepler's third law. Similarly, astronomers can find the mass of the Milky Way by treating it as a scaled-up version of a binary star in which one star, the Sun, for example, orbits under the gravitational influence of the entire galaxy (fig. 15.23). The details are given in Extending Our Reach: "Measuring the Mass of the Milky Way."

More refined methods build on this same technique but account for the mass of the entire galaxy and use the rotation speed of stars at a variety of distances (the so-called **rotation curve**). The rotation curve, shown in figure 15.24, shows how fast the galaxy spins at each distance from its core. That speed must be exactly sufficient to offset the gravity of the Milky Way that pulls orbiting stars in toward the center. The gravity in turn depends on the Milky Way's mass, and so from the rotation curve and Newton's laws of motion and gravity, astronomers can calculate the mass of our galaxy. The first task, therefore, is to measure the rotation curve.

# EXTENDING OUR REACH

## MEASURING THE MASS OF THE MILKY WAY

In our calculation, we will make use of a result from physics that, if the Milky Way were spherical, only that part of it inside the Sun's orbit would affect the Sun's motion. Moreover, the gravity of that inner part acts as if all the mass interior to the Sun's orbit were packed into a point at the orbit center. The Milky Way is not spherical, but detailed calculations using its true shape differ only slightly from what we will find with our simple model. A more serious difficulty, however, is that our method of calculation omits all the mass in the Milky Way that lies outside the Sun's orbit. That is, we can assert the Milky Way contains *at least* the amount of mass given by our calculation.

We proceed as follows. We represent the mass of the Milky Way inside the Sun's orbit as a single mass, $M$, and assume the Sun (with mass $m$) moves in a circular orbit of radius $a$ about $M$, as shown in figure 15.23. We now use the modified form of Kepler's third law described in chapter 12 for measuring the masses of binary stars. The modified form of Kepler's third law states that

$$m + M = a^3/P^2$$

$m$ = Mass of smaller object
$M$ = Mass of larger object
$a$ = Distance between objects
$P$ = Years to complete one orbit

where $m$ and $M$ are the masses of the orbiting bodies in solar masses, $P$ is their orbital period in years, and $a$ is their separation in astronomical units. The orbital period of the Sun around the Milky Way is about 240 million years (we will see below how this number is determined). Its orbital radius is about 8.5 kiloparsecs, or about $1.8 \times 10^9$ AU. If we insert these values into the equation, we find that the mass of the Sun ($m$) plus that of the Milky Way ($M$) is

$$m + M = (1.8 \times 10^9)^3/(2.4 \times 10^8)^2$$
$$= 1.8^3 \times 10^{27}/(2.4^2 \times 10^{16})$$
$$= 10^{11} \, M_\odot$$

Because $m$, the mass of the Sun, is so tiny compared with the mass of the galaxy, we can ignore the Sun's mass and conclude that the Milky Way's mass is at least $10^{11} \, M_\odot$.

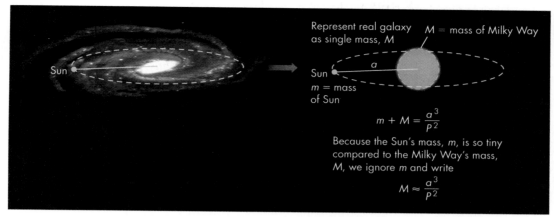

**FIGURE 15.23**

Finding the mass of the Milky Way. Represent the entire Milky Way by a single mass, $M$ (the orange blob in the figure). The Sun (mass $m$) orbits the mass $M$ at a distance $a$ and with an orbital period $P$. Apply the modified form of Kepler's third law to calculate $M + m$. Because the Sun's mass is so tiny, it can be ignored and we can set $M = a^3/P^2$.

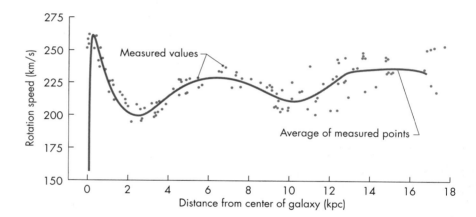

**FIGURE 15.24**

Schematic rotation curve of the Milky Way.

To determine the rotation curve, we must remember that we are observing from the Earth, which is traveling with the Sun in the Sun's orbit around the center of the galaxy. Astronomers thus have the same problem that the police in a moving patrol cruiser have in measuring the speed of a traffic violator. That is, to find the speed of the star around the Milky Way, they must measure how fast it moves with respect to the Sun and then add the Sun's motion about the galaxy. Astronomers have several methods for making this measurement, as described in Extending Our Reach Box, Measuring the Sun's Speed Around the Milky Way, page 458.

From many such measurements, astronomers conclude that the Sun is orbiting the galaxy with a velocity of about 220 kilometers per second (about 130 miles per second!). From the Sun's orbital speed, we can calculate its orbital period, $P$, by dividing the speed, V, into the circumference of its orbit.

That is

$P$ = Sun's orbital period
$R$ = Radius of Sun's orbit
$V$ = Sun's orbital speed

$$P = 2\pi R/V$$

where $R$ is the radius of the Sun's orbit, which we showed earlier is about 8.5 kiloparsecs. Inserting the values of $R$ and V determined above (in appropriate units),

## EXTENDING OUR REACH

### MEASURING THE SUN'S SPEED AROUND THE MILKY WAY

The Sun's orbital speed around the Milky Way can be found in several ways. In one method, astronomers take spectra to obtain the Doppler shift of galaxies in the Local Group, the small cluster of galaxies to which the Milky Way belongs. The Doppler shift of the galaxies is created by their own motion and that of the Sun; remember that if either the source or the observer moves, there is a Doppler shift. It turns out that the galaxies move slowly compared with the Sun's motion around the Milky Way, so almost all the Doppler shift is attributable to the Sun's motion. From that Doppler shift astronomers can find the Sun's orbital velocity.

Another method to measure the Sun's orbital velocity uses objects such as globular clusters. Because these clusters fill an approximately spherical volume around the galaxy, they, as a group, presumably revolve only slowly compared with the flat disk. Thus, the Sun's speed with respect to the globular clusters is at least approximately its true rotation speed around the Milky Way.

we find that $P$ is about $2.4 \times 10^8$ years, the number we used earlier to find the Milky Way's mass.

Unfortunately, the measurements we have described above are difficult to do accurately, so the speed of the Sun is uncertain. Because that influences the mass that we calculate for the Milky Way, our galaxy's mass is also uncertain. A far more serious problem is accounting for the large mass of our galaxy that such calculations imply. That is, the calculated mass greatly exceeds the mass of detectable stars and gas. As we saw in section 15.2, this discrepancy has led astronomers to hypothesize that our galaxy is embedded in a huge, massive halo of dark matter. The dark matter exerts a gravitational force on the detectable stars and gas, but it emits no as yet observed radiation of its own. Moreover, observations of its gravitational effects suggest that this immense halo has a radius of at least 100 kiloparsecs, so that it dwarfs the visible Milky Way. What can this dark matter be made of? Although some of it may be cool, dim stars, the nature of the rest remains unknown and is one of the greatest mysteries in astronomy today.

## 15.7 THE GALACTIC CENTER

The asterisk in Sagittarius A* is part of its name, said as "Sagittarius A star."

The Sun lies at the center of the Solar System. What lies at the center of the Milky Way galaxy? Even though the core of our galaxy is difficult to observe because interstellar dust clouds totally block the visible light emitted by objects there, radio, infrared, X-ray, and even gamma-ray telescopes allow astronomers to see through the dust and reveal a dense swarm of stars and gas swirling around the center of our galaxy (fig. 15.25). With such observations, astronomers have detected an intense source of radio waves known as Sagittarius A*, which lies at the very core of our galaxy. Measurements of this source show that it is less than 10 AU in diameter (about the size of Jupiter's orbit) and that hot gas and stars are in rapid orbit around it. Over the past few years, astronomers have tracked the motion of a number of those stars—some of which are moving at hundreds of kilometers per second (figure 15.26). From these motions, and using the modified form of Kepler's third law, astronomers calculate that the stars are orbiting an object containing about 4 million solar masses. Only such an enormous mass could keep such high-speed stars from escaping from our galaxy's core. Yet no object is visible there with any kind of telescope. Thus, the almost inescapable conclusion is that a massive black hole lies at the center of the Milky Way and holds those stars in its grip. In fact, these observations are perhaps the best evidence yet that black holes exist.

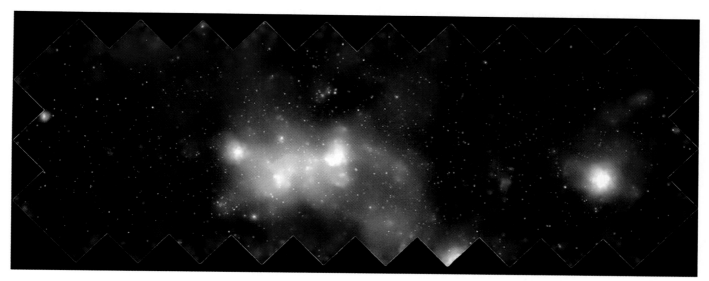

**FIGURE 15.25**

The central region of the Milky Way as observed at X-ray wavelengths by the orbiting X-ray telescope *Chandra*. The supermassive black hole at the Milky Way's core lies inside the bright white patch at the center. The image shows an area 900 light-years across. In this false-color image, the colors represent X rays of different energies. The small bright spots are white dwarfs, neutron stars, and probably gas surrounding a few black holes. The diffuse glow is from extremely hot gas surrounding our galaxy's core.

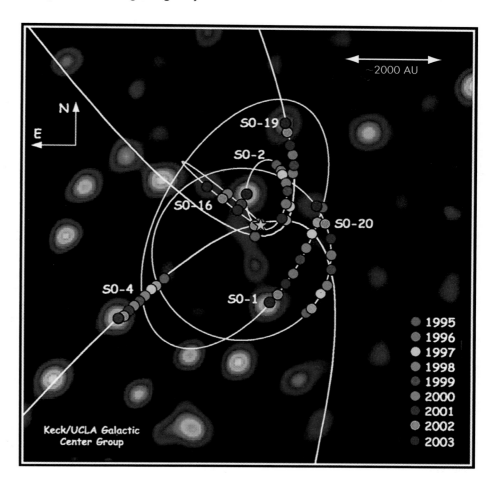

**FIGURE 15.26**

A plot of the position of several stars as they orbit the massive black hole at the core of the Milky Way. The stars are colored to show their positions in successive years. (Yellow "star" marks center of Milky Way.)

How might such a huge black hole form? We saw in chapter 14 that when a massive star reaches the end of its life, it may explode as a supernova, leaving a black hole as a remnant. Such black holes contain only 10 or so solar masses, but a black hole that formed near the core of the Milky Way during our galaxy's youth had opportunities for growth. At that long-ago time, gas naturally tended to fall inward toward the core, where any black hole lurking there might capture it. As such a black hole swallowed material, its mass and hence its gravity would increase, allowing it to capture material ever more easily. In addition, gas lost from stars by explosion, stellar wind, or tidal force stripping from neighbors in that crowded environment might also be drawn into the black hole. With so many potential sources of matter, once a black hole formed in the center of the Milky Way, it would not stop growing until it ran out of available matter. In that way, a small black hole might have grown to $10^6$ M$_\odot$ in less than 1 billion years.

The Milky Way is not alone in having a black hole at its center. Other galaxies also contain massive black holes, as we will discuss in chapter 16. For now we will simply say that, according to some hypotheses, such black holes may be a nearly inevitable by-product of galaxy formation.

## 15.8  FORMATION OF THE MILKY WAY

One of the major unsolved problems in astronomy is how galaxies form. For many years, astronomers thought that the process was simply a large-scale version of star formation; that is, instead of a small gas cloud collapsing under its own gravity to form a single star, a huge cloud collapses to form billions of stars. Although this hypothesis explains many properties of the Milky Way, it fails to explain a number of its features. Thus, astronomers have modified their hypothesis by suggesting that the formation of moderate to large galaxies (such as the Milky Way) requires several steps. In the first step, as proposed before, a gas cloud collapses and forms a midsize galaxy. But now, in a second step, the midsize galaxy grows by merging with other similar or smaller galaxies.

In trying to explain the origin of the Milky Way, astronomers draw on a variety of observations. We saw, in discussing the origin of the Solar System, that its disklike shape, the orbital regularities and common age of the planets, and the existence of two families of planets all were important clues to how it formed. Similar clues guide astronomers in studying the formation of the Milky Way. For example, our galaxy has a flattened shape and contains two main groups of stars—populations I and II—that differ greatly in their properties. In particular, Pop II stars are old, contain few heavy elements, and follow orbits that are steeply inclined to the disk. Pop I stars, on the other hand are younger, contain more heavy elements, and follow roughly circular orbits that lie in the disk. Using such information, British astronomer Donald Lynden-Bell and American astronomers Olin J. Eggen and Allan R. Sandage proposed in 1962 a two-stage model to explain the birth of the Milky Way, as described below.

### Birth of Population I and II Stars

Our galaxy probably began as a vast, slowly rotating gas cloud, perhaps a million light-years in diameter and containing about $10^{11}$ M$_\odot$ of gas. The cloud was composed of almost pure hydrogen and helium because no stars yet existed to make the heavier elements. As gravity began shrinking this immense cloud, clumps of gas within it grew more dense and gradually turned into stars (fig. 15.27A). These stars, composed of essentially pure hydrogen and helium, may have been exceptionally massive compared to stars forming today. Because of their greater mass, they evolved quickly and blew up as supernovas, which added the first heavy elements to the gas cloud even before it had finished collapsing.

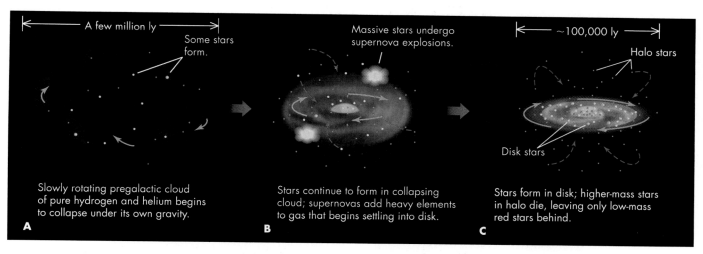

**FIGURE 15.27**
(A) Birth of the Milky Way as a gas cloud collapses. A first generation of stars (population II) forms. (B) Collapse continues with more population II stars forming and some exploding as supernovas. Heavy elements created by the stars exploding as supernovas are mixed into the gas. Gas not used in making stars settles into a rotating disk. (C) Population I stars form in the disk.

It may seem odd that stars could form and die before the cloud's collapse was complete. But collapse of such an immense cloud takes hundreds of millions of years, whereas massive stars evolve in only a few tens of millions of years—10 or so times faster.

As collapse continued, more stars formed, now with at least some heavy elements. Because these stars formed from infalling gas, they follow elongated "plunging" orbits—becoming the population II stars that we see today.

Star formation during this initial stage probably was not efficient enough to consume all the collapsing gas, and some gas survived to form a rotating disk, as sketched in figure 15.27B. That disk gas, now rich in heavy elements from the death of Pop II stars, then began to form the first population I stars. These stars followed the motion of the gas they formed from. Pop I stars, therefore, contain more heavy elements than Pop II stars and move along approximately circular orbits lying in the disk, as illustrated in figure 15.27C.

Computer simulations support this collapse model for the birth of the Milky Way. For example, figure 15.28 shows how an unstructured region of gas, of the appropriate size and rotation speed, collapses. You can see that it forms a thin disk with spiral arms, surrounded by a halo. Such simulations also show that the unseen dark matter mentioned earlier plays an important role in making galaxies, because computer models that include dark matter more closely resemble real galaxies than models that do not include dark matter.

## Building by Mergers

Although the collapse hypothesis explains many of the features of our galaxy, it fails to explain two important properties of its stars. First, according to the hypothesis, all population II stars should be about the same age, having formed during the relatively brief period of the galaxy's initial collapse. But observations of Pop II stars show that they formed over a much longer time span than the model predicts. Second, the model predicts that the low-mass stars born in that first generation that have not burned out should have virtually no heavy elements. But no such stars have ever been observed. The collapse model also omits a critical feature that astronomers now believe plays a major role in shaping galaxies: collisions with other galaxies.

**FIGURE 15.28**
Computer simulation of the birth of a galaxy similar to the Milky Way. The top three frames (A), (B), and (C) show the initial state of the gas and the development of a clump that becomes the galaxy, illustrated in the bottom two frames (D) and (E). Notice the thin disk and spiral arms. Mergers are not part of this simulation.

**FIGURE 15.29**
Two colliding galaxies known as The Mice or more formally as NGC4676.

Astronomers have recently discovered eight very dim, dwarf galaxies that are presumably companions of the Milky Way. These galaxies are just the type expected by the merger model and lend additional support to this view of how galaxies form and grow.

As galaxies move through space, they occasionally collide with one another. Sometimes these collisions lead to a merger of the two galaxies into a single larger system. Astronomers today can see collisions occurring in a number of systems (fig. 15.29), and with the Hubble Space Telescope they can see that such collisions and mergers were far more common in the past. How can the Hubble Space Telescope see into the past? Recall that light takes 5 billion years to reach us from a galaxy 5 billion light-years away. Thus, when we see distant galaxies, we observe them as they were long ago, when they were young. With this "window" back in time, astronomers see that many young galaxies are smaller than galaxies we see today, such as the Milky Way. In addition, many more of these young galaxies are colliding with neighbors than galaxies that we see today. Thus, astronomers conclude that large galaxies like the Milky Way may have begun their life with the collapse of a single gas cloud, but that they have grown to their present size by swallowing smaller galaxies.

This hypothesis is supported by the discovery of a dim, dwarf galaxy on the far side of the Milky Way and by faint streams of stars within the Milky Way's halo. The dwarf galaxy is elongated, probably because the Milky Way's gravity has stretched it and begun to pull it apart. The streams are stars that appear to have been ripped out of small galaxies that have fallen into the Milky Way and are now merging with it, as shown in figure 15.30, an artist's sketch of such an event. Even larger mergers may lie in the distant future. Some astronomers believe that two satellite galaxies of the Milky Way—the Large and Small Magellanic Clouds (see fig. 16.2)—are spiraling into our galaxy and will eventually merge with it. Thus, galaxy birth may be a dog-eat-dog event.

**FIGURE 15.30**
Artist's sketch of a small galaxy being torn apart as it falls into and is swallowed by the milky way.

## Population III

The above description of how galaxies form still leaves unanswered the question of why we see no stars formed from pure hydrogen and helium. We should expect such a composition if our galaxy formed from a cloud in which no stars had yet made heavier elements. Up to now, all stars found in the Milky Way contain at least a small percentage of heavy elements (typically at least 1/1000 the amount seen in the Sun), although one star has recently been observed with a mere $5 \times 10^{-6}$ times the heavy element abundance of the Sun.

Astronomers have devised several explanations for why they have so far seen no **population III stars,** as these hypothetical ancient pure hydrogen and helium stars are called. First, conditions in the early galaxy may have allowed only short-lived massive stars to form. For example, the gas may have been too hot or turbulent to make small stars that would have lived long enough for us to see them today. Alternatively, because massive stars can form and die before very low-mass stars even complete their contraction, Pop III stars may have had their surface layers contaminated by gas ejected at the death of their more massive brethren. Thus, Pop III stars may exist but are masquerading as Pop II stars because they contain heavy elements that were not part of their original makeup. Finally, such pure hydrogen and helium stars may simply be very rare and therefore as yet undiscovered.

## The Future of the Milky Way

We know from earlier chapters that stars change dramatically as they age. Will the Milky Way change as well? For several reasons, astronomers think the answer is yes. As our galaxy ages, it steadily uses up its interstellar gas to make stars. Although on reaching the end of their lives, stars return some of their material to space as planetary nebula shells, supernova remnants, and stellar winds, eventually that gas will also end up in stellar remnants such as white dwarfs, neutron stars, and black holes, leaving the Milky Way no gas from which to make new stars. From time to time our galaxy may acquire new gas as it captures other galaxies, but eventually that gas will also be used up. Billions of years from now our galaxy will contain only dead stars and dim, low-mass red dwarfs. Gradually even these dwarfs will burn out, and the Milky Way will fade, slowly spinning in space, a dark disk of stellar cinders.

## SUMMARY

Our galaxy, the Milky Way, consists of a disk about 100,000 light-years across embedded in a halo that extends out several times farther. Together, they contain about $10^{11}$ stars. Near the Sun, stars are a few light-years apart. Because the disk rotates, it is made flat, and it contains star groups arranged in a pattern winding out from the center to form spiral arms. Near its center, the galaxy is thicker (several kiloparsecs), forming the galactic bulge, which we can see toward the constellation Sagittarius. The Sun lies about two-thirds of the way out in the disk. From Earth, we see the Milky Way's disk as a band of dim light stretching across the night sky.

The Milky Way's stars form two main groups—population I and population II—which differ in many ways. Population I stars are found mainly in the disk. They are relatively young, tend to be blue, move in circular orbits, and are rich in heavy elements. Population II stars are found mainly in the halo and bulge. They are old, tend to be red, move in elongated and tilted orbits, and are poor in heavy elements.

About 15% of the galaxy's mass is gas and dust clouds, which orbit with the stars and dim their light. The dust makes it difficult for astronomers to see beyond 10,000 light-years in the disk. The gas, mostly hydrogen and helium, plays a crucial role in the evolution of the Milky Way because it is the source of new stars. Radio emission from the gas, both cold hydrogen emitting 21-centimeter radiation and interstellar molecules, also helps us map the galaxy and see through the dust that obscures our view of its structure.

At the very center of the Milky Way lies the galactic nucleus, a dense collection of stars and gas. Astronomers think a massive black hole, about $10^6$ $M_\odot$, lies at the center because stars orbiting there move so rapidly.

Astronomers can measure the diameter of the Milky Way and the Sun's location in it by plotting the position of globular star clusters. They measure its mass by applying the modified form of Kepler's third law to the motion of the Sun and other stars, but such measures give values of the mass that disagree with the value arrived at when stars are simply counted and their masses are added. Astronomers, therefore, infer that the Milky Way is embedded in a halo of dark matter whose nature is unknown.

The existence of two different stellar populations, I and II, leads astronomers to believe that the Milky Way formed by the collapse of a vast gas cloud about 14 billion years ago. The first stars to form became the population II stars. Some gas did not turn into stars during collapse but instead collected into a rotating disk. From supernova explosions of population II stars, heavy elements became part of the disk of gas. Stars eventually formed from this material to become the population I stars such as our Sun. Subsequently the Milky Way has captured smaller galaxies and grown larger. Eventually, the Milky Way will convert all its available gas to stars, and as the surviving stars die, our galaxy will grow dark.

### ?What It Is.

This infrared image shows the inner 2 light-years of the core of the Milky Way. The exact center of the galaxy is indicated by the two small arrows. Observations of the motion of a star near this spot are strong evidence that a black hole lies there.

## QUESTIONS FOR REVIEW

1. What does the Milky Way look like in the night sky?

2. How do we know our galaxy is a flat disk?

3. Draw a sketch of the Milky Way galaxy and label its major components. Where is the Sun in the Milky Way?

4. Roughly how big in diameter is our galaxy, and how much mass does it contain?

5. How did Shapley deduce the Milky Way's size and the Sun's position in it?

6. What are some differences between Pop I and Pop II stars?

7. How do we know interstellar matter exists?

8. How do we know that some interstellar matter is dust?

9. What is the difference between an emission and a reflection nebula?

10. What are the small, reddish blobs along the spiral arms of the galaxy in figure 15.7?

11. How does interstellar dust affect our observations of stars and the Milky Way?

12. What is the zone of avoidance?

13. What evidence makes astronomers believe the Milky Way has spiral arms?

14. What are some ways that we know cold clouds of gas exist in space?

15. How can we determine the Milky Way's diameter?

16. How can we determine the Milky Way's mass?

17. What is meant by dark matter? Why do astronomers conclude the Milky Way may contain such as yet unobserved material?

18. What is the evidence for a black hole at the center of the Milky Way?

19. Describe one model for the origin of the Milky Way. How does this model explain the difference between Pop I and Pop II stars?

20. What is meant by Pop III stars?

## THOUGHT QUESTIONS

1. What would the Milky Way look like in the night sky to an observer on a planet at the very edge of the galaxy? What would it look like to an observer at the galactic nucleus?

2. How could we measure the mass of the Milky Way using a distant orbiting globular cluster?

3. Suppose the Sun were a population II star in the halo of the Milky Way. In what ways would that make studying the Milky Way easier?

4. Cut four circles of cardboard whose diameters are 2, 4, 6, and 8 inches. Fasten them together through their centers with a brass paper fastener. Draw a small circle near the edge of the 4-inch disk so that it cuts the other disks. Now turn the two innermost circles. Does this help you see how self-propagating star formation might make spiral arms?

## PROBLEMS

1. Given that the Sun moves in a circular orbit of radius 8.5 kiloparsecs around the center of the Milky Way and that its orbital speed is 220 kilometers per second, work out how long it takes the Sun to complete one orbit of the galaxy. NOTE: 1 kiloparsec is about $3 \times 10^{16}$ kilometers, and 1 year is about $3 \times 10^7$ seconds.

2. Gas is observed to be orbiting in a circle at 200 km/s about a dark central mass. If the gas's orbit has a radius of 1 parsec, how much mass must be present to hold it in orbit?

3. Suppose the mass in problem 2 were packed into a sphere whose radius is 20 AU. Would it be a black hole?

4. According to figure 15.24, the Milky Way's rotation speed is about 230 km-s$^{-1}$ at a distance of about 16 kiloparsecs from its center. Use the method discussed in the box on page 456 to find the mass of the Milky Way implied by these values. Hint: These values correspond to a distance of about $3.3 \times 10^9$ AU and an orbital period of about $4.3 \times 10^8$ yrs.

## TEST YOURSELF

1. One way astronomers deduce that the Milky Way is a disk is that they
   (a) see stars arranged in a circular region around the north celestial pole.
   (b) see far more stars along the band of the Milky Way than in other directions.
   (c) see a large dark circle silhouetted against the Milky Way in the Southern Hemisphere.
   (d) see the same number of stars in all directions in the sky.
   (e) None of the above

2. Astronomers think the Milky Way has spiral arms because
   (a) they can see them unwinding along the celestial equator.
   (b) radio maps show that gas clouds are distributed in the disk with a spiral pattern.
   (c) young star clusters, HII regions, and associations outline spiral arms.
   (d) globular clusters outline spiral arms.
   (e) Both (b) and (c) are correct.

3. A young blue star moving along a circular orbit in the disk is a (a) Pop I or (b) Pop II star.

4. The modified form of Kepler's third law allows astronomers to determine the Milky Way's

   (a) mass.

   (b) age.

   (c) composition.

   (d) shape.

   (e) number of spiral arms.

5. Astronomers know that interstellar matter exists because

   (a) they can see it in dark clouds and clouds that absorb light.

   (b) the matter creates narrow absorption lines in the spectra of some stars.

   (c) they can detect radio waves coming from atoms and molecules in the cold gas.

   (d) spacecraft have sampled clouds near Orion.

   (e) All the above except (d).

## FURTHER EXPLORATIONS

Basri, Gibor. "The Discovery of Brown Dwarfs." *Scientific American* 282 (April 2000): 76.

Berendzen, Richard, Richard Hart, and Daniel Seeley. *Man Discovers the Galaxies.* New York: Science History Publications, 1976.

Binney, James. "The Evolution of Our Galaxy." *Sky and Telescope* 89 (March 1995): 20.

Bromm, Volker. "Out of the Dark Ages: The First Stars." *Sky and Telescope* 111 (May 2006): 30.

Combes, Francoise. "Ripples in a Galactic Pond." *Scientific American* 293 (October 2005): 42.

Hoskin, Michael. "William Herschel and the Making of Modern Astronomy." *Scientific American* 254 (February 1986): 106.

Reynolds, Ronald J. "The Gas between the Stars." *Scientific American* 286 (January 2002): 34.

Scoville, Nick, and Judith S. Young. "Molecular Clouds, Star Formation, and Galactic Structure." *Scientific American* 250 (April 1984): 42.

Trimble, Virginia, and Samantha Parker. "Meet the Milky Way." *Sky and Telescope* 89 (January 1995): 26.

Wakker, Bart P., and Philipp Richter. "Our Growing, Breathing Galaxy. *Scientific American* 290 (January 2004): 38.

**Website**

Please visit the *Explorations* website at **http://www.mhhe.com/arny** for additional online resources on these topics.

## KEY TERMS

 ## PROJECT

Find a spot where the sky is dark and you can see most of the sky. Sketch the path of the Milky Way across the sky, as best you can see it. Do you notice any particularly bright stars along the Milky Way or elsewhere? Can you see any dark regions along the Milky Way where the number of stars looks small? Mark them on your sketch. Then compare your sketch to one of the star maps inside the covers of this book.

# Galaxies

# 16

The spiral galaxy NGC 4414. The galaxy's disk is about 56,000 light-years across. The system lies about 62 million light-years from Earth. Image made with the Hubble Space Telescope.

A cluster of galaxies.

## Chapter Objectives

- Galaxies have a variety of shapes, but three types dominate:
  - Spiral
  - Elliptical
  - Irregular
- Many galaxies have a massive black hole at their center.
- All but the nearest galaxies are receding from us and show a redshift in their light.
- The farther a galaxy is from us, the faster it recedes (Hubble law).
- Galaxies occur in groups called galaxy clusters.

**?What Is This?**
(See chapter summary for the answer.)

## Concepts and Skills to Review

- Inverse square law (12.2)
- Standard candle method (12.1 and 12.8)
- Red light has a longer wavelength than blue (3.1)
- Newton's law of gravity (2.5)
- Modified form of Kepler's third law (2.7)

**B**eyond the edge of the Milky Way and filling the depths of space are billions of other star systems similar to our own. These remote, immense star clouds are called "external galaxies," and their properties are the subject of this chapter. Many external galaxies are flattened systems with spiral arms, similar to the Milky Way in size and shape. Other external galaxies differ dramatically from our own. For example, many galaxies are not disk-shaped at all. Rather, they are egg-shaped, with their stars distributed in a vast, smooth cloud surrounding a dense central core of stars. Others are neither disks nor smooth elliptical shapes but are completely irregular in appearance.

Galaxies differ in more than shape; they also differ in their contents. Some contain mostly old stars, but others contain predominantly young stars. Some are rich in interstellar matter, but others have hardly any at all. Some galaxies, despite having as much total mass as the Milky Way, have formed far fewer stars, leaving them dim and scarcely visible to us. Others have a small but tremendously powerful energy source at their core that emits as much energy as the entire Milky Way but from a region about 0.001% its size, less than a few light-years in diameter. Many of these so-called active galactic nuclei eject gas in narrow jets at nearly the speed of light.

Galaxies are not spread uniformly across the sky. Just as stars often lie in clusters, so too galaxies often lie in galaxy clusters and galaxy clusters themselves lie in clusters of galaxy clusters. These immense groups are not stationary. They are separating from one another, caught up in the expansion of the Universe, like snowflakes carried on the wind.

Astronomers can see all this diversity in galaxies, but as yet they understand only poorly how it arises. Unlike stars, whose structure and evolution are well understood, many aspects of galaxies remain a mystery. Astronomers do not even agree on why some galaxies are disks, whereas others are egg-shaped. Although this lack of knowledge is frustrating to astronomers, it does offer us a chance to see how scientists work, in particular, how astronomers begin with basic observations and go on to construct hypotheses to explain what they observe. For example, many galaxies appear to be undergoing collisions with their neighbors. Could past interactions of this kind account for the different galaxy shapes?

## 16.1 DISCOVERING GALAXIES

A **galaxy** is an immense swarm of hundreds of millions to hundreds of billions of stars and vast clouds of interstellar gas. Each star moves along its own orbit, held within the galaxy by the combined gravitational force of the rest of the matter within the galaxy. Each galaxy (apart from those undergoing collisions) is therefore an independent and isolated star system.

### Early Observations of Galaxies

All galaxies (except the Milky Way) are extremely distant from Earth, with the nearest being more than 150,000 light-years away. Their great distance makes galaxies look dim to us, and only a few can be seen without a telescope. For example, on clear, dark nights in autumn and winter in the Northern Hemisphere, you can see M31, the Great Galaxy in Andromeda (fig. 16.1) with the naked eye. It looks like a pale elliptical smudge on the sky. (See "Looking Up at M31 & Perseus," on p. 469.) Early observers such as the tenth-century Persian astronomer Al-Sufi noted this object, and it appeared on early star charts as the "Little Cloud." If you live in the Southern

# LOOKING UP AT...
# M31 & PERSEUS

The galaxy M31 lies in the constellation Andromeda, near the constellations Perseus and Cassiopeia. Northern hemisphere viewers can see M31 dimly with the naked eye, but easily with binoculars. It is about 2.9 million ly from us. The best times to see it are August through December.

**Andromeda**

~150,000 ly

## M31 Andromeda Galaxy

~200 ly

## The Double Cluster

If you scan with binoculars from M31 toward the space between Perseus and Cassiopeia, you will see the Double Cluster — two groups of massive, luminous but very distant stars. The Double Cluster is best seen with binoculars. The two clusters are about 7000 ly away and a few hundred light years apart.

**Perseus**

Perseus is easy to identify: it looks a little like the Eiffel tower.

## California Nebula

An emission nebula

## M45 Pleiades

## Capella

The brightest star in the constellation Auriga, the Charioteer. A binary star

**Auriga**

## Perseus in 3-D

1300 ly ✷

1 light year (ly) ≈ 10 trillion km ≈ 6 trillion miles

250 ly ✷
230 ly ✷
34 ly ✷
110 ly ✷
500 ly ✷
100 ly ✷
500 ly ✷

560 ly ✷

700 ly ✷

To Earth

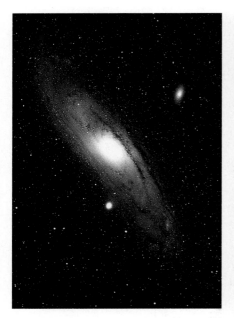

FIGURE 16.1
Photograph of M31, the Andromeda galaxy, the nearest spiral galaxy to us.

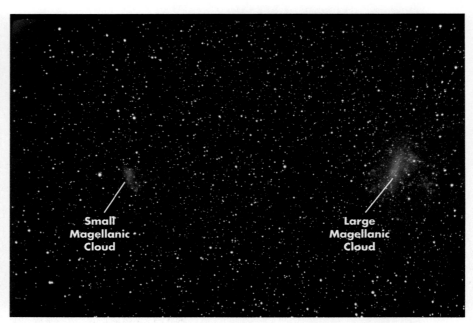

FIGURE 16.2
Photograph of the Large and Small Magellanic Cloud.

Hemisphere, you can easily see with the naked eye the Large and Small Magellanic Clouds, two satellite galaxies of the Milky Way (fig. 16.2). Today, astronomers have studied millions of galaxies besides the Magellanic Clouds and M31, and they estimate that the visible Universe contains tens of billions more.

The study of other galaxies began in the eighteenth century, when the French astronomer Charles Messier (pronounced *mess-yáy*) accidentally discovered many galaxies during his searches for new comets. He noticed a large number of faint, diffuse patches of light, and to avoid confusing them with comets, he assigned them numbers and made a catalog of their positions. Although many of Messier's objects have since been identified as star clusters or glowing gas clouds in the Milky Way, several dozen are galaxies. These and the other objects in Messier's catalog are still known by their Messier, or M number, such as M31, mentioned above.

In the nineteenth century, other astronomers, such as Sir William Herschel,* began systematically to map the heavens, finding and cataloging numerous astronomical objects, including galaxies. Herschel's work, revised and added to by his son and John Dreyer, is now known as the New General Catalog (or NGC, for short), and it gives many galaxies their name, such as NGC 1275, a peculiar, active galaxy. But many galaxies appear in more than one catalog and so bear several names. For example, Messier's M82 is the same galaxy as NGC 3034.

We know today that galaxies are immense star systems like our Milky Way, but this idea was not widely accepted by astronomers until about 1920. Before that time, astronomical opinion was strongly divided as to whether galaxies were huge star systems like the Milky Way or merely small satellites of our own galaxy. This controversy was aired in detail during a series of debates, organized by the National Academy of Sciences, in which two American astronomers, Harlow Shapley and Heber Curtis, took opposing sides. Shapley argued that the Milky Way is huge and that other galaxies (at that time called nebulas) are merely small, nearby companions. Curtis, on the other hand, argued that the Milky Way is smaller than Shapley claimed, and that the nebulas are star systems like our own and are immensely distant from it.

---

*You may recall from chapter 9 that Herschel also discovered the planet Uranus.

Historians differ on who "won" the debate. Although Shapley was more nearly correct about our galaxy's size, Curtis was right that the nebulas are distant systems like the Milky Way. In any case, once the distance and true nature of galaxies were understood, astronomers could more successfully study galaxies' properties.

## Types of Galaxies

Even early nineteenth-century observers noticed that not all galaxies look the same. However, it was Edwin Hubble, an American astronomer working first as a graduate student at the University of Chicago and later at Mount Wilson in California in the 1920s, who demonstrated that galaxies could conveniently be divided on the basis of their shape into three main types: spirals, ellipticals, and irregulars. The first type, **spiral galaxies,** have two or more arms winding out from the center. We saw in chapter 15 that the Milky Way is of this type, and figure 16.3 shows photographs of several others.

The second type shows no signs of spiral structure. These galaxies have a smooth and featureless appearance and a generally elliptical shape, as can be seen in figure 16.4. Accordingly, astronomers call them **elliptical galaxies.**

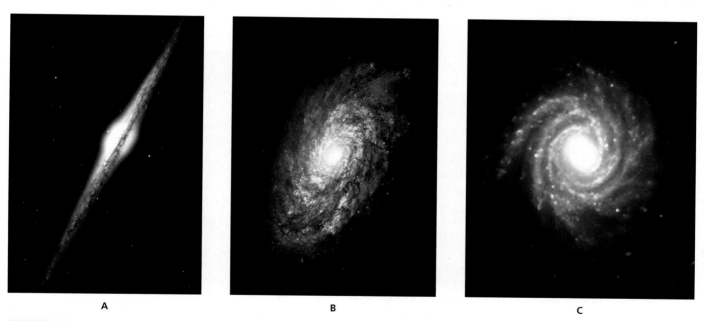

A B C

**FIGURE 16.3**
Pictures of typical spiral galaxies. (A) NGC 4565; (B) NGC 4414; and (C) NGC 1288. These examples show spiral galaxies with their disks oriented from nearly edge-on (A) to nearly face-on (C).

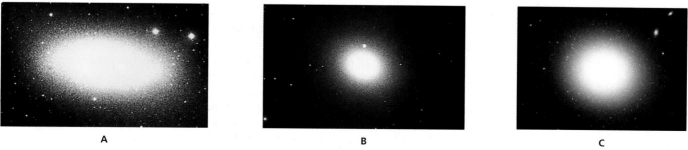

A B C

**FIGURE 16.4**
Pictures of a few elliptical galaxies: NGC 205, M49, and M87.

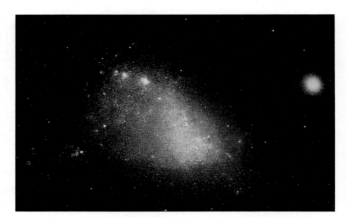

**FIGURE 16.5**
Photographs of two irregular galaxies (NGC 6822 and the Small Magellanic Cloud).

**FIGURE 16.6**
Barred spirals: NGC 1365 and NGC 5236 (M83).

Galaxies of the third major type show neither arms nor a smooth uniform appearance. In fact, they generally have stars and gas clouds scattered in random patches. For this reason, they are called **irregular galaxies** (fig. 16.5).

Spiral, elliptical, and irregular galaxies are sometimes denoted by *S*, *E*, and *Irr*, respectively. However, in addition to these three main types, astronomers recognize two additional galaxy types closely related to spiral systems. The first of these are **barred spiral galaxies,** which have arms that emerge from the ends of an elongated central region, or bar, rather than from the core of the galaxy (fig. 16.6). It is this bar that gives them their name, and they are denoted SB galaxies to distinguish them from

A

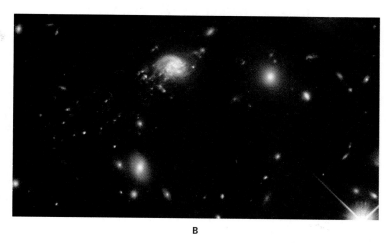

B

**FIGURE 16.7**

(A) An S0 galaxy, NGC 1201. (B) As the spiral galaxy at the upper left of this image orbits within its galaxy cluster, gas is being stripped from the galaxy. The blue blobs to the left of the galaxy are regions where stars are forming in the stripped gas. The cluster, Abell 2667, is a little over 3 billion ly distant.

normal spirals. Although Hubble treated SB galaxies as a separate class, astronomers today are less certain that they are a distinct type of galaxy. For example, computer models of galaxies show that most disk-shaped systems will form a bar if they undergo a gravitational disturbance, perhaps as the result of a close encounter with a neighboring galaxy. Even our own Milky Way has a weak bar. The bar forms because a gravitational disturbance alters the orbits of the stars, causing them to bunch up in an elliptical pattern that we see as the bar. Some models also suggest that over hundreds of millions of years, the orbits spread out again and the bar dissolves. Thus, bars may appear and disappear, and so an ordinary spiral galaxy may change into a barred spiral and then back again.

Hubble also identified another kind of galaxy similar to spirals. These are disk systems with no evidence of arms (fig. 16.7A), and they are called S0 galaxies. Hubble thought they were intermediate in type between S and E type systems. Today, however, astronomers have several theories to account for S0 galaxies. According to one theory, they are spiral galaxies whose gas has been stripped by their motion through the hot, tenuous gas in intergalactic space. The impact of such gas from outside might "sweep" gas and dust out of the moving spiral, much like a leaf-blower cleans debris from a sidewalk. Such stripped gas is clearly visible down and to the left of the blue spiral galaxy (top-left), illustrated in figure 16.7B. (The blue blobs in the stripped gas are regions of forming stars). Alternatively, the gas originally in a disk galaxy might all have turned into stars early in the galaxy's youth, perhaps because of a collision or other sort of interaction with a neighboring galaxy that compressed the gas and made it collapse into stars in one large burst. In either case, without gas, the disk galaxy can make no new stars and hence it shows no spiral structure.

Hubble took his classification system one step further, seeking a pattern that might lead to an understanding of why galaxies exhibit such diversity in their appearance. He noticed that both spiral and elliptical galaxies can be subdivided into additional classes (fig. 16.8) that, when properly ordered, show a smooth transition from spherical E type of galaxies to flatter E galaxies, to spirals, and to irregulars. Hubble went on to hypothesize that, as a galaxy aged, its type might change from E to S to Irr. This hypothesis seems very plausible if you look at the "tuning

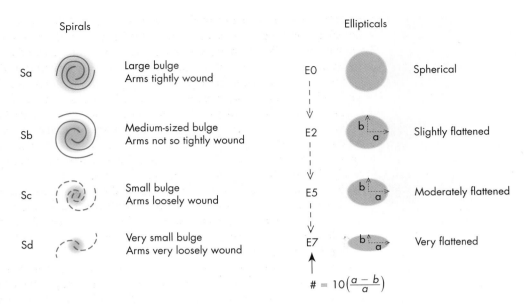

**FIGURE 16.8**
Sketch of the galaxy subclasses of spiral and elliptical galaxies.

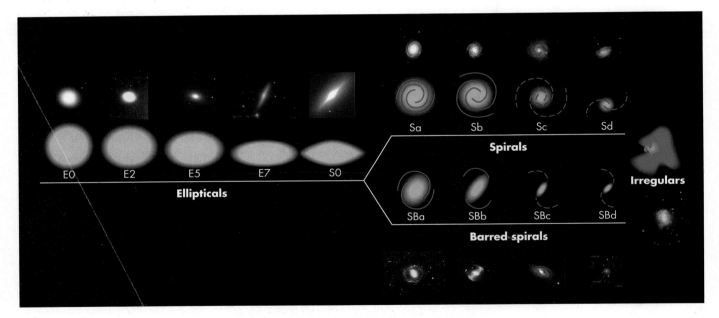

**FIGURE 16.9**
The Hubble tuning-fork
diagram.

fork" diagram that Hubble proposed (fig. 16.9), but astronomers now think it is incorrect. Nevertheless, the diagram is still a useful way to organize the galaxy types and subtypes that Hubble described.*

## Differences in the Stellar and Gas Content of Galaxies

Since Hubble's time, astronomers have discovered that spiral, elliptical, and irregular galaxies differ not only in their shape but also in the type of stars they contain. For example, spiral galaxies contain a mix of young and old stars (Pop I and II), but

---

*Although Hubble's classification system covers well the most conspicuous galaxy types, it omits two forms that astronomers today think are important: low-surface-brightness galaxies (dim, often large-diameter systems with far less star formation than ordinary galaxies) and blue dwarf systems (perhaps the building blocks of ordinary galaxies).

elliptical galaxies contain mostly old (Pop II) stars. This difference in the kind of stars composing S and E galaxies is understandable in terms of their different gas content. To make young (Pop I) stars, a galaxy must contain dense clouds of gas and dust. Spiral systems have about 15% of the mass of their disk in the form of such interstellar clouds, and so they can easily make the young stars of their spiral arms. On the other hand, ellipticals contain much less interstellar matter. In fact, astronomers used to think that the E type of galaxies contained virtually no interstellar gas or dust. However, more recent observations made with X-ray telescopes show that E galaxies often contain very low-density but very hot ($10^7$ K) gas. Such hot gas does not readily form stars, and so it is no surprise that elliptical galaxies rarely contain young, blue stars like the ones we see in the disks of spiral systems.

Although young blue stars are rare in elliptical systems, such stars are common in many irregular galaxies. These Irr systems also often contain large amounts of interstellar matter. In fact, they sometimes have more gas than stars. Accordingly, astronomers think that these galaxies, though possibly very old, have simply not yet used up much of their gas in making stars.

Apart from their different star and interstellar matter content, S, E, and Irr galaxies have few other differences to generalize about. For example, if we look at the mass and radius of galaxies, we do not find that all ellipticals are huge and all spirals are small. Instead, we discover that elliptical galaxies range enormously in size. Some are not much larger than globular clusters and contain only $10^7$ solar masses of material; others are monster galaxies, 10 to 100 times the mass of the Milky Way. This spread is easy to see in figure 16.10, which shows a group of galaxies in which most of the members are elliptical systems.

What fraction of galaxies make up each type? Even that question has no clear answer. If we make a census of the galaxies near the Milky Way, we find that most galaxies are dim dwarf E and Irr systems, sparsely populated with stars. These dim galaxies are undetectable at greater distances, where we can see only the more luminous galaxies. If we make a census of those luminous objects, we find that the majority of galaxies are spirals, but the percentage depends on whether we look at galaxies in clusters or in more sparsely populated regions. In clusters, only about 10% of the members are spirals, whereas outside of clusters that proportion is about 80%.

The diversity of galaxy form and content requires an explanation, and much of astronomical research today is devoted to answering the question of why galaxies come in such distinct and different forms. With the Hubble Space Telescope, astronomers have

**FIGURE 16.10**

Photograph of the central region of the Virgo galaxy cluster, illustrating the range in sizes of elliptical galaxies.

Q⋆ What types of galaxies can you identify in this photograph?

A⋆ A spiral galaxy lies near the center of the bottom edge of the image. An elliptical galaxy is near the upper left corner.

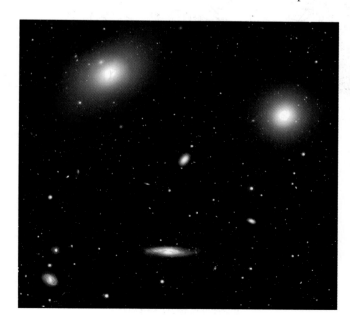

at last been able to observe young galaxies well enough to begin developing a theory of how galaxies form and evolve. As we described in chapter 15, astronomers think that galaxies form by the collapse of gas clouds, much like the process that formed our Sun and other stars but on a vastly larger scale. As the cloud collapses, stars form within it, evolve, and—if massive enough—explode as supernovas. The supernova debris adds heavy elements to the gas, which are then incorporated into later generations of stars, ultimately to become planets and living beings.

Although astronomers have not witnessed, even with the Hubble Space Telescope, the initial collapse of primordial gas clouds into galaxies, that instrument's ability to see fine details in distant (and thereby ancient) galaxies has allowed astronomers to observe galaxies just past the collapse stage. Surprisingly, these presumably very young galaxies are much smaller than systems like the Milky Way. Moreover, they are far more numerous than ordinary galaxies today. Thus, these young galaxies must somehow disappear. How can a galaxy disappear? Deducing from what is currently known, astronomers tend to think that they collide and merge into bigger systems. This would account both for the formation of large systems like the Milky Way and for the smaller number of galaxies we see today. But a major question still remains: what makes one set of merging systems become a spiral galaxy, while another becomes an elliptical galaxy?

## The Cause of Galaxy Types: Collisions and Mergers?

Astronomers are still uncertain about what causes the difference between S, E, and Irr galaxies. For many decades, they have thought that rotation plays a role. We know that rotation has shaped the Solar System. Why not galaxies, too? In fact, when they measure the rotation speed of galaxies, astronomers find that the disks of spirals rotate relatively rapidly compared to ellipticals of the same size. On the other hand, the halos and bulges of the spirals do *not* rotate especially fast. Thus, some factors other than rotation must also play a role in determining a galaxy's type.

In the section on stellar and gas content of galaxies, we discussed observations that suggest that the large galaxies we see today were formed by the merger of many smaller star systems. We also saw in chapter 15 that our own galaxy, the Milky Way, shows signs of past mergers. For example, astronomers can see streams of stars in our galaxy's halo, which appear to be debris from small galaxies that have fallen into the Milky Way and been torn apart by its gravity. Given this sort of evidence, could mergers also explain the differences between spirals and ellipticals? The answer seems to be yes. For example, figure 16.11 shows two spiral galaxies whose gravitational force has drawn them together and merged them into a single new galaxy that looks much like an elliptical system.

Observational evidence such as this and computer simulations have all encouraged astronomers to propose new hypotheses for the origin of spiral and elliptical galaxies. The basic idea is that most galaxies start out as small disklike and irregular systems. Subsequently, collisions and mergers change these systems, as shown schematically in figure 16.12. If the collision/merger leaves little gas remaining in the system, an elliptical galaxy results. If the collision/merger leaves moderate to large amounts of gas in the system, a large spiral results. Such interactions also allow some ellipticals to be incorporated back into spirals. This can happen if an elliptical captures an especially massive disk system. The disk then becomes the disk of a "new" spiral, while the elliptical becomes its bulge.

This hypothesis to explain the origin of spiral and elliptical systems has an added attraction to astronomers. It explains several puzzling features of the great spiral galaxies such as our own Milky Way. For example, astronomers at one time thought that the stars in our galaxy's bulge formed in one single, massive event at the

**FIGURE 16.11**
A merger in which two spirals are colliding and forming an elliptical galaxy.

Gas clouds collapse and form → Disk galaxies with surrounding globular clusters. → Some disks collide and form → Elliptical galaxies. Other disks remain unchanged.

**FIGURE 16.12**
Elliptical galaxies are created by the collision and merger of disk systems.

*Galaxies with tails*

galaxy's birth. But stars in the bulge have a wide range of ages, suggesting that the bulge grew gradually over time. Such gradual growth is easy to explain by mergers. Each time a large galaxy (such as the Milky Way) swallowed a smaller companion, the gas that was in the companion sank to the center of the larger system, where it formed new stars. Successive mergers led to successive generations of stars so that the bulge stars have the wide age range that is observed.

In this discussion of the factors that might shape a galaxy, we have said nothing about how spiral arms form. In chapter 15, we discussed some hypotheses about how the Milky Way may have formed its spiral arms. For example, we saw that spiral arms might form as a density wave travels around a galaxy's disk, compressing gas clouds and triggering star formation in a spiral pattern. Astronomers assume that this mechanism probably operates in other galaxies and forms their arms, too. We also discussed in chapter 15 a different model in which spiral arms form as the galaxy's rotation stretches regions of star formation into elongated arcs. This mechanism probably also forms arms in other galaxies.

Galaxy collisions and mergers may explain not only the differences among the galaxy types; they can also explain the appearance of many strongly distorted galaxies. For example, when two galaxies pass near each other, the gravitational force of one on the other alters the orbits of their stars and thereby changes the shape of each galaxy. Figure 16.13 shows a computer simulation of such a close passage, and you can see the stars torn from each galaxy by their mutual gravitational force and flung outward in long, luminous arcs. The fifth picture in the sequence shows an image of two real galaxies with just such plumes of stars.

Although interactions of this sort are disruptive to galaxies, they rarely harm the individual stars. Because stars are so far apart within a galaxy, most stars move harmlessly past each other. Thus, galaxy interactions are much like tossing two handfuls of sand together: most of the grains simply pass by one another without hitting. Dust and gas clouds in the galaxies are not so lucky. Being much larger than stars, clouds do collide and their impact may compress them enough to trigger a burst of star formation within them. Such "star burst galaxies," as these collisionally* stimulated galaxies are called, are among the most luminous galaxies known.

---

*Many astronomers use the word *collision* to describe any close interaction between galaxies, even if the galaxies do not directly hit each other.

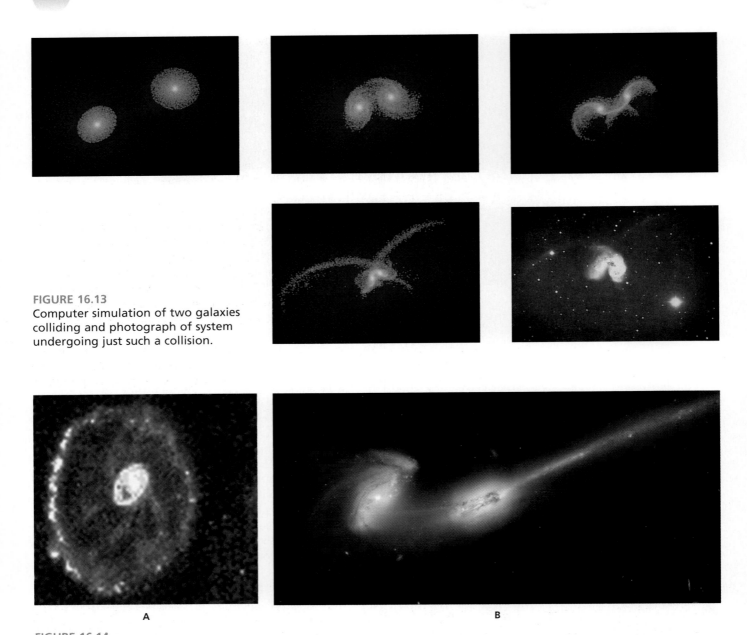

**FIGURE 16.13**
Computer simulation of two galaxies colliding and photograph of system undergoing just such a collision.

A

B

**FIGURE 16.14**
(A) A ring galaxy. (Note: The white central "blob" is the core of the ring galaxy. The galaxy that "punched" the hole is not visible in this image.) (B) The Mice—a tidally distorted system.

Galaxy interactions can create other bizarre forms, as figure 16.14 illustrates. These pictures show what can happen when a small galaxy plows directly into a large spiral galaxy. As you can see in figure 16.14A, the small galaxy has "punched a hole" through the larger one, creating what astronomers call a ring galaxy. The hole results not because the smaller galaxy destroyed the stars in the other galaxy's disk. Rather it made the hole by shifting the orbits of stars that were near the center of the larger galaxy into larger orbits farther from the center. Figure 16.14B shows a pair of galaxies—the Mice—distorted by each other's gravity.

Mergers may in fact affect just about every galaxy sometime during its life. If a large galaxy collides with a small galaxy, the small one is captured and absorbed by the larger one in a process called **galactic cannibalism.** Repeated cannibalism by a galaxy may turn it into a giant, which perhaps explains why some galaxy groups have an abnormally large elliptical galaxy at their center. Cannibalism may also account for much of the activity observed in some galaxies (see section 16.4). Even our own galaxy is devouring smaller neighbors, as we discussed in chapter 15, where we saw that astronomers have detected a small galaxy on the far side of the Milky Way that is even now being ripped apart and captured. Moreover, it seems likely that in the future our galaxy will capture and swallow the Magellanic Clouds. Our neighboring galaxy, M31, also appears to have cannibalized a small galaxy. With the Hubble Space Telescope, astronomers have detected a small bright clump of stars near the center of M31 that may be the remains of a galaxy it has absorbed.

The idea that collisions and mergers play an important role in shaping galaxies has been strikingly confirmed by pictures of more distant systems made with the Hubble Space Telescope. Such pictures show first of all that distant clusters contain a higher proportion of spirals than nearby clusters do, implying that as clusters age, some spiral galaxies change their form. Second, distant clusters contain more galaxies within a given volume, implying that collisions were more likely in the past than now. Finally, the galaxies seen in the most remote clusters show more signs of disturbance (fig. 16.15) by neighboring galaxies—odd shapes, bent arms, and twisted disks—what some astronomers call "harassment."

In all the above discussion of galaxy types and their possible evolution, we have ignored the methods by which the size and mass of a galaxy are found. An even more important point that we have overlooked is how to find the distance to a galaxy. Without knowing how far away a galaxy is, we can learn little of its properties. Thus, before we look further at galaxies, we need to discuss how their properties are measured.

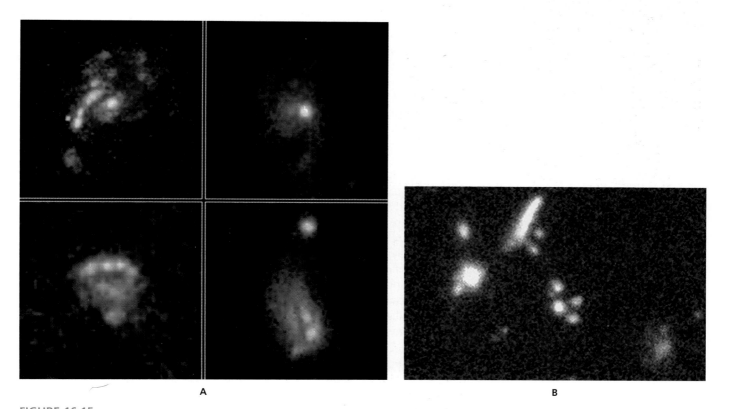

A                                                                            B

**FIGURE 16.15**
(A) Picture of young, forming galaxies taken with the Hubble Space Telescope. (B) Young galaxies possibly merging to form a larger system.

## 16.2    MEASURING THE PROPERTIES OF GALAXIES

Our ability to study galaxies is hindered by their great distance from us. In fact, the extent of their distance from us was for centuries a major controversy among astronomers. For example, in 1755, the German philosopher Immanuel Kant, whose views on the origin of the Solar System were so ahead of their time, suggested that galaxies were remote star systems—Island Universes—similar to the Milky Way. On the other hand, as we saw in section 16.1, the American astronomer Harlow Shapley argued as recently as the 1920s that galaxies were merely small satellites of the Milky Way. The reason for such disagreement comes from the difficulty of *measuring* how far away from us galaxies are.

### Galaxy Distances

Galaxies are separated from each other by enormous distances. Even the nearest galaxy to us, the Large Magellanic Cloud, is more than 150,000 light-years from the Sun. Astronomers cannot use parallax to measure such vast distances because the angle by which the galaxy's position changes as we move around the Sun is too tiny to be detected. Astronomers must therefore use other ways to find distances, such as the method of standard candles.

We discussed the method of standard candles in detail in chapter 12. We saw there that if we know the luminosity of a light source, we can use the inverse-square law to deduce its distance from how bright it looks to us. This was the technique we used in chapter 15 to find the size of the Milky Way, and we can apply it here to find the distance to galaxies.

Astronomers use many different astronomical objects as standard candles, but one of the most reliable is the class of luminous, pulsating variable stars known as the "Cepheids" (described in chapter 13). Cepheids have a luminosity about a million times the luminosity of the Sun, and because of their brilliance, they are easy to see in nearby galaxies. Furthermore, because their brightness changes as they pulsate, they are easy to distinguish. Far more important, however, is that the time it takes these variable stars to go from bright to dim to bright again (their period) is related to their luminosity. (Recall our discussion of the period-luminosity relation in chapter 13.) This then allows astronomers to determine a Cepheid's luminosity by simply measuring how long it takes the star to change its luminosity over one cycle. Once the luminosity of a variable star is known, astronomers can find its distance from a simple measurement of how bright the star appears as seen from Earth and an application of the inverse-square law. (See Extending our Reach Measuring the Distance of a Galaxy Using Cepheid Variables, page 481).

Cepheids are luminous enough that astronomers can use them to measure the distance to nearby galaxies, but, despite their luminosity, Cepheids are not observable in very distant galaxies. Therefore, astronomers use other standard candles such as supergiant stars, planetary nebulas, or supernovas when determining the distance to such remote systems. Such objects are not ideal standard candles, however, because their luminosity may not be accurately known. Or, they may be rare, and therefore not present, in the galaxy whose distance one wants to measure. Astronomers therefore have sought other ways to measure the distance to galaxies, as we will discuss in the next section.

### The Redshift and the Hubble Law

We saw in earlier chapters that we can learn much about a star from its spectrum. Likewise, we can learn much about a galaxy from its spectrum.* One of the most important such

---

*We need to keep in mind, however, that a galaxy's spectrum is the spectrum of all of its stars, gas clouds, and other contents added together.

# EXTENDING OUR REACH

## MEASURING THE DISTANCE OF A GALAXY USING CEPHEID VARIABLES

Suppose we detect a Cepheid in a galaxy and find that the Cepheid appears 100 million ($10^8$) times dimmer than a Cepheid of the same pulsation period in the Milky Way. Because the stars have the same period, we assume that they have the same luminosity. Let us denote the brightness and distance of the Cepheid in the Milky Way as $B_*$ and $d_*$, respectively. Similarly, we will denote the brightness and distance of the Cepheid in the distant galaxy by $B_g$ and $d_g$. According to the inverse-square law, if two objects have the same luminosity, their apparent brightnesses $B_*$ and $B_g$, are related to their respective distances, $d_*$ and $d_g$, by

$$\frac{B_*}{B_g} = \left(\frac{d_g}{d_*}\right)^2$$

$B_*$ = Brightness of Cepheid in Milky Way
$B_g$ = Brightness of Cepheid in distant galaxy
$d_*$ = Distance to Cepheid in Milky Way
$d_g$ = Distance to Cepheid in distant galaxy

To find their distances, we take the square root of both sides, getting

$$\sqrt{\frac{B_*}{B_g}} = \frac{d_g}{d_*}$$

In other words, the square root of the ratio of their brightness equals the ratio of their distances. Because the distant Cepheid in our problem looks $10^8$ times dimmer from Earth than the Cepheid in the Milky Way, $B_*/B_g = 10^8$. Thus, according to the inverse-square law, the ratio of the distance of the Cepheid in the galaxy to the distance of the Cepheid in the Milky Way must be

$$\frac{d_g}{d_*} = \sqrt{\frac{B_*}{B_g}} = \sqrt{10^8} = 10^4$$

Thus $d_g = 10^4 d_*$

That is, the Cepheid in the remote galaxy is $10^4$ times farther away than the one in our galaxy.

Suppose we know the Cepheid in our own galaxy is 1000 light-years away. Then the Cepheid in the other galaxy must be $10^4$ times 1000 light-years away. The distant Cepheid, and hence the galaxy in which it is located, is therefore $10^7$ light-years distant (box fig. 16.1).

$$\frac{B_*}{B_g} = \left(\frac{d_g}{d_*}\right)^2 \quad \text{or} \quad \frac{d_g}{d_*} = \sqrt{\frac{B_*}{B_g}}$$

**BOX FIGURE 16.1**
Finding the distance to a galaxy using Cepheid variables as standard candles. $B_*$ and $B_g$ are the apparent brightness of the Cepheid in the Milky Way and in the Cepheid distant galaxy, respectively. Their distances are $d_*$ and $d_g$, respectively. From a measurement of the brightness ratio of the Cepheids, astronomers can calculate the ratio of their distances and, knowing the distance of the Cepheid in the Milky Way, they can therefore find the distance to the other galaxy.

discoveries was made in the early 1900s by the American astronomer Vesto Slipher. Slipher found that of the galaxies he observed, most had their spectrum lines strongly shifted toward longer wavelengths—the red end of the spectrum. This so-called **redshift** is seen in nearly all galaxies. (The exceptions are a few nearby galaxies, whose motion is influenced by the gravitational attraction of the Milky Way and its neighbors.)

Galaxy redshifts are *not* simply Doppler shifts. Rather, as astronomers know today, they result from the expansion of the Universe, as we will discuss in chapter 17. In Slipher's time, however, the cause of galaxy redshifts was a major puzzle. This puzzle became even more mysterious when Edwin Hubble and others noticed that the redshift was larger for dimmer (and therefore presumably more distant) galaxies. This implied

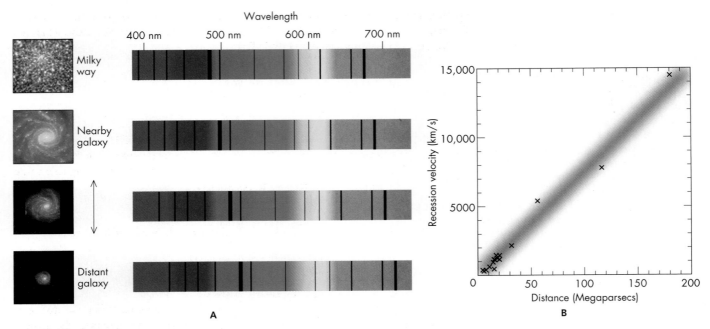

**FIGURE 16.16**

Redshift and recession velocity. (A) The more distant a galaxy, the larger its redshift. The spectrum of the total light of the Milky Way (which has zero redshift) is shown for comparison. (B) The Hubble law: Galaxies farther away recede faster.

that the speed with which a galaxy moves away from us—its **recession velocity**—increases with distance, as figure 16.16 shows. That is, *the faster a galaxy recedes, the farther away it is*. This, then, becomes a powerful tool for measuring galaxy distances. In 1929, Edwin Hubble discovered that a simple formula related the recession velocity, $V$, and the distance, $D$. He found that

$$V = HD$$

where H is a constant. Because of his discovery, this relation is called the **Hubble law,** and H is called the **Hubble constant.**

The value of the Hubble constant depends on the units used to measure $V$ and $D$. Astronomers generally measure $V$ in kilometers per second and $D$ in **megaparsecs,** where 1 megaparsec (abbreviated Mpc) is a million parsecs (= 3.26 million light-years). In these units, H is about 70 kilometers per second per megaparsec. For many years astronomers disagreed about the value of H. The reason for this controversy was that, to determine H, one had to know *both* the distance and the velocity of at least a few galaxies in order to calibrate the law. H was then simply the average value of each calibration galaxy's velocity divided by its distance. But as we have seen, it is difficult to accurately measure the distance to even a nearby galaxy. A new technique for determining H has essentially resolved this problem—establishing that H = 70 kilometers per second per megaparsec.*

This new method uses information about irregularities in the brightness of radiation emitted by the young Universe, which we will discuss briefly in chapter 17. For now, we will simply say that the method appears to give the most precise value yet for H.

Once a value for H is known, we can turn the method around and use the Hubble law to find the distance to a galaxy. The method is as follows: Take a spectrum of the galaxy whose distance you want. From the spectrum, we measure the shift of the

---

*To be more precise, H = 72 kilometers per second per megaparsec.

spectral lines. From the Doppler-shift formula, we calculate the recession velocity. Finally, we use that velocity in the Hubble law to find the distance, as we show below.

We begin by writing out the Hubble law

$$V = HD$$

| |
|---|
| $V$ = Recession velocity of galaxy |
| $D$ = Distance of galaxy |
| $H$ = Hubble constant |

To find $D$, we divide both sides by H, obtaining

$$D = V/H$$

Finally, we insert the measured value of $V$ and our choice of H and solve for the distance. For example, suppose we find from a galaxy's spectrum that its recession velocity is 56,000 kilometers per second. We find its distance by inserting this value for $V$ in the expression above and then divide by H.

$$D = (56{,}000 \text{ km/s})/H$$

For H = 70 kilometers per second per megaparsec, the value of $D$ is then

$$D = \frac{56{,}000 \text{ km/s}}{70 \text{ km/s/Mpc}} = 800 \text{ Mpc}$$

Thus, the galaxy is 800 megaparsecs away from us.

In applying this method, care must be used about the units. When we divide $V$ in kilometers per second by H in kilometers per second per megaparsec, the answer we get will be in megaparsecs. Thus, the 800 found in this example is not in parsecs or light-years but in megaparsecs. Moreover, the Hubble law applies only to galaxies that are reasonably far away. The velocity of nearby galaxies is strongly affected by the gravitational attraction of their neighbors, and so the method fails when applied to such systems. (Of course, it also does not apply to stars or other objects in the Milky Way.)

With the distance to a galaxy known, we can now measure its radius and mass as we will now show.

## Measuring the Diameter of a Galaxy

Astronomers measure the diameter of a galaxy using the same method we discussed in chapter 1 for measuring the diameter of the Moon or the Sun. There, we showed that an object's true diameter, $d$, is given by

| |
|---|
| $A$ = Angular size of object |
| $d$ = Diameter of object |
| $D$ = Distance to object |

$$d = \frac{2\pi A D}{360}$$

where $A$ is its angular size on the sky in degrees and $D$ is its distance (fig. 16.17). The units of $d$ are the same as $D$; for example, in measuring galaxy sizes, we might express both in light-years or both in parsecs.

To illustrate the method, we will measure the diameter of M31, a close neighbor of the Milky Way. M31 is a spiral galaxy, so when we speak of its diameter, we are referring to the diameter of its disk. Even that is somewhat ambiguous because galaxies do not have sharp edges—the stars just thin out gradually—and it is hard to say exactly where the galaxy ends and intergalactic space begins.

The angular diameter, $A$, of M31 is about 5° and its distance, $D$, is about 670,000 parsecs. Using the expression for $d$ above and inserting these quantities, we find that the diameter, $d$, of M31 is

$$d = \frac{2\pi A D}{360}, \quad \text{or}$$

$$d = \frac{2\pi \times 5 \times 670{,}000 \text{ pc}}{360}$$

$$= \text{about } 58{,}500 \text{ pc}$$

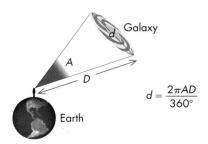

**FIGURE 16.17**

Measuring the diameter of a galaxy from its angular diameter.

## EXTENDING OUR REACH

### OTHER WAYS TO MEASURE A GALAXY'S DISTANCE

The Hubble law is a powerful tool for finding a galaxy's distance, but in some cases it cannot be applied. As we have said earlier, the Hubble Law does not work for nearby galaxies because their speed is determined by their gravitational interaction with one another rather than by the expansion of the Universe. The Hubble law is also difficult to apply to very remote galaxies. The light from such distant systems has taken billions of years to reach us, and so we see them as they were billions of years ago. At that distant time, the Universe's rate of expansion was different from its rate today, and thus the value of the Hubble constant was not the same as it is now. With these difficulties in mind, astronomers have sought other ways to find a galaxy's distance.

One method is similar to the simple observation that you have made if you look closely at a picture printed in a newspaper or comic book. Holding such a picture near your eyes, you can see that it consists of many tiny separate dots. As you move the picture farther from your eyes, the dots disappear and merge together to form the picture. Similarly, in a galaxy near us, we can easily see the separate spots of its more luminous stars. However, in more distant galaxies, the stars blend together into a uniform blur of light. Careful measurement of the graininess of a galaxy's picture can therefore give an indication of its distance.

Another method widely used for finding galaxy distances is based on the standard-candles method but uses the galaxies themselves as the standard candle. This method relies on the observation that the rotation velocity of a spiral galaxy is related to its luminosity. In particular, the more luminous a galaxy, the more rapidly it rotates. We might anticipate this result because if a galaxy rotates very rapidly, it must be very massive to hold itself together gravitationally. Moreover, other things being equal, a massive galaxy contains more stars and thus emits more light than a small galaxy. Thus, the faster a galaxy rotates, the more luminous it is.

Astronomers use this relation to find a galaxy's distance by first obtaining a spectrum of the galaxy at optical or radio wavelengths. If the galaxy is rotating, one side of it is moving toward us and the other side is moving away from us, producing a Doppler shift to the blue on one side and to the red on the other. From the Doppler shift law, astronomers can find the galaxy's luminosity and its apparent brightness in the inverse-square law to calculate the galaxy's distance. This procedure is known as the "Tully-Fisher method" after the two astronomers who devised it.

Keep in mind that most methods for finding the distance to a galaxy have *many* underlying assumptions. For example, many methods require knowing the distance to remote stars or other objects in our own galaxy, and those distances depend in turn on the distance to nearby stars. Astronomers refer to such multistep distance determinations as the distance ladder, and they must account for the inevitable errors and uncertainties at each step of the ladder. As a result, the final answer to how far away is a given galaxy can be quite uncertain.

### Measuring the Mass of a Galaxy

We can measure the mass of a galaxy the same way we measure the mass of a star or planet: by an application of the modified form of Kepler's third law. As you may recall, this method uses the principle that the orbital motion of one object around another is set by their mutual gravity. Their gravity is, in turn, set by their combined mass. Thus, from the orbital motion we can find the mass.

We went through the details of determining the mass of our own galaxy, the Milky Way, in chapter 15. There, we explained that we begin by measuring the orbital speed of a star far out in the disk of the Milky Way. With the star's speed determined and knowing the radius of its orbit, we then found its orbital period (see section 15.6 for details). Finally, we inserted the orbital period and radius in the modified form of Kepler's third law to give us the Milky Way's mass.

To measure the mass of another galaxy, we follow the same procedure. We use the Doppler shift of the galaxy's spectrum lines to find the orbital speed of a star* near the galaxy's rim. Then, from the size of the star's orbit and the modified form of Kepler's third law, we calculate the galaxy's mass.

---

*In using this method, astronomers do not in general get a spectrum of a single star but rather the spectra of groups of stars or gas clouds moving together around the galaxy. Moreover, the spectra do not need to be at visible wavelengths. Many of the most accurate observations of galaxy masses come from radio telescope observations of the Doppler shift.

The above method is an excellent and accurate way to measure a galaxy's mass, yet a puzzling discrepancy arises when astronomers compare that mass with the mass of the visible material in the galaxy. Almost without exception, the mass calculated from the modified form of Kepler's law is *much larger* than the mass of stars and interstellar matter detectable by optical, infrared, and radio telescopes. Thus, galaxies seem to contain matter that cannot be seen, what astronomers therefore call "dark matter."

# 16.3   DARK MATTER

**Dark matter** is the material that astronomers believe accounts for the discrepancy between the mass of a galaxy as found from the modified form of Kepler's third law and the mass observable in the form of stars and gas. The amount of dark matter required to resolve this discrepancy is very large, amounting in typical galaxies to 10 times their visible mass. That is, when we look at a galaxy, we may in some cases be seeing only one-tenth of its matter.

The strongest evidence that dark matter exists comes from galaxy-rotation curves. We discussed such curves in chapter 15 when we described the rotation of the Milky Way, but let us review them here briefly. A galaxy's rotation curve is a plot of the orbital velocity of the stars and gas moving around it at each distance from its center (curve A in fig. 16.18). The rotation curve is useful for studying dark matter because the orbital velocity depends on the mass, visible or not, that lies between the star and the galaxy's center. We can understand how that dependence arises by using the formula for orbital velocity that we developed in chapter 2 or by simply thinking about what keeps a star in orbit.

According to Newton's first law, if an object does not move in a straight line (for example, a star moving in a circular orbit around a galaxy), a force must be acting on it. We know that for a star orbiting in a galaxy, that force is supplied by gravity. We also know that the strength of the gravitational force depends on the mass of the attracting body (in this case, the galaxy). All other things being equal, a greater mass therefore implies a greater force. It turns out, however, that to a good approximation, *only the material between the star's orbit and the galaxy's center contributes* to the gravitational force. Stars near the center of a galaxy will therefore have only a small force acting on them because there is little mass between them and the center. Because the force acting on them is small, such stars cannot be orbiting very rapidly, or they would have flown out of the galaxy. On the other hand, stars orbiting far from the center should also feel a small gravitational force acting on them from the galaxy. They feel the effect of the galaxy's full mass, but the gravitational force on them is nevertheless weak because they are so far from the main mass (recall that the force of gravity grows rapidly weaker at greater distances from a mass). These outermost stars must also be orbiting slowly if they are not to fly out of the galaxy. Thus, both stars near the center and stars far from the center should have low orbital velocities. Accordingly, the galaxy's rotation curve should have a peak, as indicated in curve B of figure 16.18.

Detailed mathematical models bear out this analysis and show that a galaxy's rotation curve should start at small velocities near the core. The curve should then rise to higher velocities at middle distances and finally drop to lower velocities again far from the center. However, when astronomers *measure* the rotation curves of galaxies, they essentially *never* find this behavior. The rotation curves do rise near the center, but they then flatten out and virtually never drop off to low velocities again. What can explain such a major discrepancy between our hypothesis and observation? If our laws of physics are correct, only one thing can explain such behavior: galaxies must contain large amounts of unseen mass in their outer parts. As we saw in chapter 15, astronomers find evidence of such unseen mass in the Milky Way and call it dark matter. The dark matter exerts a gravitational force on stars in the outer parts of a galaxy, holding them in orbit despite their large velocity. The large gravitational force needed

INTERACTIVE

*Dark matter*

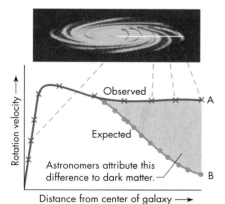

**FIGURE 16.18**
A schematic galaxy-rotation curve. The line with dots represents the curve expected if the galaxy's mass comes only from its luminous stars and gas. The line with crosses represents the observed curve, implying that the galaxy contains dark matter.

to bind such stars in a galaxy implies a huge mass of dark matter (many times the galaxy's luminous mass). Thus, we deduce that each galaxy is embedded in its own dark matter halo, vastly more massive and larger than the "visible" galaxy.

What can this dark matter be? Astronomers are puzzled. They can rule out ordinary dim stars because such objects would emit at least some infrared radiation that would be detectable. They can rule out cold gas because that would be detectable by radio telescopes. They can rule out hot gas because that would be detectable by optical, radio, or X-ray telescopes. Objects they cannot rule out are tiny planetesimal-sized bodies, extremely low-mass cool stars, dead white dwarfs or neutron stars, black holes, or sub-atomic particles such as neutrinos. Other possibilities include subatomic particles predicted by theory but not observed as yet in the laboratory, such as photinos or WIMPS (weakly interacting massive particles). The nature of the dark matter remains one of the central mysteries in astronomy today. Moreover, as we will see in section 16.6 and in chapter 17 as well, dark matter is not confined to galaxies. It also permeates galaxy clusters and perhaps the Universe itself.

We should not be surprised that something as vast as our Universe still has mysteries. In fact, dark matter is but one of many aspects of galaxies that astronomers do not fully understand. Another mystery about galaxies is the violent activity in the centers of many of them.

## 16.4    ACTIVE GALAXIES

**Active galaxies** are galaxies whose centers (nuclei) emit abnormally large amounts of energy from a tiny region in their core. Not only is the emitted radiation intense, but it also usually fluctuates in intensity. In many instances, active galaxies exhibit intense radio emission and other activity outside their main body. At least 10% of all known galaxies are active, and they can be divided into three overlapping classes: radio galaxies, Seyfert galaxies, and quasars. The centers of these active systems are sometimes called AGNs (active galactic nuclei).

### Radio Galaxies

**Radio galaxies,** as their name indicates, emit large amounts of energy in the radio part of the spectrum. They are generally elliptical galaxies, and their radio emission differs in two important ways from that of ordinary galaxies. First, it comes not so much from the main body of the galaxy but from the core and from regions outside the galaxy on either side of it, as you can see in figure 16.19A. This figure was made by aiming a radio telescope at the galaxy and recording those places where the radio signal was strong. The strength is indicated by color, with blue being weak emission and red being relatively strong emission.

The second important feature of the radio emission from such galaxies is its strength, amounting in some cases to millions of times more radio energy than normal galaxies emit. You can get some sense of these differences by comparing figure 16.19A with figure 16.20, which shows a map of the radio emission from the galaxy M81 (a system similar to the Milky Way). The radio energy shown is particularly strong along the spiral arms but some emission also comes from the entire body of the galaxy. Even so, the radio emission from M81 is puny compared to that of a radio galaxy.

The radio-emitting regions visible outside the radio galaxy in figure 16.19 are called "radio lobes." These lobes, often located symmetrically on opposite sides of the galaxy, may span hundreds to millions of light-years. Although most of the radio emission comes from the lobes, some comes from within the visible body of the galaxy, primarily from a tiny region at its core. The core source is typically less than a light-month across, which is a tiny fraction of the several-thousand-light-year diameter of the central region of the galaxy.

A

B

**FIGURE 16.19**
(A) False color map of the radio emission from the radio galaxy, Cygnus A. Notice the two radio lobes on either side of the radio galaxy. (B) A similar map of a different radio galaxy (NGC 1265) showing how its motion through gas around it has swept the radio lobes back.

**FIGURE 16.20**
False color map of the radio emission at 21 centimeters from an ordinary spiral galaxy, M81. Notice how clearly the radio intensity shows the spiral arms.

What causes this intense radio emission? Astronomers can tell from its spectrum that the emission is synchrotron radiation, created by electrons traveling at nearly the speed of light and spiraling around magnetic field lines. We discussed this mechanism in chapter 14, where we described radio emission from pulsars. Pulsars generate high-speed electrons at their surface, but radio galaxies generate them in their core. The electrons are then shot out of the core in narrow **jets** (such as the one shown in fig. 16.21) that generally point at the radio lobes. Streaming away from the end of the jet, the electrons eventually collide with the gas around the galaxy. There, they spread out to form the lobes, all the while emitting synchrotron radiation. Although the lobes are often on opposite sides of the radio galaxy, they are sometimes swept into arcs or plumes, presumably the result of the radio galaxy's motion through the gas around it (fig. 16.19B). We will discuss below one hypothesis of how the jet is formed, but before we do, we need to describe the other types of active galaxies: Seyfert galaxies and quasars.

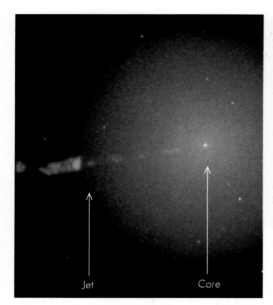

Jet                    Core

**FIGURE 16.21**
Picture of the jet in the core of the radio galaxy M87.

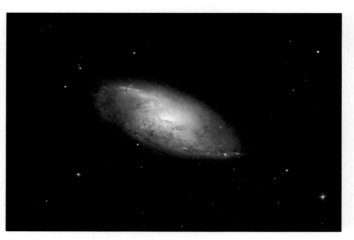

**FIGURE 16.22**
The Seyfert galaxy M106, also known as NGC 4258.

## Seyfert Galaxies

A **Seyfert galaxy** is a spiral galaxy whose nucleus is abnormally luminous (fig. 16.22). These peculiar galaxies are named for the American astronomer Carl Seyfert (pronounced see-firt), who first drew attention to their optical peculiarities. The core luminosity of a typical Seyfert galaxy is immense, amounting to the entire radiation output of the Milky Way, but coming from a region far less than a light-year across. Moreover, the radiation is at many wavelengths: optical, infrared, ultraviolet, and X-ray. However, despite its immense luminosity, the radiation from a Seyfert galaxy's core fluctuates rapidly in intensity, sometimes changing appreciably in a few minutes.

In addition to the intense source of radiation in their nuclei, Seyfert galaxies also contain gas clouds moving at speeds of about 10,000 kilometers per second or more. Such speeds are a non-negligible fraction of the speed of light! In some instances, the core is ejecting gas in small jets. This rapidly moving gas and the small, bright nuclear region make Seyfert galaxies similar to radio galaxies, and, in fact, some Seyfert galaxies are radio galaxies. Nevertheless, most Seyfert systems are spiral galaxies, whereas most strong radio galaxies are ellipticals. However, the similarity in their activity is suggestive of a similar cause, a process that probably also acts in quasars, as we shall see in the next section.

## Quasars

**Quasars** are extremely luminous, extremely distant, active galaxies. Some are powerful radio sources, similar to the radio galaxies. Others are similar both to radio and Seyfert galaxies in that they are ejecting hot gas from their centers, as deduced from the extremely strong emission lines in their spectra. (Recall our discussion in chapter 3 that a hot, low-density gas produces a spectrum consisting of emission lines.)

Quasars get their name by contraction of the term *quasi-stellar radio source*, where *quasi-stellar* (meaning "almost starlike") refers to their appearance in the first photographs taken of them (fig. 16.23), in which they look like stars (that is, like points of light). Early photographs also revealed that some quasars have small wispy clouds near them or tiny jets of hot gas coming from their cores. However, more recent images reveal that quasars often lie in faint, fuzzy-looking objects that appear to be ordinary galaxies.

The jets observed in some quasars (and at least one Seyfert galaxy) exhibit a very curious property: they contain clumps that appear to move faster than the speed of light. Astronomers refer to such seeming faster-than-light motion as superluminal motion. According to relativity theory, no material object can travel faster than light, but the theory does allow objects to *appear* to move faster than light if their true speed

A

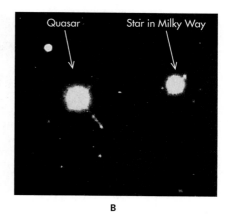
B

**FIGURE 16.23**

Two quasars: (A) PG 0052+251 as viewed from space by the Hubble Space Telescope and (B) 3C273 viewed from the ground. In the ground-based photo, you can see only a smear of light. The clearer view with the HST shows the galaxy surrounding the active core. ("PG" stands for *Palomar Green* survey and "3C" stands for *Third Cambridge Catalog of Radio Sources*.)

Q* Do you see the "jet" in the right picture?

Q* The jet points out of the blob labeled "Quasar" at an angle of about "5 o'clock."

**FIGURE 16.24**

Plot of the light variation of a quasar.

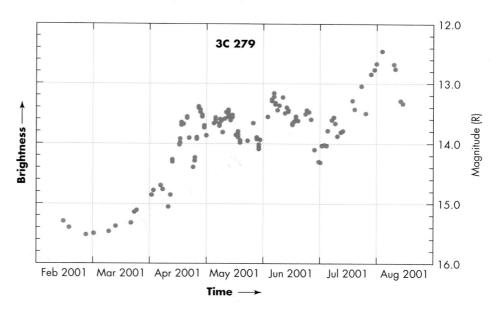

is only slightly less than the speed of light. Thus, apparent superluminal motions are additional evidence for extremely high-speed motions in active galaxies.

Pictures of quasars lack detail because these objects are so far away. We deduce their huge distance from the huge redshift of their spectra (recall that a galaxy's distance can be found from its redshift using the Hubble law). In fact, quasars have the largest redshifts known for any kind of astronomical object. On the basis of its redshift, the most distant one yet observed is about 10 billion light-years from Earth. To be visible at such an immense distance, an object must be immensely luminous, and a typical quasar turns out to be about 1000 times more luminous than the Milky Way. Yet, despite this huge power output, quasars, like Seyfert galaxies, fluctuate in brightness. Slow changes occur over months (fig. 16.24), but short-term flickerings occur in times as brief as

# EXTENDING OUR REACH

## MEASURING THE DIAMETER OF ASTRONOMICAL OBJECTS BY USING THEIR LIGHT VARIABILITY

Brightness changes give an important clue to the size of a quasar's or active galaxy's core. Suppose we have a ball of glowing material that we can turn on and off. Suppose further that when we turn it on, it begins to glow everywhere at the same time, and when we turn it off, it stops glowing everywhere at the same time. If we now look at this ball as it begins to glow, light from its nearest part, A, will reach us before light from its farthest part, B (box fig. 16.2). An exactly similar effect occurs with sound produced from a lightning stroke. The lightning stroke lasts only fractions of a second, but the sound it generates lasts many seconds. The reason the sound is drawn out for such a long time is that sound travels relatively slowly (only about 1000 feet/second). Thus, if the stroke is 5000 feet long (a typical size for a bolt of lightning), sound from the near part reaches us 5 seconds before sound from the far part does. So, too, even though the ball begins to glow everywhere on its surface at exactly the same moment, the light we see does not turn on instantly. Instead, it gradually increases in brightness over a time interval equal to the time required for light to travel across the ball. That is, if the ball has a diameter of one light-second (the distance light travels in 1 second = $3 \times 10^5$ kilometers), the light reaching an observer does not reach full brilliance in less than 1 second. Likewise, if it is a light-month across (about $10^{12}$ kilometers) and it is turned off instantly, it does not dim away to darkness in less than 1 month.

From this, we can conclude that if a quasar varies in brightness in a time as short as a few days (and assuming simultaneous illumination and dimming), then the emitting region can be no more than a few light-days across. A core a few light-days across is tiny compared to the size of a typical galaxy (100,000 light-years across). In fact, such a core is about 10 million times smaller than a galaxy, or about 200 AU in diameter, about 5 times the size of Pluto's orbit around the Sun.

$$\text{Distance} = \text{velocity} \times \text{time}$$
$$D = ct$$
$$\frac{D}{c} = t$$

**BOX FIGURE 16.2**
Sketch illustrating that when a light source turns on, we see the light from its near side before we see the light from its far side, and so we do not see its light turn on instantly. For the objects that we encounter in everyday life, the time delay is so short we do not notice it. However, for astronomical-sized bodies, the effect can create delays ranging from days to months, from which we can deduce that the emitting body is light-days or light-months across.

hours. Rapid changes of this sort give clues to the size of the emitting object, as discussed in Extending Our Reach: "Measuring the Diameter of Astronomical Objects. . . ."

However, astronomers need not rely solely on light variations to determine the size of an active galaxy's core. Observations with radio interferometer telescopes show that the active cores are extremely small (only a fraction of a light-year, in some cases).

## Cause of Activity in Galaxies

Because all three kinds of active galaxies (radio galaxies, Seyfert galaxies, and quasars) share many features in common (for example, tiny cores ejecting gas at tremendous speed), astronomers have sought a model for their activity that could explain all three. In fact, astronomers now realize that in addition to the three classes of active galaxies that we have described above, many other kinds of active galaxies exist. For example, BL Lacertae objects (BL Lac, for short) are luminous objects that are probably distant radio galaxies in which the jet of ejected matter is pointing almost straight at the Solar

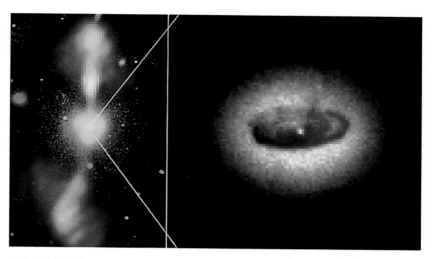

**FIGURE 16.25**

An image (on far right) made with the Hubble Space Telescope of an accretion disk in the center of the radio galaxy NGC 4261. The inset image on the left shows a wider field view of this galaxy's central region and shows the jets of hot gas being ejected from the central regions of the disk.

System. BL Lac objects, like quasars, look starlike and, in fact, BL Lac was originally believed to be a peculiar variable star within the Milky Way.

Any successful model for active galaxies must explain how such a small central region can emit so much energy over such a broad range of wavelengths. For example, at least one quasar is a powerful source of both radio waves and gamma rays. Astronomers therefore hypothesize that the core must contain something very unusual. No ordinary single star could be so luminous. No ordinary group of stars could be packed into so small a region. One kind of object that is very small but that can also emit intense radiation is the accretion disk around a black hole. We saw in chapter 14 that such accretion disks can emit intense X-ray radiation. Could the center of an active galaxy contain an immense black hole? Could the radiation we observe from the core be hot gas in an accretion disk around such a black hole? This is precisely the model for active galaxies most widely favored by astronomers today. According to this model, active galaxies have in their cores a black hole about the size of the Earth's orbit around the Sun that contains a hundred million or more solar masses. Around the black hole swirls an accretion disk of gas with a radius of tens to hundreds of AU. Gas within the disk spirals toward its center, drawn inward by the black hole's gravity, ultimately to be swallowed. Orbiting under the black hole's fatal attraction, the gas is heated to incandescence by frictional forces in the disk and by the gravitational energy converted into heat (thermal energy) as it falls inward.

Figure 16.25, an image made with the Hubble Space Telescope, supports this model. The image shows just such a disk in the central regions of the radio galaxy NGC 4261. Gas falling into the disk releases gravitational energy, which heats the material to temperatures as hot as several million K. Although most of the mass ultimately falls into the black hole, some material from the accretion disk boils off into space and escapes. If a strong magnetic field exists in the disk, it may channel such escaping gas into narrow jets from the central regions of the top and bottom surfaces of the disk (fig. 16.26), as shown by computer simulations.

Left to itself, the accretion disk would gradually drain into the black hole and activity would cease. But the disk may be replenished with matter from at least two sources. Sometimes stars within the galaxy may approach too closely to the black hole and be torn apart by its tidal forces. Their gaseous debris then falls into the accretion disk as a new source of material. Alternatively, from time to time, collisions with neighboring

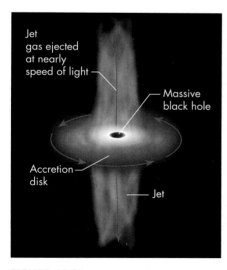

**FIGURE 16.26**

Sketch of jets formed by an accretion disk.

galaxies may add new material as the active galaxy cannibalizes smaller neighbors. Each such event gives the accretion disk a new burst of energy, causing the active galaxy's core to increase in brightness.

How might such a massive black hole form? We saw in chapter 15 that astronomers have strong evidence that our own galaxy has a black hole at its center. But that black hole has a mass of only a few million solar masses, much smaller than the one we are describing here. According to one theory, our own galaxy's black heart formed in several steps. First, a massive star in the central region of the Milky Way reached the end of its life, exploded as a supernova, and formed a black hole perhaps as small as 5 $M_\odot$. Next, this small black hole grew as interstellar matter happened to fall into it. As gas fell in, the hole's mass and radius increased, making it easier for the hole to attract and swallow yet more material. Eventually, the hole became large enough to swallow entire stars, at which point it grew rapidly to a million or more solar masses. During its growth, the hole was surrounded by an accretion disk, and our Milky Way was probably an active galaxy. As the Milky Way aged, its black hole consumed the gas around it, and the activity subsided to the low level we see today. Other galaxies may have had more mass at their cores to begin with, or they may have gained mass by collisions with small, neighboring galaxies. This extra mass thereby allowed their central black holes to grow far larger than the black hole in the Milky Way. In either case, we expect that the central black hole is active mainly in young galaxies during the time the hole is growing in mass.

Astronomers today have evidence that many (perhaps even most) galaxies have a massive black hole at their core. They come to this conclusion from the observation that stars and gas at the core of many galaxies are orbiting the core at extremely high speed. To prevent such rapidly moving material from flying out of the galaxy's core requires a huge mass. In fact, to bind the stars and gas in such tight but high-speed orbits requires (in many cases) more than a million times the mass of the Sun. Yet this huge mass seems to emit no light or other radiation, and astronomers can think of few objects that are both massive and dark. Thus, by elimination, they are led to think that many galaxies have a massive black hole at their core.

We described above how such a huge black hole might have formed at the center of the Milky Way by the steady accumulation of mass into an initially small black hole. But another recent discovery has led astronomers to wonder if massive black holes might form some other way. Observations they have made with the Hubble Space Telescope show a relationship between the mass of the black hole and the mass of its host galaxy's central stellar bulge. They find that the more massive the black hole, the more massive the bulge. Why does such a relation exist? Does the black hole form first and shape the bulge, or vice versa? Astronomers do not yet know.

## SCIENCE AT WORK

### EXPLAINING JETS

Astronomers recently proposed a revised theory to explain the jets seen in active galaxies. In this model, the massive central black hole has a strong magnetic field that is twisted as the black hole spins. Such fields (as you may recall) trap charged particles that exist in very hot gas. The spinning and twisting of the magnetic field can, therefore, capture hot gas from the accretion disk and pump it outward along the field. This model, according to some, better explains the narrowness of the jets and their kinks and turns. Moreover, it provides a more powerful supply of energy for the jet—the rotational energy of the spinning black hole.

If black holes are so common in galaxies, why are none of the galaxies close to us as active as quasars? The youth of quasars may be the key. Because quasars are so distant, we see such objects the way they were in the distant past, not the way they are "now." For example, when we look at a quasar that is 2 billion light-years away, we are seeing light that has taken 2 billion years to reach us. Thus, we are seeing the quasar as it was 2 billion years ago. The fact that most quasars are billions of light-years away implies that the phenomenon causing them occurred billions of years ago, when the Universe was younger. Galaxies whose cores were once active quasars probably surround us, but their activity has long since died away. They still harbor huge black holes, but they have used up the gas that powered their activity.

Some astronomers have argued that the black-hole theory of active galaxies is unbelievable and propose that quasars are not truly at the immense distance implied by their redshift and are in reality much closer. Astronomers favoring the black-hole model counter with the question, "What then causes their immense redshift and why do they look like Seyfert galaxies, which we *know* are far away from Earth?" Attempts to explain these extraordinary objects as powered by something other than a huge black hole have generally failed. For example, at one time, astronomers thought that the activity might be produced by a single star of $10^8$ solar masses that had managed to form in the center of the active galaxy. Astronomers have also considered more exotic explanations. For example, a few scientists proposed a new class of objects called "white holes." These would be the opposite of black holes in the sense that they would spew material out into space rather than draw it in, but no observational evidence for them exists, and therefore, using them to explain active galaxies is hardly more plausible than using black holes. So, despite the reliance of the model on an object as peculiar as a black hole, the model is consistent with observations of active galaxies and quasars. But consistency is not proof, and the history of astronomy is littered with the wrecks of consistent models (Ptolemy's epicycles, for example).

## 16.5    QUASARS AS PROBES OF INTERGALACTIC SPACE

Although interesting in their own right, quasars are also useful to astronomers as probes of intergalactic space. Because quasars are so distant, their light has a high probability of passing through other galaxies or intergalactic gas clouds on its way to Earth. Even if the intervening object is too dim to see, its matter may imprint its signature on the quasar's light, allowing us to detect these otherwise invisible objects.

For example, many quasars show absorption lines in their spectra that have very different Doppler shifts from those of the emission lines in the quasars themselves. These absorption lines are created by cool gas clouds lying between us and the quasar, much as similar lines in stars are created by cool interstellar clouds, as we discussed in chapter 15. The cool gas responsible for the absorption lines may be within galaxies, or it may lie in clouds that have not yet formed into galaxies. Matter within a galaxy cluster between us and a quasar may also affect the quasar's light in an even more remarkable way, creating a gravitational lens.

### Gravitational Lenses

About half a dozen quasars seem to have nearby companion quasars with essentially identical spectra but slightly different brightnesses and shapes. The existence of two so nearly identical objects so close together is unlikely, and astronomers now think that the "companions" are actually images of a single quasar created by a gravitational lens (fig. 16.27A).

An ordinary lens forms an image because light bends as it passes through the lens's curved glass. A **gravitational lens** forms an image because light bends as it

INTERACTIVE

*Gravitational lensing*

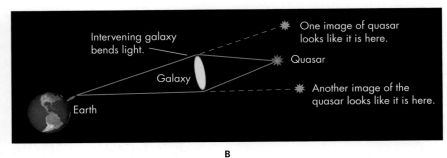

A

B

**FIGURE 16.27**
(A) An Einstein cross, a complex image of a quasar created by a gravitational lens. (B) Sketch of how a gravitational lens forms an image.

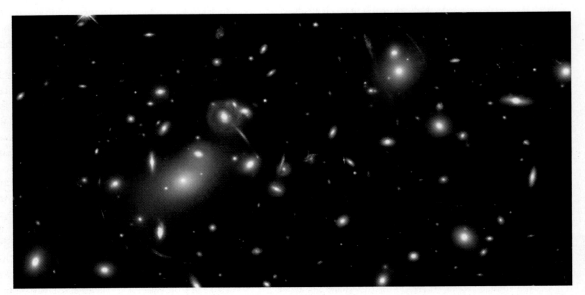

**FIGURE 16.28**
The arcs in this picture of the galaxy cluster Abell 2218 are the image of a distant galaxy distorted by the gravitational lens effect created by the galaxy cluster.

passes through the curved space around a massive object such as a galaxy.* The galaxy's gravitational force bends the space around it so that light rays that would otherwise travel off in other directions and never reach the Earth are bent so that they do, as depicted in figure 16.27B. The existence of gravitational lenses thus supports Einstein's theory that massive bodies can bend space, as we described in chapter 14 in the discussion of black holes. Although we cannot yet directly see the extreme bending of space at a black hole, we can "see" the more modest bending caused by gravitational lenses. Moreover, we will learn in the next chapter that such bending may occur on such an enormous scale that it affects the shape of the Universe.

Gravitational lenses may create forms other than multiple images of quasars. For example, figure 16.28 shows arcs created when light from a distant galaxy was bent into elongated shapes as it passed through a galaxy cluster. Because the shape of these arcs depends on the mass of material within the cluster, astronomers can use the arcs to measure the cluster's mass. The masses so found confirm yet again that galaxy clusters contain large amounts of dark matter.

---

*Recall our discussion in chapter 7 of gravitational lenses caused by otherwise invisible planets around distant stars.

## 16.6 GALAXY CLUSTERS

We have mentioned several times earlier in this chapter that galaxies are often found in groupings called **galaxy clusters.** A galaxy cluster is a group of galaxies held together by their mutual gravity. Each galaxy moves along its own orbit, just as within a galaxy each star moves on its own orbit. Galaxy clusters are huge. A typical cluster is more than several million light-years across and may contain anywhere from a handful to a few thousand member galaxies (fig. 16.29).

### The Local Group

Our own galaxy, the Milky Way, belongs to a very small cluster called the **Local Group.** The Local Group contains about forty members and is sketched in figure 16.30. Its three largest members are spiral galaxies: the Milky Way, M31, and M33. The smaller members include the satellite galaxies orbiting M31 and the Milky Way, the most famous of which are the Large and Small Magellanic Clouds, which we have referred to many times before.

The Magellanic Clouds (figs. 16.2 and 16.31) are about one-tenth the size of the Milky Way. They appear to be dwarf irregular galaxies, although the large cloud has a faint, barlike structure. Both galaxies are approaching the Milky Way along very elliptical orbits and are being pulled into a distorted shape by the Milky Way's gravity.

The Local Group is particularly interesting to astronomers not only because it is our home galaxy cluster but also because it allows us to study the "demographics" of galaxies. That is, we can see the true population of galaxies in this region of space "close" to home. For example, most of the galaxies in the Local Group are faint, small "dwarf" galaxies. Figure 16.32 shows a typical dwarf galaxy. It is a dim, small group of stars with little or no interstellar gas. Such galaxies are too faint to be visible except in clusters close by. Thus, without the Local Group to let us see the true abundance of these small systems, we might greatly underestimate their true numbers.

**FIGURE 16.29**
Photograph of the Coma Cluster of galaxies.

### Rich and Poor Galaxy Clusters

The largest groups of galaxies are called **rich clusters** because they contain hundreds to thousands of member galaxies (fig. 16.33). The great mass of such a galaxy cluster draws its members into an approximately spherical cloud with the most massive galaxies near the center.

In contrast, some galaxy clusters contain relatively few members (sometimes only a dozen or so). These so-called **poor clusters** (also called simply "groups") have so little mass that the gravitational attraction they create is much weaker than that of the rich clusters. Thus, their member galaxies are not held in so tight a grouping, and they generally have a ragged, irregular appearance with no central concentration. The Local Group (fig. 16.30) is a typical poor cluster.

M31

Dwarf
Galaxies

Milky Way

Magellanic
Clouds

M33

4 million light-years

**FIGURE 16.30**
Artist's view of the Local Group.

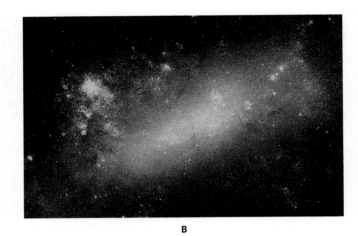

A

B

**FIGURE 16.31**
Photographs of (A) the Small Magellanic Cloud and (B) the Large Magellanic Cloud.

Q* What are the irregular reddish regions visible in the pictures?

A* HII regions—clouds of hot, ionized hydrogen.

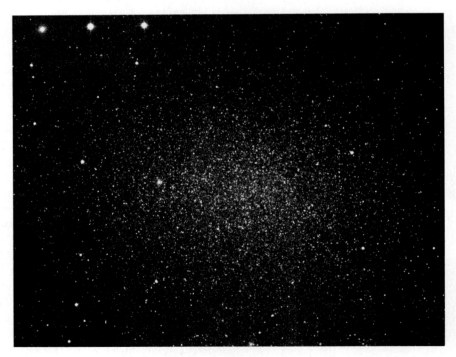

**FIGURE 16.32**
The dwarf galaxy Leo I, a member of the Local Group of galaxies.

Not only do rich and poor clusters differ in their number of galaxies, but they also tend to contain different types of galaxies. Rich clusters contain mainly elliptical and S0 galaxies. Moreover, the few spiral galaxies that they contain tend to be found in the outer parts of the cluster. On the other hand, poor clusters tend to have a high proportion of spiral and irregular galaxies. Spirals are rare in the inner regions of rich clusters, probably because of galaxy collisions. In the core of a rich cluster, the galaxies are close together, and so collisions between them are frequent. As we saw earlier, collisions can change

**FIGURE 16.33**
The Hercules Cluster, a rich cluster of galaxies.

↳⋆ What types of galaxies can you identify in this photograph? Notice the interacting galaxy pair near the center. Suppose one of those galaxies is 100,000 light-years in diameter. What is the approximate diameter of the part of the cluster visible in this picture?

↳⋆ You can see two interacting spirals in the center of the image. Slightly left of center at the bottom is an elliptical. A barred spiral is just above the central interacting pair. The cluster is about 20 times bigger across than the size of a medium-big galaxy. Thus, the cluster is about 100,000 × 20 = 2 million light-years across.

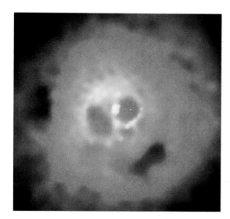

**FIGURE 16.34**
A false-color X-ray image of the Perseus galaxy cluster. The orange glow is hot gas in the cluster. The picture shows a region about 600,000 light-years across. The dark spots near the center are "holes" in the gas blown out by jets from the active galaxy at the cluster's core.

spiral galaxies into ellipticals. Thus, although the rich clusters may have once contained many spirals, today they contain few, and those are the lucky ones that escaped running into a neighbor. The many collisions in the crowded inner part of these clusters also create giant elliptical galaxies by "cannibalism," as we described in section 16.1.

Rich and poor clusters differ in yet another way. Observations with X-ray telescopes show that rich clusters often contain large amounts ($10^{12}$ to $10^{14}$ $M_\odot$) of extremely hot gas that emits X rays (fig. 16.34). In fact, some clusters contain ten times more hot gas than they do stars in their member galaxies. Some of this gas may be left over from the cluster's formation. Some may have been ejected from the galaxies within the clusters. For example, when a supernova explodes, it blasts material to such a high velocity that the material can escape a galaxy's gravity and be driven into intergalactic space.

Alternatively, gas may escape from galaxies when they collide, as they move in the crowded cluster environment. Such collisions affect the stars hardly at all. Because they are so small compared to the distance between them, the vast majority of stars in colliding galaxies simply shoot harmlessly by one another. The gas, on the other hand, because it is spread widely in clouds throughout the galaxies, cannot escape collisions. Thus, in a collision the gas in one galaxy is driven at high speed into the gas of the other, compressing the gas and heating it to millions of degrees. In addition, as the stars of one galaxy move past the stars of the other, much of the gas may remain behind. Such gas, stripped from galaxies by collisions, generally contains heavy elements synthesized in stars within the galaxies. It then mixes into the otherwise pure hydrogen and helium gas from which the galaxies formed. But the essential point is that, regardless of its source, clusters contain huge amounts of hot gas.

In addition to the X-ray-emitting gas described earlier, galaxy clusters also contain large amounts of the dark matter we discussed earlier. In fact, some of the first evidence for dark matter came from observations of the motion of galaxies within clusters. Just as stars within galaxies seem to orbit too rapidly for the gravitational force that can be attributed to their luminous mass, so too galaxies in clusters orbit too rapidly for their luminous mass. Additional evidence for dark matter in galaxy clusters comes from the hot gas they contain. To hold such hot gas within the cluster and prevent its expansion, the cluster must exert a strong inward gravitational pull on the gas. The gravity needed to confine the gas depends on the cluster's total mass, and, once again, the calculated mass greatly exceeds the mass directly observable—a signature of dark matter. Combining the evidence from these very different methods leads astronomers to deduce that the amount of dark matter in some galaxy clusters is so immense that it amounts to perhaps 100 times the amount of mass contained in stars.

Figure 16.35 shows how the dark matter is distributed in a galaxy cluster that is forming from the merger of two smaller clusters. Although the dark matter is, of course, invisible, its location can be mapped by its gravitational lensing effect on the light of background galaxies and is shown in blue. In this collision, the gas from the two clusters has collided and been compressed. The compression has heated the gas

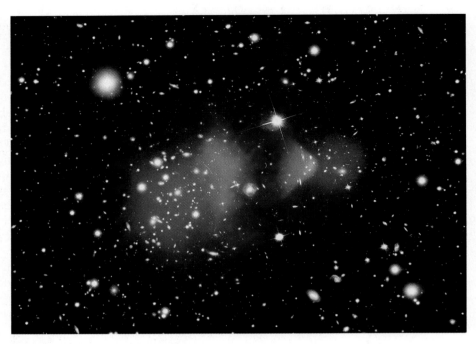

**FIGURE 16.35**
Dark matter (the blue blobs) revealed in the collision of two galaxy clusters. The image has been made by superposing three images: a picture made at optical wavelengths (showing the galaxies in the two clusters); an image made at X-ray wavelengths (showing in red the hot gas in each cluster); and a map of the dark matter in the two clusters (the blue blobs) as deduced from the gravitational lensing of the background galaxies. In the collision, the dark-matter blobs have passed through each other because the dark matter interacts with itself and other matter only gravitationally. The gas in the two clusters, on the other hand, collides and has clumped up in a bullet-like form. The merged cluster is officially known as 1E 0657-56 and more simply as *The Bullet Cluster.*

to millions of degrees, so that it emits X rays (shown here in red). The separate masses of dark matter in the two clusters did not affect each other as they collided. Rather, they simply passed harmlessly through one another. This figure also shows an example of how clusters form, according to current astronomical thinking. That is, large, rich clusters grow by the collision and merging of smaller clusters. Further evidence for this hypothesis of how clusters form comes, once again, from our ability to "see back in time" by looking out into space. The galaxies in nearby clusters are generally fairly smoothly spread within the clusters. On the other hand, galaxies in remote clusters sometimes have a clumpy distribution in the cluster, as if they have recently joined it and have not had time to spread themselves smoothly.

From such evidence, astronomers, thus, think that galaxy clusters form by galaxies attracting one another into small groups. Over time, these small groups attract other galaxies and even other entire clusters to make yet larger clusters, eventually growing into the rich clusters we see today.

## Superclusters

The great mass of galaxy clusters creates gravitational interactions between them and draws them together into larger structures called "superclusters." A **supercluster** contains a half dozen to several dozen galaxy clusters spread throughout a region of space tens to hundreds of millions of light-years across. That is, a supercluster is a cluster of galaxy clusters. For example, the Local Group is part of the Local Supercluster, which contains the Virgo galaxy cluster and more than a dozen

**FIGURE 16.36**
Artist's view of the Local Supercluster.

others spanning about 150 million light-years (about 50 mega-parsecs). It is sketched in figure 16.36.

Superclusters have irregular shapes and are themselves often part of yet larger groupings, but astronomers find it difficult to study the structure of such immense systems. To map the location of galaxies accurately, astronomers need to know not only how far away they are but also their distances relative to each other. Unfortunately, determining the distance to very remote galaxies is extremely hard. However, one feature is clear: clusters often appear to form chains or shells surrounding regions nearly empty of galaxies, as you can see in figure 16.37. These empty spaces are called *voids*, and although they are not totally empty, the few galaxies they contain are small and dim. Whether dark matter exists within voids is not yet known. At the vast distances involved, even the largest telescopes now available allow us to see only the brightest galaxies. Thus, clusters of superclusters and voids mark the end of the structure of the Universe that we can currently see. But although we may have difficulty *seeing* distant large-scale structures, our galaxy and others can "feel" their presence by their gravitational attraction. That attraction pulls on the Local Group and many nearby galaxies, altering their velocity from what is predicted by the Hubble law. From such observations, astronomers have discovered the "Great Attractor." The Great Attractor is probably an assemblage of superclusters about 160 million light-years (50 megaparsecs) away from us and about 260 million light-years (80 megaparsecs) in diameter. It may contain about $3 \times 10^{16}$ M$_\odot$, the equivalent of 10,000 or more galaxies like the Milky Way.

**FIGURE 16.37**
A map of how galaxies are distributed in space. The shape of the region shown in the figure is like two slices of pizza, tip-to-tip, as illustrated in the upper sketch. Each dot in the lower figure represents one of some 100,000 galaxies that lie in the wedge-shaped regions that extend out to about 2 billion light-years from Earth. Note the nearly empty regions (voids), the stringy structures, and the overall cobwebby appearance. The distances to the galaxies were measured from their redshifts. The angles refer to a set of coordinates on the sky.

**FIGURE 16.38**

A computer simulation of the growth of structure as the Universe expands. You can see how an initially smooth distribution of matter becomes stringy, much like the features in figure 16.37. Each box spans about 140 million light-years. The first box on the left represents a time when the Universe was roughly one-tenth its present age. The last box on the right shows that same piece of the model of the Universe as it is today.

The existence of such immense concentrations of galaxies and their effect on their neighbors is also revealed in computer simulations of the large-scale structure of the Universe. In such simulations, astronomers write out the equations that govern the motion of matter (basically Newton's laws) and then solve the equations assuming that the material is expanding in a way that mimics the expansion of the Universe. They then watch to see what happens as the gravitational force of the material acts within the simulated volume. Figure 16.38 shows a series of images from such a simulation. You can see that matter that initially had a relatively smooth distribution is drawn by gravity into a network of strings. Astronomers identify the place where strings meet as locations where galaxies form. By adjusting the amount of ordinary and dark matter and the initial lumpiness, astronomers then "experiment" to find what mix best creates structures similar to what they actually observe. The outcome of many such simulations by many different astronomers confirms that without dark matter, the large-scale distribution of galaxies fails to match what we observe—which is yet additional evidence of the existence of that mysterious matter.

But at least one larger structure remains yet for us to explore: the Universe itself.

## SUMMARY

Astronomers classify galaxies into three main types: spirals, ellipticals, and irregulars. Spirals are disklike, with arms of young stars and gas winding out from the nucleus. They contain a mixture of population I and II stars, and about 15% of their mass is in the form of interstellar gas and dust. Elliptical galaxies are spherical or egg-shaped and contain hardly any population I stars because they have essentially no dense interstellar matter from which new stars could form. Irregular galaxies lack a symmetry to their shape but are often extremely rich in interstellar matter and thus contain many young, population I stars.

Galaxies form from the collapse of primordial gas clouds. Astronomers do not yet understand why some clouds become spiral and others become elliptical galaxies. The initial rotation speed probably plays a role, with rapidly spinning clouds flattening into disk systems. Subsequently, mergers of these disk galaxies "scramble" the orbits of their stars and turn the merged galaxy into an elliptical.

Many galaxies have extremely small but very energetic and luminous nuclei. These include radio and Seyfert galaxies as well as quasars. According to current models, the activity results from matter falling in toward a supermassive black hole at the galactic center. These black holes have masses that are as large as several billion solar masses packed into a region not much larger than the Solar System. An accretion disk forms around the hole as matter falls in, and hot gas boils off the disk, forming jets and clouds of radio-emitting gas. The amount of energy generated may dwarf the light output from the galaxy's stars and, for a quasar, outshine the Milky Way by a factor of a thousand.

Galaxies are usually grouped into clusters a few million light-years across. Small (poor) clusters show little symmetry and tend to contain spiral galaxies. The Milky Way belongs to a poor cluster called the Local Group, which contains about thirty member galaxies. Rich clusters contain thousands of galaxies, which are primarily ellipticals. Clusters themselves appear to be clustered into superclusters that are many tens of millions of light-years across and are

arranged into even larger groups (shells or chains) surrounding voids.

## ?What It Is.

This is not a comet, but a radio picture of gas being stripped out of the Large and Small Magellanic Clouds by the gravitational (tidal) pull of the Milky Way. The Magellanic Clouds are the two bright blobs on the left of the image. The main arc of hydrogen gas is about 300,000 ly long, three times the diameter of the Milky Way.

## QUESTIONS FOR REVIEW

1. What are the three main types of galaxies?
2. How do the basic galaxy types differ in shape, stellar content, and interstellar matter?
3. Why is it not surprising that elliptical galaxies contain only old stars?
4. Which of the basic galaxy types has the smallest proportion of Pop I stars?
5. What happens to the stars when two galaxies collide?
6. Why are galaxy collisions of interest?
7. How do astronomers measure the distance to nearby galaxies?
8. What is the Hubble law?
9. Why does the Hubble law *not* apply to nearby galaxies?
10. How do astronomers measure the mass of a galaxy?
11. What is meant by dark matter?
12. What are the three main types of active galaxies?
13. Make a sketch of a radio galaxy illustrating its structure and where its radio emission comes from. Why do astronomers call these objects "radio" galaxies?
14. What process generates the radio emission in radio galaxies?
15. What is a Seyfert galaxy?
16. What is a quasar?
17. Why do some astronomers think that quasars are very distant?
18. How is it known that active galaxies have small core regions?
19. What mechanism has been suggested to power active galaxies?
20. How might a large black hole form in a galaxy's core?
21. What is a gravitational lens?
22. What are some differences between rich and poor clusters?
23. What is the Local Group?
24. What is the Local Supercluster?
25. What is a void?

## THOUGHT QUESTIONS

1. Hubble proposed that his tuning-fork diagram might represent evolutionary changes with E galaxies gradually changing to S galaxies. What evidence based on the amount of interstellar matter and kind of stars in E and S galaxies indicates that this is very unlikely?

2. If you wanted to look at your mother when she was a baby, how far out in space would you need to be? (Assume you have a sufficiently powerful telescope and pretend that you can travel instantly to any distance.) How does this relate to astronomers studying young galaxies?

## PROBLEMS

1. A Cepheid variable in a nearby galaxy looks $10^6$ times fainter than an identical Cepheid in the Milky Way. The Cepheid in the Milky Way is 1000 parsecs away. How far away is the nearby galaxy?

2. A galaxy has a recession velocity of 13,000 kilometers per second. What is its distance in megaparsecs?

3. A galaxy has a rotation speed of 200 kilometers per second at a distance from its center of 10 kiloparsecs. What is its mass within this distance?

4. A galaxy has an angular diameter of 0.1 degree. Its distance is 10 megaparsecs. What is its true diameter?

## TEST YOURSELF

1. A large galaxy contains mostly old (Pop II) stars spread smoothly throughout its volume, but it has little dust or gas. What type galaxy is it most likely to be?

   (a) Irr

   (b) S

   (c) SB

   (d) E

   (e) All of the above are possible.

2. A galaxy's spectrum has a redshift of 28,000 kilometers per second. If the Hubble constant is 70 kilometers per second per megaparsec, how far away from Earth is the galaxy?

   (a) $10^6$ megaparsecs

   (b) 1000 megaparsecs

   (c) 28 megaparsecs

   (d) 50 megaparsecs

   (e) 400 megaparsecs

3. The Local Group

   (a) contains about forty member galaxies.

   (b) is a poor cluster.

   (c) is the galaxy cluster to which the Milky Way belongs.

   (d) is a rich cluster.

   (e) is all of the above except (d).

4. A spiral galaxy has a small bright central region, and its spectrum shows that it contains hot, rapidly moving gas. It is most likely a _____ galaxy.

   (a) barred spiral

   (b) Seyfert

   (c) radio

   (d) BL Lac

   (e) quasar

5. Astronomers believe that dark matter exists because

   (a) they can detect it with radio telescopes.

   (b) the outer parts of galaxies rotate faster than expected on the basis of the material visible in them.

   (c) the galaxies in clusters move faster than expected on the basis of the material visible in them.

   (d) it is the only way to explain the black holes in active galaxies.

   (e) Both (b) and (c) are correct.

## FURTHER EXPLORATIONS

Barnes, Joshua, Lars Hernquist, and Francois Schweizer. "Colliding Galaxies." *Scientific American* 265 (August 1991): 40.

Bartusiak, Marcia. *Through a Universe Darkly: A Cosmic Tale of Ancient Ethers, Dark Matter, and the Fate of the Universe.* New York: HarperCollins, 1993.

Beck, Sara C. "Dwarf Galaxies and Starbursts." *Scientific American* 282 (June 2000): 66.

Bothun, Gregory D. "Beyond the Hubble Sequence." *Sky and Telescope* 99 (May 2000): 36.

———. "The Ghostliest Galaxies." *Scientific American* 276 (February 1997): 56.

Disney, Michael. "A New Look at Quasars." *Scientific American* 278 (June 1998): 52.

Dressler, Alan Michael. *Voyage to the Great Attractor: Exploring Intergalactic Space.* New York: Knopf/Random House, 1994.

Ferris, Timothy. *Galaxies.* San Francisco: Sierra Club Books.

Ford, Holland, and Zlatan I. Tsvetanov. "Massive Black Holes in the Hearts of Galaxies." *Sky and Telescope* 91 (June 1996): 28.

Henry, J. Patrick, Ulrich G. Briel, and Hans Böhringer. "The Evolution of Galaxy Clusters." *Scientific American* 279 (December 1998): 52.

Hodge, Paul. "Our New! Improved! Cluster of Galaxies [The Local Group]." *Astronomy* 22 (February 1994): 26.

Kauffmann, Guinevere, and Frank Van den Bosch. "The Life Cycle of Galaxies." *Scientific American* 286 (June 2002): 46.

Kraan-Kortewey, Renée C., and Ofer Lahav. "Galaxies Behind the Milky Way." *Scientific American* 279 (October 1998): 50.

Macchetto, F. Duccio, and Mark Dickinson. "Galaxies in the Young Universe." *Scientific American* 276 (May 1997): 92.

Naeye, Robert. "Ghosts in the Cosmic Machine: The Answer to the Dark Matter Mystery May Lie in the Murky Realm of Ghostly Elementary Particles." *Astronomy* 24 (October 1996): 48.

West, Michael. "Galaxy Clusters: Urbanization of the Cosmos." *Sky and Telescope* 93 (January 1997): 30. See also related articles in the same issue.

### Website

Please visit the *Explorations* website at **http://www.mhhe.com/arny** for additional online resources on these topics.

## KEY TERMS

active galaxy, 486

barred spiral galaxy, 472

dark matter, 485

elliptical galaxy, 471

galactic cannibalism, 479

galaxy, 468

galaxy cluster, 495

gravitational lens, 493

Hubble constant, 482

Hubble law, 482

irregular galaxy, 472

jets, 487

Local Group, 495

megaparsec, 482

poor cluster, 495

quasar, 488

radio galaxy, 486

recession velocity, 482

redshift, 481

rich cluster, 495

Seyfert galaxy, 488

spiral galaxy, 471

supercluster, 498

## PROJECT

If the season allows, use a pair of binoculars or small telescope to find M31, the great galaxy in Andromeda (see "Looking Up at M31 & Persevs). What shape does it appear to have? Use the information in the text to estimate how long it took the light you are seeing to reach you. Note: M31 is visible from most of North America, Europe, and Asia in the evening from August through February. The nebula lies about mid-way along a line from the eastern edge of the Great Square in Pegasus to Cassiopeia. (Check a detailed star chart for its exact location.)

# Cosmology

The Hubble Ultra Deep Field. This image shows a portion of the sky in the constellation Fornax that you could easily cover with a pencil eraser held at arm's length. The image shows about 10,000 galaxies. Some of them are so far away that we are seeing them the way they appeared nearly 13 billion years ago, when the Universe was only about one-twentieth of its present age.

## Concepts and Skills to Remember

- The Universe is expanding.
- From the speed of its expansion, we can deduce its age.
- The Universe began in a hot, dense state.
- We detect the remnant radiation of its beginnings as the cosmic microwave background.
- Observations of small variations in the brightness of the cosmic microwave background allow astronomers to deduce the age of the Universe and its shape.
- The Universe is 13.7 billion years old.
- The Universe is flat and will expand forever.
- Measurement of the speed of recession of the most distant galaxies shows that the Universe's expansion is speeding up.
- Dark energy may be the cause of that acceleration.
- The Universe is composed overwhelmingly of dark energy and dark matter.
- Shortly after its formation, the Universe went through a very brief, very fast expansion stage known as inflation.

**?What Is This?**
(See chapter summary for the answer.)

**C**osmology is the study of the structure and evolution of the Universe as a whole. Cosmologists seek answers to questions such as, How big is the Universe? What shape is it? How old is it? How did it form? What will happen to it in the future?

## Concepts and Skills to Review

- Hubble law (16.2)
- Density (5.1)
- Newton's law of gravity (2.5)
- Red light has a longer wavelength than blue (3.1)
- Forces (P. 12)

Given our insignificant size in the cosmos, such questions may seem futile or even arrogant, but most cultures have tried to answer them. Many of those attempted answers have become part of humanity's religious heritage. For example, the King James version of the Bible, in the book of Genesis (1:1–3), states that

> In the beginning God created the heaven and the earth; and the earth was without form and void; and darkness was upon the face of the deep and the spirit of God moved upon the face of the waters. And God said, "Let there be light," and there was light. And God saw the light, that it was good.

These simple but powerful words, rich in symbols, give one picture of the birth of the Universe. But despite their power, they offer few specifics of what our Universe was like at its birth. In this final chapter, we will learn how contemporary scientists view the creation and evolution of our Universe, beginning with the observational discoveries that the Universe is expanding and that it is filled with a very low-energy background radiation. This expansion and radiation imply that the Universe was born about 14 billion years ago in a hot, dense, violent burst of matter and energy called the **Big Bang**. Within the last few decades, cosmologists have sought to extend our knowledge of the Universe to the earliest moments of the Big Bang and have discovered that the entire Universe we see today may have expanded, during an instant known as the "inflationary stage," out of a volume smaller than a proton.

## 17.1    OBSERVATIONS OF THE UNIVERSE

In the early years of the twentieth century, most astronomers thought that the Universe consisted of the Milky Way galaxy and a number of small companions. Immanuel Kant, the German natural philosopher, suggested in the eighteenth century that the Milky Way was merely one of a multitude of galaxies, but his views were not widely accepted until the 1920s, when astronomers showed that other galaxies were millions of light-years from the Milky Way and comparable to it in size. About the same time, astronomers such as Hubble showed that galaxies were moving away from one another—in short, the Universe is expanding.

### Distribution of Galaxies

Pictures of the night sky show that distant galaxies are scattered across it approximately uniformly, as shown in figure 17.1A. This is in sharp contrast to how stars are distributed in our Milky Way: its stars are concentrated in a narrow band from which we deduce that our galaxy is a disklike system. The smooth distribution of galaxies leads astronomers to conclude that the Universe is filled approximately uniformly with galaxies, in much the way that raisins are distributed uniformly through a fruitcake.* Thus our first deduction is that the Universe is approximately the same in all directions, as shown in figure 17.1B. In addition to this uniformity in space, the Universe also exhibits a uniformity of motion.

---

*We ignore here the clumping of galaxies into groups and clusters.

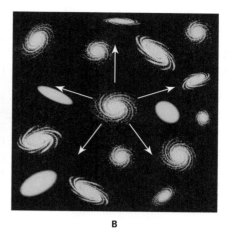

A                                                                                              B

**FIGURE 17.1**

An all-sky plot of the positions of over a million galaxies detected by 2MASS, an infrared survey of the sky. Each white dot represents a galaxy, or in clusters and superclusters of galaxies, they merge to form brighter areas. The blue shaded portion of the diagram shows where the plane of the Milky Way crosses through this map.

## Motion of Galaxies

In chapter 16, we learned that nearly all galaxies have their spectral lines redshifted, and their shift to a longer wavelength implies that most of the galaxies are moving away from us. Moreover, the redshift increases with distance, $D$, in such a way that the recession velocity, $V$, of a galaxy obeys the Hubble law:

$$V = HD$$

where H is the Hubble constant.

The motion of galaxies away from us is caused by the expansion of the Universe. This expansion is the result of space itself expanding. It is *not* like a bomb burst flinging fragments in all directions from a single central location. Instead, it is like the motion of buttons glued on a balloon when the balloon is inflated (fig. 17.2). Or, returning to the fruitcake analogy, it is like a giant, raisin-filled fruitcake expanding as it bakes. However, unlike a fruitcake, the Universe has no center to its expansion. Every point moves away from every other point.

The motion of both the buttons on the balloon* and the raisins in the cake is caused by the motion of the medium in which they are set. So too for galaxies: space itself is expanding, and galaxies are carried apart by that expansion, not by any motion of their own. However, although space expands, the objects themselves—galaxies, stars, planets, and so forth—do not. Gravity and other forces maintain such material bodies at a constant size. On the other hand, light waves *are* altered by the expansion of space, and that expansion creates the redshift of the galaxies that we observe. As a light wave from a galaxy moves through space, it is stretched as the space expands, just as a kinked curl of carrot in a piece of Jell-O is stretched if the ends of the Jell-O are pulled. If you put a series of marks on a rubber band and stretch it, you will see a similar effect—the space between the marks increases.

## Age of the Universe

**FIGURE 17.2**

Representations of the expansion of the Universe.

The above model of the Universe as an expanding space filled uniformly with galaxies allows cosmologists to estimate its age and to calculate conditions near the time of its

---

*You might wonder in the balloon analogy what lies inside of the balloon and what it is expanding into. Because we are describing space-time and because the balloon's surface represents the three dimensions of *space*, the balloon is expanding in *time*. Inside it lies the past. Outside it lies the future.

birth. Alexander Friedmann, a Russian scientist, made such calculations in the 1920s, but his work attracted little attention. Not until 1927, when Abbé Georges Lemaître, a Belgian cosmologist and priest, independently made similar calculations, did astronomers appreciate that conditions at the birth of the Universe could be deduced from what is observed today.

Lemaître argued that since galaxies are moving apart now they must in the past have been closer together. He went on to propose that at one time in the distant past, all matter in the Universe was packed into a region approximately the size of the present Solar System. With so much mass in such a relatively small volume, the density of the Universe was enormous. In fact, at that remote time in the past, the density was so immense that the protons and neutrons that now compose the far-flung stars and galaxies were touching, just as today the protons and neutrons within a single atom touch. For that reason, Lemaître called this dense stage of the early Universe the Primeval Atom.

Lemaître argued that the Primeval Atom "decayed" much like an ordinary radioactive atom such as uranium decays. Moreover, he identified this decay with the birth of the Universe, and he calculated how long ago that occurred. Astronomers no longer accept Lemaître's theory of the birth of the Universe, but they do think that it was created from a highly compressed state in the distant past, and they measure its age from that time. To determine that age, we can make a simple calculation, as explained below, but an analogy may help us better understand this calculation.

Suppose a car leaves a starting point in a long race and drives steadily at 60 miles per hour. Suppose the driver looks at the mileage and notes that the car has traveled 120 miles. How long ago did the car leave the starting point? You can easily figure that out by dividing the distance traveled (120) by the speed (60), to get 2 hours. So, too, with measuring the age of the Universe. Astronomers pick some galaxy at a known distance from the Milky Way, D. If they know the galaxy's speed, V (from the Hubble law, for example), then they can determine that the two galaxies began moving apart at a time, $t = D/V$. We will take that time as the beginning of the Universe. A little math (worked out in Extending Our Reach: "Estimating the Age of the Universe" and sketched in figure 17.3) then shows that

$$t = 1/H.$$

$D = Vt$

Therefore, $t = \dfrac{D}{V}$

But according to the Hubble law,
$$V = DH$$

Therefore, $t = \dfrac{D}{DH} = \dfrac{1}{H}$

**FIGURE 17.3**
Estimating the age of the Universe from the recession of one galaxy from another galaxy.

## EXTENDING OUR REACH

### ESTIMATING THE AGE OF THE UNIVERSE

To calculate the age of the Universe, $t$, we proceed as follows: let two representative galaxies be a distance, $D$, apart and separating with a velocity, $V$. For simplicity, we will assume that $V$ has remained constant. Then in the time $t$, the age of the Universe, they will have separated by a distance given by

$$D = Vt$$

$D$ = Distance between two galaxies
$V$ = Velocity at which galaxies are separating
$t$ = Age of the universe

We can find how long ago they began moving apart by dividing both sides by $V$. This gives

$$t = \frac{D}{V}$$

Thus, in this simple model, the age of the Universe is $D/V$. The Hubble law, however, relates a galaxy's velocity to its distance by $V = DH$. Therefore, substituting this expression for $V$ into the expression for the age of the Universe, we find

$$t = \frac{D}{V}$$
$$= \frac{D}{DH}$$
$$= \frac{1}{H}$$

where H is the Hubble constant.

To evaluate $t$ in years, we must know H, the Hubble constant. However, we cannot simply insert the value of H in the expression for $t$ because H is expressed in units of kilometers/second/megaparsec. If we were to use H in these units, we would get an answer for the age of the Universe in equally odd units. We must therefore express H in units of 1/seconds by converting megaparsecs to kilometers so that the distance units in H cancel out. One megaparsec is $3.1 \times 10^{19}$ km, so if the Hubble constant is 70 km/second/Mpc, this gives

$$H = 70 \text{ km/second}/(3.1 \times 10^{19} \text{ km})$$

$$= \frac{70}{3.1} \times 10^{-19} \text{ seconds}^{-1}$$

$$= 23 \times 10^{-19} \text{ seconds}^{-1}$$

$$= 2.3 \times 10^{-18} \text{ seconds}^{-1}$$

Therefore, $1/H = 4.3 \times 10^{17}$ seconds. Because there are about $3.2 \times 10^7$ seconds per year, $1/H = 1.4 \times 10^{10}$ years. Thus, we find that the age of the Universe is about 14 billion years.

Astronomers find this value for the age of the Universe reassuring because it is much older than the Earth (4.6 billion years, as we saw in chapter 5) and is also older than the oldest known stars (about 13 billion years). Something would be seriously wrong with our model if the Universe turned out to be younger than its stars, a clearly impossible situation.

## The Cosmic Horizon

The age of the Universe limits the distance we can see. Because light requires time to move from one point to another, we can see no farther than the distance light travels in a time span equal to the age of the Universe. If the Universe did not expand, calculating the distance that light travels in that time span would be simple. For example, if the Universe were not expanding and were 14 billion years old, we could see nothing farther than 14 billion light-years away (fig. 17.4). The Universe may well extend

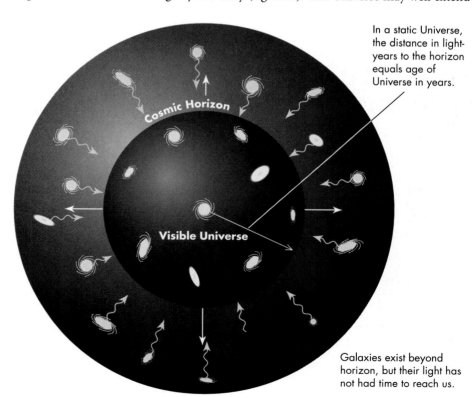

In a static Universe, the distance in light-years to the horizon equals age of Universe in years.

Cosmic Horizon

Visible Universe

Galaxies exist beyond horizon, but their light has not had time to reach us.

**FIGURE 17.4**

We cannot, even in theory, see beyond the cosmic horizon. Light takes more time than the age of the Universe to reach us from there. (The straight arrows indicate that the horizon moves outward as the Universe ages.)

beyond that distance, but the maximum distance that light can travel in the Universe's age therefore defines a **cosmic horizon.** Astronomers call the space within the horizon the visible Universe.

However, because the Universe is expanding, it turns out that there is no simple expression for the distance to the cosmic horizon, or even a distant galaxy, for that matter. A simple example will show the problem. Recall that when we see a galaxy, we see it the way it was at the time when its light left it, perhaps millions or even billions of years ago. But when it emitted its light, the galaxy was closer to us than it is now because the Universe was smaller. Furthermore, since the time when the light we now receive was emitted, the galaxy has moved yet farther from us, carried by the expansion of the Universe. Thus, which distance do we apply to the galaxy? Its distance when the light was emitted? Its distance when the light was received? Some other distance? To avoid such confusion (and for other reasons), astronomers often use the term *"light travel time distance"* when speaking of the distance to remote galaxies. The light travel time distance measures how far the light from the galaxy (or other object) would have traveled if the Universe were not expanding. This distance is simply how far the galaxy's light has traveled in the time since it was emitted. With that definition of distance, we can then say that a galaxy, which we observe today and whose light was emitted 10 billion years ago, is 10 billion light-years away from us. Defining distance in this way allows us to be more precise when we say that the cosmic horizon is approximately 14 billion light-years away. Thus, we will take the size (radius) of the visible Universe to be approximately 14 billion light-years.

## Are We at the Center of the Universe?

The recession of distant galaxies from the Milky Way often leads to the mistaken belief that the Milky Way is at the center of the Universe. Because space itself is expanding, no matter where you are in the Universe, galaxies will be moving away from you. A simple example illustrating this is shown in figure 17.5, showing galaxies distributed along a line. Each galaxy is separated from its nearest neighbor by 10 megaparsecs. Suppose the Milky Way is the third galaxy in the line and that galaxies A, B, and C obey the Hubble law, with each receding at a velocity given by $V = 70D$. That is, A recedes from us at 700, B at 1400, and C at 2100 kilometers per second. Next, suppose we could communicate with an alien on galaxy C and ask what it sees. It would see us receding at 2100 kilometers per second, A receding at $2100 - 700 = 1400$ kilometers per second, and B receding at $2100 - 1400 = 700$ kilometers per second. Thus, the alien would see galaxies receding from it exactly the same way we

**FIGURE 17.5**
A line of galaxies illustrating that in a Universe that obeys a Hubble law of expansion ($V = HD$), an observer on any galaxy sees the same law.

see galaxies receding from us. Furthermore, the galaxies seen by the alien will obey the alien's version of the Hubble law.

A similar argument can be made for an observer on any of the galaxies. Thus, for a Hubble law in which the velocity increases proportional to the distance, *every observer, no matter where in the Universe he/she/it is, will see the Universe expanding the same way.* The Universe has no preferred "center" of expansion if it expands uniformly. In fact, observers in another galaxy will have their own cosmic horizon and see some parts of the universe that we cannot see. Astronomers describe this lack of a preferred location as the **cosmological principle.** It is a statement of cosmic modesty: there are no special positions. We are not at the center of the Universe, nor is anyone else. In fact, the notion of a "center" to the Universe is meaningless, as we will discuss further in section 17.3.

## Olbers' Paradox

The Universe's finite age can also be demonstrated by the simple experiment of stepping outside at night and noting that the sky is dark. This relation between the darkness of the night sky and the structure of the Universe is known as **Olbers' paradox,** after the German astronomer Heinrich Olbers, who in 1823 presented an argument that if the Universe extends forever and has existed forever, the night sky should be bright. The paradox arises from following reasonable premises (or so it seemed at that time) to a clearly nonsensical conclusion. The argument is as follows:

Suppose you stand in a small grove of trees and look out between them to the surrounding landscape. If the grove is small, your line of sight will be blocked by trees in a few directions but will pass between the trees in most directions (fig. 17.6 A–C). If the grove

Although the paradox bears Olbers' name, more than two centuries earlier, Kepler pointed out that in an infinite Universe, the night sky should be bright, and he therefore concluded the Universe was finite in size.

In a small grove of trees, only a few block your view. Lots of space to view between trees.

**A**

In a larger forest, more distant trees block your view. No open space visible between trees.

**B**

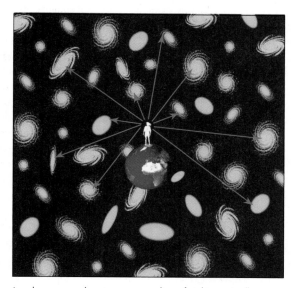

In a large enough universe, every line of sight eventually encounters a star within a galaxy

**C**

**FIGURE 17.6**

(A) An observer in a small grove of trees can see out through gaps between them to the surrounding countryside. (B) In a larger forest, distant trees block the gaps so that no matter where you look your line of sight ends on a tree. (C) So, too, in a universe that extends forever and has existed forever, your line of sight will always end on a star. Therefore, the night sky should be bright.

is larger, more distant trees in it will block your view in what were previously clear gaps. Then, no matter what direction you look, your line of sight is intercepted by a tree.

Now suppose that, rather than looking between trees, you look out into space through the galaxies and stars that compose the Universe. If space extends sufficiently far and is populated with galaxies and stars such as we see around us in the Local Group, then no matter in what direction you look, your line of sight will ultimately intercept a star just as in a sufficiently large forest your line of sight ultimately hits a tree. Therefore, the sky should be covered with stars, each glowing brilliantly, with no dark spaces between them. Even though such distant stars are very faint to us, there are so many of them that their collective brightness should be very large. In other words, the night sky should not be dark. This argument must be fallacious because the most obvious observation one can make about the sky is that it is dark at night. That is the paradox: the night sky should be bright, but it is not. Where, therefore, is the error in our reasoning?

The paradox is avoided if there are no stars beyond some distance. That is, just as we can see out of the woods if there are only a few trees, so too we will see a dark night sky if there is only a limited number of stars. The paradox is also avoided if we account for the time it takes light from the stars to reach us. Light from a star 1 billion light-years away takes a billion years to get here. Thus, to see such light requires that the Universe has existed long enough for the light to arrive. That is, if stars have not been shining for billions of years, the night sky will be dark. A final way to avoid the paradox is that light from remote stars is redshifted by the expansion of the Universe. The redshift reduces the light's energy (recall that longer-wavelength radiation carries less energy) and thereby reduces the brightness of the most distant stars and galaxies. Thus, an expanding Universe can make the night sky dark.

Today, astronomers think that the finite *age* of the Universe implies a finite cosmic horizon, making the night sky dark. Stars have simply not existed long enough for their radiation to flood space and make it bright. Moreover, they will never be able to make the night sky bright because they will burn up all their fuel first. In another way, however, the darkness of the night sky results from the limits of our senses. If we could see radio waves, the night sky would glow uniformly, as we shall see below.

## The Cosmic Microwave Background

Lemaître's idea that the young Universe was extremely dense led George Gamow, a Russian astrophysicist who fled to the United States before World War II, to realize that it must also have been extremely hot. His argument was based on the observation that compression heats a gas, and so the enormous compression of the early Universe suggests that it too was enormously hot—in fact, at least 10 billion Kelvin. Gamow went on to note that conditions in the young Universe would have been similar to those in the core of stars, and so he theorized that heavy elements might have been created in the young Universe just as they are in stars. Gamow proposed that at its birth, the Universe was composed only of neutrons, which rapidly decayed into protons and electrons. Some of the protons then fused to form helium and heavier elements. Gamow's ideas were developed further by two of his collaborators, the American physicists Ralph Alpher and Robert Herman. In 1948, they predicted that as the young Universe expanded and cooled from its hot dense state, the radiation it emitted would survive to the present, though in altered form. To understand how that alteration occurred, we need to look at the properties of the radiation in greater detail.

According to Wien's law, a dense gas radiates most intensely at a wavelength determined by its temperature, as discussed in chapter 3. In particular, hot matter radiates most strongly at short wavelengths. Because the early Universe was so hot, the wavelength of maximum emission turns out to be very short, less than a millionth of a nanometer. Such short-wavelength radiation, however, is typical only

**FIGURE 17.7**

Spectrum of the cosmic microwave background. The shape of the spectrum matches perfectly a blackbody with a temperature of 2.7 K. Notice the wavelength at which it peaks. Red dots show data points schematically.

in the young, hot, dense Universe. As space expands, it stretches the wavelength of radiation traveling through it, increasing its wavelength. Recall the analogy used earlier of a carrot curl in a block of stretched Jell-O. Calculations show that as the Universe expanded to its present size, the wavelength of the radiation emitted at its birth was stretched from fractions of a nanometer to millimeters, a radio wavelength. Thus, Alpher and Herman showed that the Universe today should be filled with faint, long-wavelength radiation created when the Universe was very young. Their prediction was confirmed in 1965, when Arno Penzias and Robert Wilson of the Bell Telephone Laboratories accidently discovered cosmic microwave radiation having just that property.

Penzias and Wilson were trying to identify sources of background noise on telephone satellite links and noticed a weak radio signal with the unusual property that its strength was constant no matter in which direction of the sky they pointed their detector. They soon demonstrated that the signal came not from isolated objects such as stars or galaxies but rather from all space. Initially mystified by the signal, they soon discovered that scientists at Princeton University had repeated Alpher, Herman, and Gamow's calculations and were searching for the predicted radiation. Once aware of that work, Penzias and Wilson realized that the background interference they had found was radiation created in the young Universe. Their discovery of the **cosmic microwave background (CMB)**, as it is now called, won for them the Nobel Prize in physics in 1978.

Figure 17.7 shows measurements of the intensity of the cosmic microwave background (CMB, for short) and demonstrates that it has a perfect blackbody spectrum, as theory predicts. Moreover, the radiation is most intense at about 1 millimeter ($10^6$ nanometers), in the microwave part of the radio spectrum. From this wavelength and using Wien's law ($\lambda = 3 \times 10^6/T$, where $\lambda$ is the wavelength in nm where the emission is strongest and $T$ is the temperature in K), the temperature that describes the radiation turns out to be about 3 K, close to the value predicted nearly 20 years earlier. A more precise measurement of $\lambda$ shows that the temperature that describes the CMB is 2.725 K, only a little warmer than absolute zero, the lowest temperature anything can have.

To understand how expansion of the Universe has cooled the CMB, we again use the relation between a body's temperature and the wavelength at which it radiates most intensely: Wien's law. We know from Wien's law that cool objects emit most strongly at long wavelengths and hot objects emit most strongly at short wavelengths.

We described earlier that as the Universe expands, it stretches the wavelengths of the radiation and makes them longer (fig. 17.8). Thus, the radiation becomes more like that emitted by a cool body rather than a hot one. We therefore say that the radiation has cooled from the high temperature at which it was born in the early Universe to the mere 2.7 K we observe today.

It is ironic that the existence of this radiation—such important proof that the Universe was born in a hot, dense state—was originally only a minor part of Alpher, Herman, and Gamow's theory of the young Universe. Their original goal was to explain the formation of the heavy elements. Today, however, we know that most heavy elements are made in stars as they evolve. Nevertheless, the idea that conditions in the early Universe might allow the synthesis of some elements is still important because it allows astronomers to determine conditions in the early Universe in yet another way.

As space expands it "stretches" the light waves moving through it, increasing their wavelength, $\lambda$.

$\lambda_0$

$\lambda_1$

Short wavelength implies *hot*.      Long wavelength implies *cool*.

**FIGURE 17.8**

As space expands, it stretches radiation moving through it, making the wavelength longer. Because longer wavelengths are associated with cooler objects, the stretching of the radiation has the effect of cooling it.

## Composition of the Oldest Stars

Gamow's hypothesis that helium could form in the young Universe is supported by more recent theory as well as by observations. According to modern theory, however, the young Universe consisted of protons, neutrons, and electrons, rather than just neutrons, as Gamow believed. Nevertheless, his basic idea is correct: in its youth, the Universe was hot and dense enough for nuclear reactions to fuse some of the protons and neutrons into helium. Moreover, current calculations show that the amount of helium formed depends on how hot and dense the Universe was and how long those conditions lasted. These conditions are determined by how rapidly the Universe expands. Too rapid an expansion cools the Universe quickly and allows too much helium to form. Too slow an expansion leaves the Universe hot too long, and too little helium forms. Current estimates of the expansion rate imply that when the Universe had expanded and cooled to about one billion kelvin, about 24% of the matter should have been transformed into helium.

This prediction can be tested by measuring the amount of helium in the oldest stars in the Milky Way and other galaxies. Those stars formed from the gas left over from the Big Bang, and therefore they have the same abundance of helium as was produced in the Big Bang. Although some helium has been produced deep in their core by the fusion of hydrogen during their long life, the composition of the outer layers of these ancient stars has not been altered. Thus, the abundance of helium we measure in those layers is essentially the helium abundance produced by the Big Bang. What is that abundance? It turns out to be 24% helium, in excellent agreement with the Big Bang model. This is not, however, the abundance of helium in stars like the Sun, which formed well after the Milky Way's birth. These later-generation stars have formed from interstellar gas that has been enriched in helium from earlier generations of stars that have exploded or lost mass in other ways. Consequently, the Sun—born some 9 billion years after the Big Bang—contains about 27% helium.

Helium is not the only element produced during the Big Bang. Tiny amounts of the extremely light elements lithium and beryllium are produced too and their abundance also agrees well with predictions based on the Big Bang model, providing further strong support that the model is correct.

## Deductions from Basic Observations of the Universe

The recession of galaxies and their uniform distribution leads us to conclude that the Universe was born in a cataclysmic event about 14 billion years ago. At that time, the Universe must have been much hotter and denser than now.

Confirmation that this hot, dense state existed comes from (1) the cosmic microwave background—the radiation created during that hot stage but cooled by the Universe's subsequent expansion—and (2) from the hydrogen-helium mix found in old stars. The high temperature and rapid expansion from a hot, dense state led the English astronomer Sir Fred Hoyle to jokingly refer to this theory of the birth of the Universe as the "Big Bang," a name that has stuck despite his sarcasm. Hoyle long championed an alternative "steady-state" cosmology, proposed by him and two other British astronomers, Hermann Bondi and Thomas Gold, in the early 1950s. In the steady-state universe, new matter forms continually out of nothing to create new galaxies in the empty, expanding space between older galaxies. This keeps the average density of the Universe constant so the Universe on the average looks the same at all times. This unchanging appearance (which gives the steady-state Universe its name) is unlike that of the Big Bang theory, where space becomes progressively emptier as the galaxies move apart. But many observations (especially the cosmic microwave background) now persuade most astronomers that the Big Bang model for the Universe and its predictions match our observations of the Universe far better than do the predictions of the steady-state theory.

## 17.2 EVOLUTION OF THE UNIVERSE: EXPANSION FOREVER OR RECOLLAPSE?

INTERACTIVE

*Cosmology and the cosmos' fate*

The Universe is currently expanding. But will it expand forever, or will it sometime stop expanding and contract? This question is fundamental to the nature of the Universe, for according to the Big Bang theory, all the hydrogen in the Universe was created in the Big Bang. Thus, if the Universe expands forever, its stars will one by one consume their hydrogen and burn out, leaving the Universe a black, empty space—a dismal prospect. However, recollapse of the Universe means that all the objects and atoms (and us) within it will be compressed into what some astronomers have called the "Big Crunch." The Big Crunch might lead to a rebirth of the Universe, perhaps leading to a new Big Bang. Such a universe might oscillate, expanding and collapsing repeatedly.

To determine which fate awaits the Universe, we need to understand what factors affect its expansion or recollapse. But before we look at details, a simple analogy may help. Suppose you throw a stone into the air. As the stone moves upward, Earth's gravitational force slows its rise and eventually stops it so the stone falls back to Earth. On the other hand, if you could throw the stone faster than the escape velocity, it would rise forever. Thus, the stone's behavior depends on the strength of gravity and the upward impulse given to it. So, too, with the Universe.

Near the time of its birth, some process began the Universe's expansion. Since that time, the gravitational force between its galaxies and other contents have generally acted to slow its expansion. If the gravitational force is weak or if the impulse of expansion is strong, we might expect the Universe to expand forever. On the other hand, if the gravitational force is strong or the initial impulse is weak, we might expect that the expansion will stop and the Universe will collapse.

Is there some way we can measure the relative strength of these effects for the Universe? The answer is yes. We can compare the gravitational energy working to hold the Universe together with the energy of its expansion. If gravity dominates, the expansion will stop. If the expansion energy dominates, the expansion will continue forever (fig. 17.9). Thus, we can learn (at least in theory) whether the Universe will expand forever or recollapse by seeing if the Universe has a gravity strong enough to stop its expansion.

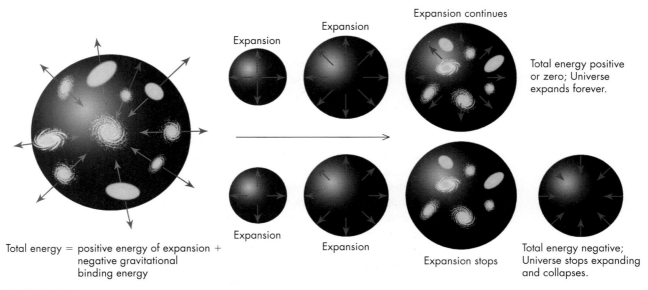

Total energy = positive energy of expansion + negative gravitational binding energy

Expansion

Expansion

Expansion continues

Total energy positive or zero; Universe expands forever.

Expansion

Expansion

Expansion stops

Total energy negative; Universe stops expanding and collapses.

**FIGURE 17.9**
A sketch of how gravity and the energy of expansion determine the behavior of the Universe.

We saw in earlier chapters that the gravitational force produced by an object depends on its mass—a result we can extend to the Universe itself. In particular, if astronomers know the Universe's mass, they can calculate the gravitational force it exerts to slow its expansion. However, to make such a calculation, it is easier to use the Universe's density (its mass divided by its volume) rather than its mass alone. What density do astronomers find? We will answer that question in the next section and also see there what effect this density has on the Universe's expansion.

## The Density of the Universe

To find the density of the Universe, astronomers choose a volume of space and add up the mass of all the material in it. In doing so, they must be sure to include the mass of dark matter as well as the visible matter—including dark matter is important because, as we have seen, it greatly outweighs the visible matter. They then divide the mass by the volume to obtain the density. When such a calculation is made, the density of the combined visible and dark matter in the Universe turns out to be about $3 \times 10^{-30}$ gm/cm$^3$.

To determine whether the Universe will expand forever or recollapse, astronomers next compare the observed density with a theoretically calculated **critical density.** If the observed density is greater than the critical density, the Universe will recollapse; if it is less, the Universe will expand forever (fig. 17.9). It turns out mathematically that the critical density $\rho_c = 3H^2/(8\pi G)$, where H is the Hubble constant, and G is the gravitational constant. Notice that the critical density depends on the Universe's rate of expansion (the Hubble constant) because if the expansion rate is very large, the critical density must also be large in order to have enough gravity to stop the expansion.

When we compare the observed and critical densities, what do we find? If we use the value of H determined by observation (70 kilometers per second per megaparsec), we find that the critical density is about $10^{-29}$ grams per cubic centimeter, the equivalent of six hydrogen atoms per cubic meter. Thus, the actual density of matter is approximately one-third the critical density, which implies that the Universe should expand forever.

Although astronomers were pleased that the question of whether the Universe would expand forever or recollapse appeared to have been settled, they were troubled by the value found for the Universe's density. According to a leading model for how the Big Bang began, the actual density should *exactly* equal the critical density, rather than be approximately one-third of that value, as was measured. Was this model, therefore, wrong? Or has some component of the Universe been overlooked? Before concluding that the model was flawed, astronomers wanted to confirm their conclusion that the Universe will expand forever. One way to do that would be to compare the present expansion speed of the Universe to its speed in the past. If the speed has not changed, it would strengthen the case for endless expansion. But how might such a comparison be made?

## Dark Energy

Astronomers are fortunate in that they, in fact, have such a way to compare present and past expansion speeds. When we look at very distant galaxies, for example, galaxies billions of light-years away from us, we see light that has taken billions of years to reach us. That is, we see the galaxies as they were billions of years ago. We can, therefore, measure how fast those distant galaxies were separating from each other compared to galaxies now. However, observing such distant galaxies is difficult because their great distance makes them very faint. But over the past decade astronomers have observed supernova blasts in a number of remote galaxies from which they can find both the distance and speed of these systems. This allows them to plot a distance–velocity diagram, as shown in figure 17.10. The nearby galaxies, which are receding at low velocity (lower left-hand part of the diagram), lie along a straight line, as predicted by the ordinary Hubble law. However, the more remote galaxies (upper right) clearly do not follow that law. Rather, the distant galaxies lie *below* the line followed by nearby galaxies. This shows that expansion in the past was slower than it is

Astronomers use a shorthand notation to indicate how close the observed density is to the critical density with a constant, $\Omega$, defined as the value of the actual density divided by the critical density. With this notation, if $\Omega > 1$, the Universe will recollapse. If $\Omega \leq 1$, the Universe will expand forever.

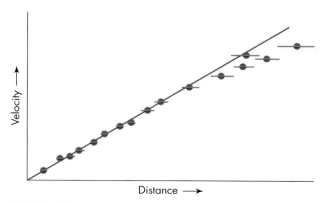

**FIGURE 17.10**

A graph of the distance and recession velocity of distant galaxies. The red line shows how the speed should change if the Universe has expanded at a constant speed. The thin blue lines show the distance uncertainty for each galaxy. The graph is schematic and based on observations by S. Perlmutter et al. Note that the distant galaxies tend to lie below the red line.

now. That is, the expansion of the Universe seems to have *sped up!* How is this possible? Gravity's pull of one galaxy on another should slow the expansion.

One possible explanation* of this seemingly odd result comes from Einstein's early efforts to understand the structure of the Universe. In his general theory of relativity, Einstein proposed that matter and energy create the force of gravity and bend space to give the Universe its structure. From his theory, he devised and solved equations to describe this structure. His solution involved a constant whose value he had no way of knowing. However, if he made the constant zero, the Universe would either collapse under gravity's influence or expand forever.

At the time of this work, astronomers had not yet discovered that the Universe is expanding. Einstein thus thought the Universe must be at rest over all. To achieve that, his equations needed a cosmic repulsive force to offset gravity's attraction. Making the constant different from zero had exactly this effect, and the term in his equations became known as the **cosmological constant.**

According to theory, the repulsion effect of the cosmological constant is far too weak to detect in the laboratory or even on the scale of the Solar System. Only over distances of millions of light years does it become powerful enough to offset gravity and prevent the Universe's collapse.

Within a decade of Einstein's introduction of the cosmological constant, astronomers discovered that the Universe was in fact expanding and so Einstein thought the constant was not needed and that he should have set it equal to zero all along. The discovery that the Universe's expansion is speeding up makes some astronomers think once more that the cosmological constant is not zero. If its value is not zero, it has major implications for our understanding of the Universe. For example, if the cosmological constant is not zero, it contributes an energy to the Universe. We cannot directly detect this energy, and so astronomers call it **dark energy,** analogously with dark matter. But although it is invisible, dark energy's effect on the Universe appears to be immense. In addition to creating the repulsive force that speeds up the recession of galaxies and will cause the Universe to expand forever, dark energy contributes a colossal mass to the Universe. Using the relation $E = mc^2$ of relativity theory, astronomers find that dark energy contributes approximately 70% of the Universe's total mass—two and a half times more than that of all the other mass in the Universe, dark and observable, combined. Thus, the Universe appears to be *mostly* dark energy.

## SCIENCE AT WORK

### THE UNIVERSE'S FATE

For many years astronomers have used the terms *open* and *closed* to describe the shape of the Universe and whether its expansion will continue forever or not. The discovery that the cosmological constant ($\Lambda$) is probably not equal to zero greatly complicates the use of these two seemingly simple terms.

If $\Lambda = 0$, then an open universe expands forever and a closed universe recollapses. If $\Lambda \neq 0$, that simple linkage does not hold. *Open* and *closed* still describe the curvature of the Universe but not its fate. An open universe may recollapse and a closed universe may expand forever. It is too bad that we lose that easy way to describe the Universe's behavior, but the Universe doesn't care about our convenience.

*There are other explanations involving exotic and untested theories of physics.

## 17.3     THE SHAPE OF THE UNIVERSE

*The future of expansion*

What shape is the Universe? Here, too, the Universe's density plays an important role, as Einstein showed with his General Theory of Relativity. According to that theory, the mass and energy in the Universe bend its space. Such bending is not just science fiction. For example, we saw in chapter 14 that the intense gravity of a black hole bends space to prevent light from escaping. Similarly, we saw in chapter 16 that the gravity of a galaxy cluster bends space in the cluster and may create a gravitational lens. So, too, gravity bends and shapes the Universe.

How can we picture such **curved space?** Our home planet Earth offers a simple example. Athletic fields and parking lots appear flat and, if we ignore hills, the surface of the Earth certainly *looks* flat. We know, however, that if we could walk in a "straight" line along this "flat" surface, we would return to our starting point. We are therefore like an ant that accidently crawls into a garden hose. The unbent hose represents a flat, uncurved space, and so, if the ant crawls in a straight line, it will reach the hose's end and emerge into the garden again. However, suppose we pick up the hose with the ant still in it and bend the hose so that its ends meet. Now the ant could crawl forever and would never reach the hose's end.

A similar situation exists for a Universe curved around on itself. Such a space looks as if it extends forever in all directions, but if we could travel in a spaceship along a "straight" line, we would eventually return to our starting position. Thus, we would have moved through space forever without coming to an end point. A universe with this property is said to have **positive curvature** and is "closed" because it has no "outside." We can infer that the Earth has a "positively curved surface" because if we travel in a straight line on it, we return to our starting point. Positively curved surfaces have other properties as well: lines that start out parallel eventually meet if they extend far enough, and the sum of the interior angles of a triangle drawn on such surfaces is greater than 180°, as illustrated in figure 17.11A.

The Universe might have other types of curvature, however. It might be **flat** (that is, have no curvature) or have **negative curvature.** Flat space is what most people picture when they think of space. In it, parallel lines do not meet, and a triangle's interior angles add up to exactly 180°, as illustrated in figure 17.11B. Negatively curved space is harder to visualize, but you can think of it as being bent into a saddle-like shape, as shown in figure 17.11C. Such curvature corresponds to bending space one way along one line and the opposite way along another line. In negatively curved space, the sum of the interior angles of a triangle is less than 180°.

**FIGURE 17.11**

Figures illustrating how the shape of a surface determines the geometry of a triangle on it. (A) The interior angles of a triangle on the surface of the Earth add up to more than 180°. (B) The interior angles of a triangle on a flat surface add up to 180°. (C) The interior angles of a triangle on a saddle-shaped surface add up to less than 180°. The top row of figures illustrates that a distant object will therefore appear larger if the curvature is positive (A) and smaller if the curvature is negative (C).

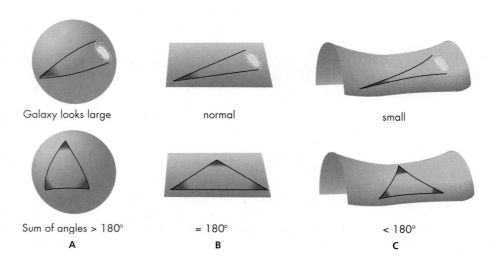

Galaxy looks large          normal          small

Sum of angles > 180°          = 180°          < 180°

A          B          C

Can we measure the curvature of space and determine whether it is positive, negative, or flat? Yes. After nearly half a century of attempts that gave only inconclusive results, astronomers at last have a method for measuring the curvature of the Universe that seems to work with good accuracy. This method uses observations of the small temperature fluctuations of the cosmic background radiation. Moreover, it gives information not only about the curvature of the Universe but also about the Universe's age and the amount of ordinary matter, dark matter, and dark energy. To understand the method, it may help to think of a simple experiment.

Imagine blowing across the mouth of an empty bottle. As you blow, you make the air in the bottle vibrate and so generate sound waves. The pitch (tone) of sound waves is set by their wavelength, with shorter wavelengths having a higher pitch. You can demonstrate that effect by putting some water into the bottle and blowing across it again. The pitch of the sound will now be higher. This happens because the water gives the sound waves less space in the bottle and makes their wavelength shorter. Thus, the length (pitch) of the sound waves gives us a clue to the amount of air in the bottle.

In earlier chapters, we saw how scientists used waves to probe objects. For example, seismic waves allow us to probe the interior of the Earth and waves in the Sun's atmosphere allow us to determine properties of the Sun's interior. We can use similar techniques to probe the Universe.

As the Universe expanded in earliest times, waves formed in its hot gas. Those waves took the form of variations in its density and temperature. We can observe traces of those waves today as variations in the brightness of the cosmic background radiation across the sky. You can see those variations as the patchiness in figure 17.12, which shows a map of the sky made by a satellite orbiting the Earth. By analyzing the size of these brightness variations, astronomers can deduce many of the properties of the young Universe, including its shape (flat or curved), its rate of expansion (the Hubble constant), and its density. These observations show that, to high accuracy, the Universe is flat. This in turn implies that the Universe must have an overall density equal to the critical density. In addition, the observations show that the age of the Universe is 13.7 billion years (to within a few percent). Finally, when combined with other measurements, the cosmic microwave background data allows astronomers to deduce the mix of the various components of the Universe: ordinary matter (stars and gas); nonluminous ordinary matter (dead stars, black holes, gas as yet unseen); nonordinary dark matter (as-yet-unknown subatomic particles such as WIMPs); radiation; and dark energy. Figure 17.13 shows the fraction of the Universe composed of each of these components.

**FIGURE 17.12**

A map of the cosmic microwave background, made by the *Wilkinson Microwave Anisotropy Probe* satellite (WMAP). The map shows the brightness distribution of the cosmic microwave background across the entire sky, much as a map of the Earth depicts its features on a flat surface. Red regions are brighter (warmer) and blue regions are fainter (colder). They show irregularities in the temperature of the Universe at the time of recombination, when the Universe was about 380,000 years old. The satellite that gathered the data for the map was named for David T. Wilkinson, a pioneer in the study of the cosmic microwave background. The term *anisotropy* refers to the lack of uniformity from point to point across the sky.

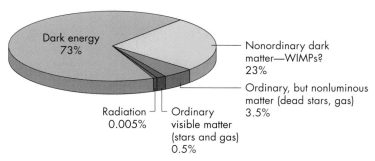

**FIGURE 17.13**

The makeup of the Universe as deduced from observations of the brightness variations in the cosmic microwave background and other data. The percentages have been rounded off, so they do not add up to exactly 100%.

## 17.4    THE ORIGIN OF THE UNIVERSE

Where did the Universe come from and what was it like in its early stages? Cosmologists study that remote time by extrapolating the present conditions in the Universe backward in time. That is, they use the laws of physics and present conditions to infer what the Universe was like at the time of its birth. We have already described one important conclusion of such extrapolations: the early Universe was hot and dense (as confirmed by the cosmic background radiation). Figure 17.14 shows how its temperature has changed over the last approximately 14 billion years.

You can see from the figure that the young Universe was extremely hot—trillions of degrees in its earliest moments. That very high temperature implies that in its youth, the Universe was a mix of gas and radiation, much like the core of our Sun, although far hotter. Such high-temperature gases turn out to obey relatively simple laws. For example, the temperature and density depend in known ways on the Universe's size. This allows astronomers to deduce the Universe's past properties from its present properties, which is how figure 17.14 was worked out.

The young Universe's enormous temperature also implies that its matter and radiation mingled in a manner unlike their sharp distinction today. The mingling occurred because, in a high-temperature gas, radiation can be so energetic that it can be converted into matter. Thus, in the early Universe, matter and radiation behaved almost as a single entity.

### Radiation, Matter, and Antimatter in the Early Universe

Einstein's equation $E = mc^2$ equates an amount of energy, $E$, to an amount of mass, $m$, times the speed of light, c, squared. It tells us that stars can obtain their energy from mass lost when, for example, hydrogen is converted to helium, as discussed in chapters 11 and 13. In the early Universe, the implication of this energy-mass relation was reversed: a quantity of energy $E$ may have turned into particles whose mass equaled $E/c^2$. Scientists know that matter can be created this way: they can observe high-energy radiation (gamma rays) forming particles in laboratory experiments with atomic accelerators. Given that energy can change into matter, it seems likely that high-energy radiation in the early Universe must do the same. Such transformations of energy into mass, however, always create a *pair* of particles, as illustrated in

*How matter goes to energy and vice versa*

FIGURE 17.14
A plot of how the temperature of the Universe changes with time according to the Big Bang theory.

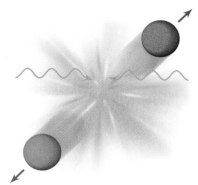

**A**    Radiation creates particle and antiparticle.

**B**    Particle and antiparticle annihilate, creating radiation.

**FIGURE 17.15**

(A) The energy of electromagnetic radiation can be converted into mass with the formation of a particle and its antiparticle. (B) The collision of a particle and its antiparticle leads to their annihilation and conversion of their mass back into energy.

figure 17.15A. Moreover, the particle pair has two special properties: the particles must have opposite charge (or no charge), and one of the pair must be made of ordinary matter, whereas the other must be made of what is called **antimatter.**

We briefly discussed antimatter in chapter 11 where we saw that the electron has an antimatter companion particle, the positron, with identical mass but opposite electric charge. Protons also have a corresponding antiparticle called the antiproton, and all other subatomic particles are similarly paired.

Antimatter has the important property that a particle and its antiparticle are annihilated on contact, leaving only high-energy radiation, as depicted in figure 17.15B. Thus, radiation (electromagnetic energy) can become matter, and matter can become radiation. In the early Universe, this shifting between mass and energy happened continually as particles formed and were annihilated.

Energetic radiation cannot turn into just any pair of particles: the pair's combined mass multiplied by $c^2$ must be less than the radiation's energy. For example, to become an electron-positron pair, the radiation needs an energy of about $1.6 \times 10^{-13}$ joules, equivalent to a wavelength of about 0.0012 nanometers (a gamma ray). To become a proton-antiproton pair the radiation needs roughly 1800 times more energy, making it a very high-energy gamma ray. Radiation with such large energies occurs only in very hot gas. For example, to have enough energy to make an electron-positron pair, the gas temperature must be about $10^{10}$ Kelvin. To make a proton-antiproton pair, the gas temperature must be about $10^{13}$ Kelvin. Such high temperatures are not reached in ordinary stars, but they occurred in the Big Bang.

## History of Matter and Radiation in the Early Universe

Astronomers hypothesize that approximately 1 microsecond after the Big Bang, the Universe had a temperature of about $10^{13}$ K, hot enough for its radiation to be converted into not only protons and antiprotons but even quarks and antiquarks, the particles that make up protons and neutrons. At that time, the Universe we now see was packed into a volume smaller than the Solar System. This volume, an inferno jammed with quarks and antiquarks created from the high-energy radiation, was expanding at a fantastic rate. As space expanded, this dazzlingly hot matter and radiation cooled. The cooler radiation was no longer energetic enough to create new quark-antiquark pairs. Thus, as quarks annihilated antiquarks, they were no longer re-created, and most of the quarks therefore disappeared from the Universe. This self-annihilation was not total, however. Theories of subatomic particles predict that a tiny excess of quarks, literally, *one in a billion,* survived with no antiquark to annihilate. These survivors then combined with one another to make the protons and neutrons of the Universe around us. Without that tiny imbalance in the annihilation of quarks, our Universe would have been essentially empty. Every particle would have had an antiparticle with which it could combine and be annihilated into radiation, leaving the Universe filled only with the cosmic background radiation and containing no matter at all.

The next landmark in the history of the early Universe occurred about 3 minutes after the Big Bang, when the expansion of space had cooled the Universe to a few hundred million degrees. Such conditions are similar to those in the core of a very hot star, and much as a star can fuse protons into helium, so did the Universe. Roughly one-quarter of the Universe's protons and neutrons fused into helium at that time. As we saw earlier, that helium, mixed with the more abundant hydrogen, permeates all space, explaining why nearly all first generation stars have the same proportion of hydrogen to helium in their outer layers.

For the next half million years, little happened in the young Universe apart from further expansion and cooling. Eventually, the temperature dropped to about 3000 K, and electrons and protons had low enough energy to bind together to form hydrogen atoms. This period (known as the recombination era because of the combination of electrons with nuclei) marks the time when the Universe's radiation and matter ceased to interact

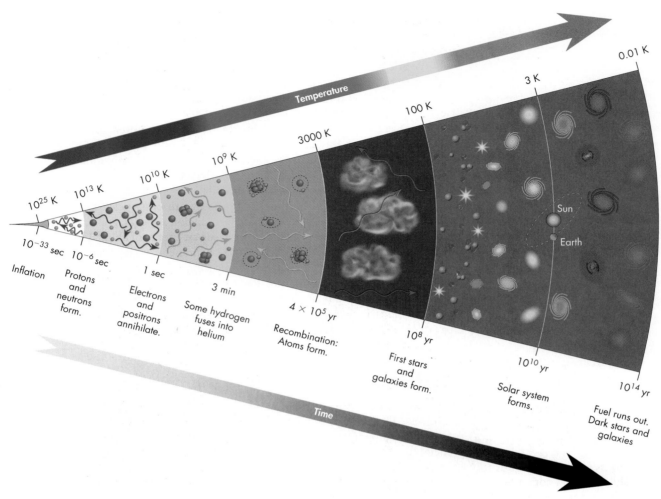

**FIGURE 17.16**
A sketch depicting the history of matter and radiation in the Universe.

strongly. Before recombination, electrons and radiation collided violently enough to exchange energy and maintain the same temperature. However, when electrons became attached to nuclei and formed hydrogen and helium atoms, their interactions with radiation became much weaker because the atoms have no overall electric charge. Thus, before recombination, the Universe behaved as a single blend of matter and radiation, but after recombination, its matter and radiation acted as separate ingredients.

The changing behavior of the early Universe's matter and radiation is summarized in figure 17.16. The diagram shows how the temperature of the Universe has changed as the Universe has expanded and cooled and changed from quarks to atoms of relatively cool hydrogen and helium. One vital stage remains to link that ancient time to the present: the formation of galaxies.

## The Formation of Galaxies

From the ages of the stars within galaxies, astronomers deduce that the Milky Way and other galaxies are about 13 billion years old. In fact, the first galaxies probably began to form soon after protons and electrons recombined to make the first hydrogen atoms in the early Universe.

Astronomers think that galaxies formed by a process similar to the way stars form, but on a much larger scale. That is, gravity drew the gas created in the Big Bang into

**FIGURE 17.17**
The Hubble Ultra Deep Field. This image shows a portion of the sky in the constellation Fornax that you could easily cover with a pencil eraser held at arm's length. The image shows about 10,000 galaxies. Some are so far away that we are seeing them the way they appeared nearly 13 billion years ago, when the Universe was only about one-twentieth of its present age.

gigantic clouds, which eventually collapsed under gravitational attraction and formed stars, as we discussed in describing the birth of the Milky Way in chapter 15. But to observe this process occurring in the first galaxies is extremely difficult because it happened so long ago. That is, to look back 13 billion years in time we need to look out into space 13 billion light-years. At such enormous distances even objects the size of galaxies appear very small and dim. Nevertheless, with the Hubble Space Telescope, astronomers have captured images (fig. 17.17) of that remote time.

In addition to such images, astronomers rely heavily on computer simulations for their understanding of how galaxies form. Such simulations show how small irregularities in the Universe's matter grow into galaxies (see figure 16.38), and they nicely create structures similar to the observed distribution of galaxies and voids, as we discussed in chapter 16. The simulations also show that dark matter plays an important role in creating a gravity that is strong enough to allow galaxies to form quickly while the Universe is still relatively young.

The simulations also suggest that the large-scale structure of the Universe—the voids and superclusters of galaxies that we see today—probably began from tiny variations of density that arose at the very earliest moments of the Big Bang.

## 17.5   THE INFLATIONARY UNIVERSE

The classic Big Bang theory allows cosmologists to understand the Universe back to about 1 microsecond after its birth. But what happened earlier? Current theories suggest that during that first microsecond, the Universe was even hotter, denser, and more energetic than we have yet described. Cosmologists hypothesize that in its infancy, the visible Universe was packed into a volume smaller than a proton and that its temperature was trillions of times hotter than trillions of degrees—about $10^{27}$ K. It was so hot, in fact, that the Universe was simply intensely high energy. Some theories imply that under such energetic conditions there can be a very large cosmological constant, creating a violent expansion, which cosmologists call **inflation.**

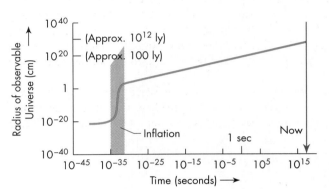

According to these theories, the inflation stage of the early Universe began about $10^{-35}$ seconds after its birth and lasted until about $10^{-32}$ seconds. During this time, space expanded explosively in size, its radius swelling by a factor of $10^{25}$, as shown in figure 17.18. If the period at the end of this sentence expanded by a similar factor, it would end up the size of the Local Group of galaxies. During that expansion, the Universe cooled (relatively), and the forces within it evolved to their present form. The inflationary stage "launched" the Big Bang.

Inflationary models of the first moments of the Universe mark the frontier of our understanding of the cosmos and offer tentative explanations of some of its most mysterious features. For example, the models suggest that the Universe may have been created literally from nothing. Some models imply the existence of other universes forever separated from our own. Still others suggest that space has more dimensions than the four we are familiar with: three of space and one of time. In fact, according to some versions of these models, space has as many as 10 or 11 dimensions. Inflationary models, however, also deal with less exotic questions, such as why (at least as far as we can tell) the Universe is all ordinary matter with essentially no antimatter. Perhaps the most intriguing aspect of these theories is that they hypothesize links between the Universe at large, the properties of subatomic particles, and the fundamental forces of nature. These theories are extremely complex and not well understood, so here we will deal with only two of the curious puzzles of our Universe that inflation neatly explains: why the Universe has just the critical density, making it "flat" and not curved, and why the Universe (especially the cosmic microwave background) is so uniform and not lumpy. Astronomers refer to the first of these puzzles as the "flatness problem," and they refer to the second one as the "horizon problem."

**FIGURE 17.18**
A sketch showing how the radius of the Universe suddenly inflates.

### The Flatness Problem

We saw in section 17.3 that space can be curved. Yet when astronomers measure the curvature of our Universe, they find that it is flat, implying that the density of the Universe equals the critical density that we referred to earlier. Inflation solves this problem in the following elegant way. Recall that during the time of inflation, the observable Universe expanded by a colossal factor—from smaller than a proton to a few centimeters in size. Such an enormous expansion of an object flattens any curvature that it has. To see why, imagine inflating a balloon. While the balloon is small, the curvature of its surface is very obvious. However, if the balloon could be inflated to the size of the Earth, its surface, for all practical intents, would be flat, just as the surface of the Earth is "flat." The rapid inflation of the early Universe does the same to space, taking an initially curved space and effectively flattening it, which is just the "shape" that astronomers deduce from the tiny variations of the cosmic microwave background.

## The Horizon Problem

We saw in section 17.1 that the brightness of the cosmic microwave background is extremely uniform across the sky. Although very small variations exist (as we discussed in the previous section), they are only about 1 part in 100,000. Such uniformity was difficult to explain prior to inflation theory. For example, if we look at the radiation coming from a given point in the sky and then look at radiation coming from exactly the opposite direction in the sky, we are seeing radiation that is coming from locations that are roughly the diameter of the observable Universe—about 28 billion light years—apart. But the Universe is only 14 billion years old, so no signal can possibly have traveled from one of these regions to the other in order to smooth out any variation. Using the terminology we used earlier, these regions lie outside each other's cosmic horizon. But the cosmic horizon is far smaller during the time when the Universe "inflates"—far smaller even than a proton when inflation begins. Because of this tiny size, energy moving at the speed of light has plenty of time to cross it many times, which ensures that each part has essentially the same physical conditions as every other part. That in turn makes the cosmic microwave background almost exactly the same in every direction.

Finally, inflation creates tiny irregularities in the density of the young Universe, which gravity will ultimately draw into galaxies. Thus, the first steps toward the formation of galaxies and their intervening voids may have been taken when the Universe was smaller than an atom.

## EXTENDING OUR IMAGINATION

### OTHER UNIVERSES?

The inflationary universe has yet another curious property: it allows adjacent regions of space to be completely separated from each other with no possible contact between them. Such isolated regions might well be considered separate universes, independent of ours and completely unobservable by us. Furthermore, according to inflationary theories, these other universes may be either expanding or contracting.

Inflation creates such sections of space much as bubbles form in a pot of boiling water. As water begins to boil, regions of the liquid suddenly change from liquid to gas (steam) and begin expanding. We see that region of steam as a bubble. Some bubbles expand, rise to the surface, and burst, whereas others collapse. So, too, in inflationary theories, entire universes may form and dissolve, entirely unknowable to us (box fig. 17.1). It would be amazing indeed if all the wonderful intricacy and beauty of our Universe is merely a single bubble in an even vaster sea of space, a space that we can never see.

"Bubbles" grow.    Expansion cools Universe.    "Bubbles" become separate universes?

**BOX FIGURE 17.1**
A sketch depicting bubbles that might become separate universes forming in an inflating universe.

# SUMMARY

Because distant galaxies are receding from the Milky Way and each other, astronomers believe that the Universe in its youth was much smaller, hotter, and denser than it is today. By extrapolating the present recessional motion of galaxies back in time, astronomers deduce that the Universe was created approximately 14 billion years ago in the Big Bang, a release of energy that initiated the expansion of space and the creation of matter. This age is consistent with that of other astronomical objects, in that it is greater than the age of the Earth and the age of the oldest stars.

Additional evidence for a hot, dense early Universe comes from the cosmic microwave background. This weak radio-wavelength radiation was originally emitted as gamma rays and X rays by the Universe itself long before galaxies formed and when the Universe was a glowing fireball of matter and energy. As the Universe expanded and cooled, so did the radiation; now its temperature is only 3 degrees above absolute zero.

Whether the Universe will expand forever or collapse depends on its density and rate of expansion. If the density is less than a critical value, not enough mass is present for gravity to stop the cosmic expansion, and the Universe will expand forever. Observations, in fact, imply that this is the case.

Astronomers have discovered that the expansion of the Universe is speeding up. To explain this acceleration, astronomers hypothesize that the Universe contains dark energy, which causes the speedup. Dark energy appears to make up approximately 70% of the Universe, but its nature is almost completely unknown to us.

Analysis of tiny brightness variations in the cosmic microwave background allows astronomers to measure the shape of the Universe and its age to much higher accuracy than was previously possible. They deduce from these measurements that the Universe is flat and 13.7 billion years old—both estimates accurate to a few percent.

The flatness of the Universe and the extreme uniformity of the cosmic microwave background lead astronomers to deduce that almost immediately after its formation, the Universe underwent a very brief but very rapid increase in volume—a process called inflation.

## ? What It Is.

This dim, thin red line in the white outline is an image of one of the most distant galaxies known. Another image of the galaxy is dimly visible in the small circle at the top of the picture. The galaxy itself lies behind a massive cluster of galaxies, and we see the distant galaxy's light distorted into the line shape by the gravitational force of the cluster that lies between it and us. The galaxy cluster is about 2 billion light-years away. The remote galaxy itself is about 13 billion light-years from us.

# QUESTIONS FOR REVIEW

1. Why do astronomers think that the Universe is expanding?
2. What is meant by a cosmic horizon?
3. What is Olbers' paradox?
4. What is the cosmic microwave background? What is its origin?
5. Why do astronomers think that the early Universe was hot and dense?
6. What is meant by the age of the Universe? How old is it? How is its age found?
7. How was helium formed from hydrogen in the early Universe?
8. What explanation do astronomers offer for why the relative abundance of hydrogen and helium is uniform?
9. Is our cosmic horizon the same as that of beings in another galaxy? Why?
10. What is meant by inflation in cosmology?
11. What is the Big Bang? What are some of its properties?
12. What problems of the Big Bang theory are resolved by inflationary models?
13. What determines whether the Universe will expand forever or recollapse?
14. What do astronomers mean when they say that the Universe is flat?
15. What is meant by dark matter?
16. What role does the cosmological constant play in cosmology? What evidence makes astronomers think it may exist?

# THOUGHT QUESTIONS

1. M31, the great galaxy in Andromeda and a member of the Local Group of galaxies, is approaching us. Does this mean that the Hubble law is wrong? Why?

2. Does it bother you that the Universe may expand forever?

3. Why do most stars, including our own Sun, contain more helium than the 24% formed during the Big Bang?

# TEST YOURSELF

1. From what evidence do astronomers deduce that the Universe is expanding?
   (a) They can see the disks of galaxies getting smaller over time.
   (b) They see a redshift in the spectral lines of distant galaxies.
   (c) They detect cosmic background X-ray radiation.
   (d) They can see distant galaxies dissolve, pulled apart by the expansion of space.
   (e) All of the above

2. Why do astronomers think the Universe was much hotter at its birth than it is now?
   (a) They detect cosmic microwave background radiation.
   (b) The amount of helium observed in all stars implies that the early Universe was hot enough to fuse some of its hydrogen into helium.
   (c) They detect intense X-ray emission from gas in galaxy clusters.
   (d) All of the above
   (e) Only (a) and (b)

3. What is meant by inflation in the early Universe?
   (a) The force of gravity suddenly grew stronger in the distant past.
   (b) Protons expanded to the size of stars, which was how our Sun formed.

   (c) The Universe increased dramatically in size in an extremely brief period of time.
   (d) The number of galaxies that we see at large distances is much greater than the number we can see near us.
   (e) The diameter of distant galaxies is much greater than the diameter of galaxies near us.

4. What is meant by the cosmic microwave background?
   (a) It is radiation from distant quasars.
   (b) It is radiation from hot gas in intergalactic space.
   (c) It is radiation from the first stars formed when the Universe was young.
   (d) It is radiation created during the early days of the Universe.
   (e) It is the explanation of Olbers' paradox.

5. What would happen if the density of the Universe were greater than the critical density?
   (a) Matter would never have formed.
   (b) Galaxies would not form.
   (c) Too many galaxies would form.
   (d) All the matter formed would be neutrons.
   (e) The Universe would recollapse.

# FURTHER EXPLORATIONS

Aguirre, Anthony. "Where Did It All Come From?" *Sky and Telescope* 112, (December 2006): 36.

Arkani-Hamed, Nima, Savas Dimopoulos, and Georgi Dvali. "The Universe's Unseen Dimensions." *Scientific American* 283 (August 2000): 62.

"Brave New Cosmos." *Scientific American* 284 (January 2001): 37. [Five articles on recent cosmological developments.]

Carroll, Sean. "Dark Energy and the Preposterous Universe." *Sky and Telescope* 109 (March 2005): 32.

Conselice, Christopher J. "The Universe's Invisible Hand." *Scientific American* 296 (February 2007): 34.

Dvali, Georgi. "Out of the Darkness." *Scientific American* 290 (February 2004): 68.

Freedman, Wendy L., and Michael S. Turner. "Cosmology in the New Millennium." *Sky and Telescope* 106 (October 2003): 30.

Guth, Alan H. "The Inflationary Universe." *Scientific American* 250 (May 1984): 116.

Hu, Wayne, and Martin White. "The Cosmic Symphony." *Scientific American* 290 (February 2004): 44.

Krauss, Lawrence M. *Quintessence: The Mystery of the Missing Mass in the Universe.* New York: Basic Books, 2000.

Lineweaver, Charles H., and Tamara M. Davis, "Misconceptions About the Big Bang." *Scientific American* 292 (March 2005): 36.

Livio, Mario. *The Accelerating Universe.* New York: John Wiley, 2000.

Nadis, Steve. "Sizing Up Inflation." *Sky and Telescope* 110 (November 2005): 32.

Naeye, Robert. "Delving into Extra Dimensions." *Sky and Telescope* 105 (June 2003): 38.

Osterbrock, Donald E., Joel A. Gwinn, and Ronald S. Brashear. "Edwin Hubble and the Expanding Universe." *Scientific American* 269 (July 1993): 84.

Rees, Martin. *Just Six Numbers: The Deep Forces That Shape the Universe.* New York: Basic Books, 2000.

Riess, Adam G., and Michael S. Turner. "From Slowdown to Speedup." *Scientific American* 290 (February 2004): 62.

Rowan-Robinson, Michael. *Nine Numbers of the Cosmos.* Oxford, England: Oxford University Press, 1999.

Singh, Simon. *Big Bang: The Most Important Scientific Discovery of All Time and Why You Need to Know About It.* Fourth Estate Press, 2005.

Strauss, Michael A. "Reading the Blueprint of Creation." *Scientific American* 290 (February 2004): 54.

Tegmark, Max. "Parallel Universes." *Scientific American* 288 (May 2003): 40.

**Website**

Please visit the *Explorations* website at **http://www.mhhe.com/arny** for additional online resources on these topics.

# PLANETARIUM EXERCISE

Center the Starry Night planetarium software on a constellation, such as Orion. Set the magnitude limit to 8 using the Settings header. Systematically place your cursor on each of the stars within the area of that constellation. In each case record the distance and the apparent magnitude. Are the brightest stars always the closest ones? How much variability is there? Make a rough sketch of the brightest stars in the constellation and their distances relative to each other. Try to imagine this constellation in 3 dimensions, adding depth to the picture. Recognize that constellations only appear the way they do from our perspective. If we were to travel to a nearby star system and look at Orion, for instance, it would look much different from that perspective.

# KEY TERMS

antimatter, 521

Big Bang, 506

cosmic horizon, 510

cosmic microwave background (CMB), 513

cosmological constant, 517

cosmological principle, 511

cosmology, 506

critical density, 516

curved space, 518

dark energy, 517

flat universe, 518

inflation, 524

negative curvature, 518

Olbers' paradox, 511

positive curvature, 518

# Life in the Universe

The evolution of the Universe from Big Bang to galaxies, stars, and planets has along the way created us. This evolutionary path from raw quarks to thinking beings who seek to understand how they fit into this grand scheme—what eighteenth-century scientists called "the great chain of being"—is one of many marvels of the Universe and is the theme of this section.

In earlier chapters, we described astronomical evolution and how the Earth and Sun fit into the cosmos. Now we turn to life on Earth and how it might have formed and developed. We will see that the existence of life on our own planet is suggestive of the possibility of its existence elsewhere, and so we will also speculate about whether we are alone, leading to the question, "Do other intelligent civilizations exist?" Astronomers are strongly divided about such possibilities. Some argue that such life is highly probable, whereas others argue that it is highly improbable. One goal of this section is to describe the evidence on both sides of this debate.

We will begin our discussion by examining the long history of life on Earth, looking at the factors thought to play a role in its development here. We will also discuss the likelihood of similar conditions elsewhere, discovering that planets suitable for life are probably numerous. In addition, we will describe how some astronomers have been motivated by such arguments to search for radio signals from other civilizations. Finally, we will speculate about the role of life in the history of our planet—the so-called Gaia hypothesis—and whether our existence may even say something about the nature of the Universe itself.

## LIFE ON EARTH

### History of Life on Earth

Life has existed on Earth for nearly the entire history of our planet. For example, fossil algae and bacteria such as those illustrated in figure E3.1 occur in rocks 3.5 billion years old. In fact, one of the most ancient life forms, algal colonies called "stromatolites," look identical to the stromatolites that exist today in shallow, warm, salty bays along the coast of western Australia.

Given that Earth formed about 4.5 billion years ago and that its surface was probably molten rock for almost 1 billion years, life began remarkably quickly. How quickly, we can perhaps appreciate from figure E3.2, which shows a time line for the history of life on Earth. This line shows the epoch at which

$\leftarrow$ 10 μm $\rightarrow$
(= 0.01 mm)

**FIGURE E3.1**
Photographs of fossil bacteria. Note: 10μm ≈ 0.004".

various life forms first appear in the fossil record. Their dates have been measured by radioactivity in their associated rock, as described in chapter 5, or other means. Along with the ages, the figure also illustrates some of these ancient life forms.

Scientists conclude from such evidence that only extremely simple life forms such as algae and single-celled animals existed for roughly three-fourths of Earth's history. Then, approximately 600 million years ago, more complex forms such as invertebrates developed, and about 500 million years ago, only a little more than one-tenth of the Earth's lifetime, shells and crustaceans appeared. Mammals and dinosaurs both appeared roughly 250 million years ago, but the latter were wiped out about 65 million years ago in the great Cretaceous extinction, probably caused by a totally chance event—an asteroidal collision, as described in chapter 10. Early hominids, our immediate ancestors, appeared roughly 5.5 million years ago, but our own species, *Homo sapiens*, evolved only about 500,000 years ago. Earth has therefore existed as a planet roughly 10,000 times longer than we have as a species. Thus, if all Earth's history were compressed into a year, humans would appear only in the last hour and civilization in the last few minutes.

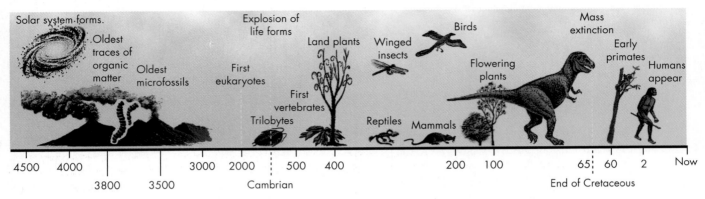

**FIGURE E3.2**
A time line illustrating the history of life on Earth.

**FIGURE E3.3**
Some of Earth's life forms.

## Unity of Living Beings

Life on Earth shows a dazzling variety of forms from butterflies to birds, worms to whales, and mushrooms to maples, as shown in figure E3.3. Despite that variety, all living beings on Earth have an underlying unity of structure, reproduction, and metabolism. For example, all living organisms use the same kinds of atoms for their structure and function (hydrogen, oxygen, carbon, nitrogen, and iron, for example). These atoms are not only widely used, but they are also abundant throughout the Universe. Thus, life on Earth generally is very economical in its use of the same chemical substances in widely differing organisms. Moreover, life tends to draw on the materials with which our planet is most richly

endowed: we are made primarily of the same substances that make up our ocean and atmosphere.

The chemical elements that compose living things are linked into long-chain molecules for their structure and function. The primary structural units of many of these chain molecules are about 20 amino acids. **Amino acids** are organic molecules containing hydrogen, carbon, nitrogen, and oxygen atoms arranged in a certain pattern (fig. E3.4). A few amino acids also contain one or more sulfur atoms.

Moving up to the next level of complexity, we find that all living things use amino acids as the structural units of more complicated molecules called **proteins.** Protein molecules give living things their structure. For example, the scales of

OH
|
CH₂
|
H₂N — C — C — OH
|    ‖
H    O

Glycine

CH₃
|
H — C — OH
|
H₂N — C — C — OH
|    ‖
H    O

Serine

Threonine

Asparagine

Glutamine

**FIGURE E3.4**

Diagrams illustrating the structure of several amino acids. H, C, N, and O represent atoms of hydrogen, carbon, nitrogen, and oxygen. The basic building block (shown in yellow) of amino acids is the same.

reptiles and fish are composed of the protein keratin, and the stiff cartilage of our own bodies is composed of the protein collagen. Proteins not only give cells their structure but also supply their energy needs. That energy supply reveals yet another underlying unity among living things: all cells use the same kind of molecule, adenosine triphosphate (abbreviated as ATP) to supply energy for action and growth.

This unity of structure and energy supply is echoed in the reproduction of life forms. All single-celled organisms divide to create new cells; all multicelled organisms produce egg cells, which divide to create new cells. Moreover, in all cases, "parents" pass on genetic information to their offspring by the same molecule, **DNA.**

DNA stands for deoxyribonucleic acid. It is a massive molecule consisting of two extremely long chains of smaller molecules wound around each other and linked by molecules called "base pairs." Thus, DNA has the appearance of a twisted ladder with the role of the connecting rungs played by the base pairs (fig. E3.5). When a cell reproduces, the DNA separates into two individual chains. Each chain then adds a new base pair at each point along its length to match the one previously present on the other chain. Thus, each strand makes a copy of the other so that after division and replication, a new DNA molecule exists identical to the original one.* That all terrestrial organisms use DNA for their reproduction gives life a truly remarkable unity.

Unity can also be seen in the way higher organisms function. For example, plants derive their food from carbohydrates that they manufacture during photosynthesis. The chlorophyll they use in this process is structurally very similar to the hemoglobin that transports oxygen in the blood of animals. These two molecules differ primarily in that chlorophyll is built around a magnesium atom, whereas hemoglobin is built around an iron atom. Likewise, cellulose, the structural material of plants, has a molecular structure very similar to chitin, the structural material in insect bodies and crustacean shells.

Finally, living systems exhibit a unity in their functions, such as the chemical processes that move biologically important molecules into and out of cells.

**FIGURE E3.5**

A diagram illustrating the structure of the DNA molecule. The blue horizontal structures in this model are the base pairs.

## DEDUCTIONS FROM THE UNITY OF LIFE AND THE TIME LINE

What conclusions can we draw from the chemical similarities of living beings and their ancient history as deduced from the time line? First, the chemical similarity almost inescapably suggests that all life on Earth had a common origin. Second, the antiquity of life suggests that under suitable conditions, life develops rapidly from complex molecules. Thus, we might infer that wherever such conditions occur, so too will life.

---

*Copying errors—mutations—sometimes occur. Most are harmful and lead to the organism's death. Nonlethal ones make evolution possible.

We should treat these conclusions with great caution, however, for they are based on only a single case: life on Earth. For example, if the first time you hit a golf ball you make a lucky hole-in-one, it does not mean that you will do so again anytime soon. Thus, the fact that life exists on Earth may say next to nothing about its chances elsewhere. Supposing, however, we do choose to speculate, what can we say about the origin of life and its likelihood elsewhere?

## ORIGIN OF LIFE ON EARTH

Most scientists today think that terrestrial life originated from chemical reactions among complex molecules present on the young Earth. This belief dates back to at least the time of Charles Darwin, but it was strongly bolstered in 1953 by an experiment performed by Stanley Miller—then a graduate student at the University of Chicago—and his professor, Harold Urey. Miller and Urey filled a sterile glass flask with water, hydrogen, methane, and ammonia—gases thought to be present in the Earth's early atmosphere (fig. E3.6). They then passed an electric spark through this mixture. The spark generated both visible and ultraviolet radiation, which triggered reactions in the gas-water mixture. At the end of a week, they analyzed the mixture, finding in it a variety of organic molecules as well as five of the amino acids used to make proteins. They therefore concluded that at least some of the ingredients necessary for life could have been generated spontaneously on the early Earth.

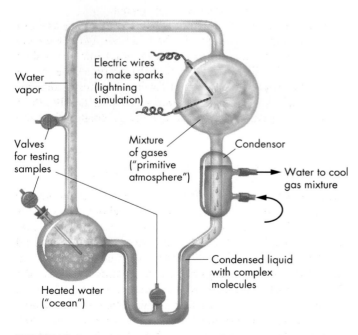

**FIGURE E3.6**
The Miller-Urey Experiment simulated conditions on the early Earth and showed that complex organic molecules could form then.

The success of Miller and Urey's experiment motivated other researchers to perform similar but more complex experiments. For example, the U.S. chemist Sydney Fox took complex organic molecules such as those made by Miller and Urey and by repeatedly heating and cooling them in water was able to create short strands of proteins called "proteinoids." More interesting, however, Fox discovered that these proteinoids spontaneously formed droplets reminiscent of cells. These droplets have no power to reproduce, nor do they show any of the complex structure commonly found in living cells, but they do demonstrate that organic material can spontaneously give itself structure and wall itself off from its environment, a step probably taken by the first living things.

Even the energy needs of living things can be produced in Miller-Urey type of experiments. For example, the Sri Lankan-American biochemist Cyril Ponnamperuma and the U.S. astronomer Carl Sagan showed that ATP can be readily synthesized from compounds similar to those generated by Miller and Urey, demonstrating one additional step in creating living from nonliving material.

Why then do we not observe similar processes occurring today on Earth? One reason is that our atmosphere now contains oxygen, which quickly attacks organic molecules, breaking them down. Thus, the absence of oxygen earlier in the Earth's history—an absence shown by analysis of ancient rocks—was probably crucial for the development of life. Another reason was given by Charles Darwin in his speculations on life's origin: today's living things immediately consume such organic material.

The importance of low oxygen abundance to the origin of life has led some researchers to speculate that life may have originated deep in the ocean near sea floor vents. These environments are rich in minerals and energy, and today they support a variety of life forms. In ancient days, before oxygen began to accumulate in our atmosphere, these same environments might have been especially congenial to the assembly of complex molecules. Even now, hot, oxygen-poor environments harbor life forms such as the bacteria that live in boiling hot springs at places such as Yellowstone National Park.

Even without the success of the Miller-Urey experiment in creating organic material, scientists studying the origin of life on Earth are confident that such material would be present on the young Earth. One source of organic material is interstellar matter. We discussed in chapter 15 that many interstellar clouds contain a rich mix of organic molecules, including many that are precursors to amino acids. Because the Solar System and Earth formed from such clouds, complex organic molecules must have been present on or near the Earth from the time of its birth. Even if these molecules were destroyed during the Earth's formation, fresh molecules may have been carried to its surface by planetesimals toward the end of the heavy bombardment of its surface by these planetesimals. In fact, carbonaceous chondritic meteors—probably surviving fragments of the ancient planetesimals—are rich in organic material and some even contain amino acids, as we saw in chapter 10. Finally, some scientists have demonstrated experimentally that the compression and heating of gases resulting from planetesimal bombardment may be as fruitful as the electric discharges used by Miller and Urey to form complex organic materials. Thus, astronomers are confident that organic molecules were common on the ancient Earth.

On the other hand, if one assumes that organic matter existed on the early Earth (whether created by lightning, ultraviolet radiation, or planetesimal bombardment), many steps are needed to take it from amino-acid-filled droplets in a primordial ocean to living cells. For example, the structures created must be able to reproduce, a feature not yet seen in any origin-of-life experiments.

Cells replicate today using the double-chained molecule DNA. Some researchers have speculated, however, that the earliest life forms used the slightly simpler molecule, ribonucleic acid, or RNA for short. In fact, biochemists have shown that if the proper chemicals are present (especially certain enzymes), RNA can replicate in the absence of living cells. Moreover, they have also discovered several other molecules far simpler in structure than RNA that can replicate but, so far, only in a "mechanical" sense. That is, if such molecules are put in a solution containing the necessary raw materials, they will make copies of themselves. In fact, one particular mix of molecules not only replicates but also can "mutate" into a slightly better replicator. Even so, far more is needed to produce life as we know it. For example, simple life forms must be able to form cells that work cooperatively.

## ORIGIN OF COMPLEX ORGANISMS

The original cellular life on Earth was probably far simpler than the cells we see today. For example, the earliest cells probably had no nucleus, a condition found now only in bacteria and certain algae. **Prokaryotes,** as these cells without nuclei are known, may accidentally have merged with other cells, a merger that benefited both cells by allowing them to grow and reproduce more rapidly. For example, a cell good at storing energy might have combined with a cell good at reproducing, thereby creating a more complex cell. From such simple beginnings, **eukaryotes** (cells with nuclei) probably evolved and, after them, multicelled organisms. Today, most cells have a complex membrane that holds them together and allows food to enter and waste products to leave. Moreover, most cells other than bacteria and some algae have a nucleus in which genetic information is stored and, in addition, mitochondria, which are tiny bodies within the cell that convert food into energy that the cell utilizes.

The greater complexity of multicelled organisms would in some cases allow them to exploit resources unavailable to simpler beings. For example, an organism that can move from nutrient-poor to nutrient-rich environments may be able to reproduce more prolifically than a stationary creature. Such reproductive advantages combined with changing environments have led to the incredible diversity of life that we see around us today. In fact, such reproductive advantage (natural selection) and random mutations are crucial to evolution and the development of higher life forms, as Charles Darwin explained in 1859 in his monumental work *On the Origin of Species by Means of Natural Selection.*

The scenario for the origin of life that we have discussed above, though hypothetical, persuades many scientists that life originated on Earth spontaneously. Not all scientists, however, accept this idea. A few favor the idea that life originated elsewhere and was transported to Earth either accidentally or deliberately.

## PANSPERMIA

The belief that terrestrial life descended from organisms created elsewhere in the Universe is sometimes called **panspermia.** According to this idea, simple life forms (perhaps bacteria) from some other location drifted from their place of origin across space to Earth. For example, bacteria spawned on some distant planet may have stuck to tiny dust particles that were lofted by winds or volcanic eruptions to the top of the planet's atmosphere. From there, they were subsequently lost to space, where, driven by radiation pressure from stars or stellar winds, they might drift almost anywhere. Although space is highly hostile to advanced life forms, simple organisms may be able to survive long passages through it if embedded in dust grains or otherwise protected. Eventually, the dust may have arrived in the Solar System, where it was captured by the Earth's gravity. Sifting down to the surface, the dust infected the virgin Earth, bringing life to our planet.

Panspermia has a certain appeal in that it suggests our roots spring from elsewhere in the Universe. The theory was very popular in the 1920s, but critics point out that it does not really simplify the problem of the origin of life, it just shifts it elsewhere. Moreover, it further requires getting the life to Earth, a perilous voyage even for a bacterium. For these reasons, the panspermia hypothesis has gradually fallen into disfavor.

## LIFE ELSEWHERE IN THE UNIVERSE

Humans have speculated for thousands of years about life elsewhere in the Universe. For example, about 300 B.C., the ancient Greek philosopher Epicurus wrote in his "Letter to Herodotus" that "there are infinite worlds both like and unlike this world of ours." Epicurus went on to add that "in all worlds there are living creatures and plants." Likewise, the Roman scholar Lucretius wrote about 50 B.C. that "it is in the highest degree unlikely that this Earth and sky is the only one to have been created." Such views were not universal, however. For example, Plato and Aristotle argued that the Earth was the only abode of life, a view that prevailed through most of the Middle Ages. Today, however, influenced by science-fiction movies, television, and books, many people find no difficulty at all in believing that extraterrestrial life exists. In fact, supermarket tabloids make silly claims almost weekly about encounters with aliens. If we look at the facts, however, there is not a shred of evidence that life exists elsewhere. This does not mean there is no extraterrestrial life; it simply means we do not *know* if there is. Thus, with no evidence, what can we say meaningfully about life elsewhere in the Universe?

## ARE WE ALONE?

Do other life forms exist in the Universe? Is there somewhere hundreds of light-years from Earth a "student alien" reading a book discussing the possibility of life on planets such as Earth? Scientists do not know, and, in fact, they are strongly divided into two groups. One group of scientists (let us call them "many-worlders") believes millions of planets with life exist in the Milky Way and many of these have advanced civilizations. The other group (let us call them "loners") argues that we are the only intelligent life in the galaxy. We will discuss below how these two radically different points of view arise.

## ARGUMENTS FOR MANY WORLDS

Many-worlders argue as follows: Earthlike planets are common. Life has formed on Earth. Therefore, it would be surprising if life did not exist elsewhere. In fact, even if only a small fraction of such planets have life, many advanced civilizations probably exist. Such arguments may actually vastly *underestimate* the number of habitable planets because life forms elsewhere may be able to live in environments intolerable to terrestrial organisms. To see how such numbers are deduced, let us begin by estimating the number of Earthlike planets in the Milky Way.

We start by asking how many Sunlike stars are in the Milky Way. We limit ourselves to such stars because luminous blue ones burn their fuel so rapidly that they will burn out and explode as supernovas before life has much chance to evolve, whereas dim red stars are so cool that planets must be very close to them for conditions to be warm enough for a terrestrial type of life. Given those restrictions on star types, we turn to censuses of stars, which show that stars like the Sun make up about 10% of our galaxy's stars, for a total of about $10^{10}$ such bodies.

Of these $10^{10}$ stars, we next ask, How many of them have Earthlike planets? Our own system has two planets out of a total of eight that might harbor life as we know it: Earth and Mars. Unfortunately, although astronomers have detected many planets around other stars, Earth-size planets are still very hard to detect and so the probability of their occurrence is still unknown. If we are conservative, therefore, and say that other systems might not be quite so lucky as we are in the Solar System, we might choose the probability of a life-sustaining planet around a solar type of star to be 1 in 10, or 10%. Taking 10% of the $10^{10}$ Sunlike stars leaves us with $10^9$ habitable planets.

How many of these actually have life on them? That will depend on how easy it is for life to form, a probability that the Miller-Urey experiments indicate may be high but that some astronomers have thought to be very low indeed. In fact, the British astrophysicist Fred Hoyle compared the likelihood of life arising spontaneously to the likelihood of a box containing several hundred tons of aluminum being shaken and by chance assembling itself into a 747 jumbo jet. Although many scientists find such a comparison silly, it illustrates the divergence of opinion on the likelihood of life forming. The difficulty in assessing that probability is that we have only the single case of life here as our basis. If, however, we use Earth as the pattern, the rapidity with which life developed on Earth and the Miller-Urey experiment's success in making molecular precursors to life indicate that the chance of life starting is fairly high, say, 1 in 100 rather than 1 in a million. Given that probability, we end up with $10^9 \times 0.01 = 10^7 = 10$ million worlds with life in our galaxy.

We now go to the next step and ask, On how many of these life-bearing planets do technical civilizations form? The chances of higher life evolving may be very large or very small. Again, scientists have no idea. Moreover, even if life succeeds in starting, perhaps it will be annihilated by asteroidal impact or, if it develops a high technology and is careless or thoughtless, life may destroy itself by pollution, nuclear war, or some other disaster. Suppose we are very conservative and say that if life forms, it has a one in a thousand chance of developing a technical civilization. Thus, we find that the number of advanced civilizations in our galaxy is $10^7 \times 0.001 = 10,000$. When we further consider that there are billions of other galaxies, it seems very likely that we are not alone.

The above argument was first presented by the U.S. astronomer Frank Drake, who pointed out that one can estimate the number of civilizations by multiplying together the probability of each condition necessary for it, a result now called the "Drake equation." Unfortunately, the accuracy of the probabilities is completely unknown. Perhaps, for example, Earthlike planets are far rarer than we have estimated. Perhaps evolution on other planets produces life forms that are so totally unlike those of Earth that we cannot recognize them as being alive. Furthermore, even if life forms similar to us do evolve elsewhere, the time period during which we could communicate with them might be so narrow that the opportunity passes. For example, if a civilization existed on Mars even as recently as 100 years ago, we could not have communicated with them by radio, and if in the meantime they destroyed themselves somehow, the chance would have been lost forever. Thus, the answer given by the Drake equation is filled with uncertainties. Nevertheless, even with pessimistic estimates, the equation suggests that *many* civilizations other than ours exist. Given their existence, we might then ask, How close is the nearest one? We can estimate the distance as follows.

Suppose the Milky Way contains $N$ civilizations scattered randomly throughout its disk, as illustrated in figure E3.7. We can calculate their average separation, $d$, as follows.

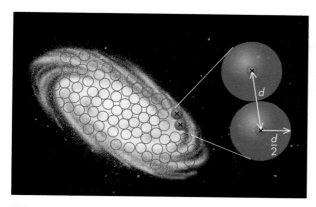

**FIGURE E3.7**
A sketch illustrating how to estimate the distance
between civilizations in the Milky Way.

Imagine drawing a sphere of radius $d/2$ around each civilized
world. Then $N$ such spheres must cover the disk of the Milky
Way. That is,

$$N\pi(d/2)^2 = \pi R^2$$

where we assume that the Milky Way's disk has radius $R$. We
can now cancel $\pi$ to obtain

$$Nd^2/4 = R^2$$

Moving the $N$ and 4 to the right-hand side of the equation
gives

$$d^2 = 4R^2/N$$

Taking the square root of both sides gives

$$d = 2R/\sqrt{N}$$

To evaluate $d$ and thereby find how far away our neighbors
are, we set $R$ equal to the Milky Way's radius (taken to be
50,000 light-years) and $N = 10,000$, the value we deduced
for the number of technical civilizations. These choices then
give a value for $d$ of 1000 light-years. Thus, even if such a
large number of civilizations exist in our galaxy, they are all
extremely far from us. Such immense separations also make
it very unlikely that we will be able to exchange information
with other civilizations.

## LONERS

Loners suggest we are the only advanced life in the Milky
Way. Their argument is based on the rapidity with which
our species has spread across Earth and the potential for its
rapid spread across the Milky Way. Loners claim that once
a civilization masters space flight, it will take it at most a few
thousand years to colonize planetary systems near their home
system and that within a few million years they will colonize
the entire galaxy. Moreover, they assume that once such colo-
nization occurs, a civilization will quickly contact other civi-

lizations, either by direct means or by radio communication.
Because no other civilization has contacted us, it must mean
that no other civilization is out there anywhere near us. Thus,
we are probably alone in the Milky Way, or nearly alone with
at most only a handful of other civilizations. However, this
argument, like that of the many-worlders for a multiplicity
of civilizations, rests on several unproved assumptions. For
example, it assumes that (1) most civilizations are driven to
colonize, (2) they seek rather than avoid contact, and (3) they
have mastered interstellar flight and communication.

## SEARCHING FOR LIFE ELSEWHERE

Is there any evidence for life elsewhere in the Universe? Not
really, although in the summer of 1996, two groups of scientists suggested that traces of organic chemicals that they
detected in meteorites from Mars may indicate that life once
existed there (see chapter 8 for more details). The absence of
hard evidence may seem discouraging, but as British astronomer Sir Martin Rees has said, "Absence of evidence is not
the same as evidence of absence." Even if the Universe were
teeming with life, we have so far examined only a tiny portion
of it: some half dozen spots on the Moon, five on Mars (by
robot space-landers), and some asteroidal fragments picked
up on Earth. The Moon's lack of atmosphere and water makes
it so inhospitable that astronomers did not expect to find life
there. Mars seemed more promising because its environment,
though harsh, is more like the Earth's than that of any other
planet, and laboratory experiments showed that some terrestrial organisms (bacteria and lichens) could in fact survive
Martian conditions, though they did not reproduce or grow.
Moreover, pictures show that its environment was less harsh
in the past. For example, Mars once had flowing water on
its surface, an indication that a warmer climate and denser
atmosphere existed long ago. Nevertheless, the two *Viking*
spacecraft that landed there in 1976 found no trace of life or
even of organic carbon compounds, although, as we mentioned earlier, some meteorites from Mars do contain organic
carbon compounds. Similarly, the *Mars Pathfinder, Spirit,* and
*Opportunity* rovers, which landed on Mars between July 1997
and January 2004 found no signs of Martian life.*

## RADIO SEARCHES

Given the failure to find life near to us, and the difficulty
of directly sampling distant locales, a few astronomers have
searched for radio signals from other civilizations (fig. E3.8).
Such signals might be either broadcast to us deliberately or
communications directed elsewhere that we might simply
"overhear." Astronomers began "listening" for extraterrestrial radio signals in the 1960s with Project Ozma, which
monitored radio emission from several nearby star systems,
hoping to detect emission from other than natural processes.

*See chapter 8 for more details.

**FIGURE E3.8**
One of the radio telescopes used to search for
signals from other civilizations.

More recent projects, called **SETI** (Search for *Extra*terrestrial *I*ntelligence), use receivers that automatically scan billions of radio wavelengths to search for signals from other civilizations.*

Searching for signals is not easy because even if they exist, they are probably very weak and buried in cosmic static. Thus, astronomers use considerable care when choosing the wavelengths to monitor. For example, at very long wavelengths, interstellar gas in the galaxy emits strongly and will overwhelm all but the most powerful signals. On the other hand, at very short wavelengths, molecules in our atmosphere block the signals. It turns out that the optimum wavelengths are near 21 centimeters, the wavelength emitted by neutral hydrogen in the cold gas clouds of our galaxy. This is also near the wavelength of OH (hydroxyl) molecules, which emit strongly at 18 centimeters. These two wavelengths are very distinctive and any other intelligent civilization would probably have discovered their significance and have equipment that could receive or broadcast them. Thus, several searches

have been made for signals in this wavelength region, which is sometimes called the "waterhole" because it is bracketed on one side by emission from H and on the other by emission from OH. However, Ozma and all subsequent searches have proved fruitless so far.

One possible drawback to such searches is that other civilizations might not be transmitting signals. Like us, they may be listening. However, if other civilizations are like ours, they inadvertently broadcast a wide range of signals. For example, television stations and defense radar transmit extremely powerful signals that would be detectable many light-years from Earth. Moreover, on at least one occasion, a ceremony in 1974 celebrating the renovation of the huge Arecibo radio telescope in Puerto Rico, astronomers deliberately broadcast a signal toward the globular cluster M13. This cluster is more than 20,000 light-years away from us, so the signal is still on its way.

The many difficulties associated with radio searches have led scientists to consider other ways to search for signs of extraterrestrial life. One interesting method involves looking for unexpected gases in the atmosphere of a planet, gases that might be produced by living things.

In the late 1960s, James Lovelock, a British chemist, noted that several of the gases making up the Earth's atmosphere would disappear if Earth had no living beings to replenish them. For example, oxygen is very reactive and readily combines with surface rock. Lovelock went on to argue that the presence of such gases in a planet's atmosphere is an indication of life there. Thus, rather than searching for signals or sending robot spacecraft to explore a planet's surface (fig. E3.9),

40 feet
(~ 12 meters)

**FIGURE E3.9**
Cliffs on the inner wall of Endurance Crater, which was visited by *Opportunity,* one of the Mars rovers. The layers (strata) that you can see here suggest that this area on Mars once was covered with water. Because water is vital to life here on Earth, scientists are anxious to explore regions on other planets and moons where water is, or was once, present—in the hope of finding traces of present or past life on other worlds.

---

*Such searches for life elsewhere require huge amounts of computer time. For example, the data in one project is e-mailed from the central search telescope to a network of millions of volunteers who allow their home computers to process the information when they are not using their machines.

astronomers might find evidence of life by analyzing the light from a planet's atmosphere. Unfortunately, we cannot yet make such observations for any planets other than those in our own Solar System. However, Lovelock's idea has since led to a very different speculation.

## THE GAIA HYPOTHESIS

In 1974, Lovelock and the U.S. microbiologist Lynn Margulis suggested that life creates a single "larger entity" with a planet, a "symbiosis" of life and planet, which they called the **Gaia hypothesis.** They chose this name from the Greek goddess of the Earth, Gaia (pronounced *Guy-uh* in this context but *Jee-uh* elsewhere).

According to the Gaia hypothesis, life does not merely respond to its environment but actually alters its planet's atmosphere and temperature to make it more hospitable. As we discussed in chapter 5, plant life has almost certainly done just that on Earth. For example, by photosynthesis, plants have created an oxygen-rich atmosphere on Earth that shields them from dangerous ultraviolet radiation. Similarly, plants may alter a planet's temperature through the greenhouse effect. If the planet is too cold, plant metabolism is slowed, and less $CO_2$ is converted to oxygen. The greater $CO_2$ abundance warms the planet and enhances plant growth. Conversely, if the planet gets too warm, plant growth, according to the hypothesis, runs amok and produces lots of oxygen, reducing the $CO_2$ abundance and therefore also reducing the greenhouse warming.

Lovelock and Margulis have also proposed that many other facets of the environment, such as humidity, salinity of the oceans, sea level, and even plate tectonics, may be controlled by living organisms so as to optimize their reproduction. Such an intimate linkage between the living and nonliving is a very appealing idea. It almost makes the planet "alive." But is the Gaia hypothesis correct? Some aspects certainly are, but many scientists are skeptical of its more extreme claims (for example, that life can control plate tectonics). Nevertheless, living beings do exert a remarkable control over their local environment, and many unsuspected links between life and the physical world surely remain to be found. One such possible connection has its roots in cosmology, a connection called the "anthropic principle."

## THE ANTHROPIC PRINCIPLE

In 1961, Robert Dicke, a physicist at Princeton University, wrote a paper concerning some remarkable cosmological coincidences. Dicke noted, for example, that the age of the Universe is very close to the lifetime of stars like the Sun and went on to argue that this "coincidence" is in fact not remarkable at all. Rather, it is a *necessity* if cosmologists are to exist to note it. Dicke pointed out that life requires elements such as carbon, silicon, and iron that are made in massive stars. Thus, for life to form, enough time must pass for massive stars to evolve and make the heavy elements and then eject them into space with a supernova explosion. Moreover, additional time is needed for the ejected material to be incorporated back into interstellar clouds and form new stars. Only with this second generation of stars does life become likely. Dicke therefore concluded that intelligent life is impossible in a Universe only a few thousand years old. He likewise argued that life requires stars, and so it can exist only in a Universe young enough that stars have not all burnt out. Thus, the very existence of an intelligent observer in the Universe requires that the age of the Universe fall within certain limits.

In 1974 the English physicist Brandon Carter took Dicke's idea further and proposed what he called the **anthropic principle.** According to the anthropic principle, "what we can expect to observe must be restricted by the conditions necessary for our presence as observers." Since Carter's work, many scientists have shown how "fine-tuned" the Universe must be for life to exist. Thus, no matter how unlikely some aspects of the Universe may appear to us, if those aspects are necessary for life, then that is what we will observe. For example, we might argue how truly marvelous it is that conditions on Earth are just right for life. According to the anthropic principle, the rude reply is, "Of course conditions are just right for life. If they were not, there would be no life here to do the marveling."

## SUMMARY

Life has existed on Earth for at least 3.8 billion years, a goodly portion of our planet's existence. That long duration and the underlying similarities of all life forms in terms of their structure, energy supply, and methods of reproduction indicate that life had a common origin relatively soon after Earth's formation. Experiments such as Miller and Urey's show that the precursors to life can be easily created under natural conditions. Thus, many astronomers believe that life has formed in numerous other places in the Universe. All searches for life, however, have found nothing, but given the vast distances between stars, that negative result is not surprising.

Once life develops on a planet, it may subtly alter its environment to its own ends, a proposition known as the "Gaia hypothesis." Such alterations may be detectable more readily than signals from other civilizations and may therefore be signatures of extraterrestrial life.

## QUESTIONS FOR REVIEW

1. What is the earliest evidence for life on Earth?
2. How old are the earliest life forms thought to be?
3. What features of life are suggestive of a common origin?
4. What is meant by panspermia?

5. What is the Drake equation?

6. Why is 21 centimeters believed to be a good wavelength region to search for signals from other civilizations?

7. What is meant by the Gaia hypothesis?

8. What is meant by the anthropic principle?

## THOUGHT QUESTIONS

1. Would you be upset if it turned out that we were the only intelligent life in the Universe? Why or why not?

2. Suppose astronomers did detect signals from another civilization. What effect do you think it would have on our society? What effect would it have on you?

3. Do you think it is a good idea to send out signals in the hope that aliens might detect them? Suppose, for example, they were more technically advanced than us and could do to us what we have done to less developed societies on Earth.

4. What message would you send to an alien civilization?

5. Use a method similar to the Drake equation to estimate the number of people getting a haircut at some given moment. That is, multiply the probabilities of each such event times the number of people in, for example, the United States.

## TEST YOURSELF

1. Evidence that life on Earth is very ancient comes from
   (a) fossil algae in rocks more than 3 billion years old.
   (b) fossils buried thousands of feet below the surface.
   (c) fossil algae in rocks a few million years old.
   (d) the discovery of silicon-based life forms in ancient rocks.
   (e) the discovery of silicon-based life forms in a Miller-Urey type of experiment.

2. A planet on which life forms create an atmosphere that makes the planet more hospitable for life is an example of
   (a) the Drake equation.
   (b) the Gaia hypothesis.
   (c) SETI.
   (d) the anthropic principle.
   (e) the Miller-Urey experiment.

3. The Miller-Urey experiment demonstrated that
   (a) very simple bacteria can be created from the chemicals present in the atmosphere of the ancient Earth.
   (b) Earth's early atmosphere was too harsh for life to have formed until recently.
   (c) conditions on the early Earth were suitable for the creation of many of the complex organic molecules found in living things.
   (d) if simple life forms landed on the early Earth, they could have survived.
   (e) early life forms were silicon-based.

4. According to the anthropic principle
   (a) we are the only civilization in the Universe.
   (b) there are $10^{12}$ civilizations more advanced than us in the Milky Way.
   (c) human beings are the highest form of life.
   (d) the Universe could not differ radically from its present properties, or we would not be here to observe it.
   (e) the greenhouse effect is beneficial to us.

5. The Drake equation attempts to determine
   (a) the conditions under which life originated on Earth.
   (b) the optimum wavelength for communicating with extraterrestrials.
   (c) the age of life on Earth.
   (d) the number of other technically advanced civilizations.
   (e) the lifetime of our own civilization.

## FURTHER EXPLORATIONS

Crowe, M. J. *The Extraterrestrial Life Debate: 1750–1900.* New York: Cambridge University Press, 1986.

Diamond, Jared. "Alone in a Crowded Universe." *Natural History* 99 (June 1990): 30.

Dick, S. J. *Plurality of Worlds.* New York: Cambridge University Press, 1982.

———. *The Biological Universe.* New York: Cambridge University Press. 1996.

Drake, Frank, and Dava Sobel. *Is Anyone Out There? The Scientific Search for Extraterrestrial Intelligence.* New York: Delacorte Press, 1992.

Falk, Dan. "A Universe Fine-tuned for Life? The Anthropic Principle's Surprising Resurgence." *Sky and Telescope* 107 (March 2004): 42.

Gould, Stephen J. "The Evolution of Life on the Earth." *Scientific American* 271 (October 1994): 84.

———. "War of the Worldviews." *Natural History* 105 (December 1996): 22.

Knoll, A. H. "End of the Proterozoic Eon." *Scientific American* 265 (October 1991): 64.

LePage, Andrew J. "Habitable Moons." *Sky and Telescope* 96 (December 1998): 50.

LePage, Andrew J., and Alan M. MacRobert. "SETI Searches Today." *Sky and Telescope* 96 (December 1998): 44.

Lovelock, James E. *The Ages of Gaia: A Biography of Our Living Earth.* New York: Norton, 1988.

———. *Gaia, A New Look at Life on Earth.* New York: Oxford University Press, 1979.

Lovelock, James E., and Lynn Margulis. "The Gaia Hypothesis." *Tellus* 26 (1974): 1.

Orgel, Leslie E. "The Origin of Life on the Earth." *Scientific American* 271 (October 1994): 76.

Schilling, Govert. "The Chance of Finding Aliens: Reevaluating the Drake Equation." *Sky and Telescope* 96 (December 1998): 36.

Shklovski, I. S., and Carl Sagan. *Intelligent Life in the Universe.* New York: Dell Publishing Co., 1966.

Shostak, Seth. "Listening for Life." *Astronomy* 20 (October 1992): 26.

Soffen, Gerald A. "Life in the Solar System." In *The New Solar System,* edited by J. Kelly Beatty, Carolyn Collins Petersen, and Andrew Chaikin. 4th ed. Cambridge, England: Sky Publishing Co. and Cambridge University Press, 1999.

Ward, Peter D., and Donald Brownlee. *Rare Earth: Why Complex Life Is Uncommon in the Universe.* New York: Copernicus, 2000.

Weinberg, Steven. "Life in the Universe." *Scientific American* 271 (October 1994): 44.

## KEY TERMS

amino acids, 530

anthropic principle, 537

DNA, 531

eukaryotes, 533

Gaia hypothesis, 537

panspermia, 533

prokaryotes, 533

proteins, 530

SETI, 536

# Space Travel and Special Relativity

A favorite theme in science fiction is space travel. At present such travel is limited to flights orbiting the Earth, although in the 1960s and 1970s astronauts traveled to the Moon, landed there, and explored some of its surface. The Moon is figuratively on our doorstep, however, so can we expect to ever be able to travel the vastly longer distances to other stars or galaxies? In science fiction stories, these immense interstellar distances are crossed by craft using faster-than-light travel. Scientists have excellent reasons for believing that faster-than-light travel is impossible and travel even at near light speed requires long spans of time. For example, it takes light about 4 years to reach us from even the nearest star beyond the Sun and tens to hundreds of thousands of years to reach us from nearby galaxies. This seems to mean that for astronauts to conquer interstellar distances, they would have to be prepared to live in space for decades, perhaps having the journey completed by their descendants. But it might surprise you to learn that astronauts could in principle travel *millions* of light-years to another galaxy and return to Earth in their own lifetime! The science that explains how such a seemingly immense journey could be made is the theme of this essay. Along the way, we will learn a little about one of science's most important and intriguing theories, the theory of special relativity. But before we deal with travel at near the speed of light, we need to look at the far simpler problem of describing motion.

## REST FRAMES

Astronomical objects are in constant motion: the Earth moves through space around the Sun; the Sun moves through space within the Milky Way galaxy; and so forth. To describe such motion we need a frame of reference, or **rest frame.** In everyday life we often use the ground as our rest frame. For example, we drive a car at 60 mph (~100 km/hr) along the freeway or walk to class across the campus at 4 mph (~6 km/hr) with respect to the ground.

Suppose, however, we are traveling in an airplane and we walk down the aisle of the plane from the back to the front. How fast are we moving? With respect to the plane, we are moving at a walking pace, say a few miles per hour. However, we could also measure our speed in the plane with respect to the ground, in which case our speed would be that of the plane plus our walking speed with respect to the plane, or hundreds of miles per hour. In other words, our measurement of an object's motion depends on the rest frame that we use for our observation.

Describing an object's motion in one rest frame when it is viewed from another rest frame is not difficult. If we consider again a person walking in a plane, we simply add the speed of the person to the speed of the plane. For another example, suppose a person is running at 10 mph and throws a javelin forward at 40 mph in the direction of his or her motion—the javelin will move across the ground at the speed 10 mph + 40 mph = 50 mph. Addition of speeds in this fashion (thrower plus javelin) (fig. E4.1) is an example of what is sometimes called **Galilean relativity,** in honor of Galileo, one of the first scientists to recognize how motions combine with one another. But Galilean relativity fails for light.

**FIGURE E4.1**
The speed of a javelin with respect to the ground is found by adding the speed of the thrower across the ground to the speed with which the javelin is thrown. That is, the velocities add.

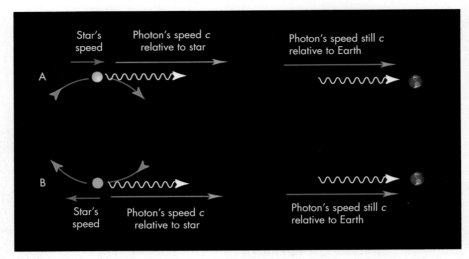

**FIGURE E4.2**
Light from orbiting stars reaches Earth at the same speed whether the star is moving toward or away from us.

## THE SPEED OF LIGHT FROM MOVING OBJECTS

Suppose we have two stars, A and B, in orbit around each other, with the orbit seen edge-on (fig. E4.2). At some point in the orbit, one of the stars, say star A, will be moving toward us and the other, say star B, will be moving away from us. Light leaves the surface of each star at the speed of light, c. We might then expect that light from the approaching star A will travel toward us a little faster than c, while light from star B, the one swinging away from us, will travel toward us a little slower than c. However, when scientists tried to measure this effect, they found no change in the light's speed: the light arrived at the same speed, c, regardless of the motion of the stars. This is not so odd as it might first seem. We see a similar effect in waves generated by a boat moving across a lake. The waves travel through the water at the same speed irrespective of how the boat moves. Such arguments led scientists in the 1800s to conclude that light moves through space like the waves on the water. But ordinary waves require some material to move through. What substance played the role of the water for light? This led scientists at that time to conclude that space was filled with some transparent substance, which they call the **æther.**

Calculations showed that the æther must have very special properties. It must be very rigid for light waves to have their high observed velocity (the stiffer a substance, the faster waves travel in it). At the same time, it must freely allow astronomical objects to plow through it with no resistance. In particular, it must flow past the Earth as we orbit the Sun.

## THE MICHELSON-MORLEY EXPERIMENT

In 1887 the American scientists Albert Michelson and Edward Morley designed an apparatus to search for the æther. They predicted that light traveling along the direction of the Earth's motion should move at a different speed than light traveling perpendicular to the Earth's motion. But when Michelson and Morley conducted their experiment, they detected no difference in the speed of light along the perpendicular paths. At first they thought this might just be bad luck: that during the experiment, the combination of the Earth's motion around the Sun and the Sun's motion through space might just have happened to make the Earth stationary relative to the æther at that point in its orbit. So they repeated their experiment many times over the course of a year. At some point they should have easily detected the motion of Earth relative to the æther, but they found no sign of the Earth's motion at all.

The Michelson-Morley experiment has been called the most famous "failed" experiment in history because it led to a revolution in physics. The results implied that there was no æther regulating the speed of light. However, if light was not moving relative to the æther, then physicists could not explain the constancy of the speed of light reaching us from sources moving at different speeds toward or away from us.

All experiments made then and now show that no matter how fast the source generating the light is moving, and no matter how fast the observer measuring the light is moving, the speed of light is always measured to be c = 299,792,458 meters per second. How can this be?

One explanation offered in the late 1800s was that motion through space somehow caused matter to contract* in the direction of motion. If matter contracts when it is moving, this would change our perception of length so that we might be tricked into thinking the speed of light had not changed. For example, Michelson and Morley were searching for a

---

*The idea of a contraction caused by motion was first proposed by Irish physicist George Fitzgerald. The Dutch physicist Hendrik Lorentz developed a model for explaining how the contraction might arise.

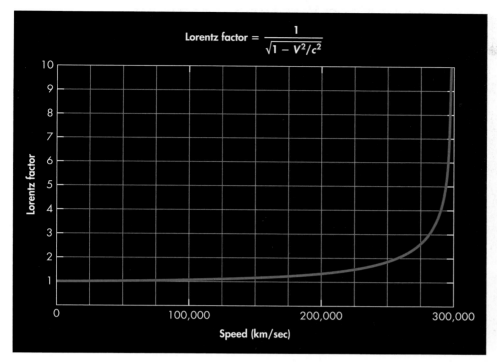

$$\text{Lorentz factor} = \frac{1}{\sqrt{1 - V^2/c^2}}$$

**FIGURE E4.3**
The Lorentz factor expresses the amount by which objects appear to compress in the direction of their motion. It is also the factor by which a moving clock appears to slow down, and the factor by which a moving object's mass appears to increase.

**FIGURE E4.4**
Albert Einstein (1879–1955).

difference in the speed of light in two perpendicular directions. If the apparatus were compressed in the direction it was moving through space but not in the perpendicular direction, the path the light traveled would be shorter. Such a contraction could potentially cancel out the effect that Michelson and Morley were searching for. The factor by which the apparatus would need to contract is known today as the **Lorentz factor.** The Lorentz factor, usually denoted by the Greek letter gamma ($\gamma$), hypothesizes that an object's length shortens in the direction that it is moving by a factor equal to

$$\gamma = \frac{1}{\sqrt{1 - V^2/c^2}}$$

where $V$ is the speed of the object and c is the speed of light. The value of $\gamma$ for different speeds $V$ is plotted in figure E4.3.

The Lorentz factor is close to 1 at small speeds. For example, at the 30 km/sec speed at which the Earth is orbiting the Sun, the Earth would contract by only a few centimeters in its direction of motion. However, at high speeds the contraction factor becomes very large, growing to infinity if the speed $V$ were to reach the speed of light. The Lorentz contraction factor can explain the Michelson-Morley experiment, but it cannot explain a variety of other conflicting results that were found. It contains, however, an important idea that grew into a whole new concept about the nature of motion.

## EINSTEIN'S THEORY OF SPECIAL RELATIVITY

In 1905 a 26-year-old graduate student named Albert Einstein (fig. E4.4) took on the problem of the seemingly inexplicable measurements of the speed of light. He was completing his physics degree while working in the Swiss patent office and supporting a new family. Yet in that one year alone, he completed his doctorate degree and wrote four papers in several areas of physics. Physicists widely agree that three of these papers were each worthy of a Nobel prize! He was little known at the time and had few colleagues with whom to discuss his ideas; but nonetheless in one of these papers he came up with a brilliant new approach to the question of the motion of light.

Einstein began by concentrating on the finding that light travels at the same speed no matter what the speed of its source or of the observer measuring the light. Even though many experiments had come to this conclusion, most physicists had assumed it was an impossibility and so were seeking other explanations—such as errors in the experiments or the Lorentz contraction. Einstein, instead of thinking of a constant speed of light for all observers as an impossibility, accepted this notion as correct, and proceeded to work out its consequences. He found that this led inevitably not only to a Lorentz-like contraction of space, *but also to a stretching of time by the same factor.* Even more important, he found that this contraction was not relative to some imagined æther

**FIGURE E4.5**
When a spacecraft travels by us, its length is contracted by the Lorentz factor, and clocks on board run more slowly by the same factor. From on board the ship it appears that the Earth is compressed and time is running slowly on Earth.

filling space, but that these effects depended on just the relative motions of any two objects.

These alterations of our perception of space and time affect everything that we see that has any speed relative to us and underlie what is known as the **special theory of relativity.** According to this theory, if a star or a rocket moves by us at high speed, we will see it squashed in its direction of motion by the Lorentz factor (fig. E4.5). If we could watch a clock tick or measure the speed of an astronaut's heart in the spaceship, we would discover that all these processes occur more slowly by the Lorentz factor.

What is remarkable is that the mathematics Einstein worked out showed that the situation is exactly symmetrical for the astronaut moving by us at high speed. That astronaut will sense herself being the one who is stationary and she will see us as moving by her at high speed. She will see us and the Earth contracted in the direction of "our" motion, and she will see our clocks and our hearts running slowly (fig. E4.5). So what

two observers in relative motion see is parallel—each would find that the other was the one undergoing the distortions of space and time. Thus, there is no preferred rest frame.

An especially important feature of Einstein's work is the behavior of light. Suppose we or the passing astronaut shine a light beam at each other. We will each measure that the light is moving past us at the same speed, c. However, as we watch each other making these measurements, we will each think that the other is measuring a shorter distance and using a clock that runs too slow.

This theory of special relativity is far-reaching in its implications. The theory is the basis for Einstein's discovery of the relationship between energy and mass: $E = mc^2$. It also predicts that other fundamental quantities change for moving objects. For instance, a moving object grows more massive by the same Lorentz factor that describes how lengths grow shorter and time slows.

These effects mean that nothing can reach the speed of light in our rest frame, let alone exceed it. The rocket ship traveling by us can fire its rockets and accelerate forever, but even though the ship goes ever faster, its mass grows ever greater, making it harder and harder to accelerate. The ship may approach the speed of light, but because the Lorentz factor goes to infinity, it would require infinite energy to reach the speed of light.

## SPECIAL RELATIVITY AND SPACE TRAVEL

The theory of special relativity may seem strange, but it has been tested by a century of high-precision experiments. There is not a single verified contradiction to it, and its predictions about such things as the slowing of time have been verified directly.

Special relativity is also about more than just perceived differences in space, time, and mass. For example, when atomic clocks (the most accurate clocks, used for establishing time worldwide) are flown between locations, it is found that their travel in airplanes leaves them a little bit slow relative to the network of fixed clocks maintained around the world. The moving clocks must be readjusted after each trip. The Lorentz factor for traveling at airplane speeds of about 1000 km/hour (600 mph) is just $\gamma = 1.0000000000004$, so every tick of the clock on an airplane takes about 4 ten-trillionths of a second longer than a tick of the clock in the ground-based network; but after several hours of flying the effect is measurable.

More intriguing is what happens at such high speed that the Lorentz factor is large. Experiments have demonstrated, for example, that a subatomic particle called a *muon* normally has a lifetime of about 2 millionths of a second before it decays. However, when muons are traveling at 99% of the speed of light (and therefore have a Lorentz factor of about 7), they live about 14 millionths of a second. This longer lifetime allows them to travel distances that would be impossible for them within their normal lifetimes.

When speaking of such brief times, this change in the rate at which time passes seems minor; but the same slowing factor would apply for a human traveling at 99% of the speed of light. If a spaceship could be built that traveled that fast, time would effectively run seven times more slowly on board compared with time here on Earth. If astronauts had food and air supplies for a 10-year trip, the Lorentz factor of $\gamma \approx 7$ (corresponding to V = 99% the speed of light) would mean they could travel for $\gamma \times 10$ years $\approx 70$ years in Earth time. At their speed relative to Earth of 0.99 c they would be able to reach a distance of almost 70 light-years (fig. E4.6) before they ran out of supplies. At speeds even closer to c the Lorentz factor becomes even larger (see table E4.1) and the potential distances greater.

From the perspective of astronauts on a craft traveling at 0.99 c, it would not seem that time was passing any more slowly than normal. Nor would they feel that they or their ship was foreshortened or more massive. From their perspective the Earth and the star they are visiting and the distance between

| TABLE E4.1 | |
| --- | --- |
| **The Lorentz Factor at High Speeds** | |
| **Speed** | **Lorentz Factor** |
| 0.87 c | 2.0 |
| 0.97 c | 4.1 |
| 0.99 c | 7.1 |
| 0.999 c | 22.4 |
| 0.9999 c | 70.7 |
| 0.99999 c | 223.6 |

the two are *contracted* by the Lorentz factor (fig. E4.6B). The distance would look like 70 light-years when they were stationary (in the rest frame of Earth); but at their high speed, the distance would look only one-seventh as large!

These marvelous "tricks" of relativity open up possibilities for traveling distances far greater than we might once have imagined. Theoretically, one could travel a million light-years within a human lifetime—although it is far beyond current technologies. It is also important to realize that such travel would have major challenges beyond simply reaching such high speeds. From the perspective of the spaceship traveling among the stars at near the speed of light, every atom and every dust particle in space along the ship's path has a mass increased by the Lorentz factor, and it is heading toward the ship at nearly the speed of light! This can give a pebble the impacting force of a ship-destroying asteroid.

Also, as intriguing as these possibilities are, they offer no time savings for the rest of us back on Earth. Consider again the astronauts traveling at 99% of the speed of light for 10 years to visit a distant star. If they then turned around and came home in another 10 years (by their reckoning), they would find that 140 years had passed on Earth. Everyone they knew would have grown old and died. They might even be younger than their great-great-grandchildren!

## THE TWIN PARADOX

Something may seem wrong with the description of space travel in the previous section. When we first introduced relativity we noted that the time stretching appears symmetrical. That is, from the rest frame of the Earth it appears that the astronauts' time is running slowly, while from the rest frame of the astronauts it appears that time on Earth is running slowly. This is sometimes presented as the **twin paradox:** If one of the astronauts left a twin back on Earth, how can we say that one ages more than the other?

The explanation of this seeming paradox is that the situations are *not* in fact symmetric. The astronauts experience accelerations as their ship speeds up and then slows down at their destination. Furthermore, when they turn their ship around to return to Earth, they experience yet more accelerations as the ship again speeds up and later slows down as it

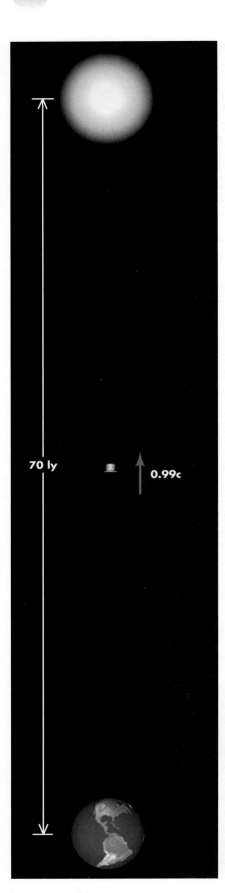

70 ly

0.99c

**FIGURE E4.6**
(A) From the rest frame of Earth and a star 70 light-years away, it appears that a spaceship traveling at 99% of c is shortened to one-seventh of its original length. (B) On the spaceship, once it is up to speed, it appears that the Earth and star are both moving at 99% of c. They and the distance between them are shortened to one-seventh of their original length.

B

0.99c

10 ly

0.99c

reaches Earth again. But people remaining back on Earth experience none of these accelerations because they always remain in the same rest frame. Thus, their progression of time remains constant. By contrast, the astronauts' rest frame keeps changing, and in the end they return to the rest frame of the Earth.

Imagine, for example, that one of the astronaut twins sends messages once each day to her twin back on Earth, and meanwhile the twin sends messages once each day to his astronaut sister on the ship. As they part, they each receive the other's messages much more slowly—both because of the Lorentz factor *and* because the separation between the ship and Earth is growing larger and the messages take longer to reach each other. When the astronaut twin reaches the distant star, she sends her 10th-year message. Until this time, the situations are symmetrical. Both have received only a small fraction of each other's messages because of the Lorentz factor and growing separation.

The astronaut twin turns around and starts back, receiving the messages from her brother that were sent years earlier and have been on their way to her across the 70 light-year gap. They come more quickly now because she is approaching Earth, cutting the distance each message has to travel. She reads about her brother getting older and older, now seemingly very rapidly. In the meantime, her brother back on Earth is still reading her messages from the outgoing trip. He would not even receive the 10th-year message announcing that the ship had reached the other star before he dies. At the speed of light in Earth's rest frame, that 10th-year message would take 70 years to reach Earth. In fact, the spaceship, traveling at 99% of the speed of light, returns to Earth just shortly after the message arrives at Earth saying that the astronauts reached the star.

In the movies, space travel is fast and everyone ages at the same rate. In reality, traveling at high speeds means that the travelers must leave behind not only their homes but their own times and the people in them.*

---

*With everyone's time running at different speeds, can you imagine a story line that would make a good movie if it portrayed space travel accurately?

## SUMMARY

Experiment shows that the speed of light is a constant and is not affected by the motion of its source or the observer. This puzzling result became the basis for Einstein's special theory of relativity. A consequence of that theory is that an observer moving past an object sees its length shrunk (the so-called Lorentz contraction) and the rate of passage of its time slowed (time dilation).

## QUESTIONS FOR REVIEW

1. What is Galilean relativity? Give an example of how it is used.
2. Describe what the Michelson-Morley experiment was trying to detect.
3. What is the Lorentz factor?
4. How are length, time, and mass altered according to special relativity?

## TEST YOURSELF

1. When a spaceship is traveling at 99% of the speed of light (Lorentz factor = 7), an astronaut on board the ship will find that
   (a) everything in the ship weighs seven times more.
   (b) the ship has become very small—only one-seventh its original length.
   (c) everyone on the ship is talking seven times more slowly than normal.
   (d) All of the above
   (e) None of the above. Everything seems normal to the astronaut on board.

2. Suppose Tom and Molly are both flying in spaceships toward each other at half the speed of light (0.5 c). If Tom shines a light toward Molly, what speed will Molly measure for the light coming toward her?
   (a) 0.25 c
   (b) 0.5 c

   (c) 1.0 c
   (d) 1.5 c
   (e) 2.0 c

3. If Bob travels at close to the speed of light to another star and then returns, he will find that his twin sister Alice who remained on Earth is
   (a) younger than him.
   (b) older than him.
   (c) the same age as him.
   (d) He cannot return to Earth because it would violate special relativity.

## THOUGHT QUESTION

1. Would you be willing to travel to a nearby galaxy if it meant you would return to Earth one million years in the future?

## FURTHER EXPLORATIONS

Gamow, George. *Mr. Tompkins in Paperback.* Cambridge: Cambridge University Press, 1993.

[This book, a reprint of Gamow's superb Mr. Tompkins books, explores the ideas of special relativity and quantum physics at a level accessible to children. Originally published in the 1940s, this reprint contains *Mr. Tompkins in Wonderland* and *Mr. Tompkins Explores the Atom.*]

## KEY TERMS

| | |
|---|---|
| æther, 542 | rest frame, 541 |
| Galilean relativity, 541 | special relativity, 544 |
| Lorentz factor, 543 | twin paradox, 545 |

| | |
|---|---|
| Preview | 1–b; 2–c; 3–e; 4–c; 5–e |
| Chapter 1 | 1–d; 2–b; 3–a; 4–e; 5–a |
| Essay 1 | 1–c; 2–b; 3–a; 4–d; 5–e |
| Chapter 2 | 1–e; 2–d; 3–d; 4–a; 5–c |
| Chapter 3 | 1–e; 2–c; 3–d; 4–e; 5–a |
| Chapter 4 | 1–d; 2–c; 3–d; 4–d; 5–a |
| Chapter 5 | 1–e; 2–a; 3–d; 4–c; 5–d |
| Essay 2 | 1–b; 2–c; 3–e; 4–e; 5–a |
| Chapter 6 | 1–d; 2–c; 3–d; 4–d; 5–b |
| Chapter 7 | 1–c; 2–c; 3–e; 4–d; 5–a |
| Chapter 8 | 1–d; 2–b; 3–c; 4–d; 5–a |
| Chapter 9 | 1–c; 2–b; 3–a; 4–d; 5–d |
| Chapter 10 | 1–e; 2–d; 3–b; 4–d; 5–d |
| Chapter 11 | 1–d; 2–c; 3–a; 4–d; 5–d |
| Chapter 12 | 1–d; 2–e; 3–e; 4–a; 5–b |
| Chapter 13 | 1–e; 2–c; 3–b; 4–b; 5–e |
| Chapter 14 | 1–d; 2–c; 3–b; 4–e; 5–e |
| Chapter 15 | 1–b; 2–e; 3–a; 4–a; 5–e |
| Chapter 16 | 1–d; 2–e; 3–e; 4–b; 5–e |
| Chapter 17 | 1–b; 2–e; 3–c; 4–d; 5–e |
| Essay 3 | 1–a; 2–b; 3–c; 4–d; 5–d |
| Essay 4 | 1–e; 2–c; 3–b |

# POWERS-OF-TEN NOTATION

Powers-of-ten notation is a shorthand method for expressing and working with very large or very small numbers. With this method, we express numbers as a few digits times ten to a power, or exponent. The power indicates the number of times that ten is multiplied by itself. For example, $100 = 10 \times 10 = 10^2$. Similarly $1,000,000 = 10 \times 10 \times 10 \times 10 \times 10 \times 10 = 10^6$. Note that we do not always need to write out $10 \times \ldots \ldots$ . Instead, we can simply count the zeros. Thus, 10,000 is 1 followed by 4 zeros, so it is $10^4$.

We can write 1 and 10 itself in powers-of-ten notation as well: $1 = 10^0$ and $10 = 10^1$.

To write a number like 300, we break it into two parts, $3 \times 100 = 3 \times 10^2$. Similarly, we can write $352 = 3.52 \times 100 = 3.52 \times 10^2$.

We can also write very small numbers (numbers less than 1) using powers of ten. For example, $0.01 = 1/100 = 1/10^2$. We can make this even more concise, however, by writing $1/10^2$ as $10^{-2}$. Similarly, $0.0001 = 10^{-4}$. Note that for numbers less than 1, the power is 1 more than the number of zeros.

We can write a number like 0.00052 as $5.2 \times 0.0001 = 5.2 \times 10^{-4}$.

Suppose we want to multiply numbers expressed in powers of ten. The rule is simple: we add the powers. Thus, $10^3 \times 10^2 = 10^{3+2} = 10^5$. Similarly $2 \times 10^8 \times 3 \times 10^7 = 2 \times 3 \times 10^8 \times 10^7 = 6 \times 10^{15}$. In general, $10^a \times 10^b = 10^{a+b}$.

Division works similarly, except that we subtract the exponents. Thus, $10^5/10^3 = 10^{5-3} = 10^2$. In general, $10^a/10^b = 10^{a-b}$.

The last operations we need to consider are raising a power-of-ten number to a power and taking a root. In raising a number to a power, we multiply the powers. Thus, $(10^3)^4 = 10^{3 \times 4} = 10^{12}$. Care must be used if we have a number like $(4 \times 10^2)^3$. Both the 4 and the $10^2$ are raised to a power, so the result is $4^3 \times (10^2)^3 = 64 \times 10^{2 \times 3} = 64 \times 10^6$.

Taking a root is equivalent to raising a number to a fractional power. Thus, the square root of a number is the number to the $\frac{1}{2}$ power. The cube root is the number to the $\frac{1}{3}$ power, and so forth. For example, $\sqrt{100} = (100)^{1/2} = (10^2)^{1/2} = 10^1 = 10$.

# SOME USEFUL FORMULAS

Distance = velocity $\times$ time $(D = Vt)$

Circumference of circle = $2\pi r$

Area of a circle = $\pi r^2$

Surface area of a sphere = $4\pi r^2$

Volume of a sphere = $\dfrac{4\pi r^3}{3}$

# SOLVING DISTANCE, VELOCITY, TIME (*D*, *V*, *t*) PROBLEMS

A large number of problems in this book (and in science in general) involve the motion of something. In such problems, we often know two of the three quantities (*D*, *V*, *t*) and want to know the third. For example, we have something moving at a speed V and want to know how far it will travel in a time *t*. Or we know that something travels with a speed V and want to find out how long it takes for the object to travel a distance *D*. We all can solve such problems in our heads if the motion involves automobiles. For example, if it is 150 miles to the city and we travel at 50 miles per hour, how long does it take to get there? Or how far can we drive in 2 hours if we are traveling at 45 miles per hour? Because we all do such problems routinely, you might find it easier to think of astronomical problems in terms of cars. Regardless of your approach, the method of solution is simple.

Begin by making a simple sketch of what is happening. Draw some arrows to indicate the motion. Label the quantities known and put question marks beside the things you want to find. Then write out the basic relation $D = Vt$. If you want to find *D* and know V and *t*, just multiply them for the answer. If you want to find the time and are given V and *D*, solve for *t* by dividing both sides by V to get $t = D/V$. If you want the velocity, divide *D* by *t*. In some problems, the motion may be in a circle of radius *r*. In that case, the distance traveled will be related to the circumference of the circle, $2\pi r$. For such cases, you may need to use the expression $D = 2\pi r$.

In most problems, you will find that it is extremely helpful to write the units of the quantities in the equation. For example, suppose you are asked how long it takes to travel 1500 km at a velocity of 30 kilometers per hour. First write out $D = Vt$. Then rewrite it as

$$D/V = t$$

Next insert the quantities so that

$$t = \frac{1500 \text{ km}}{30 \text{ km/hr}} = 50 \text{ hr}$$

Note that the units of km cancel out and leave us with units of hours, as the problem requires.

---

## APPENDIX TABLE 1

### Physical and Astronomical Constants

**Physical Constants**

| | |
|---|---|
| Velocity of light (c) | $= 2.99792458 \times 10^8$ m/s = approximately $3 \times 10^5$ km/s |
| Gravitational constant (G) | $= 6.67259 \times 10^{-11}$ m$^3$-kg$^{-1}$-s$^{-2}$ (or newton $-$ m$^2 -$ kg$^{-2}$) |
| Planck's constant (h) | $= 6.62608 \times 10^{-34}$ joule-s |
| Mass of hydrogen atom ($M_H$) | $= 1.6735 \times 10^{-27}$ kg |
| Mass of electron ($M_e$) | $= 9.1094 \times 10^{-31}$ kg |
| Stefan-Boltzmann constant ($\sigma$) | $= 5.6705 \times 10^{-8}$ watts-m$^{-2}$-deg$^{-4}$ |
| Constant in Wien's law ($\lambda_m T$) | $= 0.28978$ cm/deg |

**Astronomical Constants**

| | |
|---|---|
| Astronomical unit (AU) | $= 1.495978706 \times 10^{11}$ m $\approx 1.5 \times 10^8$ km |
| Light-year (ly) | $= 9.4605 \times 10^{15}$ m $= 9.4605 \times 10^{12}$ km $= 6.324 \times 10^4$ AU |
| Parsec (pc) $= 3.26$ ly | $= 3.085678 \times 10^{16}$ m $= 3.085678 \times 10^{13}$ km $= 206,265$ AU |
| Sidereal year | $= 365.256366$ days $= 3.155815 \times 10^7$ s |
| Mass of Earth | $= 5.974 \times 10^{24}$ kg |
| Mass of Sun ($M_\odot$) | $= 1.989 \times 10^{30}$ kg |
| Equatorial radius of Earth | $= 6,378.0$ km |
| Radius of Sun ($R_\odot$) | $= 6.96 \times 10^8$ m $= 6.96 \times 10^5$ km |
| Luminosity of Sun ($L_\odot$) | $= 3.83 \times 10^{26}$ watts |

## APPENDIX TABLE 2
### Conversion Between American and Metric Units*

**Length**

| | | | |
|---|---|---|---|
| 1 km | = 1 kilometer | = 1000 meters | = 0.6214 mile |
| 1 m | = 1 meter | = 1.094 yards | = 39.37 inches |
| 1 cm | = 1 centimeter | = 0.01 meter | = 0.3937 inch |
| 1 $\mu$m | = 1 micrometer | = $10^{-6}$ meter | = $10^{-4}$ cm = $3.93 \times 10^{-5}$ inch |
| 1 nm | = 1 nanometer | = $10^{-9}$ meter | = $10^{-7}$ cm |
| 1 mile | = 1.6093 km | | |
| 1 inch | = 2.5400 cm | | |

**Mass**

| | | | |
|---|---|---|---|
| 1 metric ton | = $10^6$ grams | = 1000 kg | = $2.2046 \times 10^3$ lb |
| 1 kg | = 1000 grams | = 2.2046 lb | |
| 1 g | = 1 gram | = 0.0022046 lb = 0.0353 oz | |
| 1 lb | = 0.4536 kg | | |
| 1 oz | = 28.3495 g | | |

*You will find it easier to use the metric system if you remember the following prefixes that denote very small to very large quantities:

| | | | | | | | | |
|---|---|---|---|---|---|---|---|---|
| nano | = | $10^{-9}$ | = | 1 billionth | kilo | = | $10^3$ | = 1 thousand |
| micro | = | $10^{-6}$ | = | 1 millionth | mega | = | $10^6$ | = 1 million |
| milli | = | $10^{-3}$ | = | 1 thousandth | giga | = | $10^9$ | = 1 billion |

## APPENDIX TABLE 3
### Physical Properties of the Planets*

| Name | Radius (Eq) (Earth units) | Radius (Eq) (km) | Mass (Earth units) | Mass (kg) | Average Density (g/cm³) |
|---|---|---|---|---|---|
| Mercury | 0.382 | 2,439 | 0.055 | $3.30 \times 10^{23}$ | 5.43 |
| Venus | 0.949 | 6,051 | 0.815 | $4.87 \times 10^{24}$ | 5.25 |
| Earth | 1.00 | 6,378 | 1.00 | $5.97 \times 10^{24}$ | 5.52 |
| Mars | 0.533 | 3,397 | 0.107 | $6.42 \times 10^{23}$ | 3.93 |
| Jupiter | 11.19 | 71,492 | 317.9 | $1.90 \times 10^{27}$ | 1.33 |
| Saturn | 9.46 | 60,268 | 95.18 | $5.68 \times 10^{26}$ | 0.69 |
| Uranus | 3.98 | 25,559 | 14.54 | $8.68 \times 10^{25}$ | 1.32 |
| Neptune | 3.81 | 24,764 | 17.13 | $1.02 \times 10^{26}$ | 1.64 |
| Pluto | 0.181 | 1,160 | 0.00216 | $1.29 \times 10^{22}$ | 2.05 |

*Pluto, now officially a dwarf planet, is retained here for historical reasons.

## APPENDIX TABLE 4
### Orbital Properties of the Planets*

| Name | Distance from Sun† (AU) | Distance from Sun† ($10^6$ km) | Period Years | Period Days | Orbital Inclination‡ | Orbital Eccentricity |
|---|---|---|---|---|---|---|
| Mercury | 0.387 | 57.9 | 0.2409 | (87.97) | 7.00 | 0.206 |
| Venus | 0.723 | 108.2 | 0.6152 | (224.7) | 3.39 | 0.007 |
| Earth | 1.00 | 149.6 | 1.0 | (365.26) | 0.00 | 0.017 |
| Mars | 1.524 | 227.9 | 1.8809 | (686.98) | 1.85 | 0.093 |
| Jupiter | 5.203 | 778.3 | 11.8622 | (4,332.59) | 1.31 | 0.048 |
| Saturn | 9.539 | 1,427.0 | 29.4577 | (10,759.22) | 2.49 | 0.056 |
| Uranus | 19.19 | 2,869.6 | 84.014 | (30,685.4) | 0.77 | 0.046 |
| Neptune | 30.06 | 4,496.6 | 164.793 | (60,189) | 1.77 | 0.010 |
| Pluto | 39.53 | 5,900 | 247.7 | (90,465) | 17.15 | 0.248 |

*Pluto, now officially a dwarf planet, is retained here for historical reasons.   †Semimajor axis of the orbit.   ‡With respect to the ecliptic.

## APPENDIX TABLE 5

### Satellites of the Solar System*

| Planet Satellite | Radius[†] (km) | Distance from Planet ($10^3$ km) | Orbital Period (days) | Mass ($10^{20}$ kg) | Density (g/cm³) |
|---|---|---|---|---|---|
| **Earth** | | | | | |
| Moon | 1,738 | 384.4 | 27.322 | 734.9 | 3.34 |
| **Mars** | | | | | |
| Phobos | $13 \times 11 \times 9$ | 9.38 | 0.319 | $1.3 \times 10^{-4}$ | 2.2 |
| Deimos | $8 \times 6 \times 5$ | 23.5 | 1.263 | $1.8 \times 10^{-5}$ | 1.7 |
| **Jupiter[‡]** | | | | | |
| Metis | 20 | 127.96 | 0.295 | $9 \times 10^{-4}$ | — |
| Adrastea | $13 \times 10 \times 8$ | 128.98 | 0.298 | $1 \times 10^{-4}$ | — |
| Amalthea | $135 \times 82 \times 75$ | 181.3 | 0.498 | $8 \times 10^{-2}$ | — |
| Thebe | $55 \times ? \times 45$ | 221.9 | 0.6745 | $1.4 \times 10^{-3}$ | — |
| Io | 1,821 | 421.6 | 1.769 | 893.3 | 3.57 |
| Europa | 1,565 | 670.9 | 3.551 | 479.7 | 2.97 |
| Ganymede | 2,634 | 1,070.0 | 7.155 | 1,482 | 1.94 |
| Callisto | 2,403 | 1,883.0 | 16.689 | 1,076 | 1.86 |
| Leda | ~8 | 11,094 | 238.7 | $4 \times 10^{-4}$ | — |
| Himalia | 92.5 | 11,480 | 250.6 | $8 \times 10^{-2}$ | — |
| Lysithea | ~18 | 11,720 | 259.2 | $6 \times 10^{-4}$ | — |
| Elara | ~38 | 11,737 | 259.7 | $6 \times 10^{-3}$ | — |
| Ananke | ~15 | 20,200 | 631R | $4 \times 10^{-4}$ | — |
| Carme | ~20 | 22,600 | 692R | $9 \times 10^{-4}$ | — |
| Pasiphae | ~25 | 23,500 | 735R | $1.6 \times 10^{-3}$ | — |
| Sinope | ~18 | 23,700 | 758R | $6 \times 10^{-4}$ | — |
| **Saturn[‡]** | | | | | |
| Pan | 10 | 133.58 | 0.574 | $4.2 \times 10^{-6}$ | — |
| Atlas | $19 \times ? \times 14$ | 137.64 | 0.602 | $1.6 \times 10^{-4}$ | — |
| Prometheus | $74 \times 50 \times 34$ | 139.35 | 0.613 | $1.4 \times 10^{-3}$ | 0.27 |
| Pandora | $55 \times 44 \times 31$ | 141.7 | 0.629 | $1.3 \times 10^{-3}$ | 0.42 |
| Epimetheus | $69 \times 53 \times 53$ | 151.42 | 0.694 | $5.6 \times 10^{-3}$ | 0.63 |
| Janus | $99 \times 95 \times 76$ | 151.47 | 0.695 | $2.0 \times 10^{-2}$ | 0.65 |
| Mimas | $210 \times 197 \times 193$ | 185.52 | 0.942 | 0.370 | 1.12 |
| Enceladus | $256 \times 247 \times 244$ | 238.02 | 1.370 | 0.65 | 1.00 |
| Tethys | $536 \times 528 \times 526$ | 294.66 | 1.888 | 6.17 | 0.98 |
| Calypso | $15 \times 8 \times 8$ | 294.66 | 1.888 | $4 \times 10^{-5}$ | — |
| Telesto | $15 \times 13 \times 8$ | 294.67 | 1.888 | $6 \times 10^{-5}$ | — |
| Dione | 559 | 377.4 | 2.737 | 10.8 | 1.49 |
| Helene | $18 \times ? \times 15$ | 377.4 | 2.737 | $1.6 \times 10^{-4}$ | — |
| Rhea | 764 | 527.04 | 4.518 | 23.1 | 1.24 |
| Titan | 2,575 | 1,221.85 | 15.945 | 1,345.5 | 1.88 |
| Hyperion | $185 \times 140 \times 112$ | 1,481.1 | 21.277 | 0.28 | — |
| Iapetus | 720 | 3,561.3 | 79.331 | 15.9 | 1.0 |
| Phoebe | $115 \times 110 \times 105$ | 12,952 | 550.4R | 0.1 | — |

(continued)

| Planet Satellite | Radius† (km) | Distance from Planet ($10^3$ km) | Orbital Period (days) | Mass ($10^{20}$ kg) | Density (g/cm³) |
|---|---|---|---|---|---|
| **Uranus‡** | | | | | |
| Cordelia | 13 | 49.75 | 0.336 | $1.7 \times 10^{-4}$ | — |
| Ophelia | 16 | 53.77 | 0.377 | $2.6 \times 10^{-4}$ | — |
| Bianca | 22 | 59.16 | 0.435 | $7 \times 10^{-4}$ | — |
| Cressida | 33 | 61.77 | 0.465 | $2.6 \times 10^{-3}$ | — |
| Desdemona | 29 | 62.65 | 0.476 | $1.7 \times 10^{-3}$ | — |
| Juliet | 42 | 64.63 | 0.494 | $4.3 \times 10^{-3}$ | — |
| Portia | 55 | 66.1 | 0.515 | $1 \times 10^{-2}$ | — |
| Rosalind | 29 | 69.93 | 0.560 | $1.5 \times 10^{-3}$ | — |
| Belinda | 33 | 75.25 | 0.624 | $2.5 \times 10^{-3}$ | — |
| Puck | 77 | 86.00 | 0.764 | $5 \times 10^{-3}$ | — |
| Miranda | $240 \times 234 \times 233$ | 129.8 | 1.413 | 0.66 | 1.26 |
| Ariel | $581 \times 578 \times 578$ | 191.2 | 2.520 | 13.5 | 1.65 |
| Umbriel | 584.7 | 266.0 | 4.144 | 11.7 | 1.44 |
| Titania | 788.9 | 435.8 | 8.706 | 35.2 | 1.59 |
| Oberon | 761.4 | 582.6 | 13.463 | 30.1 | 1.50 |
| Caliban | | | | | |
| Sycorax | | | | | |
| Prospero | | | | | |
| Setebos | | | | | |
| Stephano | | | | | |
| **Neptune‡** | | | | | |
| Naiad | 29 | 48.2 | 0.296 | $1.4 \times 10^{-3}$ | — |
| Thalassa | 40 | 50.0 | 0.312 | $4 \times 10^{-3}$ | — |
| Despina | 74 | 52.5 | 0.333 | $2.1 \times 10^{-3}$ | — |
| Galatea | 79 | 62.0 | 0.396 | $3.1 \times 10^{-2}$ | — |
| Larissa | $104 \times ? \times 89$ | 73.6 | 0.554 | $6 \times 10^{-2}$ | — |
| Proteus | $218 \times 208 \times 201$ | 117.6 | 1.121 | 0.6 | — |
| Triton | 1,352.6 | 354.59 | 5.875R | 214 | 2.0 |
| Nereid | 170 | 5,588.6 | 360.125 | 0.31 | — |
| **Pluto (Dwarf Planet)** | | | | | |
| Charon | 635 | 19.6 | 6.38718 | 17.7 | 1.83 |

*Note: Authorities differ substantially on many of the values above. "R" means orbit is retrograde.

†a × b × c values for the radius are the approximate lengths of the axes for irregular moons.

‡Astronomers have recently found many new moons orbiting each of the outer planets. All are small and many are as yet unnamed. Their orbital properties and sizes are uncertain. They may be captured asteroids. A relatively up-to-date list for each planet is available at the website solarsystem.nasa.gov/planets. Click on the planet of interest and then click on the tab labeled "moons."

## APPENDIX TABLE 6

### Properties of Some of the Solar System's Dwarf Planets

| Name | Distance from Sun (semi-major axis of orbit in AU) | Orbital Period (years) | Inclination of Orbit to Ecliptic (degrees) | Eccentricity of Orbit | Diameter (km) | Mass (kg) |
|---|---|---|---|---|---|---|
| Ceres | 2.77 | 4.599 | 10.6 | 0.08 | 975 × 909 | $9.5 \times 10^{20}$ |
| Pluto | 39.53 | 247.7 | 17.1 | 0.25 | 2300 | $1.3 \times 10^{22}$ |
| Eris | 67.7 | 557 | 44.2 | 0.44 | 2400 | $1.66 \times 10^{22}$ |
| Sedna* | 532 | 12,260 | 11.9 | 0.86 | ~1500 | $\sim4 \times 10^{21}$ |
| Orcus* | 39.42 | 247.5 | 20.6 | 0.23 | ~1300 | $7 \times 10^{20}$ |
| Quaoar* | 43.4 | 286 | 8.0 | 0.03 | ~1300 | $\sim2 \times 10^{21}$ |

*These objects are not yet officially designated as dwarf planets, but they (and at least a dozen others) appear to meet the criteria.

## APPENDIX TABLE 7

### Meteor Showers

| Shower | Dates | Hourly Rate | Radiant R.A.* | Radiant Dec.* | Associated Comet |
|---|---|---|---|---|---|
| Quadrantids | Jan. 2–4 | 30 | 15$^h$ 24$^m$ | 50° | ? |
| Lyrids | Apr. 20–22 | 8 | 18$^h$ 4$^m$ | 33° | 1861 I |
| η Aquarids | May 2–7 | 10 | 22$^h$ 24$^m$ | 0° | Halley |
| δ Aquarids | July 26–31 | 15 | 22$^h$ 36$^m$ | −10° | ? |
| Perseids | Aug. 10–14 | 40 | 3$^h$ 4$^m$ | 58° | Swift-Tuttle |
| Orionids | Oct. 18–23 | 15 | 6$^h$ 20$^m$ | 15° | Halley |
| Taurids | Nov. 1–7 | 8 | 3$^h$ 40$^m$ | 17° | Enke |
| Leonids | Nov. 14–19 | 6 | 10$^h$ 12$^m$ | 22° | Temple-Tuttle |
| Geminids | Dec. 10–13 | 50 | 7$^h$ 28$^m$ | 32° | Phaeton (an asteroid) |

*R.A. = right ascension; Dec. = declination.

## APPENDIX TABLE 8

## The Constellations*

| Constellation Name | Genitive | Description | Abbreviation | Position in Sky† | |
|---|---|---|---|---|---|
| | | | | R.A. | Dec. |
| Andromeda | Andromedae | The Princess of Ethiopia | And | $1^h$ | $+40°$ |
| Antlia | Antliae | The Air Pump | Ant | $10^h$ | $-35°$ |
| Apus | Apodis | The Bird of Paradise | Aps | $16^h$ | $-75°$ |
| Aquarius | Aquarii | The Water Bearer | Aqr | $23^h$ | $-15°$ |
| Aquila | Aquilae | The Eagle | Aql | $20^h$ | $+5°$ |
| Ara | Arae | The Altar | Ara | $17^h$ | $-55°$ |
| Aries | Arietis | The Ram | Ari | $3^h$ | $+20°$ |
| Auriga | Aurigae | The Charioteer | Aur | $6^h$ | $+40°$ |
| Boötes | Boötis | The Herdsman | Boo | $15^h$ | $+30°$ |
| Caelum | Caeli | The Sculptor's Chisel | Cae | $5^h$ | $-40°$ |
| Camelopardus | Camelopardalis | The Giraffe | Cam | $6^h$ | $+70°$ |
| Cancer | Cancri | The Crab | Cnc | $9^h$ | $+20°$ |
| Canes Venatici | Canum Venaticorum | The Hunting Dogs | CVn | $13^h$ | $+40°$ |
| Canis Major | Canis Majoris | The Big Dog | CMa | $7^h$ | $-20°$ |
| Canis Minor | Canis Minoris | The Little Dog | CMi | $8^h$ | $+5°$ |
| Capricornus | Capricorni | The Sea Goat | Cap | $21^h$ | $-20°$ |
| Carina | Carinae | The Keel of Argo‡ | Car | $9^h$ | $-60°$ |
| Cassiopeia | Cassiopeiae | The Queen of Ethiopia | Cas | $1^h$ | $+60°$ |
| Centaurus | Centauri | The Centaur | Cen | $13^h$ | $-50°$ |
| Cepheus | Cephei | The King of Ethiopia | Cep | $22^h$ | $+70°$ |
| Cetus | Ceti | The Sea Monster (Whale) | Cet | $2^h$ | $-10°$ |
| Chamaeleon | Chamaeleontis | The Chameleon | Cha | $11^h$ | $-80°$ |
| Circinus | Circini | The Compasses | Cir | $15^h$ | $-60°$ |
| Columba | Columbae | The Dove | Col | $6^h$ | $-35°$ |
| Coma Berenices | Comae Berenices | Berenice's Hair | Com | $13^h$ | $+20°$ |
| Corona Australis | Coronae Australis | The Southern Crown | CrA | $19^h$ | $-40°$ |
| Corona Borealis | Coronae Borealis | The Northern Crown | CrB | $16^h$ | $+30°$ |
| Corvus | Corvi | The Crow | Crv | $12^h$ | $-20°$ |
| Crater | Crateris | The Cup | Crt | $11^h$ | $-15°$ |
| Crux | Crucis | The Southern Cross | Cru | $12^h$ | $-60°$ |
| Cygnus | Cygni | The Swan | Cyg | $21^h$ | $+40°$ |
| Delphinus | Delphini | The Dolphin | Del | $21^h$ | $+10°$ |
| Dorado | Doradus | The Swordfish | Dor | $5^h$ | $-65°$ |
| Draco | Draconis | The Dragon | Dra | $17^h$ | $+65°$ |
| Equuleus | Equulei | The Little Horse | Equ | $21^h$ | $+10°$ |
| Eridanus | Eridani | The River | Eri | $3^h$ | $-20°$ |
| Fornax | Fornacis | The Furnace | For | $3^h$ | $-30°$ |
| Gemini | Geminorum | The Twins | Gem | $7^h$ | $+20°$ |
| Grus | Gruis | The Crane | Gru | $22^h$ | $-45°$ |
| Hercules | Herculis | Hercules | Her | $17^h$ | $+30°$ |
| Horologium | Horologii | The Clock | Hor | $3^h$ | $-60°$ |
| Hydra | Hydrae | The Water Snake (female) | Hya | $10^h$ | $-20°$ |
| Hydrus | Hydri | The Water Snake (male) | Hyi | $2^h$ | $-75°$ |
| Indus | Indi | The American Indian | Ind | $21^h$ | $-55°$ |
| Lacerta | Lacertae | The Lizard | Lac | $22^h$ | $+45°$ |
| Leo | Leonis | The Lion | Leo | $11^h$ | $+15°$ |
| Leo Minor | Leonis Minoris | The Little Lion | LMi | $10^h$ | $+35°$ |
| Lepus | Leporis | The Hare | Lep | $6^h$ | $-20°$ |
| Libra | Librae | The Scales | Lib | $15^h$ | $-15°$ |
| Lupus | Lupi | The Wolf | Lup | $15^h$ | $-45°$ |

*(continued)*

| Constellation Name | Genitive | Description | Abbreviation | Position in Sky[†] | |
|---|---|---|---|---|---|
| | | | | R.A. | Dec. |
| Lynx | Lyncis | The Lynx | Lyn | 8$^h$ | +45° |
| Lyra | Lyrae | The Lyre | Lyr | 19$^h$ | +40° |
| Mensa | Mensae | Table (Mountain) | Men | 5$^h$ | −80° |
| Microscopium | Microscopii | The Microscope | Mic | 21$^h$ | −35° |
| Monoceros | Monocerotis | The Unicorn | Mon | 7$^h$ | −5° |
| Musca | Muscae | The Fly | Mus | 12$^h$ | −70° |
| Norma | Normae | The Square | Nor | 16$^h$ | −50° |
| Octans | Octantis | The Octant | Oct | 22$^h$ | −85° |
| Ophiuchus | Ophiuchi | The Serpent Bearer | Oph | 17$^h$ | 0° |
| Orion | Orionis | The Hunter | Ori | 5$^h$ | +5° |
| Pavo | Pavonis | The Peacock | Pav | 20$^h$ | −65° |
| Pegasus | Pegasi | The Winged Horse | Peg | 22$^h$ | +20° |
| Perseus | Persei | The Hero | Per | 3$^h$ | +45° |
| Phoenix | Phoenicis | The Phoenix | Phe | 1$^h$ | −50° |
| Pictor | Pictoris | The Easel | Pic | 6$^h$ | −55° |
| Pisces | Piscium | The Fishes | Psc | 1$^h$ | +15° |
| Piscis Austrinus | Piscis Austrini | The Southern Fish | PsA | 22$^h$ | −30° |
| Puppis | Puppis | The Stern of Argo[‡] | Pup | 8$^h$ | −40° |
| Pyxis | Pyxidis | The Compass of Argo[‡] | Pyx | 9$^h$ | −30° |
| Reticulum | Reticuli | The Net | Ret | 4$^h$ | −60° |
| Sagitta | Sagittae | The Arrow | Sge | 20$^h$ | +10° |
| Sagittarius | Sagittarii | The Archer | Sgr | 19$^h$ | −25° |
| Scorpius | Scorpii | The Scorpion | Sco | 17$^h$ | −40° |
| Sculptor | Sculptoris | The Sculptor | Scl | 0$^h$ | −30° |
| Scutum | Scuti | The Shield | Sct | 19$^h$ | −10° |
| Serpens | Serpentis | The Serpent | Ser | 17$^h$ | 0° |
| Sextans | Sextantis | The Sextant | Sex | 10$^h$ | 0° |
| Taurus | Tauri | The Bull | Tau | 4$^h$ | +15° |
| Telescopium | Telescopii | The Telescope | Tel | 19$^h$ | −50° |
| Triangulum | Trianguli | The Triangle | Tri | 2$^h$ | +30° |
| Triangulum Australe | Trianguli Australis | The Southern Triangle | TrA | 16$^h$ | −65° |
| Tucana | Tucanae | The Toucan | Tuc | 0$^h$ | −65° |
| Ursa Major | Usrae Majoris | The Great Bear | UMa | 11$^h$ | +50° |
| Ursa Minor | Ursae Minoris | The Little Bear | UMi | 15$^h$ | +70° |
| Vela | Velorum | The Sails of Argo[‡] | Vel | 9$^h$ | −50° |
| Virgo | Virginis | The Maiden | Vir | 13$^h$ | 0° |
| Volans | Volantis | The Flying Fish | Vol | 8$^h$ | −70° |
| Vulpecula | Vulpeculae | The Fox | Vul | 20$^h$ | +25° |

*Constellations with declinations between −50° and −90° are below the horizon
from most of the United States.
[†]R.A. = right ascension; Dec. = declination.
[‡]Carina, Puppis, Pyxis, and Vela were once a single large constellation called Argo, the Ship.

# APPENDIX TABLE 9

## The Brightest Stars*

| Star | Name | Apparent Visual Magnitude | Spectral Type | Luminosity Class | Distance (ly) | Absolute Visual Magnitude | Visual Luminosity ($L_\odot$) |
|------|------|---------------------------|---------------|------------------|---------------|---------------------------|-------------------------------|
| α CMa A | Sirius | −1.44 | A1 | V | 8.6 | 1.4 | 26 |
| α Car | Canopus | −0.74 | F0 | Ib/II | 310 | −5.65 | 14,000 |
| α Cen | Rigel Kentaurus | −0.01 | G2 | V | 4.4 | 4.3 | 1.6 |
| α Boo | Arcturus | −0.05 | K2 | III | 36.7 | −0.31 | 110 |
| α Lyr | Vega | 0.03 | A0 | V | 25.3 | 0.58 | 50 |
| α Aur | Capella | 0.08 | G8 | III | 42 | −4.8 | 130 |
| β Ori | Rigel | 0.10 | B8 | Ia | 770 | −6.8 | 40,000 |
| α CMi | Procyon | 0.40 | F5 | IV/V | 11.4 | 2.7 | 7 |
| α Ori | Betelgeuse | 0.87 | M2 | Iab | 430 | −4.7 | 10,000 |
| α Eri | Achernar | 0.45 | B3 | V | 143 | −2.8 | 1,100 |
| β Cen AB | Hadar | 0.61 | B1 | III | 500 | −5.4 | 12,000 |
| α Aql | Altair | 0.77 | A7 | IV/V | 16.8 | 2.2 | 11 |
| α Tau A | Aldebaran | 0.87 | K5 | III | 65 | −0.63 | 4,000 |
| α Cru | Acrux | 0.78 | B2 | IV | 300 | −4.19 | 150 |
| α Vir | Spica | 0.98 | B1 | V | 260 | −3.55 | 2,200 |
| α Sco A | Antares | 1.02 | M1 | Ib | 600 | −5.3 | 11,000 |
| α PsA | Fomalhaut | 1.17 | A3 | V | 25 | 1.74 | 17 |
| β Gem | Pollux | 1.16 | K0 | III | 33.7 | 1.09 | 30 |
| α Cyg | Deneb | 1.26 | A2 | Ia | 2,600 | −8.7 | 250,000 |
| β Crux | Beta Crucis | 1.25 | B0.5 | IV | 350 | −3.9 | 3,000 |

*Luminosity and distance are uncertain for many of these stars, especially distant giants.

## APPENDIX TABLE 10

## The Nearest Stars

| Name | Distance (ly) | Spectral Type | Absolute Visual Magnitude | Visual Luminosity ($L_\odot$) | Apparent Visual Magnitude |
|---|---|---|---|---|---|
| Sun | | G2 | 4.83 | | −26.8 |
| Proxima Centauri | 4.24 | M5.5V | 15.5 | 0.00005 | 11.09 |
| α Cen A | 4.36 | G2V | 4.4 | 1.6 | 0.01 |
| B | 4.35 | K0V | 5.7 | 0.5 | 1.34 |
| Barnard's Star | 5.96 | M4V | 13.2 | 0.0004 | 9.53 |
| Wolf 359 | 7.78 | M6V | 16.6 | 0.00002 | 13.4 |
| Lalande 21185 | 8.29 | M2V | 10.4 | 0.005 | 7.47 |
| Sirius A | 8.58 | A1V | 1.5 | 26.0 | −1.43 |
| B | 8.58 | DA2 (white dwarf) | 11.3 | 0.003 | 8.44 |
| Luyten 726-8 A | 8.73 | M5.5V | 15.3 | 0.00006 | 12.41 |
| B | 8.73 | M6V | 16.1 | 0.00003 | 13.2 |
| Ross 154 | 9.68 | M3.5V | 13.1 | 0.0005 | 10.43 |
| Ross 248 | 10.3 | M5.5V | 14.8 | 0.0001 | 12.29 |
| ε Eridani | 10.5 | K2V | 6.2 | 0.3 | 3.73 |
| Lacaille 9352 | 10.7 | M1.5V | 9.75 | 0.01 | 7.34 |
| Ross 128 (FI Virginis) | 10.9 | M4V | 13.5 | 0.0003 | 11.13 |
| Luyten 789-6 ABC (EZ Aquarii) | 11.3 | M5V | 14.7 | 0.0001 | 12.33 |
| Procyon A | 11.4 | F5IV–V | 2.7 | 7.0 | 0.38 |
| B | 11.4 | DA (white dwarf) | 13.0 | 0.0005 | 10.7 |
| 61 Cygni A | 11.4 | K5V | 7.5 | 0.08 | 5.22 |
| B | 11.4 | K7V | 8.3 | 0.04 | 6.03 |
| BD + 59° 1915 A (Struve 2398) | 11.5 | M3V | 11.2 | 0.003 | 8.90 |
| B | 11.5 | M3.5V | 12.0 | 0.001 | 9.68 |
| Groombridge 34 A (GX Andromedae) | 11.6 | M1.5V | 10.4 | 0.006 | 8.08 |
| B (GQ Andromedae) | 11.6 | M3.5V | 13.4 | 0.0004 | 11.07 |
| ε Indi | 11.8 | K5Ve | 7.0 | 0.13 | 4.68 |
| GJ 1111 (DX Cancri) | 11.8 | M6.5V | 17.0 | 0.00001 | 14.79 |
| τ Ceti | 11.9 | G8Vp | 5.8 | 0.4 | 3.50 |
| GJ 1061 | 12.0 | M5.5V | 15.2 | 0.00007 | 13.03 |

Data based mainly on RECONS (Research Consortium on Nearby Stars) website.

Note: Many of the stars in the above table are known by several names. In a few instances, such alternate names are in parentheses. BD stands for Bonner Durchmusterung catalog and GJ for Gliese-Jahreiss catalog; p indicates a peculiar spectrum; e indicates that the star shows emission lines in its spectrum.

## APPENDIX TABLE 11

## Properties of Main-Sequence Stars*

| Spectral Type | Luminosity ($L_\odot$) | Temperature (K) | Mass ($M_\odot$) | Radius ($R_\odot$) |
|---|---|---|---|---|
| O5 | 790,000 | 44,500 | 60 | 12.0 |
| B0 | 52,000 | 30,000 | 17.5 | 7.4 |
| B5 | 830 | 15,400 | 5.9 | 3.9 |
| A0 | 54 | 9,520 | 2.9 | 2.4 |
| A5 | 14 | 8,200 | 2.0 | 1.7 |
| F0 | 6.5 | 7,200 | 1.6 | 1.5 |
| F5 | 3.2 | 6,440 | 1.3 | 1.3 |
| G0 | 1.5 | 6,030 | 1.05 | 1.1 |
| G5 | 0.79 | 5,770 | 0.92 | 0.92 |
| K0 | 0.42 | 5,250 | 0.79 | 0.85 |
| K5 | 0.15 | 4,350 | 0.67 | 0.72 |
| M0 | 0.08 | 3,850 | 0.51 | 0.6 |
| M5 | 0.01 | 3,240 | 0.21 | 0.27 |
| M8 | 0.001 | 2,640 | 0.06 | 0.1 |

*Authorities differ substantially on many of the above values, especially at the upper and lower mass values. Note also that the values are generally not consistent with the Stefan-Boltzmann law.

The data in the preceding tables come from many sources, including

Lang, Kenneth R. *Astrophysical Data: Planets and Stars.* New York: Springer-Verlag, 1992. NOTE: Be sure to get a sheet of corrected errors, however.

Landolt-Bornstein: *Zahlenwerte und Funktionen aus Naturwissenschaften und Technik/Numerical Data and Functional Relationships in Science and Technology.* Neue Serie/New Series. Group VII: Astronomy, Astrophysics, and Space Research, vols. 1–2, Astronomy and Astrophysics. Berlin, New York: Springer-Verlag, 1965–1982.

Donald K. Yeomans of JPL for satellite dimensions and masses.

## APPENDIX TABLE 12

## Known and Suspected Members of the Local Group of Galaxies

| Name of Galaxy | Right Ascension (hours and minutes) | Declination (degrees and minutes) | Galaxy Type | Distance (Mpc) | Diameter (kpc) | Visual Apparent Magnitude | Approximate Luminosity (millions of solar luminosities) |
|---|---|---|---|---|---|---|---|
| WLM | 0 02 | −15 28 | Irr | 1.0 | 3 | 11 | 50 |
| IC10 | 0 20 | +59 18 | Irr | 0.7 | 1 | 11 | 300 |
| Cetus | 0 26 | −11 02 | E | 0.8 | — | 14 | 1 |
| NGC 147 | 0 33 | +48 30 | E5 | 0.7 | 2 | 10 | 100 |
| Andromeda III | 0 35 | +36 30 | E5 | 0.8 | 1 | 15 | 1 |
| NGC 185 | 0 39 | +48 20 | E3 | 0.7 | 2 | 9 | 150 |
| NGC 205 | 0 40 | +41 41 | E5 | 0.8 | 5 | 8 | 300 |
| Andromeda VIII | 0 42 | +40 37 | E | 0.8 | 14 | 9 | 150 |
| M32 | 0 43 | +40 52 | E2 | 0.8 | 2 | 8 | 300 |
| M31 | 0 43 | +41 16 | Sb | 0.8 | 40 | 3 | 25,000 |
| Andromeda I | 0 46 | +38 02 | E3 | 0.8 | 0.5 | 14 | 4 |
| SMC | 0 53 | −72 50 | Irr | 0.06 | 5 | 2 | 600 |
| Sculptor | 1 00 | −33 42 | E3 | 0.09 | 1 | 9 | 1 |
| Pisces | 1 04 | +21 53 | Irr | 0.8 | 0.5 | 18 | 1 |
| IC 1613 | 1 05 | +02 07 | Irr | 0.7 | 3 | 9 | 100 |
| Andromeda V | 1 10 | +47 38 | E | 0.8 | — | 15 | 1 |
| Andromeda II | 1 16 | +33 25 | E2 | 0.7 | 0.7 | 13 | 4 |
| M33 | 1 34 | +30 40 | Sc | 0.8 | 16 | 6 | 3,000 |
| Phoenix | 1 51 | −44 27 | Irr | 0.4 | 0.6 | 12 | 1 |
| Fornax | 2 40 | −34 27 | E3 | 0.14 | 0.7 | 9 | 15 |
| UGCA 92 | 4 32 | +63 36 | Irr | 1.44 | 0.6 | 14 | 5 |
| LMC | 5 23 | −69 45 | Irr | 0.05 | 0.6 | 1 | 2,000 |
| Carina | 6 42 | −50 58 | E4 | 0.10 | 0.7 | 16 | 1 |
| Canis Major | 7 35 | −28 00 | Irr | 0.08 | 220 | — | — |
| Leo T | 9 35 | +17 03 | Irr | 0.420 | 0.34 | 16 | 0.04 |
| Leo A | 9 59 | +30 45 | Irr | 0.7 | 1 | 12 | 3 |
| Sextans B | 10 00 | +05 20 | Irr | 1.41 | 1.6 | 12 | 30 |
| NGC 3109 | 10 03 | −26 09 | Irr | 1.38 | 5 | 10 | 100 |
| Antlia | 10 04 | −27 20 | E3 | 1.41 | 0.6 | 15 | 2 |
| Leo I | 10 08 | +12 18 | E3 | 0.2 | 0.7 | 10 | 4 |
| Sextans A | 10 11 | −04 43 | Irr | 1.60 | 1.8 | 12 | 35 |
| Sextans | 10 13 | −01 37 | E | 0.09 | 1 | 12 | 1 |
| Leo II | 11 13 | +22 09 | E0 | 0.2 | 0.7 | 12 | 1 |
| DDO 155 | 12 59 | +14 13 | Irr | 2.4 | 0.4 | 14 | 8 |
| Ursa Minor | 15 09 | +67 13 | E5 | 0.06 | 0.5 | 11 | 1 |
| Draco | 17 20 | +57 55 | E3 | 0.08 | 0.8 | 10 | 1 |
| Milky Way | 17 46 | −29 00 | SBc | 0.01 | 40 | — | 20,000 |
| Sagittarius | 18 55 | −30 29 | Irr | 0.03 | — | 10 | 30 |
| SagDIG | 19 30 | −17 41 | Irr | 1.2 | 1.5 | 15 | 5 |
| NGC 6822 | 19 45 | −14 48 | Irr | 0.5 | 2 | 9 | 200 |
| Aquarius | 20 47 | −12 51 | Irr | 1.0 | 0.6 | 14 | 2 |
| Tucanae | 22 42 | −64 25 | E4 | 0.9 | 0.5 | 15 | 1 |
| UGCA 438 | 23 26 | −32 33 | Irr | 1.4 | 0.5 | 14 | 5 |
| Cassiopeia | 23 27 | +50 42 | E | 0.7 | 0.5 | 13 | 5 |
| Pegasus | 23 29 | +14 45 | Irr | 0.8 | 1 | 13 | 7 |
| Pegasus II | 23 52 | +24 35 | E | 0.8 | 0.9 | 14 | 3 |

## APPENDIX TABLE 13

## The Brightest Galaxies Beyond the Local Group*

| Name of Galaxy | Right Ascension (hours and minutes) | Declination (degrees and minutes) | Galaxy Type | Distance (Mpc) | Diameter (kpc) | Visual Apparent Magnitude | Approximate Luminosity (millions of solar luminosities) |
|---|---|---|---|---|---|---|---|
| NGC0253 | 00 48 | −25 17 | Sc | 2.6 | 21 | 7.3 | 17,000 |
| NGC0300 | 00 55 | −37 41 | Scd | 2.2 | 13 | 8.2 | 2,700 |
| NGC1068 (M77) | 02 43 | −00 01 | Sb | 13.8 | 28 | 8.7 | 60,000 |
| NGC1291 | 03 17 | −41 06 | S0-a | 9.2 | 27 | 8.7 | 25,000 |
| IC 342 | 03 47 | +68 06 | SBc | 2.5 | 15 | 8.5 | 51,000 |
| NGC2403 | 07 37 | +65 36 | Sc | 3.4 | 23 | 8.4 | 6,800 |
| NGC3031 (M81) | 09 56 | +69 04 | Sab | 3.7 | 23 | 6.9 | 33,000 |
| NGC3034 (M82) | 09 56 | +69 41 | Irr | 3.7 | 10 | 8.3 | 15,000 |
| NGC4258 (M106) | 12 19 | +47 18 | Sb | 7.4 | 41 | 8.4 | 34,000 |
| NGC4472 (M49) | 12 30 | +08 00 | E | 16.3 | 43 | 8.3 | 110,000 |
| NGC4486 (M87) | 12 31 | +12 24 | E | 15.6 | 32 | 8.6 | 81,000 |
| NGC4594 (M104) | 12 40 | −11 37 | Sa | 9.2 | 24 | 8.1 | 58,000 |
| NGC4736 (M94) | 12 51 | +41 07 | Sab | 5.2 | 11 | 8.1 | 14,000 |
| NGC4826 (M64) | 12 57 | +21 41 | Sab | 7.4 | 17 | 8.4 | 30,000 |
| NGC4945 | 13 05 | −49 28 | SBc | 4.6 | 29 | 8.6 | 37,000 |
| NGC5055 (M63) | 13 16 | +42 02 | Sbc | 7.7 | 21 | 8.6 | 25,000 |
| NGC5128 (Cen A) | 13 25 | −43 01 | S0 | 3.7 | 32 | 6.7 | 32,000 |
| NGC5194 (M51) | 13 30 | +47 12 | Sbc | 8.0 | 26 | 8.3 | 33,000 |
| NGC5236 (M83) | 13 37 | −29 52 | Sc | 4.6 | 15 | 7.3 | 28,000 |
| NGC5457 (M101) | 14 03 | +54 21 | Sc | 7.4 | 47 | 7.8 | 34,000 |
| NGC6744 | 19 10 | −63 51 | Sb | 10.7 | 65 | 8.4 | 62,000 |

*Galaxies in this table were selected from the HyperLeda database (http://leda.univ-lyon1.fr) based on their total visual apparent magnitude.

# A

**absorption** the process in which light or other electromagnetic radiation gives up its energy to an atom or molecule. For example, ozone in our atmosphere absorbs ultraviolet radiation.

**absorption-line spectrum** a spectrum showing dark lines at some narrow color regions (wavelengths). The lines are formed by atoms absorbing light, which lifts their electrons to higher orbits.

**acceleration** a change in an object's velocity (either its speed or its direction).

**accretion** the addition of matter to a body. Examples are gas falling onto a star and asteroids colliding and sticking together.

**accretion disk** a nearly flat disk of gas or other material held in orbit around a body by its gravity.

**active galaxy** a galaxy whose central region emits abnormally large amounts of electromagnetic radiation from a small volume. Examples are radio galaxies, Seyfert galaxies, and quasars.

**adaptive optics** a technique for adjusting a telescope's mirror or other optical parts to compensate for atmospheric distortions, such as seeing, thereby giving a sharper image.

**æther** a substance at one time proposed to be the substance that made it possible for light waves to travel through otherwise empty space.

**AGN** an active galactic nucleus. The core of an active galaxy.

**alpha particle** a helium nucleus: two protons plus (usually) two neutrons.

**altitude** an object's angular distance above the horizon.

**amino acid** a carbon-based molecule used by living organisms to build protein molecules.

**angstrom unit** a unit of length used in describing wavelengths of radiation and the sizes of atoms and molecules. One angstrom $= 10^{-10}$ meters.

**angular momentum** a measure of an object's tendency to keep rotating and to maintain its orientation. Mathematically, it depends on the object's mass, $M$, radius, $r$, and rotational velocity, V, and is proportional to $MVr$.

**angular size** a measure of how large an object *looks* to you. It is defined as the angle between lines drawn from the observer to opposite sides of an object. For example, the angular diameter of the Moon is about $\frac{1}{2}°$.

**annular eclipse** an eclipse in which the body in front does not completely cover the other. In an annular eclipse of the Sun, a bright ring of the Sun's disk remains visible around the black disk of the Moon. We therefore see a ring (annulus) of light around the Moon.

**anthropic principle** the principle that the properties we observe the Universe to possess are limited to those that make our existence possible.

**antimatter** a type of matter that, if brought into contact with ordinary matter, annihilates it, leaving nothing but energy. The positron is the antimatter analog of the electron. The antiproton is the antimatter analog of the proton. Antimatter is observed in cosmic rays and can be created from energy in the laboratory.

**aphelion** the point in an orbit where a body is farthest from the Sun.

**Apollo asteroids** those asteroids whose orbits cross the Earth's.

**association** a loose grouping of young stars and interstellar matter, generally consisting of several star clusters.

**asterism** an easily identified grouping of stars, often a part of a larger constellation. For example, Orion's belt.

**asteroid** a small, generally rocky, solid body orbiting the Sun and ranging in diameter from a few meters to hundreds of kilometers.

**asteroid belt** a region between the orbits of Mars and Jupiter in which most of the Solar System's asteroids are located.

**astronomical unit (AU)** a distance unit based on the average distance of the Earth from the Sun.

**atmospheric window** a wavelength band in which our atmosphere absorbs little radiation. For example, on Earth the visible window ranges from about 300 to 700 nm, allowing the light we can see with our eyes to pass through the atmosphere.

**atom** a submicroscopic particle consisting of a nucleus and orbiting electrons. The smallest unit of a chemical element.

**aurora** the light emitted by atoms and molecules in the upper atmosphere. This light is a result of magnetic disturbances and appears to us as the northern or southern lights.

**autumnal equinox** the autumn equinox in the Northern Hemisphere. Fall begins on the autumnal equinox, which is on or near September 23.

**averted vision** looking slightly to one side of a dim object so that you see it slightly away from the center of your field of view. This allows you to see a faint object better, although at a sacrifice of sharpness.

**azimuth** a coordinate for locating objects on the sky. Azimuth is the angle measured from north along the horizon to the point below the object.

# B

**barred spiral galaxy** a galaxy in which the spiral arms wind out from the ends of a central bar rather than from the nucleus.

**Big Bang** the event that, according to many astronomical theories, created the Universe. It occurred about 15 billion years ago and generated the expanding motion that we observe today.

**binary star** two stars in orbit around each other, held together by their mutual gravity.

**bipolar flow** the narrow columns of high-speed gas ejected by a protostar in two opposite directions.

**blackbody** an object that is an ideal radiator when hot and a perfect absorber when cool. It absorbs all radiation that falls upon it, reflecting no light; hence, it appears black. Stars are approximately blackbodies. The radiation emitted by blackbodies obeys Wien's law and the Stefan-Boltzmann law.

**black dwarf** a collapsed star that has cooled to the point where it emits little or no visible radiation.

**black hole** an object whose gravitational attraction is so strong that its escape velocity equals the speed of light, preventing light or any radiation or material body from leaving its "surface."

**BL Lac object** a type of active galaxy named for the peculiar galaxy BL Lac. These objects generally are strong radio sources, and their visible light varies rapidly and erratically.

**blueshift** a shift in the wavelength of electromagnetic radiation to a shorter wavelength. For visible light, this implies a shift toward the blue end of the spectrum. The shift can be caused by the motion of a source of radiation toward the observer or by the motion of an observer toward the source. For example, the spectrum lines of a star moving toward the Earth exhibit a blueshift. *See also* **Doppler shift.**

**Bode's law** a numerical expression for the approximate distances of most of the planets from the Sun.

**Bok globule** small, dark, interstellar cloud, often approximately spherical. Many globules are the early stages of protostars.

**brown dwarf** a star that has a mass too low for it to begin nuclear fusion.

**bulge** the dense, central region of a spiral galaxy.

# C

**carbonaceous chondrite** a type of meteorite containing many tiny spheres (chondrules) of rocky or metallic material stuck together by carbon-rich material.

**CCD** charged-coupled device: an electronic device that records the intensity of light falling on it. CCDs have replaced film in most astronomical applications.

**celestial equator** an imaginary line on the celestial sphere lying exactly above the Earth's equator. It divides the celestial sphere into northern and southern hemispheres.

**celestial pole** an imaginary point on the sky directly above the Earth's North or South Pole.

**celestial sphere** an imaginary sphere surrounding the Earth. Ancient astronomers pictured celestial objects as attached to it.

**cepheid** a class of yellow-giant pulsating stars. Their pulsation periods range from about a day to about 70 days. Cepheids can be used to determine distances. *See also* **standard candle.**

**Chandrasekhar limit** the maximum mass of a white dwarf. Approximately 1.4 solar masses. Named for the astronomer who first calculated that such a limit exists.

**chondritic meteorite** a meteorite containing small spherical bodies called chondrules.

**chondrule** a small spherical body embedded in a meteorite.

**chromosphere** the lower part of the Sun's outer atmosphere that lies directly above the Sun's visible surface (photosphere).

**cluster** a group of objects (stars, galaxies, and so forth) held together by their mutual gravitational attraction.

**CNO cycle/process** a reaction involving carbon, nitrogen, and oxygen (C, N, and O) that fuses hydrogen into helium and releases energy. The process begins with a hydrogen nucleus fusing with a carbon nucleus. Subsequent steps involve nitrogen and oxygen. The carbon, nitrogen, and oxygen act as catalysts and are released at the end of the process to start the cycle again. The CNO cycle is the dominant process for generating energy in main-sequence stars that are hotter and more massive than the Sun.

**coma** the gaseous atmosphere surrounding the head of a comet.

**comet** a small body in orbit around the Sun, consisting of a tiny, icy core and a tail of gas and dust. The tail forms only when the comet is near the Sun.

**compact stars** very dense stars whose radii are much smaller than the Sun's. These stars include white dwarfs, neutron stars, and black holes.

**condensation** conversion of free gas atoms or molecules into a liquid or solid. A snowflake forms in our atmosphere when water vapor condenses into ice. (Technically, the word *deposition* is used to describe the conversion of gas to solid.)

**conjunction** the appearance of two astronomical objects in approximately the same direction on the sky. For example, if Mars and Jupiter happen to appear near each other on the sky, they are said to be in conjunction. *Superior conjunction* refers to a planet that is approximately in line with the Sun but on the far side of the Sun from the Earth. *Inferior conjunction* refers to a planet that lies approximately between the Sun and the Earth.

**conservation of angular momentum** a principle of physics stating that the angular momentum of a rotating body remains constant unless forces act to speed it up or slow it down. Mathematically, conservation of angular momentum states that $MVR$ is a constant, where $M$ is the mass of a body moving with a velocity $V$ in a circle of radius $R$. One extremely important consequence of this principle is that if a rotating body shrinks, its rotational velocity must increase.

**conservation of energy** a principle of physics stating that energy is never created or destroyed, although it may change its form. For example, energy of motion may change into energy of heat.

**constellation** a grouping of stars on the night sky. Astronomers divide the sky into 88 constellations.

**continuous spectrum** a spectrum with neither dark absorption nor bright emission lines. The intensity of the radiation in such a spectrum changes smoothly from one wavelength to the next.

**convection** the rising and sinking motions in a liquid or gas that carry heat upward through the material. Convection is easily seen in a pan of heated soup on a stove.

**convection zone** the region immediately below the Sun's visible surface in which its heat is carried by convection.

**Coriolis effect** a deflection of a moving object caused by its motion across the surface of a rotating body. The Coriolis effect makes storms on Earth spin, generates large-scale wind systems, and creates cloud belts on many of the planets.

**corona** the outer, hottest part of the Sun's atmosphere.

**coronal hole** a low-density region in the Sun's corona. The solar wind may originate in these regions.

**cosmic horizon** the maximum distance one can see out into the Universe at a given time. The horizon lies at a distance in light-years approximately equal to the age of the Universe in years.

**cosmic microwave background (CMB)** the radiation that was created during the Big Bang and that permeates all space. At this time, the temperature of this radiation is 2.73 K.

**cosmic rays** extremely energetic particles (protons, electrons, and so forth) traveling at nearly the speed of light. Some rays are emitted by the Sun, but most come from more distant sources, perhaps exploding supernovas.

**cosmological constant** a term in the equations that Einstein developed to describe the structure of the Universe. The cosmological constant has the effect of a repulsive "force" opposing gravity.

**cosmological principle** the hypothesis that on average, the Universe looks the same to every observer, no matter where he or she is located in it.

**cosmology** the study of the structure and evolution of the Universe.

**Crab Nebula** a supernova remnant in the constellation Taurus. Astronomers in ancient China and the Far East saw the supernova explode in A.D. 1054. A pulsar lies in the middle of the nebula.

**crater** a circular pit, generally with a raised rim and sometimes with a central peak. Crater diameters on the Moon range from centimeters to several hundred kilometers. Most craters on bodies such as the Moon are formed by the impact of solid bodies, such as asteroids.

**critical density** the density necessary for a closed Universe. If the density of the Universe exceeds the critical density, the Universe will stop expanding and collapse. If the density is less than the critical density, the Universe will expand forever.

**crust** the solid surface of a planet, moon, or other solid body.

**curvature of space** the bending of space by a mass, as described according to Einstein's general theory of relativity. Black holes bend the space around them, curving it so that the region within the black hole is cut off from the rest of the Universe. The Universe too may be curved in such a way as to make its volume finite.

**curved space** space that is not flat. *See also* **curvature of space.**

# D

**dark adaption** the process by which the eye changes to become more sensitive to dim light.

**dark energy** a form of energy detected by its effect on the expansion of the Universe. It causes the expansion to speed up. The nature and properties of dark energy are unknown.

**dark-line spectrum** a spectrum in which certain wavelengths are darker than adjacent wavelengths. The missing light is absorbed by atoms or molecules between the source and observer. Also called an *absorption-line spectrum.*

**dark matter** matter that emits no detectable radiation but whose presence can be deduced by its gravitational attraction on other bodies.

**dark nebula** a dense cloud of dust and gas in interstellar space that blocks the light from background stars.

**daughter atoms** the atoms produced by the decay of a radioactive element. For example, uranium decays into lead. Thus, lead atoms are daughter atoms.

**daylight saving time** the time kept during summer months by setting the clock ahead one hour. This gives more hours of daylight after the workday.

**declination** one part of a coordinate system for locating objects in the sky north or south of the celestial equator. Declination is analogous to latitude on the Earth's surface.

**degeneracy pressure** the pressure created in a dense gas by the interaction of its electrons. Degeneracy pressure does not depend on temperature.

**degenerate gas** an extremely dense gas in which the electrons and nuclei are tightly packed. The pressure of a degenerate gas does not depend on its temperature.

**density** the mass of a body or region divided by its volume.

**density-wave model of spiral structure** a theory to account for the spiral arms of galaxies. According to the theory, waves traveling through the disk of a galaxy sweep stars and interstellar gas into a spiral pattern.

**differential gravitational force** the difference between the gravitational forces exerted on an object at two different points. The effect of this force is to stretch the object. Such forces create tides and, if strong enough, may break up an astronomical object. *See also* **Roche limit.**

**differentiation** the separation of previously mixed materials inside a planet or other object. This is the same separation that occurs when a heavy material, such as iron, settles to the planet's core, leaving lighter material on the surface.

**diffraction** a disturbance of light (or other electromagnetic waves) as it passes through an opening or around an obstacle. Diffraction limits the ability to distinguish fine details in images.

**disk** the flat, round portion of a galaxy. The Sun lies in the disk of the Milky Way.

**dispersion** the spreading of light or other electromagnetic radiation into a spectrum. A rainbow is an example of dispersion of light caused by raindrops.

**DNA** deoxyribonucleic acid. The complex molecule that encodes genetic information in all organisms here on Earth.

**Doppler shift** the change in the observed wavelength of radiation caused by the motion of the emitting body or the observer. The shift is an increase in the wavelength if the source and observer move apart and a decrease in the wavelength if the source and observer approach. *See also* **redshift** and **blueshift.**

**dwarf** a small dim star.

**dwarf planet** an object that orbits the Sun and is massive enough that its gravity compresses it into an approximately spherical shape. However, it has not swept its orbital region clear of other objects.

**dynamo** a physical process for generating magnetic fields in astronomical bodies. In many cases, the process involves the generation of electric currents from an interaction between rotation and convection.

# E

**Earth's magnetic field** the magnetism generated by the Earth. The magnetic field is what exerts the magnetic force on a compass needle.

**eclipse** the blockage of light from one astronomical body caused by the passage of another between it and the observer.

The shadow of one astronomical body falling on another. For example, the passage of the Moon between the Earth and Sun can block the Sun's light and cause a solar eclipse.

**eclipse seasons**  the times of year, separated by about 6 months, when eclipses are possible. At any given eclipse season, both a solar eclipse and lunar eclipse generally occur.

**eclipsing binary**  a binary star pair in which one star periodically passes in front of the other, totally or partially blocking the background star from view as seen from Earth.

**ecliptic**  the path followed by the Sun around the celestial sphere. The path gets its name because eclipses can only occur when the Moon crosses the ecliptic.

**electric charge**  the electrical property of objects that causes them to attract or repel one another. A charge may be either positive or negative.

**electric force**  the force generated by electric charges. It is attractive between unlike charges ($+-$) but repulsive between like charges ($++$ or $--$).

**electromagnetic force**  the force arising between electrically charged particles or between charges and magnetic fields. Forces between magnets are a special case of this force. This force holds electrons to the nucleus of atoms, makes moving charges spiral around magnetic field lines, and deflects a compass needle.

**electromagnetic radiation**  a general term for any kind of electromagnetic wave.

**electromagnetic spectrum**  the assemblage of all wavelengths of electromagnetic radiation. The spectrum includes the following wavelengths, from long to short: radio, infrared, visible light, ultraviolet, X rays, and gamma rays.

**electromagnetic wave**  a wave consisting of alternating electric and magnetic energy. Ordinary visible light is an electromagnetic wave, and the wavelength determines the light's color.

**electron**  a low-mass, negatively charged subatomic particle. Electrons orbit the atomic nucleus but may at times be torn free. *See also* **ionization.**

**electroweak force**  one of the fundamental forces of nature. At one time, it was considered to be two separate forces: electromagnetism and the weak force. The weak force causes radioactive decay.

**element**  a fundamental substance, such as hydrogen, carbon, or oxygen, that cannot be broken down into a simpler chemical substance. Approximately 100 elements occur in nature.

**ellipse**  a geometric figure related to a circle but elongated in one direction.

**elliptical galaxy**  a galaxy in which the stars smoothly fill an ellipsoidal volume. Abbreviated E galaxy. The stars in such systems are generally old (Pop II).

**emission**  the production of light, or more generally, electromagnetic radiation by an atom or other object.

**emission-line spectrum**  a spectrum consisting of bright lines at certain wavelengths separated by dark regions in which there is no light.

**emission nebula**  a hot gas cloud in interstellar space that emits light.

**energy**  a quantity that measures the ability of a system to do work or cause motion.

**energy level**  any of the numerous levels that an electron can occupy in an atom, roughly corresponding to an electron orbit.

**epicycle**  a fictitious, small, and circular orbit superimposed on another circular orbit and proposed by early astronomers to explain the retrograde motion of the planets.

**equator**  the imaginary line that divides the Earth (or other body) symmetrically into its northern and southern hemispheres. The equator is perpendicular to a body's rotation axis.

**equinox**  the time of year when the number of hours of daylight and night are approximately equal. The spring and fall (vernal and autumnal) equinoxes mark the beginning of the spring and fall seasons.

**escape velocity**  the speed an object needs to move away from another body in order not to be pulled back by its gravitational attraction. Mathematically, the escape velocity, V, is defined as

$$\sqrt{\frac{2GM}{R}}$$

where $M$ is the body's mass, $R$ is its radius, and G is the gravitational constant.

**eukaryotes**  cells with nuclei. Most cells in current terrestrial organisms have nuclei and are thus eukaryotes.

**Evening star**  any bright planet, often Venus, seen low in the western sky after sunset.

**event horizon**  the location of the "surface" of a black hole. An outside observer cannot see in past the event horizon.

**excited**  the condition in which the electrons of an atom are not in their lowest energy level (orbit).

**exclusion principle**  the condition that no more than two electrons may occupy the same energy state in an atom. This limitation leads to degeneracy pressure.

**exoplanet**  a planet not orbiting our Sun.

## F

**false-color picture/photograph**  a depiction of an astronomical object in which the colors are not the object's real colors. Instead, they are colors arbitrarily chosen to represent other properties of the body, for example, the intensity of radiation, which we cannot see.

**fission**  the splitting of a single body, such as an atom, into two or more pieces.

**flare**  an outburst of energy on the Sun. *See also* **solar flare.**

**flat universe**  a universe that extends forever with no curvature. Its total energy is zero.

**fluorescence**  the conversion of ultraviolet light (or other short-wavelength radiation) into visible light.

**focus**  (1) one of two points within an ellipse used to generate the elliptical shape. The Sun lies at the focus of each planet's elliptical orbit. (2) a point in an optical system in which light rays are brought together. The location where an image forms in such systems.

**frequency**  the number of times per second that a wave vibrates.

**fundamental forces**  the three basic forces of nature: gravitation, the electroweak force, and the strong force. According to some modern theories (*see also* **GUTs**), these three forces are different forms of a single, more fundamental, unified force.

## G

**Gaia hypothesis**  the hypothesis that life does not merely respond to its environment but actually alters its planet's atmosphere and temperature to make the planet more hospitable. For example, by photosynthesis, plants have created an oxygen-rich atmosphere on Earth, which shields the plants from dangerous ultraviolet radiation. Gaia is pronounced *Guy-uh* in this context.

**galactic cannibalism** the capture and merging of one galaxy into another.

**galaxy** a massive system of stars held together by their mutual gravity. Typical galaxies have a mass between about $10^7$ and $10^{13}$ solar masses. Our galaxy is the Milky Way.

**galaxy cluster** a group of galaxies held together by their mutual gravitational attraction. The Milky Way belongs to the Local Group galaxy cluster.

**Galilean relativity** a method for determining the relative speeds of motion seen by observers moving with respect to each other. This method works satisfactorily for motions at low speed, but fails when the speed becomes an appreciable fraction of the speed of light.

**Galilean satellites** the four moons of Jupiter discovered by Galileo: Io, Europa, Ganymede, and Callisto.

**geocentric theories** theories centered on the Earth. Many of the earliest attempts to describe the Solar System were geocentric in that they supposed that the planets moved around the Earth rather than around the Sun.

**giant** a star of large radius and large luminosity.

**glitches** abrupt changes in the pulsation period of a pulsar, perhaps the result of adjustments of its crust.

**globular cluster** a dense grouping of old stars, containing generally about $10^5$ to $10^6$ members. They are often found in the halos of galaxies.

**globule** *see* **Bok globule.**

**granulation** texture seen in the Sun's photosphere. Granulation is created by clumps of hot gas that rise to the Sun's surface.

**grating** a piece of material that creates a spectrum by reflecting light from or passing it through many very fine and closely spaced parallel lines.

**gravitational lens** an object that bends space (and thereby the light passing through the space) by its gravitational attraction and focuses the light to create an image of a more distant object. *See also* **curvature of space.**

**gravitational lensing** the bending of light from a distant object to form an image (usually strongly distorted) by the gravitation

of a mass between the distant object and the observer. See *gravitational lens.*

**gravitational redshift** the shift in wavelength of electromagnetic radiation (light) created by a body's gravitational field as the radiation moves away from the body. Only extremely dense objects, such as white dwarfs, produce a significant redshift of their radiation.

**gravitational waves** a wavelike bending of space generated by the acceleration of massive bodies.

**gravity** the force of attraction that is between two bodies and is generated by their masses.

**greatest elongation** the position of an inner planet (Mercury or Venus) when it lies farthest from the Sun on the sky. Mercury and Venus are particularly easy to see when they are at greatest elongation. Objects may be at greatest eastern or western elongation according to whether they lie east or west of the Sun.

**greenhouse effect** the trapping of heat by a planet's atmosphere, making the planet warmer than would otherwise be expected. Generally, the greenhouse effect operates if visible sunlight passes freely through a planet's atmosphere, but the infrared radiation produced by the warm surface cannot escape readily into space.

**Gregorian calendar** the calendar devised at the request of Pope Gregory XIII and essentially the civil calendar used throughout the world today. It omits the leap year for century years not divisible evenly by 400.

# H

**halo** the approximately spherical region surrounding spiral galaxies that contains mainly old stars, such as the globular clusters. The halo may also contain large amounts of dark matter.

**Hawking radiation** radiation that black holes are hypothesized to emit as a result of quantum effects. This radiation leads to the extremely slow evaporation of black holes.

**heliocentric models** models centered on the Sun. Models of the Solar System in which the planets orbit the Sun.

**helium flash** the beginning of helium fusion in a low-mass star. The fusion begins explosively and causes a major readjustment of the star's structure.

**Herbig-Haro objects** clumps of gas seen near some young stars.

**highlands** the old, heavily cratered regions on the Moon.

**horizon** the line separating the sky from the ground. *See also* **cosmic horizon.**

**H-R diagram** a graph on which stars are located according to their temperature and luminosity. Most stars on such a plot lie along a diagonal line, called the main sequence, which runs from cool, dim stars in the lower right, to hot, luminous stars in the upper left.

**Hubble constant** the multiplying constant H in the Hubble law, $V = HD$. The reciprocal of the Hubble constant (in appropriate units) is approximately the age of the Universe.

**Hubble law** a relation between a galaxy's distance, $D$, and its recession velocity, V, which states that more distant galaxies recede faster than nearby ones. Mathematically, $V = HD$, where H is the Hubble constant.

**hydrogen burning** nuclear fusion of hydrogen into helium. It is not "burning" like ordinary fire but is instead the transformation of one kind of atom into another accompanied by the release of energy.

**hydrostatic equilibrium** the condition in which pressure and gravitational forces in a star or planet are in balance. Without such balance, bodies will either collapse or expand.

**hypothesis** an explanation proposed to account for some set of observations or facts.

**HII region** a region of ionized hydrogen. HII regions generally show a luminous pink/red glow and often surround luminous, hot, young stars.

# I

**ideal gas law** a law relating the pressure, density, and temperature of a gas. This law states that the pressure is proportional to the density times the temperature. Also called the *perfect gas law.*

**inclination** the tilt angle of an astronomical object or its orbit.

**inertia** the tendency of an object at rest to remain at rest and of a body in motion to continue in motion in a straight line at a constant speed. *See also* **mass.**

**inferior conjunction** *see* **conjunction.**

**inferior planet** a planet whose orbit lies between the Earth's orbit and the Sun. Mercury and Venus are inferior planets.

**inflation** the rapid expansion of the early Universe by an enormous factor.

**infrared** a wavelength of electromagnetic radiation longer than visible light but shorter than radio waves. We cannot see these wavelengths with our eyes, but we can feel many of them as heat. The infrared wavelength region runs from about 1000 nm to 1 mm.

**inner core** the innermost part of a planet, also called the *solid core*. The Earth's inner core is a mixture of solid iron and nickel.

**inner planet** a planet orbiting in the inner part of the Solar System. Sometimes taken to mean Mercury, Venus, Earth, and Mars.

**instability strip** a region in the H-R diagram in which stars pulsate.

**interferometer** a device consisting of two or more telescopes connected together to work as a single instrument. Used to obtain a high resolving power, the ability to see small-scale features. Interferometers may operate at radio, infrared, or visible wavelengths.

**international date line** an imaginary line from the Earth's north to south pole, running approximately down the middle of the Pacific Ocean. It marks the location on Earth at which the date changes.

**interstellar cloud** a cloud of gas and dust in between the stars. Such clouds may be many light years in diameter.

**interstellar grains** microscopic solid dust particles in interstellar space. These grains absorb starlight, making distant stars appear dimmer than they truly are.

**interstellar matter** matter in the form of gas or dust in the space between stars.

**inverse-square law** (1) any law in which some property varies inversely as the square of the distance, $d$. Mathematically, as $1/d^2$. (2) the law stating that the apparent brightness of a body decreases inversely as the square of its distance.

**ionization** the removal of one or more electrons from an atom, leaving the atom with a positive electric charge. Under some circumstances, an extra electron may be attached to an atom, in which case the atom is described as negatively ionized.

**ionized** a condition in which the number of an atom's electrons does not equal the number of its protons. Typically, this means the atom is missing one or more electrons.

**irregular galaxy** a galaxy lacking a symmetric structure.

## J

**jets** narrow streams of gas ejected from any of several types of astronomical objects. Jets are seen near protostars and in many active galaxies.

**jet stream** a narrow stream of high-speed wind that blows in the atmosphere of a planet. Such winds occur on Earth and many other planets.

**joule** a unit of energy. One joule per second equals one watt.

**Jovian planet** one of the giant, gaseous planets: Jupiter, Saturn, Uranus, and Neptune. The name *Jovian* was chosen because Jupiter's structure is representative of the other three.

**Julian calendar** a 12-month calendar devised under the direction of Julius Caesar. It includes a leap year every four years.

## K

**Kepler's three laws** laws that describe the motion of planets around the Sun. The first law states that planets move in elliptical orbits with the Sun off center at a focus of the ellipse. The second law states that a line joining the planet and the Sun sweeps out equal areas in equal times. The third law relates a planet's orbital period, $P$, to the semimajor axis of its elliptical orbit, $a$. Mathematically, the law states that $P^2 = a^3$, if $P$ is measured in years and $a$ in astronomical units.

**Kirkwood gaps** regions in the asteroid belt with a lower-than-average number of asteroids. Some of the gaps result from the gravitational force of Jupiter removing asteroids from orbits within the gaps.

**Kuiper belt** a region from which some comets come. The region appears to extend from the orbit of Neptune, past Pluto, out to approximately 55 AU.

## L

**law of gravity** a description of the gravitational force exerted by one body on another. The gravitational force is proportional to the product of their masses and the inverse square of their separation. If the masses are $M$ and $m$ and their separation is $d$, the force between them, $F$, is $F = GMm/d^2$, where G is a physical constant.

**light** electromagnetic energy.

**light-gathering power** a measure of the ability of a telescope (or other device) to collect light. It is generally proportional to the *area* of the telescope's mirror or lens. For example, a telescope with a 6-inch diameter lens has a larger light gathering power than a telescope with a 3-inch diameter lens.

**light-year** a unit of distance equal to the distance light travels in one year. A light-year is roughly $10^{13}$ km, or about 6 trillion miles.

**liquid core** the molten interior of a planet, also called the *outer core*.

**lobes** regions lying outside the body of a radio galaxy where much of its radio emission comes from.

**Local Group** the small group of about 30 galaxies to which the Milky Way belongs.

**Local Supercluster** the cluster of galaxy clusters in which the Milky Way is located. The Local Group is one of its member clusters.

**Lorentz factor** a term that designates how much an object's time, space, and mass are altered as a result of its motion. The factor is very close to 1 except at speeds approaching the speed of light.

**luminosity** the amount of energy radiated per second by a body. For example, the wattage of a light bulb defines its luminosity. Stellar luminosity is usually measured in units of the Sun's luminosity (approximately $4 \times 10^{26}$ watts).

**lunar eclipse** the passage of the Earth between the Sun and the Moon so that the Earth's shadow falls on the Moon.

## M

**Magellanic Clouds** two small companion galaxies of the Milky Way.

**magnetic field** a representation of the means by which magnetic forces are transmitted from one body to another.

**magnetic lines of force** fictitious lines used to visualize the orientation and strength of a magnetic field.

**magnetosphere** the extreme upper regions of the Earth's atmosphere in which the motion of the gas is controlled by the Earth's magnetic field.

**magnitude** a unit for measuring stellar brightness. The smaller the magnitude, the brighter the star.

**main sequence** the region in the H-R diagram in which most stars, for instance the Sun, are located. The main sequence runs diagonally across the H-R diagram from cool, dim stars to hot, luminous ones. Stars on the main sequence fuse hydrogen into helium in their cores. *See also* **H-R diagram.**

**main-sequence lifetime** the time a star remains a main sequence star, fusing hydrogen into helium in its core.

**mantle** the solid, outer part of a planet. This part is immediately below the crust.

**mare** a vast, smooth, dark, and congealed lava flow filling a basin on the Moon and on some planets. Maria often have roughly circular shapes.

**maria** plural of *mare.*

**maser** an intense radio source created when excited gas amplifies some background radiation. *Maser* stands for *microwave amplification by stimulated emission of radiation.*

**mass** a measure of the amount of material an object contains. A quantity measuring a body's inertia.

**mass–luminosity relation** a relation between the mass and luminosity of stars. Higher-mass stars have higher luminosity.

**Maunder minimum** the time period from about A.D. 1600 to 1740 during which the Sun was relatively inactive. Few sunspots were observed during this period.

**mean solar day** the standard 24 hour day. The mean solar day is based on the average day length over a year. (The time interval from sunrise to sunrise varies slightly throughout the year.)

**megaparsec** a distance unit equal to 1 million parsecs and abbreviated Mpc.

**metal** astronomically, any chemical element more massive than helium. Thus, carbon, oxygen, iron, and so forth are metals.

**meteor** the bright trail of light created by small solid particles entering the Earth's atmosphere and burning up. A "shooting star."

**meteor shower** an event in which many meteors occur in a short space of time, all from the same general direction in the sky. The most famous shower is the Perseids in mid-August.

**meteorite** the solid remains of a meteor that falls to the Earth.

**meteoroid** the technical name for the small, solid bodies moving within the Solar System. When a meteoroid enters our atmosphere and heats up, the trail of luminous gas it leaves is called a meteor. When the body lands on the ground, it is called a meteorite. ("A meteoroid is in the void. A meteor above you soars. A meteorite is in your sight.")

**method of standard candles** *See* **standard candle.**

**Mid-Atlantic ridge** an underwater mountain range on Earth created by plate-tectonic motion and running approximately north–south down the center of the Atlantic Ocean.

**Milky Way galaxy** the galaxy to which the Sun belongs. Seen from Earth, the galaxy is a pale, milky band in the night sky.

**Miller-Urey experiment** an experimental attempt to stimulate the conditions under which life might have developed on Earth. Miller and Urey discovered that amino acids and other complex organic compounds could form from the gases that are thought to have been present in the Earth's early atmosphere, if the gases are subjected to an electric spark or ultraviolet radiation.

**millisecond pulsar** a pulsar whose rotation period is about a millisecond.

**minute of arc** a measure of angle equal to one-sixtieth of a degree.

**model** a theoretical representation of some object or system.

**molecule** two or more atoms bonded into a single particle, such as water, $H_2O$ (two hydrogen atoms bonded to one oxygen) or carbon dioxide, $CO_2$ (one carbon atom bonded to two oxygen atoms).

**Moon illusion** the illusion that the Moon appears larger when near the horizon than when seen high in the sky.

**Morning Star** any bright planet visible in the eastern sky before dawn.

# N

**nanometer** a unit of length equal to 1 billionth of a meter ($10^{-9}$ meters) and abbreviated nm. Wavelengths of visible light are several hundred nanometers. The diameter of a hydrogen atom is roughly 0.1 nm.

**neap tide** the abnormally small tide occurring when the Sun's and Moon's gravitational effects on the ocean partially offset each other.

**nebula** a cloud in interstellar space.

**negative curvature** a form of curved space sometimes described as being open in that it has no boundary. Negative curvature is analogous to a saddleshape.

**neutrinos** tiny neutral particles with little or no mass and immense penetrating power. These particles are produced in great numbers by the Sun as it fuses hydrogen into helium and also by some supernova explosions.

**neutron** a subatomic particle of nearly the same mass as the proton but with no electric charge. Neutrons and protons comprise the nucleus of the atom.

**neutron star** a very dense, compact star composed primarily of neutrons.

**Newton's first law of motion** a body continues in a state of rest or uniform motion in a straight line unless made to change that state by forces acting on it. *See also* **inertia.**

**Newton's second law of motion** $F = ma.$ In words, the amount of acceleration, $a$, that a force, $F$, produces depends on the mass, $m$, of the object being accelerated.

**Newton's third law of motion** when two bodies interact, they exert equal and opposite forces on each other.

**nonthermal radiation** radiation emitted by charged particles moving at high speed in a magnetic field. The radio emission from pulsars and radio galaxies is nonthermal emission. More generally, nonthermal means "not due to high temperature."

**north celestial pole** the point on the celestial sphere sky directly above the Earth's North Pole. Objects on the sky appear to circle around this point.

**North Star** any star that happens to lie very close to the north celestial pole. Polaris has been the North Star for about 1000 years, and it will continue as such for about another 1000 years, at which time a star in Cepheus will be nearer the north celestial pole.

**nova** a process in which a surface layer of hydrogen builds up on a white dwarf and then fuses explosively into helium. Nova explosions do not destroy the star and may be recurrent.

**nuclear force** the force that holds protons and neutrons together in the atomic nucleus. Also called the *strong force.*

**nuclear fusion** the binding of two light nuclei to form a heavier nucleus, with some nuclear mass converted to energy. For example, the fusion of hydrogen into helium. This process supplies the energy of most stars and is commonly called "burning" by astronomers.

**nucleosynthesis** the formation of elements, generally by the fusion of lighter elements into heavier ones. For example, the formation of carbon by the fusion of three helium nuclei.

**nucleus** the core of an atom around which the electrons orbit. The nucleus has a positive electric charge and comprises most of an atom's mass.

**nucleus of a comet** the core, typically a few kilometers across, of frozen gases and dust that make up the solid part of a comet.

**nucleus of galaxy** the central region of a galaxy.

# O

**obscuration** the blocking of light of background stars by interstellar matter.

**occultation** the covering up of one astronomical body by another. For example, the Moon passes directly between Earth and the planets from time to time, covering them up and causing their occultation.

**Olbers' paradox** an argument that the sky should be bright at night because of the light from many distant stars and galaxies.

**Oort cloud** a vast region in which comet nuclei orbit. This cloud lies far beyond the orbit of Pluto.

**opacity** the blockage of light or other electromagnetic radiation by matter.

**open cluster** a loose cluster of stars, generally containing a few hundred members.

**opposition** the configuration of a planet when it is opposite the Sun in the sky. If a planet is in opposition, it rises when the Sun sets and sets when the Sun rises.

**orbit** the path in space followed by a celestial body; also a description of an electron's position in an atom.

**outer core** the molten interior of a planet; also called the *liquid core.*

**outer planet** a planet whose orbit lies in the outer part of the Solar System. Jupiter, Saturn, Uranus, and Neptune are outer planets.

**ozone** a form of oxygen consisting of three oxygen atoms bonded together. Its chemical symbol is $O_3$. Because it absorbs ultraviolet radiation, ozone in our atmosphere shields us from the Sun's harmful ultraviolet radiation.

# P

**panspermia** a theory that life originated elsewhere than on Earth and came here across interstellar space either accidently or deliberately.

**parallax** the shift in an object's position caused by the observer's motion. A method for finding distance based on that shift.

**parsec** a unit of distance equal to about 3.26 light-years ($3.09 \times 10^{13}$ km) defined as the distance at which an observer sees the maximum angle between the Sun and the Earth to be one arc second.

**perfect gas law** a law relating the pressure, density, and temperature of a gas. It states that the pressure is proportional to the density times the temperature. This is also called the *ideal gas law.*

**perihelion** the point in an orbit closest to the Sun.

**period** the time required for a repetitive process to repeat. For example, orbital period is the time it takes a planet or star to complete an orbit. Pulsation period is the time it takes a star to expand and then contract back to its original radius.

**period–density relation** a relation that states that the period of a variable star is inversely proportional to the square root of its average density. That is, high-density stars pulsate more rapidly than low-density stars.

**period–luminosity law** a law stating that the longer the period of a pulsating variable star, the more luminous it is.

**phases** the changing illumination of the Moon or other body that causes its apparent shape to change. The following is the cycle of lunar phases: new, crescent, first quarter, gibbous, full, gibbous, third quarter, crescent, new.

**photo dissociation** the breaking apart of a molecule by intense radiation.

**photon** a particle of visible light or other electromagnetic radiation.

**photosphere** the visible surface of the Sun. When we look at the Sun in the sky, we are seeing its photosphere.

**planet** a body in orbit around a star.

**planetary nebula** a shell of gas ejected by a low-mass star late in its evolutionary lifetime. Planetary nebulas typically appear as a glowing gas ring around a central star.

**planetesimal** one of the numerous small, solid bodies that, when gathered together, form a planet.

**plate tectonics** the hypothesis that the crust of the Earth (or some other planet) is divided into large regions (plates) that move very slowly over the planet's surface. Interaction between plates at their boundaries creates mountains and activity such as volcanoes and earthquakes.

**polarity** the property of a magnet that causes it to have a north and south pole.

**poor cluster** a galaxy cluster with a small number of members. The galaxy cluster to which the Milky Way belongs, the Local Group, is a poor cluster.

**population (Pop) I** the younger stars, some of which are blue, that populate a galaxy's disk, especially its spiral arms.

**population (Pop) II** the older, redder stars that populate a galaxy's halo and bulge.

**population (Pop) III** a hypothetical stellar population consisting of the first stars that formed in a galaxy, composed of only hydrogen and helium.

**positive curvature** bending of space leading to a finite volume. A space that is "closed." A universe with positive curvature is analogous to a spherical shape.

**positron** a subatomic particle of antimatter with the same mass as the electron but a positive electric charge. It is the electron's antiparticle.

**precession** the slow change in direction of the pole (rotation axis) of a spinning body.

**pressure** the force exerted by a substance such as a gas on an area divided by that area. That is, pressure × area = force.

**prokaryotes** cells without nuclei. Presumably the first life forms on Earth were prokaryotes.

**prominence** a cloud of hot gas in the Sun's outer atmosphere. This cloud is often shaped like an arch.

**protein** any of many complex organic molecules composed of a chain of amino acids. Proteins serve many functions in cells, including structure and metabolism.

**proton** a positively charged subatomic particle. One of the constituents of the nucleus of the atom.

**proton–proton chain** the nuclear fusion process that converts hydrogen into helium in stars like the Sun and thereby generates their energy. This is the dominant energy-generation mechanism in cool, low-mass stars.

**protostar** a star still in its formation stage.

**pulsar** a spinning neutron star that emits beams of radiation that happen to sweep across the Earth each time the star spins. We observe the radiation as regularly spaced pulses.

**pulsate** to expand and contract regularly. For example, pulsating variable stars swell and shrink in a predictable, regular fashion.

# Q

**quantized** the property of a system that allows it to have only discrete values.

**quasar** a peculiar galaxy characterized by a large redshift, high luminosity, and an extremely small, active core. Quasars are among the most luminous and most distant objects known to astronomers.

# R

**radial velocity** the velocity of a body along the line of sight. That is, the part of its motion directly toward or away from the observer.

**radiant** the point in the sky from which meteors in showers appear to come. *See also* **meteor showers.**

**radiation pressure** the force exerted by radiation on matter.

**radiative zone** the region inside a star where its energy is carried outward by radiation (that is, photons).

**radio galaxy** a galaxy, generally an elliptical system, that emits abnormally large amounts of radio energy. *See also* **lobes.**

**radioactive decay** the breakdown of an atomic nucleus by the emission of subatomic particles.

**radioactive element** an element that undergoes radioactive decay and breaks down into a lighter element.

**rays** long, narrow, light-colored markings on the Moon or other bodies that radiate from young craters. Rays are debris "splashed" out of the crater by the impact that formed it.

**recession velocity** the velocity of an external galaxy (or other object) away from the Sun.

**recombination** the reattachment of one (or more) electrons to an atom following its removal.

**reddening** the alteration in a star's color as seen from Earth as the star's light passes through an intervening interstellar dust cloud. The dust preferentially scatters the blue light from the beam, leaving the remaining light redder.

**red giant** a cool, luminous star whose radius is much larger than the Sun's.

**redshift** a shift in the wavelength of electromagnetic radiation to a longer wavelength. For visible light, this implies a shift toward the red end of the spectrum. The shift can be caused by a source of radiation moving away from the observer or by the observer moving away from the source. For example, if a star is moving away from Earth, its spectrum lines exhibit a redshift. *See also* **Doppler shift.**

**reflection nebula** an interstellar cloud in which the dust particles reflect starlight, making the cloud visible.

**reflector** a telescope that uses a mirror to collect and focus light.

**refraction** the bending of light when it passes through one substance and enters another.

**refractor** a telescope that uses a lens to collect and focus light.

**regolith** the surface rubble of broken rock on the Moon or other solid body.

**resolving power** the ability of a telescope or instrument to discern fine details. Larger-diameter telescopes have greater (that is, better) resolving power.

**resonance** a condition in which the repetitive motion of one body interacts with the repetitive motion of another so as to reinforce the motion. Sliding back and forth in a bathtub to make a big splash is an example.

**rest frame** a system of coordinates that appear to be at rest with respect to the observer.

**retrograde motion** the drift of a planet westward against the background stars. Normally, planets shift eastward because of their orbital motion. The planet does not actually reverse its motion. The change in its direction is caused by the change in the position from which we view the planet as the Earth overtakes and passes it.

**rich cluster** a galaxy cluster containing hundreds to thousands of member galaxies.

**rifting** the breaking apart of a continental plate.

**right ascension** a coordinate for locating objects on the sky, analogous to longitude on the Earth's surface. Measured in hours and minutes of time.

**rilles** narrow canyons on the Moon or other body.

**ring galaxy** a galaxy in which the central region has an abnormally small number of stars, causing the galaxy to look like a ring. Caused by the collision of two galaxies.

**Roche limit** the distance from an astronomical body at which its gravitational force breaks up another astronomical body.

**rotation axis** an imaginary line through the center of a body about which the body spins.

**rotation curve** a plot of the rotation velocity of the stars or gas in a galaxy at different distances from its center.

**RR Lyrae stars** a type of white, giant, pulsating variable star with a period of about one day or less. They are named for their prototype star, RR Lyrae.

# S

**satellite** a body orbiting a planet.

**scattering** the random redirection of a light wave or photon as it interacts with atoms or dust particles.

**Schwarzschild radius** the radius of a black hole. The distance from the center of a black hole to its event horizon.

**scientific method** the process of observing a phenomenon, proposing a hypothesis on the basis of the observations, and then testing the hypothesis.

**scientific notation** a shorthand way to write numbers using ten to a power. For example, $1,000,000 = 10^6$. Also called "powers-of-ten notation."

**scintillation** the twinkling of stars.

**seeing** a measure of the steadiness of the atmosphere during astronomical observations. Under conditions of bad seeing, fine details are difficult to see. Bad seeing results from atmospheric irregularities moving between the telescope and the object being observed.

**seismic waves** waves generated in the Earth's interior by earthquakes. Similar waves occur in other bodies. Two of the more important varieties are S and P waves. The former can travel only through solid material; the latter can travel through either solid or liquid material.

**selection effect** an unintentional selection process that omits some set of the objects being studied and leads to invalid conclusions about the objects.

**self-propagating star formation** a model that explains spiral arms as arising from stars triggering the birth of other stars around them. The resulting pattern is then drawn out into a spiral by the galaxy's rotation.

**semimajor axis** half the long dimension of an ellipse.

**SETI** *S*earch for *E*xtra*t*errestrial *I*ntelligence. Some such searches involve automatic "listening" to millions of radio frequencies for signals that might be from other civilizations.

**Seyfert galaxy** variety of active galaxy with a small, abnormally bright nucleus containing hot gas. Named for the astronomer Carl Seyfert, who first drew attention to these objects.

**shell source** a region in a star where the nuclear energy generation occurs around the core rather than in it.

**shepherding satellite** one or more satellites that by their gravitational attractions prevent particles in a planet's rings from spreading out and dispersing. Saturn's F-ring is held together by shepherding satellites.

**short-period comet** a comet whose orbital period is shorter than 200 years. For example, Halley's comet has a period of 76 years.

**sidereal day** the length of time from the rising of a star until it next rises. The length of the Earth's sidereal day is 23 hours 56 minutes.

**sidereal period** the time it takes a body to turn once on its rotation axis or to revolve once around a central body, as measured with respect to the stars.

**sidereal time** a system of time measurement based on the motion of stars across the sky rather than the Sun.

**silicates** material composed of silicon and oxygen, and generally containing other substances as well. Most ordinary rocks are silicates. For example, quartz is silicon dioxide.

**solar cycle** the cyclic change in solar activity, such as sunspots and solar flares.

**solar day** the time interval from one sunrise to the next sunrise or from one noon to the next noon. That time interval is not always exactly 24 hours but varies throughout the year. For that reason, we use the mean solar day (which, by definition, is 24 hours) to keep time.

**solar eclipse** the passage of the Moon between the Earth and the Sun so that our view of the Sun is partially or totally blocked. *See also* **total eclipse.**

**solar flare** a sudden increase in brightness of a small region on the Sun. This flare is caused by a magnetic disturbance.

**solar nebula** the rotating disk of gas and dust from which the Sun and planets formed.

**solar nebula theory** the hypothesis that the Solar System formed from a rotating cloud of gas and dust, the solar nebula.

**Solar System** the Sun, planets, their moons, and other bodies that orbit the Sun.

**solar wind** the outflow of low-density, hot gas from the Sun's upper atmosphere. It is partially this wind that creates the tail of a comet by blowing dust and gas away from the comet's immediate surroundings.

**solid core** the inner iron-nickel core of the Earth or other planet. Despite its high temperature, the core is solid because it is under great pressure. Also called the *inner core.*

**solstice** (winter and summer) the beginning of winter and summer. Astronomically the solstice occurs when the Sun is at its greatest distance north (June) or south (December) of the celestial equator.

**south celestial pole** the imaginary point on the celestial sphere directly over the Earth's South Pole.

**special theory of relativity** a theory developed by Einstein to explain why the speed of light is always the same, regardless of the motion of the source of the light or the observer. The theory shows (and experiment has confirmed) that an object's mass, length, rate of time passage, and other quantities change depending on the speed of the observer relative to the object.

**spectral class** an indicator of a star's temperature. A star's spectral class is based on the appearance of its spectrum lines. The fundamental classes are, from hot to cool, *O, B, A, F, G, K,* and *M.*

**spectrograph** a device for making a spectrum.

**spectroscopic binary** a type of binary star in which the spectrum lines exhibit a changing Doppler shift as a result of the orbital motion of one star around the other.

**spectroscopy** the study and analysis of spectra.

**spectrum** electromagnetic radiation (for example, visible light) spread into its component wave-lengths or colors. The rainbow is a spectrum produced naturally by water droplets in our atmosphere.

**spicule** a hot, thin column of gas in the Sun's chromosphere.

**spiral arm** a long, narrow region containing young stars and interstellar matter that winds outward in the disk of spiral galaxies.

**spiral galaxy** a galaxy with a disk in which its bright stars form a spiral pattern.

**spring tide** the abnormally large tides that occur at new and full moon.

**standard candle** a type of star or other astronomical body in which the luminosity has a known value, allowing its distance to be determined by measuring its apparent brightness and applying the inverse-square law: for example, Cepheid variable stars, supernovas, and so forth.

**standard time** a uniform time kept within a given region so that all clocks there agree.

**star** a massive, gaseous body held together by gravity and generally emitting light. Normal stars generate energy by nuclear reactions in their interiors.

**star cluster** a group of stars numbering from hundreds to millions held together by their mutual gravity.

**steady-state theory** a model of the Universe in which the overall properties of the Universe do not change with time.

**Stefan-Boltzmann law** the amount of energy radiated from one square meter in one second by a blackbody of temperature T is proportional to $T^4$.

**strong force** the force that holds protons and neutrons together in the atomic nucleus. Also called *nuclear force*.

**subatomic particles** particles making up an atom, such as electrons, neutrons, and protons, or other particles of similar small size.

**subduction** the sinking of one crustal plate where it encounters another.

**sunspot** a dark, cool region on the Sun's visible surface created by intense magnetic fields.

**supercluster** a cluster of galaxy clusters. Our Milky Way belongs to the Local Group, a small galaxy cluster that is but one of many galaxy groups making up the Local Supercluster.

**supergiant** a very large-diameter and luminous star, typically at least 10,000 times the Sun's luminosity.

**superior conjunction** *see* **conjunction.**

**superior planet** a planet orbiting farther from the Sun than the Earth. Mars, Jupiter, Saturn, Uranus, and Neptune are superior planets.

**supernova** an explosion marking the end of some star's evolution. Astronomers identify two main kinds of supernovas: Type I and II. Type I occurs in a binary system in which one star is a white dwarf. The explosion is triggered when mass from a companion star falls onto the white dwarf, raising its mass above the Chandrasekhar limit and causing the star to collapse. Collapse heats the white dwarf so that its carbon and oxygen fuse explosively, destroying the star and leaving no remnant. Type II probably occurs when a massive star's iron core collapses. A Type II supernova leaves either a neutron star or a black hole, depending on the mass of the collapsing core.

**supernova remnant** the debris ejected from a star when it explodes as a supernova. Typically, this material is hot gas, expanding away from the explosion at thousands of kilometers or more per second.

**surface gravity** the acceleration caused by gravity at the surface of a planet or other body.

**synchronous rotation** the condition that a body's rotation period is the same as its orbital period. The Moon rotates synchronously.

**synchrotron radiation** a form of nonthermal radiation emitted by charged particles spiraling at nearly the speed of light in a magnetic field. Pulsars and radio galaxies emit synchrotron radiation. The radiation gets its name because it was first seen in synchrotrons, a type of atomic accelerator.

**synodic period** the time between successive configurations of a planet or moon. For example, the time between oppositions of a planet or between full moons.

# T

**tail** the plume of gas and dust from a comet. The plume is produced by the solar wind and radiation pressure acting on the comet. The tail points away from the Sun and gets longer as the comet approaches perihelion.

**terrestrial planet** a rocky planet similar to the Earth in size and structure. The terrestrial planets are Mercury, Venus, Earth, and Mars.

**tidal braking** the slowing of one body's rotation as a result of gravitational forces exerted on it by another body.

**tidal bulge** a bulge on one body created by the gravitational attraction on it by another. Two tidal bulges form, one on the side near the attracting body and one on the opposite side.

**tides** the rise and fall of the Earth's oceans created by the gravitational attraction of the Moon. Tides also occur in the solid crust of a body and its atmosphere.

**time zone** one of 24 divisions of the globe every 15 degrees of longitude. In each zone, a single standard time is kept. Most zones have irregular boundaries.

**total eclipse** an eclipse in which the eclipsing body totally covers the other body. Only at a total solar eclipse can we see the Sun's corona.

**transit** the passage of a planet directly between the observer and the Sun. At a transit, we see the planet as a dark spot against the Sun's bright disk. From Earth, only Mercury and Venus can transit the Sun.

**trans-Neptunian objects (TNOs)** numerous small, icy objects orbiting in the outer part of the Solar System beyond the orbit of Neptune.

**triangulation** a method for measuring distances. This method is based on constructing a triangle, one side of which is the distance to be determined. That side is then calculated by measuring another side (the base line) and the two angles at either end of the base line.

**triple alpha process** the fusion of three helium nuclei (alpha particles) into a carbon nucleus. This process is sometimes called helium burning, and it occurs in many old stars.

**T Tauri star** a type of extremely young star that varies erratically in its light output.

**tuning-fork diagram** a diagram devised by Hubble to classify the various forms of spiral, elliptical, and irregular galaxies. The diagram is named for its shape.

**turnoff point** the location on the main sequence where a star's evolution causes it to move away from the main sequence toward the red giant region. The location of the turnoff point can be used to deduce the age of a star cluster.

**21-cm line (or radiation)** a spectrum line at radio wavelengths produced by un-ionized (neutral) hydrogen.

**twin paradox** a supposed paradox in special relativity arising from the difference in time measurement of two observers moving relative to each other. It is usually expressed in terms of a twin who travels to a distant star and returns to Earth to find that the other twin has aged much more than the first twin. The paradox is explained because the motion of the twins is not in symmetric. The twin traveling away from Earth experiences accelerations that the twin remaining on Earth does not experience.

**type I supernova** *see* **supernova**

# U

**ultraviolet** a portion of the electromagnetic spectrum with wavelengths shorter than that of visible light but longer than that of X rays. By convention, the ultraviolet region extends from about 10 to 300 nm.

**Universal time** the time kept at Greenwich, England. Universal time is the same as Greenwich mean time. Most local times (in the United States, these are Eastern, Central, Mountain, Pacific, and so forth) differ from it by an even number of hours.

**Universe** the largest astronomical structure we know of. The Universe contains all matter and radiation and encompasses all space.

# V

**Van Allen radiation belts** doughnut-shaped regions surrounding the Earth containing charged particles trapped by the Earth's magnetic field.

**variable star** a star whose luminosity changes in time.

**vernal equinox** the spring equinox in the Northern Hemisphere. Spring begins on the vernal equinox, which is on or near March 21.

**visible spectrum** the part of the electromagnetic spectrum that we can see with our eyes. It consists of the familiar colors red, orange, yellow, green, blue, and violet.

**visual binary star** a pair of stars held together by their mutual gravity and in orbit about each other, and which can be seen with a telescope as separate objects.

**visual double star** two stars that appear to lie very close together on the sky but in reality are at greatly different distances.

## W

**wavelength** the distance between wavecrests. It determines the color of visible light and is generally denoted by the Greek letter $\lambda$.

**wave-particle duality** the theory that electromagnetic radiation may be treated as either a particle (photon) or an electromagnetic wave.

**weak force** the force responsible for radioactive decay of atoms. Now known to be linked to electric and magnetic forces and therefore called the electroweak force.

**white dwarf** a dense star whose radius is approximately the same as the Earth's but whose mass is comparable with the Sun's. White dwarfs burn no nuclear fuel and shine by residual heat. They are the end stage of stellar evolution for stars like the Sun.

**white light** visible light exhibiting no color of its own but composed of a mix of all colors. Sunlight and many artificial light sources are "white."

**Wien's law** a relation between a body's temperature and the wavelength at which it emits radiation most intensely. Hotter bodies radiate more intensely at shorter wavelengths. Mathematically, the law states that $\lambda_m = 3 \times 10^6 / T$, where $\lambda_m$ is the wavelength of maximum emission in nanometers and $T$ is the body's temperature on the Kelvin scale.

## X

**X-ray binaries** a binary star system in which one of the stars, or the gas associated with a star, emits X rays intensely. Such systems generally contain a collapsed object such as a neutron star or a black hole. *See also* **X-ray burster.**

**X-ray burster** a stellar system producing repetitive outbursts of X-ray radiation. X-ray bursters are thought to consist of a neutron star and a normal star in a close binary system. Mass from the normal star spills onto the neutron star, where it slowly accumulates and heats. Eventually, the temperature becomes large enough to initiate nuclear fusion. The released energy heats the material further, causing more fusion and leading to an explosion that we observe as the X-ray burst.

## Y

**year** the time that it takes the Earth to complete its orbit around the Sun; that is, the period of the Earth's orbit.

## Z

**Zeeman effect** the splitting of a single spectrum line into two or three lines by a magnetic field. A method for detecting magnetic fields in objects from their radiation.

**zenith** the point on the celestial sphere that lies directly overhead at your location.

**zodiac** a band running around the celestial sphere in which the planets move.

**zone of avoidance** a band running around the sky in which few galaxies are visible. It coincides with the Milky Way and is caused by dust that is within our galaxy. This dust blocks the light from distant galaxies.

# CREDITS

## PHOTOGRAPHS

Star Charts: From *Patterns in the Sky* by Ken Hewitt-White, © New Track Media, LLC.

### Preview Chapter

Page 1(top): © The Bridgeman Art Library/Getty; (bottom): Courtesy of NASA, ESA, S. Beckwith (STScI) and the HUDFT Team; P.1: Courtesy of NASA; P.2: Courtesy of Stephen E. Strom; P.3: Planets courtesy of NASA/JPL; P.4: Jupiter: Courtesy of ASA/JPL/University of Arizona; Earth: Courtesy of NASA; Sun: Courtesy of SOHO, ESA & NASA; P.7(top): Atlas Image courtesy of 2Mass/UMass/IPAC-Caltech/NASA/NSF; (bottom): Image adapted from NASA; P.8a: Courtesy of David Malin, Anglo-Australian Telescope Board; P.8b: Courtesy of Anglo-Australian Telescope Board; P.11: Milky Way and Earth photos courtesy of NASA.

### Chapter 1

Page 19(top): © English Heritage Photo Library; (bottom): © CNES/Jean-Pierre Haignere, 1999; 1.1A: Courtesy of Tony Stone/Rob Talbot; 1.1B: © English Heritage Library; 1.3A(both): © From Roger Ressmeyer, digitally enhanced by Jon Alpert; 1.3B(both): Courtesy of Eugene Lauria; 1.13B: © English Heritage Photo Library; 1.16B: Courtesy Tenc Tezel; 1.17: Courtesy of Antonio Cidadao; 1.18: Stephen Schneider; 1.19: John Walker; 1.21B: Rod Somerville; 1.28: © Art Resource/Erich Lessing; 1.31: © Corbis-Bettmann; 1.32: © Art Resource/Erich Lessing; 1.35: Courtesy of the Bridgeman Art Library/Getty; 1.37: Courtesy of Bausch and Lomb Optical Co.; E1.4: © Hermann Eisenbeiss/Photo Researchers, Inc.; E1.6A,B: © Roger Ressmeyer/Corbis; E1.10: Courtesy of Dominique Dierick.

### Chapter 2

Page 75(both): Courtesy of NASA; Figure 2.9: Courtesy of NASA.

### Chapter 3

Page 91(top): Courtesy of Bausch and Lomb Optical Co.; (bottom): Courtesy of T. Arny;

3.3(left): Photo © Ewing Galloway, Inc.; 3.5(Pulsar): Courtesy of NASA/CXC /SAO; (Sun): SOHO (ESA & NASA); (Stars): Courtesy of NOAO and Anglo-Australian Observatory, photograph by David Malin; (Cloud): Courtesy of Anglo-Australian Telescope Board, photo by David Malin; (Cosmic microwave): Courtesy of NASA/WMAP Science Team; (Galaxy): NRAO/AUI/NSF; 3.7A,B: © McGraw-Hill Companies/photographer Joe Franek.

### Chapter 4

Page 119(top): © Roger Ressmeyer/Corbis; (bottom): © Scala/Art Resource, NY; Box Figure 4.1: Tom Arny; Box Figure 4.3C: Photo courtesy of Patrick Watson; 4.3: Photo courtesy of Yerkes Observatory; 4.5A: Courtesy of Roger Ressmeyer/Corbis; 4.5B: US Gemini Project/AURA/NOAO/NSF; 4.7: W. M. Keck Observatory; 4.8A: © E.R. Degginger; 4.8B: Tom Arny; 4.9: Courtesy of Steve Criswell; 4.10A,B: Courtesy of Andrea Ghez, UCLA; 4.11: © European Southern Observatory; 4.12: California Institute of Technology; 4.13A: Courtesy of NRAO/AUI; 4.13B: Courtesy of NASA/CXC/SAO/Rutgers/J. Hughes; Box Figure 4.5A: Courtesy of Richard Wainscoat; Box Figure 4.5B: Courtesy of NRAO; Box Figure 4.5C: Courtesy of NASA/CXC/SAO; 4.15(top left): Courtesy of NASA/JPL; 4.15(bottom left): Courtesy of NASA/JPL – Caltech; 4.15(top right, bottom right): Courtesy of NASA, ESA, and Max Planck Institute for Extraterrestrial Physics, Germany; 4.16(top): Courtesy of NASA and The Hubble Heritage Team; 4.16(bottom left): Courtesy of NASA and The Hubble Heritage Team; 4.16(bottom right): NASA and Jeffrey Kenney and Elizabeth Yale (Yale University); 4.18: Courtesy of USAF; 4.19A: Courtesy of Kitt Peak National Observatory; 4.19B: Courtesy Astronomical Society of the Pacific; 4.19C: W.T. Sullivan, © 1993.

### Chapter 5

Page 145(top left): Courtesy of NASA; (top right): Tom Pfeiffer/VolcanoDiscovery; (bottom): Courtesy of MSSS, JPL, NASA; 5.1A,B: Courtesy of NASA; 5.2B: Courtesy of NASA; 5.5: © Taxi/Getty Images; 5.7: Photo courtesy of NASA; Box Figure 5.1: Copyright

by Marie Tharp 1977/2003. Reproduced by permission of Marie Tharp Oceanographic Cartographer, One Washington Ave., South Nyack, New York 10960; 5.16A: Tom Pfeiffer/VolcanoDiscovery; Page 164: Steve Schneider; 5.20A: Courtesy Paul Schneider; 5.20B: Courtesy of NASA; 5.24: Courtesy of NOAA; 5.25A-C: Courtesy of NASA/JPL; 5.26: Courtesy of Martin Hackworth; Page 177: © The Museum of Modern Art/Licensed by SCALA/Art Resource, NY.

### Chapter 6

Page 187(top left): UC Regents; UCO/Lick Observatory; (top right): Stefan Seip - www.astromeeting.de; Bottom: Laurent Laveder - PixHeaven.net; 6.1: Courtesy of Stephen E. Strom; 6.2: UC Regents; UCO/Lick Observatory image; 6.3A,B & 6.4: UC Regents; UCO/Lick Observatory image; 6.5(both) & 6.6: Courtesy of NASA; 6.7: © John Gillmoure/CORBIS; 6.8: UC Regents; UCO/Lick Observatory image; 6.9: Courtesy of NASA; 6.10A,B: Courtesy of NASA/National Geographic Society; 6.20A: Photo by Dennis di Cicco/Sky and Telescope; 6.21A: 2002 South African solar eclipse composite photo by Wendy Carlos and Jonathan Kern Copyright 2002 Serendip LLC. All Rights Reserved; 6.21B: Courtesy of Richard Wainscoat.

### Chapter 7

Page 213(top: Planets & Sun): courtesy of NASA/JPL; (bottom): Mark McCaughren (Max-Planck-Institute for Astronomy), C. Robert O'Dell (Rice University), and NASA; 7.1(left): Courtesy of NASA, JPL; 7.1(right): Courtesy SOHO/ESA/NASA; 7.4: Courtesy of Alan Dyer; 7.5(both): courtesy SOHO/ESA/NASA; 7.9: Courtesy of Anglo-Australian Telescope Board, photo by David Malin; 7.11A: Courtesy of Space Telescope Science Institute; 7.11B: NASA, ESA, D. Golimowski (Johns Hopkins University), D. Ardila (IPAC), J. Krist (JPL), M. Clampin (GSFC), H. Ford (JHU), and G. Illingworth (UCO/Lick) and the ACS Science Team; 7.12: © Fundamental Photographs/Diane Schiumo; 7.14(Mercury): Courtesy NASA's Planetary Geology and Geophysics Program/Arizona State University; (Mars, Earth, Tethys, Dione, Enceladus &

Titania): Courtesy of NASA; 7.15A: Courtesy of A.M. Lagrange, D. Mouillet, and J.L. Beusit, Grenoble Observatory; 7.15B: B. Smith (University of Hawaii), G. Schneider (University of Arizona), E. Becklin and A. Weinberger (UCLA and NASA).

## Chapter 8

Page 237(Mercury, Venus, Earth & Mars): NASA; Page 237(Bottom): Courtesy of NASA/JPL/Malin Space Science Systems; 8.1(all): Courtesy of NASA/JPL/USGS; 8.2(left): Courtesy NASA/JPL/Northwestern University; 8.2(right): Courtesy of NASA/JPL; 8.3: Courtesy of JPL/California Institute of Technology/NASA; 8.4 – 8.6 & 8.10: Courtesy of NASA; 8.11: Courtesy of Alan Dyer; 8.12, 8.13, 8.14(all): Courtesy of NASA/JPL; 8.15: Courtesy of Vernadsky Institute, Russian Academy of Sciences; 8.16: Courtesy of HST; 8.17: Courtesy of A.S. McEwen, USGS; 8.18A: Courtesy of A.S. McEwen, USGS; 8.18B,C: Courtesy of NASA/JPL; 8.19: Courtesy of NASA; 8.20B: Courtesy of A.S. McEwen, USGS; 8.21: Courtesy of NASA; 8.22 – 8.23B: Courtesy of NASA/JPL; 8.24A,B: Courtesy of NASA; 8.25: Courtesy of NASA/JPL; 8.26: Courtesy of NASA; 8.27: Courtesy of Mars Exploration Rover Mission, JPL, NASA; 8.28: Courtesy of NASA/JPL/CORNELL; 8.29: Courtesy of NASA/JPL/CORNELL/ARC; 8.30: Courtesy of Malin Space Science Systems/NASA; 8.31: Courtesy of NASA/JPL/Malin Space Science Systems.

## Chapter 9

Page 269(Top): Courtesy of Cassini Imaging Team/SSI/JPL/ESA/NASA; (Bottom): Courtesy Hubble Space Telescope Comet Team and Nasa; 9.1: Courtesy of NASA; 9.4: Courtesy of NASA/JPL; 9.5A: Courtesy of J. Clarke and G. Ballester, University of Michigan; J. Trauger and R. Evans, Jet Propulsion Laboratory; and NASA; 9.5B: Courtesy of NASA/JPL; 9.6 (Jupiter, Neptune, Saturn): NASA/JPL; 9.6 (Uranus): courtesy of E. Karkoschka, University of Arizona Lunar & Planetary Lab and NASA; 9.7: Courtesy of NASA/JPL; 9.8(both): Courtesy of NASA/JPL; 9.9A,C: Courtesy of NASA/JPL; 9.10: Courtesy of Cassini Imaging Team/SSI/JPL/ESA/NASA; 9.12: Courtesy of NASA/JPL; 9.13: Courtesy of NASA/JPL/Space Science Institute; 9.14: Courtesy of NASA/JPL; 9.15: Courtesy of NASA; 9.16: Courtesy of STScI/HST; 9.17: Courtesy of NASA/JPL/Univ. of Arizona; 9.18B&D: NASA/JPL; 9.19(Titan): NASA Jet Propulsion Laboratory (NASA-JPL); 9.19(all others): Courtesy of NASA/JPL; 9.20(both), 9.23(both), 9.25(both), 9.26: Courtesy of NASA/JPL; 9.27(top middle): UC Regents; UCO/Lick Observatory image; 9.27(bottom): Courtesy of NASA; 9.28: Courtesy of STScI; 9.30(Titan):

National Space Science Data Center; 9.30(Pluto & Charon): Images courtesy of Marc W. Buie/Lowell Observatory; 9.30(Earth): © Stocktrek/Getty Images RF; 9.30(Orcus, Sedna): NASA Jet Propulsion Laboratory (NASA-JPL); 9.30(Moon): © Digital Vision RF/Punchstock.

## Chapter 10

Page 297(top): Courtesy of Mike Skrutskie, University of Virginia; (middle): Courtesy of NASA/JPL; (bottom): Courtesy of Juan Carlos Casado and Isabel Graboleda; 10.1: Courtesy of Ronald A. Oriti, Santa Rosa Junior College, Santa Rosa, Calif.; 10.3A,B: Stephen Schneider; 10.5: Courtesy of NASA/JPL; 10.6A: Courtesy of NASA/JPL; 10.6B: NASA, ESA, J. Parker (Southwest Research Institute), P. Thomas (Cornell University), L. McFadden (University of Maryland, College Park), and M. Mutchler and Z. Levay (STScI); 10.7: ISAS/JAXA; 10.10: © Robert McNaught/Photo Researchers, Inc.; 10.12A: Copyright 1986, Max Planck Institute for Aeronomy; courtesy of Harold Reitsema, Ball Aerospace; 10.12B: NASA Jet Propulsion Laboratory (NASA - JPL); 10.16: Courtesy of Mike Skrutskie, University of Virginia; 10.19A: Courtesy of David Roddy Meteor Crater, Northern Arizona, USA; 10.20: Courtesy of Landsat/EOS.

## Chapter 11

Page 321(Top): Courtesy of LMSAL and NASA; (Bottom): Courtesy of Stefan Seip/www.astromeeting.de; 11.4: Courtesy of Royal Swedish Academy of Sciences; 11.5: Courtesy of Jacques Guertin, Ph.D.; 11.6: Courtesy of NOAO/AURA/NSF; 11.7: Photo by Dennis di Cicco/Sky and Telescope; 11.12: Courtesy of Ernest Orlando, Lawrence Berkeley National Laboratory; 11.13: Courtesy of NOAO/AURA/NSF; 11.14: Courtesy of MDI/SOHO, ESA, NASA; 11.17(both), 11.18(both): Courtesy of NOAO/AURA/NSF; 11.19: Courtesy of SOHO - EIT Consortium, ESA, NASA; 11.20: Courtesy of Eugene Lauria; Box 11.1: Courtesy of NOAO/AURA/NSF.

## Chapter 12

Page 347(top): Courtesy of STScI; (bottom): Courtesy of Dean Ketelsen; 12.4D: © Getty Images; 12.6: Courtesy of Mullard Radio Observatory; 12.7: NOAO/AURA/NSF; 12.9: Courtesy of Harvard College Observatory; 12.10: Courtesy of Katherine Haramundanis; 12.11: A computer-generated depiction, courtesy of Roger Bell and Michael Briley.

## Chapter 13

Page 383(top): Courtesy of Anglo-Australian Observatory; photograph by David Malin; (bottom): © Akira Fujii/DMI; 13.4: Courtesy

of M.J. McCaughrean, MPIA; C.R. O'Dell, Rice University; NASA; 13.5(top): Courtesy of Karen Strom, Mark Heyer, Ron Snell, FCAD, and FCRAO; (bottom): Five College Radio Astronomy Observatory (FCRAO); 13.6(both): Courtesy of Patrick Hartigan, NOAO, and STScI; 13.7: Courtesy of Ronald Snell, FCRAO; 13.8(both): Courtesy of Anglo-Australian Observatory; photographs by David Malin; 13.9A: Courtesy of Anglo-Australian Observatory; photograph by David Malin; 13.16A: Courtesy of Anglo-Australian Observatory; photograph by David Malin; 13.16B: Courtesy of Hubble Heritage Team (AURA STScI/NASA; 13.16C: Bruce Balick, University of Washington; Vincent Icke, Leiden University, Netherlands; Garrelt Mellema, Stockholm University/NASA; 13.16E: Courtesy of NASA, ESA, and Valentin Bujarrabal-Observatorio Astronomico Nac'ional, Spain; 13.19: Pablo Candia; 13.20A: Courtesy of Anglo-Australian Observatory; photograph by David Malin; 13.20B: Courtesy of Richard Wainscoat; 13.20C: Courtesy of NASA/CXC/SAO/Rutgers/J. Hughes; 13.22: Courtesy of Anglo-Australian Observatory; photograph by David Malin.

## Chapter 14

Page 413(top): Courtesy of Anglo-Australian Observatory; photograph by David Malin; (bottom): Courtesy of Paresce, R. Jedrzejewski (STScI), NASA, ESA; 14.9: Photos courtesy of NOAO.

## Chapter 15

Page 433(top): Atlas Image courtesy of 2MASS/UMass/IPAC - Caltech/NASA/NSR; (bottom): Courtesy of Rainer Schodel (MPE) et al., NAOS-CONICA, ESO; 15.1: Courtesy of Steward Observatory and NOAO; 15.2: Courtesy of Anglo-Australian Observatory; photograph by David Malin; 15.5: Courtesy of Hubble Heritage Team (AURA/STScI/NASA); 15.7: Courtesy of Anglo-Australian Observatory; photograph by David Malin; 15.11: Courtesy of NASA and The Hubble Heritage Team (STScI/AURA); 15.13: Courtesy of Anglo-Australian Observatory; photograph by David Malin; 15.15: Atlas Image courtesy of 2MASS/UMass/IPAC-Caltech/NASA/NSF; 15.16B: Courtesy of Anglo-Australian Observatory; photograph by David Malin; 15.17(all): Courtesy of Anglo-Australian Observatory; photograph by David Malin; 15.20: Courtesy of NOAO; 15.25: Courtesy of NASA, Umass, D. Wang et al.; 15.28(all): Courtesy of Neal Katz, University of Massachusetts, and James E. Gunn, Princeton University; 15.29: Courtesy of Neal Katz, University of Massachusetts, and James E. Gunn, Princeton University; 15.30: David Law/University of Virginia.

## Chapter 16

Page 467(top): Courtesy of Hubble Heritage Team, AURA/STScI/NASA; (Middle): NASA, ESA, and The Hubble Heritage Team (STScI/AURA); (bottom): Courtesy Mary E. Putman, University of Michigan; Lister Staveley-Smith, Australia Telescope National Facility; Kenneth C. Freeman, Australian National University; and Brad K. Gibson and David G. Barnes, both Swinburne University of Technology. The Parkes Telescope is operated by CSIRO; 16.1: Courtesy of George Greany; 16.2: Courtesy of William C. Keel; 16.3A: Courtesy of Bruce Hugo and Leslie Gaul, Adam Block, NOAO, AURA, ASF; 16.3B: Courtesy of Hubble Heritage Team, AURA/STScI/NASA; 16.3C: © European Southern Observatory; 16.4A: California Institute of Technology: 16.4B: Courtesy of NOAO; 16.4C, 16.5(both), 16.6(both): Courtesy of Anglo-Australian Observatory, photographs by David Malin; 16.7A: Courtesy of J. Young and L. Allen, University of Massachusetts, and J. Kenney, Yale University; 16.7B: NASA, ESA, and J. P. Kneib (Laboratorie d'Astrophysique de Marseille); 16.9(EO): Courtesy of Anglo-Australian Observatory; photograph by David Malin; (E2 & E5): NOAO/AURA/NSF; (E7): David W. Hogg, Michael R. Blanton, and the Sloan Digital Sky Survey Collaboration. © 2006 Michael R. Blanton & David W. Hogg; (S0): John Kormendy (University of Texas at Austin); (Sa): Adam Block/Steve Mandel/Jim Rada and Students/NOAO/AURA/NSF; (Sb): © European Southern Observatory; (Sc): Lauren Grodnicki; (Sd): © European Southern Observatory; (SBa): University of Arizona SEDS; (SBb): Roger Lynds/NOAO/AURA/NSF; (SBc): NOAO/AURA/NSF; (SBd): The NGC/IC Project; (Irregulars): Local Group Galaxies Survey Team, NOAO, AURA, NSF; 16.10: Courtesy of NOAO; 16.11: Courtesy of F. Schwizer, Carnegie Institution of Washington; 16.13(all): Courtesy of Joshua Barnes, University of Hawaii; 16.14A: Courtesy of STSci and NOAO; 16.14B: Courtesy of NASA; 16.15A,B: Courtesy of STScI; 16.19A: Courtesy of NRAO; 16.19B & 16.20: Courtesy of NRAO/AUI; 16.21: Courtesy of NASA; 16.22: Courtesy Bernie and Jay Stlonick, Adam Block, AOP, NOAO, AURA, NSF; 16.23A: Courtesy of John Bahcall, Institute for Advanced Study; M. Disney, University of Wales; and NASA; 16.23B: Courtesy of NOAO; 16.25A: Courtesy of W. Jaffe, Leiden Observatory, and H. Ford, Johns Hopkins University, Space Telescope Science Institute; and NASA; 16.25B: Courtesy of NRAO and California Institute of Technology; 16.27A: Courtesy of NASA; 16.28: Courtesy of STScI; 16.29: Courtesy of California Institute of Technology; 16.31A,B: Courtesy of Anglo-Australian Observatory; photographs by David Malin; 16.32: © Anglo-Australian Observatory; photograph by David Malin; 16.33: Courtesy of NOAO; 16.34: Courtesy of A. Fabian (IoA Cambridge) et al., NASA; 16.35: NASA/STScI; Magellan/U. Arizona/D. Clowe et al.; 16.38: Numerical simulations performed at the National Center for Supercomputing Applications, NCSA, Urbana-Champaign, Illinois; by Andrey Kravtsov (The University of Chicago; and Anatoly Klypin, New Mexico State University.

## Chapter 17

Page 505(both): Courtesy of S. Beckwith and the HUDF Working Group (STScI, HST, ESA, NASA); 17.1A: Two Micron All Sky Survey, a joint project of the University of MA and the Infrared Processing and Analysis Center/CA Institute of Technology, funded by NASA/National Science Foundation/T.H. Jarrett, J. Carpenter, & R. Hurt. Perha: 17.12: Courtesy of NASA/WMAP Science Team; 17.17: Courtesy of S. Beckwith and the HUDF Working Group (STScI), HST, ESA, NASA; Figure E3.1: Courtesy of Andrew H. Knoll; E3.3(bird): © Visuals Unlimited; (whale): © Brandon Cole/Visuals Unlimited; (butterfly): © Corbis RF; (beetle): © Michael Fogden/Corbis; (giraffe): © DV/Getty; E3.5: Courtesy of Ken Eward/BioGrafx; E3.8: Courtesy of Seth Shostak; E3.9: Courtesy of NASA/JPL/Cornell; E4.4: © Hulton-Deutsch Collection/Corbis.

# LOOKING UP FEATURES

## Orion

Background: © Akira Fujii/DMI; Horsehead Nebula: Courtesy of Anglo-Australian Observatory; photograph by David Malin; close-up of Horsehead; Nebula: Courtesy NOAO/AURA/NSF; Betelgeuse: Courtesy A. Dupree (CFA), NASA, ESA; Orion Nebula: Courtesy of Carol B. Ivers; Rigel: Courtesy Gary Bernstein (U. Pennsylvania); Copyright U. Michigan, Lucent; Protoplanetary Disk: Courtesy of A.M. Langrange, D. Mouillet, and J.L. Buzit, Grenoble Observatory.

## Summer Triangle

Background: © Akira Fujii/DMI; M57: Courtesy of H. Bond et al., Hubble Heritage Team (STScI/AURA); Alberio: Courtesy of Randy Brewer; M27: © IAC/RGO/Malin.

## Taurus

Background: © Akira Fujii/DMI; M1: Courtesy of Richard Wainscoat; M45: Courtesy of Anglo-Australian Observatory; photographs of David Malin; Hyades: © Akira Fujii/DMI.

## Northern Circumpolar Constellations

Background (both): © Akira Fujii/DMI; M52: Hartmut Frommelt; M97: Gary Whie and Verlenne Monroe/Adam Block/NOAO/AURA/NSF; M101: Adam Block/NOAO/AURA/NSF.

## Centaurus and Crux, The Southern Cross

Background: Courtesy of Anglo-Australian Observatory; photograph by David Malin; Centaurus A: Courtesy Peter Ward, 2004; Omega Centauri: Al Kelly; Eta Carinae: J. Hester/Arizona State University/NASA; The Jewel Box: Research School of Astronomy and Astrophysics, the Australian National University.

## Ursa Major

Background: © Akira Fujii/DMI; M81&M82: © Robert Gendler; Mizar and Alcor: Courtesy of DSS/Processing by Coelum (www.coelum.com); M51: © Tony and Daphne Hallas; 12.21: Based on a figure courtesy of Roger Bell and Michael Briley.

## Sagittarius

Background: Courtesy of Till Credner, AlltheSky.com; M16 Close-up: Courtesy of NASA, HST, J. Hester & P. Scowen (ASU); M16 with Gas Cloud: Courtesy of Bill Schoening/NOAO/AURA/NSF; M20: Courtesy of Jason Ware; M22: Courtesy of N.A. Sharp, REU program/NOAO/AURA/NSF.

## M31 & Perseus

All Photos: Courtesy Mary E. Putman, University of Michigan; Lister Staveley-Smith, Australia Telescope National Facility; Kenneth C. Freeman, Australian National University; and Brad K. Gibson and David G. Barnes, both Swinburne University of Technology. The Parkes Telescope is operated by CSIRO.

# LINE ART

## Chapter 3

Table 3.4: Reprinted from *Gerchim Cosmochim Acta* 53 (1989): 1987, by Anders and Grevesse. Copyright 1989, with permission from Elsevier; 3.17: Courtesy of Stephen M. Larson, University of Arizona; 3.18A: Courtesy of Doug McGonagle, FCRAO; Figure 3.18B: Courtesy of P.F. Winkler, Middlebury College.

## Chapter 6

6.15A: Courtesy of Robin Canup, Southwest Research Institute.

## Chapter 7

7.19: Courtesy of G. Marcy at University of California at Berkeley, from SFSU.

## Chapter 8

8.8: Based on a computer simulation, courtesy of W. Benz, University of Arizona; W. Slattery, Los Alamos; A.G.W. Cameron, CFA.

## Chapter 10

10.4: Courtesy of E.L.G. Bowell, Lowell Observatory; 10.9: Courtesy of E.L.G. Bowell, Lowell Observatory.

## Chapter 14

14.6: Courtesy of M.I. Large, University of Sydney; R.N. Manchester, Australia Telescope, CSIRO; and Joseph H. Taylor, Princeton University.

## Chapter 15

15.14A: Courtesy of NOAO; 15.18: Courtesy of Y.M. and Y.P. Georgelin; 15.21: This sequence was generated by P. Seiden (deceased) and H. Gerola, *Ap. J.* 233, 56, 1979. Reproduced by permission; 15.24: Based on diagram by Dan P. Clemens, *Ap J.* 295, 422.

## Chapter 16

16.16B: Data courtesy of Robert P. Kirshner, CFA; 16.24: Courtesy of Tom Balonek and Jeyhan Kartaltepe, Colgate University.

## Chapter 17

17.7: Based on curve supplied by NASA/Goddard Spaceflight Center: COBE Science Working Group.

## Essay 3

E3.4: From Raven and Johnson, *Biology, 2/e.* 1989. Copyright © The McGraw-Hill Companies. All rights reserved.

# INDEX